Number Systems and Their Uses

Number Systems and Their Uses

Frank L. Wolf

CARLETON COLLEGE

Xerox College Publishing

WALTHAM, MASSACHUSETTS • TORONTO

CONSULTING EDITOR
Seymour Schuster
Carleton College

LIBRARY OF CONGRESS CATALOG CARD NUMBER: 75-135633
PRINTED IN THE UNITED STATES OF AMERICA.

To the Joyful one

Preface

A child said *What is the grass?* fetching it to me
with full hands; . . .

WALT WHITMAN

When I was a child, I spake as a child, I
understood as a child, I thought as a child: but
when I became a man, I put away childish
things.

CORINTHIANS I; 13.

Some knowledge of number systems and their uses must be part of the intellectual baggage to be carried by any educated person. In this book we take a careful look at the intuitive foundations, practical applications, and logical structures of various common number systems.

The two quotations given above are pertinent to the attitude that the reader should have as we consider these matters. In one sense, the reader should be as a child. He should be willing and even eager to ask questions that may seem simple and naive. He should be open to the wonder of everyday useful things. And, above all, he should not assume that he already knows all the answers to all the interesting questions. In another sense, the reader should put away childish things. He should not be surprised that some elementary questions may have rather subtle and complicated answers. And he should be willing to work hard to make sense out of proposed answers. He should want to see *why* things are as they are. He should *not* be satisfied with the answer so often given by the impatient parent: "Sorry, that's just the way it is."

The book should be especially useful to those people who are planning to teach arithmetic to elementary school children. The material covered by the book is very close to that in the year course on the number systems which was recommended for potential elementary teachers by the Committee on the Undergraduate Program in Mathematics (CUPM) of the Mathematical Association of America.

A potential reader looking at the book for the first time may be "put off" by noting, in the later portions, the use of a few terms he does not understand. But the book very deliberately builds up to these notions and the reader should realize that he will be prepared to understand later portions of the book by material in the earlier parts. Many different devices are used to help in the gradual development of these concepts.

The logic used in the book is informal. No special symbols of mathematical logic are introduced.

Basic ideas are introduced by generalizing from familiar everyday situations.

Various kinds of diagrams are used to illustrate operations, mappings, and relations.

Much use is made of tables for operations on finite sets.

Flow charts are used to clarify many different processes — and, incidentally, to introduce the reader to a technique that is very useful in this age of computers.

Natural numbers are first introduced as cardinal numbers for certain sets. The other useful number systems are built on the natural number system.

The order properties of each number system are treated rather thoroughly.

Many fundamental ideas from algebra, geometry, and analytic geometry are introduced along the way as various number systems are considered.

An optional chapter on the uses of the real numbers gives intuitive introductions to such topics as vector spaces, matrix algebra, and complex numbers.

The final optional chapter on transfinite arithmetic introduces some intriguing ideas about infinite sets and throws further light on the natural numbers.

In many sections, a more formal treatment of a topic is included as an appendix to the section. These more formal treatments may be omitted without interfering with progress through the book. The definitions and theorems in such an appendix can serve as a review of the section.

Many exercises are included to be worked by the reader. In fact, most sections contain more problems than would make up a typical "assignment." The exercises are designed to assist the reader in understanding the material presented and, in some cases, to extend this material.

Answers for approximately half of the exercises are given in the back of the book. A booklet that gives the answers to all of the exercises may be ordered from the publisher. The author will welcome suggestions for improvements in these answers or, indeed, for any part of the book.

Theorems and definitions in the book are identified by triples of numbers. The first number in a triple indicates the chapter, the second number indicates the section, and the third number indicates the position within that section. Thus, Theorem 6.3.4 is the fourth theorem in the third section of Chapter 6. Problems have a separate notation. Problem 5, Section 6.4 refers to the fifth problem in Chapter 6, Section 4.

Starred sections and chapters are optional and may be omitted without inhibiting the study of later unstarred sections.

I wish to extend my sincere thanks to Mrs. Frances L. Kirchner and G. Eric Blistain for doing such impeccable work in typing various portions of the manuscript, and to the staff at Xerox College Publishing for their sensitive assistance and guidance. Special thanks are due to all members of my family who cheerfully put up with many inconveniences (especially a geographical one) while this book was being written in its final form.

FRANK L. WOLF

Northfield, Minnesota
April 11, 1971

Contents

8. The rational number system

9. Sequences

Number Systems and Their Uses

1

Introduction

1.1 Some matters of language

Mathematics is a strange and wonderful beast. The careful use of precise language makes up her very bones and rigid support. These bones are filled out with the flesh of logical deductions. She feeds on multitudinous problems that arise in the physical and social sciences and drinks from the bottomless well of man's need to know. To her, the breath of life is the human imagination.

Our first task here is to discuss some aspects of this basic matter of the careful use of language.

Some of the simplest words in the English language are ambiguous. The word "or," for example, has at least two distinct meanings. Since we wish to use this word often, this fact must be discussed. Consider the following quotations:

A: Enrichment has usually been achieved by the following: (a) guiding students to deeper considerations of standard topics in a course of study; (b) encouraging individual research; and (c) organizing extracurricular activities in the form of mathematics clubs, or mathematics teams (for intra- or extra-mural mathematical competitions).†

B: In such cases [of independent study] the students must have the opportunity to consult other materials, his own teacher, or some other interested adult about the progress of his investigation.‡

† *Enrichment Mathematics for the Grades—Twenty-seventh Yearbook*, National Council of Teachers of Mathematics, Washington, D.C., 1963, page 2.

‡ *Ibid.*, page 3.

C: Provision for the gifted in mathematics has in the past taken the form of either enrichment or acceleration.†
D: Eat your spinach or you won't get your dessert.‡
E: Are there any additions or corrections to the minutes?§

We are interested in how the word "or" is used in each of these cases. In A, what is evidently being claimed in the part that uses "or" is that enrichment has been achieved by mathematics clubs or mathematics teams or by both of these. It is clearly not intended in A that a club and a team haven't been used together. Here we would say that "or" is being used in the **inclusive sense.** This is the sense intended by the "and/or" that is sometimes used in legal documents.

In B, the meaning to be put on "or" is no doubt again the inclusive one. That is, B is asserting that at least one of the three (materials, teacher, interested adult) must be available. It is certainly not intended that the availability of more than one of these would be inappropriate.

In C, however, we find what seems to be a different use of "or." Here, the most likely intent is that provisions for the gifted have been as enrichment or as acceleration but not as both. When "or" is used in this manner, we are taking it in the **exclusive sense.** In the exclusive sense "or" is taken to mean "this or that but not both." It is fairly common practice, as was done in C, to use the terminology "either . . . , or . . ." to signal use of "or" in the exclusive sense.

Another example of the use of "or" in the exclusive sense is the manner in which D is usually interpreted. The child would be very disappointed if he *did* eat his spinach and *still* did not get his dessert. That is, most people would take it here that the intent of the speaker of D is that it won't happen that the child will eat his spinach and *also* not get his dessert.

E is an especially clear example of the inclusive sense for "or." Clearly, *both* additions *and* corrections would be allowed.

When "or" is used in a question, it often seems to gain even a third sense different from the two above. Suppose that at the check-out counter in the grocery store the person who is getting your purchases ready to carry asks you, "Would you like a bag or a box?" If you answer, "Yes," you will get a very strange look from the questioner. No doubt, "yes" is a logically acceptable answer to the question put. But what was really intended and what most people would take the question to mean is: Do you have a preference between a bag

† *Ibid.*, page 1.

‡ Personal communication from a relative.

§ This example is suggested in the discussion on the use of "and/or" in *Modern American Usage* by Wilson Follet (Hill and Wang, New York, 1966). If you refer to this interesting analysis, you should watch for the error made at the end of the first paragraph.

and a box? If you do, let me know which you prefer. If you do not, say something such as "either will do."

A similar situation arises in some examinations when the instructions state "True or false?" and then list some assertions. There is the possibility of putting down as the answer to each question simply "Yes"—since it is indeed so that each of the assertions is either true or false. But this is clearly not what was intended by the questioner.

In mathematics, it has become customary to use "or" in the inclusive sense only. If the meaning of the exclusive "or" is needed then it is expressed by some such terminology as ". . . or . . . , but not both" or "either . . . or . . . ."

Adopting this convention, we see, for example, that the first three of the following four assertions are true, while the last one is false.

(a) $1 + 1 = 2$ or $1 + 1 = 3$.
(b) The moon is made of green cheese or $5 + 7 = 12$.
(c) $1 + 1 = 2$ or $2 + 2 = 4$.
(d) $2 = 3$ or $3 + 3 = 5$.

Let us now turn to a discussion of implications.

Since mathematics is so involved with logical deductions, sentences in the form "If . . . , then . . ." occur repeatedly in it. Exactly how is such a sentence to be taken? There *are* various possible interpretations of such a sentence, and logicians and philosophers argue about which interpretations are suitable in certain situations. In mathematics, however, it is standard practice to interpret an "If . . . , then . . ." sentence as a **material implication.** What this means is that "If . . . , then . . ." is considered false only in case the assertion made by the part following the "If" is true while the assertion made by the part following the "then" is false. In all other cases, "If . . . , then . . ." is taken to be true. Thus, "If $2 + 1 = 3$, then $2 + 1 = 4$" is false because "$2 + 1 = 3$" is true while "$2 + 1 = 4$" is false. On the other hand, "If $2 + 1 = 3$, then $2 + 1 = 3$," "If $2 + 1 = 4$, then $2 + 1 = 3$," and "If $2 + 1 = 4$, then $2 + 1 = 4$" are all true. Notice then that, on the face of it, "If . . . , then . . ." seems to have nothing to do with the logical deducibility of the "then part" from the "if part."†

It might well be asked why mathematicians choose to interpret an "If . . . , then . . ." statement as they do. The answer is, in large measure, a practical one. The adopted interpretation is simple, clear, and very useful. And it ranks higher in regard to these matters than any other of the interpretations that have been proposed by others.

We shall refer to a sentence in the form "If . . . , then . . ." as an **implication.**

† Actually there is an important result in mathematical logic (the Deduction theorem) that shows that material implication as we have defined it is intimately connected with logical proof. But this result is not obvious from what has been said here.

The assertion made by the part following the "If" is called the **hypothesis** of the implication. The assertion made by the part following the "then" is called the **conclusion** of the implication.

In this terminology, an implication is false only when its hypothesis is true and its conclusion is false. Thus, "If $2 \times 3 = 4$, then $2 \times 3 = 6$" and "If $2 \times 3 = 4$, then $2 \times 3 = 4$" are both true since in neither case is the hypothesis true.

Endless repetition in form is objectionable and makes for dull reading. And since implications occur over and over again in mathematical writing, there is a need for several different ways to express exactly the same meaning as "If . . . , then ***."† Some of these are:

(a) ***, if . . . .
(b) In order that *** it is sufficient that . . . .
(c) In order that . . . it is necessary that ***.
(d) . . . only if ***.
(e) . . . implies ***.

As a particular case, all the following mean exactly the same thing: "If my father is alive, then I have at least one living parent." "In order that I have at least one living parent it is sufficient that my father be living." "In order that my father be living it is necessary that I have at least one living parent." "My father is alive only if I have at least one living parent." "My father being alive implies I have at least one living parent."

It is true that "If today is Tuesday then tomorrow will be Wednesday."‡ It is also true that "If tomorrow will be Wednesday, then today is Tuesday." Both of these facts may be asserted together by the sentence "Today is Tuesday if and only if tomorrow will be Wednesday." Equivalently, we may say, "A necessary and sufficient condition that today be Tuesday is that tomorrow will be Wednesday." Even another way to say the same thing is: "If today is Tuesday, then tomorrow will be Wednesday, and conversely."

This last form is tied up with the notion of the **converse** of an implication. The converse is obtained by interchanging the hypothesis and the conclusion.

For example, the converse of the implication "If $3 + 2 = 5$, then $(3 + 2) + 1 = 5 + 1$" is the implication "If $(3 + 2) + 1 = 5 + 1$, then $3 + 2 = 5$." The converse of "If an animal is a crow, then it is black" is "If an animal is black, then it is a crow." Note that the converse of an implication may be false while the implication itself is true.

To take another tack at some of these problems of language, what do you see in the box at the top of the next page?

† Here we use ". . ." to stand for the hypothesis and "***" to stand for the conclusion, whatever they may be.

‡ Note that this is true (as it should be) even if it is not Tuesday when you read this.

A little green man sitting on the page.

If you see a little green man sitting on the page, something unexpected has happened to your copy of the book. If you did not see a little green man sitting on the page, what did you see? You saw the words, "A little green man sitting on the page." Now this distinction between some words and the things the words refer to is in some sense quite obvious. Yet it is a distinction that in many cases is glossed over in such a way as to cause some confusion. It is customary to use quotation marks as we have done above to indicate that we are referring to the words themselves rather than to the things to which the words refer.

To emphasize the distinction again, consider the vast difference in the meanings of the two expressions: "I saw 'men' on the door" and "I saw men on the door."

The study of the connections between words and the things words refer to is called **semantics.** It is sometimes helpful in mathematics to be very careful in making semantic distinctions between words and the objects or ideas to which the words refer. As we shall see, the distinction between an object and its name comes up in arithmetic quite often.

Much childhood humor is based on semantic confusion. All of the following are semantic riddles.

1. Q. What part of London is in France?
   A. The letter *N*.
2. Q. What is the longest word in the English language?
   A. "Smiles" because there is a mile between the first and last letters.
3. Q. What is 1 and 1?
   A. 11.
4. Q. What is half of 8?
   A. 0, the top half. (How about 3, the right half?)

There is something to be learned in these even if, because of our sophistication, they fail to amuse us.

Semantic ideas are helpful in explaining another matter of language. Most mathematicians today use the symbol "=" between two other expressions to assert that those two expressions name the same thing. "$A = B$" (which is read "A equals B") means that, whatever "A" and "B" may be, they are names for the same thing. To make the assertion "$2 + 1 = 3$" is to claim that "$2 + 1$" and "3" are names for the same number.

This usage of "=" is usually described by saying that "=" is taken to mean logical identity.

If we wish to assert that A is not equal to B we may write "$A \neq B$."

To summarize these matters of language:

(a) We shall always use "or" in the inclusive sense so that a statement consisting of two assertions connected by "or" is false only in the case that both the assertions are false.

(b) An implication is false only in the case that its hypothesis is true while its conclusion is false. There are several equivalent ways to state an implication.

(c) There is a distinction between a word or phrase and the thing the word or phrase refers to. Quotation marks are used to indicate referral to the word or phrase itself.

(d) We shall use "=" to mean logical identity. That is, $A = B$ if and only if "A" and "B" are both names for the same thing.

All of these matters will come up repeatedly in what follows and you will no doubt find it helpful to refer back to this section quite often.

Problems

1. Which of the following are true?

(a) $2 + 4 = 6$ or $2 + 4 = 5$.
(b) $19 + 1 = 20$ or $19 + 1 = 20$.
(c) $5 + 2 = 7$ or $5 + 1 = 6$.
(d) $2 + 2 = 4$ or $4 + 4 = 2$.
(e) 9 is larger than 6, or $2 + 5 = 3$.
(f) $3 + 7 = 10$ or $7 + 3 = 10$.
(g) If $6 + 1 = 7$, then $7 + 1 = 8$.
(h) If 9 is larger than 6, then 6 is larger than 9.
(i) If 9 is less than 12, then 9 is less than 15.
(j) If 9 is less than 12, then 1 is less than 3.
(k) If $3 + 2 = 5$, then 3 is less than 5.
(l) If $3 + 2 = 6$, then 3 is less than 6.
(m) If $3 + 2 = 6$, then 3 is greater than 6.
(n) If $3 + 2 = 6$, then 6 is less than 3.
(o) If $5 + 7 = 12$, then $5 = 7 + 12$.

2. Restate the following in each of the forms (a), (b), (c), (d), and (e) given in the text on page 4.

(i) If a number is even, then its square is even.
(ii) If you turn the key, the motor will start.
(iii) If $a + 1 = b$, then $a + 2 = b + 1$.
(iv) If you don't eat your spinach, you will get no ice cream.
(v) If you're not confused, then you don't understand the situation.

3. Restate each of the following in two other ways.

(a) A triangle is equilateral if and only if it is equiangular.
(b) A number is even if and only if its square is even.

(c) A necessary and sufficient condition that two sets be equal is that each be a subset of the other.

4. Which of the following are true?

(a) A "rose" is a rose.
(b) "Five" has four letters.
(c) Five has four letters.
(d) A name, by any other Rose, would swell as meat.†
(e) "1" and "1" make "11."
(f) "1" and "1" make 11.
(g) 1 and 1 make 2.
(h) 1 and 1 make "2."
(i) If "A" is a name for 17 and "B" is a name for 17, then $A = B$.
(j) "2 + 1" = "3."

5. What does "All men are created equal" mean if the "equal" is taken in the strict mathematical sense?
6. "Analyze" each of the riddles on page 5. (Spoilsport!)
7. State the converse of each implication in Problem 2.
8. What is the converse of the converse of an implication?
9. Show that "If . . . , then ***" is true if and only if "*** or not . . ." is true.
10. The assertion "Not *** implies not . . ." is said to be the *contrapositive* of the assertion ". . . implies ***."

(a) Show that the contrapositive of an implication is true if and only if the implication is true. (In other words, an implication and its contrapositive are logically equivalent.)
(b) State the contrapositive of each implication in Problem 2.

1.2 Some matters of proof and understanding

What is a proof, anyway? What it takes to "prove" something to a person depends both on the nature of the item to be proved and also on the surrounding situation. If you want to prove to someone that a certain dog has a black tongue you might well do so by showing the person the dog and his tongue. If you want to prove to someone that human beings are mammals, you might well ask that someone to observe both the characteristic behavior of human beings and also a dictionary definition of "mammal."

Quite a few mathematical proofs are similar to this last example. That is,

† Don't take this problem too seriously. Relax. There will probably be more silly problems like this along the way.

they involve some straightforward observations and some reference to one or more definitions. An illustration would be the proof of the fact that 84 is an even number. We take it to be a matter of definition that an **even number** is a whole number that is divisible by 2. And to say that one whole number is **divisible** by another is to say that the former number, when divided by the latter number, yields a remainder of zero. With these definitions in mind, the proof that 84 is even consists in observing that 84 is a whole number and that 84 divided by 2 yields 42 with a remainder of zero.

Many useful facts in mathematics have proofs that are no more complicated than this. This suggests that if we wish to construct proofs, one of the first things we should learn to do is to look very carefully at the meanings of the terms involved in the statement we wish to prove. It may develop that the proof we seek is not much more than a direct reference to the definitions of the terms.

From a different point of view, all we have just finished saying is that if we wish to prove something it is only following common sense to first be very sure exactly what it is that we want to prove. We must ask exactly what the terms mean. Even more bluntly—we must simply know what we are talking about.†

As we consider various proofs in what follows, we shall see many examples that are simple applications of definitions. But such proofs are, in a certain sense, trivial and shallow. The proof of a *deep* result will involve much more than direct reference to definitions. Indeed, the significance and interest of the result is intimately involved with its apparent "distance" from such trivial proof. It is in the proof of highly significant results that ingenuity, depth of understanding, and mathematical talent come into play.

There are several different kinds of proofs used within mathematics itself. These cover a spectrum from very informal plausibility arguments to very formal proofs in which all the applications of logical rules are made quite explicit. Since we wish to move along fairly rapidly towards our study of number systems, we shall not take the time necessary to build up a formal logic in which to express our proofs. We shall, instead, largely rely on an informal, more or less intuitive, notion of what constitutes a correct logical argument. In other words, we shall use the somewhat imprecise logic of ordinary discourse.

We shall, however, take pains to state definitions clearly and carefully. For, when this is done, the door is at least open to construct rigorous, complete proofs.

There are many stages of understanding in mathematics. They probably cannot be strictly ranked from less to more understanding because they overlap and intertwine. But one approximation to these levels of understanding is the

† A protest that all this is obvious will be made by hardly any teacher who has tried to teach some form of mathematics.

following list. As one moves along, learning a piece of mathematics, one may develop the ability to:

1. Perform certain mechanical tasks which are usually considered mathematical. (An example of this level of understanding would be the very small child who answers the question, "What is two and two?" with "Four"—whether or not the child can even count to four. Another example would be the college freshman who answers the question, "What is the derivative of x^2?" with "$2x$"—whether or not he knows what a derivative is.†)
2. Grasp the meaning of basic concepts. (An example would be the concept of addition of whole numbers.)
3. Understand fundamental logical relationships between various concepts. (An example would be the basic relationship between addition and multiplication of whole numbers.)
4. Recognize real situations where the concepts and techniques are applicable, and make efficient use of them in the real situation. (An example would be using addition and multiplication to figure the cost of building a bookcase.)
5. Understand (and recognize the validity of) proofs of simple results. (An example would be the above proof of the evenness of 84.)
6. Construct simple proofs. (See item 5.)
7. Understand (and recognize the validity of) more subtle proofs that require long chains of logical reasoning.
8. Recreate difficult proofs of deep results once one has seen the main ideas behind the proof.
9. Conjecture new, interesting, and (perhaps) useful results and relationships.
10. Create new concepts for use in applications or for other reasons.
11. Prove or disprove nontrivial conjectures about old or new concepts.

A major reason for the study of certain portions of mathematics by many people is that they want to gain the ability listed as item 4. That is, they wish to get a tool that will be useful to them in everyday life or in professional activity. On the other hand, many people seem to study mathematics simply because they enjoy it. Part of its attraction lies in the fact that mathematical questions usually have quite definite and clear answers. People find satisfaction in finding these.‡

A mathematician would be especially interested in the abilities listed as 9, 10, and 11. But it also is true that many young students, be they potential mathematicians or not, have the abilities stated in 9 and 10. And these abilities seem

† You certainly don't need to know here what derivatives are.

‡ It should also be remarked that mathematics, being rather "culture-free," can be an important device for building self-assurance in disadvantaged children.

to be present even before the student has much ability in dealing with proofs at all. What is called the "discovery approach" to teaching mathematics attempts to take advantage of this fact.

It is the main intent of this book to assist you in getting an understanding in depth of various useful number systems. The types of understanding we shall strive for are largely of the kind 1, 2, 3, and 4 above. That is, it is not a primary purpose of this book to get you to be skilled at proving theorems. Yet we *shall* be presenting many proofs as we go along. There are several reasons for this.

First of all, we are not machines. And there would be something intellectually unsatisfying if we learned some mechanical procedures for performing mathematical tasks and we *never* asked: "Why does it work this way?" And when we try to answer such a question we are usually led to give some proofs.

Secondly, understanding proofs of general results turns out to be very *efficient*. If we develop some skill in carrying through arguments of a general type we find that we can often use a single method of attack on many, many different applied problems. The alternative, *inefficient* approach is to memorize many, many *different* tricks or gimmicks which work for the individual problems but have little generality.

Finally, if we wish to develop more than a superficial understanding of the meanings of certain concepts and their interrelationships we must look at how logic forces these relations on the concepts. That is, we must again look at some proofs.

PROBLEMS

1. Let us call a whole number a **perfect square** if it may be obtained by multiplying some positive whole number by itself. Thus, 1, 4, 9, and 16 are all perfect squares since $1 \times 1 = 1$, $2 \times 2 = 4$, $3 \times 3 = 9$, and $4 \times 4 = 16$. For each of the following assertions, decide whether it is true or false and explain carefully how you know. Your explanation should be written up in correct, complete sentences.

   (a) 36 is a perfect square.
   (b) 625 is a perfect square.
   (c) 17 is a perfect square.
   (d) The sum of any two perfect squares is a perfect square.
   (e) There is a perfect square which is 5 greater than another perfect square.
   (f) If a number is a perfect square then the product of that number and 4 is also a perfect square.
   (g) The product of any two perfect squares is a perfect square.
   (h) Some perfect squares are even and some are odd.

2. In the following table put a "1" below the name of the number if the number is a perfect square and put a "0" if not.

1	2	3	4	5	6	7	8	9	10
11	12	13	14	15	16	17	18	19	20
21	22	23	24	25	26	27	28	29	30
31	32	33	34	35	36	37	38	39	40
41	42	43	44	45	46	47	48	49	50

3. Let us call a number a **pair square**† if it is the sum of two perfect squares. Thus 25 is a pair square since $25 = 16 + 9$. Find some other pair squares.

4. In view of what was done in Problem 3, suggest a definition for the phrase "trio square." Give some examples of trio squares according to your definition.

5. Notice that $1 = 1$, $1 + 3 = 4$, $1 + 3 + 5 = 9$, $1 + 3 + 5 + 7 = 16$, and $1 + 3 + 5 + 7 + 9 = 25$. State a general conjecture suggested by these results.

6. Conjecture whether or not the following are true. (No explanations are required.)

   (a) Adding 2 to a perfect square always yields a number which is not a perfect square.
   (b) There is a largest perfect square.
   (c) Multiplying a perfect square by 2 always yields a number which is not a perfect square.
   (d) If n is a perfect square, there must be another perfect square between n and $n + 1{,}000{,}000$.

1.3 The grammar of mathematical symbols

Since the careful use of language is so important in mathematics, it is natural for us to be concerned with matters of punctuation, sentence structure, and what might be called parts of speech—just as an English teacher must be concerned with these same matters.

Sentences almost always begin with capital letters and end with a period—or

† This is not a standard term. It is introduced only to be used with the problems in this section.

perhaps a question mark or an exclamation point. For some reason it has never been common practice to have lower case or capital Arabic numerals. That is, there is no lower case "2," nor a capital "2," there is just a "2." This fact may contribute to the fact that so many students seem to forget rules for capitalization when writing a mathematics paper. When we write, for example, "$5 + 7 = 12$" there is no need to worry about the use of capitals and we may or may not include a period at the end. However, if we write "if $5^2 = 25$, then $5^2 + 1 = 26$" we have committed two errors. The first letter should be a capital, and there should be a period at the end of the expression.

Concern with proper rules of punctuation is not mere pedantry. These rules help to make sense out of a string of words, showing where a "thought" ends and where a new one begins. Another matter is important in this regard. People who use mathematics love short names for things. In studying probability we might say, for example, "Let X be the number of aces obtained in drawing 3 cards at random without replacement from a well-shuffled ordinary deck of cards." If we proceed to use "X" in our subsequent discussions we are letting this single symbol stand for quite a mouthful of words. The same kind of thing is reflected in the use of "π" to stand for the ratio of the circumference to the diameter for a circle. Indeed, a strong argument could be made for the assertion that one of the main reasons that mathematics is so useful is that it enables you to express very complicated ideas in very brief language.†

Since mathematics is such a condensed language compared with, say, English, the need for correct punctuation is even more pressing than one might expect. There is so much information crammed into a short space that sometimes one error in punctuation can change immensely the meaning of a short statement. For example, "$11 = 10 + 111 + 1 = 12$ and $10 + 2 = 12$" is weird, but "$11 = 10 + 1$, $11 + 1 = 12$, and $10 + 2 = 12$" is clear (and true).

Mathematical symbols that allow us to express ideas about numbers play a very definite grammatical role. The symbol "2" is the name of a number and, hence, plays the same role as a noun. "1532" is also a noun. The equality sign "$=$" plays the role of a verb. "$2 = 3$" is a properly constructed (and false) sentence. "$2 =$" is an improperly constructed sentence. It is neither true nor false because it is meaningless. Whenever the sign "$=$" is preceded and followed by names for objects, we have a properly constructed sentence. "$\frac{1}{2} = \frac{2}{4}$" is a properly constructed (and true) sentence.

There is a handy convention used with the "$=$" sign. Suppose we wish to assert that $\frac{1}{2} = \frac{2}{4}$ and $\frac{2}{4} = \frac{3}{6}$ and $\frac{3}{6} = \frac{4}{8}$. We may make all three of these claims more compactly by writing "$\frac{1}{2} = \frac{2}{4} = \frac{3}{6} = \frac{4}{8}$." That is, a **continued equality** such as the above is to be taken as the assertion that the first equality ($\frac{1}{2} = \frac{2}{4}$) *and* the second equality ($\frac{2}{4} = \frac{3}{6}$) *and* the third equality ($\frac{3}{6} = \frac{4}{8}$) all hold.

† On the other hand, it is easy to get carried away with the brevity of mathematical expressions and to think it wrong to supply the few words in English that let the reader know what is going on. It is not!

In dealing with equalities and other mathematical assertions it is useful to have a device that plays a role similar to that of pronouns in English. Consider the sentence "It is red." This is certainly grammatically correct. Yet it is uncertain as to whether or not the statement is true. This is because of the indefinite character of the pronoun "it." The "it" evidently refers to some object. But we cannot tell which object without more information about the circumstances in which "It is red." is asserted.

In the same vein, "$x = 2$" is a mathematical sentence that is grammatically correct. But whether or not it is true we cannot tell without further information. It says almost the same as "It is equal to 2." And until we know what the "x" (or the "it") refers to, we cannot decide the matter.

A symbol such as the "x" in "$x = 2$" is called a **variable.** A **variable** is a symbol for which we may substitute certain names. If we substitute "$1 + 1$" for the "x" in "$x = 2$" then we obtain the (true) statement "$1 + 1 = 2$." If we substitute "3" for "x" in "$x = 2$" then we obtain the (false) assertion "$3 = 2$."

To illustrate how variables may be used, we consider the following solution to Problem 1(f), Section 1.2, where we are asked to show that if n is a perfect square, then so is $4n$. A proof is as follows. Suppose n is a perfect square. Then there is a positive whole number, let us call it k, such that $n = k \times k$. We now consider $4n$. We have $4 = 2 \times 2$ and therefore $4n = (2 \times 2) \times (k \times k) = (2 \times k) \times (2 \times k)$. But it then follows that $4n$ is a perfect square and we have completed the proof. We shall discuss the details involved later. But you are no doubt willing to accept here the crucial assertion that $(2 \times 2) \times (k \times k) = (2 \times k) \times (2 \times k)$.

Consider the following expression:

E: If x is a perfect square then $x + 1$ is not a perfect square.

Strictly speaking, this is not a mathematical statement since the x that occurs in it is a variable. If we substitute "3" for "x" in the E we get: "If 3 is a perfect square, then $3 + 1$ is not a perfect square." This is true since 3 is not a perfect square. If we substitute "4" for "x" in E we get: "If 4 is a perfect square, then $4 + 1$ is not a perfect square." This is also true since 5 is not a perfect square. A little thought may lead us to the correct guess that a true statement always results when we substitute the name of any positive whole number for "x." Now it is customary in mathematics to interpret the expression E as asserting that this is so. In other words, we take E to be an abbreviation for:

F: For all possible values of x, if x is a perfect square, then $x + 1$ is not a perfect square.

Such a convention may be referred to (in language which may be impressive but is certainly unnecessary for our purposes) as the convention of implicit universal quantification.

The assertion:

G: If $x = y$, then $x + 1 = y + 1$.

is taken to mean:

H: For all possible values of x and y, if $x = y$, then $x + 1 = y + 1$.

There is some small chance for confusion with this convention. Suppose we are given the problem of finding a number x for which $x + 7 = 12$. We might find, in elementary algebra, the following "solution" submitted.

$$x + 7 = 12$$

$$x = 12 - 7$$

$$x = 5.$$

There is something satisfactory going on here, because it is true that $5 + 7 = 12$. How is it that we are using (or misusing) the above convention here? If we apply the convention to the last line alone we would seem to be asserting that for all possible values of x, $x = 5$. But this is clearly false. However, if we take the entire piece of work as a condensation of

If $x + 7 = 12$, then $x = 12 - 7$; and if $x = 12 - 7$, then $x = 5$.

Then we are on satisfactory ground. This is because if we do now apply our convention we have the following true assertion:

For all possible values of x, if $x + 7 = 12$, then $x = 12 - 7$; and if $x = 12 - 7$, then $x = 5$.

Let us remark on what is required to show the falsity of an implication that uses this convention. Suppose we consider the assertion

J: If x is a perfect square, then $x - 3$ is not a perfect square.

In order to show that J, interpreted according to our convention, is false, all that is required is that we find a particular value of x for which "x is a perfect square" is true and "$x - 3$ is not a perfect square" is false. Taking $x = 4$ will do the job since 4 is a perfect square and $4 - 3$ is also a perfect square.

In other words, to show that an implication to which our convention applies is false we need only display a particular case where the hypothesis is true while the conclusion is false.

Problems

1. In each of the following cases, tell whether or not we have a properly constructed sentence. If it is properly constructed, tell (in view of the conven-

tions we discussed above) whether it is true or false. If it is false, explain how you know it.

(a) $8 + 1 = 2 + 7 = 1 + 8$
(b) $= + 5 =$
(c) If A is an even number, then so is $A + 4$.
(d) If x is an odd number, then $x + 2$ is also odd.
(e) If $x = 2$, then $x + 1 = 3$.
(f) $2x =$
(g) If x is a perfect square.
(h) If x is a perfect square, then $x + 2$ is not a perfect square.
(i) If x is even, then x is not a perfect square.
(j) $x = x$
(k) $7 = 5 =$
(l) $37 + 8 = 16 + 19 = 8 + 37$

2. With our convention about the use of variables in assertions we note that from a given statement we may easily deduce a raft of other statements by simple substitution. For example, from

P: If $x = y$, then $x + 1 = y + 1$.

We may deduce "If $3 = 2 + 1$, then $3 + 1 = (2 + 1) + 1$" by substituting "3" for "x" and "$2 + 1$" for "y." We may also deduce from P the assertion "If $a = b$, then $a + 1 = b + 1$" by substituting "a" for "x" and "b" for "y." Write several assertions we may conclude from Q by simple substitution if we have

Q: *If* $x = y + 1$, *then* $z + x = z + (y + 1)$.

3. Disprove each of the following:

(a) If x is a number, then x is even.
(b) x is a perfect square implies x is greater than 2.
(c) If x is greater than 4, then x is less than 6.
(d) If x is a student, then x studies hard.
(e) Today is February 28 only if tomorrow is March 1.
(f) A sufficient condition that x be a perfect square is that x be odd.
(g) A necessary condition that x be a properly constructed English sentence is that x end with a period.
(h) A necessary and sufficient condition that x be a name for the number 3 is that x = "3."
(i) $x = y$ or $z = w$ only if $x = y$ and $z = w$.
(j) $x = y = z$ implies $x = y$ or $y = z$, and conversely.
(k) A sufficient condition that not both $x = y$ and $z = w$ is that $x = y$ or $z = w$.

1.4 Flow charts

When we are faced with explaining a new concept to someone, we had best put our explanation in terms that that person can understand. It does little good to explain a concept in terms of notions that are themselves unfamiliar. A pattern of education that attempts to introduce the student to comprehensive and useful concepts must build towards the understanding of these concepts by careful steps that first get the student familiar with simpler ideas. This would be true in banking, political science, geology, and mathematics, or in any field that has developed a technical vocabulary of its own.

There is a similar structure in correctly telling someone how to *do* something. If the person already knows how to do certain simple things then we may tell him how to do something more complex by telling him how to do several simple things in a certain order.

To take a simple example, suppose that a young child has learned how to print the letters C, A, and T and that the child understands that when someone says, "Move one space to the right," he is to get ready to do something in the space immediately to the right of where he was just working. We may then tell the child how to perform the complex act of printing the word "CAT." We do so by asking him to follow the instructions: 1. Print C. 2. Move one space to the right. 3. Print A. 4. Move one space to the right. 5. Print T. 6. Stop.† These instructions may be summarized in what is called a **flow chart** as in Figure 1.1.

The chart consists of several rectangular boxes connected by arrows. Inside each box is an instruction to carry out a simple procedure that we assume he has already learned to do. To carry out the entire procedure given by the flow chart, the child should start with the top box, carry out the instruction given there, and then follow the arrows to find the successive simple acts he is to perform. We could include a last box with the "Stop" in it, but we shall adopt the convention that when we reach a box out of which no arrow flows, we are to stop.

Flow charts such as the above can be very useful in displaying how a complex procedure may be broken down into a sequence of simpler ones.

Suppose we now consider the procedure for deciding whether or not "*A* or *B*" is true in view of our convention as to how we shall use "or." Here we are taking "*A*" and "*B*" as symbols that stand for certain assertions. Whether "*A* or *B*" is true or not will depend on the nature of *A* and *B*. Let us suppose that we can correctly answer the questions "Is '*A*' true?" and "Is '*B*' true?". Then a procedure for deciding whether or not "*A* or *B*" is true can be given in terms of the instructions:

† If we were being more careful here we would also instruct the child to position the initial C so that there is room on the page for the rest of the word.

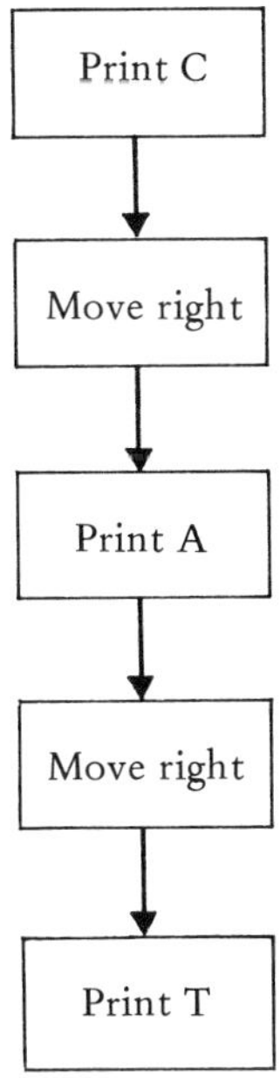

FIGURE 1.1

1. Answer the question "Is 'A' true?".
2. If the answer is "Yes," then conclude "A or B" is true.
3. If "No" then answer the question "Is 'B' true?".
4. If the answer to this second question is "Yes," then conclude that "A or B" is true.
5. If "No," then conclude "A or B" is false.

This procedure is forced on us by the convention that "A or B" is false when and only when "A" and "B" are both false. We may summarize the above procedure in a flow chart as in Figure 1.2.

Here, as before, we find simple instructions in rectangles. We also find, in diamond-shaped boxes, the questions that should be asked along the way. The path we follow out of a diamond-shaped box depends on the answer to the question for the particular case being considered. A position in a flow chart where a question must be asked to determine the proper path is called a **branch point.** Branch points are indicated by diamond-shaped boxes.

Another example will illustrate a different aspect of flow diagrams. Words and sentences and mathematical expressions are all strings of symbols. Given two strings of symbols it is important to be able to decide if they are the same strings or not. It is easy to see that "CAT" and "CAT" are the same strings of symbols. It is easy to see that "CAT" and "CAK" are not. It is perhaps a bit

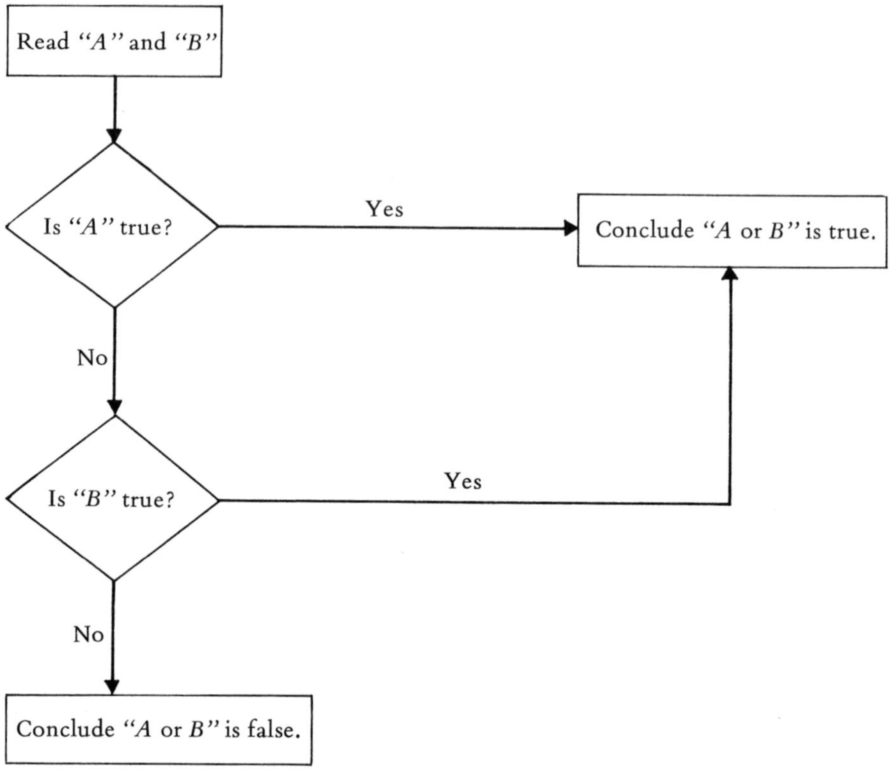

FIGURE 1.2

more difficult to see that "444444444454" and "4444444444454" are not the same strings. Suppose we want to explain carefully how to decide whether or not two strings are the same. If we assume we can tell whether or not two individual symbols are the same, the flow chart in Figure 1.3 describes how to tell whether or not two strings are the same.

The procedure the flow chart describes is as follows. First check and see if the two strings have the same number of symbols. If not, then the strings are different. If so, then compare the two first symbols. If they are different, conclude the strings are different. If they are the same, see if the two strings contain further symbols. If not then both strings have the same single symbol and, hence, they are the same strings. If there is a next symbol in each string, we determine if the two of them are the same. If not, the strings are different. If so, then we go back and ask again if we have now looked at all the symbols in the strings. If so, we conclude the strings are the same. If not, we look at the next symbol in the strings. And so forth. An indication in a flow chart that we are to go back and do certain tasks again is called a **loop.**

We may notice one more thing from this example. There are often many

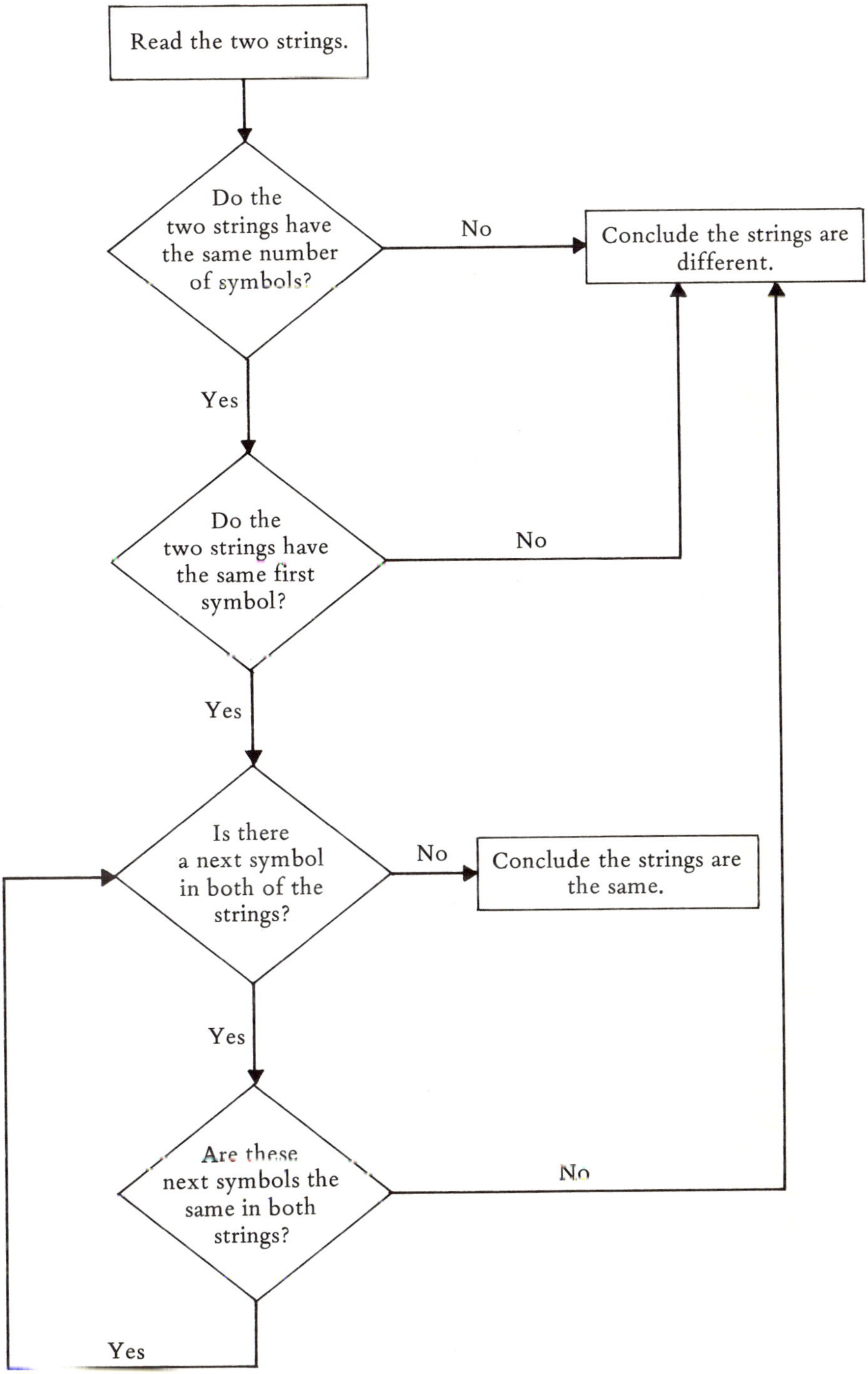
Read the two strings.
Do the two strings have the same number of symbols?
No
Conclude the strings are different.
Yes
Do the two strings have the same first symbol?
No
Yes
Is there a next symbol in both of the strings?
No
Conclude the strings are the same.
Yes
Are these next symbols the same in both strings?
No
Yes

FIGURE 1.3

different ways of reaching a desired goal. For instance, an alternative procedure for deciding if two strings of symbols are the same is given in Figure 1.4.

There are several different ways to look at flow charts. We may think of a given one as a summary of instructions to ourselves as to how to carry out (or attempt to carry out) a certain procedure. We shall often use flow charts in this way as we did above.

But we might also consider a flow chart as an indication of how we should design a machine to carry out a procedure for us. We may know how to build simple machines that will do certain elementary tasks for us and other simple machines that will answer certain elementary questions for us. Then a flow chart whose rectangular boxes contain instructions for carrying out these elementary tasks and whose diamonds contain these elementary questions can be thought of as a description of how to connect up our simpler machines to build a more complex machine. Or the flow chart can be thought of as a set of instructions telling us in what order we should use our simpler machines in order to carry out the desired task. We shall often use this approach in talking about a flow chart. In describing how a flow chart is dreamed up or explaining why one works we shall often speak of the flow chart as if it were the design for a machine which is meant to perform the task we are interested in.

A third way to think about a flow chart is an important one in this age of electronic computers. Imagine that we have a large electronic computer at our disposal. Such a device is capable of carrying out simple computational tasks and answering simple questions. And it does these things very, very quickly, makes no mistakes, and never gets tired. If we want the computer to carry out a complex task for us, we must tell it what simple tasks to carry out and what questions to ask and in what order to do these things. That is, we must *program* the machine. Flow charts are often used as a step in figuring out how to construct a program (i.e., a set of instructions) for an actual computer.

From a pedagogical point of view, flow charts are helpful in breaking down a complex procedure into simpler steps. It can then be determined what simpler skills should be developed first before we attempt to teach the complex one. And a flow chart may actually be helpful to the student (as it is to a machine) in that it makes it crystal clear in what order things should be done.

Problems

1. Make a flow chart for writing the sentence "I AM", assuming that (a) we are instructing a young child who knows how to print letters, (b) we are typing on a typewriter.

2. Make a flow chart similar to that in Figure 1.2 for deciding whether or not it is true that (a) A and B, (b) A implies B, (c) A if and only if B.

3. Using some of the questions in Figure 1.4, make a flow chart for deciding whether or not two strings of symbols have the same number of symbols in them.

4. Make a flow chart describing the following plan for studying a section of

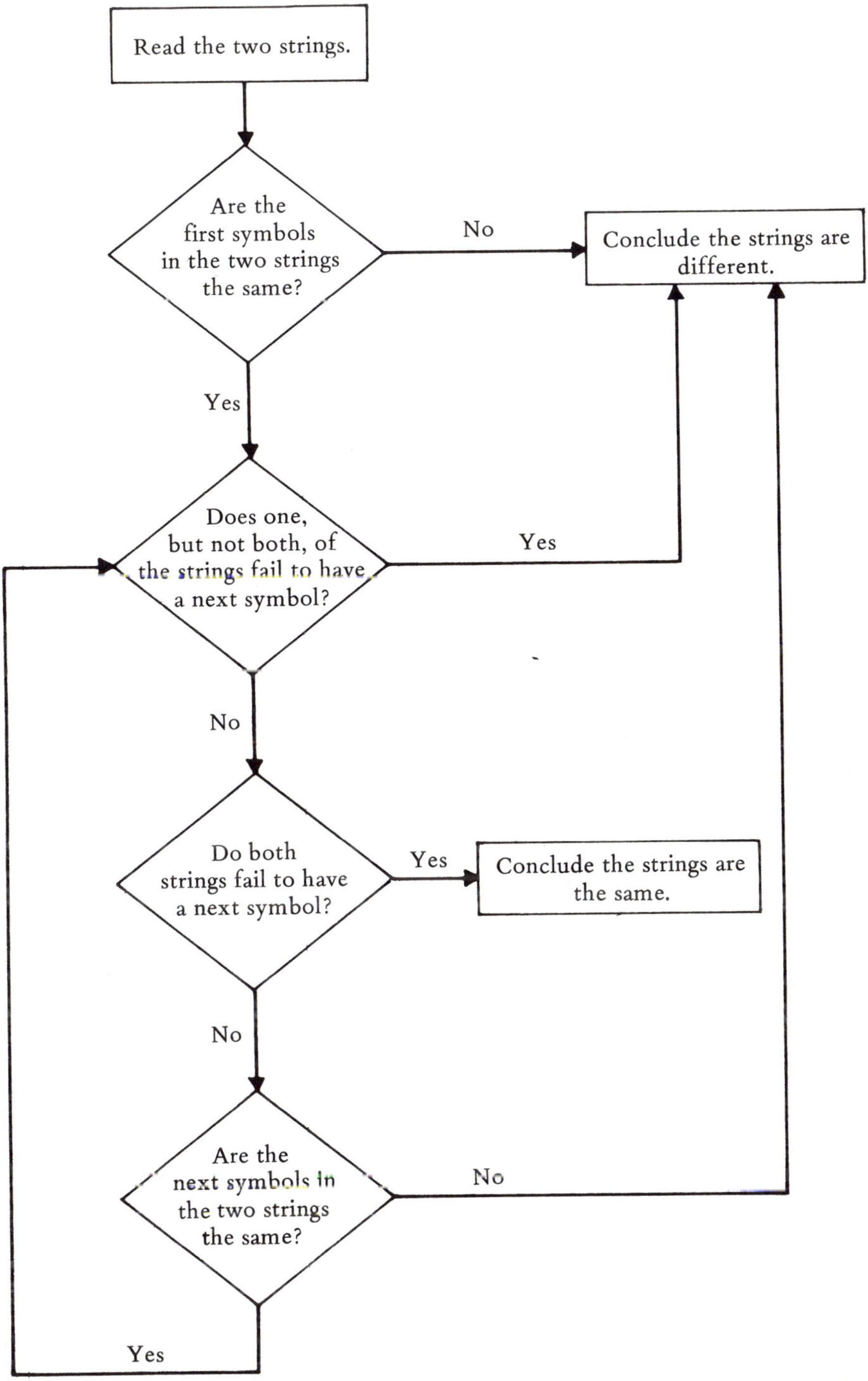
Read the two strings.
Are the first symbols in the two strings the same?
No
Conclude the strings are different.
Yes
Does one, but not both, of the strings fail to have a next symbol?
Yes
No
Do both strings fail to have a next symbol?
Yes
Conclude the strings are the same.
No
Are the next symbols in the two strings the same?
No
Yes

FIGURE 1.4

this book: Read the material in the section. Check the details of the examples in the section. Try some of the problems at the end of the section. If you have no trouble with the problems, stop working on the section. If you have trouble, reread the section, recheck the examples, and then return to the problems again. If you still have trouble with the problems, reread the section. Etc.

5. Suppose you have two large jugs, jug A and jug B, and you also have a small glass. Make a flow chart describing how you could decide if the two jugs will hold approximately the same amount of water. You have no scale and no other measuring devices, but you do have a tap from which you may draw as much water as you wish.

6. Use a flow chart to describe how to deal out the cards for a bridge game. Assume south, S, is the dealer and the other players are W, N, and E. In bridge, the 52 cards are dealt out so that each player gets 13.

7. Work out a flow chart for a better study plan than that suggested in Problem 4. In particular, a good study plan should someplace include the question "Am I too tired to get much out of further work on this topic now?".

2

Sets and set operations

2.1 Set membership

Some of the very earliest stages of human learning are involved with making classifications. The very young child learns that certain people belong to his family and others do not. He learns that certain objects are classified as red and others are not. He learns (if he is lucky) that certain actions are classified as polite in his society and others are not. In language that is becoming more and more common we could describe these situations by saying that the child learns to distinguish the **members of certain sets.** He learns to recognize the members of the set of people who make up his family. He learns to recognize members of the set of all red things. He learns which are some of the members of the set of polite actions.

A **set** is simply a collection or class of objects or ideas. "Family," "class," "collection," and "assemblage" are all words that mean the same as the word "set."

Classifying things in various ways (that is, putting the things in certain sets) is an important component of almost any logical, scientific, practical, or mathematical endeavor. Objects that belong to a set are said to be **members** or **elements** of that set.

Suppose we wish to consider the set of perfect squares that are smaller than 30. The members in this set are 1, 4, 9, 16, and 25. A handy way to state exactly the nature of this set is to give the set a name, say "A," and then write "$A = \{1,4,9,16,25\}$." We read this as "A is the set whose members are 1, 4, 9, 16, and 25." The symbols "{" and "}" are called braces. We may consider the left brace as an abbreviation for the phrase "the set whose members are." The right brace tells us when the list of (names of) members of the set stops.

If we let $B = \{1,2,3,4,5,6,7,8,9,10\}$, then B is the set whose elements are 1,2,3,4,5,6,7,8,9, and 10 and no others.

When we describe a set in this fashion we are using the **roster notation.**

Notice that 5 is a member of the set B given above. 12 is not a member of B. We introduce a short notation for such assertions as these last two. We write "$5 \in B$" to assert that 5 is a member of B. In fact, we read "$5 \in B$" as "5 is a member of B" or "5 is an element of B" or "5 is in B." Here "ϵ" is a small Greek epsilon. This Greek e can be taken to suggest the property of "elementhood."

If we wish to assert that 12 is not a member of B, we may do so quite compactly by writing "$12 \notin B$." Thus, we use a slant line through "ϵ" to give the negation.

In many cases it is laborious (or impossible) to list all the elements of a set. The device used to describe sets in such cases can be illustrated by noting that we could describe the sets A and B above as:

$$A = \{x \mid x \text{ is a perfect square smaller than } 30\}$$

$$B = \{x \mid x \text{ is a whole number between 1 and 10 inclusive}\}.$$

We read these as "A is the set of all those x such that x is a perfect square smaller than 30." and "B is the set of all those x such that x is a whole number between 1 and 10 inclusive."

In general, if ". . . x . . ." is a statement form involving x, then

$$C = \{x \mid \ldots x \ldots\}$$

is read as "C is the set of all those x such that . . . x . . . ." Such notation is often called the **set-builder notation.**

Let $D = \{x \mid x + 7 = 12\}$. What are the members of D? Evidently, the only x for which $x + 7 = 12$ is $x = 5$. Hence $D = \{5\}$.

What is the set $\{x \mid x \neq x\}$? Evidently this set has no members since $x = x$ for all x. It turns out to be very convenient in some instances to have a set with no members. We call such a set the **empty set** or **null set.** There are two names in common usage for the empty set: "ϕ" and "$\{\quad\}$." That is, we define both ϕ and $\{\quad\}$ by the equations

$$\phi = \{\quad\} = \{x \mid x \neq x\}.$$

Here the Greek letter "ϕ" is pronounced "fee." We shall usually use "ϕ" to name the empty set, but the name "$\{\quad\}$" is also helpful.

The symbol that is used right after the left brace in the set-builder notation plays a rather strange role. It is not a variable in the usual sense. For instance, if we substitute "3" for "x" in "$\{x \mid x \neq x\}$" we obtain "$\{3 \mid 3 \neq 3\}$" and this last is very strange. We are *not* at liberty to substitute the name of an object for the "x" in the set-builder notation. For this reason such an x is usually called a **dummy variable,** or, perhaps better, a **dummy.** Note, however, that we may change a dummy variable to any other symbol that has not been used to name an object. Certainly $\{x|x + 7 = 12\}$, $\{t|t + 7 = 12\}$, and $\{w|w + 7 = 12\}$ are all the same set.

If A is a set, then "A" plays the part of a noun. "ϵ" behaves like a verb. If we hope to construct a sentence that makes a true assertion we must follow "ϵ" with a name of a set, which will then be the object of the verb. The expression "7 ϵ" is neither true nor false; it is simply meaningless. If "A" is not the name of a set in "$2 \epsilon A$" then we should consider "$2 \epsilon A$" to be false. It is common practice to allow the deletion of the period at the end of the sentence "$2 \epsilon A$.".

What about the assertion "$2 \epsilon 2$"? This is false. To see the precise reason for its being false we shall need to come to grips with the question of the meaning of "2." We shall do this in Section 6.1. But right now we should be willing to admit that whatever 2 is, it does not contain 2 as a member.

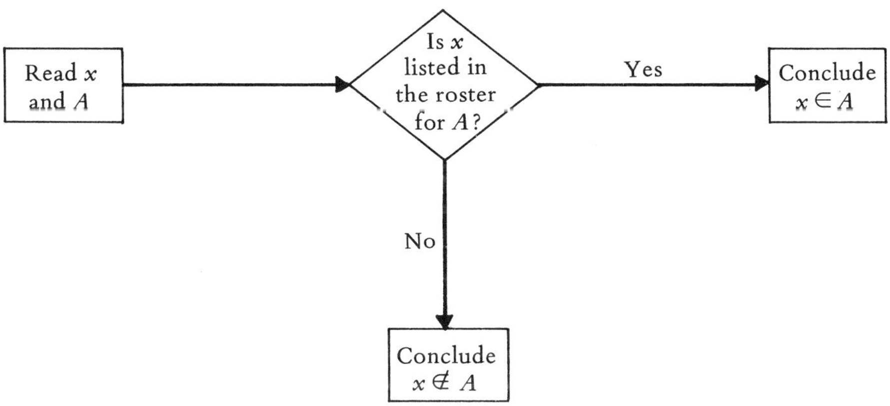

FIGURE 2.1 Determination of set membership if A is described in roster notation.

On the other hand "$2 \epsilon \{2\}$" is a correctly constructed sentence and is true. Evidently we must distinguish between a number and the set which has that number as its single element.

We may restate the correct use of set description, "ϵ," and "$\notin$" by flow charts as in Figure 2.1 and Figure 2.2.

PROBLEMS

1. True or false?

(a) $4 \epsilon \{1,2,3,4,5,6\}$
(b) $4 \notin \{1,2,3,4,5,6\}$
(c) $9 \epsilon \{1,2,3,4,5,6\}$
(d) $9 \notin \{1,2,3,4,5,6\}$
(e) $5 \epsilon \{x|2x = 10\}$
(f) $6 \epsilon \{x|2x = 10\}$
(g) $13 \epsilon \{x|x \text{ is a perfect square}\}$
(h) $25 \epsilon \{x|x \text{ is a perfect square}\}$

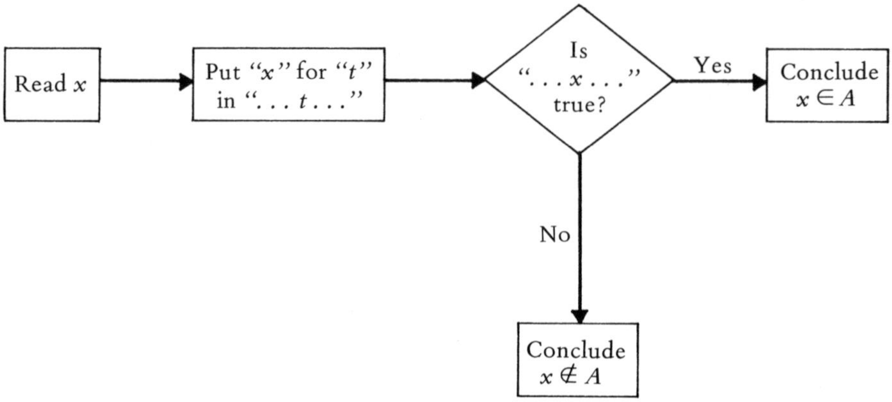

FIGURE 2.2 Determination of set membership if A is described in set-builder notation, $A = \{t | \ldots t \ldots\}$.

(i) Alaska $\in \{x | x \text{ is a state in the U.S.A.}\}$
(j) $10 \notin \{x | 5x = 10\}$ (k) $5 \in \{x | 5x = 10\}$

2. Let T be the collection of all postal zip codes in the U.S.A. Is $5505 \in T$?

3. Make up a set S whose members are seven different numbers and make several true assertions about S using "$\in$" and "$\notin$."

4. Sets whose elements are of a particular kind are often given special names. Thus, we have a **pride** of lions and a **flock** of geese. Find several other such examples.

5. In a Thesaurus look up some of the many synonyms for "set" or "assemblage." (Strangely enough, in view of common usage, the latter word yields more.)

6. We have said that "set," "class," and "collection" all mean the same thing. In everyday discourse, however, which word is used seems to depend somewhat on the kind of members involved. For example, we usually speak of a "**set** of silverware" rather than a "**class** of silverware." Fill in each blank below with "set" or a synonym for "set" that seems most natural to you.

(a) ______ of tires (b) ______ of students
(c) ______ of coins (d) ______ of dinnerware
(e) ______ of airplane models (f) ______ of jewelry
(g) ______ of books (h) ______ of tools
(i) ______ of furniture (j) ______ of automobiles
(k) ______ of drawers (l) ______ of papers
(m) ______ of cities (n) ______ of pencils
(o) ______ of people (p) ______ of schools
(q) ______ of examinations

7. Which of the following are correctly constructed sentences? Of those that are, which are true?

(a) $2 \in$	(b) $\in \{5,6\}$	(c) $5 \in \{7,8\}$
(d) $8 \in \{7,8\}$	(e) $8 \subset 8$	(f) $8 \in \{8\}$
(g) $7 \in \{\ \ \}$	(h) $\phi \in \{\ \ \}$	(i) $\phi \in \{\phi\}$

2.2 Relations between sets

Every person who is a resident of New York State is also a resident of the United States of America. Every black bird is a bird. Every multiple of 10 is also a multiple of 5. All these statements can be put in set terminology as follows. Every member of $\{x|x$ is a resident of New York State$\}$ is also a member of $\{x|x$ is a resident of the U.S.A.$\}$. Every member of $\{x|x$ is a black bird$\}$ is also a member of $\{x|x$ is a bird$\}$. Every member of $\{x|x$ is a multiple of $10\}$ is also a member of $\{x|x$ is a multiple of $5\}$.

If every member of a set A is also a member of a set B, then we say A is a subset of B and may write "$A \subseteq B$." We read "$A \subseteq B$" as "A is a **subset** of B" or "A is **contained** in B." The above assertions could have been written more concisely using this notation. For example, "$\{x|x$ is a multiple of $10\} \subseteq \{x|x$ is a multiple of $5\}$" is a way of putting the third assertion made at the beginning of this section.

The following are true: $\{1,2\} \subseteq \{1,2,3\}$, $\{5\} \subseteq \{1,5\}$, $\{3,4\} \subseteq \{1,2,3,4\}$, $\{1,2,3\} \subseteq \{1,2,3\}$, and $\{5\} \subseteq \{5\}$. If A is any set, then certainly every member of A is a member of A, so that $A \subseteq A$. It is customary and helpful to interpret our definition of "$\subseteq$" in such a way that $\phi \subseteq A$ no matter what the set A is. That is, the empty set is a subset of any set.

Suppose we know that, for certain sets A and B, we have both $A \subseteq B$ and $B \subseteq A$. This means that every member of A is a member of B and, *vice versa*, every member of B is a member of A. Then it follows that A and B have exactly the same members. In this case we say that A and B are really the same sets and assert that $A = B$. Of course, if we know that A and B have exactly the same members, then $A \subseteq B$ and $B \subseteq A$. Thus, we have a basic result about equality of sets:

$$A = B \text{ if and only if } A \subseteq B \text{ and } B \subseteq A.$$

Some examples of equalities between sets are:

$$\{x|x + 7 = 12\} = \{5\}$$

$$\{\ \ \} = \phi$$

$$\{2,3,5,7\} = \{7,3,5,2\}$$

$$\{1,2,3\} = \{3,2,1\}$$

$$\{1,2,3,3\} = \{1,2,3\}$$

$$\{5,5,6,6\} = \{5,6\}.$$

Notice that, as the third and fourth of these examples point out, it makes no difference in what order we put the names when we use the roster notation for a set. The only thing that matters is which objects are elements of the set. In the fifth and sixth examples above, this fact is again emphasized. We probably should consider "$\{1,2,3,3\}$" and "$\{5,5,6,6\}$" as misuses of the roster notation. Unless there is some comment made to the contrary we then can assume that when the roster notation is used, no member of the set is listed more than once.

Another relation between sets which is sometimes useful is **proper inclusion.** We say A is **properly included** in B (or is a **proper subset** of B) if every element of A is an element of B and yet $A \neq B$. We write "$A \subset B$" to assert that A is a proper subset of B. Thus, to claim that $A \subset B$ is to assert that $A \subseteq B$ and $A \neq B$. For example, "$\{1,2\} \subset \{1,2,3\}$" is true while "$\{1,2,3\} \subset \{1,2,3\}$" is false.

Pictures are often helpful in discussing sets. Suppose we draw a large circle and label the inside of the circle A. We may then think of the circle as being a fence inside of which we find all the elements of A. Then if we wish to represent geometrically the assertion that $B \subseteq A$ we may use a picture such as Figure 2.3. Here any element of B must be inside the smaller circle and, hence, also inside the larger. Of course, it might be the case that all the elements in A are also in B. If this is not the case so that $B \subset A$, then we must have at least one element of A which is not in B. A standard manner for showing this pictorially is given in Figure 2.4. The x which is shown in the picture is meant to show that the region in which it lies contains at least one member.

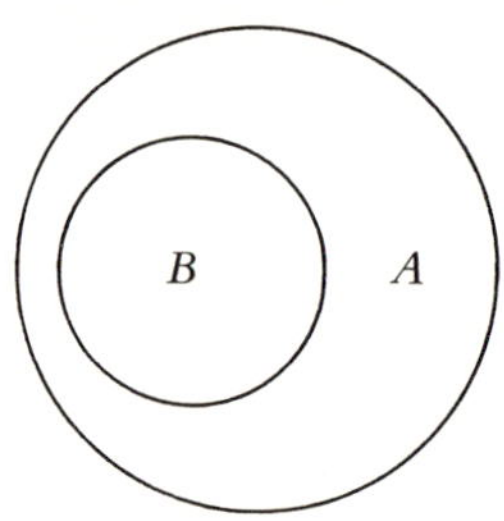

FIGURE 2.3 Geometric representation for $B \subseteq A$.

The symbols "$\subseteq$" and "$\subset$" play the role of verbs. Names of sets must occur on both sides of them if we are to have sentences that make true assertions. If "A" and "B" are not both names of sets in "$A \subseteq B$" then we should consider "$A \subseteq B$" to be false.

Inclusion, equality, and proper inclusion are relations that have some properties in common and some striking differences. An instance of their similarity is that they are all **transitive.** By this we mean that

If $A \subseteq B$ and $B \subseteq C$, then $A \subseteq C$.

If $A = B$ and $B = C$, then $A = C$.

If $A \subset B$ and $B \subset C$, then $A \subset C$.

These are all fairly obvious to the intuition. We shall return to a further discussion of transitivity in Section 3.1.

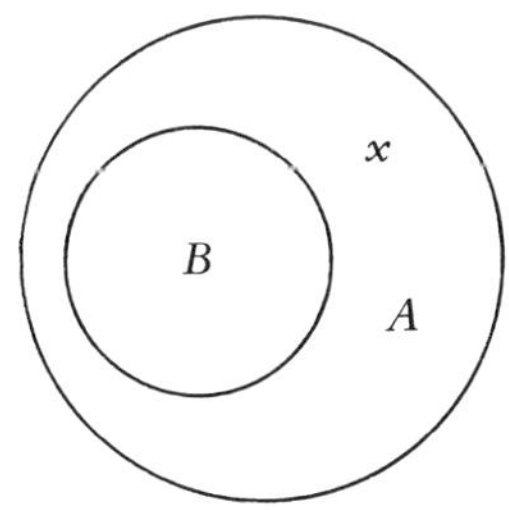

FIGURE 2.4 Geometric representation for $B \subset A$.

We have now covered the basic material you need to proceed to further topics and you should feel free to go on to the problems and then to the next section. However, we shall now present, as a kind of appendix or addendum to this section, a "more mathematical" treatment of what we have covered. This will be presented compactly and logically with no examples and with more detail than earlier in the section. Proofs will be given for all those theorems which do not follow "immediately" from the definitions or were not proved earlier in the section. You may either skip the rest of this section, or you may read the definitions and theorems (omitting proofs) as a review of this section, or you may read the entire section in detail to begin to get a feel for the kind of approach given.

You will find this same structure in many sections to follow. When you come to the separating line in any section you should always remember you have all of the above options. The problems at the ends of sections are often separated by a similar line. Those below the line are more theoretical and may make use of details which come below the line in the section.

In what follows A, B, and C are sets.

Definition 2.2.1. $A \subseteq B$ if and only if $x \in A$ implies $x \in B$ for all x. We read "$A \subset B$" as "A is a **subset** of B" or "A **is contained** in B."

Theorem 2.2.1. $A \subseteq A$

Theorem 2.2.2. $\phi \subseteq A$

Definition 2.2.2. $A = B$ if and only if A and B have exactly the same members.

Theorem 2.2.3. $A = A$

Theorem 2.2.4. $A = B$ implies $B = A$.

Theorem 2.2.5. $A = B$ if and only if $A \subseteq B$ and $B \subseteq A$.

Definition 2.2.3. $A \subset B$ if and only if $A \subseteq B$ and $A \neq B$. We read "$A \subset B$" as "A is a **proper subset** of B" or "A is **properly contained** in B."

Theorem 2.2.6. $A \subset B$ implies $A \subseteq B$.

Theorem 2.2.7. $A \subseteq B$ and $B \subseteq C$ implies $A \subseteq C$.

PROOF: Suppose $x \in A$. Since $A \subseteq B$ by hypothesis, we have $x \in B$ by Definition 2.2.1. Since $B \subseteq C$, this implies $x \in C$ by Definition 2.2.1.† Since x was any element in A we conclude from Definition 2.2.1 that $A \subseteq C$.

Theorem 2.2.8. $A = B$ and $B = C$ implies $A = C$.

PROOF: You may either consider this to be obvious or do the following. Since $A = B$ we have $A \subseteq B$ by Theorem 2.2.5. Since $B = C$ we have $B \subseteq C$ by Theorem 2.2.5. Then $A \subseteq C$ by Theorem 2.2.7. But from $A = B$ and $B = C$ we also get $B \subseteq A$ and $C \subseteq B$ by Theorem 2.2.5. Therefore by Theorem 2.2.7, $C \subseteq A$. We now have $A \subseteq C$ and $C \subseteq A$ and so by Theorem 2.2.5, we have $A = C$.

Theorem 2.2.9. $A \subset B$ and $B \subset C$ implies $A \subset C$.

PROOF: By Theorem 2.2.6, $A \subset B$ and $B \subset C$ imply $A \subseteq B$ and $B \subseteq C$. Then by Theorem 2.2.7, $A \subseteq C$. Suppose $A = C$. Then by Theorem 2.2.5, $C \subseteq A$. $B \subseteq C$ and $C \subseteq A$ give $B \subseteq A$ by Theorem 2.2.7. But $A \subseteq B$ and $B \subseteq A$ give $A = B$ by Theorem 2.2.5. Since $A \subset B$ it is impossible that $A = B$. Therefore, our supposition that $A = C$ must be wrong and we conclude $A \neq C$. Since $A \subseteq C$ and $A \neq C$, we have $A \subset C$ by Definition 2.2.3.

PROBLEMS

1. Which of the following are true?

(a) $\{1,2,3\} \subseteq \{2,3,4\}$
(b) $\{2,3,4\} \subseteq \{1,2,3\}$
(c) $\{5\} \subseteq \{5,10,15,20\}$
(d) $\{5\} \subset \{5,10,15,20\}$
(e) $\{5\} \subseteq \{x \mid 2x + 5 = 15\}$
(f) $\{5\} \subset \{x \mid x \text{ is an odd number}\}$
(g) $\{2,3,4,5\} \subseteq \{5,3,4,2\}$
(h) $\{2,3,4,5\} \subset \{5,3,4,2\}$
(i) $\{x \mid x \text{ is an odd perfect square}\} \subseteq \{x \mid x \text{ is a perfect square}\}$
(j) $\{x \mid x \text{ is an odd perfect square}\} \subset \{x \mid x \text{ is a perfect square}\}$
(k) $\{x \mid x \text{ is a black cat}\} \subset \{t \mid t \text{ is a cat}\}$
(l) $\{x \mid x \text{ is a cat}\} \subseteq \{w \mid w \text{ is a black cat}\}$

† Note that with the convention we discussed in Section 1.3, Definition 2.2.1 is taken to apply to *any* sets A and B. Thus, we may substitute "B" for "A" and "C" for "B" in Definition 2.2.1 and get "$B \subseteq C$ if and only if $x \in B$ implies $x \in C$ for all x." It is this form of the definition we are applying here.

(m) $\{x \mid x \text{ is an animal}\} \subseteq \{x \mid x \text{ is a cat}\}$
(n) $\{z \mid z \text{ is a person over 21 years of age}\} \subseteq \{z \mid z \text{ is a married person}\}$
(o) $\{w \mid w \text{ is a first grade student}\} \subset \{s \mid s \text{ is a student}\}$
(p) $\phi \subseteq \phi$
(q) $\phi \subset \phi$
(r) $\{\phi\} = \phi$

2. A complete list of all the different subsets of $\{1,2,3\}$ is: ϕ, $\{1\}$, $\{2\}$, $\{3\}$, $\{2,3\}$, $\{1,3\}$, $\{1,2\}$, $\{1,2,3\}$. Make a complete list of all the subsets of (a) ϕ, (b) $\{1\}$, (c) $\{1,2\}$, (d) $\{1,2,3,4\}$.

3. On the basis of your work in Problem 2, conjecture how many different subsets there are for (a) $\{1,2,3,4,5\}$, (b) $\{1,2,3,4,5,6\}$.

4. A **singleton** is a set with exactly one element. $\{2\}$ and $\{574\}$ are singletons; $\{1,4\}$ is not. How many different singletons are there which are subsets of $\{1,2,3,4,5,6,7,8,9,10\}$?

5. An **(unordered) pair** is a set that contains exactly two numbers. $\{5,7\}$ and $\{a,b\}$ are pairs. $\{5\}$ and $\{a,b,c\}$ are not. How many different pairs are there which are subsets of (a) ϕ, (b) $\{1\}$, (c) $\{1,2\}$, (d) $\{1,2,3\}$, (e) $\{1,2,3,4\}$? (See Problem 2.) Conjecture how many pairs would occur as subsets of (f) $\{1,2,3,4,5\}$, (g) $\{1,2,3,4,5,6\}$.

6. Let $E = \{x \mid x \text{ is a square circle}\}$. Let $T = \{x \mid x \text{ is a triangle}\}$. Surely E is empty. Since the empty set is a subset of any set, we conclude $E \subseteq T$. This may be put as "Every square circle is a triangle." The convention that $\phi \subseteq A$ for all sets A forces us to admit that this is indeed true. Introducing the proper sets, argue for the truth or falsity of (a) Every whole number between 1 and 2 is a perfect square. (b) Every four-sided triangle is red. (c) Every three-sided rectangle is green.

7. What can we conclude if in the course of an argument we prove that both $A \subset B$ and $B \subset A$?

8. In each case below, state whether the assertion is true or false and explain how you know.

(a) If $A \subseteq B$ and $B \subset C$, then $A \subset C$.
(b) If $A \subseteq B$ and $B \subseteq C$ and $C \subseteq A$, then $B = C$.
(c) There are no proper subsets of ϕ.
(d) If A has no proper subsets, it must be empty.
(e) For any two sets A and B, either $A \subseteq B$ or $B \subseteq A$.

9. May we conclude for certain that either $A \subseteq C$ or $C \subseteq A$ if we know that $A \subseteq B$ and $C \subseteq B$?

---

10. The proofs given for Theorems 2.2.7, 2.2.8, and 2.2.9 may be written in a more formal and detailed step-by-step fashion. For example, we might reconsider Theorem 2.2.9 as follows:

Theorem 2.2.9. $A \subset B$ and $B \subset C$ implies $A \subset C$.

	Step	Reason
PROOF:	1. $A \subset B$	Hypothesis
	2. $A \subseteq B$	Step 1 and Theorem 2.2.6†
	3. $B \subset C$	Hypothesis
	4. $B \subseteq C$	Step 2 and Theorem 2.2.6 with "B" for "A" and "C" for "B"‡
	5. $A \subseteq C$	Steps 2 and 4 and Theorem 2.2.7
	*6. $A = C$	Assumption§
	*7. $C \subseteq A$	Step 6 and Theorem 2.2.5 with "A" for "A" and "C" for "B"
	*8. $B \subseteq A$	Steps 4 and 7 and Theorem 2.2.7 with "B" for "A," "C" for "B," and "A" for "C"
	*9. $A = B$	Steps 2 and 8 and Theorem 2.2.5
	10. $A \neq B$	Step 1 and Definition 2.2.3
	11. $A \neq C$	Since steps 9 and 10 contradict each other, our assumption in Step 6 must be incorrect.
	12. $A \subset C$	Steps 5 and 11 and Definition 2.2.3

Write up a proof in this fashion for

(a) Theorem 2.2.7, (b) Theorem 2.2.8, (c) If $A \subseteq B$ and $B \subset C$, then $A \subset C$.

11. If A is a set that has exactly n elements, let $\binom{n}{k}$ be the number of different subsets of A that contain exactly k elements. (a) By actually listing subsets, complete the following table which gives some values of $\binom{n}{k}$.

n \ k	0	1	2	3	4	5
0	1	0	0	0	0	0
1	1	1	0	0	0	0
2	1	2	1	0	0	0
3	1	3	3	1	0	0
4	1	4	6			
5						

† Notice that in proving Theorem 2.2.9 we assume that we have already proved earlier theorems in the list so they may be used.

‡ See the footnote on page 30.

§ If you are interested in logical details here, note that we star the assumed step and all subsequent steps based on it. We do not use the starred steps except to get our contradiction.

(b) Conjecture at least part of what the next row in the table would be if we expand it to include the case $n = 6$.

2.3 Intersections and unions

Among the students in a typical class some will have had mumps, some will have had chicken pox, and some will have had both. The three sets $M = \{x \mid x$ has had mumps$\}$, $C = \{x \mid x$ has had chicken pox$\}$, and $B = \{x \mid x$ has had both mumps and chicken pox$\}$ have a rather obvious connection. The students in B are exactly those that are in both M and C. We call B the **intersection**† of M and C and write "$B = M \cap C$."

More generally, if E and F are sets, their **intersection**, which we denote by "$E \cap F$," is the collection of all those objects which are in both E and F. We read "$E \cap F$" as "E intersection F" or "E intersect F" or "E cap F." We may define this operation on sets more formally by:

$$E \cap F = \{x \mid x \in E \text{ and } x \in F\}.$$

For example, $\{1,2,3,4,5,6\} \cap \{2,3,5,7,11\} = \{2,3,5\}$, $\{2,3,5,7\} \cap \{7\} = \{7\}$, and $\{2,3,5,7\} \cap \{4,6,8,10\} = \phi$.

A flow chart for set membership in $E \cap F$ is given in Figure 2.5.

A geometric representation may be given for an intersection. In Figure 2.6, the left circle is taken to be a fence inside of which all the members of E are to be found. The right circle is to be interpreted similarly for F. The shaded region represents $E \cap F$ since objects in both E and F (if any) must be here.

Two sets which have no elements in common will have an empty intersection. Such sets are said to be **disjoint** or **nonoverlapping.** $\{1,3,5,7\}$ and $\{4,6,8,10\}$ are disjoint since $\{1,3,5,7\} \cap \{4,6,8,10\} = \phi$. To say that A and B are disjoint is to say that $A \cap B = \phi$.

Many of the basic properties of intersections are quite easy to see. For example, we certainly must have

$$A \cap B = B \cap A$$

because those objects which are in both A and B are exactly those objects which are in both B and A. Since the operation of intersection has this property, it is said to be **commutative.** We speak of "$A \cap B = B \cap A$" as the **commutative law for intersections.** To commute is to go back and forth, and this is a kind of back and forth law.

The operation of intersection is also **associative.** What this means is that the following holds:

$$(A \cap B) \cap C = A \cap (B \cap C).$$

† Some books also call B the **logical product** of M and C.

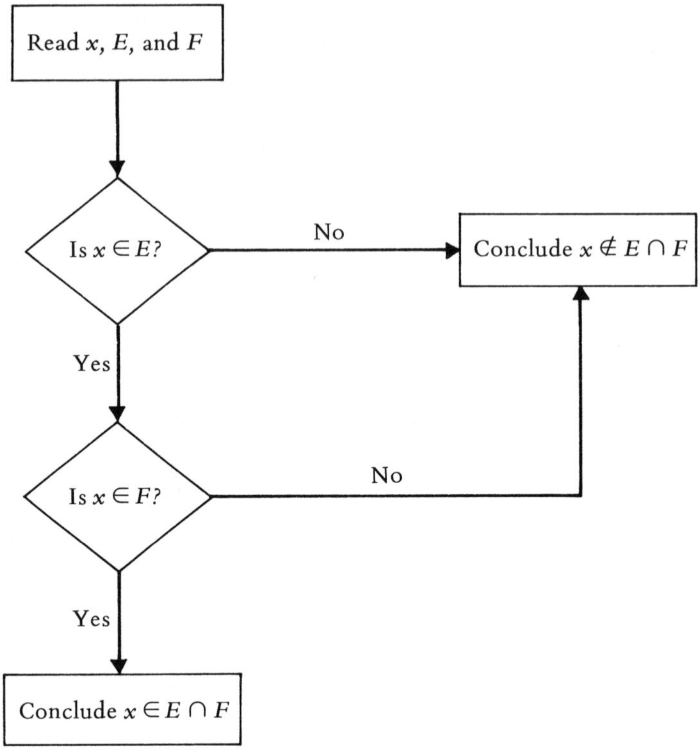

FIGURE 2.5 Determination of set memberships for an intersection.

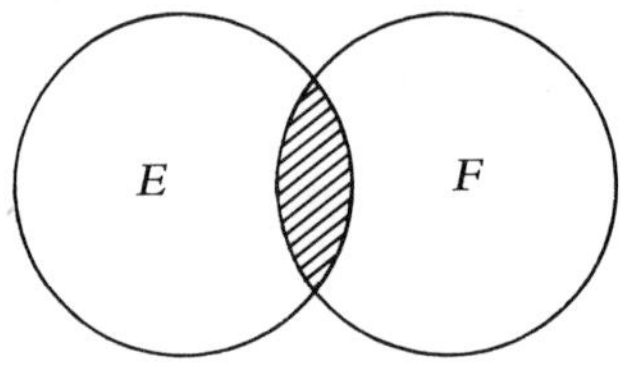

FIGURE 2.6 Geometric representation of an intersection.

This **associative law** asserts that the objects that are in the intersection of A and B and also in C are exactly those objects which are in A and also in the intersection of B and C. Clearly, $(A \cap B) \cap C$ or $A \cap (B \cap C)$ is simply the collection of all those objects that are in all of the sets A, B, C. The associative law says that we may associate (or group) the sets involved in two different ways and yet still get the same collection.

Since intersection is associative we may save ourselves some trouble and write "$A \cap B \cap C$" to mean either "$(A \cap B) \cap C$" or "$A \cap (B \cap C)$."

The associative law is illustrated geometrically in Figure 2.7.

Notice that "$\cap$" makes sense only when names of sets appear on both sides of

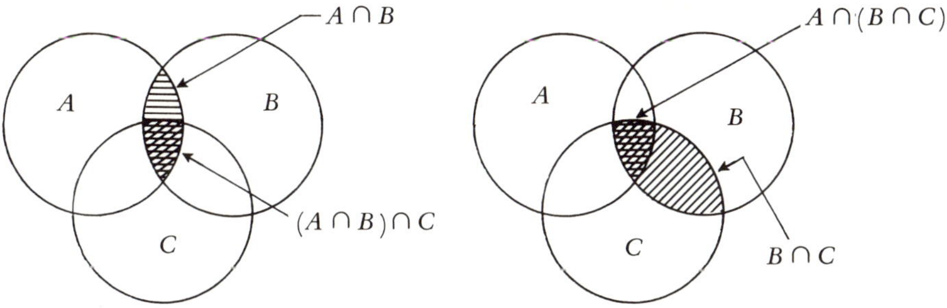

FIGURE 2.7 The associative law for intersections.

it. And when names for sets appear on both sides of it, the result is not an assertion but is the name of a set. The operation of intersection is a device for building new sets out of old ones.

The students in a class who have had mumps or chicken pox or both make up a set. We would call this set the **union** of the set of students who have had mumps and the set of students who have had chicken pox. More generally, we have the definition

$$A \cup B = \{x \mid x \in A \text{ or } x \in B\}.$$

We read "$A \cup B$" as "the union of A and B" or "A union B" or "A cup B."†

A geometric representation of the union of A and B is shaded in Figure 2.8.

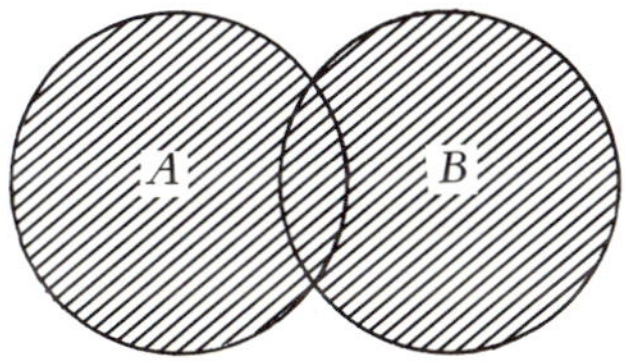

FIGURE 2.8 Geometric representation of a union.

Taking the union of two sets amounts to throwing all the elements of the two sets together to make one set.

Some examples of unions are: $\{1,2,3\} \cup \{2,3,4,5\} = \{1,2,3,4,5\}$, $\{1,2,3\} \cup \{1,2,3\} = \{1,2,3\}$, $\{1,2,3,4\} \cup \phi = \{1,2,3,4\}$, $\phi \cup \phi = \phi$, $\{4\} \cup \{1,2,3,5\} = \{1,2,3,4,5\}$.

From the way we use the word "or" it follows that the operation of taking a union is both **commutative** and **associative.** In other words, the following **commutative law** and **associative law** hold:

$$A \cup B = B \cup A$$

$$(A \cup B) \cup C = A \cup (B \cup C).$$

† The union of two sets is sometimes also called the **logical sum** of the two sets.

The second of these laws is illustrated in Figure 2.9.

Because of the associative law for unions we may write "$A \cup B \cup C$" to mean either "$(A \cup B) \cup C$" or "$A \cup (B \cup C)$."

We notice from Figures 2.6 and 2.8 that we always have

$$A \cap B \subseteq A \cup B.$$

$A \cup B$ is slant hatched
$(A \cup B) \cup C$ is horizontally hatched

$B \cup C$ is slant hatched
$A \cup (B \cup C)$ is horizontally hatched

FIGURE 2.9 The associative law for unions.

Hence, roughly speaking, $A \cup B$ holds more than $A \cap B$. This leads to an easy memory device for keeping the symbols "$\cup$" and "$\cap$" straight. A cup ("$\cup$") will hold more water than a cap ("$\cap$").

Suppose we consider taking both unions and intersections at once and look at $A \cap (B \cup C)$. In Figure 2.10 we have shown $B \cup C$ with horizontal shading and $A \cap (B \cup C)$ with slanted shading.

In Figure 2.11 we show the construction of $(A \cap B) \cup (A \cap C)$. There $A \cap B$ has horizontal shading, $A \cap C$ has vertical shading, and $(A \cap B) \cup (A \cap C)$ has slant shading. Evidently

$$A \cap (B \cup C) = (A \cap B) \cup (A \cap C).$$

We call this the **distributive law** for intersections over unions. Because this law holds we say that intersections **distribute across** unions.

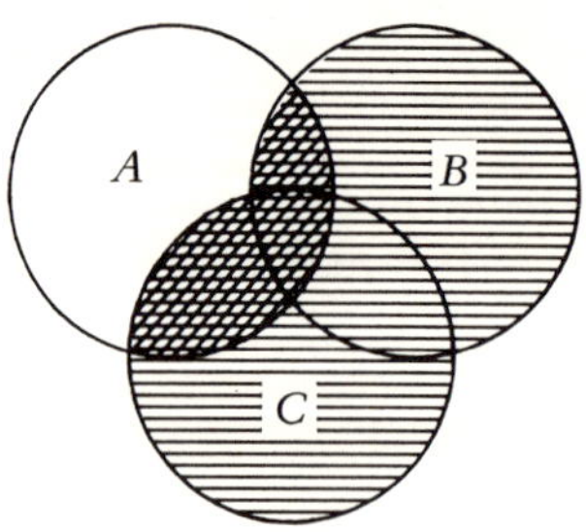

$A \cap (B \cup C)$

FIGURE 2.10

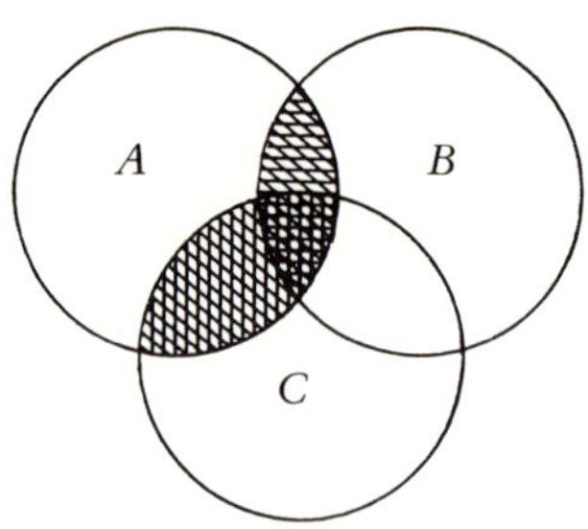

$(A \cap B) \cup (A \cap C)$

FIGURE 2.11

It can be shown in a similar fashion that

$$A \cup (B \cap C) = (A \cup B) \cap (A \cup C).$$

So we also have a **distributive law for unions over intersections** and say that unions **distribute across** intersections.

As was the case with "$\cap$", "$\cup$" makes sense only when names for sets appear on both sides of it. $A \cup B$ is a new set constructed in a special way out of the sets A and B.

Definition 2.3.1. $A \cap B = \{x \mid x \in A \text{ and } x \in B\}$. $A \cap B$ is called the intersection of A and B.

Theorem 2.3.1. (The commutative law for intersections) $A \cap B = B \cap A$

Theorem 2.3.2. (The associative law for intersections) $(A \cap B) \cap C = A \cap (B \cap C)$

Theorem 2.3.3. $A \cap A = A$

Theorem 2.3.4. $A \cap \phi = \phi$

Theorem 2.3.5. $A \cap B \subseteq A$

PROOF: Suppose $x \in A \cap B$. Then by Definition 2.3.1, $x \in A$ and $x \in B$. Therefore, $x \in A \cap B$ implies $x \in A$ and we conclude $A \cap B \subseteq A$.

Theorem 2.3.6. $A \cap B = A$ if and only if $A \subseteq B$.

PROOF: Suppose $A \subseteq B$. Let $x \in A$. Then also $x \in B$. Hence $x \in A \cap B$. This shows that $A \subseteq A \cap B$. But by Theorem 2.3.5 we have $A \cap B \subseteq A$. Hence, $A = A \cap B$ whenever $A \subseteq B$.

Suppose $A = A \cap B$. By Theorem 2.3.1 and Theorem 2.3.5, $A \cap B \subseteq B$. Therefore, $A \subseteq B$ whenever $A = A \cap B$.

Definition 2.3.2. $A \cup B = \{x \mid x \in A \textit{ or } x \in B\}$. $A \cup B$ is called the **union** of A and B.

Theorem 2.3.7. (The commutative law for unions) $A \cup B = B \cup A$

Theorem 2.3.8. (The associative law for unions) $(A \cup B) \cup C = A \cup (B \cup C)$

Theorem 2.3.9. $A \cup A = A$

Theorem 2.3.10. $A \cup \phi = A$

Theorem 2.3.11. $A \subseteq A \cup B$

PROOF: If $x \in A$, then $x \in A$ or $x \in B$. That is, if $x \in A$, then $x \in A \cup B$.

Theorem 2.3.12. $A \cup B = A$ if and only if $B \subseteq A$.

PROOF: See Problem 10.

Theorem 2.3.13. $A \cap B \subseteq A \cup B$

PROOF: Apply Theorem 2.3.5, Theorem 2.3.11, and Theorem 2.2.7.

Theorem 2.3.14. (The distributive laws for unions and intersections)

(a) $A \cap (B \cup C) = (A \cap B) \cup (A \cap C)$
(b) $A \cup (B \cap C) = (A \cup B) \cap (A \cup C)$

PROOF of (b): Suppose $x \in A \cup (B \cap C)$. Then by Definition 2.3.2, $x \in A$ or $x \in B \cap C$. If $x \in A$ then $x \in A \cup B$ and $x \in A \cup C$ (by two applications of Definition 2.3.2) and hence $x \in (A \cup B) \cap (A \cup C)$ by Definition 2.3.1. If $x \in B \cap C$, then $x \in B$ and $x \in C$ by Definition 2.3.1. Then $x \in A \cup B$ and $x \in A \cup C$ (by two applications of Definition 2.3.2) and, hence, $x \in (A \cup B) \cap (A \cup C)$ by Definition 2.3.1. Thus, we have shown that whenever $x \in A \cup (B \cap C)$ we have $x \in (A \cup B) \cap (A \cup C)$. We conclude that $A \cup (B \cap C) \subseteq (A \cup B) \cap (A \cup C)$.

Suppose $x \in (A \cup B) \cap (A \cup C)$. Then by Definition 2.3.1, $x \in A \cup B$ and $x \in A \cup C$. By Definition 2.3.2, this implies $x \in A$ or $x \in B$ and also $x \in A$ or $x \in C$. By Theorem 2.3.11, we know that $A \subseteq A \cup (B \cap C)$ so if $x \in A$ then certainly $x \in A \cup (B \cap C)$. If $x \notin A$ then we must have $x \in B$ and $x \in C$ and, consequently, $x \in B \cap C$. But by Theorem 2.3.7 and Theorem 2.3.11, this implies $x \in A \cup (B \cap C)$. We see then that whenever $x \in (A \cup B) \cap (A \cup C)$ we have $x \in A \cup (B \cap C)$. That is, $A \cup (B \cap C) \subseteq (A \cup B) \cap (A \cup C)$.

Finally by Theorem 2.2.5 we conclude $A \cup (B \cap C) = (A \cup B) \cap (A \cup C)$.

PROBLEMS

1. Let $V = \{a,e,i,o,u\}$, $A = \{a,b,c,d,e,f\}$, $B = \{u,v,w,x,y,z\}$, $D = \{d\}$, $M = \{m,a,t,h,e,i,c,s\}$, $E = \{e,a,s,y\}$, and $S = \{s,e,t\}$. Find (use the roster notation to express your answer): (a) $A \cap V$, (b) $V \cap A$, (c) $E \cup S$, (d) $S \cup E$, (e) $(E \cup E) \cup S$, (f) $S \cap E$, (g) $B \cap E$, (h) $(B \cap E) \cap M$, (i) $B \cap (E \cap M)$, (j) $D \cap (A \cup D)$, (k) $(B \cup E) \cap M$, (l) $(E \cup S) \cup D$, (m) $E \cap M$, (n) $M \cup \phi$, (o) $M \cap \phi$, (p) $D \cap A$, (q) $B \cap V$, (r) $B \cap D$, (s) $V \cap (E \cup M)$, (t) $V \cup (E \cap M)$.

2. Use the set-builder notation to describe briefly the sets V, A, and M in Problem 1.

3. Draw a flow chart (see Figure 2.5) for the determination of set membership for the union of two sets.

4. What, if anything, may be concluded about the sets A and B if $A \cup B = A \cap B$?

5. Use geometric representations of sets to verify that $A \cup (B \cap C) = (A \cup B) \cap (A \cup C)$.

6. (a) Is there a set which is disjoint from itself?
   (b) Is there a set which is disjoint from every set?

7. Give a geometric representation (see Figure 2.9) for (i) $A \cup (B \cap C)$, (ii) $A \cap (B \cup A)$, (iii) $B \cap (A \cup C)$, (iv) $(B \cap A) \cup (B \cap C)$, (v) $(A \cap B) \cup (A \cap C) \cup (B \cap C)$.

8. Construct examples of three small nonempty sets and find all possible intersections and unions (or combinations of intersections and unions) that can be constructed from them.

9. Let $L = \{x \mid x \text{ is a perfect square smaller than } 30\}$, $R = \{x \mid x \text{ is an odd number}\}$, $S = \{x \mid x \text{ is 2 greater than a perfect square}\}$. Find (in roster notation) (a) $L \cap R$, (b) $L \cap S$, (c) $L \cap R \cap S$.

---

10. Prove Theorem 2.3.12.

11. Prove or disprove each of the following.

   (a) If $A \subseteq B$, then $A \cup C \subseteq B \cup C$.
   (b) If $A \subseteq B$, then $A \cap C \subseteq B \cap C$.
   (c) If $A \subset B$, then $A \cup C \subset B \cup C$.
   (d) If $A \subset B$, then $A \cap C \subset B \cap C$.
   (e) If $A \cup C \subseteq B \cup C$, then $A \subseteq B$.
   (f) If $A \cap C \subseteq B \cap C$, then $A \subseteq B$.
   (g) If $A \cup B = \phi$, then $A = \phi$ or $B = \phi$.
   (h) If $A \cap B = \phi$, then $A = \phi$ or $B = \phi$.
   (i) If $A = \phi$ or $B = \phi$, then $A \cap B = \phi$.
   (j) If $A \cup B = \phi$, then $A = \phi$ and $B = \phi$.

2.4 Complements

The collection of all the students in a class who have *not* had mumps is determined once we know the set of students who *have* had mumps. The former set is said to be the **complement** of the latter. We use a prime on the name of a set to name the complement of that set. In the general situation we have

$$A' = \{x \mid x \notin A\}$$

and we read "A'" as "A-prime", or "A **complement**" or "the **complement** of A."

However, for this definition to make sense we must have in mind the entire collection of things which we want to consider as potential members of sets. If we do not do this we run into trouble with such questions as "Does this chair I am sitting in belong to the complement of $\{x|x + 5 = 12\}$?" So if we wish to talk about complements of sets we must have in mind the collection of all possible objects that could be considered. Such a collection we shall call the **universal set.** It corresponds roughly to what a philosopher would call the universe of discourse. We shall use "U" as a name for the universal set. The exact nature of U will depend on the particular concern we have when we mention it. In a classroom situation where we are concerned with who has not had mumps it would be natural to take the universal set to be the collection of all the students in the class.

The universal set plays a special role with unions and intersections. Since $x \in U$ for all x we have for any set A

$$A \cup U = U$$

and

$$A \cap U = A.$$

If $U = \{1,2,3,4,5,6,7,8,9,10\}$ and $A = \{1,2,3,4,5\}$, then $A' = \{6,7,8,9,10\}$. Suppose we let $U = \{x|x \text{ is a positive whole number}\}$, $E = \{x|x \text{ is even}\}$, and $O = \{x|x \text{ is odd}\}$. Then $E' = O$ and $O' = E$.

Since every object must either be in A or not in A we have

$$A \cup A' = U.$$

Since no object can be both in A and not in A we have

$$A \cap A' = \phi.$$

How does complementation interact with unions and intersections? The results that tell us the answer to this are called **De Morgan's Laws.** These two laws are

$$(A \cup B)' = A' \cap B'$$

and

$$(A \cap B)' = A' \cup B'.$$

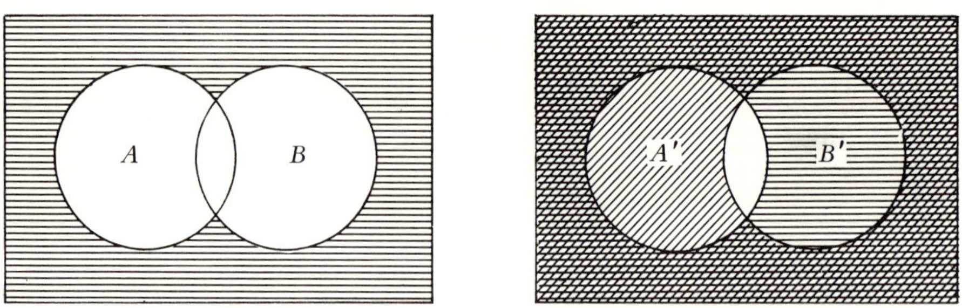

$(A \cup B)'$ is shaded

A' is horizontally hatched
B' is slant hatched

FIGURE 2.12 Verification that $(A \cup B)' = A' \cap B'$.

A pictorial argument for these is made in Figures 2.12 and 2.13. In these pictures we have constructed a rectangle to represent a fence inside of which all the elements of U are to be found.

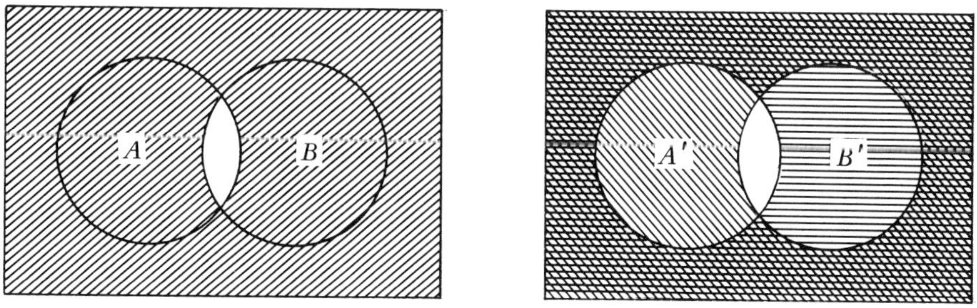

FIGURE 2.13 Verification that $(A \cap B)' = A' \cup B'$.

Complementation is an operation to be performed on a set to yield another set. A' is the name of a certain set.† An expression such as "$\cap'$" does not make sense since $\cap$ is not a set.

Definition 2.4.1. The collection of all objects being considered as possible members of sets is called the universal set and is denoted by "U."

Definition 2.4.2. $A' = \{x \mid x \notin A\}$. A' is called the **complement** of A.

Theorem 2.4.1. $A \cup U = U$

Theorem 2.4.2. $A \cap U = A$

Theorem 2.4.3. $A \cup A' = U$

Theorem 2.4.4. $A \cap A' = \phi$

Theorem 2.4.5. (De Morgan's laws)

(a) $(A \cup B)' = A' \cap B'$
(b) $(A \cap B)' = A' \cup B'$

Theorem 2.4.6. $(A')' = A$

Theorem 2.4.7. If $A \subseteq B$, then $B' \subseteq A'$

PROOF: Suppose $A \subseteq B$ and $x \in B'$. Then $x \notin B$ by Definition 2.4.2. Could $x \in A$ be true? If it were, we would have $x \in B$ since $A \subseteq B$. But $x \notin B$. So we

† Other notations are found in the literature for the complement of A. Some of these are: CA, $\tilde{A}$, and $\bar{A}$.

must have $x \notin A$. That is, $x \in A'$. We have shown that $x \in B'$ implies $x \in A'$. Therefore, $B' \subseteq A'$.

Theorem 2.4.8. $\phi' = U$

Theorem 2.4.9. $U' = \phi$

PROBLEMS

1. Let $U = \{x | x$ is letter in the alphabet we use when we write English$\}$. Let V, A, B, D, M, E, and S be defined as in Problem 1, Section 2.3. Find (in roster notation) (a) $S \cap E'$, (b) $E' \cap S$, (c) $E \cup S'$, (d) $E \cap S'$, (e) $A' \cap B'$, (f) $D' \cap M$, (g) $(E' \cap S')'$, (h) $(E' \cup S')'$.

2. Use the set-builder notation to describe briefly the sets V', M', and S' in Problem 1.

3. Let the universal set be the set of all positive whole numbers $N = \{1,2,3, \ldots\}$, and find, using roster notation

(a) $\{x | 2x = 4\}$
(b) $\{x | x + 6 = 12\}$
(c) $\{x | x + x = 2x\}$
(d) $\{y | y + 2 = 1\}$
(e) $\{z | z + 2 = 1\}$
(f) $\{x | x \neq x + 1\}$
(g) $\{x | x + 2$ is an odd number$\}$
(h) $\{x | x + 1 \in N\}$
(i) $\{x | x + 1 \in N$ and x is even$\}$
(j) $\{r | 5r - 7 = 3\}$

4. It is possible to define the **difference** of two sets by

$$A - B = A \cap B'.$$

Then $A - B$ consists of those elements of A that are not in B. Thus, $\{1,2,3\} - \{2,3,4\} = \{1\}$, $\{1,2,3\} - \{4\} = \{1,2,3\}$, and $\{1,2\} - \{1,2,3\} = \phi$. For the sets of Problem 1 find (a) $V - E$, (b) $M - V$, (c) $M - D$.

5. Simplify the following:

(a) $(A \cup A)' \cap A'$ (b) $(\phi \cup A)' \cap (A' \cup A)$ (c) $(A' \cup B') \cap (A \cup B)'$

---

6. Prove or disprove: If $A' = B'$, then $A = B$.

7. Prove or disprove: If $A \subseteq B$, then $C - A \subseteq C - B$. (See Problem 4.)

8. Prove or disprove: If $A \subseteq B$, then $C - B \subseteq C - A$. (See Problem 4.)

2.5 Cartesian products

A young woman getting ready to go skiing needs to decide which outfit to wear. She has three ski sweaters, a red one, a green one, and a purple one. She has two pairs of ski pants, grey and navy blue. The available outfits are: the red sweater with the grey pants, the red sweater with the navy blue pants, the green sweater with the grey pants, the green sweater with the navy blue pants, the purple sweater with the grey pants, and the purple sweater with the navy blue pants.

A young man planning an evening wants to go out with either Joan or Barbara to either the theater or a basketball game. The possible arrangements are: take Joan to the theater, take Joan to the game, take Barbara to the theater, take Barbara to the game.

A certain factory can produce, among other things, up to 3 large scientific computers per month and up to 4 large business computers per month. The manager must decide how many of each kind of computer to produce. His possible choices can be represented as: (0,0), (0,1), (0,2), (0,3), (0,4), (1,0), (1,1), (1,2), (1,3), (1,4), (2,0), (2,1), (2,2), (2,3), (2,4), (3,0), (3,1), (3,2), (3,3), (3,4), where we have let the first number in each pair be the number of scientific computers to be produced and the second number the number of business computers to be produced.

A man runs into two old friends in New York City. He knows that each friend is either from New York City or from Boston. The possible situations as to where his friends live are (New York, New York), (New York, Boston), (Boston, New York), and (Boston, Boston). Here we let the first city in each pair indicate the home city of the first friend, the second that of the second.

In each of the above situations we end up considering a set of **ordered pairs** which is constructed by considering all possible ordered pairs that can be made by taking the first member of the pair from one set and the second member from another. The set of ordered pairs obtained in this fashion is called the **Cartesian product** of the two given sets. More formally, we have the definition:

$$A \times B = \{(x,y) \mid x \in A \text{ and } y \in B\}.$$

We read "$A \times B$" as "the **Cartesian product** of A and B" or "A **cross** B."

For example, {red, green, purple} × {grey, navy} = {(red, grey), (red, navy), (green, grey), (green, navy), (purple, grey), (purple, navy)}, {Joan, Barbara} × {theater, game} = {(Joan, theater), (Joan, game), (Barbara, theater), (Barbara, game)}, {0,1,2,3} × {0,1,2,3,4} = {(0,0), (0,1), (0,2), (0,3), (0,4), (1,0), (1,1), (1,2), (1,3), (1,4), (2,0), (2,1), (2,2), (2,3), (2,4), (3,0), (3,1), (3,2), (3,3), (3,4)}, and {New York, Boston} × {New York, Boston} = {(New York, New York), (New York, Boston), (Boston, New York), (Boston, Boston)}.

So we now have another operation to use with sets. "×" makes sense whenever names for sets appear on both sides of it.

What are some of the basic properties of this operation? For example, does the order of the sets in a Cartesian product make any difference? Let $A = \{1\}$ and $B = \{1,2\}$, then $A \times B = \{(1,1), (1,2)\}$ and $B \times A = \{(1,1), (2,1)\}$. We see that $A \times B \neq B \times A$ since (1,2) and (2,1) are not the same ordered pairs.† Therefore, the order does matter and the Cartesian product is **not** commutative.

Is it, however, associative? The answer is in the negative. With A and B as above and $C = \{2\}$, we have $(A \times B) \times C = \{(1,1),(1,2)\} \times \{2\} = \{((1,1),2),((1,2),2)\}$ and $A \times (B \times C) = \{1\} \times \{(1,2),(2,2)\} = \{(1,(1,2)),(1,(2,2))\}$. Therefore, $(A \times B) \times C \neq A \times (B \times C)$ since $((1,1),2)$, which is an ordered pair whose first number is an ordered pair, is in $(A \times B) \times C$ but not in $A \times (B \times C)$.

From the definition, we do get, however, the following:

$$A \times \phi = \phi \times A = \phi$$

since there are no pairs with one member from the empty set.

How does the Cartesian product interact with the operation of taking a union? A partial answer to this is given by the law

$$A \times (B \cup C) = (A \times B) \cup (A \times C).$$

This law could be said to assert that Cartesian products distribute across unions.

For example if $A = \{1,2\}$, $B = \{a,b\}$ and $C = \{c\}$, then $A \times (B \cup C) = \{1,2\} \times (\{a,b\} \cup \{c\}) = \{1,2\} \times \{a,b,c\} = \{(1,a), (1,b), (1,c), (2,a), (2,b), (2,c)\}$. Also $(A \times B) \cup (A \times C) = (\{1,2\} \times \{a,b\}) \cup (\{1,2\} \times \{c\}) = \{(1,a), (1,b), (2,a), (2,b)\} \cup \{(1,c), (2,c)\} = \{(1,a), (1,b), (2,a), (2,b), (1,c), (2,c)\} = A \times (B \cup C)$.

A flow chart for deciding set membership in a Cartesian product is given in Figure 2.14.

A new kind of picture is needed if we wish a useful geometric representation for a Cartesian product. Suppose we consider the sets {red, green, purple} and {grey, navy} and want their Cartesian product. One way to display it is in the form of a table:

	grey	navy
red		
green		
purple		

Here we assign each element of the first set to a (horizontal) row in the table and each element in the second set to a (vertical) column. We may then think of

---

† As we have been assuming earlier without specifically mentioning it, we say that (a,b) and (c,d) are ordered pairs which are equal if and only if $a = c$ and $b = d$. This is the basic property of **ordered** pairs. An ordered pair is to be distinguished from an unordered pair. $\{1,2\}$ is an **unordered** pair and $\{1,2\} = \{2,1\}$. However, an **ordered** pair consists of two objects and *also* an assignment of one of these to be the first object and the other to be the second. $(1,2)$ is *not* the same **ordered** pair as $(2,1)$.

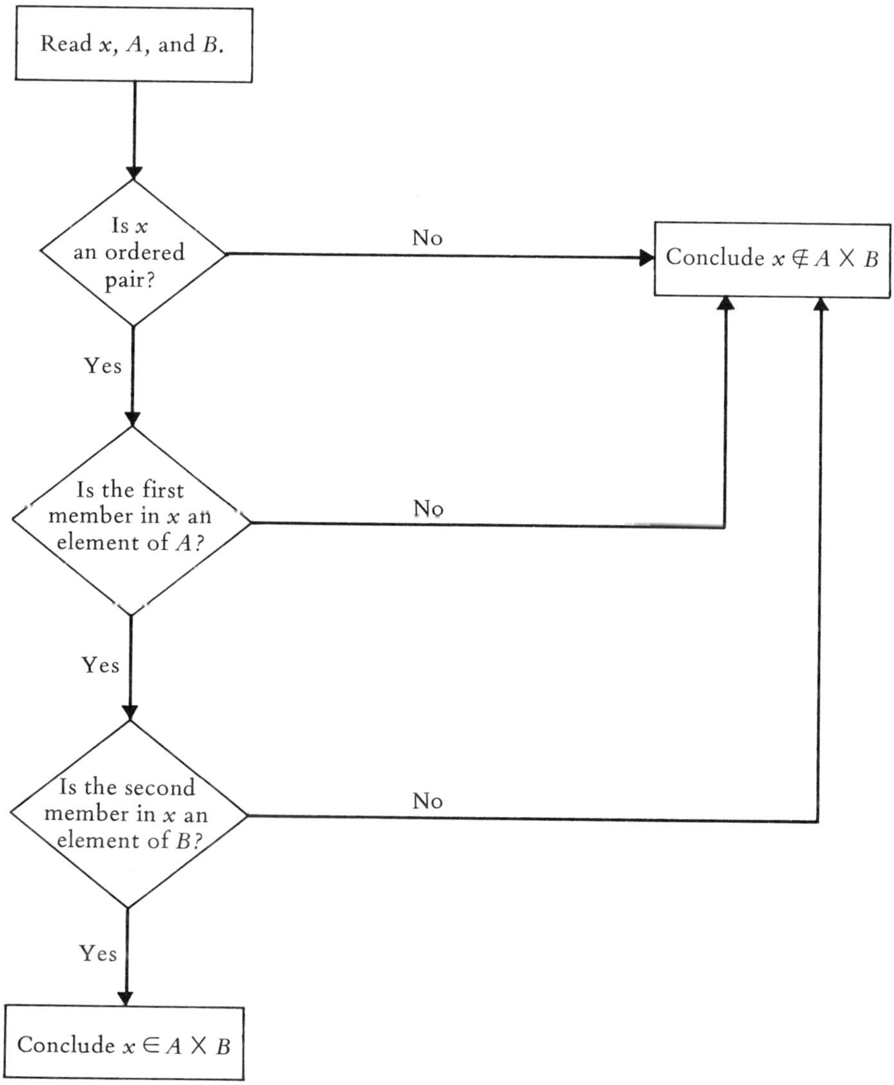

FIGURE 2.14 Determination of membership in $A \times B$.

each box (or position) in the table as representing an element in the Cartesian product. For example, the lower right box represents (purple, navy). When we use this device we shall always take the first member of a pair to indicate the row and the second member the column.

We could make the colors diagram even more schematic and simply draw a horizontal line to the right of (the name of) each member of the first set and a vertical line below each (name of a) member of the second:

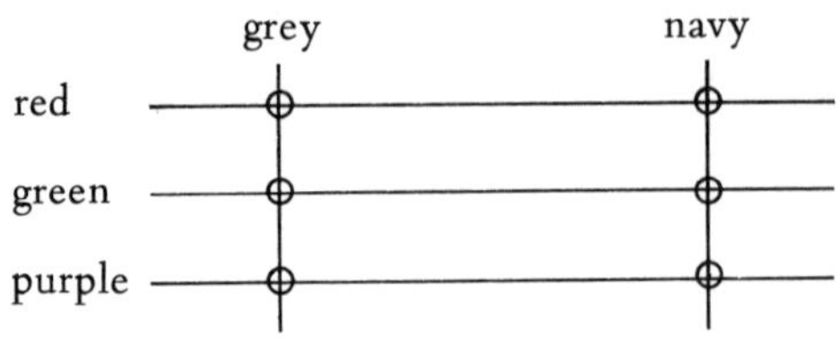

Then the points of intersection of the lines may be taken to represent the elements in the Cartesian product. For example the point at the middle left represents (green, grey).

If the sets involved happen to be sets of numbers, something very close to this last gimmick proves to be very helpful. A graphical representation of $\{1,2,3\} \times \{1,2,3\}$ is shown in Figure 2.15.

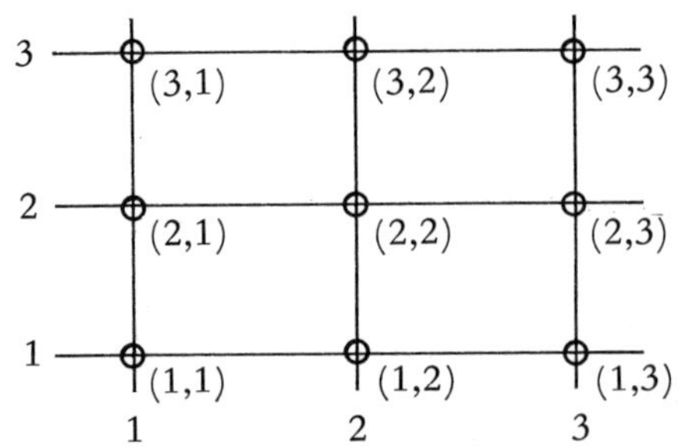

FIGURE 2.15 Geometric representation of $\{1,2,3\} \times \{1,2,3\}$.

There is another useful way to represent a Cartesian product. In Figure 2.16, we have given what is called a tree diagram for $\{1,2\} \times \{a,b,c\}$.

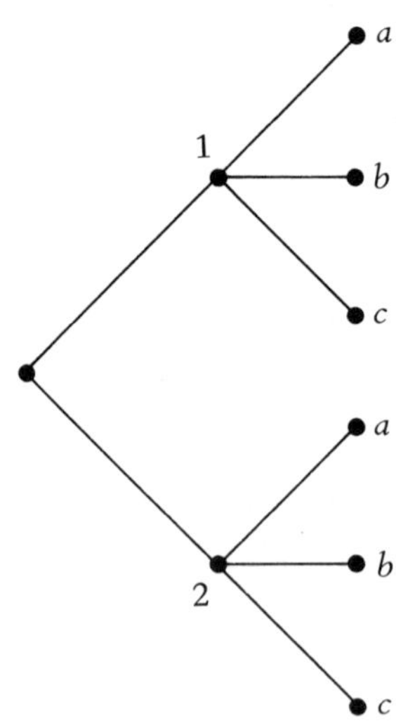

FIGURE 2.16

In this figure, each point in $\{1,2\} \times \{a,b,c\}$ is represented by a path running from the leftmost point to the right. For example, the path along the bottom represents $(2,c)$.

Definition 2.5.1. $A \times B = \{(x,y) | x \in A \text{ and } y \in B\}$. $A \times B$ is called the **Cartesian product** of A and B.

Theorem 2.5.1. $A \times \phi = \phi \times A = \phi$

Theorem 2.5.2. $A \times (B \cup C) = (A \times B) \cup (A \times C)$

PROOF: Suppose $x \in A \times (B \cup C)$. Then $x = (a,d)$ for some $a \in A$ and some $d \in B \cup C$. But then $d \in B$ or $d \in C$. If $d \in B$ we have $x = (a,d) \in A \times B$ and then certainly $x = (a,d) \in (A \times B) \cup (A \times C)$. If $d \in C$, then $x = (a,d) \in A \times C$ and again $x = (a,d) \in (A \times B) \cup (A \times C)$. So in every case $x \in (A \times B) \cup (A \times C)$. We conclude that $A \times (B \cup C) \subseteq (A \times B) \cup (A \times C)$.

Now suppose that $x \in (A \times B) \cup (A \times C)$. Then $x = (a,b)$ for some $a \in A$ and some $b \in B$, or $x = (a,c)$ for some $a \in A$ and some $c \in C$. In the former case we certainly have $b \in B \cup C$ so that $x = (a,b) \in A \times (B \cup C)$. In the latter case, we see that $c \in B \cup C$ and so $x = (a,c) \in A \times (B \cup C)$. So in every case, $x \in A \times (B \cup C)$ whenever $x \in (A \times B) \cup (A \times C)$. That is, $(A \times B) \cup (A \times C) \subseteq A \times (B \cup C)$.

Finally, since $A \times (B \cup C) \subseteq (A \times B) \cup (A \times C)$ and $(A \times B) \cup (A \times C) \subseteq A \times (B \cup C)$, we conclude that $A \times (B \cup C) = (A \times B) \cup (A \times C)$.

Theorem 2.5.3. $A \times (B \cap C) = (A \times B) \cap (A \times C)$

PROOF: See Problem 7.

Theorem 2.5.4. If $A \subseteq B$, then $A \times C \subseteq B \times C$.

PROOF: Suppose $x \in A \times C$. Then by Definition 2.5.1, $x = (a,c)$ for some $a \in A$ and some $c \in C$. But $A \subseteq B$, so $a \in B$. Hence, $x = (a,c) \in B \times C$. Since $x \in B \times C$ whenever $x \in A \times C$, we conclude $A \times C \subseteq B \times C$.

Theorem 2.5.5. If $A \times B = \phi$, then $A = \phi$ or $B = \phi$.

PROOF: Suppose $A \neq \phi$ and $B \neq \phi$. Then there is an element a such that $a \in A$ and there is an element b such that $b \in B$. Then $(a,b) \in A \times B$ and $A \times B$ is not empty. (If this proof is not convincing to you, see Problem 10, Section 1.1.)

PROBLEMS

1. Describe the following sets in roster notation: (a) $\{1,2\} \times \{3\}$, (b) $\{1,2,3\} \times \{2\}$, (c) $\{1,2\} \times \{2,3\}$, (d) $\{7,10\} \times \{1,2\}$, (e) $\{1,2,3\} \times \{1,2,3\}$.

2. By introducing the appropriate sets, use a Cartesian product to describe the collection of all possible ways for (a) two coins (say a penny and a dime) to fall when they are tossed, (b) two dice (say a white one and a red one) to fall, (c) two cards to be drawn from an ordinary deck if the first card is put back and the deck is shuffled before the second is drawn, (d) going from city E to city F and from there to city G if you may drive, fly, or go by rail from E to F and must drive or fly from F to G (assume you may rent a car in F), (e) us to decide on the wall colors for two rooms in our house if white, yellow, and ivory are the only available colors.

3. In each of the following cases, state whether the assertion is true or false and explain how you know.

(a) $A \cup (B \times C) = (A \cup B) \times (A \cup C)$
(b) $A \cap (B \times C) = (A \cap B) \times (A \cap C)$
(c) If $A \subset B$, then $A \times C \subset B \times C$.
(d) $A' \times B' = (A \times B)'$
(e) If $A \times C \subseteq B \times C$, then $A \subseteq B$.
(f) If $A \times C \subseteq B \times C$ and $C \neq \phi$, then $A \subseteq B$.
(g) If $A \subseteq B$ and $C \subseteq D$, then $A \times C \subseteq B \times D$.
(h) $(A \times A) \cap (B \times B) = (A \cap B) \times (A \cap B)$

4. If $A \times B = B \times A$, may we conclude that $A = B$?

5. Sketch a geometric representation (as in Figure 2.15) of

(a) $\{1,2,3,4\} \times \{1,2,3,4\}$,
(b) $\{1,2,3\} \times \{1,2,3,4,5,6\}$,
(c) $\{1,2,3,4,5,6\} \times \{1,2,3,4,5,6\}$.

6. Sketch a tree diagram for

(i) $\{a,b,c\} \times \{1,2\}$,
(ii) $\{1\} \times \{1,2,3,4\}$,
(iii) $(\{a,b\} \times \{1,2,3\}) \times \{c,d\}$.

---

7. Prove Theorem 2.5.3.

3

Relations

3.1 Properties of relations

All the students in a class in an elementary school were asked to write down on a piece of paper, if they were willing, the name of their best friend in that class. They were then asked to write their own name on the top of the paper and the teacher collected the papers. Perhaps the teacher was interested in the social structure in her classroom. In any event, the result of the procedure was a set of ordered pairs of names. Suppose (Patti N., Allison W.) was one of these pairs, where "Patti N." was the name at the top of the paper. Then these names would give us a true sentence when we substitute them for "x" and "y" in "x considers y to be his (her) best friend in the class." Any relationship such as friendship is somehow intimately involved with ordered pairs.

If we are interested in a certain set, C, of people and consider the statement form: "x is a friend of y," then for certain ordered pairs of names of people in C we would get a true statement when we substitute the names in the proper order for "x" and "y". The collection of all such pairs would be the set $F = \{(x,y) \mid x$ is a friend of $y\}$. On C, we can then identify the relation of "being a friend of" as being exactly this set F. We shall then have $(x,y) \in F$ if and only if x is a friend of y.

This kind of reasoning leads us to make the following definition:

A **relation** is a set of ordered pairs.

Being a set of ordered pairs, a relation may be thought of as a subset of $A \times B$ for some sets A and B. Of special interest is the case where we may take A and B to be the same set. We adopt the following terminology:

R is a **relation in the set** A if and only if R is a subset of $A \times A$.

The first relation we considered above was a relation **in the class of students.** F was a **relation in** C.

Further examples of relations in a set C of people would be:

$$M = \{(x,y) \mid x \text{ is the mother of } y\}$$

$$G = \{(x,y) \mid x \text{ is the grandfather of } y\}$$

$$B = \{(x,y) \mid x \text{ is a brother of } y\}$$

$$S = \{(x,y) \mid x \text{ is the spouse of } y\}$$

$$A = \{(x,y) \mid x \text{ is of the same age as } y\}$$

$$L = \{(x,y) \mid x \text{ loves } y\}$$

$$Y = \{(x,y) \mid x \text{ is younger than } y\}$$

$$H = \{(x,y) \mid x \text{ is heavier than } y\}$$

$$N = \{(x,y) \mid x \text{ is of the same nationality as } y\}$$

$$P = \{(x,y) \mid x \text{ has the same parents as } y\}$$

$$V = \{(x,y) \mid x \text{ is of the same sex as } y\}$$

$$E = \{(x,y) \mid x \text{ has the same color eyes as } y\}$$

$$D = \{(x,y) \mid x \text{ is a descendant of } y\}.$$

If $C = \{$John, William, Mary, Peter, Alfred, Alice$\}$ where John is William's father, Mary is William's wife, and Peter, Alfred, and Alice are the children of William and Mary; then, for even more concrete examples of some of the above relations, $M = \{$(Mary, Peter), (Mary, Alfred), (Mary, Alice)$\}$, $G = \{$(John, Peter), (John, Alfred), (John, Alice)$\}$, $B = \{$(Peter, Alfred), (Alfred, Peter), (Peter, Alice), (Alfred, Alice)$\}$, and $S = \{$(William, Mary), (Mary, William)$\}$.

A very helpful notation to use in connection with relations is the following:

If R is a relation we shall write "xRy" to assert that $(x,y) \in R$.

For example, in the above case of a family, we may write "(Mary) M (Peter)" to assert that (Mary, Peter) $\in M$ (i.e., to assert that Mary is the mother of Peter).

A relation may be represented geometrically. For instance, the relation G in the above family is diagrammed in Figure 3.1. The relation B is diagrammed in Figure 3.2.

In drawing a diagram for a relation R we draw an arrow from x to y if and only if xRy. In a classroom where xRy if and only if x considers y his (her) best friend in the class, we could get a physical representation of such a diagram by asking each child to *point* to his best friend. (Actually doing this would probably cause some embarrassment and it is not a recommended procedure.)

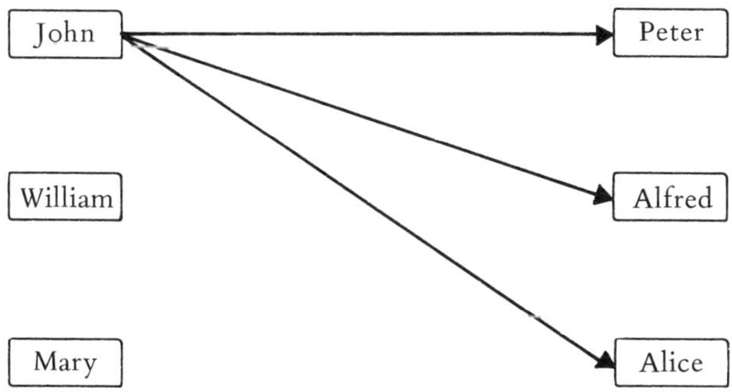

FIGURE 3.1 Geometric representation of the relation G.

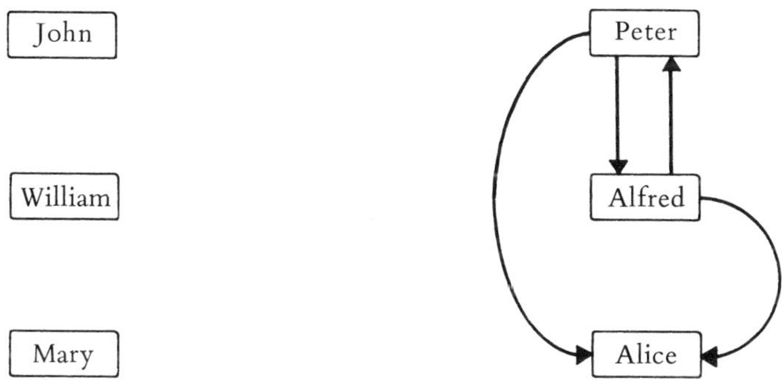

FIGURE 3.2 Geometric representation of the relation B.

In Figure 3.3 the relation V is diagrammed for the family.

Suppose we let $S = \{\phi, \{1\}, \{2\}, \{1,2\}\}$ so that S is the collection of all possible subsets of $\{1,2\}$. The relation "$\subseteq$" in S may be diagrammed as in Figure 3.4.

The relation "$=$" in S is diagrammed in Figure 3.5.

One thing we might notice about the diagram in Figure 3.3 for V is that whenever we have an arrow going from one point to another we also have an arrow going back the other way. A relation whose diagram has this property is said to be **symmetric.** Without referring to the diagram we may state the definition:

> R is said to be **symmetric** on A if for all x and y in A, xRy implies yRx.

S as defined above is symmetric since if x is the spouse of y then y is the spouse of x. B is not symmetric for the given family since x is the brother of y does not imply y is the brother of x. (See Figure 3.2.)

In Figure 3.3 and Figure 3.4 we have diagrams for relations where every point has an arrow coming from it and returning to it. Such relations are said to be **reflexive.** In other words,

> R is said to be **reflexive** on A if for all x in A, xRx.

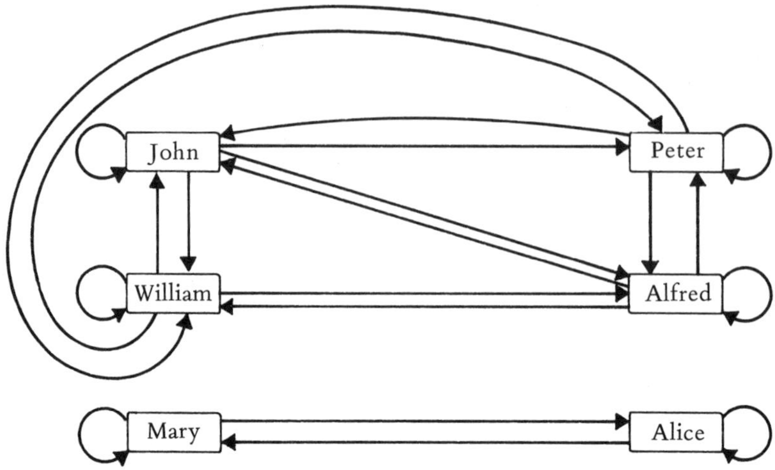

FIGURE 3.3 Geometric representation of the relation *V*.

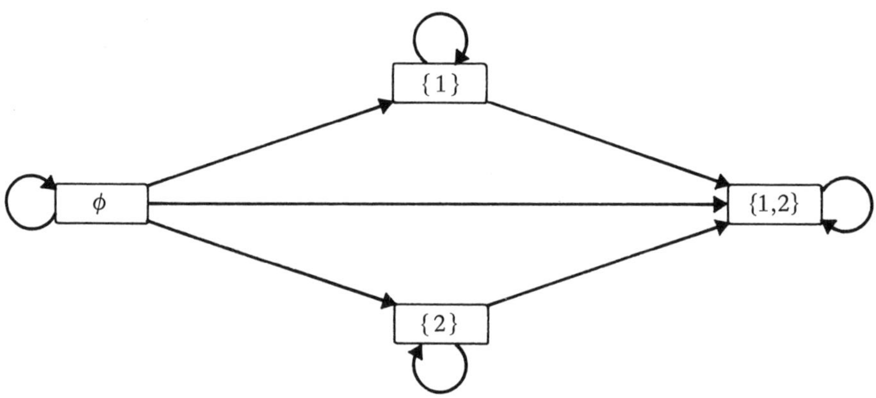

FIGURE 3.4 Geometric representation of the inclusion relation in *S*.

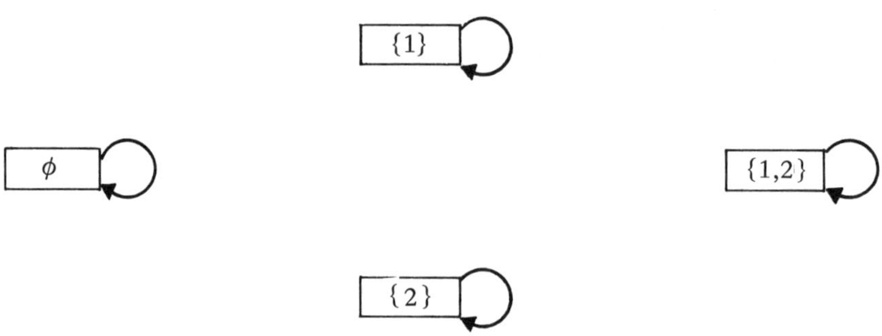

FIGURE 3.5 Geometric representation of the equality relation in *S*.

Inclusion is reflexive on any collection of sets. (See Theorem 2.2.1.) The relation M above is not reflexive on C since it is not true that everyone is his own mother.

In the diagram for set inclusion (Figure 3.4) we might notice that whenever there is an arrow going from one point to a second point and then also an arrow going from that second point to a third point, then there is also an arrow going directly from the first point to the third. A relation with this property is said to be **transitive.** Alternatively, we may state the definition:

> R is said to be **transitive** on A if for all x, y, and z in A, xRy and yRz implies xRz.

Inclusion is transitive on any collection of sets (see Theorem 2.2.7). The relation of "being the spouse of" is not transitive on the family given above since if x is the spouse of y and y is the spouse of z, it certainly does not follow that x is the spouse of z.†

There are many other properties of relations that we could discuss, but the most important properties for our purposes are the ones we have defined: symmetry, reflexivity, and transitivity.

Definition 3.1.1. A **relation** is a set of ordered pairs. If R is a relation we may write "xRy" to assert $(x,y) \in R$. R is said to be a relation **in** A if $R \subseteq A \times A$.

Definition 3.1.2. A **diagram** for a relation R in a set A is a picture where each element in A is represented by a point and an arrow is drawn from the point for x to the point for y if and only if xRy.

Definition 3.1.3. R is said to be **symmetric** on A if xRy implies yRx for all x and y in A.

Definition 3.1.4. R is said to be **reflexive** on A if xRx for all x in A.

Definition 3.1.5. R is said to be **transitive** on A if xRy and yRz implies xRz for all x, y, and z in A.

PROBLEMS

1. Consider the relation

$$L = \{(x,y) \mid x \text{ is less than } y\}$$

in the set $\{1,2,3,4,5\}$. (a) Display L in the roster notation. (b) Diagram L.

2. Diagram the relation of equality in $\{1,2,3,4,5,6\}$. Also give this relation as a set of ordered pairs in roster notation.

---

† Note that "being a mother of" *is* transitive on the family given. This is because "xMy and yMz" is always false on this family. We would say that M is transitive *vacuously* on this family.

3. Let S be the collection of all possible subsets of $\{1,2,3\}$. Diagram (a) the inclusion relation in S, (b) the proper inclusion relation in S.

4. (a) Diagram the relation P defined above in the family given in this section. (b) Do the same for the relation D.

5. Make a table in which it is shown for each of the relations M, G, B, S, A, L, Y, H, N, P, V, E, and D defined above whether it is symmetric, reflexive, or transitive on the family C given in this section. In the case of the relation L you should explain why you make the claims you do about it.

6. Let S be the collection of all students in your school. Consider the following relations in S:

$$C = \{(x,y) \mid x \text{ owns the same make of car as } y\}$$

$$D = \{(x,y) \mid x \text{ has had at least one "date" with } y\}$$

$$H = \{(x,y) \mid x \text{ and } y \text{ have the same home town}\}$$

$$Y = \{(x,y) \mid x \text{ and } y \text{ are in the same year in school}\}$$

$$R = \{(x,y) \mid x \text{ considers } y \text{ to be a radical}\}.$$

Make up a table which shows which of these are symmetric, reflexive, or transitive. You may need to give a rationale for your choice in a few cases. Be especially careful with the transitivity of C. Someone may own two cars!

7. A relation R is said to be **antisymmetric** on a set A if for all x and y in A, xRy implies that it is false that yRx. Which of the relations in Problem 5 are antisymmetric?

8. A relation R is said to be **irreflexive** on a set A if for all x in A it is false that xRx. Which of the relations in Problem 5 are irreflexive?

9. A relation R is said to be **strictly intransitive** on a set A if for all x and y and z in A, xRy and yRz implies that it is false that xRz. Which of the relations in Problem 5 are strictly intransitive?

10. Since a relation is simply a set of ordered pairs, one possible relation is the empty relation ϕ. Is this relation symmetric, reflexive, or transitive on (a) $\{1,2\}$, (b) ϕ?

11. (a) Describe in roster notation all possible relations in $\{1,2\}$. (b) Diagram each of these. (You should get 16 different relations.)

12. If R is a relation, we define the **domain** of R to be the set $\{x \mid xRy \text{ for some } y\}$. And we define the **range** of R to be the set $\{y \mid xRy \text{ for some } x\}$. Find the domain and range for (a) the relations M, G, B, and A of Problem 5, (b) the relations of Problem 6.

13. Diagram all possible relations in $\{1,2,3\}$ which are both reflexive and symmetric. Which of these are transitive?

14. Let $P = \{(x,y) \mid x \text{ and } y \text{ differ by an even number}\}$ in $\{1,2,3,4,5,6\}$. Sketch the diagram for P. (Note that we take 0 to be an even number.)

15. Let R be a relation in A. Suppose R is transitive and symmetric on A and that the domain of R (see Problem 12) is A. Prove that R must be reflexive on A.

16. Let R and S be relations in a set A and suppose $R \subseteq S$. Prove or disprove each of the following: (a) If R is reflexive on A then so is S. (b) If R is symmetric on A then so is S. (c) If R is transitive on A then so is S.

17. If R is a relation in A and B is another set such that $A \subseteq B$, then R may also be thought of as a relation in B. If R is (a) symmetric, (b) reflexive, (c) transitive on A, does it follow that R must have this same property on B? Explain.

3.2 Equivalence relations

Among the various relations considered in the preceding section, we found a few which were symmetric and reflexive and transitive. Such relations are extremely useful. They are called equivalence relations. That is:

> A relation R in a set A is said to be an **equivalence relation** on A if R is symmetric and reflexive and transitive on A.

The relation of equality is, in a way, the epitome of all equivalence relations. On any set we would have: $x = x$ (reflexivity), $x = y$ implies $y = x$ (symmetry), and $x = y$ and $y = z$ implies $x = z$ (transitivity).

The relation of set inclusion is generally not an equivalence relation since it fails to be symmetric.

The relation V, diagrammed in Figure 3.3, is an equivalence relation. Notice how in Figure 3.3 the diagram is in two pieces which have no arrows running between them. The relation V seems naturally to divide or partition the set into two parts. These two pieces are the subsets {John, William, Peter, Alfred} and {Mary, Alice}. Notice that in the first of these subsets we find all the elements that are in the relation V to John (or to William, or Peter, or Alfred). In the second subset we find all the elements which are in the relation V to Mary (or to Alice).

Let $T = \{(x,y) \mid x \text{ and } y \text{ differ by a multiple of } 3\}$ in the set $A = \{1,2,3,4,5,6\}$. T is diagrammed in Figure 3.6.

It is quite easy to see that T is an equivalence relation.† And we note again that the relation breaks up or (as we shall say) decomposes the underlying set A into different parts which are unconnected in the diagram. Here the three subsets into which A is partitioned by T are $\{1,4\}$, $\{2,5\}$, and $\{3,6\}$. The subset $\{1,4\}$ consists of all elements in A which are in the relation T to 1 (or to 4). $\{2,5\}$ is the set of all elements which are in the relation T to 2 (or to 5). We may write $\{3,6\}$ as $\{y \mid y \in A \text{ and } 3Ty\}$ or as $\{y \mid y \in A \text{ and } 6Ty\}$.

It turns out that we get this sort of partitioning whenever we have an equiva-

† Here we consider 0 to be a multiple of 3 since $0 = 0 \cdot 3$.

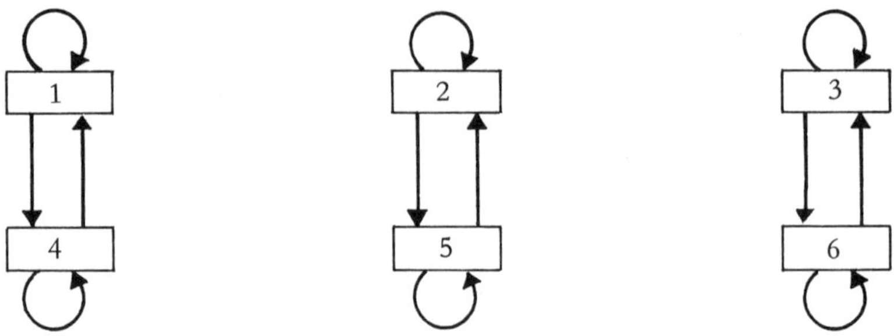

FIGURE 3.6

lence relation on a set. We shall now consider the general case. Suppose R is an equivalence relation on a nonempty set A. Let $[x] = \{y \mid y \in A \text{ and } xRy\}$. We shall call $[x]$ the **equivalence class** for x. We may also read "$[x]$" as "square brackets x" if we wish. For the equivalence relation T given alone, $[1]$ is the set of members that are in the relation T to one. That is, $[1] = \{1,4\}$. Similarly, $[2] = \{2,5\}$, $[3] = \{3,6\}$, $[4] = \{1,4\}$, $[5] = \{2,5\}$, and $[6] = \{3,6\}$ for the relation T.

For the relation V above, we have as examples [John] = {John, William, Peter, Alfred} and [Alice] = {Mary, Alice}.

For the equality relation on any set we have $[x] = \{x\}$ for each x.

Suppose R is an equivalence relation on a nonempty set A. Let us consider the nature of the equivalence classes that arise from R. We first note that since R is reflexive $x \in [x]$ for all $x \in A$. We conclude from this that every element of A is in at least one equivalence class and also no equivalence class is empty.

Suppose xRy. What can be said about $[x]$ and $[y]$ in this case? We claim that if xRy then $[x] = [y]$. We can prove this as follows. Suppose $z \in [x]$. Then by definition xRz. Since R is symmetric, xRy implies yRx. But yRx and xRz together with the transitivity of R gives yRz. Then $z \in [y]$. We see then that whenever $z \in [x]$ we have $z \in [y]$. That is, $[x] \subseteq [y]$. The symmetry of R and xRy implies yRx. So exactly the same argument with x and y interchanged will give us $[y] \subseteq [x]$. Therefore, $[x] = [y]$ whenever xRy.

Is it possible for two equivalence classes to overlap? To answer this question, suppose that $[x] \cap [y]$ is not empty. Since this set is not empty there is at least one element in it. Call this element z. Since $z \in [x]$ we have xRz. Since $z \in [y]$ we have yRz and hence, by the symmetry of R, zRy. But xRz and zRy and the transitivity of R give us xRy. And by what we just proved above, this implies $[x] = [y]$. Hence, if two equivalence classes overlap they are actually equal.

We summarize the above results as:

> *An equivalence relation on a nonempty set decomposes (or partitions) the set into nonempty disjoint equivalence classes.*

Let A be the set of students at a certain college. Let $H = \{(x,y) \mid x \text{ is from the same home town as } y\}$. It is easy to see that H is an equivalence relation on A.

H decomposes A into equivalence classes. If Johnny Bright is a student at the college and his home town is Chicago then [Johnny Bright] would be the collection of all the students who are from Chicago. Similarly, if Sue Smart is from New Orleans then [Sue Smart] would be the collection of all the students from New Orleans.

The relation V of Figure 3.3 partitions the set of people into the set of males and the set of females.

The equality relation on a set partitions the set into equivalence classes which contain exactly one element. (Equality seems to be, in some sense, the "finest" of all possible equivalence relations on a set.)

The idea of equivalence classes is a fundamental one in mathematics and crops up in many different places. We shall make repeated use of the notion as we proceed in our study of the various number systems that people find useful.

Definition 3.2.1. A relation R in a set A is said to be an **equivalence relation** on A if R is symmetric and reflexive and transitive on A.

Definition 3.2.2. If R is an equivalence relation on a nonempty set A, the set $[x] = \{y \mid y \in A \text{ and } xRy\}$ is called the **equivalence class** for x on A (with respect to R).

Lemma 3.2.1. Every element of A belongs to some equivalence class.

Lemma 3.2.2. No equivalence class is empty.

Lemma 3.2.3. If xRy, then $[x] = [y]$.

Lemma 3.2.4. If $[x] \cap [y] \neq \phi$, then xRy.

Theorem 3.2.1. An equivalence relation on a nonempty set A decomposes (or partitions) A into nonempty, disjoint equivalence classes.

PROBLEMS

1. Describe the equivalence classes for those relations that are equivalence relations in (a) Problem 2, Section 3.1, (b) Problem 5, Section 3.1, (c) Problem 6, Section 3.1, (d) Problem 13, Section 3.1, (e) Problem 14, Section 3.1.

2. Verify that each of the following is an equivalence relation on the set of all students in your school. Then describe the decomposition corresponding to each relation.

$S = \{(x,y) \mid x \text{ is from the same state as } y\}$

$B = \{(x,y) \mid x \text{ is the same as } y \text{ or } x \text{ is a sibling of } y\}$

$W = \{(x,y) \mid \text{the weight of } x \text{ to the nearest pound is the same as that for } y\}$

$E = \{(x,y) \mid x \text{ and } y \text{ have the same color eyes}\}$

3. Is the relation diagrammed in Figure 3.7 an equivalence relation on $\{a,b,c,d,e,f\}$?

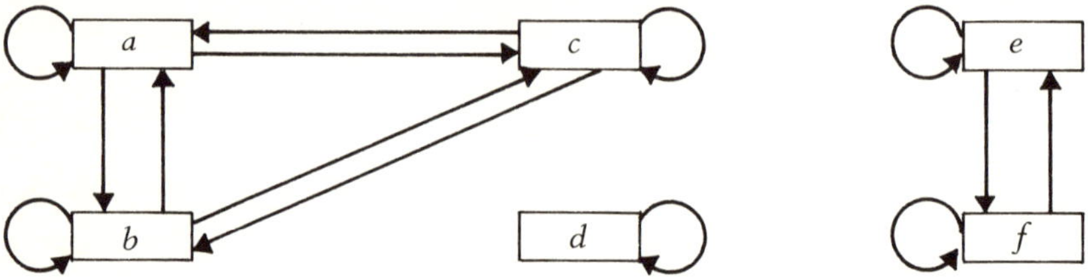

FIGURE 3.7

4. Suppose A is a nonempty set. Since a relation is simply a set of ordered pairs, the set $A \times A$ is a relation in A. Is it an equivalence relation? If so, what are the equivalence classes?

5. Let $A = \{1,2,3,4,5,6,7,8,9,10,11,12\}$ and $F = \{(x,y) \mid x$ and y differ by a multiple of $4\}$. Diagram the relation F in A and find its equivalence classes.

6. x and y are said to have the same **parity** if x and y are both odd or if x and y are both even. Find the equivalence classes for $P = \{(x,y) \mid x$ and y have the same parity$\}$ on $N = \{1,2,3,4, \ldots\}$.

7. If A is the set of all citizens of the United States, which of the following are equivalence relations on A? For each equivalence relation, describe some of its equivalence classes.

$V = \{(x,y) \mid x$ and y either both did not vote or both voted for the same presidential candidate in the last election$\}$

$K = \{(x,y) \mid x$ knows $y\}$

$T = \{(x,y) \mid x$ and y sometimes play cards together$\}$

$C = \{(x,y) \mid x$ owns exactly as many cars as $y\}$

$F = \{(x,y) \mid x$ has the same number of children as $y\}$

$W = \{(x,y) \mid x$ works in the same building as $y\}$

---

8. How many different possible equivalence relations are there on (a) $\{1,2,3\}$ (see Problem 13, Section 3.1), (b) $\{1,2,3,4\}$?

9. Suppose that P and Q are relations in a set A and that $P \subseteq Q$. Prove or disprove: If P is an equivalence relation on A, then so is Q.

10. A relation R is said to be a **partial ordering** on (or of) A if R is reflexive and transitive on A and if, for all x and y in A, xRy and yRx implies $x = y$. If R is a partial ordering of A then A is said to be **partially ordered by**

R. Given any collection A of subsets of a fixed set, set inclusion is a partial ordering.

(a) What relations on a set A can be both a partial ordering and an equivalence relation?

(b) Diagram all possible partial orderings on $\{1,2\}$. (Note that in the diagram for a partial ordering at most one arrow connects two distinct points.)

(c) Diagram all possible partial orderings on $\{1,2,3\}$.

11. If R is a partial ordering on a set A then R is called a **linear ordering** of (or on) A if for all x and y in A either xRy or yRx. If R is a linear ordering on A, then A is said to be **linearly ordered by** R.

(a) Show that set inclusion is a linear ordering on the collection $\{\phi, \{1,2\}, \{1,2,3\}\}$.

(b) Show that set inclusion is not a linear ordering on the collection of all possible subsets of $\{1,2,3\}$.

(c) Show that $\{(x,y) \mid x$ is less than or equal to $y\}$ is a linear ordering on $\{1,2,3,4,5,6\}$.

(d) Diagram all possible linear orderings on $\{1,2\}$ and $\{1,2,3\}$. (There should be two of the former and six of the latter.)

3.3 Isomorphic relations

In Figure 3.8 we find the diagrams for all possible equivalence relations on $\{1,2,3\}$.

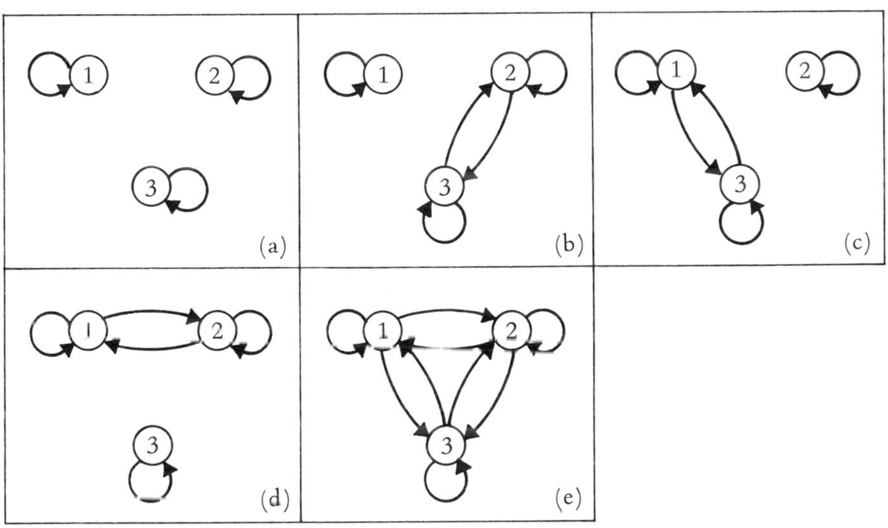

FIGURE 3.8 The equivalence relations on $\{1,2,3\}$

Note that the relations in (b), (c), and (d) all "look alike." They all have a similar structure. In fact, we see that if we interchange the "1" and the "2" in (c) we get a diagram for exactly the relation diagrammed in (b). If we change the "3" to a "1" and the "1" to a "3" in (d) we again get a diagram for the relation diagrammed in (b). We would say that the relations diagrammed in (b), (c), and (d) are all **isomorphic** meaning that they have essentially the same structure.

In Figure 3.9, we have diagrammed all the equivalence relations on $\{1,2,3,4\}$.

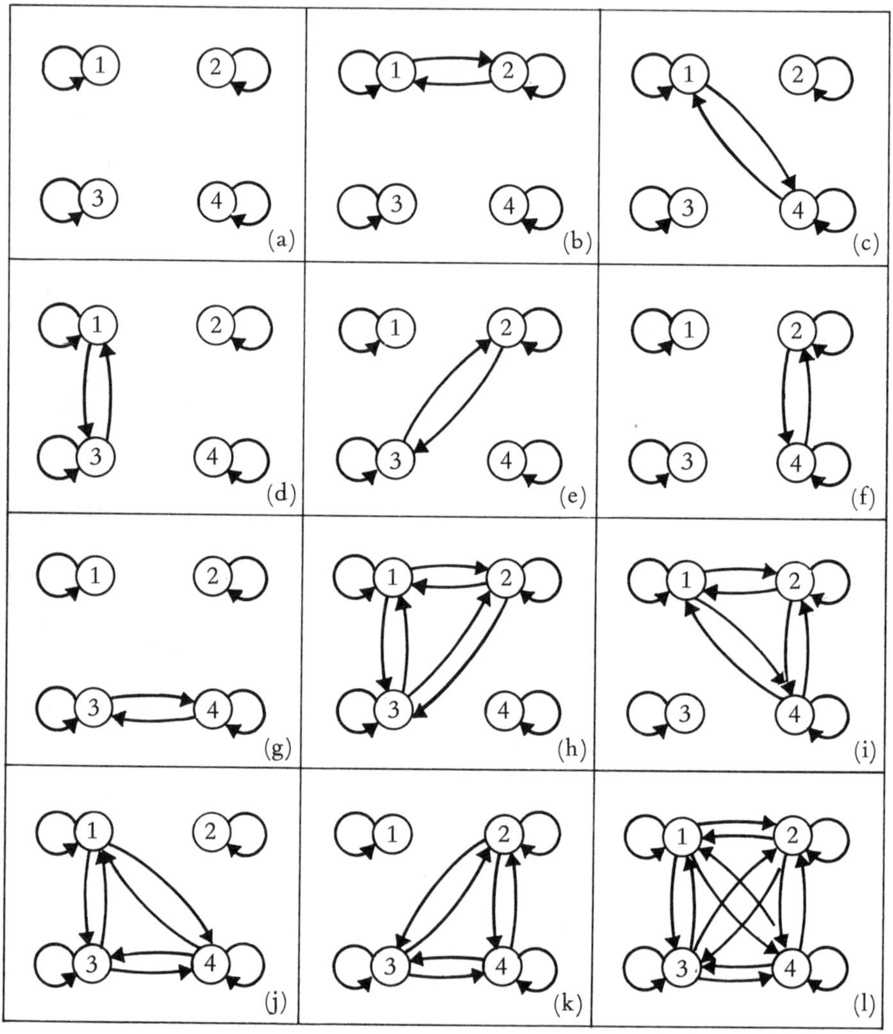

FIGURE 3.9 The equivalence relations on $\{1,2,3,4\}$.

Here, the relations diagrammed in (b), (c), (d), (e), (f), and (g) are all isomorphic.† So are those in (h), (i), (j), and (k).

Are the two relations diagrammed in Figure 3.10 isomorphic? Evidently, they are. If in (b) we change the "1" to a "5", change the "2" to a "1", change the "3" to a "2", leave the "4" as it is, and change the "5" to a "3", we get a diagram of the same relation as in (a). Of course, to make the pictures look exactly alike after making these changes in (b) we would need to rearrange the points on the page. But the relation that is being diagrammed does not depend on the positions we choose for the points on the page.

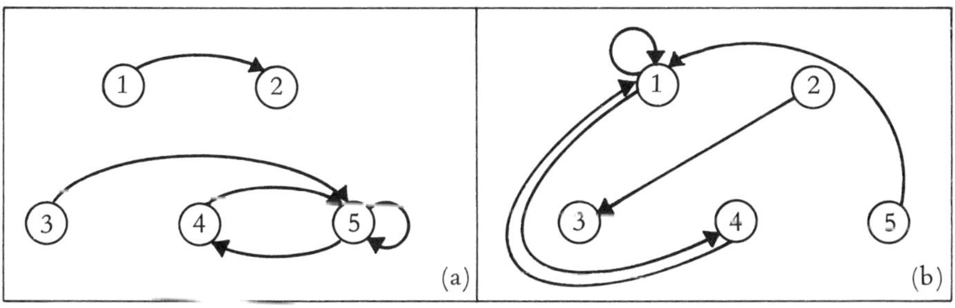

FIGURE 3.10

In this last example, the roster notation for the relations might make the situation even clearer. The relation in (a) is $\{(1,2), (3,5), (4,5), (5,4), (5,5)\}$. That in (b) is $\{(1,1), (1,4), (2,3), (4,1), (5,1)\}$. And if we make the suggested changes in the pairs for (b) we get exactly the pairs in (a).

When is it then that two relations have the same structure or are isomorphic? When we can change the names of the points in one relation so that it becomes identical to the other. We shall state the definition of "isomorphic" more precisely later, but we can operate informally at this stage with the definition given here.

PROBLEMS

1. Which of the relations are isomorphic in (a) Problem 11, Section 3.1, (b) Problem 13, Section 3.1, (c) Problem 11(d), Section 3.2.
2. Which of the relations diagrammed in Figure 3.11 are isomorphic?
3. On any collection S of relations, the relation $R = \{(x,y) \mid x$ is isomorphic

† Marshall McLuhan (as interpreted by Tom Wolfe) asserts that "The tribal man—the new TV generation—is far more apt at **pattern recognition,** which is the basis of computers." It is somewhat uncertain as to how this should make you feel if you find it easy to recognize the patterns above. See G. E. Stearn (ed.), *McLuhan: Hot and Cold* (New York: The New American Library, 1967), p. 38.

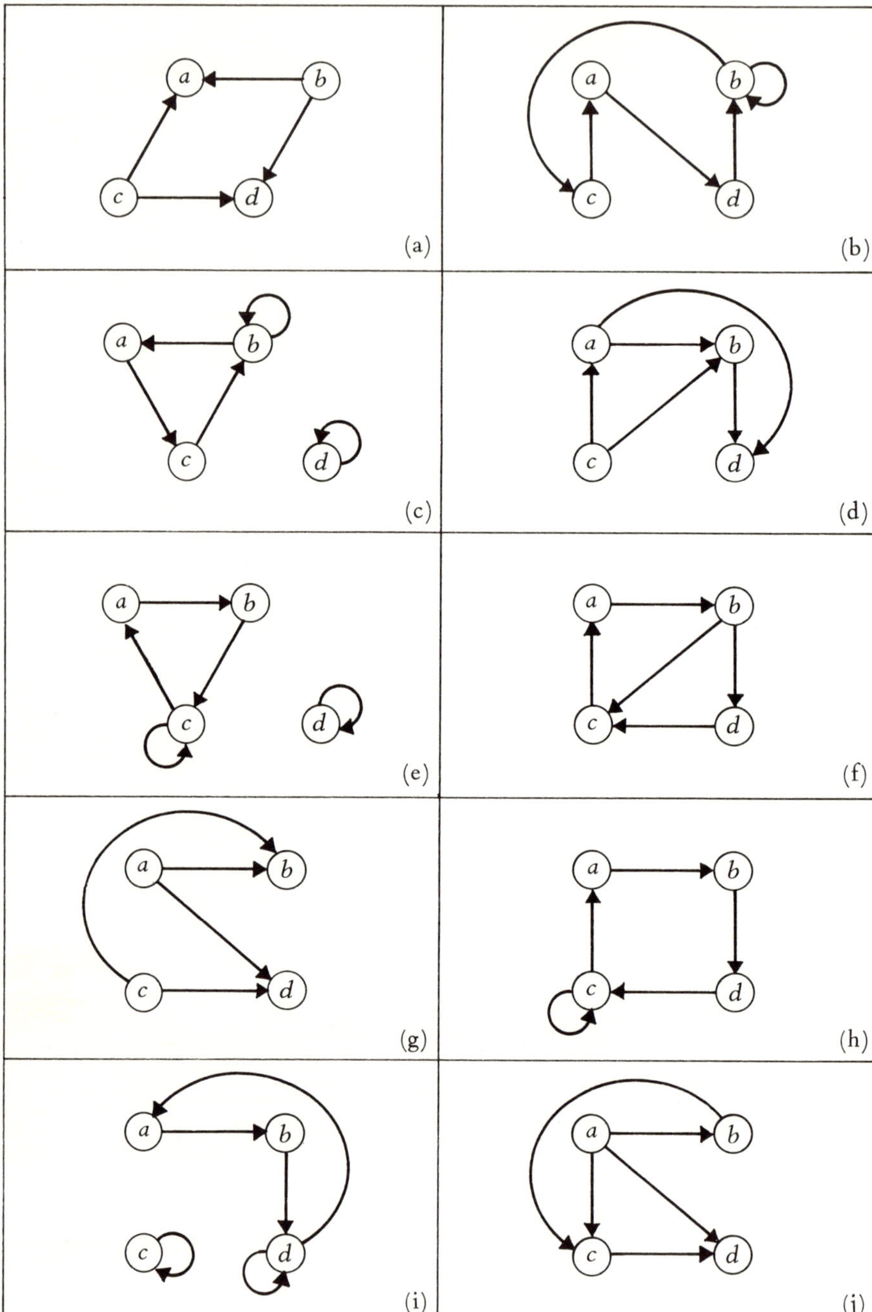

FIGURE 3.11

to $y\}$ is an equivalence relation. We shall prove this later. Meanwhile, find the equivalence classes for R if we take R as defined on the family S of all equivalence relations on (a) $\{1,2,3\}$ (see Figure 3.8), (b) $\{1,2,3,4\}$ (see Figure 3.9).

4. Conjecture a general theorem suggested by your results in Problem 1 (c).

5. Let $A = \{1,2,3,4,5\}$. Let S be the collection of all relations in A which contain exactly one pair (i.e. the diagram for any relation in S has one and only one arrow in it). Are all the relations in S isomorphic? If not, explain the situation. (See Problem 3.)

6. Let $A = \{1,2,3,4\}$. Let D be the collection of all relations in A that contain exactly two pairs. (That is, each diagram for a relation in D will have exactly two arrows.) Are all the relations in D isomorphic? If not, explain the situation. (See Problem 3.)

7. Suppose P and Q are both relations in a set A and that P and Q are isomorphic as relations in A. If $A \subseteq B$, then P and Q can also be thought of as relations in B. Must P and Q be isomorphic as relations in B? Explain your answer in terms of the diagrams we could draw for the relations.

8. Draw a triangle and label its vertices A, B, and C. Let S be the collection of all those relations in $\{A,B,C\}$ which can be diagrammed by putting in three arrow heads in our picture. Are all the relations in S isomorphic? If not, explain the situation. (See Problem 3.) (You should find 8 different relations in S.)

4

Functions

4.1 Functional notation

In 1965, the approximate populations of the ten major countries in the world were as given in the following table.†

Country	*Population in millions*
China	755
India	487
U.S.S.R.	231
U.S.A.	195
Japan	98
W. Germany	59
U.K.	55
Italy	52
France	49
Canada	20

In 1965 the approximate gross national products (GNP) per capita (in 1965 U.S. dollars) for these same countries were‡

† The data is from H. Kahn and A. J. Wiener, *The Year 2000* (New York: Macmillan, 1967), p. 157.

‡ *Ibid.*, p. 161.

Country	*GNP per capita*
China	98
India	99
U.S.S.R.	1,288
U.S.A.	3,557
Japan	857
W. Germany	1,905
U.K.	1,804
Italy	1,101
France	1,924
Canada	2,464

In each of these tables, the name of a country is paired with a number which has a certain meaning. Thus, each table gives us a set of ordered pairs and, hence, a relation. The relation in the first table could be defined as $\{(x,y) \mid$ the country x had approximately the population y (in millions) in 1965$\}$. Both the relations given by these tables have a very special property: each country in the set of ten countries is paired with **exactly one** number.

In the table in Problem 2, Section 1.2 you were asked to consider the numbers 1,2,3, . . . , 50 and to assign 1 to each number that was a perfect square, and zero to each number that was not. The table that results if you do this also gives us a set of ordered pairs since each whole number from 1 to 50 is paired with either a 1 or a 0. The relation defined by this collection of ordered pairs has the property that each number from 1 to 50 is paired with **exactly one** of the numbers in $\{0,1\}$.

The table of contents of a book may be thought of as a relation in which each section heading is paired with a page number. This relation has the property that each section heading is paired with **exactly one** page number.

All of the above relations are examples of what is called a **function.** A **function** is a relation in which each object that occurs as a first element in some pair is paired with **exactly one** object by the relation. Putting this another way, a relation F is said to be a **function** if xFy and xFz implies $y = z$. Still another way to put this is to say that a relation F is a **function** if no two distinct pairs in F have the same first member.

As far as diagrams go, a relation F is a function if *at most* one arrow runs from every point in the diagram. The relation diagrammed in Figure 4.1 (a) is a function. That in Figure 4.1 (b) is not.

In many cases, it is helpful to describe a function by means of a table as we did above for the population figures. There is a special notation that is useful in this connection and also turns out to be very helpful in many ways. If F is a function and $(x,y) \in F$ so that y (and only y) is paired with x by F, then we call

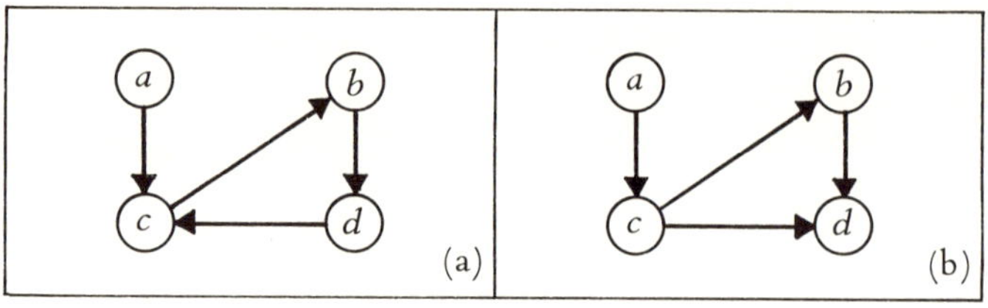

FIGURE 4.1

y the **value of** F at x and write "$y = F(x)$." We read "$F(x)$" as "F at x" or "F of x." If $y = F(x)$, then F is said to **take on the value** y **at** x.

If we let P be the name of the population function given by the table at the beginning of this section, then, for example, P(China) = 755, P(U.S.) = 195, and P(Italy) = 52. If "G" is the name of the GNP-per capita function in the second table, then G(U.S.S.R.) = 1,288 and G(U.K.) = 1,804.

If we wish to describe the function diagrammed in Figure 4.1 (a) by means of a table we may call the function h and display:

x	a	b	c	d
$h(x)$	c	d	b	c

Many tables of data may be thought of as defining functions. An instance would be a table which gives the age distribution of the population of the U.S. in 1965.†

Age group	*Population in 1000's*
Under 5	20,434
5–9	20,519
10–14	18,956
15–19	17,052
20–24	13,667
25–34	22,358
35–44	24,431
45–54	22,045
55–64	16,966
65 and over	18,156

Here, each age group is paired with one (and only one) population.

The collection of all the elements that occur as first elements in the pairs in

† *Ibid.*, p. 169.

a function is called the **domain** of the function. In other words, the domain of a function F is $\{x \mid \text{there is a } y \text{ such that } (x,y) \in F\}$. If x is in the domain of F then F is said to be **defined at** x. If x is not in the domain of F, then F is said to be **undefined at** x.

The domain of the population function in the first table above is the set of ten major countries. The domain of the function diagrammed in Figure 4.1 (a) is $\{a,b,c,d\}$.

The collection of all the values taken on by a function is called the **range** of the function. That is, the **range** of a function F is $\{y \mid \text{there is an } x \text{ such that } (x,y) \in F\}$.

The range of the population function given at the beginning of this section is $\{755, 487, 231, 195, 98, 59, 55, 49, 20\}$. The range of the function in Figure 4.1 (a) is $\{b,c,d\}$.

A function is said to be a function *on* its domain. And a function is said to **map** its domain *onto* its range. The population function above maps the set of ten major countries onto the set $\{755, 487, 231, 195, 98, 59, 55, 52, 49, 20\}$. The function in Figure 4.1 (a) maps $\{a,b,c,d\}$ onto $\{b,c,d\}$. The function defined by the table in Problem 2, Section 1.2 maps $\{1,2,3, \ldots, 50\}$ onto $\{0,1\}$.

Suppose D is the domain of a function F and R is the range of F. If we have a set S of which R is a subset, then we may say that F maps D *into* S. In other words, F is said to map its domain *into* a set S if every value of F is a member of S.

If F maps a set A onto a set B then certainly F maps A into B. But if F maps A into B, then it may not be the case that F maps A onto B. The idea of "onto B" is that B is "covered" by the values of the function. "Into B" simply tells us that the values of the function are all in B (and there may be other elements in B which are not values of the function).

This "mapping terminology" ties in with a very convenient way for thinking of functions. Suppose we take the state of Missouri and think of it as a set M of physical points. If we draw a map of the State of Missouri, each actual point in the state should correspond to some one point on the map. Thus, in mapping the state we set up a function. In the function each point in the state is paired with one (and only one) point on the map. We speak of each point on the map as the **image** of the corresponding point on the ground. A photograph of a scene gives the same kind of correspondence with each point in the picture being an image of a point in the scene.

In the general situation, if F is a function such that $y = F(x)$ we may call y the **image** of x under (or for) F and we may call x a **pre-image** of y under (or for) F. Notice that an object may have more than one pre-image. For example, a and d are both pre-images of c for the function h above.

Corresponding to this notion of a mapping, there is a geometric way to represent a function as a mapping. The procedure for doing this will now be explained. Suppose F is a function which maps A into B. We first represent the elements in A by points in a picture. Then in a different part of the picture we put in points to represent the points in B. We then draw an arrow from x in A

to y in B if and only if $y = F(x)$. For example, we find the diagram of the function h of Figure 4.1 (a) as a mapping in Figure 4.2.

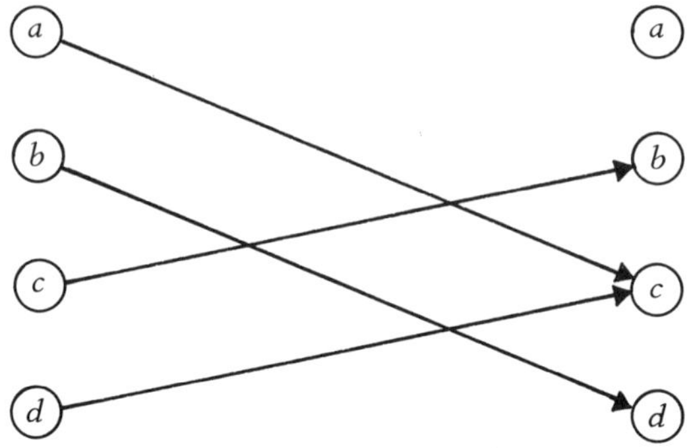

FIGURE 4.2 Diagram for the mapping h.

Notice that diagramming a function as a mapping is somewhat different from diagramming the function as a relation as we did in Section 3.1. To diagram a function as a mapping we must have two given sets—one the domain of the function and the other a set containing the range. In the diagram of the function as a mapping we insert a point for each member of the domain and in another portion of the picture we also insert a point for each point in the other set. We then draw an arrow from the point for x in the domain to the point for y in the range if and only if y is the value of the function at x.

The function g given by the table:

x	1	2	3	4	5	6
$g(x)$	2	4	6	8	10	12

is diagrammed as a mapping from its domain into $\{1,2,3,4,5,6,7,8,9,10,11,12\}$ in Figure 4.3.

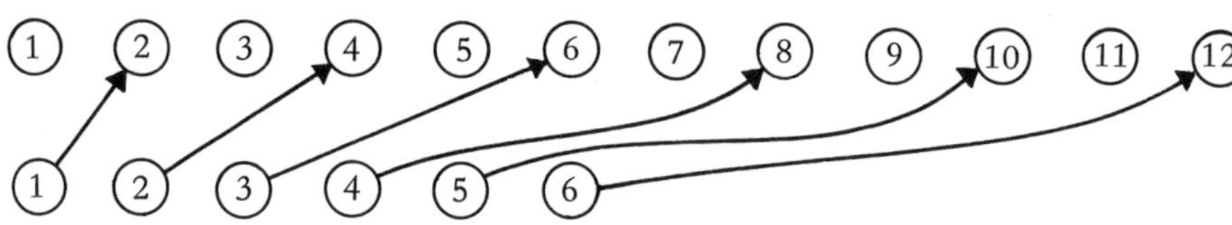

FIGURE 4.3 Diagram for the mapping g.

There is another useful way to think about functions. Suppose F is a function with domain D. Then we may think of F as a sort of machine. This machine, when given (a name of) any element in D, does some sort of maneuvers and then puts out (a name of) $F(x)$, the value of F at x. If the machine is fed a (name for an) element not in D, then the machine simply fails to operate and gives out nothing. See Figure 4.4.

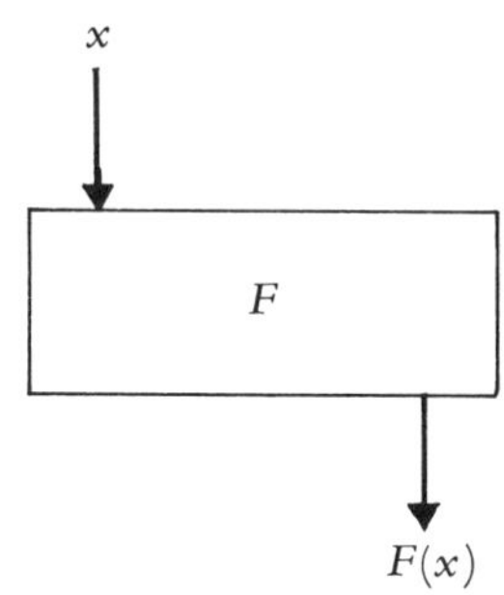

FIGURE 4.4 A function as a machine.

A function on a set A is sometimes also called a **correspondence** on A. If y is the value of the function at x, then y is said to **correspond to** x through (or by way of) the function.

We have in this section introduced quite a multitude of terms that are used in connection with functions. Fundamental notions, such as that of a function, somehow usually require (or acquire) a considerable technical vocabulary because of their many uses.

Definition 4.1.1. A relation F is said to be a **function** if for all x, y, and z, xFy and xFz implies $y = z$.

Definition 4.1.2. If F is a function and xFy, then we say that y is **the value of** F at x and write "$y = F(x)$." We read "$F(x)$" as "F at x" or "F of x."

Definition 4.1.3. The **domain** of a function F is $\{x \mid y = F(x) \text{ for some } y\}$. The **range** of a function F is $\{y \mid y = F(x) \text{ for some } x\}$. A function is said to be a **function on** its domain. A function is said to be **defined** at each point in its domain. If z is not in the domain of F, then F is said to be **undefined** at z.

Definition 4.1.4. Suppose F is a function with domain D and range R. We say that F **maps** D *onto* R. If B is a set such that $R \subset B$, we say that F **maps** D *into* B. If $y = F(x)$ we may say that y is the **image** of x under (or for) the mapping F and that x is a **pre-image** of y under (or for) the mapping F.

Definition 4.1.5. The **diagram** of a function F **as a mapping** from A into B is a geometric representation in which the elements in A are represented by certain points, the elements in B are represented by certain other points, and an arrow is drawn from the point for x to the point for y if and only if $y = F(x)$.

PROBLEMS

1. For each relation in Problem 5, Section 3.1 decide whether or not the relation is a function. For each function find the domain and range. Diagram each function as a mapping from its domain into C.

2. (a) By means of tables, list all possible functions mapping A into B, and (b) diagram each of these functions as a mapping into B if

(i) $A = \{1,2\}$, $B = \{1\}$
(ii) $A = \{1,2\}$, $B = \{1,2\}$
(iii) $A = \{1\}$, $B = \{1,2\}$
(iv) $A = \{1,2,3\}$, $B = \{1,2\}$
(v) $A = \{1,2\}$, $B = \{1,2,3\}$
(vi) $A = \{1,2,3\}$, $B = \{1,2,3\}$

3. On the basis of your results in Problem 2, conjecture how many different mappings there are (a) from $\{1,2,3\}$ into $\{1,2,3,4\}$, (b) from $\{1,2,3,4\}$ into $\{1,2,3\}$.

4. Do the problem that results if each occurrence of "into" is changed to "onto" in (a) Problem 2, (b) Problem 3.

5. Which of the following is a function?

(a) $\{(x,y)\,|\,y = 1 \text{ and } x = 1,2,3,4, \text{ or } 5\}$
(b) $\{(x,y)\,|\,x = 1 \text{ and } y = 1,2,3,4, \text{ or } 5\}$
(c) $A \times A$ where $A = \{1,2,3\}$
(d) $\{(x,y)\,|\,y$ is a parent of x, and x and y live in the U.S.$\}$
(e) $\{(x,y)\,|\,y = x$, and x and y are citizens of the U.S.$\}$
(f) $\{(x,y)\,|\,y$ is the mother of x, and x and y live in Chicago$\}$
(g) $\{(x,y)\,|\,y$ is the governor of x, and y is a person and x is a state in the U.S.$\}$
(h) $\{(x,y)\,|\,y$ is a sister of x, and x and y are students in your school$\}$
(i) $\{(x,y)\,|\,y$ is the number of sisters x has, and x is a student in your class$\}$
(j) $\{(x,y)\,|\,x$ is a student in your class, and y is the number of living grandparents that x has$\}$

6. Find the domain and range for each relation that is a function in Problem 5.

7. Diagram as mappings all possible functions mapping $\{0,1\} \times \{0,1\}$ into $\{0,1\}$. (There are 16 of them.)

8. Let the function p be defined by the following table:

x	1	2	3	4	5	6	7	8	9	10
$p(x)$	2	3	5	7	11	13	17	19	23	29

(a) If $y = p(3)$, find $p(y)$. Note that we could call this $p(p(3))$. (b) If $y = p(4)$, find $p(y)$. (c) If possible, find an x such that $p(x) = 13$. (d) If possible, find an x such that $p(x) = 14$. (e) If possible, find an x such that $p(x) = x + 2$. (f) If possible, find an x such that $p(x) = x$.

9. Answer the questions in Problem 8 if the function p is given by:

x	1	2	3	4	5	6	7	8	9	10
$p(x)$	1	2	2	3	2	4	2	4	4	4

10. If for some particular x we have $F(x) = x$, then x is said to be a **fixed point** for the function F. (That is, a **fixed point** for F is a point that is mapped onto itself.) Which of the functions have at least one fixed point in (a) Problem 2, (b) Problem 5, (c) Problem 7?

---

11. Prove or disprove each of the following: (a) if f is a function with domain A and g is a function with domain B, then $f \cup g$ (i.e. their union as a set of ordered pairs) must be a function with domain $A \cup B$. (b) If f is a function with domain A and g is a function with domain B and $A \cap B = \phi$, then $f \cup g$ is a function on $A \cup B$. (c) If f and g are functions then $f \cap g$ is a function. (d) If f maps A into B where $A \cap B = \phi$, then the diagram for f as a mapping into B is identical with its diagram as a relation in $A \cup B$. (e) If F is a function then $\{(x,y) \mid x = F(y)\}$ is also a function.

12. Suppose A is a set in a universal set U. The function F_A defined by $F_A(x) = 1$ if $x \in A$, $F_A(x) = 0$ if $x \notin A$ is called the **characteristic function** for the set A. Show that for all x (a) $F_{A'}(x) = 1 - F_A(x)$, (b) $F_{A \cap B}(x) = F_A(x)F_B(x)$, (c) $F_{A \cup B}(x) = 1 - (1 - F_A(x))(1 - F_B(x))$.

13. Argue that the total number of different subsets of $\{1,2,3,4,5,6\}$ is exactly the same as the number of different mappings from $\{1,2,3,4,5,6\}$ into $\{0,1\}$.

14. Suppose A is a collection of assertions. The function on A defined by

$$T(x) = 1 \text{ if } x \text{ is true}$$

$$T(x) = 0 \text{ if } x \text{ is false}$$

may be called the **truth value function** on A. Show that (a) $T(\text{not-}x) = 1 - T(x)$, (b) $T(x \text{ and } y) = T(x)T(y)$, (c) $T(x \text{ or } y) = 1 - (1 - T(x))(1 - T(y))$. [NOTE: not-$x$ is true if and only if x is false.]

15. Suppose R is an equivalence relation on a set A and R is also a function on A. What may we conclude about it?

16. If f is a function whose domain is a set A, we may extend f to a function defined on the collection of all subsets of A by the following device. For $B \subseteq A$ we let

$$f(B) = \{y \mid y = f(x) \text{ for some } x \text{ in } B\}.$$

In other words, $f(B)$ is the collection of the images (under f) of all the points in B. Note that the new function (for which we still used the name f) is an extension of f in the sense that $f(\{x\}) = \{f(x)\}$ for all x in A. Make a table for such an extension for the function (a) in Figure 4.2, (b) the function f on $A = \{1,2,3\}$ such that $f(x) = x$ for all x in A.

4.2 Composition of functions

Suppose F and G are functions and we think of them as machines. What happens if we connect up the machines for F and G in the manner shown in Figure 4.5? That is, what happens if we feed (the name of) an object x into the machine for G and then take the output of G and feed it into the machine for F? If F gives an output y when we do this, then we evidently have y as the value of F at $G(x)$. We may write this more briefly as "$y = F(G(x))$" which we read as "y is F of G of x."

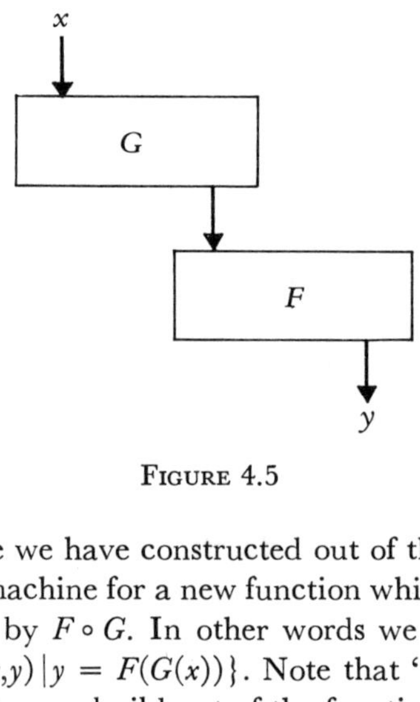

FIGURE 4.5

Now the new machine we have constructed out of the machines for F and G may be thought of as a machine for a new function which we call the **composite** of F and G and denote by $F \circ G$. In other words we define the composite of F and G by $F \circ G = \{(x,y) \mid y = F(G(x))\}$. Note that "$F \circ G$" is the name of a single function—a function we build out of the functions F and G. We call this building process **composition.**

Suppose F and G are defined by the tables

x	1	2	3	4	5	6	7	8	9	10
$G(x)$	10	9	8	7	6	5	4	3	2	1

z	1	2	3	4	5	6	7	8	9	10
$F(z)$	1	2	2	3	2	4	2	4	3	4

Then $F(G(4)) = F(7) = 2$, $F(G(5)) = F(6) = 4$, and $F(G(7)) = F(4) = 3$. Carrying through all such cases we get as a table for $F \circ G$:

x	1	2	3	4	5	6	7	8	9	10
$(F \circ G)(x)$	4	3	4	2	4	2	3	2	2	1

In Figure 4.6 we have diagrammed both F and G together as mappings and may consider the result as a diagram for $F \circ G$. To find $(F \circ G)\ (x)$ from the diagram in Figure 4.6, we start at the point for x at the top, follow the arrow from there to the point for $G(x)$ and then follow the arrow from there to the point for $F(G(x))$.

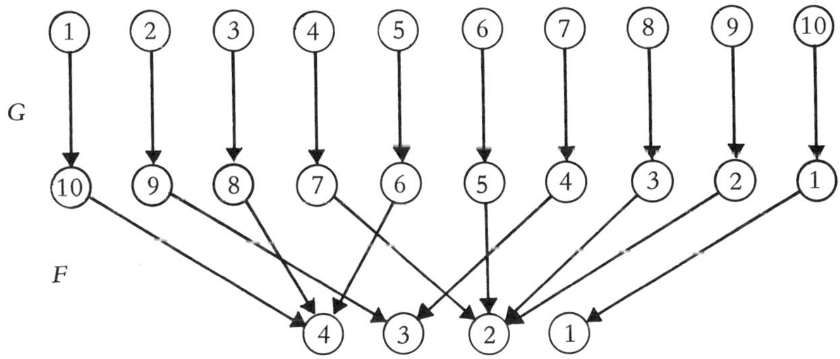

FIGURE 4.6

Notice how natural the mapping terminology is here. To find the value of $F \circ G$ at, say, 8 we refer to Figure 4.6 and see that 8 is mapped on 3 by G and then 3 is mapped on 2 by F. Thus $(F \circ G)\ (8) = 2$.

Will $H \circ K$ always be a function if H and K are functions? The answer is in the affirmative. Given any x, K pairs at most one value, $K(x)$, with that x. Then H pairs at most one value with $K(x)$. So $H \circ K$ is a set of ordered pairs in which no two distinct pairs have the same first number.

Of course, it is possible that $H \circ K$ will be empty. This will happen if no elements in the range of K happen to be in the domain for H.

With F and G as defined above what is $G \circ F$? We have $G(F(2)) = G(2) = 9$, $G(F(6)) = G(4) = 7$, and $G(F(8)) = G(4) = 7$. And if we work out all possible cases we have $G \circ F$ given by the table:

x	1	2	3	4	5	6	7	8	9	10
$(G \circ F)\ (x)$	10	9	9	8	9	7	9	7	8	7

Certainly $F \circ G \neq G \circ F$. Thus we see that the operation of composition of functions is *not* commutative.

However, composition is associative. That is, when F, G, and H are functions

$$F \circ (G \circ H) = (F \circ G) \circ H.$$

This is easy to believe after a look at Figure 4.7 where we have indicated how one would construct the machines for $F \circ (G \circ H)$ and $(F \circ G) \circ H$ out of those for F, G, and H.

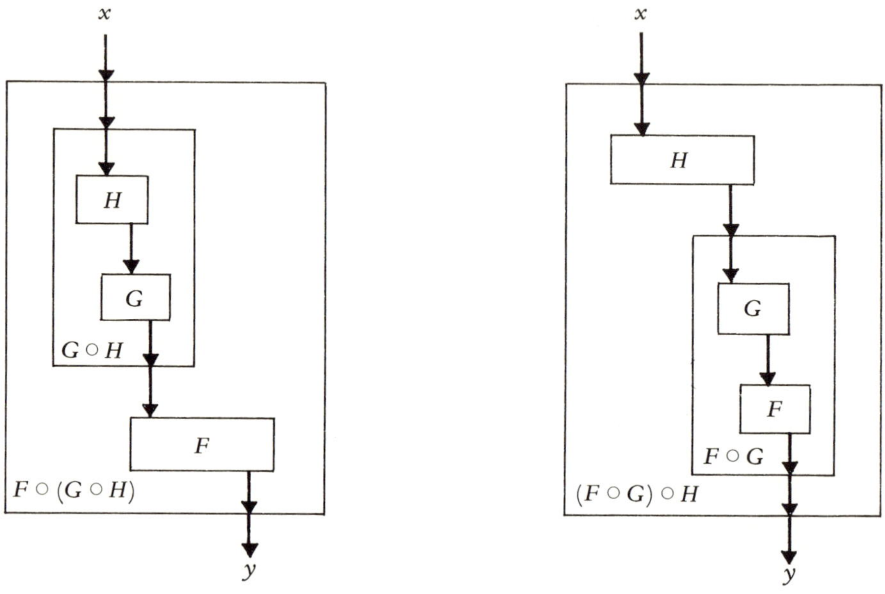

FIGURE 4.7

The associativity of composition turns out to be a fundamental fact that is used in a great many arguments although it may not be mentioned explicitly in many cases.

Sometimes, the domain of a function we wish to consider is a collection of ordered pairs. Consider, for example, the function diagrammed in Figure 4.8. The domain of this function is $\{1,2,3\} \times \{1,2,3\}$. If we call the function S, then it would be natural to use our functional notation in the fashion indicated by the equalities: $S(1,3) = 4$, $S(3,1) = 4$, $S(3,3) = 6$. That is, instead of writing $S((1,3)) = 4$ since S pairs 4 with $(1,3)$, we drop one set of parentheses and simply write $S(1,3) = 4$. More generally, $S(x,y)$ is the value paired with (x,y). When this notation is used S may be called a **function of two variables.** In other words, when the domain of a function is a set of ordered pairs we may say that the function is a function of two variables. We shall not use this "two-variable-terminology" very often here, but it is common in the literature of mathematics.

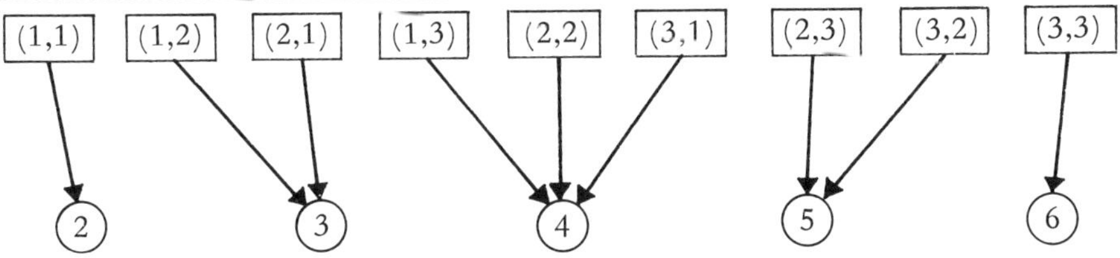

FIGURE 4.8

Definition 4.2.1. $F \circ G = \{(x,y) \mid y = F(G(x))\}$. $F \circ G$ is called the **composite** of F and G.

Theorem 4.2.1. If F and G are functions then $F \circ G$ is a function.

Theorem 4.2.2. $F \circ (G \circ H) = (F \circ G) \circ H$

PROOF: By Theorem 4.2.1, $F \circ (G \circ H)$ and $(F \circ G) \circ H$ are functions. Suppose $(x,y) \in F \circ (G \circ H)$ and $(x,z) \in (F \circ G) \circ H$. Then $y = F((G \circ H)(x)) = F(G(H(x)))$ and $z = (F \circ G)(H(x)) = F(G(H(x)))$. Therefore, $y = z$. Hence, $F \circ (G \circ H) = (F \circ G) \circ H$.

PROBLEMS

1. In each case below, if f and g are given by the table, construct a table for $f \circ g$ and $g \circ f$ and also diagram these functions as mappings as was done in Figure 4.6.

(a)

x	1	2	3	4	5	6
$f(x)$	1	2	1	2	1	2
$g(x)$	1	1	1	2	2	2

(b)

x	1	2	3	4	5	6	7	8	9
$f(x)$	6	5	3	7	9	1	2	4	8
$g(x)$	6	7	3	8	2	1	4	9	5

(c)

x	1	2	3	4	5	6	7
$f(x)$	2	2	2	3	3	3	3
$g(x)$	3	3	3	2	2	2	2

(d)

x	a	b	c	d	e
$f(x)$	e	d	c	b	a
$g(x)$	e	d	c	a	b

(e)

x	1	2	3	4	5
$f(x)$	a	b	a	d	c

(f)

y	a	b	c	d	e
$g(y)$	4	1	4	3	3

2. Make a table for $F \circ F$ and $F \circ F \circ F$ if F is the function in (a) Figure 4.1 (a), (b) Figure 4.3, (c) the population function at the beginning of Section 4.1.

3. Let F be a function on a set A whose range is a subset of A. Define a function I on A by $I = \{(x,y) \mid y = x\}$. I is called the **identity function** on A.† What can be said about $I \circ F$ and $F \circ I$?

4. Suppose F has domain A and range C. Suppose G has domain B and range D. Describe precisely in terms of A, B, C, and D what will be the domains of $F \circ G$ and $G \circ F$.

5. For the function S of Figure 4.8 find (a) $S(1,S(1,1))$, (b) $S(S(1,1),1)$, (c) $S(1,S(1,2))$, (d) $S(S(1,1),2)$, (e) $S(S(1,2),S(1,1))$, (f) $S(S(1,1),S(2,1))$.

6. Suppose P is the function diagrammed in Figure 4.9. The domain of P is $\{1,2,3\} \times \{1,2,3\}$. Find each of the following. (S is the function of Figure 4.8.) (a) $P(2,2)$, (b) $P(P(2,1),P(1,2))$, (c) $P(2,P(3,1))$, (d) $P(2,P(1,1))$, (e) $P(P(2,1),1)$, (f) $P(2,S(2,1))$, (g) $S(P(2,2),P(2,1))$.

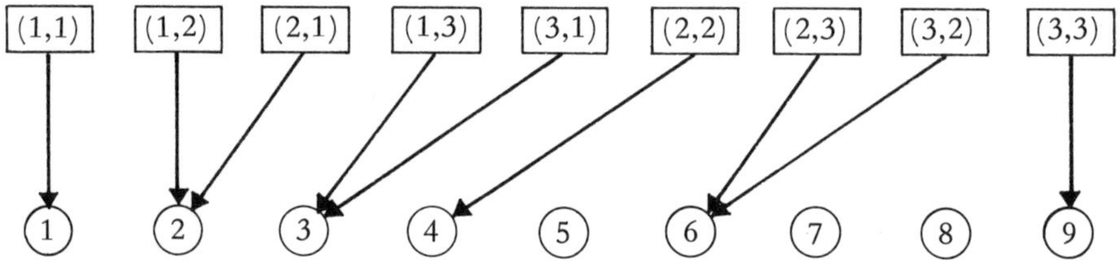

FIGURE 4.9

7. Given the function K defined by the table

x	1	2	3	4	5	6
$K(x)$	4	1	2	6	5	3

† Of course, I is just our old friend the equality relation on A, but sometimes old friends do get new names.

find a function M such that $(M \circ K)(x) = (K \circ M)(x) = x$ for all x in $\{1,2,3,4,5,6\}$.

8. Suppose B and C are subsets of A, and F is a function on A. Prove or disprove each of the following. (See Problem 15, Section 4.1.)

(a) If $B \subseteq C$, then $F(B) \subseteq F(C)$
(b) $F(B \cap C) = F(B) \cap F(C)$
(c) $F(B \cup C) = F(B) \cup F(C)$
(d) $F(B') = [F(B)]'$
(e) If $F(B) = B$ for all subsets B of A, then $F(x) = x$ for all $x \in A$, and conversely.

9. Among all the mappings from $\{1,2,3\}$ into $\{1,2,3\}$ as obtained in Problem 2 (vi), Section 4.1 (there are 27 of them), which of them are isomorphic as relations in $\{1,2,3\}$?

4.3 Inverse functions

Suppose we take the diagram for a relation in a set A and reverse all the arrows that appear in it. The result will be a diagram of another relation in A. This new relation is called the **inverse** of (or for) the original relation.†

For example, if A is a set of people and we consider the relation

$$C = \{(x,y) \mid x \text{ is a child of } y\}$$

then the inverse of C is the relation

$$C^{-1} = \{(x,y) \mid y \text{ is a child of } x\}.$$

This relation‡ would be described more naturally perhaps as

$$C^{-1} = \{(x,y) \mid x \text{ is a parent of } y\}.$$

Thus, "parent of" is the inverse of "child of." Notice that we have written "C^{-1}" as a name for the inverse of the relation C. This is standard practice. We read "C^{-1}" as "the inverse of C" or "C inverse." We could read "C^{-1}" as "C to the minus one," but this last terminology is best avoided.

In the general situation, when R is a relation, the **inverse relation** for R is given by

$$R^{-1} = \{(x,y) \mid yRx\}.$$

In Figure 4.10 we have diagrammed several relations and their inverses.

† What we call here the inverse relation is sometimes, in other books, called the **converse** relation.

‡ Note that $\{(x,y) \mid y \text{ is a child of } x\}$ and $\{(y,x) \mid x \text{ is a child of } y\}$ are the same sets. "x" and "y" are dummy variables here.

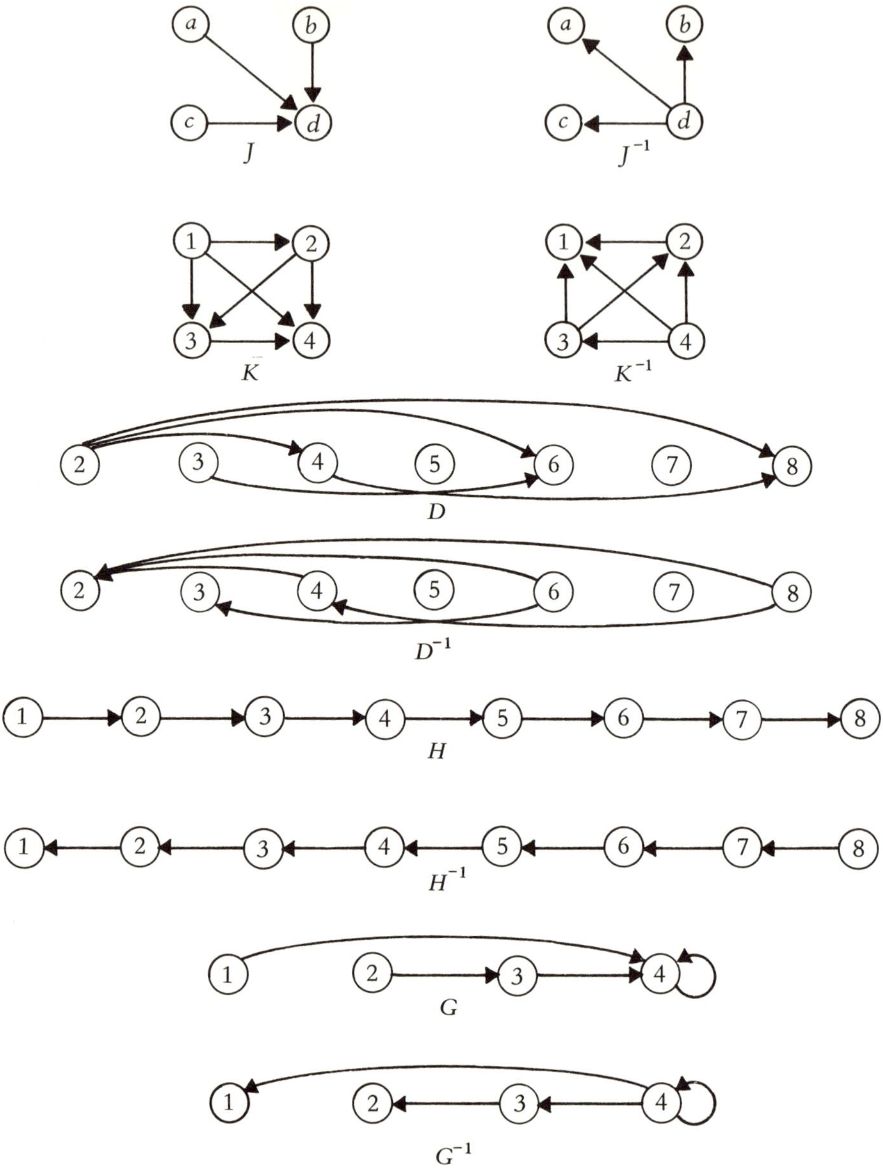

FIGURE 4.10

Notice that the relations H and G in Figure 4.10 are functions. Notice also that H^{-1} is a function but G^{-1} is *not*.

H and G are diagrammed as relations in Figure 4.10. If we diagram them as mappings we get Figure 4.11.

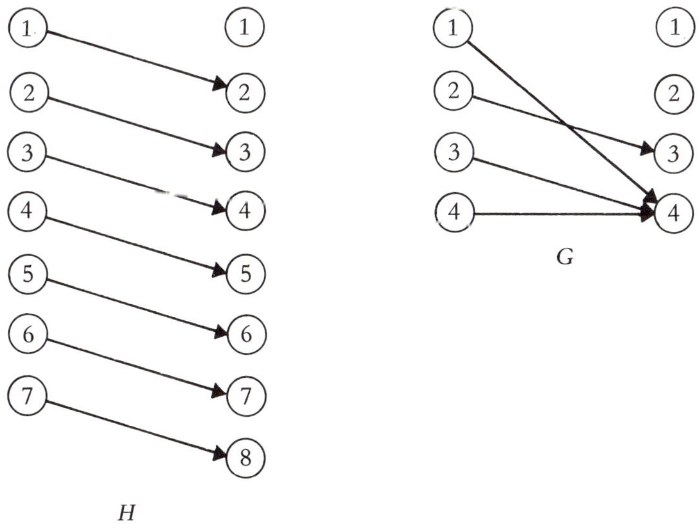

FIGURE 4.11

If we reverse all the arrows that appear in Figure 4.11, we clearly get a function from the diagram for H since each element of $\{2,3,4,5,6,7,8\}$ will then be paired with exactly one number, but not from G since there we shall have 4 paired with more than one item.

If F is any function and F^{-1} is a function we say that F^{-1} is the **inverse function** for F. We see from the above that there are some functions that do not have inverse functions.

If L is the function given by the table

x	2	3	4	5	6	7	8
$L(x)$	7	8	9	10	11	12	13

,

then L has an inverse function and it is given by the table

y	7	8	9	10	11	12	13
$L^{-1}(y)$	2	3	4	5	6	7	8

.

If S is the function given by the table

x	1	2	3	4	5	6	7	8
$S(x)$	1	4	9	16	25	36	49	64

,

then S has an inverse function and it is given by the table

y	1	4	9	16	25	36	49	64
$S^{-1}(y)$	1	2	3	4	5	6	7	8

.

If P is the function given by the table

x	1	2	3	4	5	6	7	8
$P(x)$	1	0	1	0	1	0	1	0

,

then P does not have an inverse function since 1 (as well as 0) is paired with more than one element in the domain of P.

Under what conditions does a function F have an inverse function? When for each element y in the range of F there is only one x such that $y = F(x)$. This was the case for H above, but not for G.

If a function F has an inverse function, it means that each y in the range of F has exactly one pre-image. The process of finding the pre-image x for the given y is the process of finding the value of F^{-1} at y. In other words, $y = F(x)$ if and only if $x = F^{-1}(y)$.

The machine for F^{-1} (when F has an inverse function) simply reverses whatever the machine for F does. If $x \to \boxed{F} \to y$ then $y \to \boxed{F^{-1}} \to x$, and conversely.

This brings us to another result. If F maps A onto B and F has an inverse function F^{-1}, then, as we see from Figure 4.12, $F^{-1}(F(x)) = x$ for all x in A and

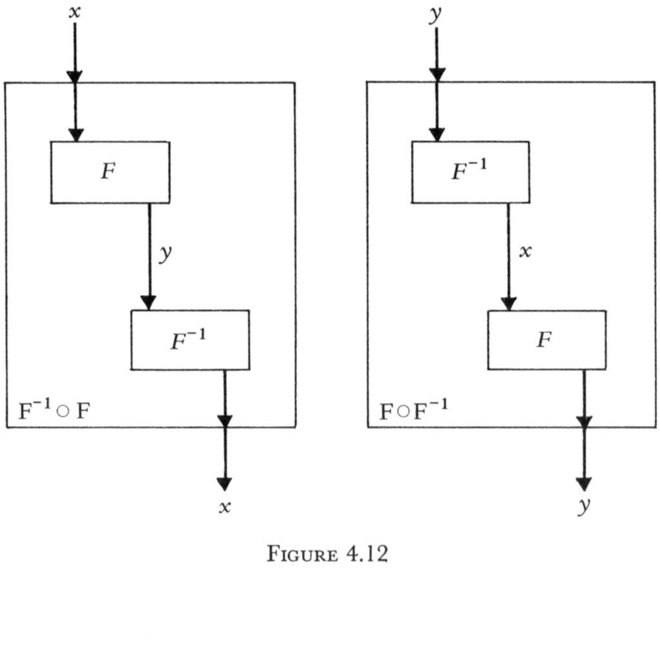

FIGURE 4.12

$F(F^{-1}(y)) = y$ for all y in B. We could express this somewhat differently by saying that $F^{-1} \circ F$ is the identity function on A and $F \circ F^{-1}$ is the identity function on B. (The **identity function** on a set is the one which maps each element onto itself. See Problem 3, Section 4.2.)

Looking at the situation from another angle, suppose F maps A onto B and G is a function such that $G(F(x)) = x$ for all x in A and $F(G(y)) = y$ for all y in B. Then we may conclude that F has an inverse function and that this inverse function is G.

A function that has an inverse function is said to be a **one-to-one** correspondence between its domain and range. The function H in Figure 4.11 is a one to one correspondence between $\{1,2,3,4,5,6,7\}$ and $\{2,3,4,5,6,7,8\}$.

In Figure 4.13 we have diagrammed as mappings all possible one to one correspondences between $\{1,2,3\}$ and $\{1,2,3\}$.

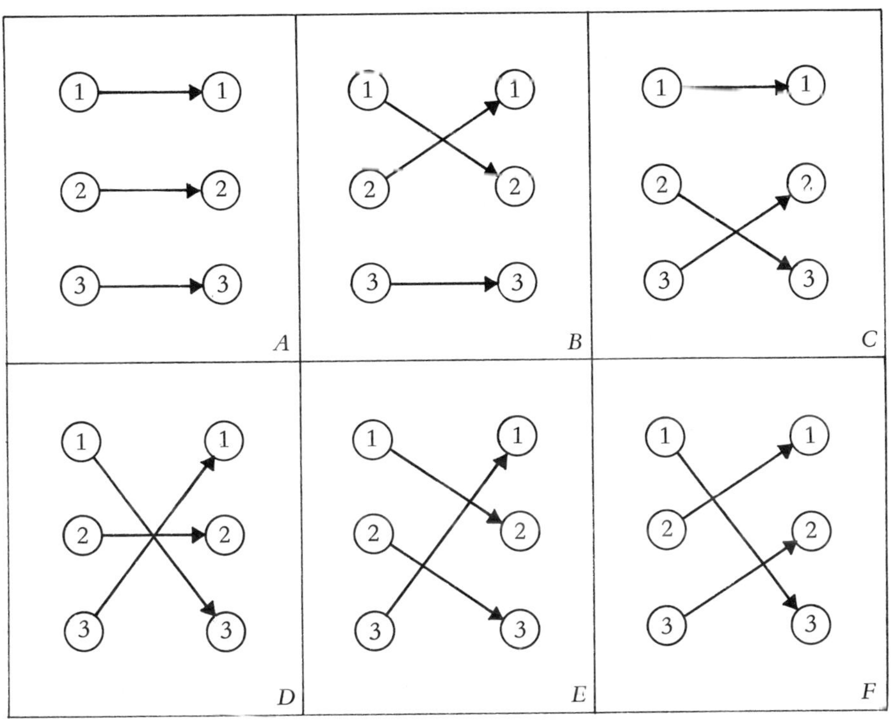

FIGURE 4.13 The one-to-one correspondences between $\{1,2,3\}$ and $\{1,2,3\}$.

If each person in a group is seated in a chair at a dinner table and there are no chairs left over, then this seating of the people sets up pairs which form a one-to-one correspondence.

If we make a list of all the states in the U.S. along with (the names of) the governors of the states, then this list sets up a one-to-one correspondence between the set of states and the set of governors. Every state has a governor and no governor is the governor of more than one state.

If we put our two hands together so that the tips of the fingers of the two hands touch, we set up a one-to-one correspondence between the two collections of fingers.

In a set of married people with both members of each couple in the set, $\{(x,y) \mid x \text{ is the husband of } y\}$ is a one-to-one correspondence between the set of men and the set of women. (We assume there is no bigamy and no polyandry in the group.)

Definition 4.3.1. If R is a relation the **inverse relation** for R is given by $R^{-1} = \{(x,y) \mid yRx\}$.

Definition 4.3.2. If F is a function and F^{-1} is a function then F^{-1} is called the **inverse function** for F.

Theorem 4.3.1. If F maps A onto B and F has an inverse function F^{-1}, then for all x in A and all y in B, $y = F(x)$ if and only if $x = F^{-1}(y)$.

Theorem 4.3.2. If F maps A onto B and F has an inverse function F^{-1}, then $F^{-1}(F(x)) = x$ for all x in A and $F(F^{-1}(y)) = y$ for all y in B.

Theorem 4.3.3. If F maps A onto B, G maps B onto A and $G(F(x)) = x$ for all x in A, then G is the inverse function for F.

Definition 4.3.3. If F is a function mapping A onto B and F has an inverse function, then F is said to be a **one-to-one correspondence** between A and B.

Problems

1. In each part of Problem 1, Section 4.2, determine if f and g have inverse functions. If they do, display them by means of tables.

2. (a) What is the inverse of the inverse of a relation R? (b) If F is a function with an inverse F^{-1}, must the function F^{-1} have an inverse function? What is it?

3. Find a natural phrase to describe the inverse of the relations described by (a) grandchild of (answer: grandparent of), (b) aunt or uncle of, (c) spouse of, (d) sibling of, (e) older than, (f) ancestor of, (g) richer than, (h) taller than, (i) brighter than, (j) heavier than, (k) longer than, (l) bigger than, (m) harder than, (n) fatter than, (o) dirtier than, (p) faster than, (q) equal to, (r) colder than, (s) ruder than.

4. Which of the functions in Problem 2, Section 4.1, have inverse functions? In each case where an inverse function exists find the sets A and B such that the function is a one-to-one correspondence between A and B.

5. Which of the functions in Problem 7, Section 4.1, have inverse functions?

6. If F^{-1} is the inverse function of F, what can be said about the fixed points for F^{-1} (see Problem 10, Section 4.1) in terms of the fixed points for F?

7. From your results in Problem 2, Section 4.1, conjecture how many different one-to-one correspondences there are mapping (a) $\{1,2,3\}$ onto $\{1,2,3,4\}$, (b) $\{1,2,3,4\}$ onto $\{1,2,3,4\}$, (c) $\{1,2,3,4\}$ onto $\{1,2,3\}$.

---

8. Suppose R is a relation in a set A. Prove that $R = R^{-1}$ if and only if R is symmetric. (Note then that any equivalence relation is its own inverse.)

9. Show that a relation R is reflexive on a set A only if R^{-1} is reflexive on A.

10. Show that the transitivity of R on a set A implies the transitivity of R^{-1} on A.

11. If R and S are relations in a set A and $R \subseteq S$, then $R^{-1} \subseteq S^{-1}$. Prove this.

12. Suppose K is the collection of all possible relations in $\{0,1\}$. On K, let "RIS" mean that the relation R is the inverse of the relation S. Is the relation I (a) reflexive on K, (b) symmetric on K, (c) transitive on K?

13. On the set K of Problem 12 let "RJS" mean that the relation R is either the same as S or is the inverse of S. Is J an equivalence relation? If so, what are the equivalence classes for J on K?

14. Suppose F is a one-to-one correspondence between A and B and G is a one-to-one correspondence between B and C. Show that $G \circ F$ is a one-to-one correspondence between A and C. [HINT: What kind of a mapping is $F^{-1} \circ G^{-1}$?]

4.4 Isomorphisms

We are now in a position to be somewhat more precise about the notion of two relations having essentially the same structure (i.e. being isomorphic). Consider the two relations diagrammed in Figure 4.14.

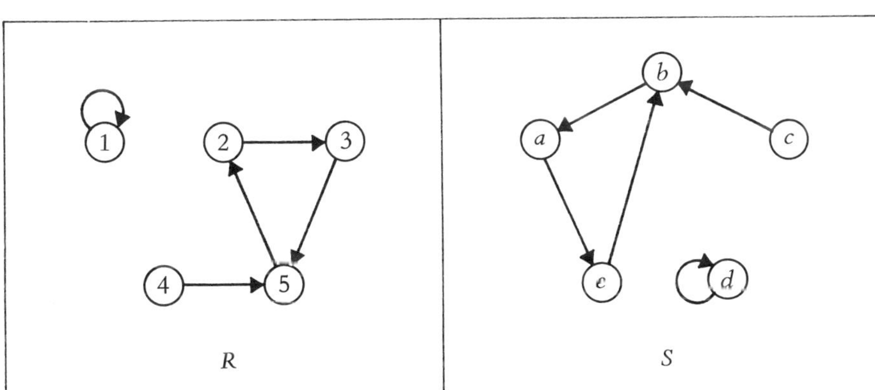

FIGURE 4.14

Let $A = \{1,2,3,4,5\}$ and $B = \{a,b,c,d,e\}$. Then R is a relation in A and S is a relation in B. Clearly, these two relations are "look alikes." In fact, if we reposition the points appropriately when we diagram S we get the diagram of Figure 4.15 for S. And if in Figure 4.15 we erase the "d" and put a "1" in its

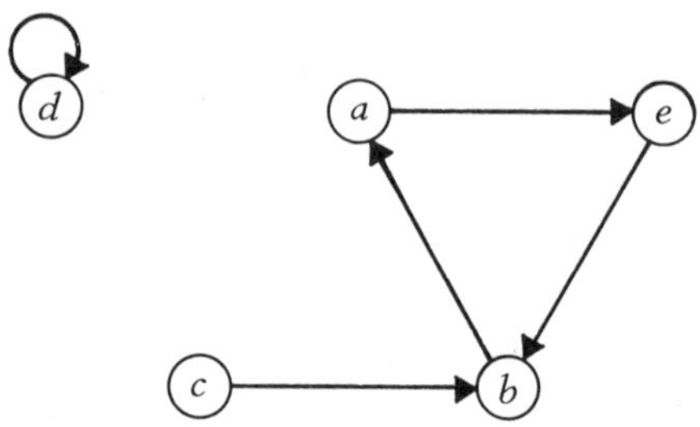

FIGURE 4.15

place, replace "a" with "2," replace "e" with "3," replace "c" with "4," and replace "b" with "5," we get exactly the diagram for R.

Now another way to put this is to say that we have discovered a one-to-one correspondence, call it f, between B and A given by the table:

x	a	b	c	d	e
$f(x)$	2	5	4	1	3

And this correspondence has the property that when, in the diagram for S, we replace each x in B by $f(x)$ we get exactly a diagram for R. That is, whenever there is an arrow connecting x and y in the diagram for S there is an arrow connecting $f(x)$ and $f(y)$ in the diagram for R, and conversely. In other words, f is a one-to-one correspondence between B and A with the property that for all x and y in B, xSy if and only if $f(x)Rf(y)$. It is this last way of putting the situation that we follow in stating our definition of "isomorphic" more precisely than we have so far. The precise definition is:

> If M is a relation in a set C and N is a relation in a set D we say that M and N are **isomorphic** if there is a one-to-one correspondence F between C and D such that, for all x and y in C, xMy if and only if $F(x)NF(y)$. Such a function F is said to be an **isomorphism** between C and D with respect to the relations M and N.

Consider the relations L and G diagrammed in Figure 4.16. L and G are isomorphic since the function g defined by the table:

x	1	2	3	4
$g(x)$	4	3	2	1

is an isomorphism between $\{1,2,3,4\}$ and $\{1,2,3,4\}$ (i.e. between $\{1,2,3,4\}$ and itself) with respect to L and G.

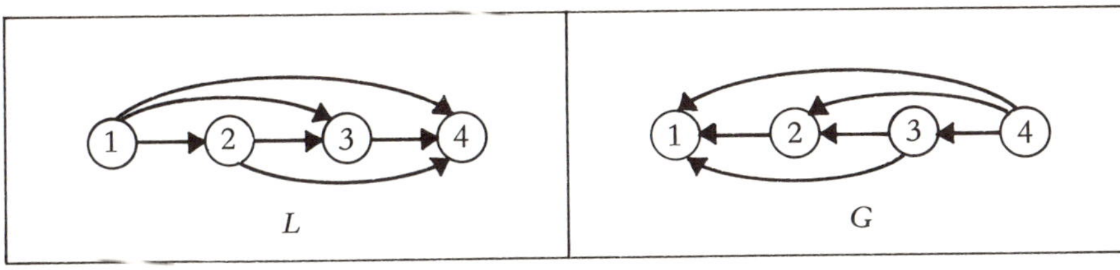

FIGURE 4.16

If M is any relation in a set C, then M is certainly isomorphic to itself since the function F defined by $F(x) = x$ is an isomorphism between C and C with respect to M and M.

If the relation M in a set C is isomorphic to the relation N in a set D then certainly N is isomorphic to M. This follows from the fact that if F is an isomorphism between C and D with respect to M and N, then F^{-1} (which exists since F is a one-to-one correspondence) is an isomorphism between D and C with respect to N and M.

Both of these last results certainly make a great deal of sense when we think of "M is isomorphic to N" as meaning that M and N have the same type of structure. Each relation should have the same kind of structure as itself. And if M has the same kind of structure that N has, then surely N should have the same kind of structure as M.

Furthermore, we would expect that if M and N have the same type of structure, and N and P have the same type of structure, then M and P do also. Stating this more precisely we have:

> Suppose M is a relation in C, N is a relation in D, and P is a relation in E. If M is isomorphic to N and N is isomorphic to P, then M is isomorphic to P.

This is indeed true as we now show. Let F be an isomorphism between C and D with respect to M and N. Let G be an isomorphism between D and C with respect to N and P. We would like to find an isomorphism between C and E with respect to M and P. The function $H = G \circ F$ is such. This is so because H is a one-to-one correspondence between C and E (see Problem 14, Section 4.3), and whenever x and y are in C we have xMy if and only if $F(x)NF(y)$, and $F(x)NF(y)$ if and only if $G(F(x))PG(F(y))$. Hence, for all x and y in C we have xNy if and only if $H(x)PH(y)$.

From the above results we see that if we consider the collection W of all relations in a certain set A, then the relation $I = \{(P,Q) \mid P \text{ is isomorphic to } Q\}$ is an **equivalence relation**† on W.

---

† Note that I is a relation between relations!

Definition 4.4.1. Let M be a relation in a set C and N be a relation in a set D. We say that M and N are **isomorphic** if there is a one-to-one correspondence F between C and D such that, for all x and y in C, xMy if and only if $F(x)NF(y)$. Such an F is said to be an **isomorphism** between C and D with respect to M and N (in that order).

Theorem 4.4.1. Any relation in any set is isomorphic to itself.

Theorem 4.4.2. Suppose M is a relation in a set C and N is a relation in a set D. If M is isomorphic to N, then N is isomorphic to M.

Theorem 4.4.3. If M, N, and P are relations in the sets C, D, and E, respectively, and M is isomorphic to N and N is isomorphic to P, then M is isomorphic to P.

Problems

1. For the relations given by Figure 3.8 prove, by displaying the isomorphism, that (i) b is isomorphic to c, (ii) b is isomorphic to d, (iii) c is isomorphic to d. (iv) Construct the composition of the isomorphisms in parts i and iii, and show that it is an isomorphism between b and d.

2. For the relations R and S of Figure 4.14, is there any other function, aside from the function f we constructed, which maps $\{a,b,c,d,e\}$ onto $\{1,2,3,4,5\}$ and is an isomorphism between these two sets with respect to S and R? Explain.

3. Verify your answers to Problem 2, Section 3.3 by actually displaying an isomorphism for each pair of isomorphic relations. (A nice way to display an isomorphism is to diagram the relations side by side and then use colored arrows to show the isomorphism mapping.)

4. Construct all possible isomorphisms between the relations diagrammed in (a) Figure 4.17 (there are two of them), (b) Figure 4.18, (c) Figure 4.19. (See the suggestion at the end of Problem 3 and use different colors for different isomorphisms if you wish.)

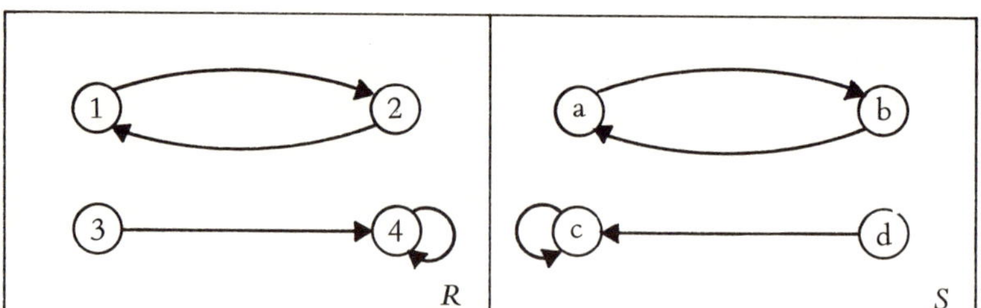

Figure 4.17

5. If R and S are relations in a set A we may say that R and S are *of the same* (structural) **type** if R and S are isomorphic. It ties in nicely with this termi-

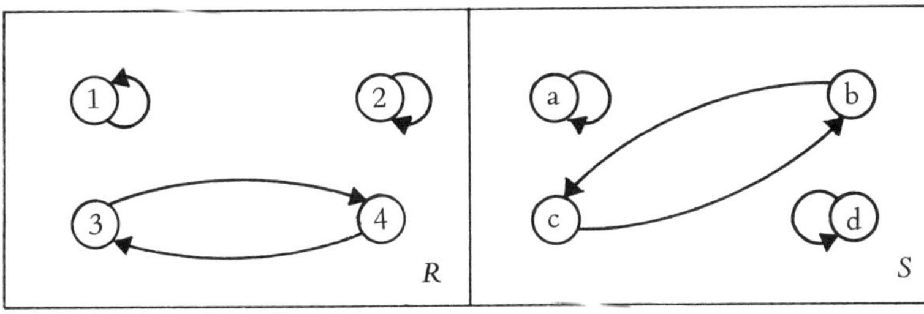

FIGURE 4.18

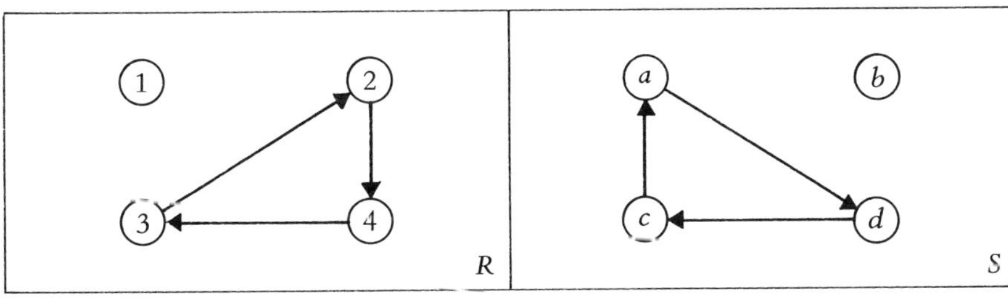

FIGURE 4.19

nology to say that the **type** of a relation is the equivalence class to which the relation belongs when we consider the equivalence relation $\{(P,Q) \mid P$ is isomorphic to $Q\}$ on the collection of all relations in A. Taking $A = \{1,2,3,4\}$, diagram all the relations that are of the same type as R if R is (a) the relation R in Figure 4.17, (b) the relation $\{(1,1)\}$, (c) the relation of equality on A, (d) the relation L in Figure 4.16 (there are 24 of them), (e) the relation G in Figure 4.16.

6. Do Problem 7, Section 3.3 using our more precise statement of what it means for two relations to be isomorphic.

5

Operations

5.1 Properties of operations

Quite early in life you learned about the operations of addition and multiplication. If we perform the operation of addition with the numbers 5 and 7 we get a result, called the sum, which is the number 12. We take a pair of numbers and perform the operation of addition to get another number we call their sum. Similarly, we can take a pair of numbers, perform the operation of multiplication, and get a number called their product. With each of these operations we take a pair of numbers and find another number that we call the result of the operation.

But this amounts to saying that an operation is a function defined on a set of ordered pairs. We simply call the value of the function by the name "result of the operation." It is in this form that we state our general definition of what an operation is:

An **operation** is a function whose domain is a set of ordered pairs.† For example, the function M defined by the table:

(x,y)	(0,0)	(0,1)	(1,0)	(1,1)
$M(x,y)$	0	1	1	1

† What we have called an operation here is also sometimes called a **binary operation.** If this terminology is used (and we shall not usually use it), then "unary operation" is synonymous with "function."

is an operation. We could have defined this operation on the given set of ordered pairs by: $z = M(x,y)$ if and only if z is the larger of the numbers x and y. We can also describe M by means of the **operation table:**

M	0	1
0	0	1
1	1	1

Here we are recognizing that the domain of the function is the Cartesian product $\{0,1\} \times \{0,1\}$ and we are making use of the tabular device for representing such a product. And in each box in the table corresponding to an ordered pair we have entered the value of M at that pair.

Thus, we are constructing a table whose use is similar to that of the multiplication table we all memorized at some time in elementary school.

As further examples, consider the operations defined by the tables below. Some of them will be familiar.

+	1	2	3	4	5	6	7	8	9	10
1	2	3	4	5	6	7	8	9	10	11
2	3	4	5	6	7	8	9	10	11	12
3	4	5	6	7	8	9	10	11	12	13
4	5	6	7	8	9	10	11	12	13	14
5	6	7	8	9	10	11	12	13	14	15
6	7	8	9	10	11	12	13	14	15	16
7	8	9	10	11	12	13	14	15	16	17
8	9	10	11	12	13	14	15	16	17	18
9	10	11	12	13	14	15	16	17	18	19
10	11	12	13	14	15	16	17	18	19	20

·	1	2	3	4	5	6	7	8	9	10
1	1	2	3	4	5	6	7	8	9	10
2	2	4	6	8	10	12	14	16	18	20
3	3	6	9	12	15	18	21	24	27	30
4	4	8	12	16	20	24	28	32	36	40
5	5	10	15	20	25	30	35	40	45	50
6	6	12	18	24	30	36	42	48	54	60
7	7	14	21	28	35	42	49	56	63	70
8	8	16	24	32	40	48	56	64	72	80
9	9	18	27	36	45	54	63	72	81	90
10	10	20	30	40	50	60	70	80	90	100

÷	1	2	3	4	5	6	7	8	9	10
1	1	–	–	–	–	–	–	–	–	–
2	2	1	–	–	–	–	–	–	–	–
3	3	–	1	–	–	–	–	–	–	–
4	4	2	–	1	–	–	–	–	–	–
5	5	–	–	–	1	–	–	–	–	–
6	6	3	2	–	–	1	–	–	–	–
7	7	–	–	–	–	–	1	–	–	–
8	8	4	–	2	–	–	–	1	–	–
9	9	–	3	–	–	–	–	–	1	–
10	10	5	–	–	2	–	–	–	–	1

$\oplus$	0	1	2
0	0	1	2
1	1	2	0
2	2	0	1

$\odot$	0	1	2
0	0	0	0
1	0	1	2
2	0	2	1

$\cap$	ϕ	{1}	{2}	{1,2}
ϕ	ϕ	ϕ	ϕ	ϕ
{1}	ϕ	{1}	ϕ	{1}
{2}	ϕ	ϕ	{2}	{2}
{1,2}	ϕ	{1}	{2}	{1,2}

$\cup$	ϕ	{1}	{2}	{1,2}
ϕ	ϕ	{1}	{2}	{1,2}
{1}	{1}	{1}	{1,2}	{1,2}
{2}	{2}	{1,2}	{2}	{1,2}
{1,2}	{1,2}	{1,2}	{1,2}	{1,2}

$\underset{+}{\sim}$	0	1
0	0	1
1	1	0

$\underset{\cdot}{\sim}$	0	1
0	0	0
1	0	1

G	0	1	2
0	0	0	0
1	1	0	0
2	1	1	0

E	0	1	2
0	1	0	0
1	0	1	0
2	0	0	1

D	0	1	2
0	0	–	–
1	1	0	–
2	2	1	0

All of these tables are to be used just as we use an ordinary multiplication table. To find the product of 5 and 7 we go to the table for $\cdot$, look in the row labeled "5" and in the column labeled "7," and there find the (name of the) product 35. To find the result of the operation G on 1 and 2 (in that order) we go to the table for G, look in the row labeled "1" and in the column labeled "2," and there find the result 0.

If we used functional notation to express our last conclusions we would write $\cdot(5,7) = 35$ and $G(1,2) = 0$. But it is customary when dealing with operations to use another notation where the operation name is written between the names of the members of the pairs.† In this notation, we would write the above results as "$5 \cdot 7 = 35$" and "$1\ G\ 2 = 0$."

In general, when F is an operation we write "$z = xFy$" to mean that $z = F(x,y)$.

Notice then that it is important for us to be aware of whether we are talking about **operations** or **relations** as we were earlier. If F is an operation then xFy is an object—the object that F pairs with (x,y). If, however, F is a relation and is being treated as such, then xFy is an assertion—the assertion that $(x,y) \in F$. It should always be clear from the context which is meant. And, of course, in many cases we use special names for the operation or relation involved so there is little chance for confusion. We would not think of "$2 = \frac{4}{2}$" as the result of an operation to be performed on 2 and 4/2. Nor would we consider "$2 + 3$" to be an assertion.

In the table for $\div$ above, we find $9 \div 3 = 3$. If we look for $9 \div 4$ we find simply a dash. In constructing this division table, we allowed ourselves only whole numbers so we left $9 \div 4$ undefined. This is the intent of the blank (or dash) in that position.

To illustrate further this operator notation we have the following from the appropriate tables: $1 \oplus 1 = 2$, $1 \oplus (1 \oplus 1) = 1 \oplus 2 = 0$, $1 \underset{+}{\sim} 1 = 0$, $1\ D\ 0 = 1$, and $1\ D\ 2$ is undefined.

Any two-way table may be thought of as defining an operation. For example, the first table on page 92 summarizes the 1966 American League baseball season‡

† We have a similar choice of devices in English where we may either speak of "the product of 5 and 7" or else "5 times 7."

‡ Data from *The New Information Please Almanac* (New York; Simon and Schuster, 1967), p. 925.

by giving in the x row and the y column the number of times that team x beat team y during the season. The second table below gives† in the x row and the y column the (observed or predicted) population in millions of continent x in the year y.

B	Baltimore	Minnesota	Detroit	Chicago	Cleveland	California	Kansas City	Washington	Boston	New York
Baltimore	–	10	9	9	8	12	11	11	12	15
Minnesota	8	–	7	14	9	7	10	14	12	8
Detroit	9	11	–	10	9	9	6	13	10	11
Chicago	9	4	8	–	11	10	13	12	7	9
Cleveland	10	9	9	7	–	8	6	9	11	12
California	6	11	9	8	10	–	9	7	9	11
Kansas City	5	8	12	5	12	9	–	9	9	5
Washington	7	4	5	6	9	11	9	–	10	10
Boston	6	6	8	11	7	9	9	8	–	8
New York	3	10	7	9	6	7	13	5	10	–

P	1965	1975	1985	2000	2020
Africa	310	398	520	779	1,320
Asia	1,889	2,343	2,863	3,701	5,100
Europe	675	732	792	886	1,040
Oceania	14	16	20	25	35
North America	294	354	431	578	850
South America	166	221	291	420	680

We might even use the operator notation in connection with such tables and, for example, read "(Asia) P (1975)" as the "Asia(n) population in 1975."

There are many things we could notice about several of the above tables. First, we might remark that in some of the tables only objects that appear as row or column headings appear as entries in the body of the table. This is true in the tables for $\div$, $\oplus$, $\cup$, $\cap$, $\underset{+}{\sim}$, $\underset{\cdot}{\sim}$, $\odot$, G, E, and D. But it is not true in the tables for $+$, $\cdot$, B, and P. This property we characterize, in general, as follows:

We say that an operation F is an operation *in* a set A if the domain of F is contained in $A \times A$ and the range of F is contained in A. Thus $\div$ as given by

† Data from H. Kahn and A. J. Wiener, *The Year 2000* (New York; Macmillan, 1967), p. 139.

the table above is an operation *in* $\{1,2, \ldots ,10\}$. Similarly, $\oplus$ as given by the table above is an operation *in* $\{0,1,2\}$. However, $+$ as given by the table above is not an operation *in* $\{1,2, \ldots ,10\}$ since some of the sums fall outside this set.

Among the above tables some have blank places in them and some do not. For operations that are operations *in* a set and whose tables have no blanks there is a special terminology.

We say that a set A is **closed under an operation** F if F is an operation in A and the domain of F is $A \times A$.

Thus, $\{0,1,2\}$ is closed under the operation $\oplus$ above and also closed under the operation $\odot$. However, $\{1,2, \ldots ,10\}$ is not closed under the operation $\div$ and $\{0,1,2\}$ is not closed under the operation D. $\{1,2, \ldots ,10\}$ is not closed under either $+$ or $\cdot$. $\{\phi, \{1\}, \{2\}, \{1,2\}\}$ *is* closed under both $\cap$ and $\cup$. The idea of a set A being closed under an operation is that the operation may be carried out on any pair of elements in A and the result will always be in A.

Some of the tables above have the property that they are symmetric about the diagonal running from upper left to lower right. That is, if we were to flip the table about this line it would look just the same. The tables for $+$, $\cdot$, $\oplus$, $\odot$, $\cap$, $\cup$, $\underset{+}{\sim}$, $\underset{\cdot}{\sim}$, and E all have this property. The tables for $\div$, G, D, B, and P do not An operation whose table has this property is said to be **commutative.** More precisely we have:

> An operation F is said to be **commutative** if $xFy = yFx$ for all x and y for which xFy or yFx is defined.

Earlier, we commented that $\cup$ and $\cap$ are commutative. We also said in Chapter 2 that these two operations are **associative.** In the general case we have the definition:

> An operation F is said to be **associative** if $(xFy)Fz = xF(yFz)$ for all x, y, and z for which either of the two sides of the equality is defined.

It is much more difficult to test for associativity by looking at the table for an operation than it is to test for commutativity. We know from what we did earlier that the operations $\cup$ and $\cap$ defined by the above tables *are* associative. $\div$ is clearly not associative since $8 \div (4 \div 2) = 8 \div 2 = 4$ whereas $(8 \div 4) \div 2 = 2 \div 2 = 1$.

A property that two operations in a set may have jointly is the following:

> If F and G are operations in a set A we say that F **distributes across** G if $xF(yGz) = (xFy)G(xFz)$ for all x, y, and z in A.

From what we said in Chapter 2 about unions and intersections it is clear that with $\cap$ and $\cup$ as defined by the above tables, $\cap$ distributes across $\cup$. On the other hand, D does not distribute across $\oplus$ since $1\ D\ (1 \oplus 1) = 1\ D\ 2$ which is undefined, whereas $(1\ D\ 1) \oplus (1\ D\ 1) = 0 \oplus 0 = 0$.

We may think of an operation as a machine into which we may feed (the names of) two objects. The machine then does its work† and puts out (a name of) the result. In Figure 5.1 we have represented the operation F as a machine.

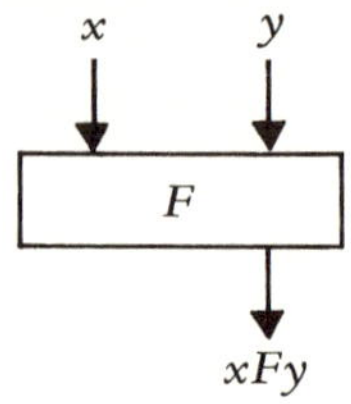

FIGURE 5.1 An operation as a machine.

To say that F is commutative is to say that for any inputs x and y the machine in Figure 5.1 always gives the same output as the machine in Figure 5.2.

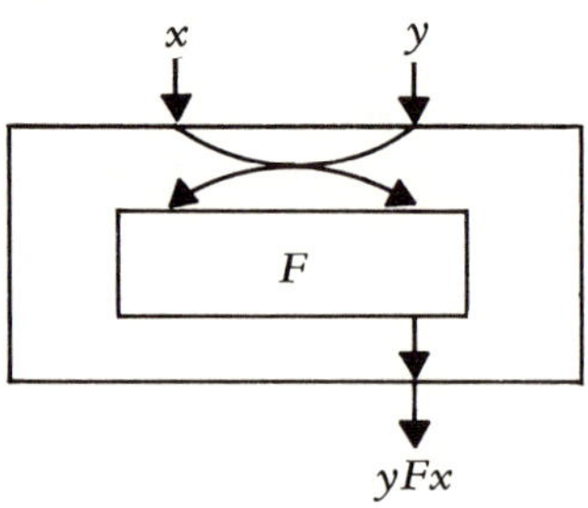

FIGURE 5.2

To say that F is associative is to say that the two machines in Figure 5.3 always give the same output (or none at all) for any inputs x, y, and z.

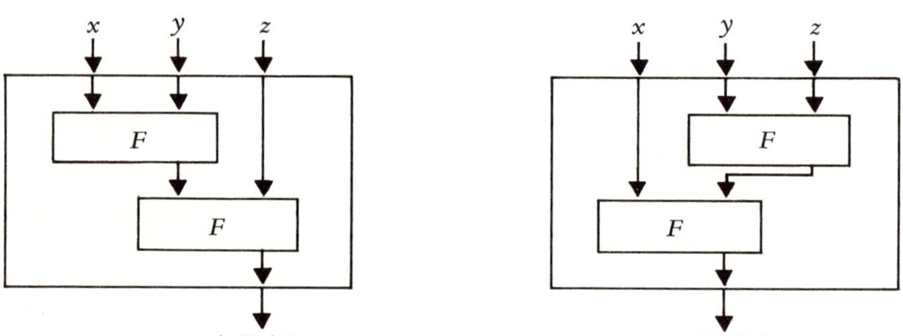

FIGURE 5.3

† We may think of a table that defines the operation as part of the machine. Given x and y (in that order), a marker moves down the left column until it finds the row labeled "x." Then the marker moves right along that row until it reaches the column labeled "y." The machine then reads the name in the position reached and gives it as output.

To say that F distributes across G is to say that the two machines in Figure 5.4 always give the same output (or none at all) for any inputs x, y, and z.

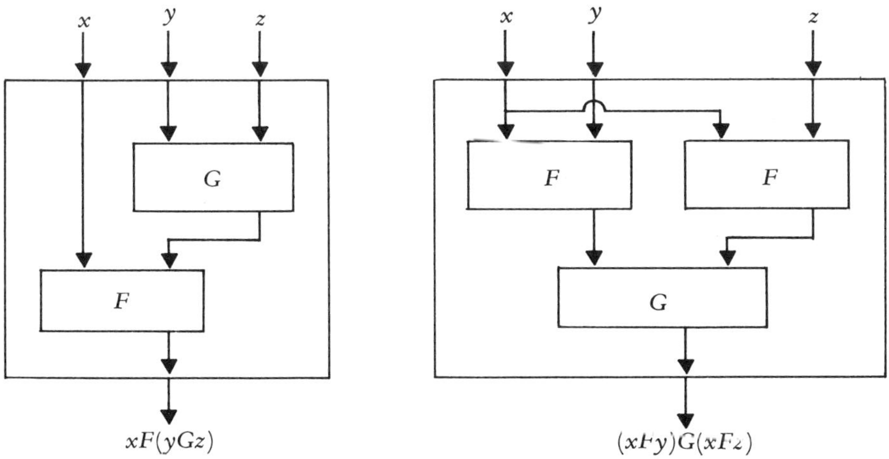

FIGURE 5.4

We may state the conditions for commutativity, associativity, and distributivity in functional notation. An operation F is commutative if $F(x,y) = F(y,x)$. It is associative if $F(x,F(y,z)) = F(F(x,y),z)$. F distributes across G if $F(x, G(y,z)) = G(F(x,y),F(x,z))$. These equalities are somewhat more difficult to fathom than the corresponding ones using the standard notation for the result of an operation.

Definition 5.1.1. An **operation** is a function whose domain is a set of ordered pairs. If F is an operation and $z = F(x,y)$ then we call z the **result** of performing the operation F on x and y (in that order).

Definition 5.1.2. A **table** for an operation F is a rectangular display such that on the row labeled "x" and in the column labeled "y" we find a name for $F(x,y)$.

Definition 5.1.3. An operation F is said to be an operation *in* a set A if the domain of F is a subset of $A \times A$ and the range of F is a subset of A.

Definition 5.1.4. If F is an operation we shall write "$z = xFy$" to mean that $z = F(x,y)$.

Definition 5.1.5. To say that a set A is **closed** under the operation F is to say that F is an operation in A and, further, that the domain of F is $A \times A$.

Definition 5.1.6. An operation F is said to be **commutative** if $xFy = yFx$ for all x and y for which xFy or yFx is defined.

Definition 5.1.7. An operation F is said to be **associative** if $(xFy)Fz = xF(yFz)$ for all x, y, and z for which either side of the equality is defined.

Definition 5.1.8. If F and G are operations in a set A then F is said to **distribute across** G if $xF(yGz) = (xFy)G(xFz)$ for all x, y, and z for which either side of the equality is defined.

PROBLEMS

1. Referring to the tables in this section where the operations are defined find (a) $1 \oplus 2$, (b) $2 \oplus (1 \oplus 2)$, (c) $(2 \oplus 1)$, (d) $(2 \oplus 1) \oplus 2$, (e) $2 \odot (1 \oplus 2)$, (f) $(2 \odot 1) \oplus (2 \odot 2)$, (g) $2 \oplus (2 \odot 2)$, (h) $(2 \oplus 2) \odot (2 \oplus 2)$, (i) $1 \underset{+}{\sim} (1 \underset{+}{\sim} 0)$, (j) $(1 \underset{+}{\sim} 1) \underset{+}{\sim} 0$, (k) $(1 \odot 0) \odot 1$, (l) $1 \odot (0 \odot 1)$, (m) $1 \odot (1 \oplus 0)$, (n) $(1 \odot 1) \oplus (1 \odot 0)$, (o) $1 \oplus (1 \odot 1)$, (p) $(1 \oplus 1) \odot (1 \oplus 1)$.

2. Consider the operations $\overset{+}{\sim}$ and $\dot{\sim}$ in $\{0,1,2,3\}$ defined by the tables:

$\overset{+}{\sim}$	0	1	2	3
0	0	1	2	3
1	1	2	3	0
2	2	3	0	1
3	3	0	1	2

$\dot{\sim}$	0	1	2	3
0	0	0	0	0
1	0	1	2	3
2	0	2	0	2
3	0	3	2	1

(a) Are $\overset{+}{\sim}$ and $\dot{\sim}$ commutative? (b) It can be shown that both $\overset{+}{\sim}$ and $\dot{\sim}$ are associative. Verify this partially by taking four different triples x, y, and z and checking that $(x \overset{+}{\sim} y) \overset{+}{\sim} z = x \overset{+}{\sim} (y \overset{+}{\sim} z)$ and $(x \dot{\sim} y) \dot{\sim} z = x \dot{\sim} (y \dot{\sim} z)$. (c) It can be shown that $\dot{\sim}$ distributes across $\overset{+}{\sim}$. Partially verify this by working out a few examples. (d) Show that $\overset{+}{\sim}$ does not distribute across $\dot{\sim}$ by evaluating $1 \overset{+}{\sim} (1 \dot{\sim} 2)$ and $(1 \overset{+}{\sim} 1) \dot{\sim} (1 \overset{+}{\sim} 2)$.

3. The operations $\oplus$ and $\odot$ as defined above can be interpreted physically in terms of a three-hour clock as shown in Figure 5.5. $x \oplus y$ can be interpreted

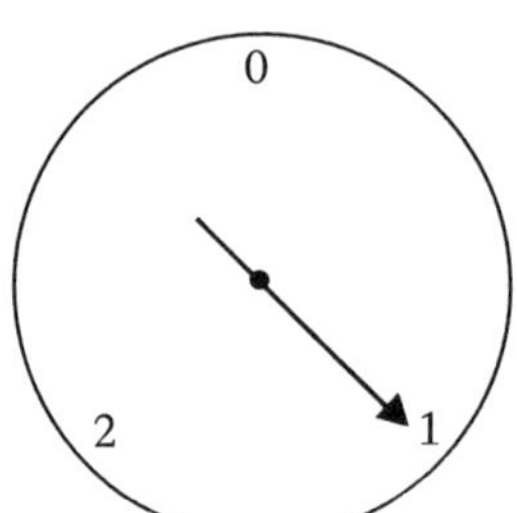

FIGURE 5.5

as the result we obtain if we set the clock hand at x, then rotate the hand in a clockwise direction through y hours, and then read the number indicated by the arrow. $\odot$ can then be interpreted as repeated operations of $\oplus$. That is, we could define $\odot$ by saying $0 \odot x = 0$, $1 \odot x = x$, and $2 \odot x = x \oplus x$ for all x in $\{0,1,2\}$. (a) Check that this does give the tables for $\oplus$ and $\odot$. (b) Write up similar interpretations for the operations (i) $\underset{+}{\sim}$ and $\underset{\cdot}{\sim}$ as defined above, (ii) $\overset{+}{\sim}$ and $\overset{\cdot}{\sim}$ as defined in Problem 2.

4. Prove that $\underset{+}{\sim}$ is indeed associative by completing the following table:

x	y	z	$(x \underset{+}{\sim} y) \underset{+}{\sim} z$	$x \underset{+}{\sim} (y \underset{+}{\sim} z)$
0	0	0	0	0
1	0	0	1	1
0	1	0		
0	0	1		
1	1	0		
0	1	1		
1	0	1		
1	1	1		

5. Does $\oplus$ distribute across $\odot$? (See the results of Problem 1.)

6. By consulting a table similar to that in Problem 4, prove or disprove that $\underset{+}{\sim}$ distributes across $\underset{\cdot}{\sim}$.

7. Make a list of all the possible triples we would need to consider if we wished to prove that $\oplus$ is associative by means of a table such as in Problem 4. (Do not construct a table.) Guess how many triples we would need to consider to use a table to show that $\overset{+}{\sim}$ is associative.

8. (a) Construct a table for subtraction (denoted by "$-$") in $\{1,2,\ldots,10\}$. Allow only positive whole numbers as results. (b) Is $-$ commutative? (c) Is $-$ associative? (d) Does $-$ distribute across $-$? (e) Is $\{1,2,\ldots,10\}$ closed under $-$? (f) Is $-$ an operation *in* $\{1,2,\ldots,10\}$?

9. **Concatenation** is the operation of writing two symbols (or strings of symbols) next to each other in a row. If we let "C" name this operation then, for example, $aCb = ab$, $(an)Cd = and$, $oCr = or$. (a) Make up an operation table for C taking its domain to be restricted to $A \times A$ where $A = \{a,b,c,d\}$. (b) Is C commutative? (c) Is A closed under C?

10. Construct the table for the Cartesian product operation in the collection S of all subsets of $\{1,2\}$. (a) Is S closed under $\times$? (b) Is $\times$ commutative? (c) Is $\times$ associative?

---

11. If F is an operation in a set A, we may extend F to get an operation in the collection of all subsets of A by making the following definition: For $D \subseteq A$ and $C \subseteq A$ we define $D\,F\,C$ by $D\,F\,C = \{z \mid z = xFy$ for some x in D and some y in $C\}$. (a) Make out a table for such an extension of $\underset{+}{\sim}$ as defined above. (b) Make out a table for such an extension of $\odot$ as defined above. (c) Does the commutativity or associativity of F in A imply the same property for the extension of F?

12. Suppose F is a commutative operation in a set A. If B is a subset of A with the property that for every x in A there is a y in A such that xFy is in B, then we call B a **complete set of results** for F in A. (a) Is $\{0\}$ a complete set of results for $\odot$? (b) Is $\{1\}$ a complete set of results for $\oplus$? (c) Is $\{1,2\}$ a complete set of results for $\odot$? (d) Is $\{\{1,2\}\}$ a complete set of results for $\cup$ in $\{\phi, \{1\}, \{2\}, \{1,2\}\}$? (e) Is $\{\{1,2\}\}$ a complete set of results for $\cap$ in $\{\phi, \{1\}, \{2\}, \{1,2\}\}$?

13. An operation whose domain is $\{1,2, \ldots ,n\} \times \{1,2, \ldots ,m\}$ is sometimes called an n by m **matrix.** For example, the operation A defined by:

A	1	2	3
1	1	3	1
2	2	1	2

is a 2 by 3 matrix. There is no chance for ambiguity if we suppress the row and column headings in this table as long as we know we are dealing with a matrix. Therefore, we may simply define the matrix A by

$$A = \begin{pmatrix} 1 & 3 & 1 \\ 2 & 1 & 2 \end{pmatrix}.$$

It is in this sense that a matrix is sometimes defined to be simply a rectangular array. If A is a matrix, the object $A(i,j)$ is called the (i,j) **th element** of the matrix and is usually denoted by a_{ij}. a_{ij} is the number named in the ith row and the jth column in the table for A. The "ij" here is a **double subscript.** It does not, of course, mean i times j. List all possible 2 by 3 matrices whose elements are in the set $\{0,1\}$ and for which $a_{11} = 1$ and $a_{23} = 0$.

14. One way to treat the situation when F is an operation in A and A is not closed under F is the following. Suppose $u \notin A$. Let us put a "u" in each blank space that occurs in the table for A. That is, we insert "u" in each position where F is undefined. Also add to the table for F a row and a column headed "u". Fill this new row and this new column with "u"s. Finally, replace the "F" in the table with "F^*." Now F^* is an operation in $B = A \cup \{u\}$. And B is **closed** under F^*. Show that F^* is (a) commutative, (b) associative on B if and only if F has the same property on A.

5.2 Identity elements and inverses

We again display here, for easy reference, the tables defining the operations $\oplus$ and $\odot$.

$\oplus$	0	1	2
0	0	1	2
1	1	2	0
2	2	0	1

$\odot$	0	1	2
0	0	0	0
1	0	1	2
2	0	2	1

Notice that in the table for $\oplus$ the column headings may simply be copied in the same order as given to obtain the first row in the body of the table. Also the row headings may simply be copied to obtain the first column in the table.

In the table for $\odot$ we may obtain the second row in the body of the table by simply copying the column headings. We may obtain the second column in the table for $\odot$ by simply copying the row headings.

Putting these observations in a more compact form we would have

$$0 \oplus x = x \oplus 0 = x \quad \text{for all } x \text{ in } \{0,1,2\}$$

and

$$1 \odot x = x \odot 1 = x \quad \text{for all } x \text{ in } \{0,1,2\}.$$

In other words, performing the operation $\oplus$ on 0 and any number x gives x as a result. Performing the operation $\odot$ on 1 and any number x leaves x unchanged. We say that 0 is the **identity element** for $\oplus$. We say that 1 is the **identity element** for $\odot$.

More generally, if F is an operation whose domain is contained in $A \times A$ and e is an element in A with the property that

$$eFx = xFe = x \quad \text{for all } x \text{ in } A,$$

then we say that e is the† **identity element** for F.

In the table given in the preceding section for $\cdot$, the ordinary operation of multiplication on $\{1,2, \ldots ,10\} \times \{1,2, \ldots ,10\}$, we see that 1 is the identity element for $\cdot$.

From the tables given on page 90 for $\underset{+}{\sim}$ and $\underset{\cdot}{\sim}$ we see that 0 is the identity element for $\underset{+}{\sim}$ and 1 is the identity element for $\underset{\cdot}{\sim}$.

On the other hand, there is no identity element for the operation E given in the preceding section since, for example, there is no element e such that $eE2 = 2$.

The existence of an identity element for an operation is important since when there is an identity element you may operate with it and cause no change.

Does the identity element 0 for $\oplus$ occur in every row in the table for $\oplus$? In every column? The answer to both questions is "Yes." Does the identity element 1 for $\odot$ occur in each row of the table for $\odot$? In each column? The answer to both these questions is "No."

In other words, for each x in $\{0,1,2\}$, it is possible to find a y such that $y \oplus x = 0$, and it is also possible to find a z such that $x \oplus z = 0$. But given an x in $\{0,1,2\}$ it may not be possible to find a y such that $x \odot y = 1$ or to find a z such that $z \odot x = 1$.

Now, since 0 is the identity element for $\oplus$, we know that if we perform the operation $\oplus$ on any number and 0 we leave that number unchanged. But in view of what we said above, if we perform the operation $\oplus$ on, say, w and x, we

† It is perhaps not obvious that there could not be two different elements with the desired property. But such is the case as we shall confirm in the appendix to this section.

know that there is a y such that $x \oplus y = 0$. Then we can undo what we did in operating on w with x because we shall have (by the associativity of $\oplus$):

$$(w \oplus x) \oplus y = w \oplus (x \oplus y) = w \oplus 0 = w.$$

It is important to be able to undo something we have done.

If we look closely at the table for $\oplus$, we see that when $y \oplus 0 = 0$ and $0 \oplus z = 0$, then we must have $y = z = 0$. Similarly, when $y \oplus 1 = 0$ and $1 \oplus z = 0$, we have $y = z = 2$. When $y \oplus 2 = 0$ and $2 \oplus z = 0$, we have $y = z = 1$. As a consequence, we say that 0 is the **inverse element** for 0, 2 is the **inverse element** for 1, and 1 is the **inverse element** for 2 with respect to the operation $\oplus$.

In general, when F is an operation with identity element e, we say that y is an **inverse element** for x with respect to F if $yFx = xFy = e$.

If an element x has exactly one inverse we shall often use "x^{-1}" as a name for this inverse. We read "x^{-1}" as "the inverse of x" or "x inverse."

We may summarize the situation for inverses with respect to $\oplus$ by the table:

x	0	1	2
x^{-1}	0	2	1

Even though an operation has an identity element, there is no guarantee that a given element will have an inverse. In the table for $\odot$, 1 is the identity element. 1 is the inverse element for 1 with respect to $\odot$, and 2 is the inverse element for 2 with respect to $\odot$, but 0 has no inverse element with respect to $\odot$.

In the table for $\cup$ in $\{\phi, \{1\}, \{2\}, \{1,2\}\}$ on page 90 we see that ϕ is the identity element. But in this table ϕ is the only element with an inverse.

In the table for E on page 91, there is no identity element, so no element has an inverse with respect to E.

It is possible for an element to have more than one inverse. For example, if we have the operation W defined by the table

W	1	2	3
1	1	2	3
2	2	1	1
3	3	1	1

then 1 is the identity element, and 2 and 3 are both inverses for 2. However, it can be shown that if an operation is associative, each element has at most one inverse.† We shall prove this later for those of you that are interested.

† Since this is so, we see that W must not be associative.

Definition 5.2.1. If F is an operation whose domain is a subset of $A \times A$, we say that e is an **identity element** for F in A if $xFe = eFx = x$ for all x in A.

Theorem 5.2.1. There is at most one identity element for an operation.

PROOF: Suppose $aFx = xFa = x$ and $bFx = xFb = x$ for all x. Then $aFb = b$ from the stated property of a. Also $aFb = a$ from the stated property of b. Hence, $a = b$.

Definition 5.2.2. If F is an operation with an identity element e, we say that y is an **inverse element** for x with respect to F if $xFy = yFx = e$.

Theorem 5.2.2. If F is associative then each element has at most one inverse.

PROOF: Suppose $xFa = aFx = e$ and $xFb = bFx = e$. Since F is associative we have $aF(xFb) = (aFx)Fb$. But also $aF(xFb) = aFe = a$ and $(aFx)b = eFb = b$. Hence $a = b$.

PROBLEMS

1. For the operations $\overset{+}{\sim}$, $\dot{\sim}$, $\underset{+}{\sim}$, $\underset{\cdot}{\sim}$, and $\cap$ as defined in the previous section find (a) the identity element, (b) the inverse for each element that has one.

2. Are there identity elements for the operations $\div$, G, D, B, or P as defined in the previous section?

3. If we consider the operation of concatenation in a set of strings of symbols (see Problem 9, Section 5.1) what string of symbols should we include to serve as an identity for this operation?

4. In the table for multiplication $(\cdot)$ on page 89 which elements have inverses?

5. For the operation M on page 89 is there an identity element? If so, find the inverses for the elements that have them.

6. Show that if e is the identity element for the operation F, then e is an inverse element for e.

7. Suppose a set A is closed under the operation F and F is associative with identity element e. Show that if a and b have inverses, then

$$(aFb)^{-1} = (b^{-1})\ F\ (a^{-1}).$$

[HINT: Start by considering $(aFb)\ F\ [(b^{-1})\ F\ (a^{-1})] = [(aFb)\ F\ (b^{-1})]\ F\ (a^{-1}) = [aF(bFb^{-1})]\ F\ (a^{-1})$.]

8. For each of the operations given below find the identity element if there is one, and find the inverse for each element, if any.

J	a	b
a	a	b
b	b	a

K	a	b	c
a	a	b	c
b	b	c	a
c	c	a	b

L	a	b	c	d
a	a	a	a	a
b	a	b	c	d
c	a	c	b	c
d	a	d	c	b

M	a	b	c	d
a	a	b	c	d
b	b	a	d	c
c	c	d	a	b
d	d	c	b	a

N	a	b	c	d
a	a	a	a	a
b	b	b	b	b
c	a	b	c	d
d	d	d	d	d

V	t	f
t	t	t
f	t	f

$\rightarrow$	t	f
t	t	f
f	t	t

$\boxplus$	odd	even
odd	even	odd
even	odd	even

$\boxdot$	odd	even
odd	odd	even
even	even	even

9. In the tables below we find all possible operations under which $\{1,2\}$ is closed. In how many of them is it the case that (a) there is an identity element, (b) both 1 and 2 have inverses, (c) the operation is commutative?

(a)	1	2
1	1	1
2	1	1

(b)	1	2
1	2	1
2	1	1

(c)	1	2
1	1	2
2	1	1

(d)	1	2
1	1	1
2	2	1

(e)	1	2
1	1	1
2	1	2

(f)	1	2
1	2	2
2	1	1

(g)	1	2
1	2	1
2	2	1

(h)	1	2
1	2	1
2	1	2

(i)	1	2
1	1	2
2	2	1

(j)	1	2
1	1	2
2	1	2

(k)	1	2
1	1	1
2	2	2

(l)	1	2
1	1	2
2	2	2

(m)	1	2
1	2	1
2	2	2

(n)	1	2
1	2	2
2	1	2

(o)	1	2
1	2	2
2	2	1

(p)	1	2
1	2	2
2	2	2

10. As a class project determine which of the operations in Problem 9 are associative.

11. Construct tables for all possible operations F under which $\{1,2,3\}$ is closed, and for which 1 is the identity element and $2F2 = 3F3 = 1$. (There are nine of them.) In how many of these is it the case that (a) the operation is

commutative, (b) all elements have inverses? [SUGGESTION: You might use Ⓐ, Ⓑ, Ⓒ . . . as names for your operations.]

12. Using the "three-hour clock" interpretation for $\oplus$, explain why the inverses with respect to $\oplus$ are what they are.

13. Verify that the operation W on page 100 is indeed not associative by computing $(3W2)W2$ and $3W(2W2)$.

14. Let A be the collection of all mappings from $\{1,2\}$ into $\{1,2\}$. Diagram each member of A and assign a name to each. Make a table for the operation of composition on A. Find the identity element and the inverse of each function that has one (with respect to the operation of composition).

15. Do Problem 14 for the collection of all mappings of (a) $\{1,2\}$ **onto** $\{1,2\}$, (b) $\{1,2,3\}$ **onto** $\{1,2,3\}$. (See Figure 4.13 for part b.)

---

16. Suppose A is a set and F is an operation in A such that (i) A is closed under F, and (ii) F is associative. Then A is said to be a **semigroup** under (or with respect to) the operation F. The study of semigroups is an area of considerable interest and research in mathematics. If F is also commutative, then A is said to be a **commutative semigroup** or an **Abelian semigroup.** (a) In view of the results of Problem 10, in how many different ways may $\{1,2\}$ be made into a semigroup? (b) Is $\{0,1,2\}$ a semigroup with respect to $\oplus$? (c) Is $\{0,1,2\}$ a semigroup with respect to $\odot$? (d) Is $\{a,b,c\}$ a semigroup with respect to K? (See Problem 8 and for the question of associativity look carefully at the situation in part (b).) (e) Is $\{t,f\}$ a semigroup with respect to V? (See Problem 8.)

17. Suppose A is a semigroup with respect to F. Prove or disprove each of the following: (a) There must be an identity element† in A. (b) If there is an identity element in A then each element must have an inverse. (c) If A has an identity element e and x has an inverse x^{-1} then for all y in A

$$\text{(i) } z = xFy \quad \text{if and only if} \quad x^{-1}Fz = y$$

and

$$\text{(ii) } w = yFx \quad \text{if and only if} \quad wFx^{-1} = y.$$

18. Let $N = \{1,2,3, \ldots\}$ be the set of positive whole numbers. If $A \subseteq N$ and $B \subseteq N$ define the operation $\tilde{+}$ by $A \mathbin{\tilde{+}} B = \{z \mid z = x + y$ for some x in A and for some y in $B\}$. Argue informally that the collection of all subsets of N is a semigroup with respect to this operation. (To make clear the nature of the operation we have, for example, $\{2,3,5\} \mathbin{\tilde{+}} \{2,7,9\} = \{4,5,7,9,10,11,12,14\}$ and $\{2,4,6,8, \ldots\} \mathbin{\tilde{+}} \{2,4,6,8, \ldots\} = \{4,6,8,10,12, \ldots\}$.)

19. Suppose that A is a semigroup with respect to F. Show that the collection

† A semigroup with an identity element is called a **monoid.**

of all subsets of A is a semigroup with respect to the extension of F as defined in Problem 11, Section 5.1.

5.3 Inverse operations

In our everyday interactions with our environment and with other people, we often find ourselves doing certain acts or operations and then, later, undoing them. We put our shoes on in the morning and we take them off at night. We put some money into the bank on Friday and then take some money out on Monday. We put the milk into the refrigerator in the morning and then take it out at lunch. We give someone a pen to use and later we take it back.

When we undo what we did earlier we end up (roughly) back where we started in the first place.

An action which undoes what another action does could be called the **inverse action** for the other. "Taking off" is the inverse action for "putting on" for shoes. "Withdrawing" is the inverse action for "depositing" at a bank. "Taking out" is the inverse action for "putting in" for milk and refrigerators. "Taking back" is the inverse action for "lending" for pens.

If we *put on* two shoes in the morning and then *take off* two shoes at night, we end up as we started, with two bare feet. If we *deposit* \$150 in a bank on Friday and then *withdraw* \$150 on the next Monday our balance will be just as it was when we started.† If we *put in* two quarts of milk in the morning and *take out* two quarts at lunch the amount of milk in the refrigerator is as it was when we started.

We follow the lead of these examples as we begin to see what it means to say that one operation is the inverse operation of another. If F is a given operation and G is to be something like an inverse operation for F, we would like $(xFy)Gy = x$ wherever this is possible. For if this is so then G in some sense undoes what F does.

Consider, for example, the operation $\underset{+}{\sim}$ defined by the table

$\underset{+}{\sim}$	0	1
0	0	1
1	1	0

We claim that $\underset{+}{\sim}$ is its own inverse in the above sense. The proof of this is displayed in the table at the top of page 105.

† At least this is the case if we ignore the possibility of the account having earned some interest between Friday and Monday.

x	y	$x \underset{+}{\sim} y$	$(x \underset{+}{\sim} y) \underset{+}{\sim} y$
0	0	0	0
0	1	1	0
1	0	1	1
1	1	0	1

We see that $(x \underset{+}{\sim} y) \underset{+}{\sim} y = x$ for all x and y in $\{0,1\}$. Thus, if we wish to undo what we have done when we perform the operation $\underset{+}{\sim}$ on x with y we need only perform the operation $\underset{+}{\sim}$ with y again.

As another example, consider the operation $\underset{\cdot}{\sim}$ defined by the table

$\underset{\cdot}{\sim}$	0	1
0	0	0
1	0	1

Suppose we want to find an operation K which is an inverse operation for $\underset{\cdot}{\sim}$. What should be the value of $0K0$? We evidently want $(0 \underset{\cdot}{\sim} 0)K0 = 0$ and $(1 \underset{\cdot}{\sim} 0)K0 = 1$ both to hold. But $0 \underset{\cdot}{\sim} 0 = 1 \underset{\cdot}{\sim} 0 = 0$ so we evidently need both $0K0 = 0$ and $0K0 = 1$. This is, of course, impossible since we want K to be an operation and an operation must assign at most one value to each pair. Evidently we should leave $0K0$ undefined.

Such a situation corresponds to that with ordinary actions where we cannot undo what we have done. An example would be capital punishment. We certainly cannot undo what we have done when we kill a man.

Moving ahead to see to what extent we *can* construct an inverse operation K for $\underset{\cdot}{\sim}$, we ask what should be the value of $0K1$. We want $(x \underset{\cdot}{\sim} 1)K1 = x$ whenever $x \underset{\cdot}{\sim} 1 = 0$. But the only x value for which $x \underset{\cdot}{\sim} 1 = 0$ is $x = 0$. Hence, we should let $0K1 = 0$.

What should be the value of $1K1$? We want $(x \underset{\cdot}{\sim} 1)K1 = x$ whenever $x \underset{\cdot}{\sim} 1 = 1$. But $x \underset{\cdot}{\sim} 1 = 1$ implies $x = 1$. Hence, we should define $1K1$ to be 1.

What should be the value of $1K0$? We want $(x \underset{\cdot}{\sim} 0)K0$ whenever $x \underset{\cdot}{\sim} 0 = 1$. But $x \underset{\cdot}{\sim} 0$ is never equal to 1. Hence, to force K to have the property we want we could evidently define $1K0$ to be anything at all or else we could leave $1K0$ undefined. We shall choose the latter of these alternatives for reasons that we shall explain subsequently.

Finally, then we have an inverse operation K for $\underset{\cdot}{\sim}$ to the extent that we can get one:

K	0	1
0	–	0
1	–	1

The procedure we have followed above can be described in the general case by means of a flow diagram. Figure 5.6 gives a process for finding the value of zGy if G is to be an inverse operation for F and F is given by a table. The process can be summarized by stating that we take zGy to be equal to x if x is the unique value such that $xFy = z$. Otherwise we leave zGy undefined. The G constructed in this fashion will, insofar as it is possible, have the property we wish. When there is a unique x such that $z = xFy$ we shall have $(xFy)Gy = zGy = x$. When we have both $xFy = z$ and $wFy = z$ with $x \neq w$, then we leave zGy undefined for if not we would evidently have $x = (xFy)Gy = zGy = (wFy)Gy = w$ which is impossible. When there is no value x such that $xFy = z$, then we opt to leave zGy undefined, but this does not affect the desired property for G.

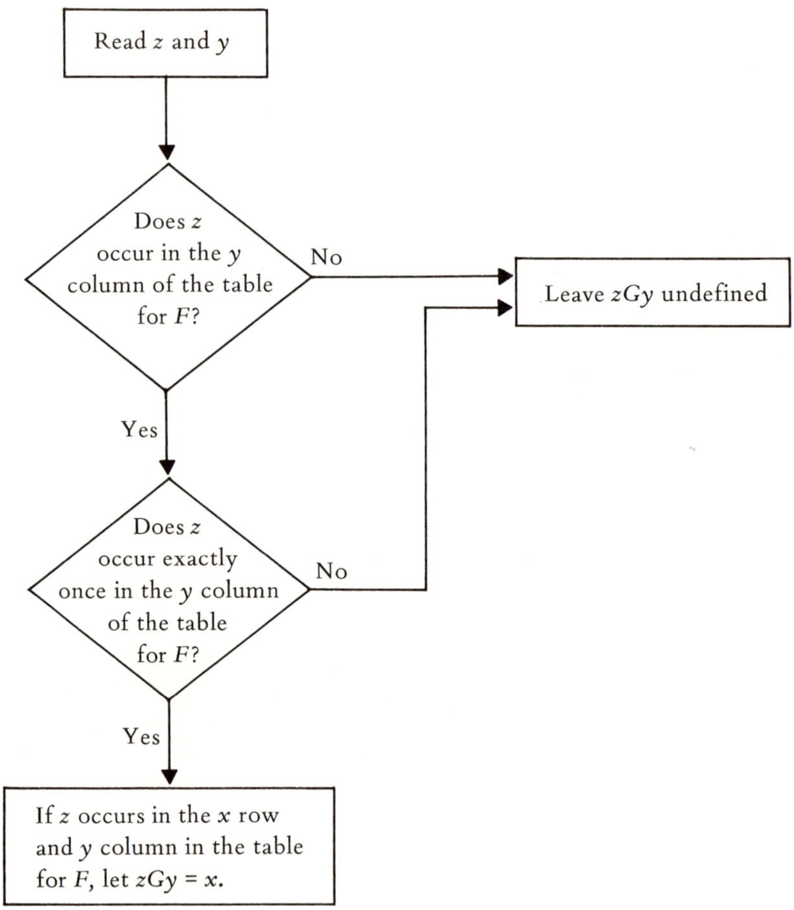

FIGURE 5.6 The construction of an inverse operation G for an operation F which is given by a table.

The reason for this last option is as follows. If G is an inverse operation for F it would be nice to have F, insofar as it is possible, an inverse operation for G. Suppose we discover that for a given y and z we have $xFy \neq z$ for all values of x and yet we define zGy to be some value, say, w. If we want F to act as much as possible as an inverse for G then we would want $wFy = (zGy)Fy$ to be equal to z. But xFy is never equal to z for any x. So it is better to leave zGy undefined.

We can now state exactly what it means for one operation to be the inverse for another:

An operation G is said to be the **inverse operation** for an operation F if (a) $zGy = x$ whenever x is the unique value such that $xFy = z$, and (b) zGy is undefined otherwise.

This definition simply spells out in detail the nature of the operation constructed by the procedure given in Figure 5.6. Application of this procedure to the operation A below yields the inverse operation B.

A	a	b	c
a	c	a	b
b	c	b	c
c	b	a	a

B	a	b	c
a	–	–	c
b	c	b	a
c	–	–	b

We can express the notion of an inverse operation in terms of solving equations. Suppose F is a given operation and y and z are given values. We may then consider the problem of solving the equation $xFy = z$. To say that we have *solved* this equation is to say that we have found all values of x which satisfy it. That is, we have found all values which, when substituted for x, make the equation true. Any x such that $xFy = z$ may be called a **solution** to (or for) the equation. The collection of all solutions is called the **solution set** for the equation. Thus, the solution set for the equation $xFy = z$ is $\{x \mid xFy = z\}$.

For example, the solution set for the equation $x + 7 = 12$ is $\{5\}$. The solution set for $2x = 8$ is $\{4\}$. In this terminology, we say that G is an inverse operation for F if for any given y and z we have (a) zGy is defined when and only when the equation $xFy = z$ has a single solution, and (b) $zGy = x$ implies $xFy = z$. In other words, G is the inverse operation for F if for a given y and z we have $zGy = x$ if and only if x is the unique solution of the equation $xFy = z$.

There is one case of special interest where the inverse operation G will have a value at every pair (z,y). This is the case where F is sufficiently "well behaved" in a set A so that A is closed under F, there is an identity element e for F on A, F is associative on A, and each element in A has an inverse element with respect to F. For such an F we know that $(xFy)Fy^{-1} = xF(yFy^{-1}) = xFe = x$ for all x and y in A. But if we want the operation G to have the property that $(xFy)Gy = x$ for all x and y, then we see that we should define G by taking $zGy = zFy^{-1}$ for all z and y in A. If we do so, we shall have $(xFy)Gy = (xFy)Fy^{-1} = x$ for all x and y in A.

Consider, for example, the operation $\oplus$ on $\{0,1,2\}$ as given by the table

$\oplus$	0	1	2
0	0	1	2
1	1	2	0
2	2	0	1

$\oplus$ is associative, $\oplus$ has the identity element 0, each element has an inverse, and $\{0,1,2\}$ is closed under $\oplus$. The inverses are: $0^{-1} = 0$, $1^{-1} = 2$, $2^{-1} = 1$. Hence, we may construct the table for the inverse operation for $\oplus$, which we shall call $\ominus$, by making such calculations as the following: $0 \ominus 0 = 0 \oplus 0^{-1} = 0 \oplus 0 = 0$, $2 \ominus 1 = 2 \oplus 1^{-1} = 2 \oplus 2 = 1$, $1 \ominus 1 = 1 \oplus 1^{-1} = 1 \oplus 2 = 0$, $1 \ominus 2 = 1 \oplus 2^{-1} = 1 \oplus 1 = 2$. The final result of such calculations is the table:

$\ominus$	0	1	2
0	0	2	1
1	1	0	2
2	2	1	0

In the special case where F is associative and each element has an inverse and we have closure, suppose we imagine that the operation F is a machine into which we feed two inputs. Suppose also that we have a machine called "Inverse" which when fed any element gives out the inverse of that element with respect to F. Then we may represent G, the inverse operation for F, as in Figure 5.7.

We may also go through this last procedure thinking of our basic interest as being the solving of equations. If F is associative and each element has an inverse in A, and A is closed under F, it can be shown (see the end of this section) that $xFy = z$ if and only if $x = zFy^{-1}$. In other words, when F has these properties and y and z are given, the equation $xFy = z$ has exactly one solution for x and that solution is given by $x = zFy^{-1}$.

Definition 5.3.1. G is said to be the **inverse operation** for an operation F if (a) $zGy = x$ whenever x is the unique value such that $xFy = z$ and (b) zGy is undefined if there is not a unique x such that $xFy = z$.

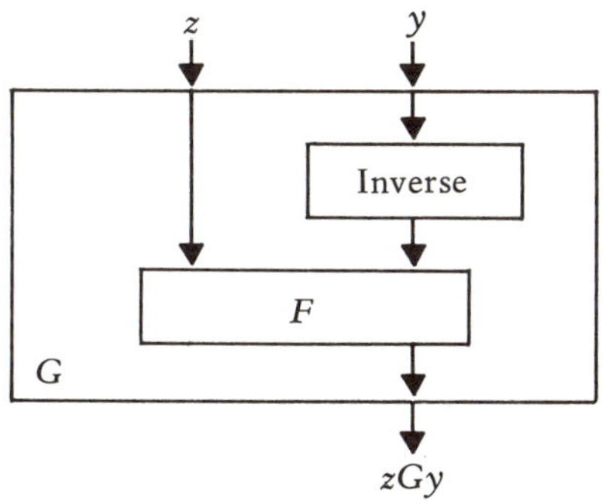

FIGURE 5.7

Theorem 5.3.1. If G is the inverse operation for F then either $(xFy)Gy$ is undefined or $(xFy)Gy = x$.

PROOF: If xFy is undefined then $(xFy)Gy$ is undefined. If xFy is defined suppose its value is z. Either there is a w such that $w \neq x$ and $wFy = z$ or there is not. If there is such a w then $(xFy)Gy = zGy$ is undefined by Definition 5.3.1. If not, then x is the unique value such that $xFy = z$. Hence, $(xFy)Gy = zGy = x$ by Definition 5.3.1.

Theorem 5.3.2. If F is associative and each element has an inverse in A and A is closed under F, then

(a) $xFy = z$ if and only if $x = zFy^{-1}$,
(b) the operation G defined by $zGy = zFy^{-1}$ is the inverse operation for F.

PROOF: Suppose F has the given properties in A. If $xFy = z$ then $(xFy)Fy^{-1} = zFy^{-1}$. But this implies by the associativity of F that $xF(yFy^{-1}) = zFy^{-1}$. But $yFy^{-1} = e$ where e is the identity element for F. Since $xFe = x$, we have $x = zFy^{-1}$.

Suppose, on the other hand, that $x = zFy^{-1}$. Then $xFy = (zFy^{-1})Fy$. The associativity of F gives from this that $xFy = zF(y^{-1}Fy) = zFe = z$. This completes the proof of (a).

The proof of (b) is immediate from (a) and Definition 5.3.1 since $xFy = z$ has the unique solution (for x) given by $x = zFy^{-1}$.

PROBLEMS

1. Use the process of Figure 5.6 to find an inverse operation for the operation (a) $+$ as given on page 89, (b) B as given in this section, (c) $\odot$ as given on page 90, (d) G as given on page 91, (e) $\cdot$ as given on page 89, (f) $\div$ as given on page 90, (g) $\overset{+}{\sim}$ in Problem 2, Section 5.1, (h) $\overset{\cdot}{\sim}$ in Problem 2, Section 5.1, (i) W as given on page 100.

2. Use the process of defining the inverse operation G by $zGy = zFy^{-1}$ to find the table for G when F is the operation (a) $\overset{+}{\sim}$ in Problem 2, Section 5.1, (b) M as given on page 89.

3. The table for $\oplus$ is given again in this section. The tables for $\overset{+}{\sim}$ and $\overset{\cdot}{\sim}$ are given in Problem 2, Section 5.1. Using these, find all the values of x (if any) for which (a) $x \oplus 1 = 2$, (b) $x \oplus 2 = 1$, (c) $x \overset{+}{\sim} 1 = 3$, (d) $x \overset{+}{\sim} 2 = 0$, (e) $3 \overset{+}{\sim} x = 2$, (f) $0 \overset{\cdot}{\sim} x = 1$, (g) $x \overset{\cdot}{\sim} 2 = 0$, (h) $2 \overset{\cdot}{\sim} x = 1$, (i) $x \overset{\cdot}{\sim} 3 = 1$.

4. Characterize the operation $\ominus$ constructed above in terms of the 3-hour clock of Problem 3, Section 5.1.

5. Characterize your result in Problem 1 (g) in terms of a 4-hour clock.

---

6. If G is the inverse operation for F, does it follow that F must be the inverse operation for G? [HINT: See Problem 1 (b).]

5.4 Isomorphic systems

Displayed below are the tables for several operations we have considered before.

$\oplus$	0	1	2
0	0	1	2
1	1	2	0
2	2	0	1

$\odot$	0	1	2
0	0	0	0
1	0	1	2
2	0	2	1

$\cap$	ϕ	$\{1\}$	$\{2\}$	$\{1,2\}$
ϕ	ϕ	ϕ	ϕ	ϕ
$\{1\}$	ϕ	$\{1\}$	ϕ	$\{1\}$
$\{2\}$	ϕ	ϕ	$\{2\}$	$\{2\}$
$\{1,2\}$	ϕ	$\{1\}$	$\{2\}$	$\{1,2\}$

$\underset{+}{\sim}$	0	1
0	0	1
1	1	0

$\underset{\cdot}{\sim}$	0	1
0	0	0
1	0	1

J	a	b
a	a	b
b	b	a

K	a	b	c
a	a	b	c
b	b	c	a
c	c	a	b

V	t	f
t	t	t
f	t	f

$\rightarrow$	t	f
t	t	f
f	t	t

$\boxed{+}$	odd	even
odd	even	odd
even	odd	even

$\boxed{\cdot}$	odd	even
odd	odd	even
even	even	even

Some of these operations look alike. For instance, $\underset{+}{\sim}$ and J are very similar. In fact, if we erase each "a" in the table for J and replace it with "0" and also replace each "b" with "1" in this table, we get exactly the table for $\underset{+}{\sim}$. We would say that the **system** $\{0,1\}$ with the operation $\underset{+}{\sim}$ and the **system** $\{a,b\}$ with the operation J have the same structure or are **isomorphic.**

In a similar fashion, the system consisting of $\{0,1,2\}$ with the operation $\oplus$ is **isomorphic** to the system consisting of $\{a,b,c\}$ with the operation K. This is because taking the table for K and replacing each "a" by "0", each "b" by "1", and each "c" by "2" will yield a table identical to that for $\oplus$.

In the general situation, suppose F is a relation in a set A and G is a relation in a set B. We then speak of A along with F as the **system** (A,F). In general, any set with an operation defined in it will be called a **system.** We shall say that the system (A,F) has the same structure as, or is **isomorphic** to, the system (B,G) if it is possible to rename the elements in A so that we get essentially the same system as (B,G). More precisely, we have:

The system (A,F) is said to **isomorphic** to the system (B,G) if there is a one-to-one correspondence f between A and B such that $xFy = z$ if and only if $f(x)Gf(y) = f(z)$ for all x, y, and z in A.

In other words, the two systems are isomorphic if it is possible to set up a one-to-one correspondence which preserves the results of the operations.

In terms of the machines for the operations, the systems are isomorphic if and only if there is a one-to-one correspondence f between A and B such that the two machines in Figure 5.8 always give the same results for all x and y in A.

If f is such a one-to-one correspondence, then the two machines in Figure 5.9 also always give the same results for any a and b in B. If we think of f (or f^{-1}) as simply a renaming function then this is saying that the systems behave exactly alike except for the names we happen to call the objects.

Returning again to the examples at the start of this section, it is not difficult to see that $\{0,1\}$ with the operation $\underset{\cdot}{\sim}$, $\{t,f\}$ with the operation V, and $\{$even, odd$\}$ with the operation $\boxed{\cdot}$ are all isomorphic.

A one-to-one correspondence which establishes that two systems are isomorphic is, of course, called an **isomorphism.** And, as we would expect, the relation of "being isomorphic to" is reflexive, symmetric, and transitive on any collection of systems.

All of the properties of systems that we have discussed are preserved under isomorphisms. For example, if F is associative on a set A and (A,F) is isomorphic to (B,G), then surely G must be associative on B. Similar statements hold for

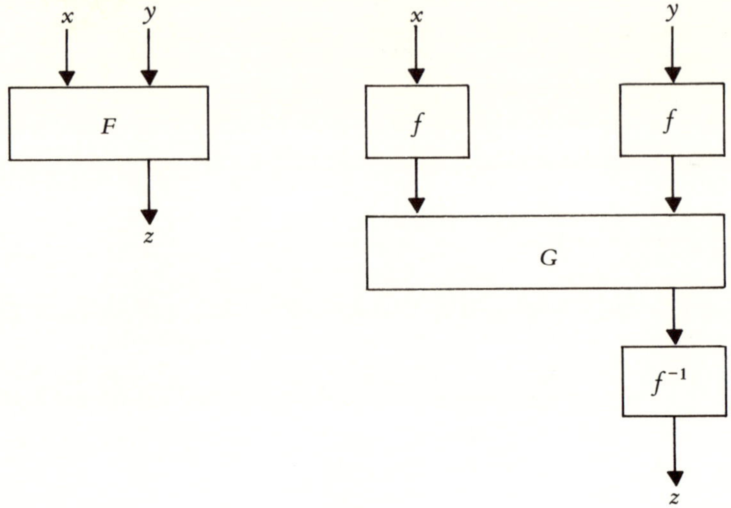

FIGURE 5.8

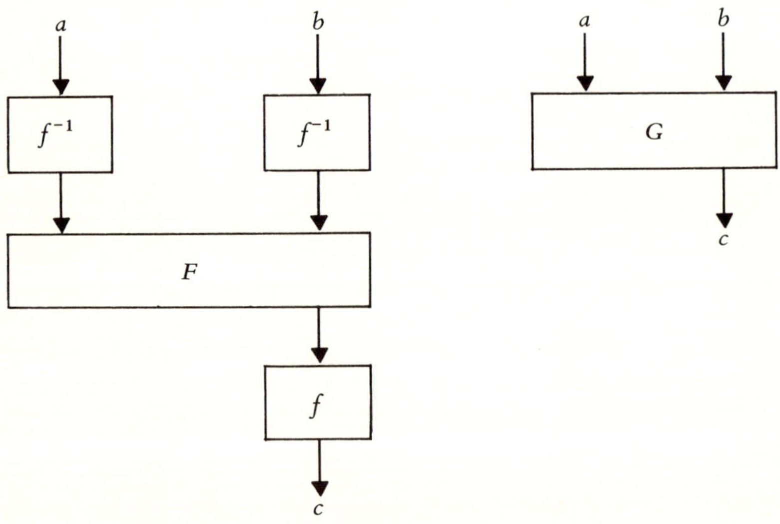

FIGURE 5.9

closure, commutativity, the existence of an identity element, and the existence of inverse elements.

This gives us one powerful reason for being interested in whether or not one system is isomorphic to another. If we establish a certain property for one system we have automatically established that same property for any system isomorphic to the given one. We may thus be able to save ourselves a tremendous amount of work. For example, once we know that $\oplus$ above is associative we know immediately that K is also associative.

The general result above is often useful the other way around to show that two systems are not isomorphic. For example, in the system $(\{0,1,2\},\oplus)$ every element has an inverse element. This is not true for $(\{0,1,2\},\odot)$. Hence, these two systems cannot be isomorphic.

Sometimes we are interested in systems where more than one operation is defined in a given set. In such situations, the above definition of isomorphism is extended in a very natural way and the results about isomorphic systems with a single operation generalize as you would expect. For example, if the system (A,F,G), where F and G are operations in the set A, is isomorphic to (B,H,J) and F distributes across G on A, then H distributes across J on B.

Definition 5.4.1. A **system** is a set with one or more operations defined in it. We shall write "(A,F)" for the system consisting of the operation F in the set A and use a similar notation if more than one operation is involved.

Definition 5.4.2. A system (A,F) is said to be **isomorphic** to a system (B,G) if there exists a one-to-one correspondence f between A and B such that $xFy = z$ if and only if $f(x)Gf(y) = f(z)$ for all x, y, and z in A. Such a correspondence f is called an **isomorphism.**

Theorem 5.4.1. Any system (A,F) is isomorphic to itself.

PROOF: $f(x) = x$ for all x in A gives the desired isomorphism.

Theorem 5.4.2. If (A,F) is isomorphic to (B,G) then (B,G) is isomorphic to (A,F).

PROOF: If f is an isomorphism between (A,F) and (B,G), then f^{-1} is an isomorphism between (B,G) and (A,F).

Theorem 5.4.3. If (A,F) is isomorphic to (B,G) and (B,G) is isomorphic to (C,H), then (A,F) is isomorphic to (C,H).

PROOF: If f is an isomorphism between (A,F) and (B,G) and g is an isomorphism between (B,G) and (C,H), then $g \circ f$ is the desired isomorphism between (A,F) and (C,H).

Theorem 5.4.4. Suppose (A,F) is isomorphic to (B,G). Then:

(a) when F is associative on A, G is associative on B,
(b) when F is commutative on A, G is commutative on B,
(c) when A is closed under F, B is closed under G,
(d) when there is an identity element for F on A, there is also an identity element for G on B,
(e) when each element in A has an inverse element for F, then each element in B has an inverse element for G.

PROOF: See Problem 6.

Definition 5.4.3. A system (A,F,G) is said to be **isomorphic** to a system (B,H,J) if there is a one-to-one correspondence f between A and B such that $xFy = z$ and $xGy = w$ if and only if $f(x)Hf(y) = f(z)$ and $f(x)Jf(y) = f(w)$ for all x, y, z, and w in A.

Theorem 5.4.5. For systems with two operations the relation of "being isomorphic to" is reflexive, symmetric, and transitive.

PROOF: See the proofs of Theorems 1, 2, and 3.

Theorem 5.4.6. If (A,F,G) is isomorphic to (B,H,J) and F distributes across G in A, then H distributes across J in B.

PROOF: See Problem 8.

PROBLEMS

1. In Problem 9, Section 5.2, which systems are isomorphic to which?
2. In Problem 11, Section 5.2, which systems are isomorphic to which?
3. Explain exactly why the system with operation $\cap$ given by the table at the beginning of this section cannot be isomorphic to the system with operation $\odot$.
4. Show that the systems $(\{t,f\},V)$ and $(\{t,f\},\rightarrow)$ as given above are not isomorphic.
5. Display isomorphisms that establish the fact that $(\{0,1\},\sim)$, $(\{t,f\},V)$, and $(\{\text{even},\text{odd}\},\boxdot)$ are all isomorphic.
6. In Theorem 5.4.4 prove part (a), (b), (c), (d), (e).
7. Prove Theorem 5.4.5.
8. Prove Theorem 5.4.6.
9. Suppose f is an isomorphism between (A,F) and (B,G). Show that (a) if e is the identity element for F in A, then $f(e)$ is the identity element for G in B, (b) if y is an inverse element for x with respect to F, then $f(y)$ is an inverse element for $f(x)$ with respect to G. (In other words, an isomorphism must map an identity element on the identity element, and it must map inverses on inverses.)

6

The natural numbers

6.1 Cardinal numbers

In any group of physically normal people there are just as many right hands as left hands. In an auditorium in which all the seats are filled and no one is standing, there are just as many people as there are chairs. There are just as many first letters on the printed pages in a book as there are last letters.

In each of these cases it is possible to tell that there are just as many objects in one set as there are in another and to do this *without referring to numbers at all*. Somehow the notion of two sets being "of the same size" is logically prior to the notion of number.

We now wish to turn to the question whose answer is a major theme of this book—What are numbers? What is the nature of numbers? What is a number, anyway? In the first chapters of this book we have been developing tools that will help us come to grips with these questions.

In previous sections, we often used numbers in constructing examples for the concepts with which we dealt. But we made this use of numbers for illustrative and pedagogical reasons, not for logical reasons. The basic concepts of sets, relations, functions, and operations do not depend on the notion of number. We may, therefore, build on these concepts to explain what numbers are. When we do this we are following not only a logically sound path, but also a *psychologically* sound path.

Before a young child can learn to deal with numbers he must learn how to group things (i.e. put things into certain sets) or at least to recognize the limits of a group someone else has made (i.e. to recognize the set that is intended). He must be able to distinguish between different objects in the group (i.e. he must know at least implicitly what is meant by "="). And, as we shall discuss in detail later, he must learn how to make a one-to-one correspondence between two sets.

Thinking of such situations as those mentioned at the very beginning of this section we make the following definition:

> A set A is said to have *exactly as many members as* a set B if there exists a one-to-one correspondence between A and B.†

Then we can see that any set A has (as it should have) exactly as many members as itself. This is because the function on A defined by $f(x) = x$ is a one-to-one correspondence between A and itself. Also, as we would hope, if A has exactly as many members as B, then B has exactly as many members as A. This follows from the fact that when f is a one-to-one correspondence between A and B then f^{-1} is a one-to-one correspondence between B and A. If A has exactly as many members as B, and B has exactly as many members as C, then we should have it the case that A has exactly as many members as C. That this is so follows from the fact that if f is a one-to-one correspondence between A and B, and g is a one-to-one correspondence between B and C, then $g \circ f$ is a one-to-one correspondence between A and C (see Problem 14, Section 4.3). This situation is illustrated in Figure 6.1 where we have shown $g \circ f$ by means of dotted arrows.

† In other places, different terminology is used for this concept. Instead of "A has exactly as many members as B" you will find "A and B are **equipollent**," "A and B are **cardinally equivalent**," "A and B are **equivalent**," "A and B are **equinumerous**," "A and B are **cardinally similar**," and "A and B **have the same power**" in various places in the literature.

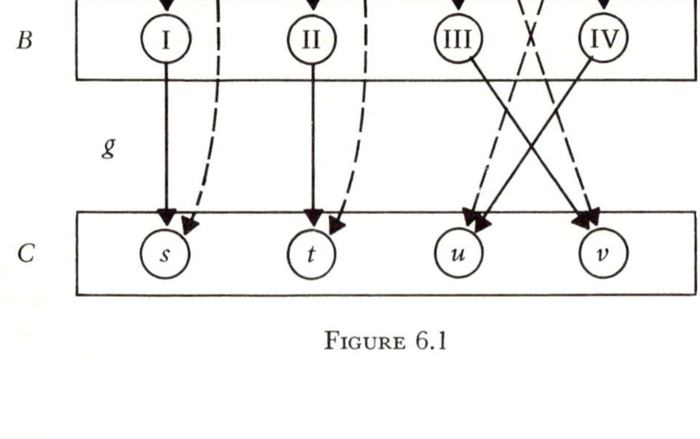

FIGURE 6.1

We could summarize the above by saying that on any collection of sets the relation $E = \{(A,B) \mid A$ has exactly as many members as $B\}$ is reflexive, symmetric, and transitive, and hence, is an equivalence relation.

Suppose $A = \{a\}$ is a set in a collection of sets. What sets are in the equivalence class $[A] = [\{a\}]$? They are the sets which can be paired up in a one-to-one fashion with $\{a\}$.

Suppose B is a set that can be put into a one-to-one correspondence with $\{a\}$ by a function f. Then B must be nonempty, since it must contain an element to be the pre-image of a under f. Also if x is in B and y is in B, then we must have $x = y$ for we must have $f(x) = f(y) = a$ and f is one to one. In other words, B is in the class $[A]$ when and only when B contains a single element. We take this sort of thing to lead us to our definition of the number 1.

We shall say that a set B **has cardinal number** 1 or *contains* exactly 1 element if it is a member of $[\{a\}]$. And the **cardinal number** 1 itself we take to be the collection $[\{a\}]$. That is, 1 is the collection of all sets that can be put into a one-to-one correspondence with $\{a\}$.

In any collection of sets, each set A has the **cardinal number** $[A]$ consisting of all those sets which have exactly as many elements as A. We shall write "$\#(A)$" for the cardinal number of A. We shall also then say that $\#(A)$ is the **number of members in** A. Our above definition of the meaning of "1" can then be stated as $1 = \#(\{a\})$. We now restate this and also assign names to some other cardinal numbers by means of the following definitions:

$$0 = \#(\phi)$$

$$1 = \#(\{a\})$$

$$2 = \#(\{a,b\})$$

$$3 = \#(\{a,b,c\}).$$

"0" is read as "zero". "1" is read as "one". "2" is read as "two". "3" is read as "three".

Thus, 2 is the number of members in any set which has exactly as many members as $\{a,b\}$. Also 2 is the cardinal number of $\{a,b\}$ and 2 is the cardinal number of any set that can be put into a one-to-one correspondence with $\{a,b\}$. And, finally, if you are pushed by someone who says, "You've told me what it means for a set to *have* the cardinal number 2. But what **is** the cardinal number 2?," the number 2 *is* the collection $[\{a,b\}]$. That is, 2 *is* the collection of all sets which have exactly as many members as $\{a,b\}$. This gives a relatively concrete answer to the question "What is 2?."

But notice that the answer is still more than somewhat abstract—as it should be for such a comprehensive concept. None of us has ever seen the number 2. This is because none of us has ever seen the collection of all possible sets which can be put into a one-to-one correspondence with $\{a,b\}$.† But we can always tell whether or not a set has the cardinal number 2. We have seen many examples of sets with cardinal number 2 or, in other words, many instances of twoness, but we cannot point to a physical spot and say there is the number 2.

This process of abstraction is not peculiar to mathematics. None of us has ever seen the concept redness, but we have all seen many instances of redness and we can tell quite well whether or not a given physical object is red or not.

We all learned as children what it meant to say that an object is red even though we never saw (and never will see) the collection of all red objects. Similarly, we all learned what it means to say that a set has 2 elements, even though we have certainly not *seen* the collection of all such sets.

Since such abstract concepts as twoness, redness, warmth, darkness, roundness, etc., are learned by most people at a very early age, it seems clear that even very young human beings are capable of thinking abstractly. A child does not need to be told each time it is warm. He learns very early the concept of warmth. He can decide properly for himself in each new instance whether it is warm or not. Indeed, the ability to carry out such abstractions is probably a basic distinction between man and the other animals.

We now have many ways to say that a set A has 2 members. All of the following mean the same thing:

(i) $\#(A) = 2$
(ii) A has cardinal number 2.
(iii) There are (exactly) 2 elements in A.
(iv) A has exactly as many members as $\{a,b\}$.
(v) A can be put into a one-to-one correspondence with $\{a,b\}$.
(vi) $A \in [\{a,b\}]$.
(vii) $A \in 2$.

In assertion (vi) we are taking it for granted that the equivalence relation involved is the one defined by "has exactly as many members as." $[\{a,b\}]$ is the equivalence class for $\{a,b\}$ with respect to this relation.

In a more formal treatment of numbers, the form used in (vii) might be a quite useful way to put the assertion, but in what follows we shall usually avoid this manner of speaking.

We now define some more cardinal numbers:

$$4 = \#(\{a,b,c,d\})$$

$$5 = \#(\{a,b,c,d,e\})$$

$$6 = \#(\{a,b,c,d,e,f\})$$

† Notice that semantic distinctions are important here. We have all seen "2" but none of us has seen 2.

$$7 = \#(\{a,b,c,d,e,f,g\})$$

$$8 = \#(\{a,b,c,d,e,f,g,h\})$$

$$9 = \#(\{a,b,c,d,e,f,g,h,i\})$$

$$10 = \#(\{a,b,c,d,e,f,g,h,i,j\}).$$

We read "4" as "four," "5" as "five," "6" as "six," "7" as "seven," "8" as "eight," "9" as "nine," and "10" as "ten."

We have chosen the above names for these cardinal numbers because they are the ones that you have, no doubt, used the most when you learned or applied the corresponding concepts. But, of course, just as the parents of a new baby have considerable freedom of choice in picking a first name for the child, we could have chosen other names. Certainly other names for these concepts have been used at other times and are being used in other geographic locations. Just as the child exists independently of his name, these concepts exist independently of the manner in which we name them. We shall temporarily call the above names we *have* chosen, **standard** names for the first few cardinal numbers. We shall return to the matter of naming numbers in a later section.

With the above cardinal numbers on hand we may begin to make use of them. In particular, we may now consider sets whose elements are cardinal numbers. How many elements are there in $\{1\}$? Clearly there is exactly 1 since the function $f(a) = 1$ is a one-to-one correspondence between $\{a\}$ and $\{1\}$ and, by definition, a set has 1 element in it when it is possible to set up a one-to-one correspondence between that set and $\{a\}$. In compact form, we have shown that $\#(\{1\}) = 1$. In a similar fashion, it is easy to display the one-to-one correspondences that prove the following:

$$\#(\{1,2\}) = 2$$

$$\#(\{1,2,3\}) = 3$$

$$\#(\{1,2,3,4\}) = 4$$

$$\#(\{1,2,3,4,5\}) = 5$$

$$\#(\{1,2,3,4,5,6\}) = 6$$

$$\#(\{1,2,3,4,5,6,7\}) = 7$$

$$\#(\{1,2,3,4,5,6,7,8\}) = 8$$

$$\#(\{1,2,3,4,5,6,7,8,9\}) = 9$$

$$\#(\{1,2,3,4,5,6,7,8,9,10\}) = 10$$

These equalities can all be thought of as theorems that follow from our definitions. But notice that we can now assert that the cardinal number of a set A is 5 (or there are 5 elements in A) if and only if we can set up a one-to-one correspondence between A and $\{1,2,3,4,5\}$. Similarly, there are 8 elements in a set B if and only if there exists a one-to-one correspondence between B and $\{1,2,3,4,5,6,7,8\}$.

It is results such as this that lead us to the process of **counting.** In learning to count, a very young child must first memorize a certain sequence of sounds in a certain order. These are the sounds which we represent in writing as "one," "two," "three," and, say, "four." It is clearly *sounds* that the child memorizes since he cannot read. The child, when prompted to, may well be able to say, "One, two, three (or free), four." But this does not mean that the child can count. All it means is that the child has learned to parrot back these sounds in the proper order.

If the child is to proceed to count the objects in a collection, he must first have reached a stage of development at which he can recognize the collection that is presented to him. He must be able to tell of any given object whether or not it is in the collection.

When he can do this he can then proceed to *count* the objects. But what is this counting? It is, most typically, a physical act of touching an object in the set and, at the same time, making the sound "One," then touching another object in the set and making the sound "Two," and so forth. The counting is done correctly if the child touches each object in the set once and only once, if he goes through his memorized sequence of sounds in the proper order, and if he emits one and only one of the sounds for each touched object.

If the child counts correctly and there are, say, 4 elements in the set being counted, then the child succeeds in setting up a one-to-one correspondence between the set and the collection of sounds {"one," "two," "three," "four"}. The cardinal number of this collection of sounds is clearly the same as $\#(\{1,2,3,4\})$ so the child could (and would after he realizes this is the proper last step) correctly conclude that there are 4 members in the set with which he started.

Once a child learns this counting process with a certain memorized sequence of sounds, it is often the case that then he can carry out the setting up of the needed one-to-one correspondence for larger sets but he runs out of sounds in his list—just as we, at this stage, have only named the numbers 0, 1, 2, . . . , 10.

If we ignore the distinctions between sounds and symbols and numbers, we can say that the process of counting a set of, say, 4 objects is the setting up of a one-to-one correspondence between the set and $\{1,2,3,4\}$. In other words, we take $\{1,2,3,4\}$ as a kind of standard set for fourness and use this set to show that a set has 4 elements.

If a set has 10 members, counting the members of the set amounts to setting up a one-to-one correspondence between that set and $\{1,2,3,4,5,6,7,8,9,10\}$.

In making the claim as we did at the beginning of this section that the basic concepts of sets, relations, and functions did not depend on numbers, we were ignoring a fine point or two about the actual way we presented these concepts. If we think of a relation as a set of **ordered pairs** don't we need the concept of 2 to define "pair"? The answer is that we do not. We may define the ordered pair (x,y) as the set $\{x,\{x,y\}\}$. Then the needed basic property of ordered pairs that $(a,b) = (c,d)$ if and only if $a = c$ and $b = d$ can be proved from basic properties of sets without any reference to numbers.

We shall not pursue the details of this here. We did not use the "number independent" definition of "ordered pair" because such a definition seems to complicate the notion unnecessarily when it is being presented for the first time.

It might also be asked whether we have used the number 1 in saying that a function is a relation that pairs exactly *one* element with each element in the domain. But a way of putting the idea of what a function is without using any numbers was given. We said a function F is a relation such that whenever $(x,y) \in F$ and $(x,z) \in F$ then $y = z$. And this statement involves no use of numbers.

Also, despite the occurrence of the word "one" in "one-to-one correspondence," we need make no use of the number 1 when defining what a one-to-one correspondence is.

Definition 6.1.1. A set A is said to have *exactly as many* **elements** *as* a set B if there exists a one-to-one correspondence between A and B.

Theorem 6.1.1. On any collection of sets the relation $\{(A,B) \mid A$ has exactly as many elements as $B\}$ is an equivalence relation.

Definition 6.1.2. The **cardinal number** of a set A is the collection of all sets which have exactly as many elements as A. We write "$\#(A)$" to stand for the cardinal number of the set A. If $\#(A) = n$, then we say that A *has* (or contains) *exactly n* **elements.**

Definition 6.1.3. $\#(\phi) = 0$, $\#(\{a\}) = 1$, $\#(\{a,b\}) = 2$, $\#(\{a,b,c\}) = 3$, $\#(\{a,b,c,d\}) = 4$, $\#(\{a,b,c,d,e\}) = 5$, $\#(\{a,b,c,d,e,f\}) = 6$, $\#(\{a,b,c,d,e,f,g\}) = 7$, $\#(\{a,b,c,d,e,f,g,h\}) = 8$, $\#(\{a,b,c,d,e,f,g,h,i\}) = 9$, and $\#(\{a,b,c,d,e,f,g,h,i,j\}) = 10$.

Theorem 6.1.2. $\#(\{1\}) = 1$, $\#(\{1,2\}) = 2$, $\#(\{1,2,3\}) = 3$, $\#(\{1,2,3,4\}) = 4$, $\#(\{1,2,3,4,5\}) = 5$, $\#(\{1,2,3,4,5,6\}) = 6$, $\#(\{1,2,3,4,5,6,7\}) = 7$, $\#(\{1,2,3,4,5,6,7,8\}) = 8$, $\#(\{1,2,3,4,5,6,7,8,9\}) = 9$, and $\#(\{1,2,3,4,5,6,7,8,9,10\}) = 10$.

Problems

1. Prove that 2 is the number of elements in $\{z,y\}$.
2. By diagramming the appropriate one-to-one correspondences, display all possible ways to count the elements of $\{x,y,z\}$.
3. Explain carefully why we are certain that, with 0 and 1 defined as in this section, we have $0 \neq 1$.

4. Describe in the language of one-to-one correspondences, several ways in which a young child may make a mistake while trying to find out how many objects there are in a certain set of 5 objects.

5. Prove that $\#(\{1,2,3,4,5,6,7,8\}) = 8$.

6. In Figure 6.2 we have displayed the flow diagram for a process which enables you to determine the cardinal number of any set A which has 0, 1, 2, 3, or 4 elements in it. Explain carefully in terms of the definitions in this section why the process works.

7. Prove that $\#(\{a,b\} \cup \{1,2,3,4\}) = 6$.

8. Prove that $\#(\{1,2\} \times \{1,2\}) = 4$.

9. Prove that $\#(\{1,2,3\} \times \{1,2,3\}) = 9$.

6.2 Addition of cardinal numbers

Let $A = \{a,b\}$ and $B = \{1,2,3\}$. Then by our previous work we know that $\#(A) = 2$ and $\#(B) = 3$. We have $A \cup B = \{a,b,1,2,3\}$. And $\#(A \cup B) = 5$ since the function f defined by $f(a) = a$, $f(b) = b$, $f(1) = c$, $f(2) = d$, and $f(3) = e$ is a one-to-one correspondence between $A \cup B$ and $\{a,b,c,d,e\}$.

We recognize here the process of addition that all of us learned very early in life. And we now state a definition for this familiar operation:

If n and m are cardinal numbers, then we define $n + m$ to be $\#(C \cup D)$ where $n = \#(C)$, $m = \#(D)$, and $C \cap D = \phi$. We call $n + m$ the **sum** of n and m. We read "$n + m$" as "the **sum** of n and m" or "n **plus** m." The operation we have thus defined is called **addition**.

This, it is hoped, is merely a careful statement of what we all learned as the meaning of addition. If we want to add 3 and 7, we find a set of 3 elements (maybe 3 fingers), find another set of 7 elements (fingers?) which does not overlap the first, lump these two sets together to get the union, and then find the cardinal number of the resulting set.

If we are being very careful we should ask ourselves if we actually have an operation defined here. If it is an operation it must yield a single result when it yields any. Maybe when we are given cardinal numbers n and m and we find some sets with these cardinal numbers, it makes a difference which sets we choose to work out the sum. Maybe if we change the sets we use for C and D, we get different sums. It can be shown (and will be shown at the end of this section) that this does not happen. This amounts to saying that it is true that if $n = p$ and $m = q$, then $n + m = p + q$.

Using the above definition of addition, we can now prove such a theorem as

$$2 + 2 = 4.$$

The proof is obtained when we observe that $\#(\{a,b\}) = 2$, $\#(\{1,2\}) = 2$, $\{a,b\} \cap \{1,2\} = \phi$, and $\#(\{a,b\} \cup \{1,2\}) = \#(\{a,b,1,2\}) = 4$ where the last

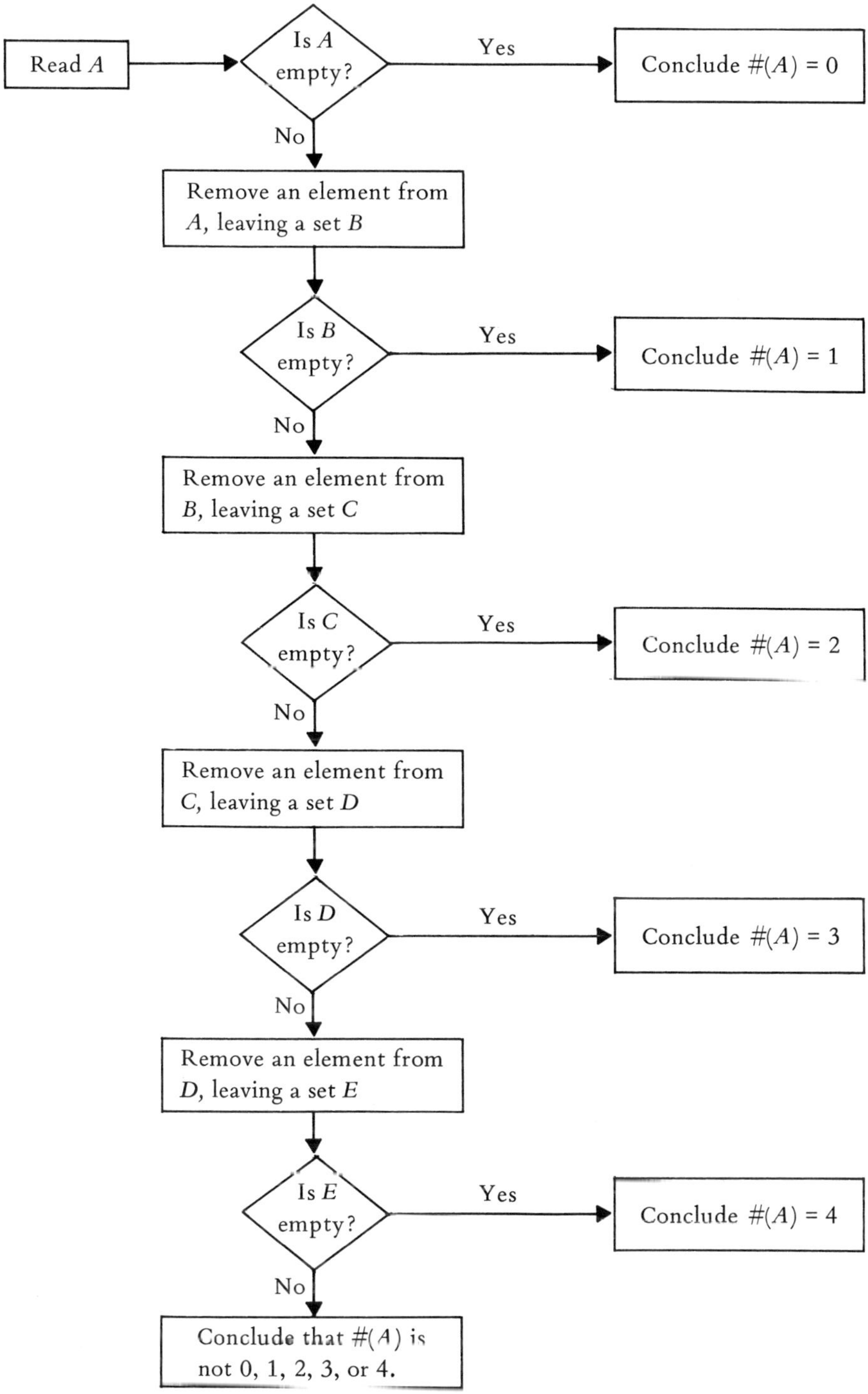
Read A
Is A empty?
Yes
Conclude #(A) = 0
No
Remove an element from A, leaving a set B
Is B empty?
Yes
Conclude #(A) = 1
No
Remove an element from B, leaving a set C
Is C empty?
Yes
Conclude #(A) = 2
No
Remove an element from C, leaving a set D
Is D empty?
Yes
Conclude #(A) = 3
No
Remove an element from D, leaving a set E
Is E empty?
Yes
Conclude #(A) = 4
No
Conclude that #(A) is not 0, 1, 2, 3, or 4.

FIGURE 6.2

equality is justified by the one-to-one correspondence f between $\{a,b,1,2\}$ and $\{a,b,c,d\}$ where f is defined by $f(a) = a$, $f(b) = b$, $f(1) = c$, $f(2) = d$.†

We can summarize a raft of such theorems by means of the table:

+	0	1	2	3	4	5
0	0	1	2	3	4	5
1	1	2	3	4	5	6
2	2	3	4	5	6	7
3	3	4	5	6	7	8
4	4	5	6	7	8	9
5	5	6	7	8	9	10

This is a partial table for the **operation of addition** on cardinal numbers.

From another point of view, the table summarizes some important results about addition, but the meaning of addition is something apart from the table. A very young child can be taught to answer the question "What is 3 plus 3?" with "Six," and yet may not know what addition is. We should not conclude from this, however, that it is not helpful to *know* the facts summarized in the table.

The table suggests that the operation of addition is commutative. We cannot however prove this by reference to the table since it does not give all possible sums of cardinal numbers. We *can* prove the commutativity of addition by noting that when A and B are sets, $A \cup B = B \cup A$, so that we must have $\#(A \cup B) = \#(B \cup A)$.

We can also easily show that addition of cardinal numbers is associative since when A, B, and C are sets, $A \cup (B \cup C) = (A \cup B) \cup C$, so that $\#(A \cup (B \cup C)) = \#((A \cup B) \cup C)$.

Since $p + (q + r) = (p + q) + r$ for all cardinal numbers p, q, and r, we may write "$p + q + r$" to mean either $p + (q + r)$ or $(p + q) + r$.

In fact, in any string of names for cardinal numbers connected by "+" signs we may leave out parentheses and cause no confusion. This result is sometimes referred to as the **generalized associative law** for addition. An example would be: $((p + q) + (r + s)) + t = ((p + q) + r) + (s + t) = p + ((q + r) + (s + t)) = p + q + r + s + t$.

0 is the identity for addition on cardinal numbers. This follows from the fact that for any set A we have $A \cup \phi = A$ and $A \cap \phi = \phi$. Then whenever $n = \#(A)$ we have $0 + n = n + 0 = \#(A \cup \phi) = \#(A) = n$.

† Note here that, as was assumed without mention earlier, we are not taking a or b to be variables which might possibly be names for numbers. We are, rather, thinking of a and b as simply distinct letters in the alphabet.

Suppose we know that n and m are cardinal numbers and $n + m = 0$. What may we conclude about n and m? Surely we must have $n = 0$ and $m = 0$. This is due to the fact that when $\#(A) = n$ and $\#(B) = m$ and $A \cup B = \phi$, we surely must have both $A = \phi$ and $B = \phi$. If more details are desired here note that $A \subseteq A \cup B$ and $B \subseteq A \cup B$ and when $A \cup B = \phi$ this implies $A = B = \phi$.

Definition 6.2.1. If n and m are cardinal numbers then $n + m$ is defined to be the cardinal number of $A \cup B$ where $\#(A) = n$, $\#(B) = m$, and $A \cap B = \phi$. We call $n + m$ the **sum** of n and m, and read "$n + m$" as "n **plus** m."

Theorem 6.2.1. If $n = p$ and $m = q$, then $n + m = p + q$.

PROOF: Suppose $n = \#(A)$ and $p = \#(C)$ and $n = p$. Then there exists a one-to-one correspondence f between A and C. If $m = \#(B)$ and $q = \#(C)$ and $m = q$, then there is a one-to-one correspondence g between B and C. Suppose $A \cap B = \phi$ and $D \cap C = \phi$. Let us define the function h on $A \cup B$ by: $h(x) = f(x)$ if $x \in A$ and $h(x) = g(x)$ if $x \in B$. Then h is a one-to-one correspondence between $A \cup B$ and $C \cup D$. In fact, $h^{-1}(y) = f^{-1}(y)$ if $y \in C$ and $h^{-1}(y) = g^{-1}(y)$ if $y \in D$. We conclude that $A \cup B$ has exactly as many members as $C \cup D$. In other words, $n + m = \#(A \cup B) = \#(C \cup D) = p + q$.

Note that if we have $\#(A) = n$ and $\#(B) = m$ it is always possible to construct, out of A and B, sets C and D such that $C \cap D = \phi$ and $\#(C) = n$ and $\#(D) = m$. Simply let $C = A \times \{a\}$ and $D = B \times \{b\}$.

Theorem 6.2.2. Addition of cardinal numbers is commutative and associative, and 0 is the identity element for addition.

Theorem 6.2.3. If n and m are cardinal numbers and $n + m = 0$, then $n = 0$ and $m = 0$.

PROBLEMS

1. Prove that (a) $3 + 4 = 7$, (b) $2 + 8 = 10$, (c) $2 + 5 = 7$.
2. Prove three of the following assertions: $0 + 1 = 1$, $1 + 1 = 2$, $2 + 1 = 3$, $3 + 1 = 4$, $4 + 1 = 5$, $5 + 1 = 6$, $6 + 1 = 7$, $7 + 1 = 8$, $8 + 1 = 9$, $9 + 1 = 10$.
3. Find the standard name for the cardinal number of each of the following sets. (a) $\phi \cup \{a,b\}$, (b) $\{\phi\}$, (c) $\{\phi,\{\phi\}\}$, (d) $\{\phi,\{\phi\},\{\phi,\{\phi\}\}\}$.
4. Prove or disprove: If m, n, p, and q are cardinal numbers and $m + n = p \mid q$ then $m = p$ and $n = q$.
5. Let $A = \{\phi,\{1\},\{2\},\{1,2\}\}$. For x and y in A let $x \star y = \phi$ if $x = \phi$ or $y = \phi$. Let $x \star y = \{1,2\}$ if $y = \{1,2\}$. Is $\star$ an operation in A?
6. Even though we have not named any cardinal numbers past 10, we can, in one sense, still work certain addition problems. For example, $5 + 6 =$

$\#(\{a,b,c,d,e,1,2,3,4,5,6\})$. Find, in this manner, the sums (a) $5 + 7$, (b) $8 + 8$, (c) $9 + 4$.

6.3 Multiplication of cardinal numbers

The definition of addition given in the preceding section was a very natural one in terms of how we all learned what addition is. The definition we now give for multiplication of natural numbers is, at first blush, perhaps not so natural.

If n and m are cardinal numbers, we define $n \cdot m$ to be the cardinal number of the set $A \times B$ where $\#(A) = n$ and $\#(B) = m$. We call $n \cdot m$ the **product** of n and m and read "$n \cdot m$" as "n **times** m" or "n **multiplied by** m." We call this operation **multiplication.**

The symbol "$\times$" is also commonly used (instead of the dot) to denote the operation of multiplication.

That the above definition is very close to our intuitive notion of multiplication becomes clear when we think of a geometric representation of a Cartesian product as a table. How many boxes (or positions) are there in the table below which represents $\{a,b,c\} \times \{1,2\}$?

	1	2
a		
b		
c		

There are three rows of two each. There are 6.

We can now prove as a theorem:

$$3 \cdot 2 = 6.$$

The proof is obtained when we note that $\#(\{a,b,c\}) = 3$, $\#(\{1,2\}) = 2$, $\{a,b,c\} \times \{1,2\} = \{(a,1),(a,2),(b,1),(b,2),(c,1),(c,2)\}$ and the function k is a one-to-one correspondence between this last set and $\{a,b,c,d,e,f\}$ if we define k by: $k(a,1) = a$, $k(a,2) = b$, $k(b,1) = c$, $k(b,2) = d$, $k(c,1) = e$, $k(c,2) = f$.

It can easily be shown that multiplication of cardinal numbers is commutative and associative. But the proofs are not quite as simple as the proofs of the corresponding theorems for addition, and they are postponed until the end of this section. Because of the associativity of multiplication we may write "$p \cdot q \cdot r$" to mean either $(p \cdot q) \cdot r$ or $p \cdot (q \cdot r)$.

1 is the identity element for multiplication. For suppose $n = \#(A)$. Then the function f defined by $f(x,a) = x$ is clearly a one-to-one correspondence between $A \times \{a\}$ and A, and, hence, $1 \cdot n = n \cdot 1 = \#(A \times \{a\}) = \#(A) = n$.

Also, since $A \times \phi = \phi \times A = \phi$ for any set A, we have $n \cdot 0 = 0 \cdot n = 0$ for any cardinal number n.

It is a straightforward matter to prove the other theorems summarized by the table below.

·	0	1	2	3
0	0	0	0	0
1	0	1	2	3
2	0	2	4	6
3	0	3	6	9

This is, of course, only a partial table for the operation of multiplication on cardinal numbers.

In a (perhaps expanded) table for multiplication, might we find some 0's that are not in the first row or in the first column? Our intuition tells us that the answer is "no." And this is, indeed, correct. This tells us that when $n \cdot m = 0$, we must have $n = 0$ or $m = 0$. We shall leave the proof of this as an exercise.

Definition 6.3.1. If n and m are cardinal numbers then $n \cdot m = \#(A \times B)$ where $\#(A) = n$ and $\#(B) = m$. We call $n \cdot m$ the **product** of n and m and read "$n \cdot m$" as "n **times** m" or "n **multiplied by** m."

Theorem 6.3.1. If $n = p$ and $m = q$, then $n \cdot m = p \cdot q$.

PROOF: Suppose $\#(A) = n$, $\#(B) = m$, $\#(C) = p$, and $\#(D) = q$. If $n = p$ there is a one-to-one correspondence f between A and C. If $m = q$ there is a one-to-one correspondence g between B and D. Define the mapping h on $A \times B$ by taking $h(x,y) = (f(x),g(y))$. Then h is a one-to-one correspondence between $A \times B$ and $C \times D$. In fact, $h^{-1}(z,w) = (f^{-1}(z), g^{-1}(w))$. We conclude that $n \cdot m = \#(A \times B) = \#(C \times D) = p \cdot q$.

Theorem 6.3.2. Multiplication of cardinal numbers is commutative and associative, and 1 is the identity element for multiplication.

PROOF: Suppose $p = \#(A)$, $q = \#(B)$, and $r = \#(C)$. We shall first show that $p \cdot q = q \cdot p$. By definition, $p \cdot q = \#(A \times B)$ and $q \cdot p = \#(B \times A)$. Define a mapping f on $A \times B$ by $f(x,y) = (y,x)$. Then f is clearly a one-to-one correspondence between $A \times B$ and $B \times A$, and, hence, $p \cdot q = q \cdot p$. To establish the associativity we define a mapping g on $A \times (B \times C)$ by $g(x,(y,z)) = ((x,y),z)$. g is clearly a one-to-one correspondence between $A \times (B \times C)$ and $(A \times B) \times C$. Hence, $p \cdot (q \cdot r) = (p \cdot q) \cdot r$. The fact that 1 is the identity element for multiplication was proved earlier in the section.

Theorem 6.3.3. If n is a cardinal number, then $0 \cdot n = n \cdot 0 = 0$.

Theorem 6.3.4. If n and m are cardinal numbers and $n \cdot m = 0$, then $n = 0$ or $m = 0$.

PROOF: See Problem 3.

PROBLEMS

1. Prove each of the following: (a) $3 \cdot 3 = 9$, (b) $2 \cdot 5 = 10$, (c) $2 \cdot 4 = 8$.
2. Prove or disprove: If m, n, p, and q are cardinal numbers and $n \cdot m = p \cdot q$, then $n = p$ and $m = q$.
3. Prove that if $n \cdot m = 0$, then $n = 0$ or $m = 0$. (See page 47.)
4. Draw a tree diagram for $\{a,b,c\} \times \{1,2,3\}$. How many different paths are there through the tree from left to right?
5. Verify that $\#(\{a,b\}) + \#(\{c,d\}) + \#(\{e,f\}) = \#(\{a,b\}) \cdot 3$.

6. Introduce certain symbols to stand for the cardinal numbers involved and then explain how we may use the operations of this section to solve each problem stated below. Assume all cash values are in a whole number of dollars. (We have given a sample solution for part (a).)

(a) How many chairs will we need in the classroom we use if we have Miss A's class and Mr. B's class meet together? (Solution: Let a be the cardinal number of Miss A's class. Let b be the cardinal number of Mr. B's class. We shall then need $a + b$ chairs.)

(b) How many people will be at our party if we invite one group and then invite some more? (Assume that all those invited will attend.)

(c) We plan to put a certain number of ounces of mouth-wash in each of a certain group of bottles. How many ounces of mouth-wash will we need?

(d) Each person in a group has normal hands. How many fingers belong to the people in the group?

(e) We wish to give a certain sized cash gift to each of our nieces and nephews. How much money will we need to do this?

(f) Suppose we have a set of water glasses and a set of juice glasses and we wish to put a fixed number of ounces of a beverage in each one of these glasses. How much beverage will we need?

(g) We have a group of workers who all get the same salary. How much money do we need to meet the payroll for this group?

(h) As in g but with two groups of workers who get two different salaries.

(i) Each team in a baseball league has the same number of players on its roster. How many players are in the league?

(j) We own shares of two different stocks with different market values. What are our total assets from these stocks?

(k) The three elementary schools in a town have various numbers of classrooms in them. How many classrooms are available in the town?

(l) Each shelf in a certain portion of a library will hold a fixed number of copies of a certain journal. How many copies of the journal may we shelve there?

6.4 The natural numbers

We now wish to characterize the numbers with which we count. Certainly 0, 1, 2, 3, 4, 5, 6, 7, 8, 9, and 10, as we have already defined them, are such numbers. We can get another by adding 1 to 10. We can then get another by adding 1 to that. And so forth. The numbers we get in this way are called the **natural numbers.** More precisely, we have the following three-part definition:

(i) 0 is a natural number.
(ii) If n is a natural number, then so is $n + 1$.
(iii) The only things that are natural numbers are those which are such as a consequence of (i) and (ii).

We see that 0 is a natural number by part (i). Then $0 + 1 = 1$ is a natural number by (ii). Then $1 + 1 = 2$ is a natural number by (ii). And so forth. Part (iii) tells us that the only way to get the things called natural numbers is by use of (i) and (ii).

We shall from this point on use "N" as a name for the collection of *all* natural numbers.

We now have a device for getting names for natural numbers beyond those we have already introduced. By successive applications of (ii) we see that $10 + 1$, $(10 + 1) + 1$, $((10 + 1) + 1) + 1$, and $(((10 + 1) + 1) + 1) + 1$ are all natural numbers. It would, of course, be convenient to have more compact names for these numbers. We shall discuss efficient naming procedures later. But we now need to investigate some of the consequences of our definition† of N.

The basic idea of N is that it contains 0 and it also contains all the other numbers (and only the other numbers) which we can generate by successive actions of adding on 1.

Is the set of natural numbers closed under addition? It is at least conceivable that for some natural numbers, say x and y, we find that $x + y$ is not a natural number. This, of course, cannot happen, as we now prove. Let x be any natural number. Given any other natural number y we wish to show that $x + y$ is also a natural number. To that end, let us look at the collection of those y values for which $x + y$ *is* a natural number. In other words, consider the set $A = \{y \mid x + y \text{ is a natural number and } y \in N\}$. Certainly 0 is in A because we know

† Whereas our set of natural numbers is $\{0,1,2,3, \ldots\}$, some books take the set of natural numbers to be $\{1,2,3, \ldots\}$ where 0 is excluded. Unfortunately, the terminology is not completely standardized.

$x + 0 = x$ and x is a natural number. Suppose k is some member of A. Then, by the definition of A, $x + k$ is a natural number. But then by part (ii) of our basic definition $(x + k) + 1$ is a natural number. By the associativity of addition, $(x + k) + 1 = x + (k + 1)$. Therefore, $x + (k + 1)$ is a natural number. This means that $k + 1 \in A$. What we have shown about the set A is that 0 is in A, and k in A implies $k + 1$ is in A. But then surely every natural number is in A. This is because the natural numbers are exactly those that are generated by starting with 0 and successively adding 1. In other words, we conclude that $A = N$. Since x was an arbitrary natural number and A had the meaning it did, we have proved:

> If x and y are natural numbers, then so is $x + y$. In other words, we have proved that N is closed under addition.

The method of proof we used above is a very important one. It is called **mathematical induction** or simply **induction** (or sometimes **complete induction**). This method relies on the fact that we made use of (and argued for) above that:

> If (i) $A \subseteq N$, (ii) $0 \in A$, and (iii) $k \in A$ implies† $k + 1 \in A$; then $A = N$.

We shall see many examples of proofs by mathematical induction as we proceed.

A brief word is appropriate here about the power of the procedure. Look at what we accomplished above! We evidently proved *infinitely many* facts in a brief paragraph. We proved at one fell swoop that $0 + 0$, $0 + 1$, $1 + 0$, $1 + 1$, $2 + 1$, $1 + 2$, $2 + 2$, $3 + 1$, $1 + 3$, $3 + 2$, $2 + 3$, $3 + 3$, etc., etc., are all natural numbers. We are now dealing with N, the set of *all* natural numbers. This set is infinite and we cannot usually construct proofs about it by listing all possible cases. Proof by mathematical induction is a powerful tool we need to help us in this situation.

We shall return, later, to a careful discussion of what the word "infinite" means. But we can now at least say that the natural numbers are sometimes also called **finite cardinal numbers** and that there are other cardinal numbers which are *not* finite.

In connection with the finite cardinal numbers (i.e. the natural numbers) there are several basic questions we should ask.

(a) Is it possible that for some natural number x we have $x = x + 1$? (If so, our process of successively adding 1 will never get us past x.)

(b) Is it possible that for some natural numbers x and y we have $x + 1 = y + 1$ and yet $x \neq y$? (If so, it would evidently mean that there are two different routes in our march of successively adding 1 which end up at the same place.)

---

† We are, of course, making use here of the convention we discussed in Chapter 1. By "$k \in A$ implies $k + 1 \in A$" we really mean "For all possible values of k, $k \in A$ implies $k + 1 \in A$."

(c) Is there a case where we have $x + y = x$ and yet $y \neq 0$? (If so, we may be going in a possibly huge circle in our march.)

(d) Do there exist natural numbers x, y, and z such that $x + z = y + z$ and yet $x \neq y$? (If so, we have two different routes of z steps in our march that start at different places and end up at the same place.)

(e) Could there be a natural number x which is different from 0 and yet for no y do we have $y + 1 = x$? (If so, there would be a natural number we could never reach in our march.)

As your intuition no doubt tells you, the answer to each of these questions is "No." That is, each of the following is true for any natural numbers x, y, and z.

(a) $x \neq x + 1$
(b) $x + 1 = y + 1$ implies $x = y$
(c) $x + y = x$ implies $y = 0$
(d) $x + z = y + z$ implies $x = y$
(e) $x \neq 0$ implies there is a w such that $w + 1 = x$.

It will afford us another example of a proof by induction if we prove (c) assuming that (b) has already been established. We leave the proofs of the other results to the appendix of this section. To construct our proof of (c), let $A = \{x \mid x + y = x \text{ implies } y = 0\}$. We first claim that $0 \in A$. This follows because $0 + y = y$ coupled with the assumption that $0 + y = 0$ clearly implies $y = 0$. Next, we assume that $k \in A$ and shall attempt to show that this implies $k + 1 \in A$. To this end, suppose that $(k + 1) + y = k + 1$. By the associative and commutative laws for addition we have $(k + 1) + y = k + (1 + y) = k + (y + 1) = (k + y) + 1$. This combined with our supposition gives $(k + y) + 1 = k + 1$. But this by (b) implies $k + y = k$. Since $k \in A$, this implies $y = 0$. We have shown that when $k \in A$ then $(k + 1) + y = k + 1$ implies $y = 0$. That is, $k \in A$ implies $k + 1 \in A$. We conclude that $A = N$ and the proof is complete.

Such a proof as the above may well strike you as difficult, but you should not let this discourage you. Proofs by induction are often *not* easy. We shall be getting much practice with such proofs in later work and, fortunately, some of the later examples are even easier to follow than the one just given.

In Figure 6.3 we have constructed a flow diagram to suggest a method of trying to find a proof that $A = N$ where A is a set of natural numbers.

If you look closely at the figure you will see that there is no guarantee that the method will get you anyplace. If you cannot find a proof that $k \in A$ implies $k + 1 \in A$ and also cannot find a natural number outside of A, you may simply go around and around in circles. 'Twas ever thus. Proofs are often not easily found.

In view of the way we have defined natural numbers, it *is* now feasible to give a flow diagram for an adding machine which will always get us the desired sum of any two natural numbers. We shall be somewhat more precise with this new flow diagram than we have been before. For example, in the past, we have

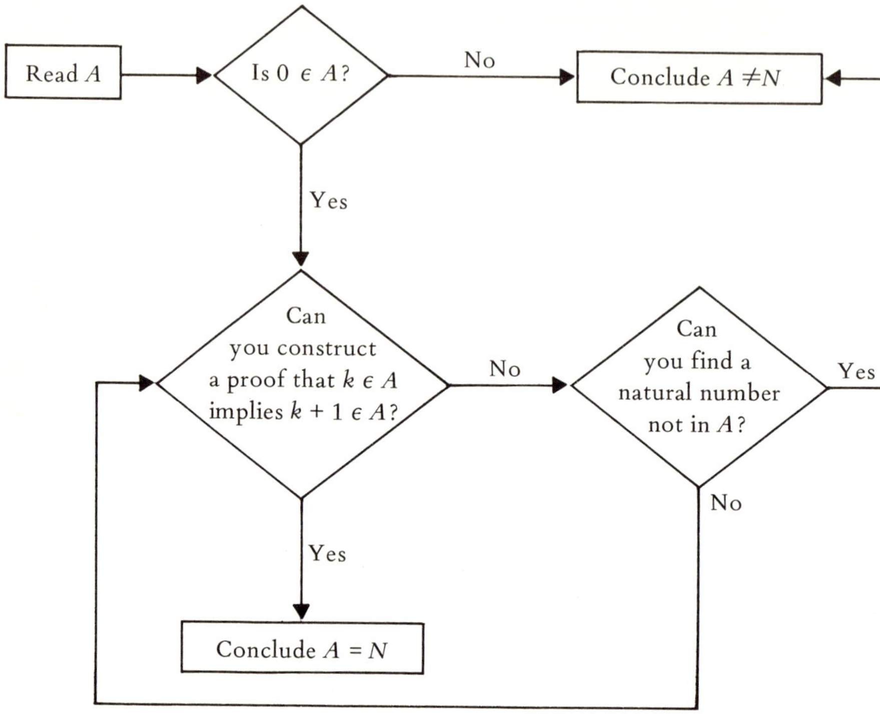

FIGURE 6.3

displayed flow diagrams where we had an instruction box [Read A] where A was a set. And we were not at all precise about what this meant. But we now want to take into account that the machine we are diagramming will need to deal with the names of numbers rather than with the numbers themselves. After all, we can't feed abstract concepts into a machine. The best we can do is feed in names for the concepts. So let us now suppose that the machine we wish to construct can recognize certain names for natural numbers. Let us also suppose that when the machine has a name for the natural number x, it can then produce a name for $x + 1$. Then Figure 6.4 gives us a flow diagram for an adding machine.

This machine is a fairly complicated one. But then it should be since it can supposedly perform infinitely many different tasks. It can add *any* two natural numbers. The diagram involves some notation we have not used before. The instruction $s \leftarrow s + 1$ means that inside the machine $s + 1$ is put in place of s. (Or, more precisely, a name for $s + 1$ is put in place of the name for s already stored in the machine.) $i \leftarrow i + 1$ means that, inside the machine, $i + 1$ is put in place of i.

Let us see how the machine works. Suppose we consider the case $y = 0$. The machine reads "x" and "0". It then sets (in its memory, if you like) $s = x$ and

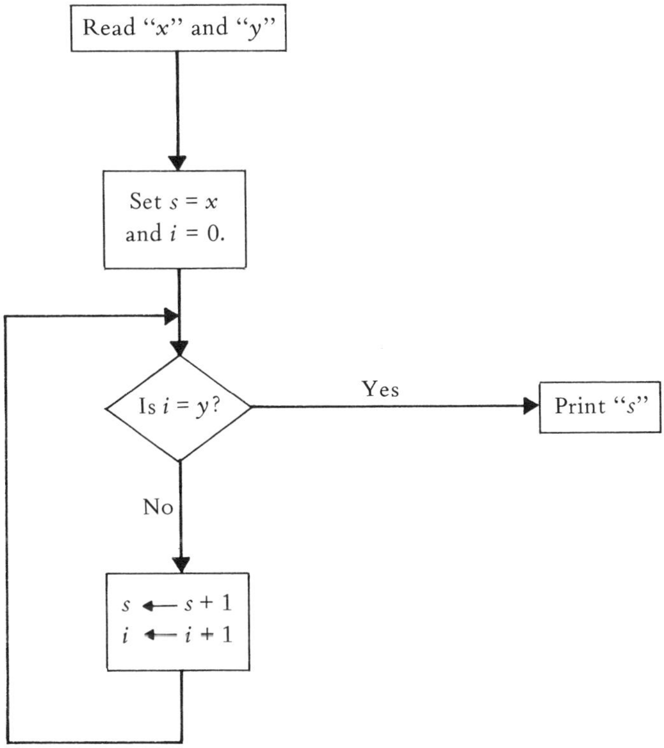

FIGURE 6.4 An adding machine.

$i = 0$. It then asks, "Is $0 = y$?" and finds the answer to be "Yes" so it prints "x". Of course, when $y = 0$, then $x + y = x + 0 = x$ so we got the correct result.

Suppose $y = 1$. Then when the machine asks "Is $0 = 1$?" it gets the answer "No." It then increases both s and i by 1 to get $s = x + 1$ and $i = 0 + 1 = 1$. It then goes back and asks again "Is $i = y$?". That is, it now asks "Is $1 = 1$?" The answer is "Yes" so it prints out "s". But now $s = x + 1 = x + y$. So again we get the correct result.

With $y = 2$ the machine makes one more loop than it did before and prints out a name for $x + 2$. And so forth.

The machine, if you like, starts with x and builds up to $x + y$ by successively adding 1's. After each act of adding 1, it asks itself, "Have I added enough 1's yet?" If it has, it prints the answer and quits. If it hasn't, it goes on to add another 1.

We close this section with a comment on a matter of notation that will be useful in the problems coming up. We have $N = \{0,1,2,3, \ldots\}$. The notation "$\{1,2,3, \ldots\}$" may be used to denote the set that we obtain by starting with the number 1 and including all the numbers we can obtain by successively adding 1. Similarly, $\{5,6,7, \ldots\}$ is the set which contains 5 and all the numbers we can generate by successively adding on 1. In the above expressions "..." is the mathematical form for "etc." We read "..." as "and so on."

Definition 6.4.1. The **natural numbers** are defined as follows:

(i) 0 is a natural number.
(ii) If n is a natural number, then $n + 1$ is also.
(iii) The only things which are natural numbers are those which are such as a consequence of (i) and (ii).

Definition 6.4.2. We let $N = \{x \mid x \text{ is a natural number}\}$.

Theorem 6.4.1. If (i) $A \subseteq N$, (ii) $0 \in A$, and (iii) $k \in A$ implies $k + 1 \in A$; then $A = N$.

Theorem 6.4.2. N is closed under addition.

Theorem 6.4.3. If x is a natural number then $x \neq x + 1$.

PROOF: Let $A = \{x \mid x \in N \text{ and } x \neq x + 1\}$. We see that $0 \in A$ because $0 + 1 = 1$ and it is impossible to set up a one-to-one correspondence between the empty set and $\{a\}$, so that $0 \neq 1$. Let $k \in A$. Assume that $k + 1 \notin A$. Then $k + 1 = (k + 1) + 1$. k is a cardinal number. Suppose B is a set such that $\#(B) = k$. There is no loss in generality if we suppose that $a \notin B$ and $b \notin B$. Then $\#(B \cup \{a\}) = k + 1$ and $\#(B \cup \{a,b\}) = k + 2 = k + (1 + 1) = (k + 1) + 1$. If $k + 1 = (k + 1) + 1$ then there is a one-to-one correspondence f between $B \cup \{a\}$ and $B \cup \{a,b\}$. Let $f(a) = c$. Define a function g on B by: $g(x) = f(x)$ if $x \in B$ and $f(x) \neq b$, $g(x) = c$ if $x \in B$ and $f(x) = b$. Then g is a one-to-one correspondence between B and $B \cup \{a\}$. This is so because g consists of either (i) exactly the pairs of f except that the pair (a,b) has been removed (this is the case where $c = b$), or else (ii) exactly the pairs of f except that the pairs (a,c) and $(f^{-1}(b),b)$ have been removed and replaced by the single pair $(f^{-1}(b),c)$ (this is the case where $c \neq b$). But then, since g is a one-to-one correspondence between B and $B \cup \{a\}$, and $\#(B) = k$ and $\#(B \cup \{a\}) = k + 1$; we have $k = k + 1$. This is a contradiction of the fact that $k \in A$. Hence, our assumption that $k + 1 \notin A$ must be wrong. Then, by induction, $A = N$ and the proof is complete.

Theorem 6.4.4. If x and y are natural numbers and $x + 1 = y + 1$, then $x = y$.

PROOF: Suppose $\#(A) = x$, $\#(B) = y$, and $x + 1 = y + 1$. Let a be an element not in A. Let b be an element not in B. Then $\#(A \cup \{a\}) = x + 1$ and $\#(B \cup \{b\}) = y + 1$. Since $x + 1 = y + 1$ there is a one-to-one correspondence f between $A \cup \{a\}$ and $B \cup \{b\}$. Let $f(a) = c$. We define a mapping g on A as follows: $g(x) = f(x)$ if $x \in A$ and $f(x) \neq b$, $g(x) = c$ if $x \in A$ and $f(x) = b$. The mapping g is a one-to-one correspondence between A and B. This follows from the fact that the pairs in g are either (i) exactly the pairs in f with (a,b) removed (this is the case where $c = b$) or else (ii) exactly the pairs in f except that the two pairs (a,c) and $(f^{-1}(b),b)$ have been replaced by the single pair $(f^{-1}(b),c)$ (this is the case where $c \neq b$). But since there is a one-to-one correspondence between A and B, we have $\#(A) = \#(B)$ and, hence, $x = y$. (Note that in this proof we needed only the fact that x and y were cardinal numbers and did not make use of the fact that they were **finite** cardinal numbers.)

Theorem 6.4.5. If x and y are natural numbers and $x + y = x$, then $y = 0$.

Theorem 6.4.6. If x, y, and z are natural numbers and $x + z = y + z$, then $x = y$.

PROOF: See Problem 4.

Theorem 6.4.7. If $x \in N$, then either $x = 0$ or there is a y in N such that $y + 1 = x$.

PROOF: Let $A = \{x | x = 0$ or there is a y such that $y + 1 = x\}$. Clearly $0 \in A$. Also $1 \in A$ since $0 \in N$ and $0 + 1 = 1$. Suppose $k \in A$ and $k \neq 0$. Then there is a y in N such that $y + 1 = k$. Then by Theorem 6.2.1 $(y + 1) + 1 = k + 1$. But $y + 1 \in N$, so $k + 1 \in A$. Hence, $A = N$ and the proof is complete.

PROBLEMS

1. Assuming the results in Problem 5, Section 6.2, use the basic definition of this section to show that 1,2,3,4,5,6,7,8,9, and 10 are all natural numbers.
2. Suppose $B \subseteq N$, $0 \notin B$, $1 \in B$, and $k \in B$ implies $k + 1 \in B$. Prove that $B = \{x | x \in N$ and $x \neq 0\}$. [HINT: Let $A = B \cup \{0\}$ and use induction. Be careful about your argument that for *all* k we have k in A implying $k + 1$ in A.]
3. Find a proper subset B of N such that $k \in B$ implies $k + 1 \in B$.
4. Fill in the blanks in the following.

Theorem. If x, y, and z are natural numbers and $x + z = y + z$, then $x = y$.

PROOF: Let x and y be given natural numbers. Let $A = \{z | x + z = y + z$ implies $x = y\}$. We first show that $0 \in A$. Suppose that $x + 0 = y + 0$. Since 0 is the identity element for addition of cardinal numbers we know that ______ and ______. Therefore, $x + 0 = y + 0$ implies ______ and we may conclude that ______.

Now suppose that $k \in A$. If $x + (k + 1) = y + (k + 1)$, then $(x + k) + 1 = (y + k) + 1$ by ______. But this last equality implies $x + k = y + k$

by ______. Furthermore $x + k = y + k$ implies $x = y$ because ______. Thus, we have shown that when $k \in A$, $x + (k + 1) = y + (k + 1)$ implies $x = y$. That is, $k \in A$ implies ______. We conclude that ______ and this completes the proof.

5. The function S defined by $S(n) = n + 1$ for $n \in N$ is called the **successor function.** What are the domain and range of S? Is S a one-to-one correspondence between certain sets? Explain how you know.

6. If $y = x + 1$ where x and y are natural numbers, then y is said to be the **immediate successor** of x and x is said to be the **immediate predecessor** of y. Restate the result (b) in this section in the "successor language." Which natural numbers have an immediate successor? Which natural numbers have an immediate predecessor? What is the immediate predecessor of the immediate successor of a natural number x?

7. Restate in "successor language" the conditions used in any argument by induction to show that $A = N$.

8. Suppose $A \subseteq N$, and $5 \in A$, and $k \in A$ implies $k + 1 \in A$. What exactly may we conclude about A?

9. Suppose $A \subseteq N$, and $0 \in N$, and $k \in A$ implies $k + 2 \in A$. May we conclude that $A = N$? Explain.

10. Suppose $A \subseteq N$, and $0 \in N$, and $k + 1 \in A$ implies $k \in A$. May we conclude that $A = N$? Explain.

11. Suppose $A \subseteq N$, and $10 \in A$, and $k + 1 \in A$ implies $k \in A$. What may we conclude about A?

12. Suppose $A \subseteq N$, and whenever $0 \in A$ and $k \in A$ it follows that $k + 1 \in A$. May we conclude that $A = N$? Explain.

13. Suppose $A \subseteq N$, and $k \in A$ if and only if $k + 1 \in A$. May we conclude that $A = N$? If not, what may we conclude about A?

---

14. Discuss the following theorem and its "proof."

Theorem. If $A = \{0,k, k + 1\}$ *then every natural number is in* A.

PROOF (?): Clearly $0 \in A$ and also $k \in A$ implies $k + 1 \in A$. Therefore, $A = N$.

15. Explain why the "counting machine" whose flow diagram is given in Figure 6.5 will correctly determine the cardinal number of any finite set if it can follow the instructions in the boxes.

6.5 The distributive law

Our definition of multiplication for cardinal numbers in terms of a Cartesian product may well have been disturbing to you because it did not seem to reflect immediately the idea that multiplication is "repeated addition." We can now show that there is no reason to be disturbed.

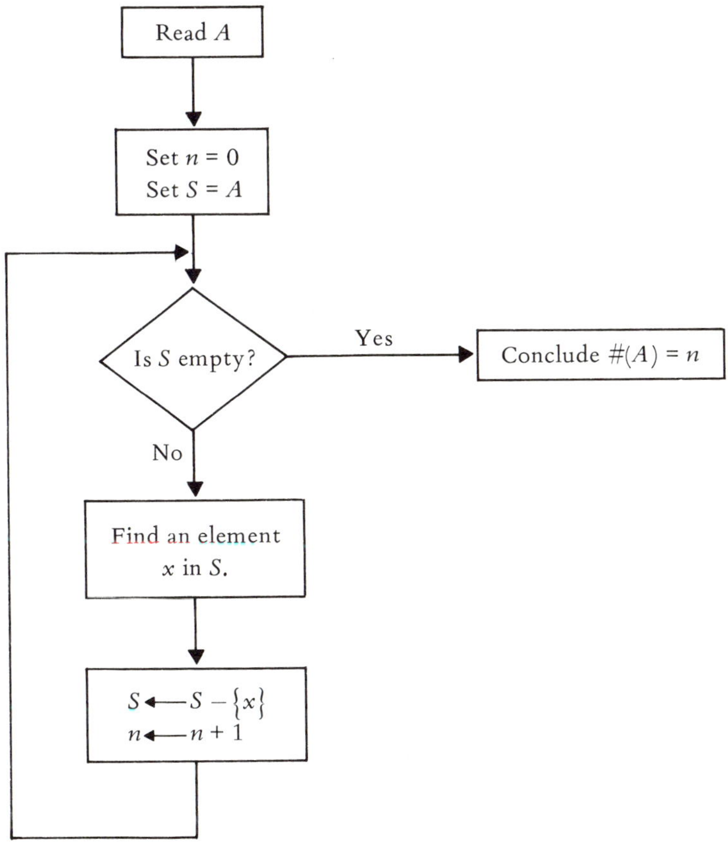

FIGURE 6.5

The basic fact we need is that, for cardinal numbers, multiplication distributes across addition. This is quite easy to prove. Suppose $x = \#(A)$, $y = \#(B)$, $z = \#(C)$ and $B \cap C = \phi$. Then, by our definitions for addition and multiplication of cardinal numbers $\#(B \cup C) = y + z$ and $\#(A \times (B \cup C)) = x \cdot (y + z)$. But by Theorem 2.5.2, $A \times (B \cup C) = (A \times B) \cup (A \times C)$. Also since $B \cap C = \phi$, $(A \times B) \cap (A \times C) = \phi$. Hence $x \cdot (y + z) = \#(A \times (B \cup C)) = \#((A \times B) \cup (A \times C)) = \#(A \times B) + \#(A \times C) = (x \cdot y) + (x \cdot z)$. We have thus proved:

> If x, y, and z are cardinal numbers then $x \cdot (y + z) = (x \cdot y) + (x \cdot z)$. This is called the **distributive law for multiplication across addition** or simply the **distributive law.**

If we make use of the commutativity of multiplication the above result also yields immediately that when x, y, and z are cardinal numbers, then $(x + y) \cdot z = (x \cdot z) + (y \cdot z)$.

If we recall that 1 is the identity for multiplication so that $1 \cdot x = x \cdot 1 = x$ for all x, then we have by the distributive law:

$$x \cdot 2 = x \cdot (1 + 1) = (x \cdot 1) + (x \cdot 1) = x + x$$

$$x \cdot 3 = x \cdot (2 + 1) = (x \cdot 2) + (x \cdot 1) = (x + x) + x$$

$$x \cdot 4 = x \cdot (3 + 1) = (x \cdot 3) + (x \cdot 1) = (x + x + x) + x.$$

And we begin to suspect that, indeed, multiplication as we defined it *is* repeated addition.

We can save ourselves some trouble in writing such equations as the above if we adopt some standard conventions about notation. First, it is customary to drop the dot which is the sign for the multiplication operation and simply use juxtaposition to indicate multiplication. Then in place of "$x \cdot y$" we write simply "xy." In place of "$(x \cdot y) + (x \cdot z)$" we write simply "$(xy) + (xz)$." It is as if we adopted the empty symbol (i.e. the absence of a symbol, not a blank) as the name for the multiplication operation. Of course, we need to be a bit careful in using this convention. When we see "10" we should interpret it as a name of $\#(\{a,b,c, \ldots, j\})$ not as an abbreviation for $1 \cdot 0$. We shall comment on this matter again later.

The second standard convention we shall adopt is that in place of "$(xy) + (xz)$" we may write simply "$xy + xz$" and mean the same thing. This convention is sometimes described by saying that we agree that multiplication takes precedence over addition. With this convention "$ab + c$" means $(ab) + c$ and not $a(b + c)$. "$5 \cdot 4 + 2$" means $(5 \cdot 4) + 2$ and not $5(4 + 2)$. The idea is that when we have some multiplications and additions to perform and parentheses do not inform us as to which should be done first, we always should perform all the possible multiplications before we do any of the additions.

"$abw + zx + y$" means $(abw) + (zx) + y$ or, even more precisely, $(((ab)w) + (zx)) + y$. When faced with nested sets of parentheses such as this we simply work our way out from the inside finding the indicated results as we include more and more of the expression. For instance, with $a = x = 2$, $b = 3$, and $w = z = y = 1$ in the above, we have $(((ab)w) + (zx)) + y = (((2 \cdot 3) \cdot 1) + (1 \cdot 2)) + 1 = ((6 \cdot 1) + (1 \cdot 2)) + 1 = (6 + (1 \cdot 2)) + 1 = (6 + 2) + 1 = 8 + 1 = 9$.

A sort of verification of the fact that multiplication is repeated addition can be put in the form of the assertion:

$$x \cdot 1 = x \text{ and } x \cdot (y + 1) = x \cdot y + x.$$

The last part of this follows easily from the first part and the distributive law. But looking at the situation from another point of view, the above result along with the fact that $x \cdot 0 = 0$ tells us that the set of natural numbers is closed under multiplication.

To see how this is so, let x be any natural number and consider the set $A = \{y \mid y \in N \text{ and } xy \in N\}$. Clearly, $0 \in A$ since $x \cdot 0 = 0$. And if $k \in A$ then we have

$x(k + 1) = xk + x$. But $xk \in N$ since $k \in A$. Also N is closed under addition. Therefore, $x(k + 1) \in N$ and it follows that $k + 1 \in A$. Hence, $A = N$ and we have shown that the product of **any** two natural numbers is a natural number.

What are sometimes referred to as algebraic rules for "adding variables" or (worse) "adding literal numbers" can be thought of as applications of the distributive law. For example

$$a + a = 2a$$

may be proved as follows:

$$a + a = 1 \cdot a + 1 \cdot a = (1 + 1)a = 2a.$$

Similarly, $5w + 4w = (5 + 4)w = 9w$.

We may make use of adding machines to construct a multiplying machine as in Figure 6.6.

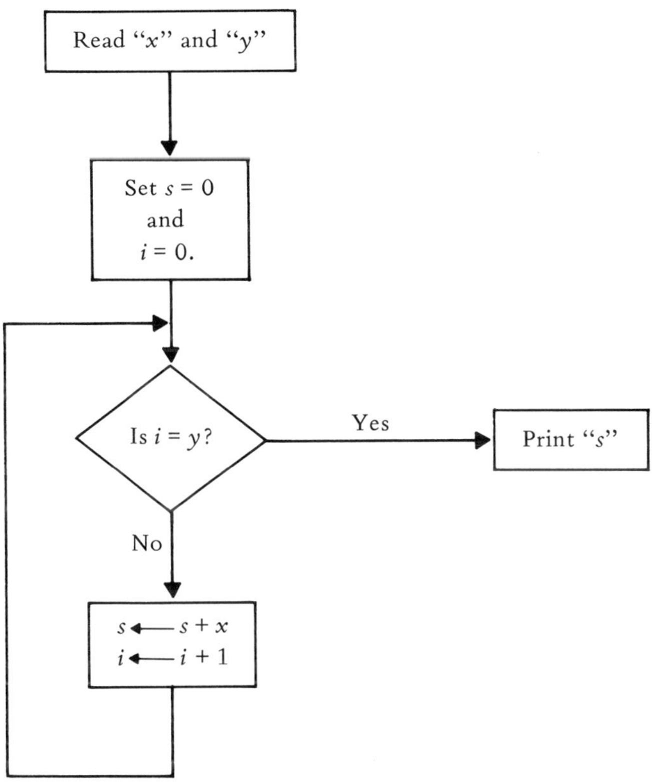

FIGURE 6.6 A multiplying machine.

Here we have assumed that we have the adding machine of Figure 6.4 built into this machine so that it may replace s with $s + x$. To compute xy the machine starts with 0 and successively adds x's until it has reached a sum of y x's. At each stage it must ask itself, "Have I added enough x's yet?"

Is it possible to find natural numbers x, y, and z such that $x \neq y$ and yet $xz = yz$? The answer is in the affirmative. For example, with $x = 2$, $y = 1$, and $z = 0$ we have $xz = 2 \cdot 0 = 0 = 1 \cdot 0 = yz$ and yet $x \neq y$.

However, if $z \neq 0$ and $xz = yz$ then we must have $x = y$. In terms of a table for multiplication this means that when $z \neq 0$, the column under z never contains two entries that are the same. In the proof of this fact (which will be given at the end of the section) we make use of the distributive law.

In terms of adding and multiplying machines, the distributive law asserts that the two machines in Figure 6.7 always give the same results when fed (the names of) the natural numbers x, y, and z.

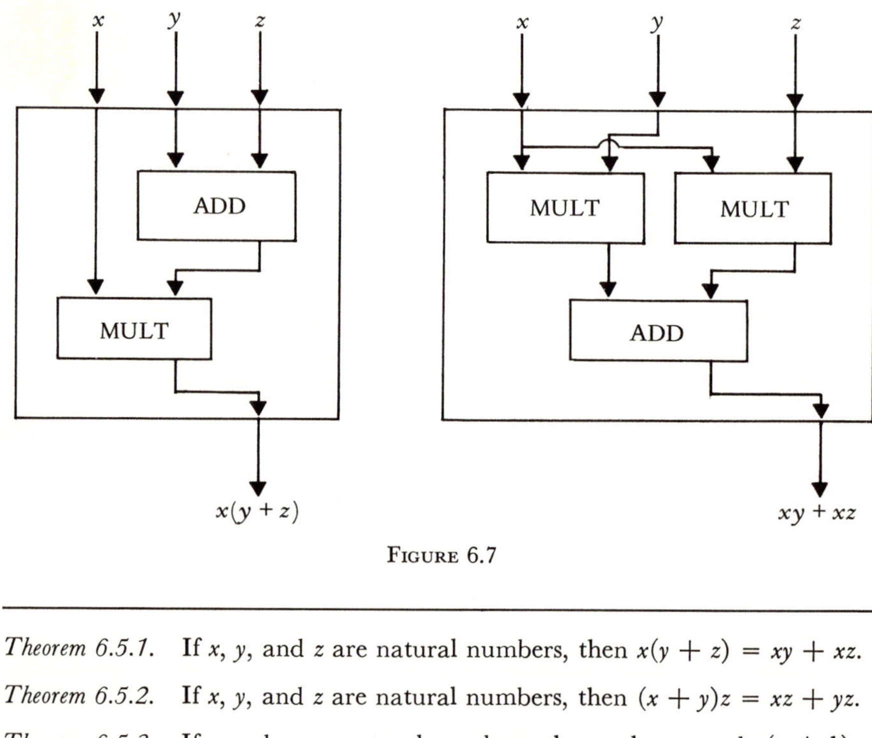

FIGURE 6.7

Theorem 6.5.1. If x, y, and z are natural numbers, then $x(y + z) = xy + xz$.

Theorem 6.5.2. If x, y, and z are natural numbers, then $(x + y)z = xz + yz$.

Theorem 6.5.3. If x and y are natural numbers, then $x \cdot 1 = x$ and $x(y + 1) = xy + x$.

Theorem 6.5.4. N is closed under multiplication.

Theorem 6.5.5. If x, y, and z are natural numbers such that $z \neq 0$ and $xz = yz$, then $x = y$.

PROOF: Suppose $z \in N$ and $z \neq 0$. Let $A = \{y \mid y \in N$ and for all x in N $xz =$

yz implies $x = y\}$. We claim that $0 \in A$. For suppose $x \in N$ and $xz = 0z$. Since $0 \cdot z = 0$ this gives $xz = 0$ which implies $x = 0$ since $z \neq 0$. (See Theorem 6.3.4.) Now suppose $k \in A$ and for some x in N we have $xz = (k + 1)z$. Since $k \in N$, $k + 1 \neq 0$, (see Theorem 6.2.3.) and therefore $x \neq 0$ (see Theorem 6.2.2). But if $x \neq 0$ there is a w such that $w \in N$ and $w + 1 = x$. (See Theorem 6.4.7.) Then we have $(w + 1)z = (k + 1)z$. By Theorem 6.5.3, this gives $wz + z = kz + z$. But then Theorem 6.4.6 gives us that $wz = kz$. Since $k \in A$, this implies $w = k$. Then $x = w + 1 = k + 1$. We have thus shown that when $k \in A$ we have $xz = (k + 1)z$ implying $x = k + 1$ for any x in N. That is, $k \in A$ implies $k + 1 \in A$ and this completes the proof.

Problems

1. Simplify each of the following: (a) $2 \cdot 3 + 4$, (b) $3 \cdot 2 + 2 \cdot 2$, (c) $7x + 2y + x + 3y$, (d) $3 \cdot 2 \cdot 1 + 4$, (e) $(2a + 1)(2 + 2)$, (f) $(2 + 1)(3 + 0) + (2 + 10) \cdot (0)$, (g) $2(2 + 1) + 2(1 + 1)$, (h) $w + z + w + z + w + z$, (i) $2w + a + 5w + 3a + w$.

2. As a consequence of the associative and commutative laws, we may add a string of numbers in any order that we choose. For example, $(a + b) + (c + d) = a + (b + c) + d = (d + a) + (b + c) = (a + c) + (b + d)$, etc. As a consequence, we may omit all parentheses in such a sum. Then, for example, we may write "$a + b + c + d$" to stand for any of the above. Prove carefully the third equality in the example.

3. Prove that $5x = x + x + x + x + x$.

4. Prove or disprove:

   (a) If x and y are natural numbers and $xy = x$, then $y = 1$.
   (b) If x and y are natural numbers and $xy = x$ and $x \neq 0$, then $y = 1$.

5. Prove that $(x + y)(z + w) = xz + yz + xw + yw$ for all natural numbers x, y, z, and w.

6. Given that $xy = 6$ simplify $(x + 2)(y + 1)$.

7. Illustrate the result in Problem 5 by constructing tables to represent certain Cartesian products when we take $x = \#(\{a,b\})$, $y = \#(\{1,2,3,4\})$, $z = \#(\{a,b,c\})$ and $w = \#(\{1\})$. (Tables representing certain subsets could be circled in different colors.)

8. Suppose we define the **area** of a square with a side of one inch to be 1 square inch. A rectangle with a height of 3 inches and a width of 2 inches may be divided up in a natural way so as to be a representation for the set $\{1,2,3\} \times \{1,2\}$. We take the area of such a rectangle to be $3 \cdot 2 = 6$ square inches. In general, we may think of $x \cdot y$ as the number of square inches of area in a rectangle of height x inches and width y inches. Use the area interpretation of products to illustrate the result in Problem 5.

9. Let $A = \{5,6,7,8, \ldots\}$. Is the system $(A,+)$ isomorphic to the system $(N,+)$? Is $(A,+,\cdot)$ isomorphic to $(N,+,\cdot)$?

6.6 Exponents

The result of multiplying a number by itself is called the *square* of the number. This ties in nicely with the geometric representation of a Cartesian product as a table. If we make the rows just as wide as the columns in a table representing $\{1,2,3\} \times \{1,2,3\}$ we end up with the geometric figure called a square.

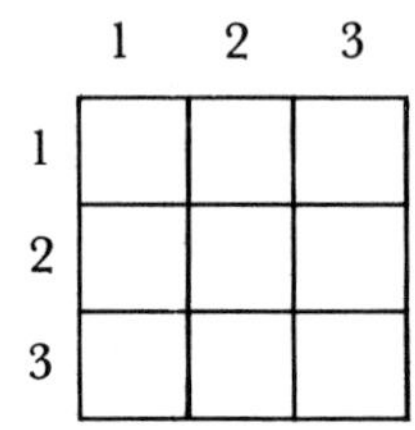

We have $3 \cdot 3 = 9$ and we write this more compactly as $3^2 = 9$.

In general, "x^2" means $x \cdot x$. We call the small "2" an **exponent** or a **power.** We may read "3^2" as "3 squared," "the square of 3," "3 to the power 2," "3 raised to the power 2," or "3 to the exponent 2."

It is easy to generalize this notion. 2^3 means $2 \cdot 2 \cdot 2$. 3^3 means $3 \cdot 3 \cdot 3$. 2^4 means $2 \cdot 2 \cdot 2 \cdot 2$. In general, x^n means the product of n x's. This can be put more precisely and completely as follows:

$$x^1 = x$$

$$x^{n+1} = x \cdot x^n.$$

This *inductive* definition enables us to evaluate x^n when x and n are natural numbers and n is different from 0. For example, $5^3 = 5 \cdot 5^2 = 5 \cdot 5 \cdot 5^1 = 5 \cdot 5 \cdot 5$.

A flow chart for evaluating nonzero powers of natural numbers is given in Figure 6.8.

Suppose, for example, we wish to find 5^2. We take $x = 5$ and $n = 2$. The machine starts by taking $p = 5$ and $i = 1$. It then asks, "Is $i = n$?" That is, "Is $1 = 2$?" Getting the answer "No," it makes $p = 5 \cdot 5$ and $i = 1 + 1 = 2$. It then asks, "Is $2 = 2$?", gets the answer "Yes," and prints out a name for $5 \cdot 5$.

In order to extend the above definition somewhat we do the customary thing and define x^0 to be 1 when $x \neq 0$. Then when $x \neq 0$, it is consistent with the above definition to assert that for all natural numbers n we have

$$x^0 = 1$$

$$x^{n+1} = x \cdot x^n.$$

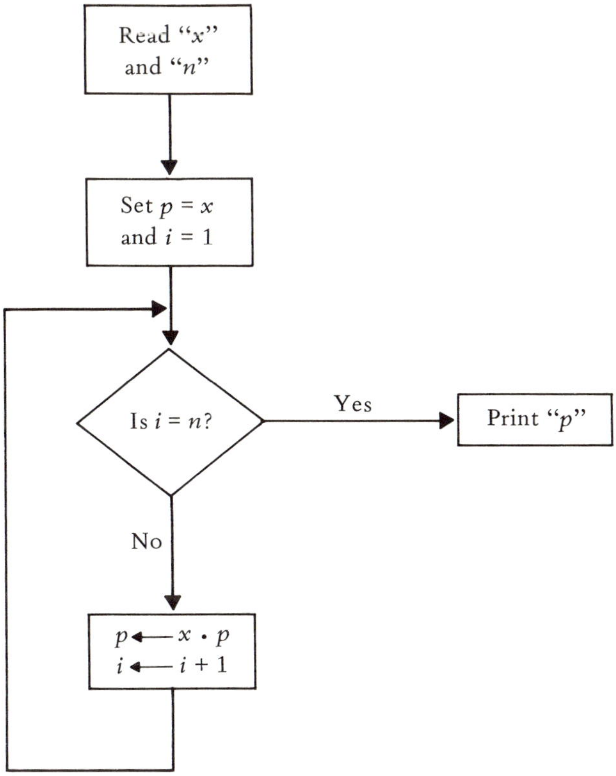

FIGURE 6.8 Flow chart for the evaluation of x^n.

This checks with the earlier definition since $x^1 = x^{0+1} = x \cdot x^0 = x \cdot 1 = x$.

The case where $x = 0$ is essentially trivial since 0^n is 0 for all natural numbers n different from 0. We leave 0^0 undefined for reasons that will become clear as we proceed.

Taking the definition which includes the exponent 0 as our starting point it is easy to prove the basic properties of exponents.† First of all, we note that when $x \neq 0$, x^m is defined for every natural number m. This is because we know what x^0 is and, since N is closed under multiplication, we can find x^{n+1} whenever we know x^n.

Notice that $(2^2)(2^3) = (2 \cdot 2)(2 \cdot 2 \cdot 2) = 2 \cdot 2 \cdot 2 \cdot 2 \cdot 2 = 2^5$. Also $3^4 \cdot 3^2 = (3 \cdot 3 \cdot 3 \cdot 3)(3 \cdot 3) = 3 \cdot 3 \cdot 3 \cdot 3 \cdot 3 \cdot 3 = 3^6$. More generally, we have

If x, n, and m are natural numbers and $x \neq 0$, then $(x^n)(x^m) = x^{n+m}$.

The proof of this is by an easy induction. Suppose x and n are given natural

† It would be better to say "the basic properties of exponentiation" but the other language is more commonly used.

numbers and $x \neq 0$. Let $A = \{m \mid m \in N \text{ and } (x^n)(x^m) = x^{n+m}\}$. Clearly $0 \in A$ since $(x^n)(x^0) = x^n \cdot 1 = x^n = x^{n+0}$. Suppose that $k \in A$. Then $(x^n)(x^{k+1}) = (x^n)(x \cdot x^k) = (x^n)(x^k \cdot x) = (x^n \cdot x^k) \cdot x$. But since $k \in A$, this last is equal to $(x^{n+k}) \cdot x = x \cdot x^{n+k} = x^{(n+k)+1} = x^{n+(k+1)}$. Therefore, $k + 1 \in A$. Hence, $A = N$ and this completes the proof.

In regard to raising products to powers we have such examples as: $(2 \cdot 3)^2 = (2 \cdot 3) \cdot (2 \cdot 3) = (2 \cdot 2) \cdot (3 \cdot 3) = 2^2 \cdot 3^2$ and $(3 \cdot 4)^3 = (3 \cdot 4) \cdot (3 \cdot 4) \cdot (3 \cdot 4) = (3 \cdot 3 \cdot 3) \cdot (4 \cdot 4 \cdot 4) = 3^3 \cdot 4^3$. More generally, if x and y are not 0,

$$(xy)^n = x^n y^n.$$

This may be proved by an easy induction on n.

What happens if we raise powers to powers? We have $(2^2)^2 = (2^2)(2^2) = (2 \cdot 2)(2 \cdot 2) = 2^4$, $(3^3)^2 = (3^3)(3^3) = (3 \cdot 3 \cdot 3)(3 \cdot 3 \cdot 3) = 3^6$, $(2^2)^3 = (2^2)(2^2)(2^2) = (2 \cdot 2)(2 \cdot 2)(2 \cdot 2) = 2^6$. More generally,

If x, n, and m are natural numbers and $x \neq 0$, then $(x^n)^m = x^{nm}$.

The proof we give is by induction on m. Suppose x and n are given natural numbers and $x \neq 0$. Let $A = \{m \mid m \in N \text{ and } (x^n)^m = x^{nm}\}$. Certainly $0 \in A$ since† $(x^n)^0 = 1$ and $x^{n \cdot 0} = x^0 = 1$. Suppose $k \in A$. Then $(x^n)^k = x^{nk}$ and $(x^n)^{k+1} = (x^n)(x^n)^k = (x^n)(x^{nk}) = x^{nk+n} = x^{n(k+1)}$. Therefore, $k \in A$ implies $k + 1 \in A$. It follows that $A = N$ and this completes the proof.

Exponents are helpful in simplifying certain expressions involving variables. For example, $(x + 2)(x + 3) = (x + 2)x + (x + 2)3 = x \cdot x + 2x + 3x + 6 = x^2 + 5x + 6$. Somewhat more generally we have for any natural numbers a, b, and x:

$$(x + a)(x + b) = x^2 + (a + b)x + ab.$$

The proof of this follows the pattern of the special case we just worked out.

Despite the fact that we have not used the usual notation for an operation, the process of raising a natural number to a power *is* an operation in N. We could use operator notation and write, say, "xEn" for x^n. Note that N is not closed under E since $0E0 = 0^0$ is undefined. It is also the case that E is not commutative and not associative. We do have $xE1 = x$ for all x in N. But $1Ex = 1$ for all x in N. Thus, 1 is a sort of a **right identity** element for E, but there is no identity element for E.

In these first sections of this chapter we have avoided using the common names for natural numbers greater than 10. The reason for this is that we have not yet explained carefully how we arrive at names for larger numbers. We now have the logical tools to do this and shall do so in the next section. In the problems

† Note that x^n cannot be 0 since $x \neq 0$.

for this section, however, we shall recognize the fact that you do know how to work with such names as "11," "100," "256," etc.

Definition 6.6.1. $0^n = 0$ for all $n \neq 0$. If $x \neq 0$, then $x^0 = 1$ and $x^{n+1} = x \cdot x^n$.

Theorem 6.6.1. If x, n, and m are natural numbers and $x \neq 0$, then $(x^n)(x^m) = x^{n+m}$.

Theorem 6.6.2. If x and y are in N and different from 0, then $(xy)^n = x^n y^n$.

Theorem 6.6.3. If x, n, and m are natural numbers and $x \neq 0$, then $(x^n)^m = x^{nm}$.

Theorem 6.6.4. If a, b, and x are natural numbers, then

$$(x + a)(x + b) = x^2 + (a + b)x + ab.$$

PROOF:

$$\begin{aligned}(x + a)(x + b) &= (x + a)x + (x + a)b \\ &= (x \cdot x + a \cdot x) + (x \cdot b + a \cdot b) \\ &= x^2 + ax + bx + ab \\ &= x^2 + (a + b)x + ab.\end{aligned}$$

PROBLEMS

1. Complete the following partial table for the operation E when $xEn = x^n$.

E	0	1	2	3	4	5
0	–	0	0	0	0	0
1	1	1	1	1	1	1
2	1	2	4			
3	1	3				
4	1	4				
5	1	5				

2. If $xEy = x^y$ show that E is not commutative and not associative.

3. Define a function T on N by $T(x) = 2^x$. What is the range of T? Is T a one-to-one correspondence between its domain and range? (If you wish to justify your answer to the last question, try proving that $2^{n+m} = 2^m$ implies $n = 0$.)

4. Prove Theorem 6.6.2.

5. Revise the flow chart of Figure 6.8 so that the resulting machine will compute x^n as long as x and n are not both 0.

6. Do there exist natural numbers n and m such that $n \neq m$ and yet $n^m = m^n$?

6.7 Decimal names for the natural numbers

10 is the cardinal number of the collection of all the fingers a person has. We can think of each of the numbers 0, 1, 2, 3, 4, 5, 6, 7, 8, 9, and 10 as cardinal numbers for easily displayed collections of a person's fingers.

We now wish to develop a system of names for the natural numbers. The system we shall first develop will be the decimal numeration system with which you are familiar. So what we shall be doing can be thought of as explaining exactly what this familiar system is.

In the logical development we are putting forth, we should think of ourselves as rather sophisticated children who understand the idea of what a natural number is, know what it means to add or multiply two natural numbers, and understand the process of raising a natural number to a natural number power. But we do not know what "124" means or even what "11" means. In other words, we know about putting sets together in certain ways to obtain sets with cardinal numbers that are sums or products or powers but we have no convenient names for natural numbers beyond 10.

We first adopt names for the powers of 10. For 10^2 we shall write "100." We read "100" as "one hundred." For 10^3 we shall write "1000." We read "1000" as "one thousand." We continue in this fashion getting the following table.†

To this number	we assign the name	which is read as
10^1	10	ten
10^2	100	one hundred
10^3	1000	one thousand
10^4	10,000	ten thousand
10^5	100,000	one hundred thousand
10^6	1,000,000	one million
10^7	10,000,000	ten million
10^8	100,000,000	one hundred million
10^9	1,000,000,000	one billion

More generally, the name we assign to 10^n is a "1" followed by n "0's."

The symbols "0," "1," "2," "3," "4," "5," "6," "7," "8," and "9" are called decimal digits. Thus the digit "7" is a name for the number 7. If we reread the first paragraph in this section it should not startle us to use this terminology since "digit" is also used to mean finger.

We now arrange it so that strings of decimal digits are used to name the natural numbers. In particular:

† In the last two columns in the table there should be, strictly speaking, quotes about each entry. But their omission should cause no confusion.

This expression	which we read as	is a name for this number
11	eleven	$10 + 1$
12	twelve	$10 + 2$
13	thirteen	$10 + 3$
14	fourteen	$10 + 4$
15	fifteen	$10 + 5$
16	sixteen	$10 + 6$
17	seventeen	$10 + 7$
18	eighteen	$10 + 8$
19	nineteen	$10 + 9$
20	twenty	$2 \cdot 10$
21	twenty-one	$2 \cdot 10 + 1$

More generally suppose we have a string of $n + 1$ decimal digits. Let us call the last (or rightmost) digit in the string "a_0." Let "a_1" be the digit immediately to the left of "a_0." Let "a_2" be the digit to the left of "a_1." . . . Let "a_n" be the leftmost digit in the string.† Then the string of digits is a name for the number

$$a_n \cdot 10^n + a_{n-1} \cdot 10^{n-1} + \cdots + a_0.$$

For example, "5312" is a name for the number

$$5 \cdot 10^3 + 3 \cdot 10^2 + 1 \cdot 10 + 2.$$

Here $a_0 = 2$, $a_1 = 1$, $a_2 = 3$, and $a_3 = 5$. We read "5312" as "five thousand three hundred twelve."‡

In the decimal name "$a_n \cdots a_1 a_0$," "a_0" is said to be the digit in the **units place,** "a_1" is said to be the digit in the **tens place,** "a_2" is said to be the digit in the **hundreds place,** etc.

When "a_n," . . . , "a_1," and "a_0" are decimal digits, the string "$a_n a_{n-1} \ldots a_0$" is taken to be a name for a number according to the above scheme. It is **not** to be taken as the product of a_n times a_{n-1} times $\cdots$ times a_0. If we wish to indicate the product of numbers written in the above notation, we must use the dot to indicate multiplication. Thus, $32 = 3 \cdot 10 + 2$ while $3 \cdot 2 = 6$.

Whereas it would be natural to read "23" as "two tens and three" it is common to have special ways to read the names for numbers between 1 and 100. Thus, "23" is read as "twenty-three." "74" is read as "seventy-four" instead of "seven tens and 4." Except for some slight changes in spelling, these special names for numbers past 19 are obtained if we treat "ty" as an abbreviation for "tens and." 84 is eight tens and 4, but we read it as "eighty-four."

† The small number name written to the right and below the "a" is called a **subscript.** "a_n" is read as "a sub n."

‡ Evidently, "five thousand*s* three hundred*s* twelve" would also be correct—but less common.

We shall assume that you are familiar with the usual devices for reading strings of decimal digits and simply give a couple of examples. "37430" or "37,430" is read as "thirty-seven thousand four hundred thirty." "101542716" or "101,542,716" is read as "one hundred one million, five hundred forty-two thousand, seven hundred sixteen." We note that it is customary but not necessary to use a comma to separate clumps of three digits to make things easier.

A name of a number is often called a **numeral.** Thus, "7" is the **numeral** for the number 7. "6312" is the **numeral** for the number six thousand three hundred twelve.

From the manner in which we have described our naming procedure it follows that "0017" is a name for 17. However, it is customary, although not essential, not to allow "0" at the beginning of a string of digits that is to name a number. The one exception is that where we wish to name 0 itself. In practice, when dealing with a desk calculator or any computer it is sometimes convenient to violate this convention and allow one or more initial "0's". It would also make sense to be patient with a child who violates the convention if this ever occurs.

The names we have introduced are called the **decimal** names for the natural numbers. When we use these names we say we are using **decimal notation** or **base ten notation** for the numbers.

If the decimal naming method is a good one, we should have each natural number getting a name, and no string of decimal digits should be the name of two different numbers. It will be shown in the appendix to this section that our method does indeed have these properties.

In what follows in this book we shall usually do what most people in the world do and use decimal notation for the natural numbers. But there are special cases where other naming systems are useful.

Definition 6.7.1. "0," "1," "2," "3," "4," "5," "6," "7," "8," and "9" are called **decimal digits.**

Definition 6.7.2. If "a_n," . . . , "a_1," and "a_0" are decimal digits, then "$a_n \ldots a_1a_0$" is the **decimal name** for the number

$$a_n \cdot 10^n + \cdots + a_1 \cdot 10 + a_0.$$

This is called the **decimal notation** or **base ten notation** for numbers.

Theorem 6.7.1. Each finite string of decimal digits names a unique natural number.

PROOF: This follows from the closure of N under addition and multiplication and the fact that the decimal digits and "10" all name natural numbers.

Theorem 6.7.2. Each natural number has a decimal name.

PROOF: Let $A = \{n \mid n \in N \text{ and } n \text{ has a decimal name}\}$. Since "0" is a decimal name for 0, $0 \in A$. Suppose $k \in A$. Let "$a_n \ldots a_1a_0$" be a decimal name for k. Then $k + 1 = a_n \cdot 10^n + \cdots + a_1 \cdot 10 + a_0 + 1$. If $a_0 + 1$ is less than 10, let "b_0" be the decimal digit that names it. Then "$a_n \ldots a_2a_1b_0$" is a decimal name of $k + 1$. If $a_0 + 1 = 10$ then $k + 1 = a_n \cdot 10^n + \cdots + a_1 \cdot 10 + 10 = a_n \cdot 10^n + \cdots + (a_1 + 1) \cdot 10 + 0$. If $a_1 + 1 = b_1$ is less than 10, then "$a_n \ldots a_2b_10$" is a decimal name for $k + 1$. If $a_1 + 1 = 10$, then $k + 1 = a_n \cdot 10^n + \cdots + a_2 \cdot 10^2 + 10 \cdot 10 = a_n \cdot 10^n + \cdots + a_3 \cdot 10^3 + (a_2 + 1) \cdot 10^2$. Now either $a_2 + 1 = 10$ or not, and we proceed as before. In each case we find a decimal name for $k + 1$. The extreme case would be where $a_n = \cdots = a_1 = a_0 = 9$ and the decimal name for $k + 1$ is a "1" followed by $n + 1$ "0's."

PROBLEMS

1. Write out in words the proper way to read the expression (a) 195, (b) 1,000,006, (c) 52,723, (d) 7,502,794, (e) 8,964,000,032.

2. Draw a table which represents $A \times A$ where $A = \{1,2,3,4,5,6,7,8,9,10\}$. How many boxes are there in the table?

3. Suppose we have many small cubical blocks each of which has a face that just fits one of the boxes in the table in Problem 2. Tell how we may get a physical representation of 10^3.

4. Discuss how we might build color names for natural numbers by assigning a different color to each of the numbers 0,1,2,3,4,5,6,7,8, and 9 and then using strings of colored dots (or strokes) as number names.

5. Suppose we assign names to the numbers 0,1,2,3,4,5,6,7,8, and 9 as in the table below:

This number	0	1	2	3	4	5	6	7	8	9
is assigned this name	0	1	11	*T*	*f*	*F*	*s*	*S*	*e*	*n*

(a) First show that if we use these names as decimal digits we cannot build a satisfactory naming system by following the method in this section. (b) Then change the name assigned to 2 to "*t*" and argue that we can now build an adequate naming system using the above table and the method of this section. (c) Finally, in this new set-up in (b) could we simply use a blank in place of "0" or would this get us into trouble?

6. Let $B = \{0,1,2,3,4,5,6,7,8,9\}$. Construct a tree diagram for $(B \times B) \times B$. (You need not draw the entire tree but give enough of it to indicate its nature.) Explain how certain paths in this tree may be thought of as representing 3-digit numbers. (Allow initial "0's" here.) How many such paths are there?

6.8 Other naming systems

One place where it turns out that decimal names are not the most convenient names to use is in the study of the foundations of mathematics where mathematical logicians ask themselves what computing machines can and cannot do. Are there some mathematical problems that no machine—no matter how large or complicated—could solve? Such a question is important and leads to the question of what computers actually are—not in terms of hardware but in terms of the principles behind them. When such questions are considered, it is helpful to have very simple methods for naming the natural numbers so that very simple machines can recognize the names. The naming method often used in such a situation is perhaps the simplest of all such systems. This could be called the **tally** system. In this system, the natural numbers are named as follows:

The number whose decimal name is	is, in the tally system, called
0	0
1	1
2	11
3	111
4	1111
5	11111
6	111111
7	1111111
8	11111111
9	111111111
10	1111111111
11	11111111111

In general, the tally system name for the natural number n consists of a string of n "1's."

It is interesting and somewhat ironic that modern mathematicians concerned with subtle questions about the limitations of computers should return to this most primitive system of names.†

Another naming system in fairly common use is the system of **Roman numerals.** This archaic system is of little practical value and will not be discussed here.‡

† In actual practice, the logician would be somewhat more likely to use "1" as a name for 0, "11" as a name for 1, "111" as a name for 2, etc. But the tally system as we have described it above is almost the same as this and will be less confusing to us.

‡ A competent discussion of how to read Roman numerals will be found in almost any encyclopedia.

We are led to some other naming systems which **are** of some practical value when we ask the question: What is so special about the number ten? Can't we construct naming systems which use powers of other numbers besides 10? The answer is that we can.

One such system is the **binary** system which is useful in working with electronic computers. Here we follow the pattern of the decimal system but use 2 in place of ten. There are two **binary digits:** "0" and "1." In the binary system we name 2^n by writing a "1" followed by *n* "0's." We obtain the following:

The number	is, in the binary system, called	which is read as
2^0	1	one
2^1	10	two
2^2	100	one quad
2^3	1000	one octet
2^4	10,000	two octet
2^5	100,000	one quad octet
2^6	1,000,000	one octet octet
2^7	10,000,000	two octet octet
2^8	100,000,000	one quad octet octet

Here, while the names in the middle are standard and used by everyone using the binary system, the way these names are read as given in the last column is not standard. This terminology is introduced here in the hope that it will be helpful to us in dealing with the binary system.

A possible alternative way of reading the binary names for the powers of two would be suggested by the units of liquid measure as follows:

The number	is, in the binary system, called	which we read as
2^0	1	one
2^1	10	two
2^2	100	one gill
2^3	1000	one cup
2^4	10000	one pint
2^5	100000	one quart
2^6	10,00000	two quarts
2^7	100,00000	one gallon
2^8	1000,00000	two gallons

But again, these verbalizations for the names given in the center are not standard.

If we are given a string of $n + 1$ binary digits, let "b_0" be the rightmost digit, "b_1" the next digit, . . . , and "b_n" the leftmost digit. Then this string of digits is taken as a name for the number $b_n \cdot 2^n + \cdots + b_1 \cdot 2 + b_0$. Thus, "10111" is a name in the binary system for $1 \cdot 2^4 + 0 \cdot 2^3 + 1 \cdot 2^2 + 1 \cdot 2 + 1 = 2^4 + 2^2 + 2 + 1$. In the decimal system we would call this number "16 + 4 + 2 + 1" or "23." When we use the binary naming system we say that we are writing the numbers in the **binary notation** or in the **base two notation.**

Using binary notation we could read "10,111" as "two octet, one quad, two, and one." We could read "1,011,001" as "one octet octet, two and one octet, and one."

If we adopt the alternative way of reading binary names as given above then we could read "10111" as "one pint, one gill, two, and one," and "10,11001" as "two quarts, one pint, one cup, and one."

So that we can distinguish between various naming systems, we shall adopt the standard practice of writing a name for the base as a subscript on the number name.† If no such subscript appears, the base ten is to be assumed. Thus, we may write

$$10{,}111_2 = 23$$

$$1011001_2 = 2^6 + 2^4 + 2^3 + 1 = 89.$$

Every natural number has a name in the binary system and any string of binary digits names a natural number. Again, it is customary, but not necessary, to use only strings of binary digits that do not start with a "0" unless we wish to refer to 0 itself.

In order to find the binary name for a natural number we must express the number as a sum of powers of two. If we wish to find the binary name for 147 we may proceed as follows:

$$\begin{aligned} 147 &= 128 + 19 = 128 + 16 + 3 = 128 + 16 + 2 + 1 \\ &= 2^7 + 2^4 + 2^1 + 2^0 = 10010011_2. \end{aligned}$$

At each step here, we find the largest power of two in the last term in the sum and then write that last term as a sum of that power of two and another number.

Now that we have seen what it means to write numerals to the base 2 or to the base 10, it is easy to see how a naming system may be built using any natural number greater than 1 as a base. The bases that are most often encountered aside from 10 and 2 are 3, 8, and 12.

In the base 3 or **ternary** naming system we would have the **ternary digits** "0," "1," and "2" and:

---

† We shall also use a subscript T to indicate a tally name. We must, of course, choose some naming system for the subscripts indicating the base. We shall use decimal names for them.

The number	is, in the base 3 system, called	which is read as
1	1	one
2	2	two
3	10	three
3^2	100	one court
3^3	1000	three courts
3^4	10000	one court court
3^5	100000	three court court

Here again the verbalizations are not standard, but in the U.S.A. "court" seems supremely justifiable as a name for 3^2.

If the base 12 is used we need, in addition to the usual decimal digits, some standard names for 10 and 11 to serve as base 12 digits. We shall use "*t*" for 10 and "*e*" for 11. Then:

The number whose decimal name is	is, in the base 12 system, called	which is read as
10	*t*	ten
11	*e*	eleven
12	10	twelve (or one dozen)
12^2	100	one gross
12^3	1000	one dozen gross
12^4	10000	one gross gross
12^5	100000	one dozen gross gross

Then "$81t0e_{12}$" is read as "eight gross gross, one dozen gross, ten gross, and eleven." And

$$81t0e_{12} = 8 \cdot 12^4 + 1 \cdot 12^3 + 10 \cdot 12^2 + 11 = 169{,}067.$$

There is an international organization which supports the universal adoption of the base 12 naming system. Such a system does have some computational advantages that we shall mention from time to time as we proceed.

The method for verbalizing the base 12 names as given above is not standard, but it may be helpful.

PROBLEMS

1. Give the decimal name for the number named by "1111" if we use the base (a) 2, (b) 3, (c) 5, (d) 8, (e) 12, (f) if we use the tally system.

2. Find the decimal name for each of the following. (a) 1000000_2, (b) 222_3, (c) eee_{12}, (d) 210212_3, (e) $9t7e_{12}$, (f) 6740_8, (g) 4432_5, (h) 1111111111111111_T.

3. (a) What is the decimal name for 999999999 + 1?
   (b) What is the base 2 name for $111111111_2 + 1$?
   (c) What is the base 12 name for $eeeeeeeee_{12} + 1$?
   (d) What is the base 3 name for $222222222_3 + 1$?

4. Describe carefully in your own words how to find the tally name for the sum of two numbers whose tally names are given.

5. Find the base 2 name for each of the following numbers: (a) 74, (b) 1346, (c) 7496, (d) 1,545,463.

6. Find the base 3 names for the numbers given in Problem 5.

7. Find the base 12 names for the numbers given in Problem 5.

8. Complete the following table to display the names for 0,1, . . . , 100 when the bases 2, 3, and 12 are used. (The idea is to work down one column at a time to get the pattern for getting the next name in the list.)

Decimal name	Base 2 name	Base 3 name	Base 12 name
0	0	0	0
1	1	1	1
2	10	2	2
3	11	10	3
⋮			
40	10100	1111	34
⋮			
99	1100011	1020	83
100	1100100	1021	84

9. Complete the following partial tables for addition and multiplication where the numbers are written to the base 2. (Here we suppress the subscript 2 for convenience since only names to the base 2 will be used.)

+	0	1	10	11	100
0	0	1	10	11	100
1	1	10	11	100	101
10	10	11	100		
11	11	100			
100	100	101			

·	0	1	10	11	100
0	0	0	0	0	0
1	0	1	10	11	100
10	0	10	100		
11	0	11			
100	0	100			

10. Construct partial tables for addition and multiplication of natural numbers expressed in base 3 notation by filling the blanks in the following:

(Here we suppress the subscript 3 since only base 3 names are involved.)

+	0	1	2	10	11	12	20
0							
1							
2							
10							
11							
12							
20							

·	0	1	2	10	11	12	20
0							
1							
2							
10							
11							
12							
20							

11. Construct partial tables for addition and multiplication of natural numbers expressed in tally notation by completing the following tables. (Here we suppress the subscript T since only tally names are involved.)

+	0	1	11	111
0				
1				
11				
111				

·	0	1	11	111
0				
1				
11				
111				

12. What is the main disadvantage of the tally naming system as compared with the decimal naming system?

13. Suppose you are told that "113" is the name of a number to some unknown base. What is the smallest number that this expression could be a properly constructed name for? Is there a largest number it could be a name for?

6.9 Inequalities

We all have a reliable idea of what it means for one natural number to be larger than another. Comparisons between numbers which indicate such things as height, distance, weight, age, and cost are common in everyday discourse.

Partly because our intuitive grasp of inequalities is so sound, it is quite easy to give a careful logical development of the basic properties of inequalities.

It is convenient to start on this development by first saying exactly what it means to say that one natural number is at least as large as another. For brevity, we shall write "$x \geq y$" to assert that x is at least as large as y. Or, in other words, "$x \geq y$" should mean that x is larger than y or equal to y. "$x \geq y$" should be true when x occurs at least as far out as y in the march of grinding out successive natural numbers. We have the following formal definition:

We say that $x \geq y$ if there is a natural number z such that $x = y + z$. We read "$x \geq y$" as "x is greater than or equal to y."

We may then define the relations $>$, $\leq$, and $<$ in terms of $\geq$ as follows:

We say $x > y$ if $x \geq y$ and $x \neq y$. In place of $x > y$ we may write $y < x$. In place of $x \geq y$ we may write $y \leq x$.

We read "$x > y$" as "x is greater than y", "$x \leq y$" as "x is less than or equal to y", and "$x < y$" as "x is less than y."

For example, $9 \geq 6$ since $9 = 6 + 3$. Also, $9 > 6$ since $9 \geq 6$ and $9 \neq 6$. And $6 \leq 9$ means the same as $9 \geq 6$, while $6 < 9$ means the same as $9 > 6$.

Note that the shape of the symbol "$>$" is helpful in remembering what it means. The large, open end of the symbol is next to the name of the number that is asserted to be larger. The small, pointed end of the symbol is next to the name of the number asserted to be smaller.

The expressions "$x < y$," "$y > x$," "$x \geq y$," and "$y \leq x$" are all called **inequalities.** "$x \geq y$" and "$x \leq y$" are said to be **weak inequalities.** "$x > y$" and "$x < y$" are said to be **strong inequalities.**

We shall sometimes write "$x < y < z$" to assert that $x < y$ *and* $y < z$. Similar interpretations are put on other **continued inequalities** such as $x < y \leq z$.

When writing continued inequalities it is customary to insist that all the inequalities run in the same direction. Thus, "$6 \leq 7 < 10$" is permissible and true. But "$6 \leq 10 > 7$" is an improper use of the notation. Violations of this convention often cause great confusion. So great care should be exercised, in using continued inequalities, to see that all the inequalities have the same sense. Then as we run through a (true) continued inequality we are always moving towards larger numbers or always moving towards smaller numbers.

In general, when $x < y < z$, we say that y is *between* x and z, or else that y is *between* z and x.

We shall now list some of the basic properties of inequalities. Our intuitive grasp of what it means to have one number larger than another is often so good that we can simply glance at some of these assertions and say immediately, "Of course. That must be so." Proofs of some of the following are to be given as exercises, but some of the more difficult proofs are given in the appendix to this section. In these assertions x, y, and z are natural numbers.

1. $x \geq 0$
2. $x \geq x$
3. If $x \geq y$ and $y \geq z$, then $x \geq z$.
4. $x > y$ and $y > z$ implies $x > z$.
5. $x \geq y$ or $y \geq x$
6. If $x \geq y$ and $y \geq x$, then $x = y$.
7. (The law of trichotomy) $x > y$ or $y > x$ or $x = y$ and only one of these may hold.
8. $x + 1 > x$
9. $x \geq y$ if and only if $x + z \geq y + z$.
10. $x > y$ if and only if $x + z > y + z$.
11. If $x \geq y$ and $z \geq w$, then $x + z \geq y + w$.

12. If $x > y$ and $z > w$, then $x + z > y + w$.
13. If $x < y$, then $x + 1 \leq y$.
14. There is no x such that $0 < x < 1$.
15. If $x \geq y$, then $xz \geq yz$.
16. If $x > y$ and $z \neq 0$, then $xz > yz$.
17. If $x \geq y$ and $z \geq w$, then $xz \geq yw$.
18. If $x > y$ and $z > w$, then $xz > yw$.
19. If $x \geq y$, then $x^2 \geq y^2$.
20. If $x \geq y$ and $z \neq 0$, then $x^z \geq y^z$.

Many of these statements are often put in other terms. For instance, assertion number 3 may be restated as: $\geq$ is transitive in N. Parts of 9 and 10 together can be put as: adding the same thing to both sides of an inequality preserves the inequality. Assertion number 14 could be put as: there is no natural number between 0 and 1.

The decimal numeration system has a wonderful property in regard to inequalities—larger names go with larger numbers. The more digits there are in a numeral (assuming we do not allow initial "0's") the larger the number. Thus, we can tell at a glance that $101{,}312 > 98{,}694$ and $2{,}631{,}000 > 996{,}543$.

If two numbers have decimal names with the same number of digits, then we may determine which is larger by the following procedure. We compare the numerals a digit at a time working from the left. We stop when we find a (number named by the) digit from one numeral larger than the (number named by the) corresponding digit from the other. And we conclude that the number contributing the larger digit is larger. For example, $999{,}611 > 999{,}498$.

This ease with which the decimal numeration system enables us to determine the relative sizes of numbers is not afforded us by such a system as Roman numerals. For example, X > VI although "VI" is longer than "X."

The proofs of some of the theorems of this section do not follow quite so immediately from the definition of $\geq$. We give these proofs below. But the theorems below are not necessarily any more important or useful than the others given above. In giving these proofs we assume that assertions earlier in the list have already been established. Let us consider assertions 1 through 20 above as Theorem 6.9.1, Theorem 6.9.2, . . . , Theorem 6.9.20.

Theorem 6.9.5. If x and y are in N, then $x \geq y$ or $y \geq x$.

PROOF: Suppose $x \in N$. Let $A = \{y \mid y \in N$, and $x \geq y$ or $y \geq x\}$. Clearly $0 \in A$ since $x \geq 0$ by Theorem 6.9.1. Suppose $k \in A$. Then $x \geq k$ or $k \geq x$. We consider these two cases separately. Case A. Suppose $x \geq k$. Then there is an n in N such that $x = k + n$. If $n = 0$, then $x = k$ and $k + 1 \geq x$ so that $k + 1 \in A$. If $n \neq 0$, then there is a natural number m such that $n = m + 1$. Hence, in this case, $x = k + n = k + (m + 1) = k + (1 + m) = (k + 1) + m$. It follows that $x \geq k + 1$ and, hence, $k + 1 \in A$.

Case B. Suppose $k \geq x$. We know that $k + 1 \geq k$. Hence, by Theorem 6.9.3, $k + 1 \geq x$ and $k + 1 \epsilon A$. We see then that $k \epsilon A$ implies $k + 1 \epsilon A$ in every case. Hence, $A = N$ and the proof is complete.

Theorem 6.9.6. If x and y are in N and $x \geq y$ and $y \geq x$, then $x = y$.

PROOF: Suppose $x \geq y$ and $y \geq x$. Then there are natural numbers n and m such that $x = y + n$ and $y = x + m$. Hence, $x = y + n = (x + m) + n = x + (m + n)$. Hence, (Theorem 6.4.5) $m + n = 0$. But this implies (Theorem 6.2.3) that $m = n = 0$. Therefore, $x = y$.

Theorem 6.9.7. If x and y are in N, then exactly one of the following must hold. (a) $x = y$. (b) $x > y$. (c) $y > x$.

PROOF: Suppose x and y are in N. Either $x = y$ or $x \neq y$. If $x = y$ then clearly neither $x > y$ or $y > x$ can hold. If $x \neq y$ we see from Theorem 6.9.5 that we must have $x > y$ or $y > x$. It remains to show that we cannot have *both* $x > y$ and $y > x$. But this follows from Theorem 6.9.6 since $x > y$ and $y > x$ imply $x \geq y$ and $y \geq x$ which imply $x = y$ and this is not the case.

Theorem 6.9.13. If x and y are natural numbers and $x < y$, then $x + 1 \leq y$.

PROOF: Suppose $x < y$. Then there is a nonzero natural number n such that $x + n = y$. But if $n \neq 0$, then there is a natural number m such that $m + 1 = n$. (Theorem 6.4.7). Hence, $y = x + n = x + (m + 1) = x + (1 + m) = (x + 1) + m$. Therefore, $x + 1 \leq y$.

Theorem 6.9.14. There is no natural number x such that $0 < x < 1$.

PROOF: Suppose x is a natural number and $0 < x$. Then, by Theorem 6.9.13, $0 + 1 \leq x$. That is, $1 \leq x$. But then, by Theorem 6.9.7, we cannot have $x < 1$.

Theorem 6.9.18. If x, y, z, and w are in N and $x > y$ and $z > w$, then $xz > yw$.

PROOF: Suppose $x > y$ and $z > w$. Then there are nonzero natural numbers n and m such that $x = y + n$ and $z = w + m$. Hence, $xz = (y + n)(w + m) = yw + ym + nw + nm$. Now $yn + nw + nm$ can be 0 only if all three terms are 0. But $nm \neq 0$ since $n \neq 0$ and $m \neq 0$. Hence $xz > yw$.

PROBLEMS

1. True or false (or improperly constructed): (a) $17 > 15$, (b) $17 > 18$, (c) $17 \geq 17$, (d) $17 \leq 17$, (e) $6 \leq 8 \leq 10$, (f) $6 \leq 12 \leq 10$, (g) $100 \leq 200 \leq 101$, (h) $75 \geq 64 \leq 73$, (i) $675 > 1000$, (j) $47 < 60 \leq 60$, (k) $36 \geq 14 > 0$, (l) $36 \geq 14 \leq 40$, (m) $10^4 < 10^8 < 10^{12}$.

2. Prove the assertions given in this section as numbers (a) 1, (b) 2, (c) 3, (d) 4, (e) 8, (f) 9, (g) 10, (h) 11, (i) 12, (j) 15, (k) 16, (l) 17, (m) 19, (n) 20. In

proving any one of these you may assume that any earlier result has already been established. One method of attack on (n) is to do an induction on z.

3. Prove or disprove each of the following. x, y, z, and w are natural numbers.

(a) If $x < y < z$ then $xw < yw < zw$.
(b) If $x < y < z$ and $w \neq 0$, then $xw < yw < zw$.
(c) If $x \leq y$ and $z < w$, then $xz < yw$.
(d) If $x \leq y$ and $z < w$ and $y \neq 0$, then $xz < yw$.
(e) If $x \leq y$ and $y \leq z$ and $z \leq x$, then $x = y = z$.
(f) There is no natural number y such that $x < y < x + 1$.

4. Is the relation $\geq$ in N isomorphic to the relation $\geq$ in the set $\{6,7,8,9, \ldots\}$?
5. Is the relation $>$ in N isomorphic to the relation $>$ in $\{6,7,8,9, \ldots\}$?

---

6. Is $\geq$ in N isomorphic to $\leq$ in N?
7. A proof that the algorithm for deciding the relative sizes of two numbers written in the decimal notation actually works may be built out of the following facts. (a) If $n \geq m$ then $10^n \geq 10^m$. (b) If $n > m$ then $10^n > 10^m$. (c) If "a" and "b" are decimal digits and $a > b$ then $a \cdot 10^n > b \cdot 10^n$. (d) If "$b_0$", "$b_1$", "$b_2$", ..., and "$b_k$" are decimal digits then $10^{k+1} > b_k \cdots b_2 b_1 b_0$. Prove each of these. [HINT FOR PART (d): Show first that $9 \cdots 999 < 10^{k+1}$.]
8. Draw a flow chart for deciding the relative sizes of two natural numbers given in (a) decimal notation, (b) binary notation.

6.10 Geometric representations

Let us take a horizontal straight line, choose a point on it, and label that point "0." Let us also choose a unit of distance such as an inch. We then find the point on the line one unit to the right of the point labeled "0." We label this new point "1." We then find the point one unit to the right of the point labeled "1" and label this point "2." And so forth. The result is shown in Figure 6.9.

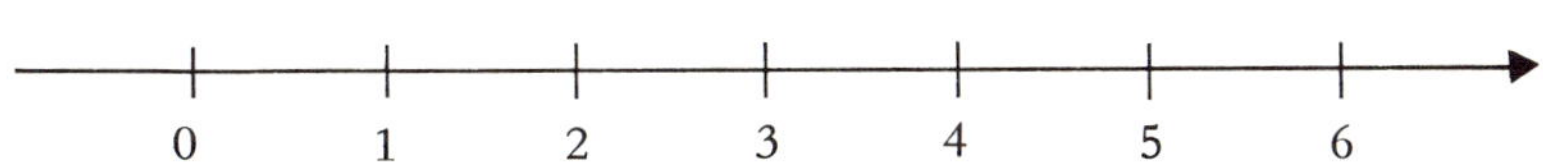

FIGURE 6.9

We see then that certain points on the line are assigned to certain natural numbers by this process. Also each natural number is assigned a point on the line. (We imagine the line as unbounded on the right.) We refer to the point labeled "0" as the **origin** or the point 0. We refer to the point labeled "1" as the point 1, the point labeled "2" as the point 2, etc.

We may interpret many of the things we discussed earlier in terms of this number line.

For example, to say that $x < y$ where x and y are natural numbers is to say that the point x is to the left of the point y. To say that $x > y$ is to say that the point x is to the right of the point y.

In Figure 6.10 we have again shown our number line, but we have added an arrow running from the point 2 to the point 5.

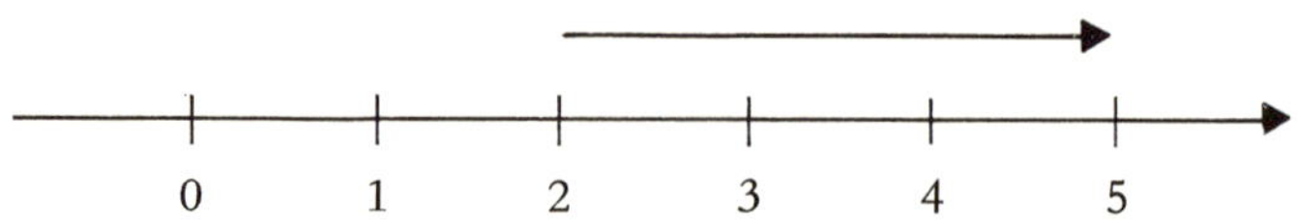

FIGURE 6.10

We have shown the arrow above the line for clarity but we should think of the arrow as actually lying along the line in the indicated position. The arrow shown in Figure 6.10 is said to have the initial point 2 and the terminal point 5. Suppose we let A be the collection of all possible arrows that run from left to right and have points labeled with a natural number for both initial and terminal points. We want to include in A the "degenerate" arrows that have the same point for both initial and terminal point. We could represent such a degenerate arrow in a picture by simply a dot. Since each arrow in A is determined by its initial point and terminal point we may think of an arrow in A as an ordered pair of natural numbers. Let us denote the arrow in Figure 6.10 by "(2,5)." In general, let (n,m) be the arrow running from the point n to the point m. If the arrow (n,m) is in A we must have $n \leq m$ since all the arrows in A run from left to right. In Figure 6.11, we have displayed several different members of A.

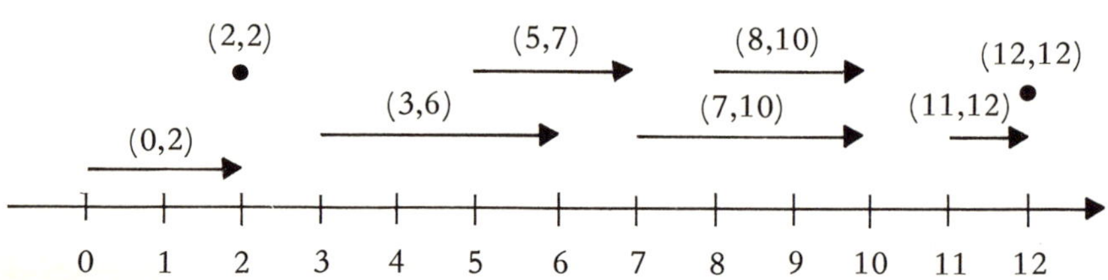

FIGURE 6.11

We notice in this diagram that the arrows (0,2), (5,7), and (8,10) all have the same length. In the collection A, let us consider the relation E defined by

> Two arrows in A are said to be in the relation E if they have the same length.

Now it is not at all difficult to believe that this relation E in A is an equivalence relation. What do the equivalence classes for E in A look like? Each one of them is a collection of arrows all having the same length. Some of these equivalence classes are:

$$\{(0,0),(1,1),(2,2),(3,3), \ldots\},$$

$$\{(0,1),(1,2),(2,3),(3,4), \ldots\},$$

$$\{(0,2),(1,3),(2,4),(3,5), \ldots\}, \ldots$$

$$\{(0,17),(1,18),(2,19),(3,20), \ldots\}, \ldots$$

$$\{(0,1000),(1,1001),(2,1002),(3,1003), \ldots\},$$

and so on. Using our usual notation for equivalence classes we could call the above $[(0,0)],[(0,1)],[(1,3)],[(2,19)]$, and $[(0,1000)]$, respectively.

Each one of these equivalence classes is called a **vector.** Thus, a vector is a collection of arrows all having the same length and direction.

The notion of a vector is a very useful one. For example, consider the idea of a trip of 3 miles to the east. If we consider the unit on our number line to be one mile, then a trip from the point 0 to the point 3 would be a particular trip of 3 miles to the east. A trip from 1 to 4 would also be a trip of 3 miles to the east. So would a trip from 2 to 5. The vector $\{(0,3),(1,4),(2,5), \ldots\}$ may be thought of then as corresponding to the *abstract* notion of a trip of 3 miles to the east. Any particular pair in this collection represents such a trip. The idea of a trip of 3 miles to the east does not depend on our point of departure. But any particular trip from 0 to 3, or from 1 to 4, or from 2 to 5, . . . represents such a trip. In the same way we may use any particular arrow (or pair) in an equivalence class to represent that class (vector).

It will be helpful to use "$\vec{n}$" as a name for the vector which is represented by the pair $(0,n)$. Notice that each vector can be represented by such a pair for some natural number n.

Let V be the class of all vectors as constructed above. We may think of V as the collection of all possible trips to the east of a natural number of miles. V contains the (non) trip of 0 miles which we call $\vec{0}$.

There is a very natural way to define an operation of addition in V. We use "$+_V$" to denote this operation and state for any natural numbers n and m

$$\vec{n} +_V \vec{m} = \overrightarrow{n + m}.$$

For example, $\vec{3} + \vec{5} = \vec{8}$, which may be thought of as asserting that a trip of 3 miles to the east followed by a trip of 5 miles to the east is equivalent to a trip of 8 miles to the east.

In terms of the equivalence classes our definition may be stated as

$$[(0,n)] +_V [(0,m)] = [(0,n + m)].$$

There are two other useful ways to look at this operation of vector addition. First, suppose we look at the diagram in Figure 6.12.

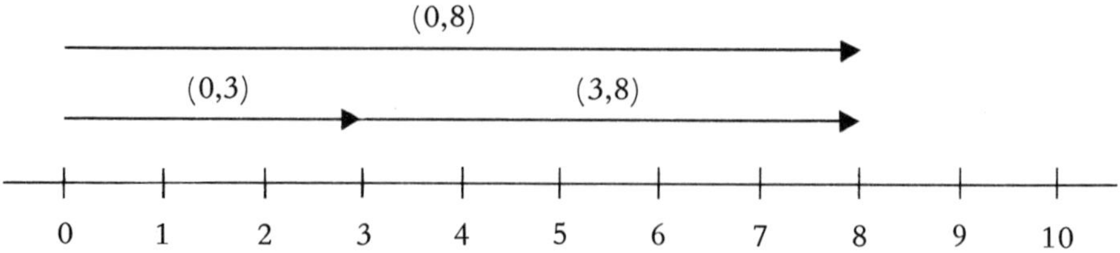

FIGURE 6.12

There we have constructed the arrow (0,3) to represent $\vec{3}$ and then the arrow (3,8) to represent $\vec{5}$. We then draw the arrow from the initial point for (0,3) to the terminal point for (3,8) and we get the arrow (0,8) which represents $\vec{8}$.

In general, to add any two vectors we may draw an arrow which represents the first vector and then draw an arrow that starts where the first one stopped and represents the second vector. An arrow which represents the sum is obtained by drawing an arrow from the initial point of the first arrow to the terminal point of the second.

Second, let us look at the diagram in Figure 6.13.

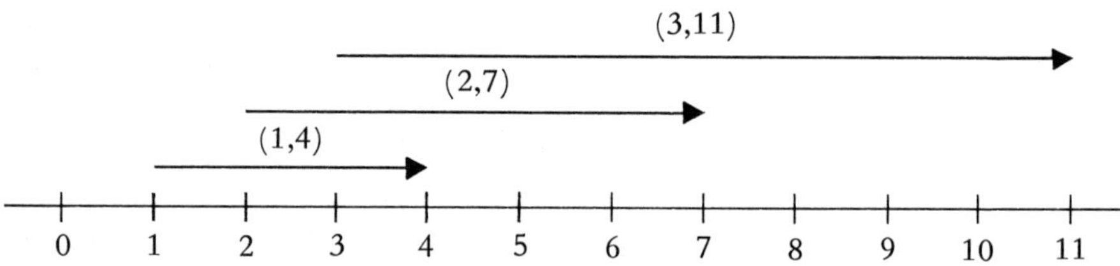

FIGURE 6.13

We note that $[(1,4)] = \vec{3}$, $[(2,7)] = \vec{5}$, and $[(1 + 2, 4 + 7)] = [(3,11)] =$ 8. But this sort of thing always works. That is, in terms of equivalence classes we have

$$[(n,m)] +_V [(p,q)] = [(n + p, m + q)].$$

We may see why this must be so as follows. Since (n,m) and (p,q) are in A we have $n \leq m$ and $p \leq q$. Hence there are natural numbers k and s such that $m = n + k$ and $q = p + s$. Therefore, $m + q = (n + k) + (p + s) = (n + p) + (k + s)$. That is, the arrow from $n + k$ to $m + q$ has the length $k + s$ and $[(n + p, m + q)] = [(n,m)] +_V [(p,q)] = \vec{k} +_V \vec{s} = \overrightarrow{k + s} = [(n + p, m + q)]$.

For each natural number n there is a corresponding vector $\vec{n}$ in V. Conversely, any vector in V has a length (which may be zero) and, hence, corresponds to the natural number which expresses that length. In fact, we claim that the system $(V, +_V)$ is isomorphic to the system $(N, +)$. This is proved by showing that the function f defined on V by $f(\vec{n}) = n$ is an isomorphism between $(V, +_V)$ and $(N, +)$. If x and y are in V then they have certain lengths. If these lengths are n and m, then $x = \vec{n}$ and $y = \vec{m}$ and $f(x +_V y) = f(\vec{n} +_V \vec{m}) = f(\overrightarrow{n + m}) = n + m = f(\vec{n}) + f(\vec{m})$. And this does establish that f is an isomorphism.

We conclude that we may think of addition of natural numbers in terms of addition of the corresponding vectors.

Many applications of the natural number system make use of this vector interpretation of addition. And we shall find that this vector approach is very helpful when we eventually move ahead to build more complicated number systems.

Definition 6.10.1. Suppose that on a horizontal line we choose a point and label it "0," choose a point one unit to the right of that and label it "1," choose a point one unit to the right of that and label it "2," etc. We call the point labeled "0" the **origin** or the **point** 0. We call the point labeled "1" the point 1, the point labeled "2" the point 2, etc.

Definition 6.10.2. Let (n,m) be the arrow running from the point n to the point m. We call the point n the initial point of the arrow and the point m the terminal point.

Definition 6.10.3. Let A be the collection of all arrows constructed as in Definition 6.10.2 and running from left to right. That is, $A = \{(n,m) \mid n \in N, m \in N, \text{ and } n \leq m\}$.

Definition 6.10.4. Two arrows in A are said to be in the relation E to one another if they have the same length.

Theorem 6.10.1. The relation E of Definition 6.10.4 is an equivalence relation.

PROOF: Any arrow has the same length as itself. Thus, E is reflexive. The symmetry and transitivity of E are equally obvious.

Definition 6.10.5. A **vector** is an equivalence class in A with respect to E. We write "$\vec{n}$" to stand for $[(0,n)]$.

Definition 6.10.6. We define addition of vectors by $\vec{n} +_V \vec{m} = \overrightarrow{n + m}$.

Theorem 6.10.2. $[(n,n + k)] +_V [(n + k,n + k + j)] = [(n,n + k + j)]$.

PROOF: Clearly $[(n,n + k)]$ has length k and $(n + k,n + k + j)$ has length j. Also $(n,n + k + j)$ has length $k + j$. Therefore, $[(n,n + k)] +_V [(n + k,n + k + j)] = [(0,k)] +_V [(0,j)] = \vec{k} +_V \vec{j} = \overrightarrow{k + j} = [(0,k + j)] = [(n,n + k + j)]$.

Theorem 6.10.3. $[(n,m)] +_V [(p,q)] = [(n + p,m + q)]$.

Theorem 6.10.4. The system $(V,+_V)$ is isomorphic to the system $(N,+)$.

PROOF: If $(n,m) \in A$ then $n \leq m$. Hence, there is a natural number k such that $n + k = m$. Clearly k is the length of (n,m). Hence $(n,m)\ E\ (0,k)$ and

$[(n,m)] = \vec{k}$. Thus, each arrow in A can be written in the form $\vec{k}$ for some natural number k. Then the function f defined by $f(\vec{k}) = k$ is a one-to-one correspondence between V and N. But also $f(\vec{n}) + f(\vec{m}) = n + m = f(\overrightarrow{n+m}) = f(\vec{n} +_V \vec{m})$. Hence, f is an isomorphism.

PROBLEMS

1. Make a sketch which shows each of the following arrows: (a) (0,4), (b) (2,6), (c) (3,3), (d) (1,9).
2. Find seven other arrows that belong to the same equivalence class (with respect to E as defined in this section) as the arrow (a) (1,9), (b) (0,1653), (c) (5795,5798), (d) (3313,3313).
3. Display a vector diagram that represents the assertion (a) $\vec{5} + \vec{7} = \overrightarrow{12}$. (b) $\vec{4} + \vec{4} = \vec{8}$.
4. In a sketch of a number line show points for 0,1,2, . . . ,15. Draw *all* the arrows which represent $\vec{3}$ and do not run off the sketch.
5. Prove in a very small number of words that V is closed under $+_V$, $+_V$ is commutative, $+_V$ is associative, and there is an identity element for $+_V$ in V. What is the identity element?

6.11 Some graphical devices

We may follow the ideas behind vector addition to build a mechanical slide rule for addition. Construct two number lines, one along the bottom edge of a piece of cardboard and the other along the top edge of a separate piece of cardboard.† Then we may perform addition of vectors (or natural numbers) by the following procedure. To add x and y slide the top number scale along the bottom until its origin falls on the point x on the lower scale. Under the point y on the top scale we shall find the point $x + y$ on the lower scale. See Figure 6.14 where we have the calculation of 4 + 5.

Another graphical device for adding natural numbers is the structure called

† Two yardsticks may be used if they are numbered properly.

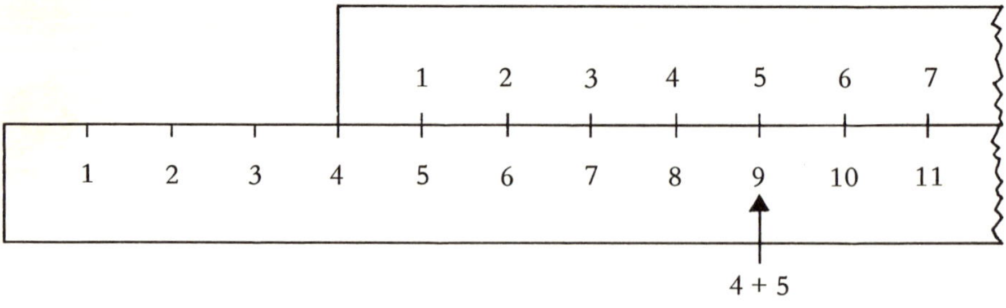

FIGURE 6.14

a **nomograph** given in Figure 6.15. We shall not discuss here the justification for the method of construction of this nomograph, but you may verify, with several examples, that it works.

If you wish to construct such a nomograph for yourself, note that the two outside scales are equidistant from the center scale, and one unit of distance on an outside scale is equal to two units on the center scale.

Another graphical device for performing additions is shown in Figure 6.16. We shall not discuss the construction of this device here. The similarity between the graph and a partial operation table for + is evident.

Products of natural numbers may be represented geometrically as areas of certain rectangles. We take it to be a matter of definition that the area of a square with length of side of one unit is one square unit. We then define the area of a rectangle with a base b units long and an altitude of h units to be bh. We do this because, in terms of our representation of a Cartesian product as a table, such a rectangle contains bh unit squares.

This area interpretation may also be put in terms of two number lines as follows. We construct a horizontal number line as before and call it the **horizontal axis.** We then construct a line perpendicular to the horizontal axis and passing through the origin of that axis. On this **vertical axis** we lay out points representing the natural numbers as before. We take the origin to be the point of intersection of the two axes and we put the point for 1 *above* the horizontal axis. The result is the system of axes shown in Figure 6.17.

If we now draw a vertical line through the point x on the horizontal axis and a horizontal line through the point y on the vertical axis we determine a rectangle whose area is xy. Two sides of the rectangle fall along the axes and two fall along the other lines we drew. In Figure 6.17 the rectangle constructed has area $2 \cdot 3 = 6$ square units.

Graphical devices, nomographs, and a slide rule may be constructed for multiplication. We shall give here a single example (Figure 6.18). We shall return to this matter in another section.

Problems

1. Check each of the following results by means of a slide rule for addition, Figure 6.15, and Figure 6.16. (a) $3 + 3 = 6$, (b) $3 + 7 = 10$, (c) $7 + 3 = 10$, (d) $0 + 10 = 10$, (e) $7 + 9 = 16$.
2. Use Figure 6.18 to find (a) $5 \cdot 6$, (b) $6 \cdot 5$, (c) $6 \cdot 4$.
3. Conjecture how the graphical device of Figure 6.18 could be modified so that it could be used in case $y = 7$ or $y = 8$.
4. Construct a table to represent $\{1,2,3,4,5,6,7\} \times \{1,2,3,4,5,6,7,8\}$. Make all the boxes in your table be squares. Number the boxes to show that $7 \cdot 8 = 56$.
5. What is the area of each of the figures in Figure 6.19?
6. Use a horizontal and a vertical axis to represent $7 \cdot 8$ as the area of a rectangle.

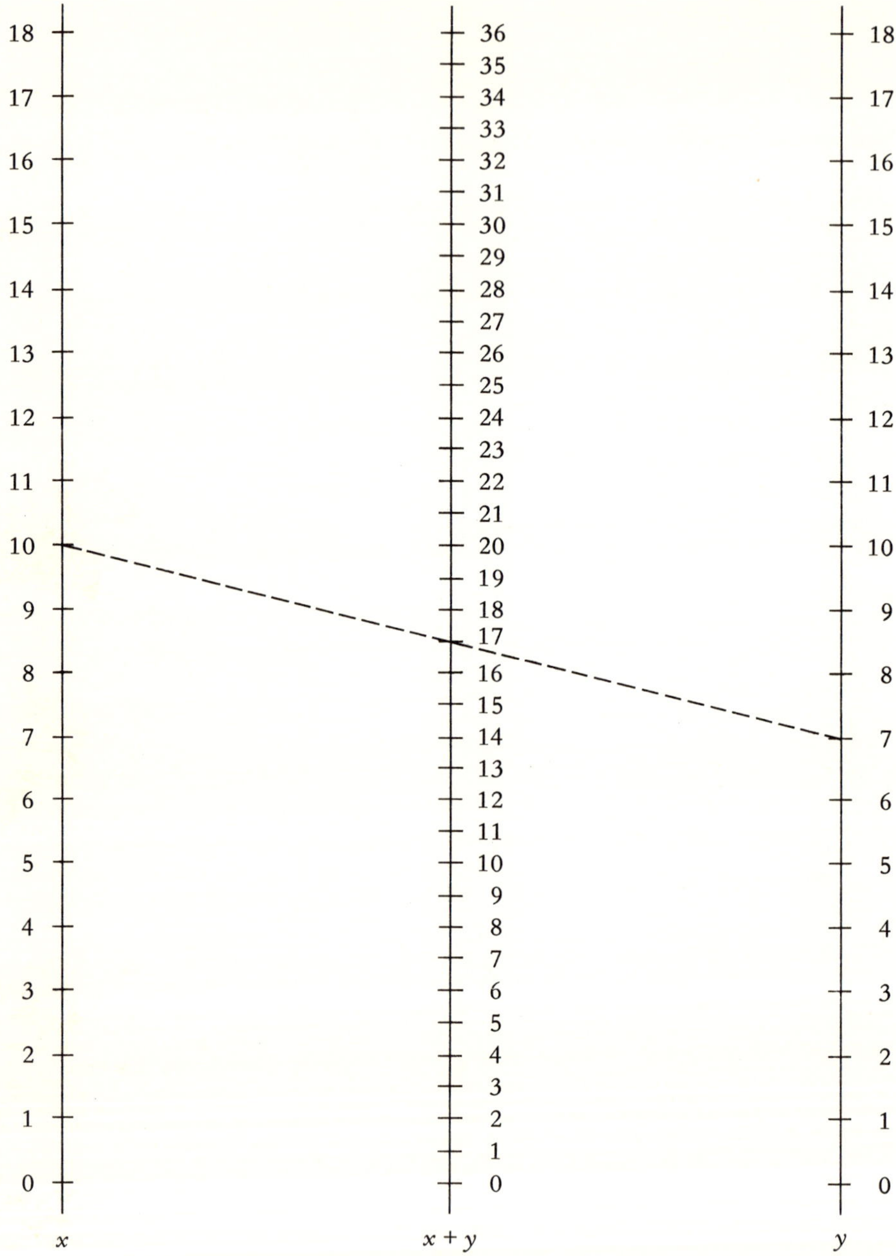

FIGURE 6.15 Nomograph for addition. To add x and y, draw a line from x on the left scale to y on the right scale. $x + y$ is the scale reading where this line crosses the middle scale. The computation of $10 + 7$ is shown.

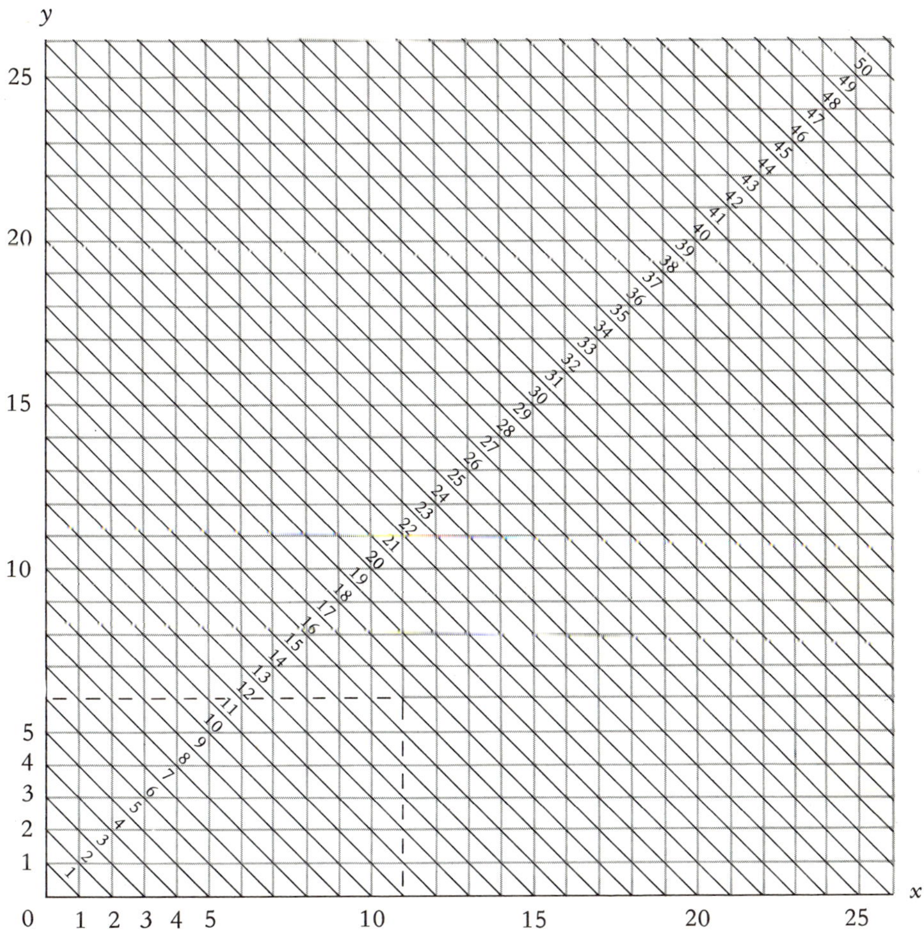

FIGURE 6.16 A graphical device for addition. To add x and y, draw a vertical line through x on the horizontal axis and a horizontal line through y on the vertical axis. The lines drawn will intersect on a slant line labeled $x + y$. The computation of $11 + 6$ is shown.

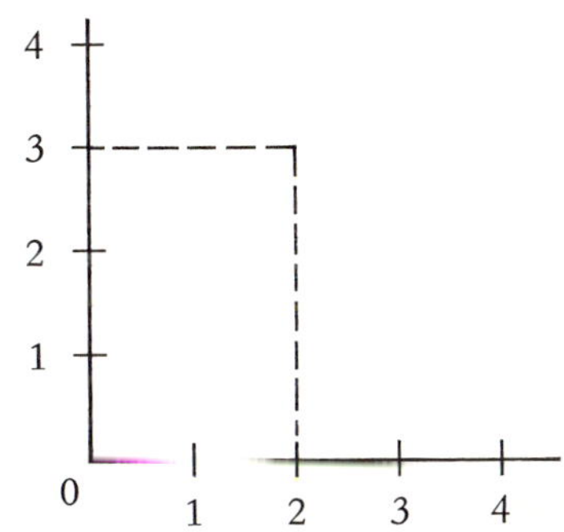

FIGURE 6.17

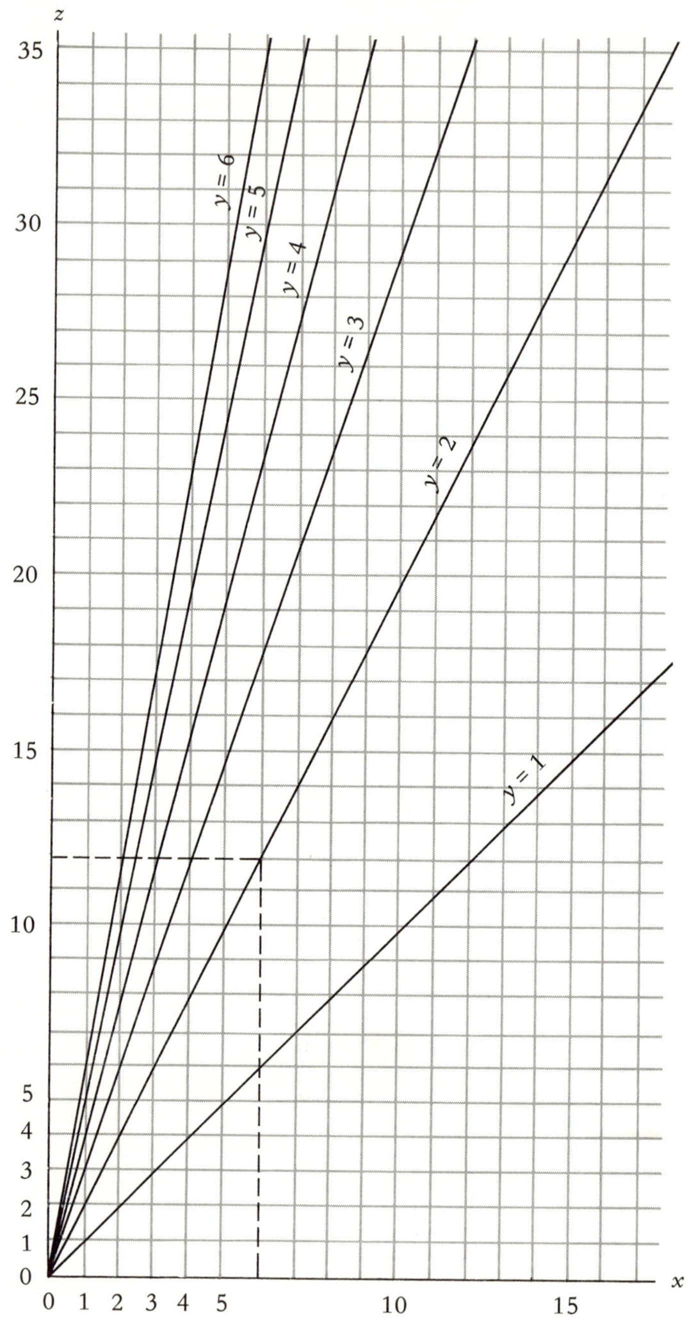

FIGURE 6.18 Graphical device for multiplication. To multiply x by y, draw a vertical line at x to the slant line for y. From there draw the horizontal to $z = xy$. The computation of $6 \cdot 2$ is shown.

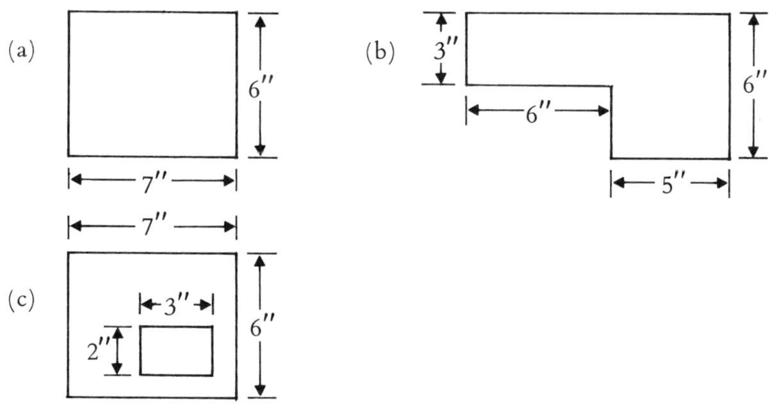

FIGURE 6.19

6.12 Maximum and minimum elements

Almost everyone is interested in both giants and midgets. It is somehow fascinating to find out which is the tallest mountain in the world or which is the longest river on the face of the earth. And who wouldn't be interested in observing the smallest bird that flies, the tiniest workable transistor radio, or the world's biggest elephant?

On a much more mundane level we are often interested in the cheapest among several items that we consider to be of about the same quality and wish to purchase. Or we may want very much to know the quickest way to get a job done.

Extremes are not only intriguing but useful.

Many of the qualities by which we judge whether or not items are extremes are measured by means of numbers. The extremes of a set of numbers are given special names.

The smallest number in a set of numbers is called the **minimum** of the set. The largest element is called the **maximum.** More precisely, if $A \subseteq N$ and $m \in A$ and $m \leq x$ for all x in A, then m is said to be the **minimum** (or the **minimum element**) of (or for) A. It will sometimes be convenient to use "min (A)" to refer to the minimum of A. If $A \subseteq N$ and $M \in A$ and $x \leq M$ for all x in A, then we say that M is the **maximum** (or **maximum element**) of (or for) A. We use "max (A)" to refer to the maximum of A.

For example, if $A = \{7,9,10,12,15,74\}$, then min $(A) = 7$ and max $(A) = 74$. Also, min $(N) = 0$.

There are two fundamental facts about the collection N of natural numbers in reference to maxima and minima.

1. N has no maximum.
2. Every nonempty subset of N has a minimum.

These are deep results that underlie much of what mathematics is all about. And yet they are relatively obvious to the intuition. Let us look at how we might prove these facts in terms of what we have done earlier.

We can prove the first by simply noting that no natural number can be the maximum for N. For suppose x is in N. Then $x + 1$ is also in N and $x < x + 1$. Therefore, x cannot be the maximum for N. Hence, N does not have a maximum.

The second result is more difficult to establish. As might be expected, we fall back on induction to construct the proof. Suppose $A \subseteq N$. We shall consider another set B which contains all the natural numbers (and only those) which are smaller than every element in A. That is, let $B = \{x \mid x < y \text{ for every } y \text{ in } A\}$. Surely $A \cap B = \phi$ because no number could be in A and at the same time less than every element in A. Now suppose that A has no minimum element. Then surely 0 is not in A. For if 0 is in A it must be the minimum of A since it is the minimum of N itself. It follows that 0 must be in B. For we know that $0 \leq y$ for all y in N and, hence, $0 \leq y$ for all y in A. And since 0 is not in A, $0 < y$ for all y in A. Let k be some member of B. Then $k < y$ for all y in A. It follows that $k + 1 \leq y$ for all y in A (Theorem 6.9.19). Suppose that $k + 1$ is in A. Then $k + 1$ must be the minimum of A. But A has no minimum. Hence, $k + 1$ cannot be in A. But when $k + 1 \leq y$ for all y in A and $k + 1 \neq y$ for all y in A, we have $k + 1 < y$ for all y in A. Therefore, $k + 1$ is in B. Thus, 0 is in B and k in B implies $k + 1$ is in B. Hence, $B = N$. Then $B \cap A = N \cap A = A$. But $B \cap A = \phi$ and A must be empty. We have shown that if A has no minimum element it must be empty. This proves the second result above.

The fact that there is no maximum among the natural numbers is intimately tied up with the fact that there are infinitely many natural numbers.

We say that a set is **finite** if its cardinal number is a natural number. That is, a set A is **finite** if $\#(A) = n$ for some natural number n. A set that is not finite is said to be **infinite.**

It is not hard to believe that if a set of natural numbers is finite then it must have a maximum. To find it we simply compare two elements from the set, pick the larger one, compare that with another element from the set, pick the larger one, compare that with another element, and so on. The process must eventually stop and yield the maximum if the set is finite.

It is perhaps not so obvious that the converse of the above is also true. That is, if a set of natural numbers has a maximum, then that set is finite. Another way to put this would be to say that no infinite set of natural numbers has a maximum.

Hence, a nonempty set of natural numbers has a maximum if and only if it is a finite set.

As these results indicate, infinite sets are quite different from finite sets. Even a very large finite set is strikingly different from an infinite set. The set of all grains of sand on a beach is, for example, very large—but finite. The collection of all the leaves on all the trees in a forest is large—but finite. Indeed, it has been asserted by some physicists that the collection of all atoms in the *universe*, while large, is finite.

If we move to geometric representations of numbers by points on a number line we might introduce the notion of **bounded** sets. A set of points on a line is said to be **bounded** if all the points in the set fall between some two points on the line. A set of points which is not bounded is said to be **unbounded.** For example, the set of points $\{2,3,4,5,6\}$ is a bounded set since all these points fall between the point 1 and the point 7. On the other hand, the set of points $\{2,4,6,8,10, \ldots\}$ is unbounded. Evidently, a set of points corresponding to natural numbers on a number line is bounded if and only if it is finite. Notice, however, that if we recognize that there are other points on the number line aside from those assigned to natural numbers then we might have a bounded set which is infinite. In fact, if we are willing to accept the fact that there are infinitely many points on the line which fall between the point 1 and the point 2, then this collection of points does constitute a set which is both bounded *and* infinite. We have not yet assigned any numbers to any points between the point 1 and the point 2. But we shall be working towards that end in later sections.

Some of the confusion that occurs about infinite sets is due to the failure to distinguish between what it means to say that a set is infinite and what it means to say that it is bounded. We shall return to this topic again in a later section.

If we continue to think of a nonempty collection of natural numbers in terms of the points on a number line, we may speak of the minimum of the set as the **first** point in the set and the maximum as the **last** point in the set. When we do this we behave as if we were thinking of moving along the line from left to right. We have seen that every nonempty set of natural numbers must have a first point in this sense. And a set of natural numbers will have a last point in this sense if and only if the set is finite. As we move on to construct other number systems, we shall find it helpful to contrast the above situation for the natural numbers with that for other systems.

Problems

1. Which of the following sets are finite?

   (a) $\{1,2,3,10^{6},10^{1,000,000}\}$
   (b) $\{x \mid x \in N \text{ and } x + 6 = 7\}$
   (c) $\{x \mid x \in N \text{ and } x + 1 = x\}$
   (d) $\{x \mid x \in N \text{ and } x = x\}$
   (e) $\{x \mid x \in N \text{ and for some } k \text{ in } N,\ x = 2k\}$
   (f) $\{x \mid x \in N \text{ and for some } k \text{ in } N,\ x = k\}$
   (g) $\{x \mid x \in N \text{ and for some } k \text{ in } N,\ x + k = 0\}$
   (h) $\{x \mid x \text{ is a person living on the earth}\}$
   (i) $\{x \mid x \text{ is an insect living on the face of the earth}\}$

(j) $\{x \mid x$ is an atom of oxygen in the earth or its atmosphere$\}$
(k) $\{x \mid x \in N$ and the decimal name for x can actually be written down using a particular standard typewriter by the year 2700$\}$
(l) $\{x \mid x \in N \text{ and } x < 10^6\}$
(m) $\{x \mid x \in N \text{ and } x > 10^6\}$.

2. For each set of numbers in Problem 1 find (if possible) the maximum and minimum.

3. Which of the following assertions are true? (Our universal set is N.)

(a) The union of two finite sets must be finite.
(b) The intersection of two finite sets must be finite.
(c) The union of two infinite sets must be infinite.
(d) The intersection of two infinite sets must be infinite.
(e) The complement of a finite set must be infinite.
(f) The complement of an infinite set must be finite.
(g) The Cartesian product of two finite sets must be finite.
(h) The Cartesian product of a finite set and an infinite set must be infinite. (Be careful here.)
(i) If the Cartesian product of two sets is infinite then one of the sets must be infinite.
(j) Every subset of a finite set is finite.
(k) Every subset of an infinite set is infinite.
(l) Every subset of an infinite set is finite.
(m) If a set has an infinite subset then it must be infinite.
(n) Among all the elephants alive in the world there must be one whose weight is a maximum among the weights of all elephants.
(o) Among all the birds in the world that fly there must be a heaviest one.
(p) Among all the natural numbers which are squares of natural numbers there must be a largest.
(q) Among all the natural numbers whose decimal names contain at least 1000 digits, there is a smallest.
(r) Among all the numbers x such that $x = x + 1$, there is a smallest.
(s) There must be a largest natural number whose square is less than 5,742,600.
(t) There must be a smallest natural number whose square is greater than 5,742,600.
(u) If A, B, and C are finite sets of natural numbers, then we must have $\max(\{\min(A), \min(B), \min(C)\}) = \min(\{\max(A), \max(B), \min(C)\})$.

4. Give examples of each of the following and argue that they must hold for any finite sets A and B of natural numbers.

(a) $\max(A \cup B) = \max(\{\max(A), \max(B)\})$

(b) $\max(A \cap B) \leq \max(A)$
(c) $\min(A \cap B) \geq \min(A)$.

6.13 Subtraction

The inverse operation for addition is called subtraction. Subtraction is the operation that *undoes* what addition *does*. We use "$-$" to denote subtraction.

From the way we defined inverse operations, we shall have $z - y = x$ whenever x is the unique value such that $z = x + y$. That is, $z - y$ is the value which we must add to y to get z. From some of our earlier work it is easy to see that if there is any x such that $z = x + y$, then there is exactly one such x. And there will be such an x when and only when $z \geq y$. It follows that $z - y$ makes sense or is defined if and only if $z \geq y$. (Of course, we are assuming that all we have to work with here are the natural numbers.)

For example, $10 \geq 7$ so $10 - 7$ must be a natural number. If we let $10 - 7 = x$ then we must have $10 = x + 7$. But we know that $10 = 3 + 7$. Therefore, $10 - 7 = 3$. In other words, we know that $10 - 7 = 3$ because $10 = 3 + 7$. Similarly, $13 - 6 = 7$ because $13 = 7 + 6$. A partial table for subtraction on the natural numbers is given below.

−	0	1	2	3	4	5	6	7	8	9	10
0	0	–	–	–	–	–	–	–	–	–	–
1	1	0	–	–	–	–	–	–	–	–	–
2	2	1	0	–	–	–	–	–	–	–	–
3	3	2	1	0	–	–	–	–	–	–	–
4	4	3	2	1	0	–	–	–	–	–	–
5	5	4	3	2	1	0	–	–	–	–	–
6	6	5	4	3	2	1	0	–	–	–	–
7	7	6	5	4	3	2	1	0	–	–	–
8	8	7	6	5	4	3	2	1	0	–	–
9	9	8	7	6	5	4	3	2	1	0	–
10	10	9	8	7	6	5	4	3	2	1	0

In terms of an adding machine as represented in Figure 6.20 we may think of a process of finding $y - z$ as that of determining what x value we must feed into the machine along with z to obtain y. As long as we have only natural numbers to work with there is no answer when $y < z$.

In terms of vector addition, subtraction amounts to finding the missing arrow in a properly constructed diagram. We draw an arrow to represent y and an arrow to represent z putting both initial points together. We then ask what vector is represented by the arrow running from the arrowhead for z to the arrowhead for y. This is the vector of length $y - z$. In Figure 6.21 we have illustrated the problem of finding $10 - 7$.

We may interpret the operation of subtraction in terms of cardinal numbers.

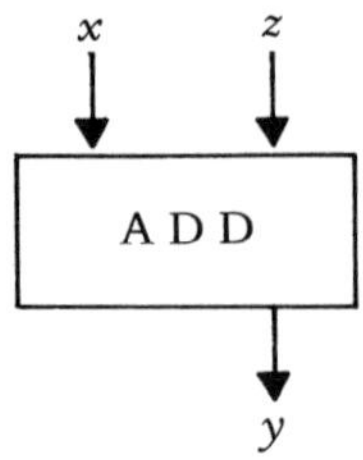

FIGURE 6.20

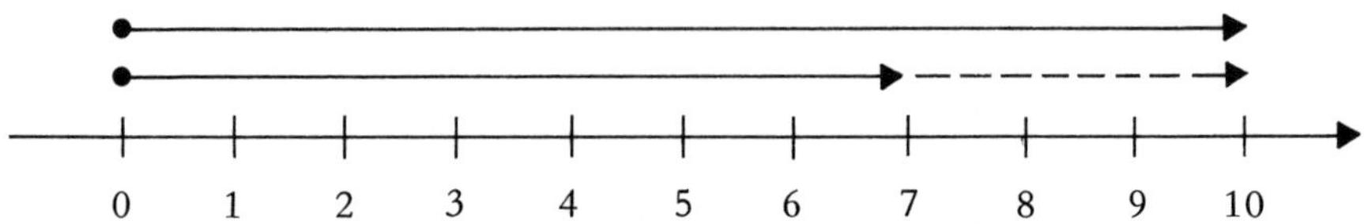

FIGURE 6.21

Suppose z and y are natural numbers and $z \geq y$. If A is a set with z elements and $B \subseteq A$ and B has y elements, then the set $A - B$ obtained by removing all the elements of B from A is a set with $z - y$ elements. This interpretation is, no doubt, the cause for reading "$z - y$" as "z take away y." We may read "$z - y$" this way if we wish, but is perhaps better to read it as "z minus y."

In view of the manner in which subtraction is defined, it is easy to give a mechanical procedure for finding $y - z$ when it is defined. Suppose we are given natural numbers y and z and we wish to find $y - z$. We first check to see if $y \geq z$. If not, then $y - z$ is undefined. If so, then there is an n such that $y = z + n$. Our problem is to find this n since $y = z + n$ implies $y - z = n$. But knowing that such an n exists we may find it by the following procedure. First, determine if $y = z$. If so then $y - z = 0$ since $y = z + 0 = z$. If $y \neq z$, then we find $z + 1$ and ask if $y = z + 1$. If it does, then $y - z = 1$. If not,then we find $z + 2$ and ask if $y = z + 2$. If so, $y - z = 2$. If not, we find $z + 3$ and ask if $y = z + 3$. If so, $y - z = 3$. If not, we continue as before. The process must eventually stop since we know that there is an n such that $y = z + n$ and this procedure is bound to find it.

A table summarizing the steps in this process for the example $z = 7$ and $y = 12$ is as follows: (We assume that we have already checked that $12 \geq 7$.)

d	$7 + d$	Is $7 + d = 12$?
0	7	No
1	8	No
2	9	No
3	10	No
4	11	No
5	12	Yes

The process leads to the conclusion that $12 - 7 = 5$.

This process may be put in terms of a flow chart as in Figure 6.22 which we may think of as a design for a subtraction machine.

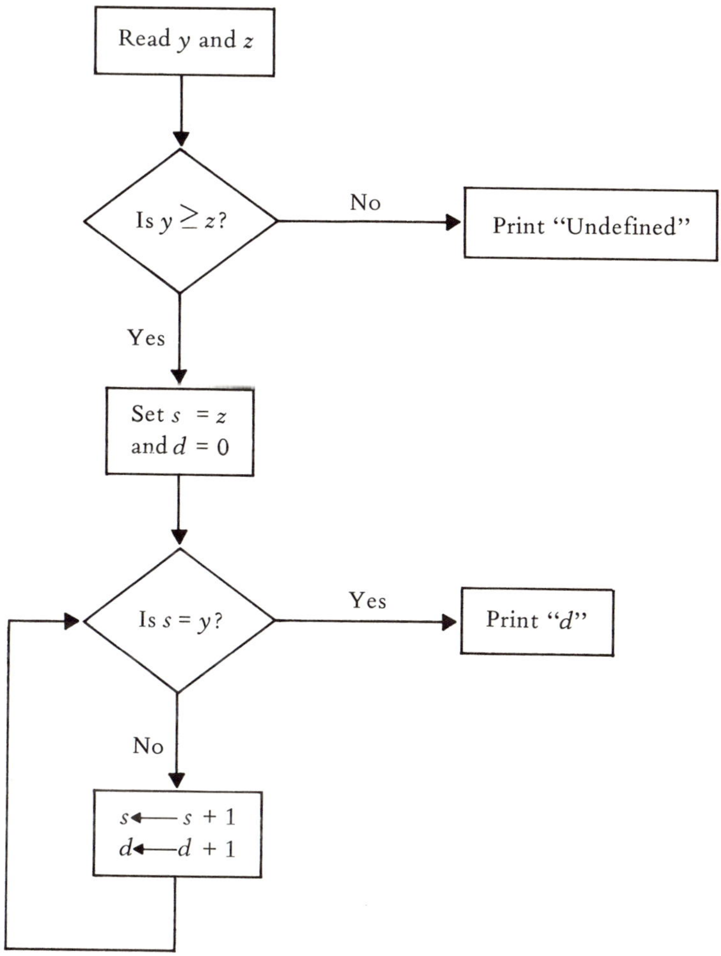

FIGURE 6.22 A subtraction machine. The machine computes $y - z$.

There is a basic property of subtraction that leads us to another mechanical procedure for subtraction. It is the equation

$$(y + x) - (z + x) = (y - z).$$

This equality holds when either side is well defined. And it is easy to believe in terms of putting in and taking out elements from sets. If $\#(A) = y$ and $\#(B) = z$ and $B \subseteq A$, then $y - z$ is the number of elements in the set obtained by removing the elements of B from A. But if $\#(C) = x$ and $A \cap C = \phi$ then $y + x = \#(A \cup C)$ and $z + x = \#(B \cup C)$. $(y + x) - (z + x)$ is the number of elements in the set obtained by removing the elements in $B \cup C$ from $A \cup C$. But surely this number is simply $y - z$ since we first put in the elements of C and then remove them all.

The above equality leads to a sort of **cancellation procedure** where the x in the $(z + x)$ "cancels out" the x in $(y + x)$. We may make use of this as in the following example.

$$\begin{aligned} 11 - 7 &= (10 + 1) - (6 + 1) = 10 - 6 = (9 + 1) - (5 + 1) \\ &= 9 - 5 = (8 + 1) - (4 + 1) = 8 - 4 = (7 + 1) - (3 + 1) \\ &= 7 - 3 = (6 + 1) - (2 + 1) = 6 - 2 = (5 + 1) - (1 + 1) \\ &= 5 - 1 = (4 + 1) - (0 + 1) = 4 - 0 = 4. \end{aligned}$$

Such a procedure makes use of the fact that $n - 0 = n$ for all n in N and assumes that when $m \neq 0$ we can find a k such that $m = k + 1$. That is, it assumes we can find $m - 1$ for any nonzero m.

Or the procedure may be thought of in terms of sets and their cardinal numbers. Suppose $\#(A) = y$ and $\#(B) = z$ and, in this case, A and B are disjoint. We start removing elements from B one at a time. And each time we remove an element from B we also remove an element from A. When B is empty, we have left in A exactly $y - z$ elements.

A flow chart for this procedure is given in Figure 6.23.

Definition 6.13.1. **Subtraction** is the inverse operation for addition in N. We use "$-$" to denote subtraction. We call $z - y$ the **difference** between z and y and read "$z - y$" as "z minus y."

Theorem 6.13.1. With x, y, and z in N, $z - y = x$ if and only if $z = x + y$.

PROOF: In view of Definition 6.13.1 the only thing to check is that if $z = x + y$ and $z = x + w$ then $y = w$. But this follows from Theorem 6.4.6.

Theorem 6.13.2. $z - y$ is defined if and only if $y \leq z$.

PROOF: $y \leq z$ if and only if there is a natural number n such that $z = y + n$. Then we may apply Theorem 6.13.1.

Theorem 6.13.3. If x, y, and z are in N, $(z + x) - (y + x) = z - y$ whenever either side of the equality is defined.

PROOF: Suppose $z - y = n$. Then $z = n + y$ and $z + x = (n + y) + x = (y + n) + x = y + (n + x) = y + (x + n) = (y + x) + n$. Hence, $(z + x) - (y + x) = n$. If $(z + x) - (y + x) = m$, then $z + x = (y + x) + m = y + (x + m) = y + (m + x) = (y + m) + x$. But by Theorem 6.4.6 this implies $z = y + m$. And, hence, $z - y = m$ by Theorem 6.13.1.

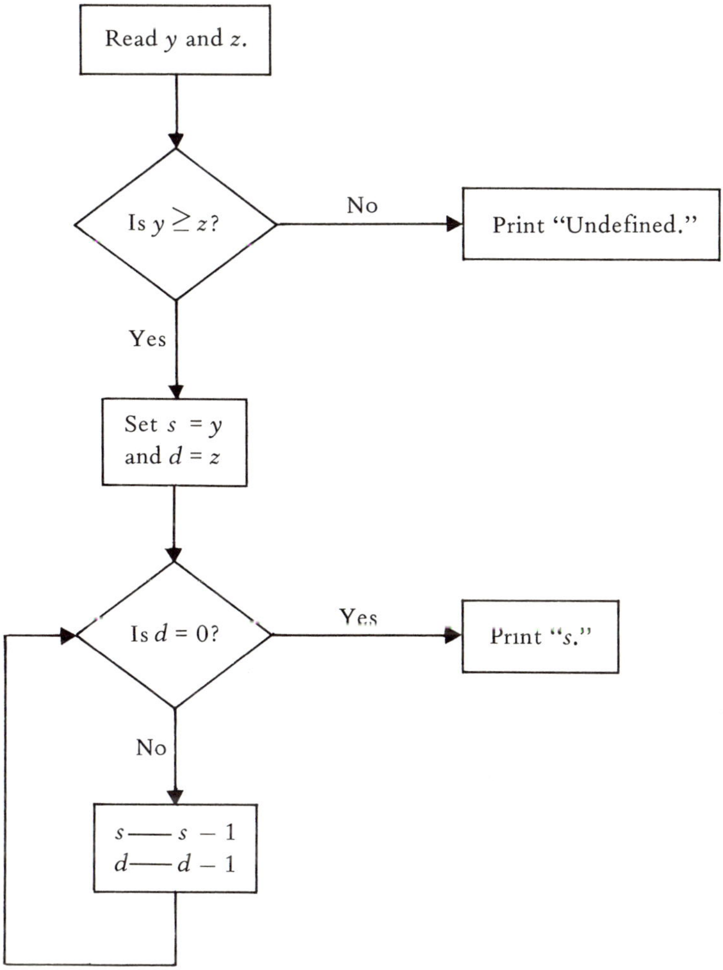

FIGURE 6.23 Another subtraction machine. The machine computes $y - z$.

PROBLEMS

1. Write up instructions for using Figure 6.15 to perform subtractions. Illustrate your procedure with an example.
2. Write up instructions for using Figure 6.16 to perform subtractions. Illustrate your procedure with an example.
3. Write up instructions for using a slide rule for addition to perform subtractions. Illustrate your procedure with an example.
4. Use a vector diagram to find (a) $8 - 5$, (b) $16 - 4$.
5. Explain the step-by-step procedure followed by the machine of Figure 6.22 in finding $4 - 2$.

6. Explain the step-by-step procedure followed by the machine in Figure 6.23 in finding $5 - 2$.

7. Construct a partial table for subtraction in which the numerals are written to the base 2 and rows and columns are included for $0, 1, 10_2, \ldots, 1000_2$.

8. Describe a mechanical procedure for subtraction when the numbers are written in the tally system. You may wish to start by writing the names of the numbers in certain convenient positions.

6.14 Properties of subtraction

It is clear that N is not closed under subtraction and that subtraction is not commutative. Is subtraction associative? We see that it is not by observing that $9 - (6 - 2) = 9 - 4 = 5$ while $(9 - 6) - 2 = 3 - 2 = 1$.

If we consider set-ups where both subtraction and addition are involved we have such examples as: $5 + (7 - 2) = 5 + 5 = 10$ while $(5 + 7) - 2 = 12 - 2 = 10$, and $13 + (6 - 1) = 13 + 5 = 18$ while $(13 + 6) - 1 = 19 - 1 = 18$. But note that $3 + (2 - 4)$ is undefined while $(3 + 2) - 4 = 5 - 4 = 1$. Such results lead to the correct conjecture that

$$x + (y - z) = (x + y) - z$$

whenever $y - z$ is defined. Note that this also may be interpreted in terms of removing elements from sets.

Since $(x - y) + z = z + (x - y)$ we also have

$$(x - y) + z = (z + x) - y$$

whenever the left side is defined.

What is the situation for repeated subtractions when they are possible? We might notice that we have such results as $(10 - 7) - 3 = 3 - 3 = 0$ and $10 - (7 + 3) = 10 - 10 = 0$, $(9 - 3) - 4 = 6 - 4 = 2$ and $9 - (3 + 4) = 9 - 7 = 2$, $(8 - 2) - 1 = 6 - 1 = 5$ and $8 - (2 + 1) = 8 - 3 = 5$. These suggest that for any natural numbers x, y, and z we have $(x - y) - z = x - (y + z)$. And this equality does indeed hold whenever either side makes sense. In fact, this is quite easy to believe if we interpret subtraction in terms of removing elements from sets. If from a set of x elements we remove y elements and then remove z elements from the resulting set, we have just as many elements left as if we had removed $y + z$ elements from the original set of x elements.

The situation is quite different with $x - (y - z)$, however. Notice that $10 - (6 - 4) = 10 - 2 = 8$ while $(10 - 6) + 4 = 4 + 4 = 8$. Also $8 - (6 - 3) = 8 - 3 = 5$ while $(8 - 6) + 3 = 2 + 3 = 5$. And $7 - (2 - 1) = 7 - 1 = 6$ while $(7 - 2) + 1 = 5 + 1 = 6$. These suggest that for any natural numbers x, y, and z we have

$$x - (y - z) = (x - y) + z.$$

This *is* true in the limited sense that whenever *both* sides are defined they must be equal. But it is possible to have either side undefined while the other makes sense. For example, $6 - (1 - 2)$ is undefined while $(6 - 1) + 2 = 5 + 2 = 7$. Also $(1 - 2) + 1$ is undefined while $1 - (2 - 1) = 1 - 1 = 0$. We see that the fact that N is not closed under subtraction badly complicates the situation for repeated subtractions.

To save ourselves some trouble, we adopt two conventions for use with subtractions. First, we shall allow ourselves to write "$x - y - z$" to mean $(x - y) - z$. Notice that this convention is *not* justified by subtraction being associative. We must realize that "$x - y - z$" does not mean $x - (y - z)$. "$x - y - z - w$" will mean $((x - y) - z) - w$, and similarly when more terms are involved. Second, we assume that multiplication takes precedence over subtraction as it does over addition. For example, "$xy - z$" means $(xy) - z$ and not $x(y - z)$.

When $x \geq y$ and $z \geq w$ we find that $(x - y) + (z - w) = (x + z) - (y + w)$. This is easy to understand in terms of sets. If we remove y elements from an x element set, and remove w elements from a z element set that does not overlap our first set, and then lump the two resulting sets together, we have just as many elements as if we have first put together our x element and z element sets and then removed $y + w$ elements.

When we consider multiplication and subtraction together, it is natural to ask if the former distributes across the latter. The answer is again complicated by the fact that N is not closed under subtraction. For instance, $0 \cdot (1 - 2)$ is undefined, yet $0 \cdot 1 - 0 \cdot 2 = 0 - 0 = 0$. However, we do have the result:

> If x, y, and z are natural numbers and $x \neq 0$, then $x(y - z) = xy - xz$ whenever either side is defined.

For example, $2(6 - 3) = 2 \cdot 3 = 6$ and $2 \cdot 6 - 2 \cdot 3 = 12 - 6 = 6$. Also $5(2 - 1) = 5 \cdot 1 = 5$ and $5 \cdot 2 - 5 \cdot 1 = 10 - 5 = 5$.

This limited distributive law may be interpreted in terms of a table for certain Cartesian products as in Figure 6.24. There we have assumed that $y \geq z$ so that the subtractions yield natural numbers. The large rectangle will contain xy boxes. And if we remove xz of these boxes we are left with the $x(y - z)$ boxes at the right. Hence, $xy - xz = x(y - z)$.

A slightly more complicated result involving subtraction and multiplication is the one dealing with $(x - y)(z - w)$. We claim that when $x \geq y$ and $z \geq w$, then

$$(x - y)(z - w) = (xz + yw) - (yz + xw).$$

To establish this we consider different cases.

If $x = y$, then $(x - y)(z - w) = 0 \cdot (z - w) = 0$ and also $(xz + yw) - (yz + xw) = (xz + xw) - (xz + xw) = 0$.

If $z = 0$, then $w = 0$ since $z \geq w$ and hence both sides of the equality are 0.

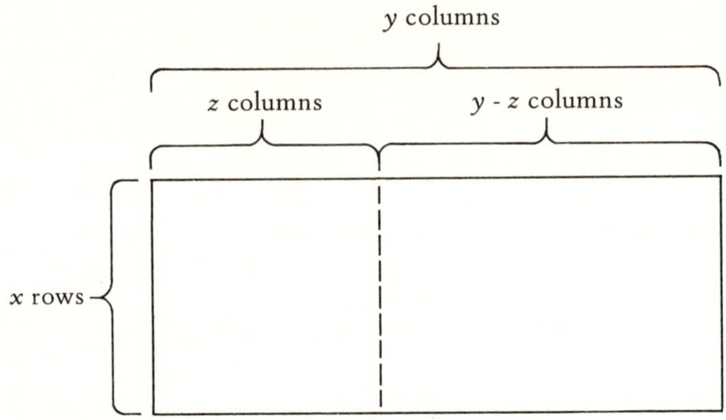

FIGURE 6.24

If $w = 0$, then $(x - y)(z - w) = (x - y)z = xz - yz$. And $(xz + yw) - (yz + xw) = (xz + 0) - (yz + 0) = xz - yz$.

If $x \neq y$ and $w \neq 0$, then we may make three uses of the above limited distributive law to get $(x - y)(z - w) = (x - y)z - (x - y)w = (xz - yz) - (xw - yw)$. But, by our results above about repeated subtraction, this last is equal to $[(xz - yz) - xw] + yw = yw + [(xz - yz) - xw] = yw + [xz - (yz + xw)] = (yw + xz) - (yz + xw) = (xz + yw) - (yz + xw)$.

We may also interpret the above result in terms of tables representing Cartesian products. In Figure 6.25 the number of boxes in region D is $(x - y)(z - w)$.

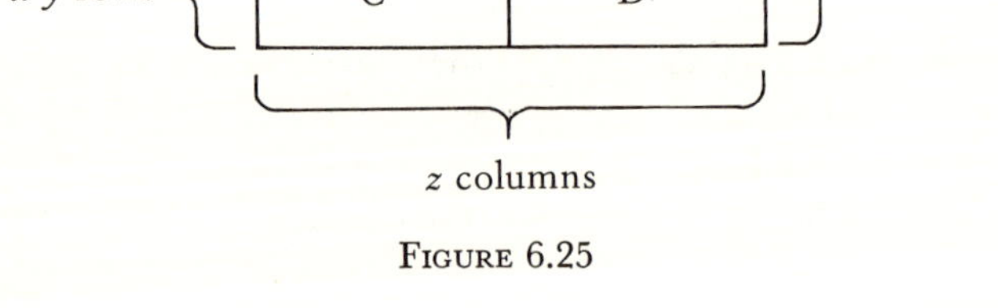

FIGURE 6.25

Region A has yw boxes. A with C has xw boxes. A with B has yz boxes. And A, B, C, and D taken together have xz boxes. The number of boxes in $A \cup B$ plus the number of boxes in $A \cup C$ is yw more than the number of boxes in $A \cup B \cup C$. Hence, $xw + yz - yw + (x - y)(z - w) = xz$. Hence, $(x - y)(z - w) + xw + yz = xz + yw$ and, therefore, $(x - y)(z - w) = (xz + yw) - (yz + xw)$.

Subtraction may be thought of as a tool for solving certain kinds of equations.

If we wish to solve the equation $x + 7 = 10$ we may claim immediately that $x + 7 = 10$ if and only if $x = 10 - 7$. And having found that $10 - 7 = 3$ we conclude that *the* solution to the equation is $x = 3$.

Another way to put the process of solving the above equation is as follows. Assume that $x + 7 = 10$. Then since subtraction is an operation $(x + 7) - 7 = 10 - 7$. But $(x + 7) - 7 = x$ and $10 - 7 = 3$, so $x + 7 = 10$ implies $x = 3$. This tells us that if there is any x value for which $x + 7 = 10$, then it must be 3. Since we know that, indeed, $3 + 7 = 10$ we have $\{3\}$ as the solution set for the equation.

Note the logic of this procedure. We find that the truth of the equation implies that x has a certain value. We then verify that that value for x actually does make the equation true. This last step is necessary since there might conceivably not be any solution for the equation even though the equality implies some fact about x.

To illustrate the need for checking, suppose we consider the equation $x + 5 = 3$. We might "solve" this equation as follows: If $x + 5 = 3$ then $(x + 5) - 5 = 3 - 5$. But $(x + 5) - 5 = x$ so $x = 3 - 5$. And if we stop here we might well conclude that we have found a natural number which is a solution for the equation, namely, $x = 3 - 5$. In other words the solution set is $\{3 - 5\}$. But $3 - 5$ is *not* a natural number. It is true that $x + 5 = 3$ implies $x = 3 - 5$, but there is *no* natural number x for which $x + 5 = 3$. That is, the correct solution set for the equation is ϕ.

We may state a general result about an equation in the same form as the above:

> In the natural number system, the equation $x + a = b$, where a and b are given natural numbers, has the solution set ϕ if $a > b$ and the solution set $\{b - a\}$ if $a \leq b$.

Theorem 6.14.1. If x and y are natural numbers, then $(x + y) - y = x$.

PROOF: Since $x + y \geq x$, $(x + y) - y$ is defined. Hence, by the definition of subtraction, $(x + y) - y = x$.

Theorem 6.14.2. If x, y, and z are in N and $y \geq z$, then $x + (y - z) = (x + y) - z$.

PROOF: Let $y - z = w$. Then $y = w + z$. It follows that $x + y = x + (w + z) = (x + w) + z$. Hence, $(x + y) - z = x + w = x + (y - z)$.

Theorem 6.14.3. If x, y, and z are in N, $(x - y) - z = x - (y + z)$ whenever either side is defined.

PROOF: Suppose $w = x - (y + z)$ where w is in N. Then, $x = w + (y + z) = (w + z) + y$. Hence, $x - y = w + z$. It follows that $(x - y) - z = w = x - (y + z)$. Here we have made repeated use of Theorem 6.13.1.

If, on the other hand, $v = (x - y) - z$, then $v + z = x - y$. And hence, $v + z + y = x$. Hence, $v = x - (z + y) = x - (y + z)$.

Theorem 6.14.4. If x, y, and z are in N, $x - (y - z) = (x - y) + z$ whenever *both* sides are defined.

PROOF: Suppose $y \geq z$ and $x \geq (y - z)$ so that the left side of the equality makes sense. Suppose also that $x \geq y$ so that the right side makes sense. Let $y - z = k$. Then $y = k + z$ and $(x - y) + z = (x - (k + z)) + z = ((x - k) - z) + z = x - k = x - (y - z)$ where the second inequality is justified by Theorem 6.14.2 and the third by Theorem 6.14.1.

Theorem 6.14.5. If x, y, z, and w are in N and $x \geq y$ and $z \geq w$, then $(x - y) + (z - w) = (x + z) - (y + w)$.

PROOF: Suppose $x - y = n$ and $z - w = m$. Then $x = n + y$ and $z = m + w$. Hence $x + z = (n + y) + (m + w) = (n + m) + (y + w)$. Therefore, $(x + z) - (y + w) = n + m$.

Theorem 6.14.6. If x, y, and z are in N and $x \neq 0$, then $x(y - z) = xy - xz$ whenever either side is defined.

PROOF: If $y \geq z$, then the left side of the equation is defined and there is an n such that $y = z + n$. Then $x(y - z) = xn$. Also $xy - xz = x(z + n) - xz = (xz + xn) - xz = (xn + xz) - xz = xn$. If $y < z$, then the left side is undefined, but since $x \neq 0$, $y < z$ implies $xy < xz$ so that in this case the right side is also undefined.

Theorem 6.14.7. If x, y, z and w are in N and $x \geq y$ and $z \geq w$, then $(x - y)(z - w) = (xz + yw) - (yz + xw)$.

PROBLEMS

1. Give three different numerical examples to illustrate each of the following equalities (which all hold when both sides are defined).

(a) $x + (y - z) = (x + y) - z$
(b) $(x - y) - z = x - (y + z)$
(c) $x - (y - z) = (x - y) + z$
(d) $(x - y) + (z - w) = (x + z) - (y + w)$
(e) $x(y - z) = xy - xz$
(f) $(x - y)(z - w) = (xz + yw) - (yz + xw)$.

2. Find the solution set in N for each of the following equations: (a) $x + 1 = 7$, (b) $x + 3 = 7$, (c) $x + 4 = 9$, (d) $x + 9 = 4$, (e) $3 + x = 5$, (f) $5 + x = 4$, (g) $x = x + 1$, (h) $(x + 2) + 1 = x + 3$.

3. Prove that if x, y, and z are in N and $x \leq y$ and $x \leq z$ and $y \leq z$, then $y - x \leq z - x$.

4. Prove that if x, y, and z are in N and $y - x \leq z - x$, then $y \leq z$.
5. Prove that if $y \leq z$, then $x + y \leq z$ if and only if $x \leq z - y$.
6. Find an inequality in the form $x \leq a$ where "a" is a decimal name for a natural number and the inequality holds if and only if (a) $x + 3 \leq 7$, (b) $x - 7 \leq 4$, (c) $x + 8 \leq 10$, (d) $2x + 1 \leq x + 7$, (e) $7x - 4 \leq 6x + 6$, (f) $4x - 3 \geq 5x - 7$, (g) $75x + 5 \leq 74x + 10$.
7. Is there an identity element for subtraction in N?
8. Does subtraction distribute across addition in N?
9. Show that if x and y are in N and $y \leq x$, then

$$(x + y)(x - y) = x^2 - y^2.$$

6.15 Division

If we have 75 sheets of paper to give out to 25 students in a class, how many sheets should we give to each student if we wish each student to have the same number of sheets? Let n be the number of sheets each student is to get. If we had all the students lined up in a row we could think of our handing out of the sheets as a physical process equivalent to building a table for a Cartesian product. In fact, if we put the sheets in a column in front of each student the sheets may be thought of as boxes in our table with n rows and 25 columns. Since there are 75 sheets available we must have $n \cdot 25 = 75$. To solve the problem we must answer the question: What number when multiplied by 25 gives 75? But it is clear that the same question can be put in terms of the inverse operation for multiplication.

We call the inverse operation for multiplication **division** and denote it by "$\div$." Hence, from our definition for inverse operations we have $y \div z = x$ if and only if x is the unique value such that $y = xz$. If there is not a unique x such that $y = xz$ then $y \div z$ is undefined.

In the above problem about handing out paper we may say that our solution is to hand out $n = 75 \div 25$ sheets to each student.

In general, to find $y \div z$ we must find that x, if there is one and only one such, for which $xz = y$.

For example, from the fact that $2 \cdot 5 = 10$ we see that $10 \div 5 = 2$. But to be very careful we should check that there is no other value say w which is different from 2 for which $w \cdot 5 = 10$. For if there is such a w, we should leave $10 \div 5$ undefined. There is no such w, of course. And we can prove it by noting that if $z \neq 0$ and $x \cdot z = w \cdot z$, then $x = w$ (Theorem 6.5.5). In general, when $z \neq 0$ there is either exactly one natural number x for which $xz = y$ or there is none at all. Hence, when $z \neq 0$ and we find an x such that $xz = y$ then we can assert that $x = y \div z$. For example, $5 \cdot 5 = 25$ implies $5 = 25 \div 5$, $3 \cdot 25$ implies $3 = 75 \div 25$, $0 \cdot 7 = 0$ implies $0 = 0 \div 7$, and $4 \cdot 2 = 8$ implies $4 = 8 \div 2$. Any multiplication fact $xz = y$ where $z \neq 0$ automatically yields the division fact $y \div z = x$.

We may make use of this result (or the method of Figure 5.6) and a partial table for multiplication to construct a partial table for division of natural numbers. See, for instance, the table on page 90. In that table we omitted the row and column for 0. We should now take care of that case.

We know that $0 \cdot z = 0$ for any natural number z. At first glance this suggests that we should have $0 = 0 \div z$ for all z, but this is not so. For suppose $z = 0$. Then we have not only that $0 \cdot z = 0 \cdot 0 = 0$ but also $5 \cdot z = 5 \cdot 0 = 0$. That is, there is *not* a unique x such that $x \cdot 0 = 0$. Hence, we must leave $0 \div 0$ undefined. We may summarize the above in the theorem:

If z is a natural number and $z \neq 0$, then $0 \div z = 0$. But $0 \div 0$ is undefined.

On the other hand, suppose $y \neq 0$ and we consider $y \div 0$. We know that $y \div 0 = x$ if and only if x is the unique value such that $y = x \cdot 0$. But $x \cdot 0 = 0$ for *all* x and $y \neq 0$. So $y = x \cdot 0$ cannot be true. Hence, when $y \neq 0$, $y \div 0$ is undefined. This, combined with part of the above results, yields:

$y \div 0$ is undefined for all y in N.

Putting this another way, in the natural number system division by 0 is impossible in all cases.

It is certainly clear that N is not closed under division.

We may conclude from our partial table that division is not commutative. It is also not associative since, for example, $8 \div (4 \div 2) = 8 \div 2 = 4$ while $(8 \div 4) \div 2 = 2 \div 2 = 1$.

In terms of a multiplication machine we may think of the problem of finding $y \div z$ as that of finding *the* x, if there is exactly one, such that when x and z are fed into the machine we get out y. (See Figure 6.26.)

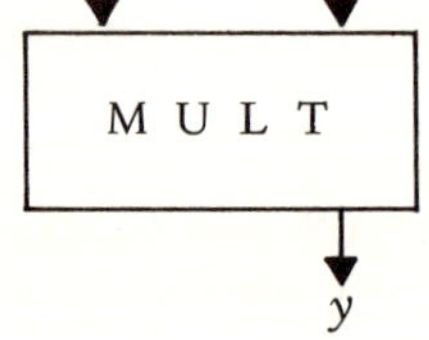

FIGURE 6.26

Since multiplication may be thought of as repeated addition, and division is the inverse operation for multiplication, and subtraction is the inverse operation for addition, it is not a surprise to discover that division may be thought of in terms of repeated subtraction. Note that

$$3 \cdot 2 = 2 + 2 + 2 = 6.$$

Then $2 + 2 = 6 - 2$, $2 = 6 - 2 - 2$ and $0 = 6 - 2 - 2 - 2$. In other words, we must subtract 2 from 6 exactly 3 times to get 0. Also $6 \div 2 = 3$.

Similarly $5 \cdot 4 = 4 + 4 + 4 + 4 + 4 = 20$ so $4 + 4 + 4 + 4 = 20 - 4$, $4 + 4 + 4 = 20 - 4 - 4$, $4 + 4 = 20 - 4 - 4 - 4$, $4 = 20 - 4 - 4 - 4 - 4$, and finally, $0 = 20 - 4 - 4 - 4 - 4 - 4$. So we must subtract 4 from 20 exactly 5 times to get 0. Also $20 \div 4 = 5$.

In the general situation, if $y \div z = x$ we have $y = xz$ and hence

$$y = \underbrace{z + z + \cdots + z}_{x \text{ terms}}.$$

Therefore, upon repeatedly subtracting z from both sides of this equation

$$y \underbrace{- z - z - \cdots - z}_{x \text{ terms}} = 0.$$

In other words, if $y \div z = x$ then we get 0 when we subtract z from y exactly x times. The converse of this also holds when $z \neq 0$.

With this interpretation in our minds we can again see why division by 0 is left undefined. If $y \neq 0$ we can never subtract 0 from y enough times to get 0. If $y = 0$ then we may subtract 0 from y any number of times we wish and get 0.

We may use the interpretation of division in terms of repeated subtraction to design a division machine. A flow chart for such a machine is given in Figure 6.27.

Suppose we feed $y = 8$ and $z = 2$ into this machine. It first checks to see if $z = 0$. Since $z \neq 0$ it sets $s = 2$ and $q = 1$. It then finds $y - s = 6$ and asks "Is $y - s = 6 = 0$?" Getting the answer "No" it sets $s = z + z = 2 + 2 = 4$ and $q = 1 + 1 = 2$. It then asks, "Is $8 - 4 = 0$?" It gets a negative answer so it sets $s = 4 + 2 = 6$ and $q = 2 + 1 = 3$. It then asks, "Is $8 - 6 = 0$?", finds it is not, and sets $s = 6 + 2 = 8$ and $q = 3 + 1 = 4$. It then asks, "Is $8 - 8 = 0$?", gets an affirmative answer, and prints out "4." And, of course, $8 \div 2 = 4$.

As we might expect since N is not closed under division, this machine will fail to give any output for certain inputs. This situation could be remedied if we assumed that the machine could tell somehow whether or not there is an x such that $y \div z = x$. We shall discuss this problem in the next section.

If there is a natural number x such that $x = y \div z$, then we say that y is **divisible** by z or that z **divides** y or z is a **divisor** of y. For instance, 10 is divisible by 5, but 10 is not divisible by 3. 5 is a divisor of 10 but 3 is not a divisor of 10. And 5 divides 10, but 3 does not divide 10.

There is another approach to the problem of division which depends on the equality:

$$(yx) \div (zx) = y \div z.$$

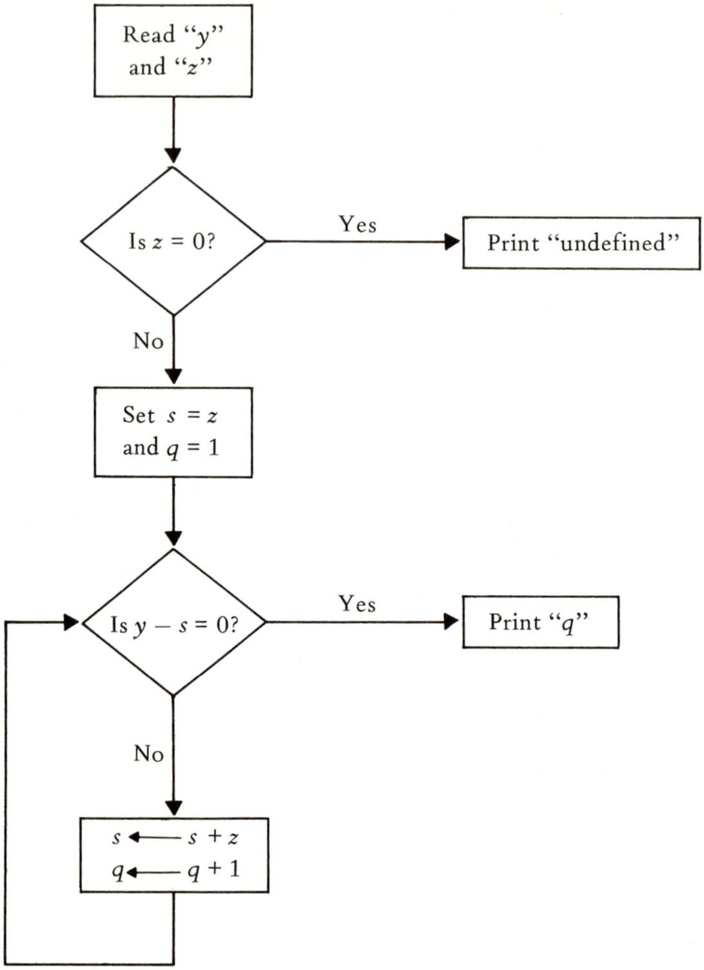

FIGURE 6.27 A division machine. The machine computes $y \div z$.

This holds whenever either side is defined. The proof is easy. Suppose $(yx) \div (zx) = w$. Then $yx = w(zx) = (wz)x$. But when $x \neq 0$ this implies $y = wz$ and, hence, $y \div z = w$. Note that $x \neq 0$ since $(yx) \div (zx)$ is defined. Suppose on the other hand that $y \div z = v$. Then $z \neq 0$ and $y = vz$. Hence $yx = (vz)x = v(zx)$. If $x \neq 0$ and $z \neq 0$, then $zx \neq 0$ and so $(yx) \div (zx) = v$. This completes the proof.

We may think of the above equality as a sort of **cancellation law.** We may use it as follows: $243 \div 27 = (81 \cdot 3) \div (9 \cdot 3) = 81 \div 9 = (27 \cdot 3) \div (3 \cdot 3) = 27 \div 3 = (9 \cdot 3) \div (3 \cdot 1) = 9 \div 1 = 9$. This procedure depends on our being able to find a number which is a divisor of both y and z.

One more example of this approach is the following:

$$\begin{aligned} 2652 \div 221 &= (204 \cdot 13) \div (17 \cdot 13) = 204 \div 17 \\ &= (12 \cdot 17) \div (1 \cdot 17) = 12 \div 1 = 12. \end{aligned}$$

This example makes it clear that there are practical difficulties in carrying out this procedure efficiently. It is difficult to recognize that 2652 may be written as $204 \cdot 13$ and that 221 may be written as $17 \cdot 13$.

However, the procedure is very helpful in some instances. For example:

$$500 \div 100 = (5 \cdot 100) \div (1 \cdot 100) = 5$$

$$81{,}000 \div 3{,}000 = (27 \cdot 3 \cdot 1000) \div (3 \cdot 1000) = 27$$

$$625^2 \div 625 = (625 \cdot 625) \div 625 = 625.$$

When we find that 2652 may be written as $204 \cdot 13$ we would say that we have **factored** 2652 into the two **factors** 204 and 13. Similar terminology is used in similar cases. The procedure for carrying out divisions that was discussed immediately above could be called **division by factoring.**

Definition 6.15.1. The inverse operation for multiplication is called division and denoted by "$\div$." We read "$y \div z$" as "y divided by z."

Definition 6.15.2. If there is an x in N such that $x = y \div z$ then we say y is a **divisor** of z or y **divides** z.

Theorem 6.15.1. If $y = xz$ and $z \neq 0$, then $y \div z = x$.

Theorem 6.15.2. If $z \neq 0$, then $0 \div z = 0$.

Theorem 6.15.3. If x is in N, $x \div 0$ is undefined.

Theorem 6.15.4. If x and y are in N and $y \neq 0$, then $(xy) \div y = x$.

PROOF: By Definition 6.15.1, $(xy) \div y = z$ if and only if z is the unique value for which $xy = zy$. But when $y \neq 0$, $xy = zy$ if and only if $x = z$.

Theorem 6.15.5. Suppose $z \neq 0$. Then $y \div z = x$ if and only if

$$y - \underbrace{z - z - \cdots - z}_{x \text{ terms}} = 0.$$

Theorem 6.15.6. $(yx) \div (zx) = y \div z$ whenever either side is defined.

Definition 6.15.3. If $z = x \cdot y$ then x or y is said to be a **factor** of z. The process of finding two numbers x and y such that $z = x \cdot y$ for a given z is called **factoring.**

PROBLEMS

1. Using an oversized piece of paper, construct a partial table for division which has a row and a column for each of the numbers 0, 1, 2, . . . , 100.

2. Use the above table or other methods to find decimal names for (a) $(75 \div 5) \div 5$, (b) $75 \div 25$, (c) $(36 \div 3) \cdot 2$, (d) $(36 \div 2) \div 3$, (e) $72 \div (3 \div 3)$, (f) $(72 \div 3) \div 3$, (g) $36 \div 6$, (h) $(4 \cdot 9) \div 6$, (i) $(36 + 12) \div 4$, (j) $(36 \div 4) + (12 \div 4)$, (k) $4 \div (2 + 2)$, (l) $(4 \div 2) + (4 \div 2)$, (m) $(81 \div 9) \div 3$, (n) $81 \div (9 \div 3)$, (o) $(5 \cdot 9) \div 9$, (p) $5 \cdot (9 \div 9)$, (q) $(49 \div 7) \cdot 7$, (r) $49 \div (7 \cdot 7)$, (s) $98 \div (7 \cdot 2)$, (t) $(98 \div 7) \cdot 2$, (u) $(7 + 8) \div 5$, (v) $(7 \div 5) + (8 \div 5)$, (w) $((100 \div 5) \div 5) \div 2$, (x) $(100 \div (5 \div 5)) \div 2$, (y) $(100 \div 5) \div (5 \div 2)$, (z) $(35 \div 5) + (20 \div 5)$, (aa) $(35 + 20) \div 5$.

3. Give instructions for using Figure 6.18 to perform certain divisions. Illustrate your method with an example.

4. Introduce symbols or names for the cardinal numbers of appropriate sets and indicate how to answer the question in terms of the operations in N we have discussed. (a) Each person in a certain club donated a fixed number of dollars to support a lobbyist. If we know the total number of dollars donated how much did each person donate? [SOLUTION: Let N be the number of people in the club. Let D be the total amount donated. Then each person donated $D \div N$ dollars.] (b) In a certain factory, there are several store rooms where parts are kept. A large batch of a certain part arrives at the plant. The parts are to be distributed evenly among the store rooms. How many of the parts go to each store room? (c) In a large pile of coconuts a few are rotten. The good coconuts are to be distributed evenly among several people. How many coconuts will each person get? (d) A length of board is to be cut into 5 pieces of equal length. How long should each piece be? (e) The government paid a total of \$100,000,000 for several jet fighters. How much did each fighter cost? (f) A person puts a fixed amount of money in a mattress on the first of each month. At the end of a year he finds he has \$600 in his mattress. How much did he put in each month? (g) A certain natural number has the property that when it is increased by 5 and then the sum is divided by 7 we get the result 6. What is the number? (h) A board of a certain length costs a certain number of dollars. The lumber yard delivers a pile of such boards and sends us a bill for the total cost. How many boards did they deliver? (i) Joan is 3 years younger than twice as old as Joy. If we know Joan's age how old is Joy? (j) If we know how many gallons our gas tank holds and how many miles we get to the gallon in our car, how many times will we need to fill up our tank to make a certain trip? (k) What is the width of a rectangle if we know its length and its area? (l) How many shares of a certain stock can we buy with a given amount of money? (m) How many gallons are there in a certain number of quarts? (n) How many miles are there in a certain large number of feet? (o) How many pounds are there in a certain large number of ounces? (p) A car travels at a constant speed for a given distance in a measured number of hours. How many miles did the car go during each hour? (q) How long will it take a space ship to reach the star Arcturus from earth if it travels at a given speed? (r) There is a puzzle which has a number of cubical blocks with painted faces. There are 8 red faces, 6 blue faces, 6 white faces, and 4 green faces. How many blocks are there?

5. Show how we may easily find $111111111_T \div 111_T$. Describe a general procedure for division (when it is possible) when we write the numbers in tally notation. [HINT: Put your description in terms of erasing or striking out appropriate strings of "1's."]

6. Find decimal names for (a) $6^0 \div 6^0$, (b) $6^3 \div 6^2$, (c) $2^5 \div 2^2$, (d) $3^4 \div 3$, (e) $10^4 \div 10^2$, (f) $10^{16} \div 10^{12}$.

7. Find the decimal name for (a) $(8\cdot 7) \div 7$, (b) $(8\cdot 7) \div (2\cdot 7)$, (c) $(9\cdot 8\cdot 7)\div(8\cdot 7)$, (d) $(1\cdot 2\cdot 3\cdot 4\cdot 5) \div (1\cdot 2\cdot 3)$, (e) $4000 \div 1000$, (f) $600{,}000 \div 10{,}000$, (g) $(3\cdot 4\cdot 5\cdot 6\cdot 7\cdot 8\cdot 9) \div (9\cdot 8\cdot 7\cdot 6\cdot 5\cdot 4)$.

8. Prove that if x is in N then $x \div 1 = x$.

9. Prove that if x is in N and $x \neq 0$, then $x \div x = 1$.

6.16 Properties of division

The cancellation law of the last section enables us to find *some* division results quite easily. For example,

$$5^3 \div 5^2 = (5\cdot 5^2) \div 5^2 = 5$$

$$10^{23} \div 10^{15} = (10^8\cdot 10^{15}) \div 10^{15} = 10^8.$$

More generally, when $x \neq 0$, we have

$$x^m \div x^n = x^{m-n}$$

whenever $m \geq n$. The proof of this is quite straightforward. When $x \neq 0$ and $m \geq n$ then $m - n$ is a natural number and $x^n \neq 0$. So we have $(x^{m-n})\cdot x^n = x^{(m-n)+n} = x^m$. This implies that $x^m \div x^n = x^{m-n}$.

This result explains why we defined x^0 to be 1 when $x \neq 0$. We should, of course, have that $x^n \div x^n = 1$ when $x \neq 0$. And we should also have, by the above, that $x^n \div x^n = x^{n-n} = x^0$.

We left 0^0 undefined when we discussed exponents. A reason for this is that it would seem proper to have $0^0 = 1$ since $x^0 = 1$ for all $x \neq 0$, but it also would seem proper to have $0^0 = 0$ since $0^y = 0$ for all $y \neq 0$. Since we can't satisfy both of these requirements, we satisfy neither. Looking at this another way, we might think of 0^0 as, say, $0^{5-5} = 0^5 \div 0^5 = 0 \div 0$ which is undefined. In a later chapter, we shall discuss an even more forceful reason for leaving 0^0 undefined.

Let us now turn to the problem of how to decide if one number is a divisor of another.

First note that if $x > y$ and $y \neq 0$ then x cannot be a divisor of y. For example, 10 can certainly not be a divisor of 7 and 147,369 cannot be a divisor of 59,430.

If $x \leq y$ then it may be that x is a divisor of y, but, of course, it *need* not be. The division process we presented in the last section suggests that if we want to find $y \div x$ when $x \neq 0$ we should try subtracting higher and higher multiples of x from y. But this brings us to an important question. With x and y given and

$x \neq 0$, is it possible that every multiple of x is less than y? The answer is that it is not possible. In other words, when $x \neq 0$ (and hence $x \geq 1$ by Theorem 6.9.2 and Theorem 6.9.20) then there is a natural number n such that $nx > y$. Putting this another way, we can say that the multiples of any nonzero natural number get arbitrarily large—given any y, no matter how large, we can find a multiple of x which exceeds y as long as x is not 0. We shall prove this at the end of this section.

So suppose we are given y and a nonzero x. We may start looking at $y - 0 \cdot x$, $y - 1 \cdot x$, $y - 2 \cdot x$, $y - 3 \cdot x$, etc. Since there is an n such that $nx > y$ there must eventually be one of these subtractions that is undefined. In fact, there must be a first one that is. Suppose this first one is $y - kx$. Note that k cannot be 0 since $y - 0$ is always simply y. Then there is a natural number q such that $q = k - 1$. Since k was chosen as it was, there must be a natural number r such that $y - qx = r$. We claim that we must have $r < x$. For suppose $r \geq x$ then $r = x + m$ for some m in N and, hence, $y - qx = x + m$ and this implies that $y - qx - x = m$. This yields $y - (qx + x) = m$ or $y - x(q + 1) = y - kx = m$. And this is impossible since $y - kx$ is not a natural number. We have thus established:

> If y and x are natural numbers and $x \neq 0$, then there are natural numbers q and r such that $r < x$ and $y = qx + r$.

Putting this another way, we may say that as we grind out the differences $y - 0 \cdot x$, $y - x$, $y - 2x$, $y - 3x$, . . . we must eventually come to a point where the difference is a natural number less than x.

If we are given x and y with $x \neq 0$ suppose we find q and r by the above technique so that $y = qx + r$ and $r < x$. If $r = 0$ then, of course, $y = qx$ and $y \div x = q$. So if $r = 0$, x is a divisor of y. If $r \neq 0$, then x cannot be a divisor of y. We shall give a proof of this in the appendix, but it is not hard to believe. In looking for $y \div x$ we are looking for something to multiply x by in order to get y. We have found that qx is too small to be equal to y, and $(q + 1)x$ is too large to be equal to y. Hence, no multiple of x can be equal to y.

We now have a mechanical procedure for finding out if x divides y. We plow through successive evaluations of $y - 0 \cdot x$, $y - x$, $y - 2x$, $y - 3x$, . . . until we find a difference less than x. If that final difference is not 0, then x does not divide y. This procedure leads to the flow chart in Figure 6.28.

In the rightmost box in this flow chart we could also have the machine print the value of i being used. We would then have the machine giving us the result $y \div x$ whenever it exists.

There is a standard pattern of terminology that goes with the above algorithm. When we are given natural numbers y and x and find the numbers q and r such that $y = qx + r$ and $r < x$, then it is said that x **goes into** y q times with a **remainder** of r. Or we say that the **dividend** y with the **divisor** x yields a **quotient** of q with a **remainder** r.

Since division and subtraction are both inverse operations of basic operations

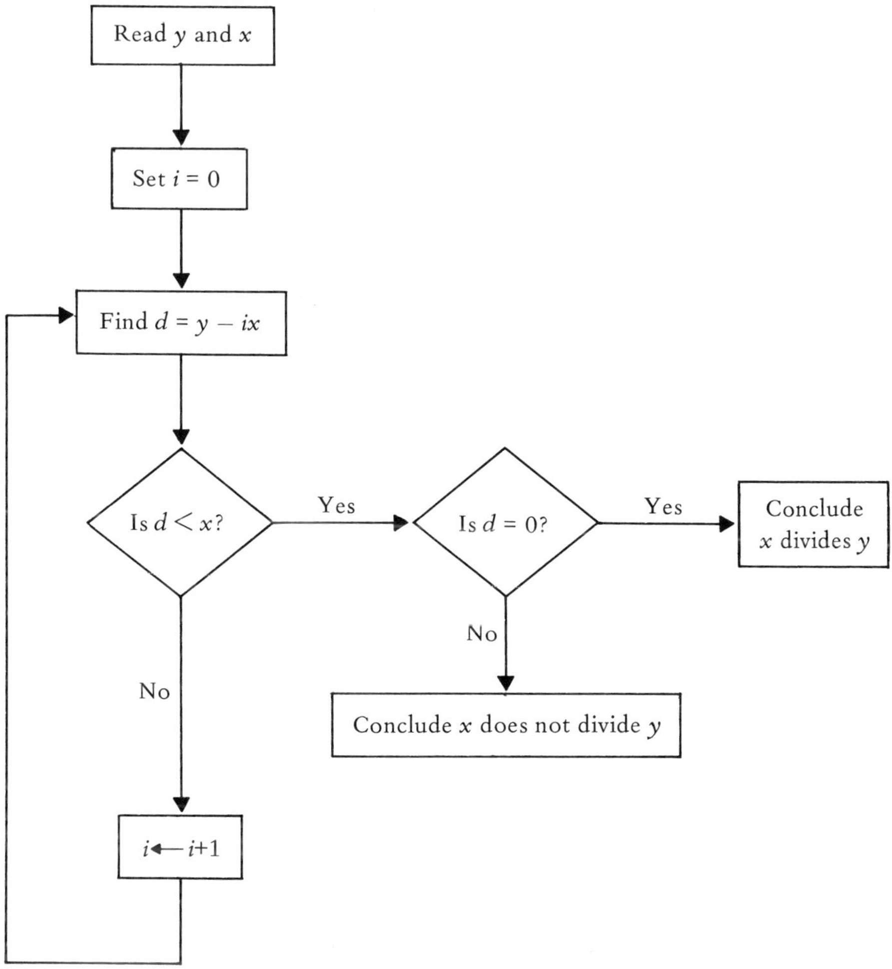

FIGURE 6.28 Flow chart for deciding if x divides y when $x \neq 0$.

we would expect some similarities between them. We shall now collect some results in pairs to emphasize such similarities. In the following x, y, and z are natural numbers.

1A. $x - y$ is defined if and only if there is a z such that $x = z + y$.
1B. When $y \neq 0$, $x \div y$ is defined if and only if there is a z such that $x = zy$.
2A. $(x + y) - y = x$.
2B. If $y \neq 0$, $(xy) \div y = x$.
3A. $(x - y) - z = x - (y + z)$ whenever either side is defined.
3B. $(x \div y) \div z = x \div (yz)$ whenever either side is defined.

As examples of 3B we have: $(36 \div 4) \div 9 = 9 \div 9 = 1$ and $36 \div (4 \cdot 9) =$

$36 \div 36 = 1$. Also $(153 \div 3) \div 17 = 51 \div 17 = 3$ and $153 \div (3 \cdot 17) = 153 \div 51 = 3$.

4A. $x + (y - z) = (x + y) - z$ whenever $y - z$ is defined.
4B. $x \cdot (y \div z) = (x \cdot y) \div z$ whenever $y \div z$ is defined.

For example, $5 \cdot (10 \div 2) = 5 \cdot 5 = 25$ and $(5 \cdot 10) \div 2 = 50 \div 2 = 25$. Also $17 \cdot (9 \div 3) = 17 \cdot 3 = 51$ and $(17 \cdot 9) \div 3 = 153 \div 3 = 51$. But notice that $9 \cdot (2 \div 3)$ is undefined while $(9 \cdot 2) \div 3 = 18 \div 3 = 6$.

5A. $x - (y - z) = (x - y) + z$ whenever both sides are defined.
5B. $x \div (y \div z) = (x \div y) \cdot z$ whenever both sides are defined.

Examples of 5B would be: $75 \div (15 \div 3) = 75 \div 5 = 15$ and $(75 \div 15) \cdot 3 = 5 \cdot 3 = 15$, $222 \div (6 \div 2) = 222 \div 3 = 74$ and $(222 \div 6) \cdot 2 = 37 \cdot 2 = 74$. But we do need both sides defined. For example, $36 \div (15 \div 5) = 36 \div 3 = 12$ while $(36 \div 15) \cdot 5$ is undefined. Also $(36 \div 9) \cdot 5 = 4 \cdot 5 = 20$ while $36 \div (9 \div 5)$ is undefined.

6A. $(x + y) - (z + y) = x - z$ whenever either side is defined.
6B. $(xy) \div (zy) = x \div z$ whenever either side is defined and $y \neq 0$.

We discussed 6B in the preceding section.

7A. $(x - y) + (z - w) = (x + z) - (y + w)$ whenever the left side is defined.
7B. $(x \div y) \cdot (z \div w) = (x \cdot z) \div (y \cdot w)$ whenever the left side is defined.

As examples of 7B, we have $(6 \div 3) \cdot (15 \div 5) = 2 \cdot 3 = 6$ while $(6 \cdot 15) \div (3 \cdot 5) = 90 \div 15 = 6$, and $(39 \div 3) \cdot (14 \div 7) = 13 \cdot 2 = 26$ while $(39 \cdot 14) \div (3 \cdot 7) = 546 \div 21 = 26$.

An important result for division and addition is the following:

$(x + y) \div z = (x \div z) + (y \div z)$ whenever the right side is defined.

For example, $(16 + 8) \div 2 = 24 \div 2 = 12$ and $(16 \div 2) + (8 \div 2) = 8 + 4 = 12$, $(36 + 18) \div 9 = 54 \div 9 = 6$ and $(36 \div 9) + (18 \div 9) = 4 + 2 = 6$. But we do need the right side defined as shown by the following example: $(3 + 5) \div 2 = 8 \div 2 = 4$ while $(3 \div 2) + (5 \div 2)$ is undefined.

Consider the following examples: $(10 \div 5) + (9 \div 3) = 2 + 3 = 5$ and $(10 \cdot 3 + 9 \cdot 5) \div (3 \cdot 5) = (30 + 45) \div 15 = 75 \div 15 = 5$. $(8 \div 2) + (10 \div 2) = 4 + 5 = 9$ and $(8 \cdot 2 + 10 \cdot 2) \div 2 \cdot 2 = (16 + 20) \div 4 = 36 \div 4 = 9$. Such results suggest the true conclusion that

$(y \div x) + (z \div w) = (yw + zx) \div (xw)$ whenever the left side makes sense.

We have not yet discussed a really efficient process for finding $y \div x$ when it is defined. We shall discuss the usual, more efficient process in Section 6.20.

Theorem 6.16.1. If $m \geq n$ and $x \neq 0$, then $x^m \div x^n = x^{m-n}$.

Theorem 6.16.2. If x and y are in N and $x \neq 0$, then there is an n in N such that $nx > y$.

PROOF: Since $x \neq 0$, we have $x \geq 1$. But this implies that $(y + 1)x \geq (y + 1) \cdot 1$. That is, $(y + 1)x \geq y + 1$. But $y + 1 > y$. So $(y + 1)x > y$. Hence, we may take $n = y + 1$ and assure the desired result.

Theorem 6.16.3. If x and y are in N and $x \neq 0$, then there are numbers q and r in N such that $y = qx + r$ and $r < x$.

Theorem 6.16.4. Suppose x and y are in N and $x \neq 0$ and $y = qx + r$ where $r < x$. Then x is a divisor of y if and only if $r = 0$.

PROOF: If $r = 0$ and $x \neq 0$, then clearly x divides y. Suppose that x divides y and $r \neq 0$. Let $y \div x = z$. Then $y = zx = qx + r$. Note we cannot have $z = q$ for this would imply $r = 0$. Also $zx - qx = r$ or $x(z - q) = r$. Since $z - q \neq 0$, we have $z - q \geq 1$. Since $x \neq 0$, this yields $x(z - q) \geq x$. Hence, $x \leq x(z - q) = r$. But this contradicts the fact that $r < x$. Hence, x cannot divide y when $r \neq 0$, and this completes the proof.

Definition 6.16.1. If $y = qx + r$ where $r < x$, then it is said that *x goes into y q times with a* **remainder** *of r*, or the **dividend** y with the **divisor** x yields a **quotient** of q and a **remainder** of r.

Theorem 6.16.5. If x, y, and z are in N, then $(x \div y) \div z = x \div (yz)$ whenever either side is defined.

PROOF: Suppose $(x \div y) \div z = w$ for some w in N. Then clearly $y \neq 0$ and $z \neq 0$. By the definition of division we then have $x \div y = wz$. But this implies $x = (wz)y = w(zy)$. Noting that $yz \neq 0$ we use the definition of division again and get $x \div (zy) = w$.

Suppose $x \div (yz) = v$ for some v in N. Then $yz \neq 0$ and this implies $y \neq 0$ and $z \neq 0$. Hence, $x = v(yz) = v(zy) = (vz)y$. This gives $x \div y = vz$ which yields $(x \div y) \div z = v$.

Theorem 6.16.6. If x, y, and z are in N and z divides y, then $x \cdot (y \div z) = (xy) \div z$.

PROOF: Suppose $y \div z = w$. Note that $z \neq 0$. Then $y = wz$ and, hence, $xy = x(wz) = (xw) \cdot z$. Then $(x \cdot y) \div z = xw = x \cdot (y \div z)$.

Theorem 6.16.7. If x, y, and z are in N, then $x \div (y \div z) = (x \div y) \cdot z$ whenever both sides are defined.

PROOF: Suppose $x \div (y \div z) = w$ and $(x \div y) \cdot z = v$ where w and v are in N. Then $z \neq 0$ and $y \neq 0$. From the definition of division we have $x = w \cdot (y \div z)$. This yields $x = (w \cdot y) \div z$ by Theorem 6.16.6. Hence, $xz = wy$.

Then $w = (xz) \div y = (zx) \div y = z \cdot (x \div y) = (x \div y) \cdot z = v$ where we have again used Theorem 6.16.6.

Theorem 6.16.8. If x, y, z, and w are in N and y divides x and w divides z, then $(x \div y) \cdot (z \div w) = (x \cdot z) \div (y \cdot w)$.

PROOF: Suppose $x \div y = n$ and $z \div w = m$. Then $y \neq 0$ and $w \neq 0$ so $yw \neq 0$. Also $x = ny$ and $z = mw$. Hence, $xz = (ny)(mw) = (yw)(nm)$. Therefore $(xz) \div (yw) = nm$.

Theorem 6.16.9. If x, y, and z are in N and z divides x and z divides y, then $(x \div z) + (y \div z) = (x + y) \div z$.

PROOF: Suppose $x \div z = w$ and $y \div z = v$ where w and v are in N. Clearly $z \neq 0$ and $x = wz$ and $y = vz$. Then $x + y = wz + vz = (w + v)z$. Therefore, $(x + y) \div z = w + v = (x \div z) + (y \div z)$.

Theorem 6.16.10. If x is a divisor of y and w is a divisor of z, then $(y \div x) + (z \div w) = (yw + zx) \div (xw)$.

PROOF: Let $y \div x = n$ and $z \div w = m$. Then $y = nx$ and $z = mw$. Note that $x \neq 0$ and $w \neq 0$ so that $xw \neq 0$. And we then have $(yw + zx) \div xw = ((nx)w + (mw)x) \div xw = (n(xw) + m(wx)) \div (xw) = (n(xw) + m(xw)) \div (xw) = ((n + m)(xw)) \div (xw) = n + m$.

PROBLEMS

1. Simplify the following: (a) $100{,}000 \div 10{,}000$, (b) $10^{17} \div 10^{6}$, (c) $2^{17} \div 2^{13}$, (d) $3^{30} \div 3^{27}$, (e) $17^5 \div 17^5$, (f) $6^3 \div 3^3$ (note that $6 = 2 \cdot 3$), (g) $27^3 \div 3^7$, (h) $144^2 \div (2^7 \cdot 3^3)$.

2. Use the method of Figure 6.28 to decide if x is a divisor of y when (a) $x = 17, y = 4$, (b) $x = 9, y = 54$, (c) $x = 100, y = 200$, (d) $x = 100, y = 201$.

3. Give four different examples that illustrate (a) 3B, (b) 4B, (c) 5B, (d) 6B, and (e) 7B as given above.

4. Give four different examples which illustrate the equality $(x \div z) + (y \div z) = (x + y) \div z$.

5. Prove that if x is a divisor of y and x is a divisor of z and $y \leq z$, then $y \div x \leq z \div x$.

6. Give four different examples to illustrate the equality

$$(y \div x) + (z \div w) = (yw + zx) \div (xw).$$

7. Suppose x divides y, w divides z, and u divides v. Show that

$$(y \div x) + (z \div w) + (v \div u) = (ywu + zxu + xwv) \div (xwu).$$

8. Suppose x divides y, x divides z, and $y \geq z$. Show that

$$(y \div x) - (z \div x) = (y - z) \div x.$$

9. Suppose that x divides y, w divides z, and $yw \geq xz$. Show that

$$(y \div x) - (z \div w) = (yw - xz) \div (xw).$$

6.17 Algorithms for addition

If we are asked to find the sum of 574 and 3967, we might answer, "The sum of 574 and 3967 is 574 + 3967." And, in a sense, this answer is correct. We know that N is closed under addition. Hence, "574 + 3967" is a name for some natural number—and that natural number is certainly the sum of 574 and 3967.

But when someone asks us for this sum, he usually would not be pleased with the answer "574 + 3967." What he usually wants is the decimal name for the requested sum. Of course, it is not a mere whim on the part of the questioner that he wants the answer in this form. In many applications and especially in communicating results to other people it is necessary or at least helpful to have our result in the form of a decimal name.

In elementary school we all learned a mechanical process of using the names "574" and "3967" in a certain way and producing the decimal name for the sum. Such a mechanical process is an example of what is called an **algorithm.** In general, an **algorithm** is a step-by-step procedure for carrying out a certain task.

We wish here to give an explanation and justification for the familiar algorithm for addition of natural numbers in decimal notation.

First note that we have all memorized the decimal name for the sum of any pair of numbers among the numbers 0, 1, 2, 3, 4, 5, 6, 7, 8, 9, and 10. In other words we all **know** the facts summarized by the table on page 89. From another point of view, such a table summarizes quite a few theorems that we can prove by falling back on simple sets (maybe of fingers and toes) and our definition of addition of cardinal numbers.

Our algorithm for addition will reduce the problem of finding the decimal name for a sum to a step-by-step process where each step is simply a use of the above table—or of our memory where we have the table's results on call.

Consider the following example where we make use of the meaning of decimal names, the associativity and commutativity of addition, and the distributive law.

$$\begin{aligned}342 + 14 &= (3\cdot 10^2 + 4\cdot 10 + 2) + (1\cdot 10 + 4)\\ &= 3\cdot 10^2 + 4\cdot 10 + 1\cdot 10 + 2 + 4\\ &= 3\cdot 10^2 + (4 + 1)\cdot 10 + (2 + 4)\\ &= 3\cdot 10^2 + 5\cdot 10 + 6\\ &= 356.\end{aligned}$$

The above string of symbols may be somewhat abbreviated and rearranged to yield

$$\begin{aligned}342 + 14 = 3\cdot 10^2 &+ 4\cdot 10 + 2\\ &+ 1\cdot 10 + 4\\ = 3\cdot 10^2 &+ (4 + 1)\cdot 10 + (2 + 4)\end{aligned}$$

$$\begin{aligned} &= 3 \cdot 10^2 + 5 \cdot 10 + 6 \\ &= 356. \end{aligned}$$

But if we are careful about lining up expressions for various powers of ten on the right, we may delete "10^2" and "10" wherever they occur and have as an abbreviation for the above

$$\begin{array}{rccc} 342 + 14 = & 3 & 4 & 2 \\ & & 1 & 4 \\ = & 3 & 5 & 6. \end{array}$$

And usually this is written as

$$\begin{array}{r} 342 \\ +\ 14 \\ \hline 356. \end{array}$$

Now consider the problem of finding the decimal name for $574 + 3967$. We have

$$\begin{aligned} 574 + 3967 &= (5 \cdot 10^2 + 7 \cdot 10 + 4) + (3 \cdot 10^3 + 9 \cdot 10^2 + 6 \cdot 10 + 7) \\ &= 3 \cdot 10^3 + (5 + 9) \cdot 10^2 + (7 + 6) \cdot 10 + (4 + 7) \\ &= 3 \cdot 10^3 + (14) \cdot 10^2 + (13) \cdot 10 + 11 \\ &= 3 \cdot 10^3 + (10 + 4) \cdot 10^2 + (10 + 3) \cdot 10 + 10 + 1 \\ &= 3 \cdot 10^3 + 10^3 + 4 \cdot 10^2 + 10^2 + 3 \cdot 10 + 10 + 1 \\ &= (3 + 1) \cdot 10^3 + (4 + 1) \cdot 10^2 + (3 + 1) \cdot 10 + 1 \\ &= 4 \cdot 10^3 + 5 \cdot 10^2 + 4 \cdot 10 + 1 \\ &= 4541. \end{aligned}$$

Notice that when we think of our addition table for 0, 1, 2, . . . , 9, and 10 as a summary of many theorems, the above is a proof of the theorem $574 + 3967 = 4541$. In the proof, we have used some of the theorems in our table as well as the commutative, associative, and distributive laws and basic properties of exponents.

The process that the above equalities indicate can be described just as well by the following array:

$$\begin{array}{rrrr} 574 + 3967 = & & 5 \cdot 10^2 & + 7 \cdot 10 + 4 \\ & + 3 \cdot 10^3 & + 9 \cdot 10^2 & + 6 \cdot 10 + 7 \\ = & & & 10 + 1 \\ & & & + (10 + 3) \cdot 10 \\ & & + (10 + 4) \cdot 10^2 & \\ & + 3 \cdot 10^3 & & \\ = & & & 1 \cdot 10 + 1 \\ & & + 1 \cdot 10^2 & + 3 \cdot 10 \end{array}$$

$$\begin{array}{rlll} & +\,1\cdot 10^3 & +\,4\cdot 10^2 & \\ & +\,3\cdot 10^3 & & \\ = & 4\cdot 10^3 & +\,5\cdot 10^2 & +\,4\cdot 10+1 \\ = & 4541. & & \end{array}$$

But if we keep our columns straight here we may suppress the names for the powers of ten and abbreviate the above as

$$\begin{array}{rcccc} 574+3967 = & & 5 & 7 & 4 \\ & 3 & 9 & 6 & 7 \\ = & & & 1 & 1 \\ & & 1 & 3 & \\ & 1 & 4 & & \\ & 3 & & & \\ = & 4 & 5 & 4 & 1. \end{array}$$

Or more compactly we could write

$$\begin{array}{r} 574 \\ +3967 \\ \hline 11 \\ 13 \\ 14 \\ 3 \\ \hline 4541 \end{array}$$

as an abbreviation for the above display.

More briefly yet, the above is sometimes changed to

$$\begin{array}{r} 111 \\ 574 \\ +3967 \\ \hline 4541 \end{array}$$

or

$$\begin{array}{r} 5\;7\;4 \\ +3_1 9_1 6_1 7 \\ \hline 4\;5\;4\;1. \end{array}$$

The above examples suggest the general procedure which we can now describe. To add $a_n a_{n-1} \cdots a_0$ and $b_m b_{m-1} \cdots b_0$ where the a's and b's are decimal digits, we proceed as follows. Find $c_0 = a_0 + b_0$. Note that we can find c_0 by use of our addition table. Also $0 \leq c_0 \leq 18$. If $c_0 \leq 9$ then we shall simply use its usual decimal name as the right digit in our sum. We call this digit "d_0." If $c_0 \geq 10$, we let $d_0 = c_0 - 10$. We now find c_1 which is simply $a_1 + b_1$ if $c_0 \leq 9$ and is $a_1 + b_1 + 1$ if $c_0 \geq 10$. If $c_1 \leq 9$, we let $d_1 = c_1$. If $c_1 > 10$, we let $d_1 = c_1 - 10$. "d_1" will be the digit in the tens-place in our sum. We now find c_2 which is $a_1 + b_1$ if $c_1 \leq 9$ and is $a_1 + b_1 + 1$ if $c_1 \geq 10$. We let $d_2 = c_2$ if $c_2 \leq 9$ and $d_2 = c_2 - 10$ if $c_2 \geq 10$. And so forth. Notice that if one numeral has more digits than the other we shall need to interpret missing digits as "0's" to carry out the process.

This process is explained in detail in the flow chart in Figure 6.29.

Let us trace through what the machine would do to find the sum of 962 and 47. It first finds that $k = 2$. It then sets $a_3 = 0$, $b_3 = b_2 = 0$, and $e = 0$. If you like, it has now rewritten our numbers as "0962" and "0047." It then sets $i = 0$ and finds $c_0 = a_0 + b_0 + e = 2 + 7 + 0 = 9$. It then asks, "Is $9 \leq 9$?", gets an affirmative answer and sets $d_0 = 9$. We assume that $d_0 = 9$ is then stored someplace in (the memory of) the machine. It then sets $e = 0$ (as it was already) and asks, "Is $0 = 3$?" Getting a negative answer it changes i to 1. It then finds $c_1 = a_1 + b_1 + e = 6 + 4 + 0 = 10$. Asking "Is $10 \leq 9$?", it now gets a "No," and so it finds $d_1 = c_1 - 10 = 0$ and sets $e = 1$. It stores $d_1 = 0$ someplace and asks, "Is $1 = 3$?" Getting a "No," it sets $i = 2$, finds $c_2 = 9 + 0 + 1 = 10$, asks, "Is $10 \leq 9$?", gets a "No," and finds $d_2 = 10 - 10 = 0$. Setting $e = 1$ again, it then asks, "Is $2 = 3$?", gets a "No," and sets $i = 3$. It then finds $c_3 = a_3 + b_3 + e = 0 + 0 + 1 = 1$, asks, "Is $1 \leq 9$?", gets a "Yes," and so finds $d_3 = c_3 = 1$. Asking "Is $3 = 3$?", it now gets a "Yes." So it asks "Is $1 = 0$?", gets a "No" and, finally, it prints out "$d_3d_2d_1d_0$." That is, it prints out "1009."

In the flow chart, the first question asked can be thought of as, "Do I have nothing to carry?" The next question is, "Have I gone far enough?" And the last question is, "Is there an initial '0' that I should delete so as to get the usual conventional name for the sum?"

When we carry through the usual process of addition with paper and pencil, we don't actually do all that the above machine does. The reason for this is that we adopt some conventions which do for us what the machine must do for itself. We treat a blank where a digit might be as if it were a "0." We do this both in terms of the names of the given numbers and in terms of the amount we carry each time. We might write the above problem as

$$\begin{array}{r} 9\,6\,2 \\ {}_{1}4\,7 \\ \hline 1\,0\,0\,9 \end{array}$$

whereas the machine, if you like, does the problem in a fashion closer to:

$$\begin{array}{r} 0\,9\,6\,2 \\ 0_{1}0_{1}4_{0}7 \\ \hline 1\,0\,0\,9. \end{array}$$

Note one thing we can learn from the machine. If we want to carry through the process on paper we had best not write our numbers too close to the left edge. We may need room there to insert another column of digits. Our having

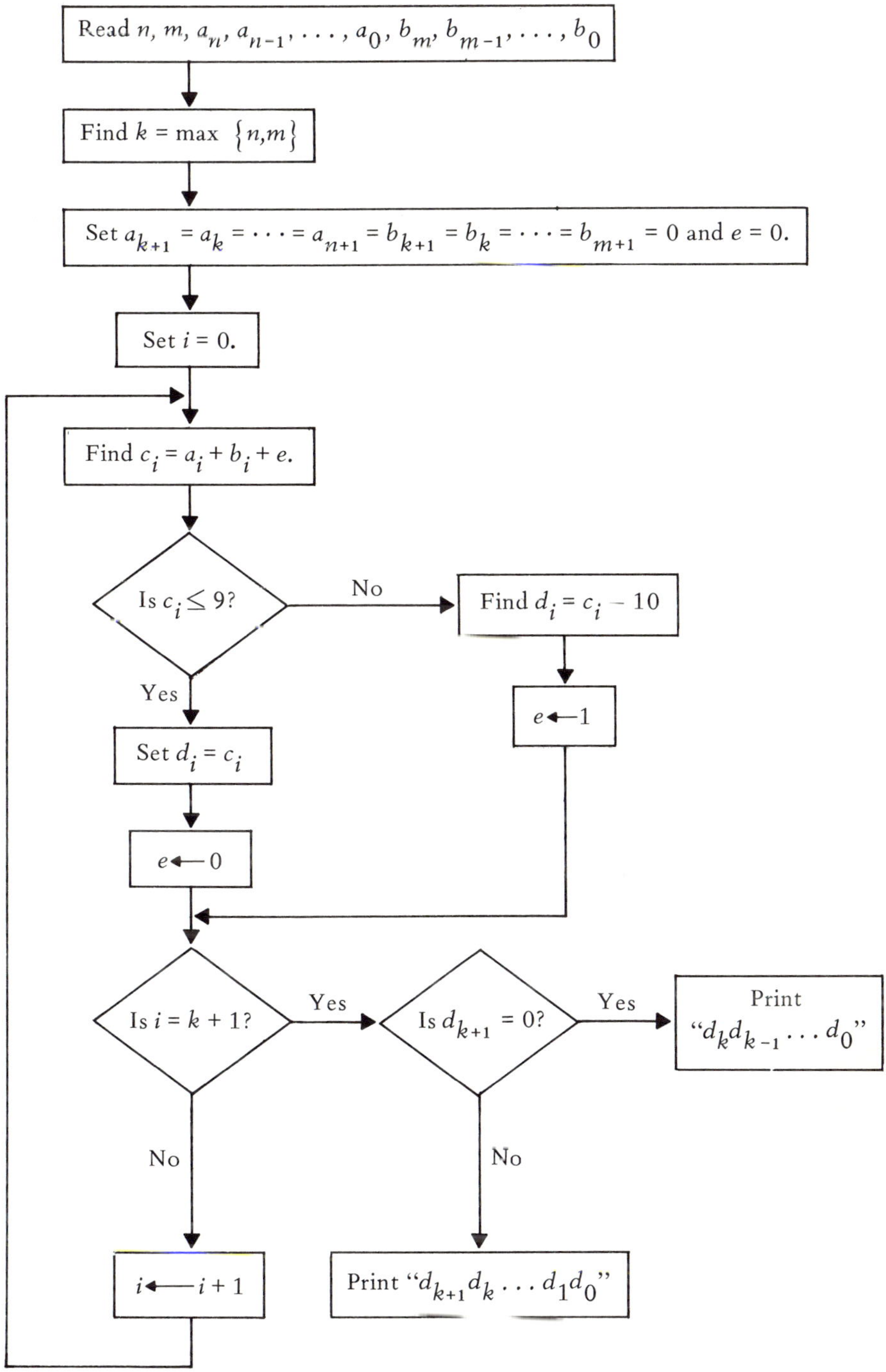

FIGURE 6.29 Flow chart for addition using decimal names.

enough space on the page corresponds to the machine's having enough storage space in its memory.

It may be recalled that the flow chart of Figure 6.29 is the second flow chart we have given for addition of natural numbers. The first one was given in Figure 6.4. What is the difference between these two? In the earlier one, the basic process was finding the name for $x + 1$ when we have the name for x. The process there is that of adding on 1 enough times. If a child can count in the sense of being able to find the decimal name for the next natural number each time he is given a natural number, then he can use that earlier process. To add 962 and 47 this way, we might make 47 marks (or tallies) on a sheet. We then add 1 to 962 and strike out one of our marks, add 1 to our result and strike out a mark, and then continue in the same fashion. When we have struck out all our marks we have the desired sum before us. The only thing that is wrong with this algorithm for addition is that it is so very inefficient in comparison with the process of Figure 6.29.

Using the process of Figure 6.4 is similar to adding on one's fingers. With enough fingers you can always do the job. But even with enough fingers, it is awfully slow.

It will perhaps throw further light on the usual algorithm for addition if we look at similar algorithms when some base different from 10 is used.

In the base 2 system we have the basic addition facts given by the table

+	0	1
0	0	1
1	1	10

$$\begin{aligned}
\text{Then } 11011_2 = 1010_2 &= (2^4 + 2^3 + 2 + 1) + (2^3 + 2) \\
&= 2^4 + 1\cdot 2^3 + 1\cdot 2^3 + 1\cdot 2 + 1\cdot 2 + 1 \\
&= 2^4 + (1 + 1)\cdot 2^3 + (1 + 1)2 + 1 \\
&= 2^4 + 2\cdot 2^3 + 2\cdot 2 + 1 \\
&= 2^4 + 2^4 + 2^2 + 1 \\
&= 2\cdot 2^4 + 2^2 + 1 \\
&= 2^5 + 2^2 + 1 \\
&= 100101_2.
\end{aligned}$$

This may be condensed to

$$\begin{array}{r}
1\,1\,0\,1\,1 \\
{}_1 1\;0_1 1\;0 \\
\hline
1\,0\,0\,1\,0\,1
\end{array}$$

where we have suppressed the subscript 2. Base 2 addition certainly has the advantage that we need to remember only a few basic addition facts to carry out the process.

In base 3, the following table summarizes the basic facts.

+	0	1	2
0	0	1	2
1	1	2	10
2	2	10	11

And we have, for example,

$$\begin{aligned}
2102_3 + 121_3 &= 2\cdot3^3 + 1\cdot3^2 + 2 + 1\cdot3^2 + 2\cdot3 + 1\\
&= 2\cdot3^3 + 1\cdot3^2 + 1\cdot3^2 + 2\cdot3 + 2 + 1\\
&= 2\cdot3^3 + (1+1)\cdot3^2 + 2\cdot3 + 1\cdot3\\
&= 2\cdot3^3 + 2\cdot3^2 + 3\cdot3\\
&= 2\cdot3^3 + 2\cdot3^2 + 3^2\\
&= 2\cdot3^3 + 3\cdot3^2\\
&= 2\cdot3^3 + 3^3\\
&= 3\cdot3^3\\
&= 3^4\\
&= 10000_3
\end{aligned}$$

which may be condensed to

$$\begin{array}{r}
2\,1\,0\,2\\
{}_1 1\,{}_1 2\,{}_1 1\\
\hline
1\,0\,0\,0\,0
\end{array}$$

where we have suppressed the subscript 3.

Logically speaking, an algorithm which gives us the decimal name for the sum of any two numbers given in decimal notation completely solves the problem of finding sums. This is because repeated sums were actually defined in terms of adding two numbers at a time. For example $a + b + c + d$ means $((a + b) + c) + d$. However, it is relatively easy to develop more efficient processes for adding several numbers together.

For example,

$$\begin{array}{rlrr}
675 + 347 + 32 = & 6\cdot10^2 & +\,7\cdot10 & +\,5\\
& 3\cdot10^2 & +\,4\cdot10 & +\,7\\
& & +\,3\cdot10 & +\,2\\
= & & & (5+7+2)\\
& & +\,(7+4+3)\cdot10 & \\
& +\,9\cdot10^2 & & \\
= & & 10 & +\,4\\
& & +\,(7+4+3)\cdot10 & \\
& +\,9\cdot10^2 & &
\end{array}$$

$$
\begin{aligned}
&= \qquad\qquad\qquad (7 + 4 + 3 + 1)\cdot 10 + 4 + 9\cdot 10^2\\
&= \qquad\qquad\qquad 15\cdot 10 + 4 + 9\cdot 10^2\\
&= \qquad\qquad\qquad (10 + 5)\cdot 10 + 4 + 9\cdot 10^2\\
&= 10^2 + 5\cdot 10 + 4 + 9\cdot 10^2\\
&= (9 + 1)\cdot 10^2 + 5\cdot 10 + 4\\
&= 10\cdot 10^2 + 5\cdot 10 + 4\\
&= 10^3 + 5\cdot 10 + 4\\
&= 1054.
\end{aligned}
$$

This entire process may be indicated more briefly by

$$
\begin{array}{r}
675\\
347\\
32\\
\hline
14\\
140\\
900\\
\hline
1054.
\end{array}
$$

Or we may suppress some of the "0's" and have

$$
\begin{array}{r}
675\\
347\\
32\\
\hline
14\\
14\\
9\\
\hline
1054.
\end{array}
$$

Even more compactly we may simply put down

$$
\begin{array}{r}
675\\
347\\
{}_1 3 {}_1 2\\
\hline
1054.
\end{array}
$$

These are simply different ways of keeping track of the process in an efficient manner.

PROBLEMS

1. In each of the following problems first find the decimal name for the sum by using the meaning of the decimal notation and basic properties of addition and multiplication. Then check your result by finding the sum in the usual manner. (a) $999 + 1$, (b) $574 + 47{,}396$, (c) $747 + 1049$, (d) $156{,}789 + 9{,}896{,}541$, (e) $736 + 554 + 1{,}532$, (f) $986 + 43 + 15{,}791$.

2. Construct a flow chart for finding the binary name of the sum of two numbers given in binary notation.

3. Find the sum in the indicated base: (Do not convert to base 10.) (a) $1100101_2 + 10111_2$, (b) $21102_3 + 11202_3$, (c) $574_8 + 1023_8$, (d) $e7t_{12} + 79te_{12}$, (e) $1111_2 + 10101_2$.

4. Check your results in Problem 3 by converting all the numerals there to base 10, performing the additions, and then converting back.

5. Verify that the machine with the flow diagram of Figure 6.30 will correctly find the binary name of the sum of any two numbers whose binary names consist of single digits. The machine is especially simple since almost the only thing it needs to be able to do is recognize "0."

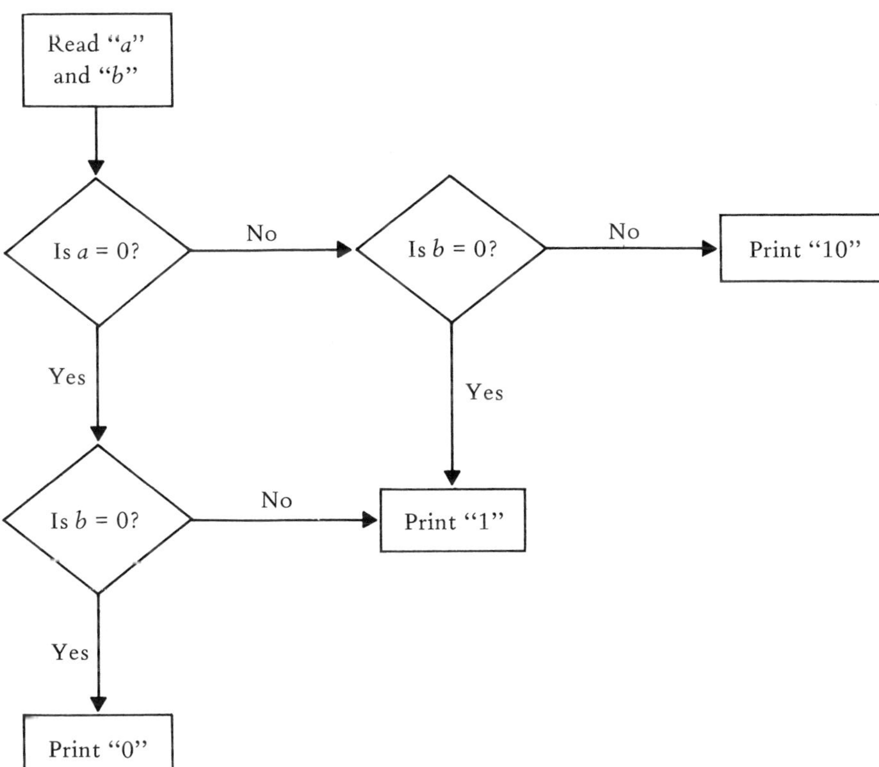

FIGURE 6.30

6.18 Algorithms for subtraction

The usual display of the subtraction problem

$$\begin{array}{r} 7\,8\,9\,4 \\ -\ 3\,2\,3 \\ \hline 7\,5\,7\,1 \end{array}$$

may be thought of as a condensation of:

$$\begin{aligned} 7894 - 323 &= 7\cdot 10^3 + 8\cdot 10^2 + 9\cdot 10 + 4 - (3\cdot 10^2 + 2\cdot 10 + 3) \\ &= 7\cdot 10^3 + (8-3)\cdot 10^2 + (9-2)\cdot 10 + (4-3) \\ &= 7\cdot 10^3 + 5\cdot 10^2 + 7\cdot 10 + 1 \\ &= 7\,5\,7\,1. \end{aligned}$$

Similarly,

$$\begin{array}{r} 7\,8\,9\,4 \\ -\ 3\,9\,7\,7 \\ \hline 3\,9\,1\,7 \end{array}$$

may be thought of as a schematic presentation of:

$$\begin{aligned} 7894 - 3977 &= \qquad 7\cdot 10^3 + 8\cdot 10^2 + 9\cdot 10 + 4 \\ &\qquad - (3\cdot 10^3 + 9\cdot 10^2 + 7\cdot 10 + 7) \\ &= 6\cdot 10^3 + (10+8)\cdot 10^2 + 8\cdot 10 + (10+4) \\ &\qquad - (3\cdot 10^3 + 9\cdot 10^2 + 7\cdot 10 + 7) \\ &= (6-3)\cdot 10^3 + (18-9)\cdot 10^2 + (8-7)\cdot 10 + (14-7) \\ &= 3\cdot 10^3 + 9\cdot 10^2 + 1\cdot 10 + 7 \\ &= 3\,9\,1\,7. \end{aligned}$$

A display that lies between the above two in regard to giving details would be

$$\begin{array}{r} 6\quad 8\ \ \ \\ \cancel{7}\,{}^{1}8\,\cancel{9}\,{}^{1}4 \\ -3\,9\,7\,7. \\ \hline 3\,9\,1\,7. \end{array}$$

As another example,

$$\begin{array}{r} 1\,4\,3\,2 \\ -\ \ 9\,7\,4 \\ \hline 4\,5\,8 \end{array}$$

may be thought of as a condensation of

$$\begin{aligned} 1432 - 974 &= 1\cdot 10^3 + 4\cdot 10^2 + 3\cdot 10 + 2 - (9\cdot 10^2 + 7\cdot 10 + 4) \\ &= 1\cdot 10^3 + 4\cdot 10^2 + 2\cdot 10 + (10+2) - (9\cdot 10^2 + 7\cdot 10 + 4) \\ &= 1\cdot 10^3 + 4\cdot 10^2 + 2\cdot 10 + (12-4) - (9\cdot 10^2 + 7\cdot 10) \end{aligned}$$

$$\begin{aligned}
&= 1\cdot 10^3 + 3\cdot 10^2 + 10\cdot 10 + 2\cdot 10 + 8 - (9\cdot 10^2 + 7\cdot 10)\\
&= 1\cdot 10^3 + 3\cdot 10^2 + (12 - 7)\cdot 10 + 8 - 9\cdot 10^2\\
&= 10\cdot 10^2 + 3\cdot 10^2 + 5\cdot 10 + 8 - 9\cdot 10^2\\
&= (10 + 3)\cdot 10 + 5\cdot 10 + 8 - 9\cdot 10^2\\
&= (13 - 9)\cdot 10^2 + 5\cdot 10 + 8\\
&= 4\cdot 10^2 + 5\cdot 10 + 8\\
&= 4\ 5\ 8.
\end{aligned}$$

In order to carry out such computations efficiently we need to **know** part of the table for subtraction on N. A portion of the table that gives the inverse operation for the operation in the table on page 89 will suffice. We give such a table below.

—	0	1	2	3	4	5	6	7	8	9
0	0	—	—	—	—	—	—	—	—	—
1	1	0	—	—	—	—	—	—	—	—
2	2	1	0	—	—	—	—	—	—	—
3	3	2	1	0	—	—	—	—	—	—
4	4	3	2	1	0	—	—	—	—	—
5	5	4	3	2	1	0	—	—	—	—
6	6	5	4	3	2	1	0	—	—	—
7	7	6	5	4	3	2	1	0	—	—
8	8	7	6	5	4	3	2	1	0	—
9	9	8	7	6	5	4	3	2	1	0
10	10	9	8	7	6	5	4	3	2	1
11	11	10	9	8	7	6	5	4	3	2
12	12	11	10	9	8	7	6	5	4	3
13	13	12	11	10	9	8	7	6	5	4
14	14	13	12	11	10	9	8	7	6	5
15	15	14	13	12	11	10	9	8	7	6
16	16	15	14	13	12	11	10	9	8	7
17	17	16	15	14	13	12	11	10	9	8
18	18	17	16	15	14	13	12	11	10	9

Of course, this table need not actually be memorized if we know the addition facts of the table on page 89 and the meaning of subtraction. Instead of remembering that $17 - 9 = 8$ we may remember $8 + 9 = 17$ and from this deduce that $8 = 17 - 9$.

Assuming that we have a machine that can perform the subtractions indicated in the above table and can tell when one number is larger than another, we may construct a flow chart for a subtraction algorithm as in Figure 6.31.

Let us see how $376 - 49$ is handled by the machine with this flow chart. The machine first checks to see that $376 \geq 49$ and finds that it is. It then finds

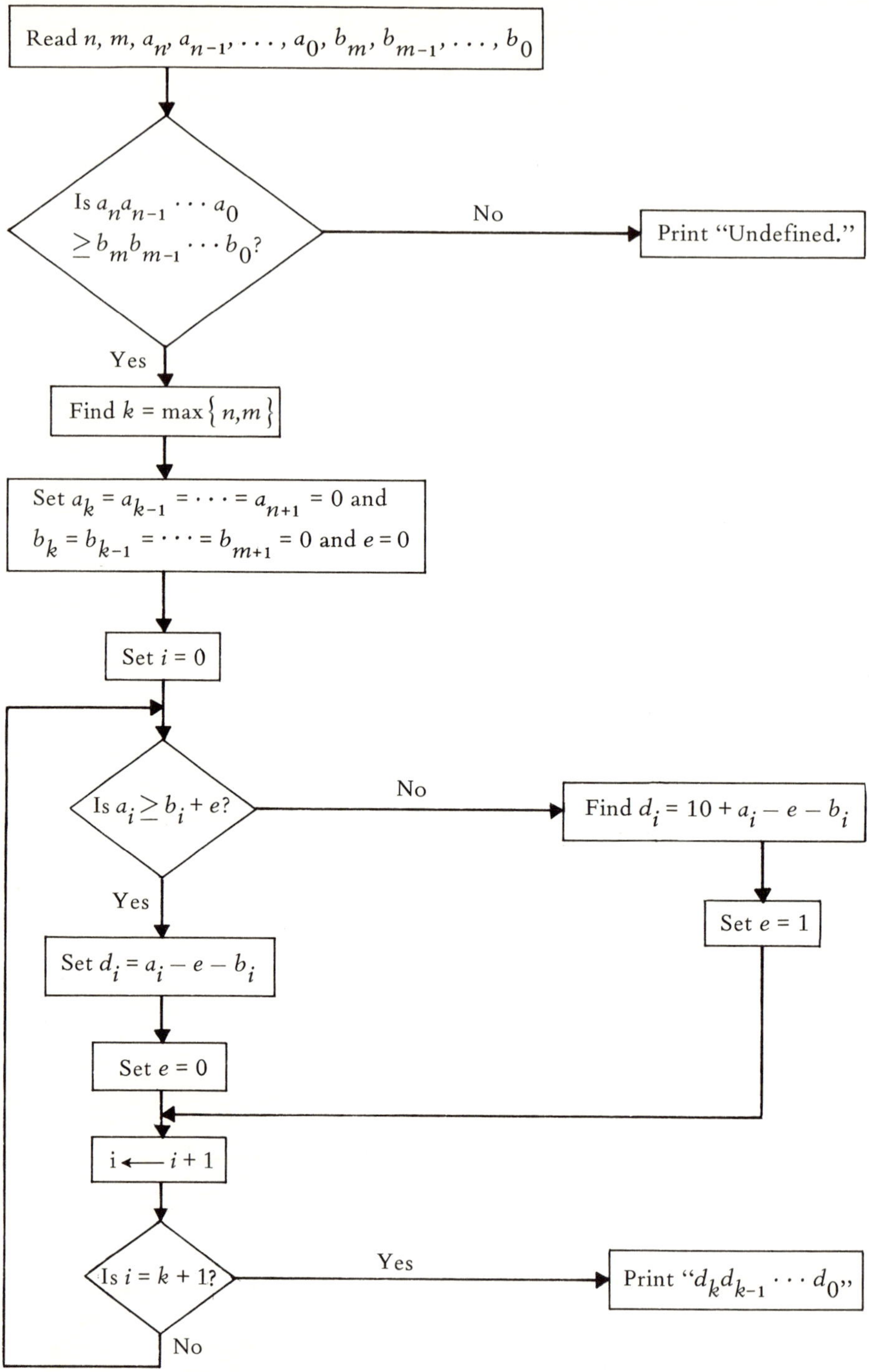

FIGURE 6.31 Flow chart for subtraction using decimal names.

$k = 2$ and sets $b_2 = 0$ and $e = 0$. It then asks, "Is $6 \geq 9 + 0$?" and finds the answer to be "No." So it computes $d_0 = 10 + 6 - 0 - 9 = 16 - 9 = 7$ and sets $e = 1$. Changing i to 1 it asks, "Is $i = k + 1 = 3$?" and finds it is not. So it goes back and asks, "Is $7 \geq 4 + 1$?" and finds it is. So it computes $d_1 = 7 - 1 - 4 = 6 - 4 = 2$ and sets $e = 0$. Setting i at 2 it asks, "Is $2 = 3$?" and finds it is not. So it goes back and asks, "Is $3 \geq 0 + 0$?" Since it is, it computes $d_2 = 3 - 0 - 0 = 3$, sets $e = 0$ and $i = 3$, and asks, "Is $3 = 3$?" Since this is so, it prints "327" and stops.

The following array may be thought of as a summary of what the machine has done:

$$\begin{array}{r} 6 \\ 3\,\not{7}\,^1 6 \\ -0\;4\;\;9 \\ \hline 3\;2\;\;7. \end{array}$$

In teaching a child to carry out this algorithm for subtraction, we must essentially train him to do what the above machine does. Of course, the child will use the positions of digits on the paper as a substitute for assigning subscripts to the digits. He will simply work from right to left instead of increasing the subscripts. He will also treat a blank space where no digit occurs as if it were a "0" instead of filling such blank spaces with "0's."

It will throw further light on the above subtraction algorithm if we look at the situation when we use other numeration systems. Using the binary system we would need to know the following subtraction table (where we have suppressed the subscript 2 on the binary numerals).

−	0	1
0	0	−
1	1	0
10	10	1

Then, in analogy with the subtraction process using decimals, we have, for example,

$$\begin{array}{r} 1 \\ \not{1}\,\not{0}\,^1 0\;1\;0\;1 \\ -\quad 1\;1\;1\;0\;0 \\ \hline 1\;0\;0\;1. \end{array}$$

This may be considered to be an abbreviation of

$$\begin{aligned} 100101_2 - 11100_2 &= 2^5 + 2^2 + 1 - (2^4 + 2^3 + 2^2) \\ &= 2^4 + 2^4 + 2^2 + 1 - (2^4 + 2^3 + 2^2) \\ &= 2^4 + 2^3 + 2^3 + 2^2 + 1 - (2^4 + 2^3 + 2^2) \end{aligned}$$

$$= 2^3 + 1$$
$$= 1\,0\,0\,1_2.$$

Note that as we have written up this last computation the basic fact used is that $2^{n+1} = 2\cdot 2^n = (1 + 1)2^n = 2^n + 2^n$. If b is any natural number we have

$$b^{n+1} = b\cdot b^n = (1 + 1 + \cdots + 1)b^n = b^n + b^n + \cdots + b^n$$

where there are b terms in each sum. But this fact is not very useful in our algorithm unless $b = 2$.

Problems

1. First find the decimal name for each result by using the meanings of the decimal names and the basic laws for addition, multiplication, and subtraction. Then check your result by the usual algorithm. (a) 47,963 − 36,462, (b) (7574 − 6943) − 427, (c) 7574 − (6943 − 427), (d) 1,797,400 − 696,001 − 4960, (e) 4675 − 3966 − 709.

2. Display a table for subtraction when base 3 numerals are used and we have rows for 0, 1, 2, 10, 11, and 12, and columns for 0, 1, and 2.

3. Find numerals for each of the following in the indicated base. (Do not convert to decimals.) (a) $1101101_2 - 111_2$, (b) $110011_2 - 1000_2$, (c) $210021_3 - 21010_3$, (d) $12210_3 - 1121_3$.

4. Check your results in Problem 3 by converting to decimals, subtracting, and then converting back.

5. Display a table for subtraction when base 12 numerals are used and we have rows for 0, 1, 2, . . . , 1*t*, and 1*e*, and columns for 0, 1, 2, . . . , and *e*.

6. Find the base 12 names for (a) $1t79e_{12} - 3471_{12}$, (b) $32114_{12} - 7ete_{12}$, (c) $t69e5_{12} - et41_{12}$. (Do not convert to decimals.)

7. Check your results in Problem 6 by converting to decimals, subtracting, and then converting back.

8. Draw a flow chart for subtraction when base 2 numerals are used.

9. Describe carefully a simple algorithm for subtraction when tally names are used for the natural numbers. (Instead of giving a flow chart, indicate what should be written down and how we should proceed if we allow ourselves to erase or strike out marks on the paper.)

6.19 Algorithms for multiplication

The table on page 89 summarizes the multiplication facts for all possible pairs of the numbers 1 to 9. The usual algorithm for multiplication using decimal names makes use of these results and basic facts about addition, multiplication, and exponents. It also assumes that we can find decimal names for sums when we need them.

Let us first note that there is a very easy algorithm for multiplying by a power of 10—simply add on the needed number of "0's" to the name. For example,

$$\begin{aligned} 576 \cdot 100 &= (5 \cdot 10^2 + 7 \cdot 10 + 6) \cdot 10^2 \\ &= 5 \cdot 10^2 \cdot 10^2 + 7 \cdot 10 \cdot 10^2 + 6 \cdot 10^2 \\ &= 5 \cdot 10^4 + 7 \cdot 10^3 + 6 \cdot 10^2 \\ &= 5\,7\,6\,0\,0. \end{aligned}$$

Similarly,

$$\begin{aligned} 127 \cdot 10{,}000 &= 1{,}270{,}000, \\ 56 \cdot 10^6 &= 56{,}000{,}000, \\ 74 \cdot 10^2 &= 7400, \end{aligned}$$

and

$$1000 \cdot 10^4 = 10{,}000{,}000.$$

Then we may use this result and those in our multiplication table to carry out such calculations as

$$\begin{aligned} 576 \cdot 4 &= (5 \cdot 10^2 + 7 \cdot 10 + 6) \cdot 4 \\ &= 4 \cdot 5 \cdot 10^2 + 4 \cdot 7 \cdot 10 + 4 \cdot 6 \\ &= 20 \cdot 10^2 + 28 \cdot 10 + 24 \\ &= 2000 + 280 + 24 \\ &= 2304. \end{aligned}$$

This may also be written without exponents as:

$$\begin{aligned} 576 \cdot 4 &= (500 + 70 + 6) \cdot 4 \\ &= 4 \cdot 500 + 4 \cdot 70 + 4 \cdot 6 \\ &= 2000 + 280 + 24 \\ &= 2304. \end{aligned}$$

In this form we are assuming the easily established fact that if the decimal name for x is a single nonzero digit "d" followed by n "0's," then we may obtain the decimal name for yx by writing down the decimal name for yd and adding on n "0's."

The usual form for displaying such a multiplication problem as that above may be thought of as simply a condensed form of either of the above displays:

$$\begin{array}{r} 5\,7\,6 \\ \times 4 \\ \hline 2\,4 \\ 2\,8\,0 \\ 2\,0\,0\,0 \\ \hline 2\,3\,0\,4. \end{array}$$

In fact, it is customary (but not necessary) to suppress some of the "0's" in this last display and write it as

$$\begin{array}{r} 576 \\ \times 4 \\ \hline 24 \\ 28 \\ 20 \\ \hline 2304. \end{array}$$

Here, we are letting blanks play the role of "0's" as we did in our addition algorithm.

The above may be condensed even further if we do some multiplying and adding in our heads. We use small numerals as we did before to indicate the amount that is "carried":

$$\begin{array}{r} 5_3 7_2 6 \\ \times 4 \\ \hline 2304. \end{array}$$

Once we have a procedure for multiplying by a number below 10, it is relatively easy to make use of it to get a procedure for multiplying by larger numbers. For example

$$\begin{aligned} 576 \cdot 24 &= 576(2 \cdot 10 + 4) \\ &= 2 \cdot 576 \cdot 10 + 4 \cdot 576 \\ &= (1152) \cdot 10 + 2304 \\ &= 11{,}520 + 2304 \\ &= 13{,}824 \end{aligned}$$

or more simply,

$$\begin{aligned} 576 \cdot 24 &= 576 \cdot 20 + 576 \cdot 4 \\ &= 11{,}520 + 2304 \\ &= 13{,}824. \end{aligned}$$

This may be presented more compactly as

$$\begin{array}{r} 576 \\ \times 24 \\ \hline 2304 \\ 11520 \\ \hline 13{,}824. \end{array}$$

Or we may suppress a "0" as is customary (but not necessary) and have

$$\begin{array}{r} 576 \\ \times 24 \\ \hline 2304 \\ 1152 \\ \hline 13{,}824. \end{array}$$

As another illustration we have

$$\begin{aligned} 47009 \times 53 &= 47009(50 + 3) \\ &= 2{,}350{,}450 + 141{,}027 \\ &= 2{,}491{,}477 \end{aligned}$$

which is written more briefly as

$$\begin{array}{r} 47009 \\ \times 53 \\ \hline 141027 \\ 235045 \\ \hline 2{,}491{,}477. \end{array}$$

If we assume that we have a machine which is able to get the decimal name for the product of any two natural numbers below 10, is able to multiply any number by a power of ten, and is able to add, then we may display a flow chart for multiplying by a one-digit number as in Figure 6.32.

The algebraic form which closely resembles the action of the machine would be similar to that in

$$5 \cdot 634 = 5 \cdot 4 \cdot 10^0 + 5 \cdot 3 \cdot 10^1 + 5 \cdot 6 \cdot 10^2.$$

If names of numbers are written in the base 2, multiplication algorithms become especially simple. For example, we have the following where we delete the subscript 2:

$$\begin{array}{r} 110101 \\ \times 101 \\ \hline 110101 \\ 0 \\ 11010100 \\ \hline 100001001. \end{array}$$

Notice that the rule for multiplying by a power of the base is just as it was in the decimal system—tack on the necessary number of "0's." And the above is a condensed version of

$$\begin{aligned} 110101 \cdot 101_2 &= 110101_2(100_2 + 1) \\ &= 11010100_2 + 110101_2 \\ &= 100001001_2. \end{aligned}$$

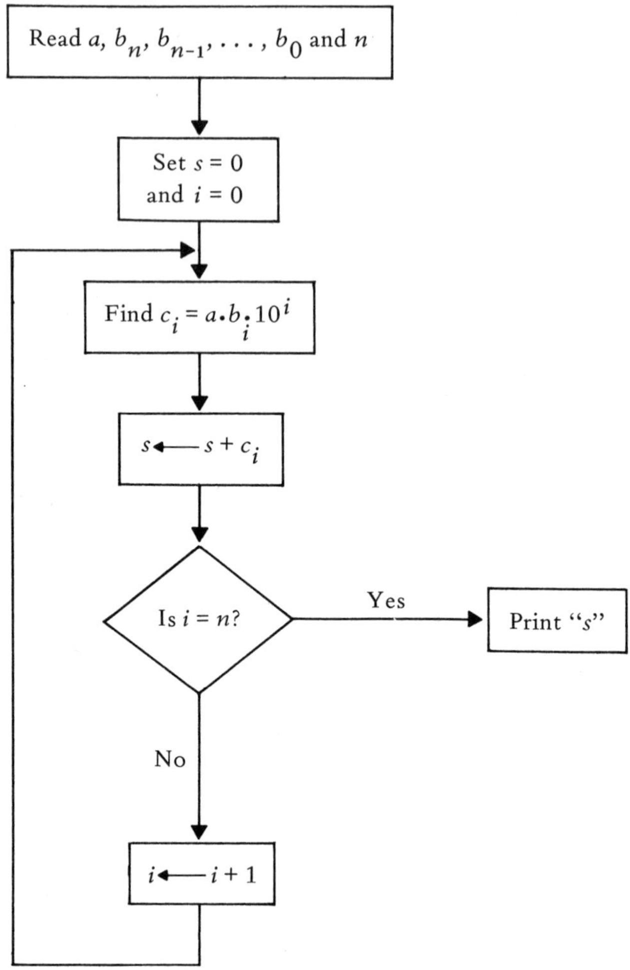

FIGURE 6.32 Flow chart for computing $a \cdot (b_n b_{n-1} \ldots b_1 b_0)$ where "a" and the "b's" are decimal digits.

Of course, we may delete some unneeded "0's" in the display and write

$$\begin{array}{r} 110101 \\ \times 101 \\ \hline 110101 \\ 110101 \\ \hline 100001001. \end{array}$$

PROBLEMS

1. First find decimal names for the following products by using the meaning of the decimal names and basic properties of exponents and multiplication

and addition, then check your result using the usual algorithm. (a) $749\cdot5$, (b) $663\cdot4$, (c) $663\cdot94$, (d) $3068\cdot116$, (e) $555\cdot111$.

2. Construct a flow chart using binary numerals for (a) multiplying a number by a one-digit number, (b) multiplying any two numbers.

3. Construct multiplication tables for numerals to the base b with columns and rows for $0, 1, \ldots, b - 1$ when (i) $b = 3$, (ii) $b = 5$, (iii) $b = 8$, (iv) $b = 12$.

4. Find the names in the indicated base for each product. (Do not convert to decimals.) (a) $101_2\cdot111_2$, (b) $10110_2\cdot1011_2$, (c) $(1010_2)^2$, (d) $1202_3\cdot210_3$, (e) $et9_{12}\cdot1et_{12}$, (f) $88e7_{12}\cdot111t_{12}$.

5. Check your results in Problem 4 by converting to base 10, multiplying, and converting back.

6. Describe in your own words (not a flow chart) a procedure involving making marks on paper and erasing them for multiplying two numbers given in tally notation.

6.20 Algorithms for division

It is very easy for us to get the result when we are asked to divide 6 by 2. This is because we remember that $6 = 3\cdot2$ and we know that this implies that $6 \div 2 = 3$. In general, we know that when x is different from 0 we have $y \div x = z$ if and only if $y = zx$. But $y = zx$ if and only if $y - zx = 0$. And the assertion that $y - zx = 0$ is logically equivalent to $y - x - x - x - \cdots - x = 0$ where there are z occurrences of "x." Thus, for $x \neq 0, y \div x = z$ when and only when we get 0 when we subtract x exactly z times from y.

Faced with the problem of finding the decimal name for $36 \div 9$, we may do so by carrying through the computations indicated by the equations $36 - 1\cdot9 = 36 - 9 = 27$, $36 - 2\cdot9 = (36 - 9) - 9 = 27 - 9 = 18$, $36 - 3\cdot9 = (36 - 9 - 9) - 9 = 18 - 9 = 9$, $36 - 4\cdot9 = (36 - 9 - 9 - 9) - 9 = 9 - 9 = 0$, and, finally, concluding from this that $36 \div 9 = 4$.

Such an approach is not very practical for the case where we wish to find the decimal name for $1256 \div 8$. We could start out computing $1256 - 1\cdot8$, $1256 - 2\cdot8$, $1256 - 3\cdot8$, etc., but we would soon get tired. And, worse than this, with so many different calculations we would be quite likely to make a mechanical mistake. A rather natural way to modify the procedure suggests itself. Let us speed up the process by subtracting off rather large, but convenient multiples of 8. It is easy to compute $1256 - 100\cdot8 = 1256 - 800 = 456$. And then we quickly get $456 - 10\cdot8 = 456 - 80 = 376$, $376 - 10\cdot8 = 296$, $296 - 10\cdot8 = 216$, $216 - 10\cdot8 = 136$, $136 - 10\cdot8 = 56$. And, indeed, these last equalities may be condensed to $456 - 5\cdot10\cdot8 = 456 - 50\cdot8 = 456 - 400 = 56$. Finally, we see that $56 = 7\cdot8$, and, combining our above work we see that $1256 - 100\cdot8 - 50\cdot8 - 7\cdot8 = 0$. That is, $1256 - 157\cdot8 = 0$. Hence, $1256 \div 8 = 157$.

The entire procedure may be summarized by the display

$$
\begin{array}{rr}
 & 1256 \\
100\cdot 8 = & 800 \\ \cline{2-2}
 & 456 \\
50\cdot 8 = & 400 \\ \cline{2-2}
 & 56 \\
7\cdot 8 = & 56 \\ \cline{1-1}\cline{2-2}
(100+50+7)\cdot 8 & 0
\end{array}
$$

where each horizontal bar on the right indicates a subtraction for the numbers given immediately above the bar.

But instead of writing the "8" and the "=" three times at the left, suppose we put down names for our multipliers for 8 in a column at the top and represent the above process by

$$
\begin{array}{r|r}
 & 7 \\
 & 50 \\
 & 100 \\ \cline{2-2}
8 & 1256 \\
 & 800 \\ \cline{2-2}
 & 456 \\
 & 400 \\ \cline{2-2}
 & 56 \\
 & 56 \\ \cline{2-2}
 & 0.
\end{array}
$$

Note the pattern here.

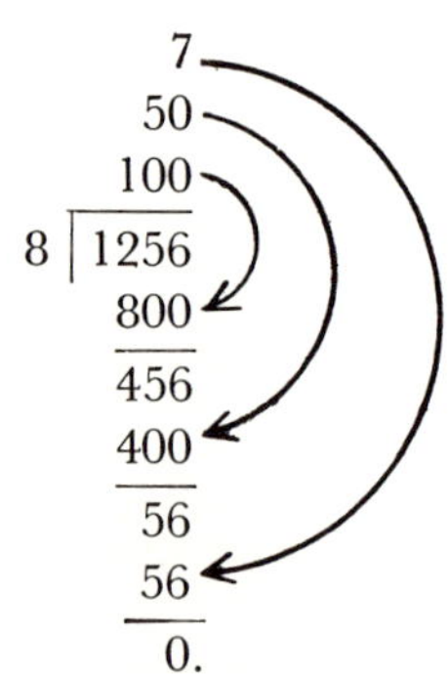

Each numeral above the "$\overline{)\qquad}$" leads to the corresponding product which we get by multiplying by 8. After each such multiplication we subtract as indicated by the bars.

A general procedure for division is easy to describe in terms of such a display. To find $y \div x$ when it exists, we write a decimal name for y under the symbol "$\overline{)\qquad}$" and a name for x at the left of this symbol. Above the symbol "$\overline{)\qquad}$" we may write the name of any number q_1. When we do, we also write a name for xq_1 below the name of y. Then we draw a bar, subtract xq_1 from y, and write the name of the result d_1 below the bar. Let us call d_1 a **partial remainder.** If $xq_1 > y$ so that we cannot subtract to get d_1, then we made a poor choice for q_1 so we start all over again with a smaller value for q_1. If $d_1 = 0$, we are through since we have found that $y - xq_1 = 0$ and, hence, $y \div x = q_1$. If $d_1 \neq 0$, then we find a number q_2 such that $xq_2 \leq d_1$. We write the decimal name for q_2 above that for q_1 in the display, and the name for xq_2 below that for d_1. Then we find the partial remainder $d_2 = d_1 - xq_2$ and write a name for it below a bar under the name for xq_2. We then proceed, as before, to find a q_3 such that $xq_3 \leq d_2$, and so forth.

By this process, we must eventually find a final **partial remainder** d_k which is less than x. If this $d_k = 0$ then we may add all the **partial quotients** q_1, q_2, . . . , q_k to get the result $y \div x$. If $d_k \neq 0$, then y is not divisible by x. But in this latter case it is still customary to add together q_1, q_2, . . . , and q_k to get a number q and say that x goes into y q times with a remainder of d_k.

Let us return to our example and see how the amount that needs to be written down may be greatly reduced. In the last display above we may simply fail to write many of the digits and cause absolutely no confusion—provided we keep our columns straight. We get as a condensation of our display:

$$\begin{array}{c|cccc}
 & & & & 7 \\
 & & & 5 & \\
 & & 1 & & \\ \cline{2-5}
8 & 1 & 2 & 5 & 6 \\
 & & 8 & & \\ \cline{2-4}
 & & 4 & 5 & \\
 & & 4 & 0 & \\ \cline{3-5}
 & & & 5 & 6 \\
 & & & 5 & 6 \\ \cline{4-5}
 & & & & 0.
\end{array}$$

But this may be condensed even further if we perform some of the additions and subtractions "in our heads" and simply record the crucial digits. The final result is usually written in one of the standard *short division forms* as

$$\begin{array}{r|l}
 & 1\;5\;7 \\ \hline
8 & 1\;2_4 5_5 6
\end{array}$$

or

$$\underline{8\,|}\;1\;2_4 5_5 6\;\underline{|\,1\;5\;7.}$$

Let us consider another numerical example. Suppose we wish to find $35{,}518 \div 7$. Using the general procedure we described above we might write

$$\begin{array}{r|r}
 & 4 \\
 & 70 \\
 & \cancel{100} \\
 & 1000 \\
 & 4000 \\ \cline{2-2}
7 & 35{,}518 \\
 & \underline{28{,}000} \\
 & 7\,518 \\
 & \underline{7\,000} \\
 & 518 \\
 & \cancel{700} \\
 & \underline{490} \\
 & 28 \\
 & \underline{28} \\
 & 0.
\end{array}$$

We conclude that $35{,}518 \div 7 = 4000 + 1000 + 70 + 4 = 5074$. In the display we have indicated a procedure for handling the choice of a partial quotient which is too large at a certain stage. We simply strike out a poor choice. Of course, the above process is not as efficient as it might be. We might have recognized that $5000 \cdot 7$ is less than 35,518 and, hence, used 5000 as our first partial quotient q_1. But no actual harm is done in the expanded form by choosing our first partial quotient to be 4000.

The above description of the general procedure *is* somewhat vague as to how the partial quotients *are* to be chosen. As a matter of fact, it doesn't matter logically *how* they are chosen—the procedure will still work. But we are interested not only in "getting the answer" but also in doing so in a fashion that is efficient. If we can keep down the number of steps to be carried out in the process, we shall also keep down our chances for making mechanical errors.

So it is appropriate to give some suggestions as to how the partial quotients should be chosen to get the desired result efficiently. The usual prescription is: (a) choose the partial quotients so they may be written as a single nonzero digit followed by a (possibly empty) string of "0's," and (b) at each step, choose the largest possible partial quotient satisfying condition (a). Condition (a) assures us of a partial quotient q_i which has the property that it will be easy to find the decimal name for $q_i x$. Condition (b) assures us that we will not include any unnecessary steps once we have decided also to satisfy condition (a).

In the two examples above, we did actually choose our partial quotients to satisfy condition (a). And we violated condition (b) only at the first step in the second example.

Although, in our examples so far, we have only considered cases where the divisor was less than 10, the general procedure we outlined (as well as the admonition as to how the partial quotients might be chosen) apply more generally. Suppose we wish to find 1,805,276 ÷ 5071. Our procedure yields:

$$
\begin{array}{r|r}
 & 6 \\
 & 50 \\
 & 300 \\
 & \sout{500} \\
 & \sout{1000} \\
\cline{2-2}
5071 & 1805276 \\
 & \sout{5071000} \\
 & \sout{2535500} \\
 & 1521300 \\
\cline{2-2}
 & 283976 \\
 & 253550 \\
\cline{2-2}
 & 30426 \\
 & 30426 \\
\cline{2-2}
 & 0.
\end{array}
$$

Hence, 1,805,276 ÷ 5071 = 300 + 50 + 6 = 356. We have included in the display some typical false starts at choosing the partial quotients. The above may be written somewhat more concisely by omitting some "0's" and some digits that are copied repeatedly, but in this particular case the gain in doing so is very slight.

The algorithms that are usually taught for "long division" and "short division" are honed to a fine edge in that almost nothing is written down unless it is absolutely necessary, and the number of steps is minimized. It is a good question as to whether or not children need to be taught these *very* efficient routines. Perhaps procedures which are not quite as efficient are better for the child to learn. The very efficiency of a procedure may disguise what is actually going on. Also, the child may well discover his own short cuts if he is left to do so and is not immediately presented with a paragon of efficiency. And, finally, in this age of desk calculators, slide rules, and electronic computers, maybe it would be best to leave the question of the *most* efficient procedure to the designers of these machines.

In the display above, several subtractions must be carried out. And it is important that our columns of digits be kept straight for these. One device that would assist us in this regard would be to do our work on squared paper and write at most one digit in each square. Rather large squares would be appropriate for younger students. Since "long division" is essentially a mechanical process, we should certainly make use of a mechanical aid such as squared paper if it is helpful.

A flow chart for a fairly efficient version of a division algorithm for decimal numerals is given in Figure 6.33. We shall have $a_i < 10$ for each a_i generated by the process given in the figure. And "$a_0a_1 \ldots a_i$" is the decimal numeral constructed out of the digits "a_0," "a_1," . . . , "a_i." Of course, in actual practice we would not bother to write initial "0's."

Let us see what the machine described in Figure 6.33 would do to evaluate $157 \div 17$. It would first find $k = 2$ and $n = 3$. Note that k is simply the number of digits in the decimal name for x and n is the number of digits in the decimal name for y. The machine computes $j = 3 - 2 = 1$. It sets $i = 0$ and $d_0 = 157$. It then finds the largest a_0 such that $17 \cdot a_0 \cdot 10^1 \leq 157$. Clearly $a_0 = 0$. Then it computes $157 - 17 \cdot a_0 \cdot 10^1 = 157 - 0 = 157 = d_1$. Since $d_1 \geq 17$, it sets $i = 1$ and finds the largest a_1 such that $x \cdot a_1 \cdot 10^{j-1} = 17 \cdot a_1 \cdot 10^0 = 17 \cdot a_1 \leq 157$. Clearly, $a_1 = 9$. It then computes $d_2 = d_1 - xa_1 \cdot 10^{j-1} = 157 - 17 \cdot 9 \cdot 1 = 157 - 153 = 4$. Since $d_2 < 17$ and $d_2 \neq 0$, the process concludes that 17 goes into $157\, a_0a_1 \cdot 10^{j-1} = 09 = 9$ times with a remainder of 4.

Of course, the entire process may be summarized more briefly on paper as

$$\begin{array}{r|l} & \;\;09 \\ \hline 17 & 157 \\ & 153 \\ \cline{2-2} & \;\;\;\;4 \end{array}$$

where the "0" may be omitted.

For the result $112500 \div 375$, the machine would find $k = 3$, $n = 6$, and $j = 3$. Setting $i = 0$ and $d_0 = 112500$, it would find that the largest a_0 such that $x \cdot a_0 \cdot 10^j = 375 \cdot a_0 \cdot 10^3 \leq 112500$ is $a_0 = 0$. Computing $d_1 = 112500 - 0 = 112500$ it would find $d_1 \geq x$. So it sets $i = 1$ and finds that the largest a_1 such that $x \cdot a_1 \cdot 10^{3-1} = 375 \cdot a_1 \cdot 10^2 \leq d_1 = 112500$ is $a_1 = 3$. Computing $d_2 = 112500 - 375 \cdot 3 \cdot 10^2 = 0$, it finds $d_2 < 375$ and indeed $d_2 = 0$. So it concludes that $112500 \div 375 = a_0a_1 \cdot 10^{j-1} = 03 \cdot 10^2 = 300$.

The process of Figure 6.33 is essentially the standard process of long division. The a_0, a_1, . . . are successive digits in our quotient. The d_0, d_1, . . . are the partial remainders.

However, when we carry out a "long division" on paper we usually rely on the position on the paper to indicate the power of ten that is actually involved in the multiplication. And it is the usual practice not to bother to copy digits that we know will not be affected at a given stage because they are bound to have "0's" below them in a given subtraction.

Feed the problem $537{,}696 \div 126$, the machine would compute successively, $k = 3$, $n = 6$, $j = 3$, $i = 0$, $d_0 = 537{,}696$, $a_0 = 4$, $d_1 = 33696$, $i = 1$, $a_1 = 2$, $d_2 = 8496$, $i = 2$, $a_2 = 6$, $d_3 = 936$, $i = 3$, $a_3 = 7$, and $d_4 = 54$. On paper we would write the process as indicated on page 220:

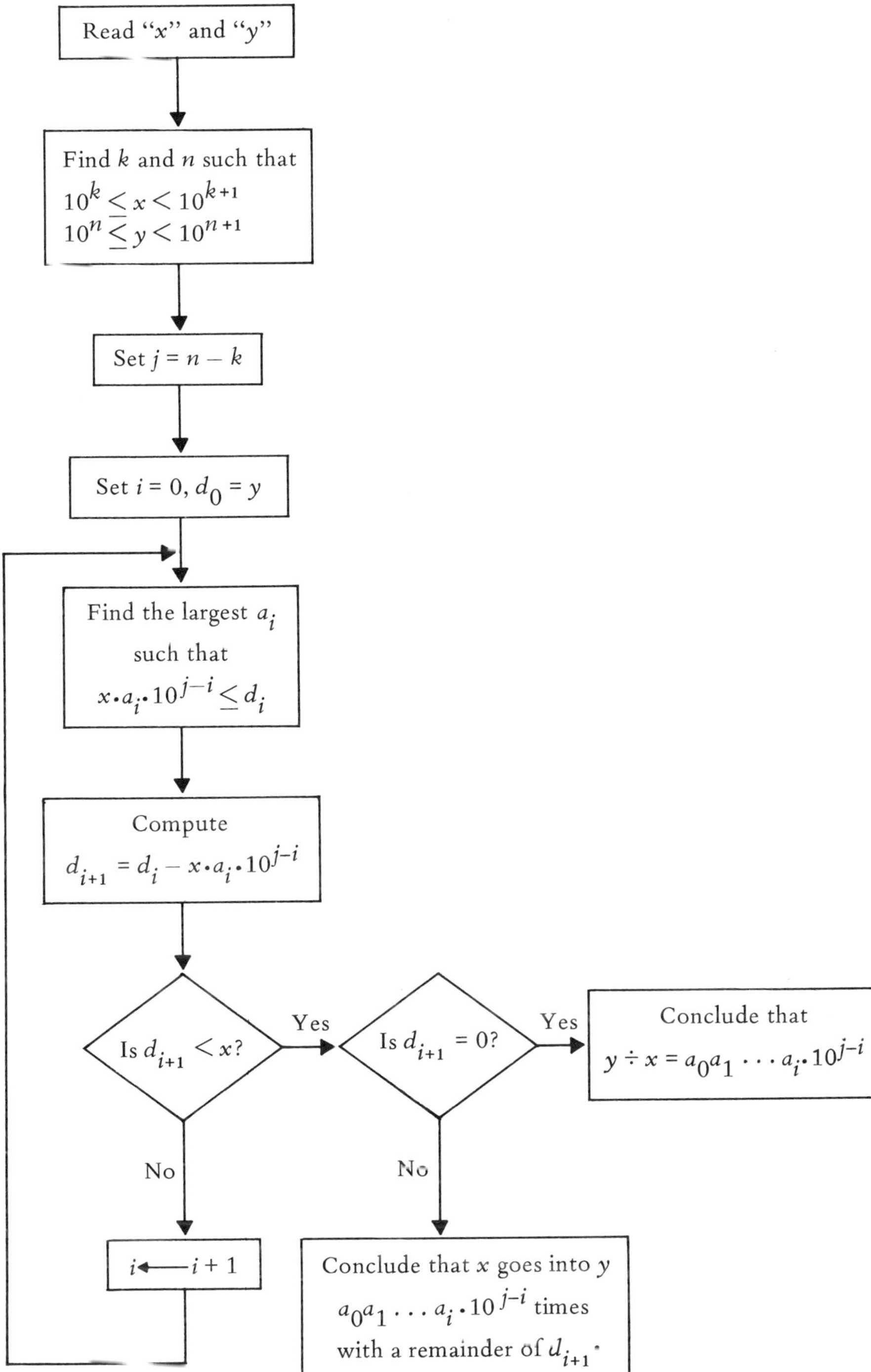

FIGURE 6.33 A flow chart for division. The process yields $y \div x$ when it is defined. It is assumed that $y \geq x$.

```
            7
           60
          200
         4000
     --------
126 | 537696
      504000
      ------
       33696
       25200
       -----
        8496
        7560
        ----
         936
         882
         ---
          54
```

or, making the customary deletions, as:

```
         4267
     --------
126 | 537696
      504
      ---
       336
       252
       ---
        849
        756
        ---
         936
         883
         ---
          54.
```

In any event, we conclude that 126 goes into 537,696 4267 times with a remainder of 54.

As a reminder that division was defined to be the inverse operation for multiplication, we shall look at how a division problem might be written up in a different way. Suppose we wish to find $25016 \div 472$. We might think of the problem as a search for the proper thing to put in the blank in order for the multiplication

```
  472
×
-----
25016
```

to be correct. If we leave room for the intermediate results in the multiplication process we might fill in the blank by the process indicated below:

$$\begin{array}{l}
472 \\
\underline{\times \cancel{100} + 50 + 3} \\
1416 \\
23600 \\
\underline{\cancel{47200}} \\
25016.
\end{array}$$

Here we have tried a partial quotient of 100 and found it too large. Then we used a partial quotient of 50 and subtracted—working from the bottom up. Then we found that the additional partial quotient of 3 gave us just the partial remainder already obtained.

Note that this layout is almost a simple physical rearrangement of the numerals in the more usual display:

$$\begin{array}{r|r}
 & 53 \\
\cline{2-2}
472 & 25016 \\
 & \underline{23600} \\
 & 1416 \\
 & \underline{1416} \\
 & 0.
\end{array}$$

The division algorithm we have developed may easily be modified to work with numerals written to other bases. The situation with binary numerals is especially simple as the following example demonstrates:

$$\begin{array}{r|r}
 & 100 \\
 & 10000 \\
\cline{2-2}
1001 & 10111011 \\
 & \underline{10010000} \\
 & 101011 \\
 & \underline{100100} \\
 & 111.
\end{array}$$

Here we have suppressed the subscript 2. We see that 1001_2 goes into 10111011_2 10100_2 times with a remainder of 111_2.

PROBLEMS

1. Use the method of finding $y \div x$ by finding $y - x$, $y - 2x$, $y - 3x$, . . . to evaluate each of the following: (a) $40 \div 10$, (b) $27 \div 9$, (c) $90 \div 15$, (d) $3760 \div 752$, (e) $50{,}000 \div 25{,}000$, (f) $79{,}536 \div 39{,}768$.

2. Use the general procedure described on page 214 to find each of the following. Display your work in a form similar to the example on page 217. (a) $861 \div 7$, (b) $50{,}000 \div 5$, (c) $52{,}290 \div 498$, (d) $52{,}290 \div 105$, (e) $3{,}955{,}056 \div 50064$, (f) $999{,}999 \div 111$, (g) $584{,}696 \div 764$.

3. Rewrite your work for each part of Problem 2 so that it appears in the condensed standard form.

4. Explain in detail the action of the machine of Figure 6.33 when it is faced with evaluating (a) $1407 \div 3$, (b) $29{,}509 \div 7503$.

5. If we carry through some form of algorithm and conclude that x goes into y q times with a remainder of r, how may we check our work using algorithms for multiplication and addition? Illustrate your method by checking your results in Problem 2.

6. Is it possible to distribute 3410 pieces of paper into 95 boxes so that all the boxes have the same number of pieces in them?

7. If a_1, a_2, . . . , and a_n are n numbers then the number $(a_1 + a_2 + \ldots + a_n) \div n$ is said to be the **arithmetic mean** (or sometimes simply the **mean,** or the **average**) of the numbers. Find the arithmetic mean of each of the following lists of numbers. (a) 3, 4, 5, 6, 7, 8, 9; (b) 367, 496, 637; (c) 46, 460, 4600, 16262.

8. What is the smallest number which is greater than 401 and is a divisor of 16,452?

9. Find the decimal name for $(((76{,}245 \div 39) \div 17) \div 23)$.

10. Find all the divisors of (a) 36, (b) 144.

11. Find x if we know that $((343 \div x) \div x) \div x = 1$.

12. Check the division example in the base 2 as given in this section by converting all the numbers to decimal notation, carrying out a division algorithm, and then converting back to binary notation.

13. Evaluate each of the following. Do not convert to decimals. (a) $11101100_2 \div 100_2$, (b) $11101111_2 \div 101_2$, (c) $220011_3 \div 102_3$, (d) $10_{12} \div 2_{12}$, (e) $100_{12} \div 3_{12}$, (f) $1000_{12} \div 4_{12}$, (g) $10_{12} \div 6_{12}$, (h) $100_{12} \div 8_{12}$, (i) $100_{12} \div t_{12}$.

14. Check each of your results in Problem 13 (a) by carrying out a multiplication (and, where necessary, an addition) in the given base, (b) by converting to decimals, dividing, and then converting back.

15. Give an argument to show that

(a) A natural number n is divisible by 2 if and only if the decimal name for n ends with "0," "2," "4," "6," or "8."

(b) A natural number n is divisible by 3 if and only if the sum of the decimal digits in the numeral for n is divisible by 3.

(c) A natural number n is divisible by 5 if and only if the decimal name for n ends with "0" or "5."

(d) A natural number n is divisible by 9 if and only if the sum of the decimal digits in the numeral for n is divisible by 9.

16. Discuss divisibility criteria similar to those in Problem 15 when other numeration systems are used.

6.21 Prime numbers

Looking at a multiplication table for N (see page 89), we see† that every number except 1 appears in the body of the table at least twice. For if n is a natural number and $n \neq 1$, then n will appear in the body of the table as the product of 1 and n, and also as the product of n and 1. Certain numbers appear *exactly twice* in the body of the table. These numbers are called **prime numbers.** Putting this another way, we say that n is a **prime number** if $n \neq 1$ and the only divisors of n are 1 and n.

The first sixteen prime numbers are 2, 3, 5, 7, 11, 13, 17, 19, 23, 29, 31, 37, 41, 43, 47, and 53.

A number which is not prime and is different from 1 and 0 is said to be **composite.**

The only even prime number is 2. 9 is not a prime since 3 is a divisor of 9.

Is 113 a prime number? To find out the answer, we may simply check by the division algorithm to see if 2, 3, 5, 7, or 11 is a divisor of 113. The reason that we need to consider only 2, 3, 5, 7, and 11 as possible divisors is as follows. We need not check for any composite divisors, because if 113 has a composite divisor d, then it certainly has a prime divisor—namely any divisor of d. We need not check any numbers larger than 11 since $11^2 = 121$ and $121 > 113$. If there is a divisor d of 113 which is larger than 11, then there certainly must be a divisor smaller than 11. For if $113 \div d = k$ and $k \geq 11$ and $d > 11$, then we must have $dk > 121$ which is impossible since $dk = 113$. Since we work up to 11 by going through all the smaller prime numbers first, we will find a smaller divisor if it exists. Since direct calculations show that no one of the numbers 2, 3, 5, 7, or 11 is a divisor of 113, we see that 113 is prime.

Is 1137 a prime number? When we start to try 2, 3, 5, 7, 11, 13, 17, 19, 23, 29, 31, and 37 as possible divisors, we need not go far. We find that $1137 \div 3 = 379$. And so 1137 is not prime.

These two examples suggest one method for finding out if a given number is prime or not.

Another process for finding prime numbers is the **sieve of Eratosthenes** which is illustrated in Figure 6.34. To construct such a table we first write names for all the natural numbers from 1 to, say, 150. We plan to circle all the names of nonprimes. We circle "1" because 1 is not a prime. 2 is a prime. But every multiple of 2 except for 2 itself is composite, so we circle all these multiples of 2. 3 is a prime, but we should circle all multiples of 3 except 3. The next uncircled numeral is "5." 5 is prime, but we circle all the other multiples of 5. We continue in this fashion until we have circled all the multiples of 11 (but not 11 itself). We may stop at 11 since the next prime after 11 is 13 and $13^2 > 150$. And if some composite number does not exceed 150 it must have a prime

† Note that the table on page 89 does not include the natural number 0. But if a row and a column were put in for 0, they would be filled with "0's."

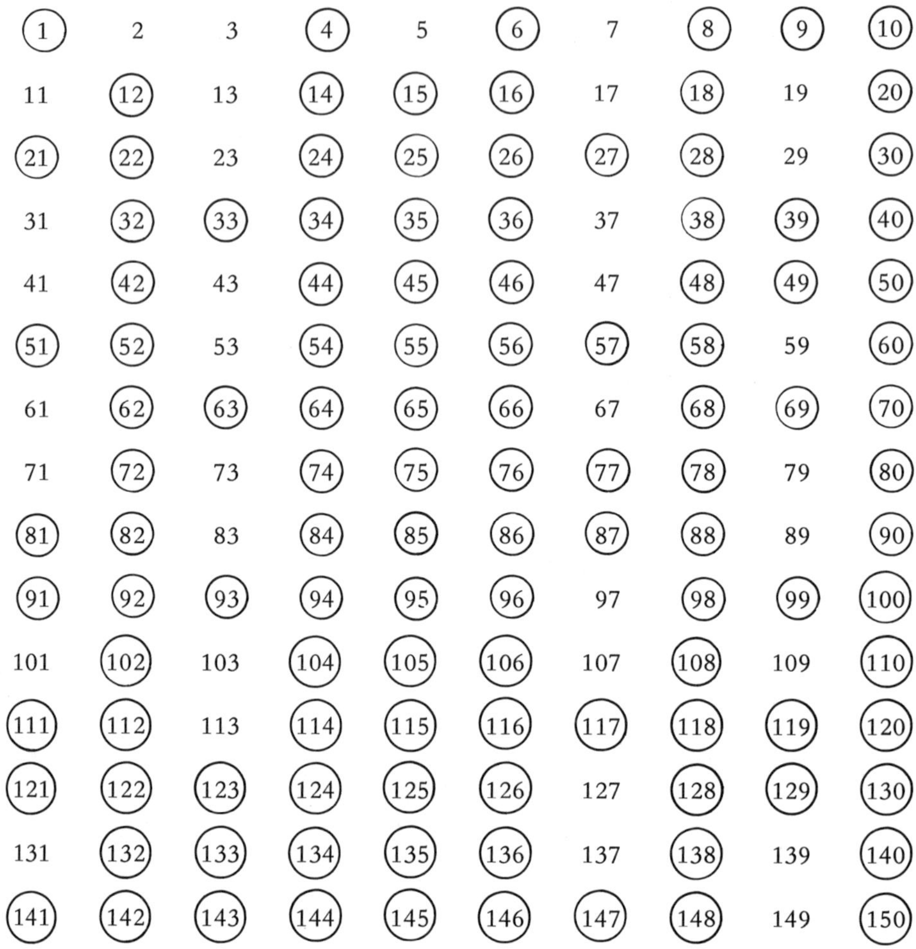

FIGURE 6.34

factor less than 13, and already would have been circled. Thus, all the uncircled numerals in the figure are names of primes.

While we found that 1137 is not prime, we may write it as a product of primes. We have $1137 = 3 \cdot 379$ and 379 is prime.

144 is not prime. But we may express 144 as a product of primes. We have $144 = 2 \cdot 72 = 2 \cdot 2 \cdot 36 = 2 \cdot 2 \cdot 2 \cdot 18 = 2 \cdot 2 \cdot 2 \cdot 2 \cdot 9 = 2 \cdot 2 \cdot 2 \cdot 2 \cdot 3 \cdot 3 = 2^4 \cdot 3^2$.

In fact, any natural number greater than 1 can be expressed as a product of primes. And, except possibly for the order in which these factors are written, this product is uniquely determined. In other words, when n is in N and $n \neq 0$

and $n \neq 1$, then there is one and (except for the order of the factors) only one way to express n as a product of primes.†

This fact is of such great importance that it is called **the fundamental theorem of arithmetic.**

Rather than give a proof of this fact here, let us get some insight into the theorem by looking at a system where it doesn't hold. We shall consider the operation $(\cdot_6)$ defined by the following table:

$(\cdot_6)$	0	1	2	3	4	5
0	0	0	0	0	0	0
1	0	1	2	3	4	5
2	0	2	4	0	2	4
3	0	3	0	3	0	3
4	0	4	2	0	4	2
5	0	5	4	3	2	1

This operation is clearly commutative and it may be shown that it is also associative. 1 is the identity element for $(\cdot_6)$. Let us call $x(\cdot_6)y$ the *product* of x and y. An element different from 1 in $\{0,1,2,3,4,5\}$ would be called a **prime** in the system if it appears exactly twice in the body of the operation table. Then 5 is the only prime in this system.

Can we write 4 as a product of primes in this system? We have $4 = 5(\cdot_6)2$ where 5 is prime. And since 2 is not prime we may factor this further and get $4 = 5(\cdot_6)2 = 5(\cdot_6)(5(\cdot_6)4)$. And 4 is not prime so we may factor this further and get $4 = 5(\cdot_6)(5(\cdot_6)(5(\cdot_6)2))$. And so forth. It is clear that we can *never* express 4 as a product of primes in this system.

It is also interesting to note that, in this system, we find that $1 = 5(\cdot_6)5$ so that 1 can be written as a product of primes. Of course, we also have that that $1 = 1(\cdot_6)1$.

PROBLEMS

1. Check our list of the first sixteen prime numbers by noting that each of them appears exactly twice or not at all in the body of the partial multiplication table on page 89.

2. Which of the following are prime? (a) 1139, (b) 111,111, (c) 11,111, (d) $10^{14} - 1$, (e) 12345, (f) 3047, (g) 10201.

3. Find the next three larger primes after 149.

4. Write each of the following as a product of primes. (a) 125, (b) 1728, (c) 4900, (d) 3759, (e) 3737, (f) 2400.

† Of course, we consider a prime standing by itself to be a product of primes.

5. Show that if two natural numbers have a common divisor different from 1 (i.e., there is a number different from 1 that is a divisor of both of them), then they must have a *prime* common divisor.

6. Let the function p be defined on the nonzero natural numbers by $p(n) =$ the number of distinct divisors of n. Thus, $p(1) = 1$, $p(2) = 2$, $p(3) = 2$, $p(4) = 3$, $p(5) = 2$, and $p(6) = 4$. (a) Find $p(n)$ for $n = 7,8,9, \ldots ,25$. (b) Show that n is prime if and only if $p(n) = 2$. (c) Does $p(n) =$ the number of times that n appears in the body of the multiplication table for N?

7. 6 is said to be a **perfect** number since it is the sum of all the numbers which are divisors of 6 and less than 6. Are there any other perfect numbers?

8. Use the fundamental theorem of arithmetic to prove that a natural number is even if and only if its square is even.

9. Two prime numbers are said to be **twin primes** if they differ by 2. 3 and 5 are twin primes. Find all the pairs of twin primes less than 150. (It is a famous unsolved problem as to whether or not there are infinitely many pairs of twin primes.)

10. Show that each number in the set $\{4,6,8,10, \ldots ,30\}$ can be written as the sum of two primes. (It is a famous unsolved problem as to whether or not every even number greater than 4 can be expressed as the sum of two odd primes. The guess that every even number greater than 4 can be so expressed is called **Goldbach's conjecture.**)

11. In analogy with the prime numbers for multiplication, let us define what we shall call **primary** numbers for addition in N. A natural number n is said to be **primary** if it appears exactly twice in the body of the table for addition on N. Which numbers (if any) are **primary** in this sense? Can every natural number except 0 be written as a *sum* of primary numbers? Is this sum unique?

12. What are the "primes" for the system (a) $(\{0,1,2\},\odot)$ given on page 90, (b) $(\{a,b,c,d\}, L)$ given on page 101, (c) $(\{\phi,\{1\}, \{2\}, \{1,2\}\}, \cup)$?

13. Find a necessary and sufficient condition in terms of the powers that appear in the prime factorization of n in order that n be a perfect square.

14. We know that 17 is a prime number. Also $17 = 10001_2$. Is 10001_2 a prime "in the base 2"? Explain why this is a silly question.

*6.22 Counting mappings

The notion of a one-to-one correspondence is basic to our understanding of natural numbers. And the notion of a function or mapping is basic to our understanding of one-to-one correspondences. We may now use our knowledge of the natural numbers to answer some very natural questions about mappings. It will develop that the answers to these questions will be very useful.

The primary question we wish to consider is this: If we are given two sets, say A and B, how many different mappings are there from A into B?

First taking A to be as simple as possible we ask: How many mappings are there from ϕ into B? The answer evidently is 1. The empty mapping is the only possible mapping whose domain is ϕ.

Suppose we take A so that $\#(A) = 1$. How many mappings from A into B are there? In order to give a mapping from A into B we must tell what object in B is the image of the single element in A. Evidently then there are $\#(B)$ different mappings from A into B when $\#(A) = 1$. Note that this answer is correct even if B is empty. If there are no elements in B then we cannot construct a mapping from A into B when $\#(A) = 1$.

To illustrate the above take $A = \{a\}$ and $B = \{1,2,3\}$. Then there are 3 mappings from A into B. We could call these mappings f, g, and h and define them by $f(a) = 1$, $g(a) = 2$, and $h(a) = 3$.

If $\#(A)$ is greater than one the situation becomes more complicated. Taking $A = \{a,b\}$ and $B = \{1,2,3\}$, we may diagram all possible mappings from A into B as in Figure 6.35. We see that there are 9 mappings from $\{a,b\}$ into $\{1,2,3\}$.

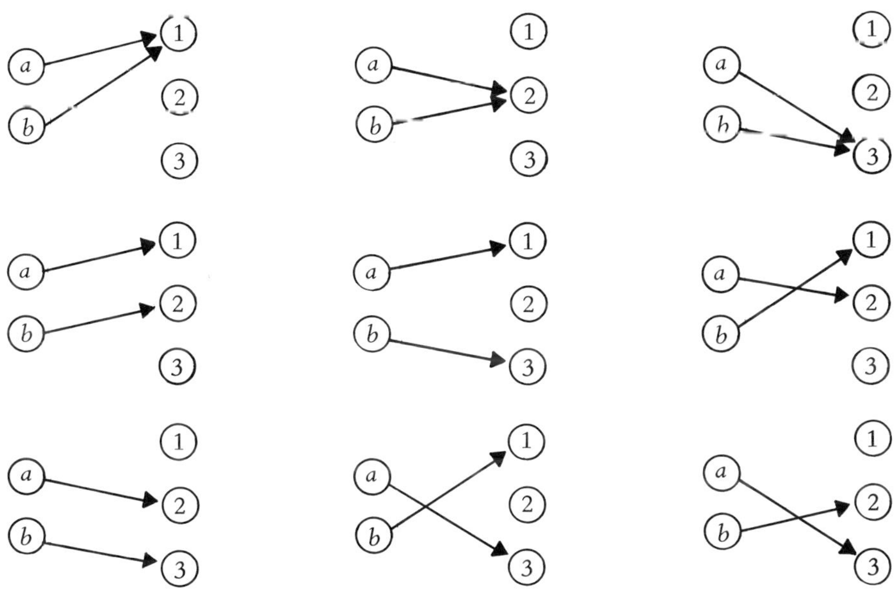

FIGURE 6.35

The possible mappings from $\{a,b,c\}$ into $\{1,2\}$ are shown in Figure 6.36. There are 8 of these.

We observe that the number of mappings of ϕ into B is $1 = (\#(B))^0$. If $\#(A) = 1$ then the number of mappings from A into B is $\#(B) = (\#(B))^1$. The number of mappings from $\{a,b\}$ into $\{1,2,3\}$ is $9 = 3^2$. The number of mappings from $\{a,b,c\}$ into $\{1,2\}$ is $8 = 2^3$. These results suggest the conjecture:

The number of mappings from A into B is $(\#(B))^{\#(A)}$.

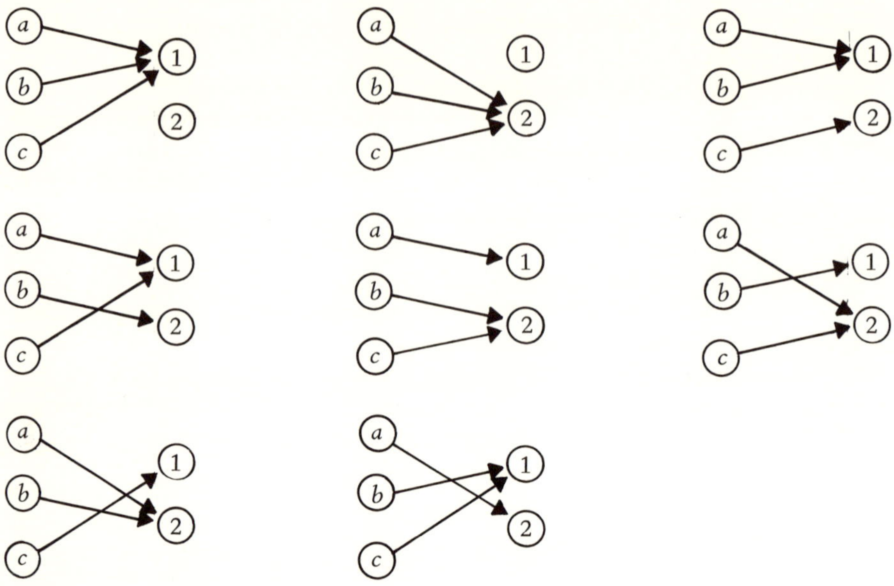

FIGURE 6.36

And this is indeed correct. When A and B are finite sets a general proof may be given by induction. (See the appendix to this section.)

A standard notation that ties in nicely with the above result is the following. When A and B are sets we write "B^A" for the collection of all mappings from A into B. Thus, Figure 6.35 is a display of the members of $\{1,2,3\}^{\{a,b\}}$ and Figure 6.36 is a display of the members of $\{1,2\}^{\{a,b,c\}}$.

Using this terminology we may state the main result above as:

$$\#(B^A) = (\#(B))^{\#(A)}.$$

Suppose we now change our basic question a bit and ask: How many mappings are there from A **onto** B?

If $A = \phi$ and $B \neq \phi$, then there are *no* mappings from A **onto** B.

If $\#(A) = 1$ and $\#(B) \neq 1$ then there are again **no** mappings from A onto B. If $\#(A) = 1$ and $\#(B) = 1$ then there is exactly one mapping from A onto B.

We see from Figure 6.35 that there are no mappings of $\{a,b\}$ onto $\{1,2,3\}$.

We see from Figure 6.36 that there are 6 mappings of $\{a,b,c\}$ onto $\{1,2\}$.

Evidently, there are no mappings from A onto B whenever $\#(A) < \#(B)$.

When $\#(A) > \#(B)$ there *are* some mappings from A onto B but the determination of the number of them is somewhat complicated.

When $\#(A) = \#(B)$, then the number of mappings of A onto B is fairly easy to find and this case is also of great practical value.

In Figure 6.37 we have diagrammed all mappings of $\{1,2, \ldots, n\}$ **onto** $\{1,2, \ldots, n\}$ for $n = 1,2$, and 3. All that we have shown for the case $n = 4$

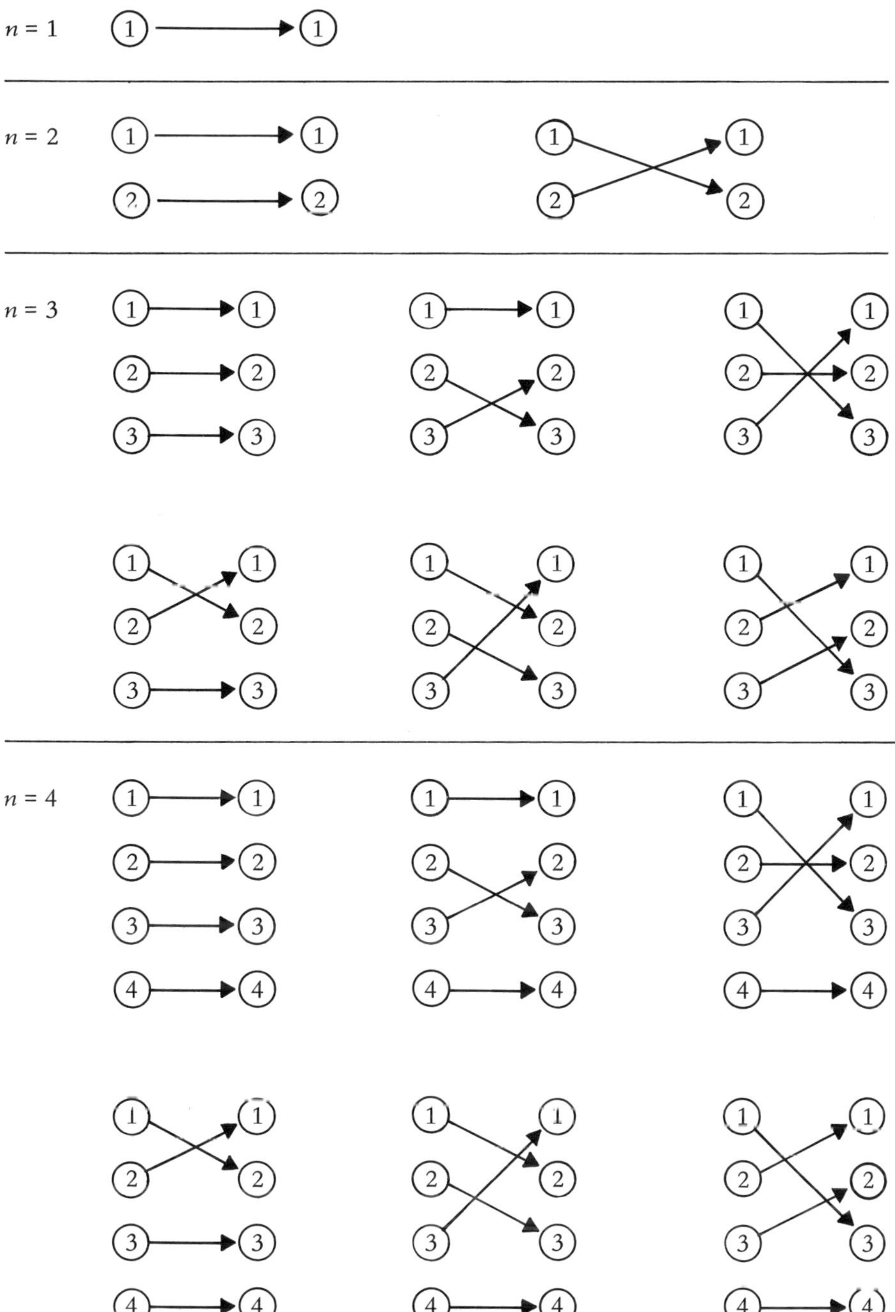

FIGURE 6.37

are the mappings in which 4 is mapped on 4. If, in the case $n = 4$, we were to draw in the mappings where 4 is mapped on 3 how many would there be? There would evidently be 6 since when we have decided that 4 maps onto 3 we must then map $\{1,2,3\}$ onto $\{1,2,4\}$ and we see from the case $n = 3$ that there are 6 ways to do this.

There are also, in the case $n = 4$, 6 mappings we did not draw where 4 maps on 2, and also 6 where 4 maps on 1. We thus see that Figure 6.37 shows only 6 of the total of $6 + 6 + 6 + 6 = 24$ mappings for the case $n = 4$.

How many mappings of $\{1,2,3,4,5\}$ onto $\{1,2,3,4,5\}$ are there? There are evidently 24 which map 5 onto 5, 24 which map 5 onto 4, . . . , and 24 which map 5 onto 1. Thus, there are $5 \cdot 24 = 120$ such mappings.

What we have found so far is that the number of mappings from $\{1\}$ onto $\{1\}$ is 1. From $\{1,2\}$ onto $\{1,2\}$ it is $2 = 1 \cdot 2$. From $\{1,2,3\}$ onto $\{1,2,3\}$ it is $6 = 1 \cdot 2 \cdot 3$. From $\{1,2,3,4\}$ onto $\{1,2,3,4\}$ it is $24 = 1 \cdot 2 \cdot 3 \cdot 4$. And from $\{1,2,3,4,5\}$ onto $\{1,2,3,4,5\}$ it is $120 = 1 \cdot 2 \cdot 3 \cdot 4 \cdot 5$. We are led to the conjecture that:

> The number of mappings from $\{1,2, \ldots, n\}$ *onto* $\{1,2, \ldots, n\}$ is $1 \cdot 2 \cdot 3 \cdot \ldots \cdot n$.

This correct conclusion may be proved by an induction which follows the lines of our argument for the case of $\{1,2,3,4,5\}$ above.

A function that is helpful to us in such cases as our question about onto mappings is the **factorial function** defined on N by:

$$0! = 1, \ (k + 1)! = (k + 1)(k!).$$

We read "$k!$" as "k factorial." The conventions of writing the function symbol "!" on the right in this instance and of omitting the parentheses that would be used in ordinary functional notation are standard ones. The above inductive definition allows us to find $n!$ for any natural number n. We have $0! = 1$, $1! = 1 \cdot 0! = 1 \cdot 1 = 1$, $2! = 2 \cdot 1! = 2$, $3! = 6$, $4! = 24$, $5! = 120$, $6! = 720$, $7! = 5040$, $8! = 40320$, ...

Using this terminology we may state our conclusion about the number of *onto* mappings as:

> There are $n!$ mappings of $\{1,2, \ldots, n\}$ onto itself.

One way to display a mapping of $\{1,2,3,4\}$ onto itself is by means of a table. For example, one such mapping f is given by

x	1	2	3	4
$f(x)$	4	3	2	1

But if we realize that the domain of the function is $\{1,2,3,4\}$ and that we shall take the elements in this set in the natural order, then we need only give the four entries in the second line in this table to identify the function. That is, we may use simply (4,3,2,1) to indicate the mapping we want. This same device can be used to identify any mapping of $\{1,2, \ldots, n\}$ onto itself.

As a consequence, we may think of any mapping of $\{1,2, \ldots, n\}$ onto itself as an ordered arrangement of the members of this set. When such mappings are considered in this fashion, they are called **permutations.** Thus, we have another way to state the result discovered above:

There are $n!$ permutations of any group of n objects.

The ideas we have introduced so far in this section turn out to be very much involved in many counting problems that, at first glance, have nothing to do with mappings. We shall now consider some illustrations of this fact.

How many different strings of symbols are there that contain 3 symbols from the list "0," "1," "2," . . . , "9"? To construct a string of symbols which satisfies the given condition we must decide which will be the first symbol, which will be the second symbol, and which will be the third symbol. But we notice that constructing such a string amounts to defining a mapping from $\{1,2,3\}$ *into* {"0," "1," "2," . . . , "9"}. By the first general result in this section there are $10^3 = 1000$ such mappings (strings). Of course we may think of these strings as the 1000 numerals "000," "001," "002," . . . , "998," "999."

Three different assignments are to be handed out to three different students so that each student gets an assignment. In how many ways may this be done? If we number the assignments 1, 2, and 3 and assign the numbers 1, 2, and 3 to the students, we see that the problem may be thought of as: How many mappings of $\{1,2,3\}$ on itself are there? The answer is $3! = 6$.

How many ways can the 52 cards in an ordinary deck of cards be arranged? We may think of the cards as numbered from 1 to 52 (by rank, say). To put the cards in a certain order we may construct 52 boxes numbered from 1 to 52 and distribute the cards one to a box. But each time we do this we determine a mapping of $\{1,2, \ldots, 52\}$ **onto** itself. Hence, there are 52! different arrangements (mappings) possible. It can be shown that 52! is roughly $8 \cdot 10^{67}$.

Definition 6.22.1. $B^A =$ the collection of all mappings from A into B.

Theorem 6.22.1. If $\#(A) = n$ and $\#(B) = m$ where n and m are natural numbers and $m \neq 0$, then $\#(B^A) = m^n$.

PROOF: Let m be a nonzero natural number and suppose $B = \{1,2, \ldots, m\}$. Let $S = \{x \mid x$ is a natural number such that $\#(A) = x$ implies $\#(B^A) = m^x\}$. We see that $0 \in S$ since the empty mapping is the only member of B^ϕ and $m^0 = 1$. Suppose that $k \in S$. Let $\#(A) = k + 1$. There is no loss in generality if we suppose $A = \{1,2, \ldots, k, k+1\}$. Let $C = \{1,2, \ldots, k\}$. Since $k \in S$, $\#(B^C) = m^k$. For each f in B^C define m functions $f_1, f_2, \ldots, f_m$ on A by taking $f_i(j) = f(j)$ if $j \neq k + 1$ and $f_i(k + 1) = i$. Clearly the collection of all functions that can be constructed in this way is B^A. But each f in B^C leads to m such functions so $\#(B^A) = m\#(B^C) = m \cdot m^k = m^{k+1}$. Therefore, $S = N$ and the proof is complete.

Definition 6.22.2. $0! = 1$, $(n + 1)! = (n + 1) \cdot n!$. We read "$n$!" as "$n$ factorial."

Theorem 6.22.2. There are $n!$ mappings of $\{1,2, \ldots, n\}$ **onto** itself.

PROOF: Let $S = \{x \mid \text{there are } x! \text{ mappings of } \{1,2, \ldots ,x\} \text{ onto itself}\}$. Clearly $0 \in S$ since $0! = 1$ and there is exactly one mapping (the empty mapping) of ϕ onto ϕ. Suppose $k \in S$. Let T be the collection of all mappings of $\{1,2, \ldots, k\}$ onto itself. For each f in T define $k + 1$ mappings of $\{1,2, \ldots ,k,k + 1\}$ onto itself as follows. Let $f_0(j) = f(j)$ for $j \neq k + 1$ and let $f_0(k + 1) = k + 1$. Let $f_i(j) = f(j)$ if $j \neq i$ and $j \neq k + 1$. Let $f_i(i) = k + 1$ and $f_i(k + 1) = i$. Here, $i = 1,2, \ldots ,k$. The collection of all the functions that can be constructed in this fashion is the set of all mappings of $\{1,2, \ldots ,k + 1\}$ onto itself. On the other hand, each f in T leads to $k + 1$ such functions and since $k \in S$ we have $\#(T) = k!$ Hence, the total number of mappings of $\{1,2, \ldots ,k + 1\}$ onto itself is $(k + 1) \cdot k! = (k + 1)!$. This implies $S = N$ and completes the proof.

PROBLEMS

1. Diagram all possible mappings of $\{a,b\}$ into $\{1,2,3,4\}$. How many of these are mappings *onto* $\{1,2,3,4\}$?
2. Diagram all possible mappings of $\{1,2,3,4\}$ into $\{a,b\}$. How many of these are mappings *onto* $\{a,b\}$?
3. Calculate 9!, 10!, and 11!.
4. Draw a flow chart for the calculation of $n!$.
5. In how many different orders is it possible to arrange the symbols *A*, *B*, *C*, and *D*?
6. Five people are to be assigned to 5 different classes with one student going to each class. In how many ways may the assignments be made?
7. In Problem 6 assume that there are no restrictions as to how many students may be assigned to each class. Then, in how many different ways may the students be assigned to the classes?
8. Each of the 10 students in a small class is to receive a grade of either *A*, *B*, *C*, *D*, or *F*. How many different possible assignments of grades are there?
9. A teacher wants to plan the order in which to present 9 different, independent topics to his class. How many different orders does he have to choose from? Should the teacher list all possible orders and then choose one?
10. How many different seating arrangements are possible in a classroom with 52 students and 52 chairs? (Assume the chairs are in fixed positions.) About how long would it take to list all of these if it takes a minute to write a description for any given arrangement?

11. Which is larger, $n!$ or n^n? Explain how you know.

12. In a class with k students, we plan to find out the month and day of birth for each student. How many different conceivable outcomes are there for our data?

13. Diagram all possible one-to-one correspondences between $\{1,2, \ldots ,k\}$ and a subset of $\{1,2, \ldots ,n\}$ for (a) $k = 1, n = 3$, (b) $k = 2, n = 3$, (c) $k = 2, n = 4$, (d) $k = 3, n = 2$, (e) $k = 2, n = 5$.

14. Referring to your results in Problem 13, partially verify that there are $n(n - 1) \cdots (n - k + 1) = n!/(n - k)!$ different one-to-one correspondences between $\{1,2, \ldots ,k\}$ and a subset of $\{1,2, \ldots ,n\}$.

15. In Problem 12, how many possible outcomes are there in which no two students in the class have the same birthday? [HINT: Make use of the result in Problem 14.]

16. Suppose p is a prime number. The number $p!$ is divisible by all the prime numbers up to and including p. Is $p! + 1$ divisible by p or any prime less than p? Use your conclusion to prove that there are infinitely many prime numbers. [HINT: Suppose there is only a finite number of primes. Then there is a last prime. Call it p. What can be said about the number $p! + 1$?]

17. Define the function g as follows: $g(1) = 1!$, $g(2) = (2!)!$, $g(3) = ((3!)!)!, \ldots ,g(n) = (\ldots((n!)!)\ldots)!$ where there are n occurrences of "!" in the expression defining $g(n)$. Is $g(4)$ a large number?

18. In stud poker, five cards from an ordinary 52 card deck are dealt to each player. The order in which the cards are dealt to a player makes a difference as to how the hands will be played. How many different stud poker hands are there? (See Problem 14.)

19. A red, a green, and a white die are all tossed together. How many different outcomes are possible? (A die is a cubical solid with six numbered faces. The "result" on a particular die is the number (named) on the side that is uppermost when the die comes to rest.) How many of these outcomes give a total of all three results which is equal to (a) 3, (b) 4, (c) 5, (d) 6?

*6.23 Counting subsets

In the game of bridge the 52 cards in the deck are dealt out evenly to 4 players. How many different bridge hands are there? Since the order in which the cards are dealt does not matter in bridge, we may reword this question as: How many different 13 element subsets of a 52 element set are there? We shall now develop the machinery that will enable us to answer this question. It helps if we first consider the general situation.

Let us use "$\binom{n}{k}$" to stand for the number of different k element subsets in a set of n elements. We see that $\binom{0}{0} = 1$ since there is one subset of an empty set—the empty set itself. In fact, $\binom{n}{0} = 1$ for all n since the empty subset is always the unique subset with 0 elements. $\binom{1}{1} = 1$ since the only 1 element subset of $\{a\}$ is $\{a\}$ itself. In fact, $\binom{n}{n} = 1$ for all n since the only n element subset of an n element set is the set itself. To investigate other cases, let us list all possible subsets for $\{a\}$, $\{a,b\}$, $\{a,b,c\}$, $\{a,b,c,d\}$, and $\{a,b,c,d,e\}$. A compact way to do so is by means of a table which uses the elements of the set as column headings and has a row for each subset. In a given row we shall put a "1" under "x" if x is a member of the subset corresponding to that row and a "0" if not. This has been done in Figure 6.38. For example, the sixth row under "$a - b - c$" indicates the subset $\{a,c\}$, and the first row under "$a - b - c - d$" indicates the empty subset ϕ.

By simply counting subsets as indicated in this table we may determine several values of $\binom{n}{k}$. We summarize these results in the following table which also includes the case $n = 0$.

n \ k	0	1	2	3	4	5
0	1	0	0	0	0	0
1	1	1	0	0	0	0
2	1	2	1	0	0	0
3	1	3	3	1	0	0
4	1	4	6	4	1	0
5	1	5	10	10	5	1

(Note that $\binom{n}{k}$ may be thought of as the result of an operation to be performed on n and k, but it is not often thought of this way.)

We might notice from this table that we have such results as:

$$\binom{2}{1} = \binom{1}{1} + \binom{1}{0},$$

$$\binom{3}{1} = \binom{2}{1} + \binom{2}{0},$$

$$\binom{3}{2} = \binom{2}{2} + \binom{2}{1},$$

$$\binom{4}{2} = \binom{3}{2} + \binom{3}{1},$$

$$\binom{5}{3} = \binom{4}{3} + \binom{4}{2},$$

$$\binom{5}{4} = \binom{4}{4} + \binom{4}{3}.$$

a
0
1

a	*b*
0	0
1	0
0	1
1	1

a	*b*	*c*
0	0	0
1	0	0
0	1	0
0	0	1
1	1	0
1	0	1
0	1	1
1	1	1

a	*b*	*c*	*d*
0	0	0	0
1	0	0	0
0	1	0	0
0	0	1	0
0	0	0	1
1	1	0	0
1	0	1	0
1	0	0	1
0	1	1	0
0	1	0	1
0	0	1	1
1	1	1	0
1	1	0	1
1	0	1	1
0	1	1	1
1	1	1	1

a	*b*	*c*	*d*	*e*
0	0	0	0	0
1	0	0	0	0
0	1	0	0	0
0	0	1	0	0
0	0	0	1	0
0	0	0	0	1
1	1	0	0	0
1	0	1	0	0
1	0	0	1	0
1	0	0	0	1
0	1	1	0	0
0	1	0	1	0
0	1	0	0	1
0	0	1	1	0
0	0	1	0	1
0	0	0	1	1
1	1	1	0	0
1	1	0	1	0
1	1	0	0	1
1	0	1	1	0
1	0	1	0	1
1	0	0	1	1
0	1	1	1	0
0	1	1	0	1
0	1	0	1	1
0	0	1	1	1
1	1	1	1	0
1	1	1	0	1
1	1	0	1	1
1	0	1	1	1
0	1	1	1	1
1	1	1	1	1

FIGURE 6.38

This would lead us to conjecture that, in general:

$$\binom{n+1}{k+1} = \binom{n}{k+1} + \binom{n}{k}.$$

And this is true. A proof will be given in the appendix to this section.

If we leave out the zeroes in the table on page 234 for the values of $\binom{n}{k}$, we obtain the array of numbers called **Pascal's triangle.** The basic rule of formation of the triangle is the equality displayed above. With this rule it is easy to determine subsequent rows in the triangle. For instance, the next row (for $n = 6$) would be 1, 6, 15, 20, 15, 6, 1.

In a sense, then, we have solved the question we asked at the beginning of this section. We know that the number of different bridge hands is $\binom{52}{13}$. And now all we need to do is build up Pascal's triangle until we get part of the 52nd row. Did we say that's *all* we need to do? It sounds like a week's work—and such a stupendous task that we would be almost certain to make some mistakes. There must be some way to get the value of $\binom{52}{13}$ without going through so much labor. There is.

It can be shown that whenever $n \geq k$, then

$$\binom{n}{k} = n! \div (k!\,(n-k)!).$$

(We shall give a proof of this in the appendix to this section.) Since $\binom{n}{k}$ is the number of different k element subsets in an n element set, it is, of course, always a natural number. But then the above tells us a rather surprising fact. When $n \geq k$, we always have $k!\,(n-k)!$ as a divisor of $n!$.

We may now make the following calculation which, while still lengthy, is *much* easier than building up 52 rows in Pascal's triangle.

$$\begin{aligned}
\binom{52}{13} &= 52! \div (13! \cdot 39!) \\
&= (52 \cdot 51 \cdot 50 \cdot 49 \cdot 48 \cdot 47 \cdot 46 \cdot 45 \cdot 44 \cdot 43 \cdot 42 \cdot 41 \cdot 40) \\
&\qquad \div (13 \cdot 12 \cdot 11 \cdot 10 \cdot 9 \cdot 8 \cdot 7 \cdot 6 \cdot 5 \cdot 4 \cdot 3 \cdot 2 \cdot 1) \\
&= (\overset{4}{\cancel{52}} \cdot \overset{17}{\cancel{51}} \cdot \overset{5}{\cancel{50}} \cdot 49 \cdot \overset{4}{\cancel{48}} \cdot 47 \cdot \overset{23}{\cancel{46}} \cdot \overset{5}{\cancel{45}} \cdot \cancel{44} \cdot 43 \cdot \cancel{42} \cdot 41 \cdot \cancel{40}) \\
&\qquad \div (\cancel{13} \cdot \cancel{12} \cdot \cancel{11} \cdot \cancel{10} \cdot \cancel{9} \cdot \cancel{8} \cdot \cancel{7} \cdot \cancel{6} \cdot \cancel{5} \cdot \cancel{4} \cdot \cancel{3} \cdot \cancel{2} \cdot \cancel{1}) \\
&= 4 \cdot 17 \cdot 5 \cdot 49 \cdot 4 \cdot 47 \cdot 23 \cdot 5 \cdot 43 \cdot 41 \\
&= 2^4 \cdot 5^2 \cdot 7^2 \cdot 17 \cdot 23 \cdot 41 \cdot 43 \cdot 47 \\
&= 635{,}013{,}559{,}600.
\end{aligned}$$

The number of different bridge hands is impressively large. No wonder it's a long wait for that one hand (out of only 4 possible) that has all 13 cards of the same suit!

We may make use of the display of subsets in Figure 6.38 to discover another useful fact. Notice that there are 1 or 2^0 subsets of ϕ, 2 or 2^1 subsets of $\{a\}$, 4 or 2^2 subsets of $\{a,b\}$, 8 or 2^3 subsets of $\{a,b,c\}$, 16 or 2^4 subsets of $\{a,b,c,d\}$, and 32 or 2^5 subsets of $\{a,b,c,d,e\}$. We are certainly led to conjecture that there are 2^n subsets of an n element set. And, indeed, Figure 6.38 even suggests a proof of that fact. To pick out a subset of a set we simply assign a 1 to each object to be included in the subset and assign a 0 if not. Then each subset of a set A corresponds (in a one-to-one fashion) to a mapping from A into $\{0,1\}$. There are 2^n such mappings of $\#(A) = n$. And, hence, there are 2^n different subsets.

Another way to put this conclusion would be:

$$\binom{n}{0} + \binom{n}{1} + \binom{n}{2} + \cdots + \binom{n}{n-1} + \binom{n}{n} = 2^n.$$

This is so because each subset of an n element set must have 0 members, or 1 member, or 2 members, . . . , or n members.

The numbers $\binom{n}{k}$ arise in many applied and theoretical problems in many places in mathematics. In algebra, these numbers are called **binomial coefficients.** We shall see the reason for this terminology in a later problem. Let us now give some examples to suggest the variety of uses of these numbers.

From a class of 20 students a teacher wants to choose a group of 3 to form a discussion group. In how many ways may she do this? Since choosing the group is simply a matter of choosing a subset of 3 students, there are $\binom{20}{3}$ ways. And

$$\binom{20}{3} = 20! \div (17! \cdot 3!) = (20 \cdot 19 \cdot 18) \div (3 \cdot 2 \cdot 1) = 20 \cdot 19 \cdot 3 = 1140.$$

We have 7 red and 2 green blocks. Blocks of the same color are indistinguishable. We wish to arrange the blocks in a row. In how many (distinguishable) ways can we do this? If we think of nine positions in a row as elements of a set, the question is simply one of how many ways we can choose the two places to put the green blocks. For we must fill the other places with red ones. Hence, there are $\binom{9}{2}$ different ways. And

$$\binom{9}{2} = (9!) \div (7! \cdot 2!) = (9 \cdot 8) \div (2 \cdot 1) = 36.$$

In how many different ways may a^7b^2 be written as a product of a's and b's (without exponents)? Such a product is a string of 7 a's and 2 b's in some order. But we see that this is simply the preceding problem in a rather algebraic disguise. So there are 36 such products.

10 distinguishable coins are tossed and it is noted how each comes to rest—with the "head" or the "tail" uppermost. How many different possible outcomes are there? Of these, how many have exactly 5 heads? An "outcome" is an assignment to each of the coins of either "head" or "tail." Hence, the number of possible outcomes is the number of mappings from $\{1,2,\ldots,10\}$ into $\{\text{head, tail}\}$. There are 2^{10} such mappings. $2^{10} = 1024$. To construct an outcome with exactly five heads, we need only choose the 5 coins which should come up heads and then have the others come up tails. Hence, there are $\binom{10}{5}$ outcomes with exactly 5 heads. And

$$\binom{10}{5} = (10!) \div (5!\cdot 5!) = (10\cdot 9\cdot 8\cdot 7\cdot 6) \div (5\cdot 4\cdot 3\cdot 2\cdot 1) = 3\cdot 2\cdot 7\cdot 6 = 252.$$

Definition 6.23.1. $\binom{n}{k}$ = the number of different k element subsets in an n element set.

Theorem 6.23.1. $\binom{n}{0} = 1$ and $\binom{n}{n} = 1$ for all n.

Theorem 6.23.2. If $k > n$, then $\binom{n}{k} = 0$.

Theorem 6.23.3. $\binom{n+1}{k+1} = \binom{n}{k+1} + \binom{n}{k}$

PROOF: Let $A = \{1,2,\ldots,n,n+1\}$. Each $k+1$ element subset of A either contains the element $n+1$ or not. To construct a $k+1$ element subset of A that contains $n+1$, we must select a k element subset of $\{1,2,\ldots,n\}$ to go with $n+1$. We may do this in $\binom{n}{k}$ ways. To construct a $k+1$ element subset of A which does not contain the element $n+1$, we must select a $k+1$ element subset from $\{1,2,\ldots,n\}$. We may do this in $\binom{n}{k+1}$ ways. Hence, $\binom{n+1}{k+1} = \binom{n}{k+1} + \binom{n}{k}$.

Theorem 6.23.4. When $k \le n$, $\binom{n}{k} = (n!) \div ((n-k)!k!)$.

PROOF: We first note that $(0!) \div ((0-0)!0!) = (1!) \div ((1-0)!0!) = (1!) \div ((1-1)!1!) = 1$ so that the formula gives the correct values for $\binom{0}{0}$, $\binom{1}{0}$, and $\binom{1}{1}$. But once we know these values for $\binom{n}{k}$ and the formula of Theorem 6.23.3 we may calculate $\binom{n}{k}$ for all other cases. In other words, once one row of Pascal's triangle is given, the next row is determined. Hence, it will complete the proof if we show that $(n+1)! \div (((n+1)-(k+1))!(k+1)!) = (n!) \div ((n-(k+1))! \cdot (k+1)!) + (n!) \div ((n-k)!k!)$. But we have $(n!) \div ((n-(k+1))(k+1)!) + (n!) \div ((n-k)!k!) = (n!) \div ((n-k-1)!(k+1)!) + (n!) \div ((n-k)!k!) = ((n-k)(n!)) \div ((n-k)!(k+1)!) + ((k+1)(n!)) \div ((n-k)!(k+1)!) = ((n-k)(n!) + (k+1)(n!)) \div ((n-k)!(k+1)!) = ((n-k+k+1)(n!)) \div ((n-k)!(k+1)!) = (n+1)! \div ((n+1-k-1)!(k+1)!) = (n+1)! \div (((n+1)-(k+1))!(k+1)!)$. This completes the proof.

Theorem 6.23.5. There are 2^n different subsets of any n element set.

Theorem 6.23.6. $\binom{n}{0} + \binom{n}{1} + \binom{n}{2} + \cdots + \binom{n}{n} = 2^n.$

PROBLEMS

1. List all possible 2 element subsets of $\{1,2,3,4,5,6\}$. Calculate $\binom{6}{2}$ and check that your list is complete.

2. List all possible 4 elements subsets of $\{1,2,3,4,5,6\}$. Calculate $\binom{6}{4}$ and check that your list is complete.

3. In draw poker a 5 card hand from an ordinary 52 card deck is dealt face down to each player. How many different poker hands are there?

4. In a class of 20 students, the teacher wants to split the class into two groups of 10 to work on two different projects. In how many ways may this be done?

5. Argue that for any $k \leq n$, we must have $\binom{n}{k} = \binom{n}{n-k}$. [HINT: See Problems 1 and 2.]

6. Complete Pascal's triangle to include the row for $n = 10$.

7. Eight coins are tossed together. How many ways can they fall? How many of these results have exactly 4 heads?

8. How many different mappings are there from $A = \{1,2,3, \ldots ,n\}$ into $\{0,1,2\}$ such that (a) no element of A is mapped on 2, (b) no element of A is mapped on 1 or 2, (c) the range of the mapping is $\{0,1,2\}$?

9. A committee of four is to be selected from 50 people. In how many ways may this be done?

10. It is very natural to ask how raising to powers interacts with the process of addition. (a) Show that it is not true, in general, that $(a+b)^n = a^n + b^n$. (b) Consider the following equalities where it is actually convenient for our purposes to forget that multiplication is commutative.

$$\begin{aligned}(a+b)^2 = (a+b)(a+b) &= a(a+b) + b(a+b)\\ &= aa + ab\\ &\quad + ba + bb\end{aligned}$$

$$\begin{aligned}
(a+b)^3 &= (a+b)(a+b)^2 \\
&= (a+b)(aa+ab \\
&\qquad\qquad + ba+bb) \\
&= \qquad a(aa+ab \\
&\qquad\qquad + ba+bb) \\
&\quad + \; b(aa+ab \\
&\qquad\qquad + ba+bb) \\
&= aaa+aab \\
&\quad + aba+abb \\
&\quad + baa+bab \\
&\qquad + bba+bbb \\
(a+b)^4 &= (a+b)(a+b)^3 \\
&= a(aaa+aab \\
&\qquad + aba+abb \\
&\qquad + baa+bab \\
&\qquad\quad + bba+bbb) \\
&\quad + b(aaa+aab \\
&\qquad + aba+abb \\
&\qquad + baa+bab \\
&\qquad\quad + bba+bbb) \\
&= aaaa+aaab \\
&\quad + aaba+aabb \\
&\quad + abaa+abab \\
&\qquad + abba+abbb \\
&\quad + baaa+baab \\
&\qquad + baba+babb \\
&\qquad + bbaa+bbab \\
&\qquad\quad + bbba+bbbb.
\end{aligned}$$

Check from these calculations that $(a+b)^2 = a^2+2ab+b^2$, $(a+b)^3 = a^3+3a^2b+3ab^2+b^3$, and $(a+b)^4 = a^4+4a^3b+6a^2b^2+4ab^3+b^4$. (c) The **binomial theorem** of algebra asserts that

$$(a+b)^n = \binom{n}{0}a^n + \binom{n}{1}a^{n-1}b + \binom{n}{2}a^{n-2}b^2 + \cdots + \binom{n}{n-1}ab^{n-1} + \binom{n}{n}b^n.$$

Use part (b) to verify this theorem for the cases $n = 2,3$, and 4. (d) Supply the details in the following proof by induction for the binomial theorem.

$$\begin{aligned}
(a+b)^{n+1} &= (a+b)(a+b)^n \\
&= a\left(\binom{n}{0}a^n + \binom{n}{1}a^{n-1}b + \cdots + \binom{n}{k}a^{n-k}b^k + \cdots\right.
\end{aligned}$$

$$+ \binom{n}{n} b^n \Big) + b \left(\binom{n}{0} a^n + \binom{n}{1} a^{n-1} b + \cdots \right.$$

$$\left. + \binom{n}{k-1} a^{n-(k-1)} b^{k-1} + \cdots + \binom{n}{n} b^n \right)$$

$$= \binom{n}{0} a^{n+1} + \left(\binom{n}{1} + \binom{n}{0} \right) a^n b + \cdots$$

$$+ \left(\binom{n}{k} + \binom{n}{k-1} \right) a^{n+1-k} b^k + \cdots$$

$$+ \left(\binom{n}{n} + \binom{n}{n-1} \right) a b^n + \binom{n}{n} b^{n+1}$$

$$= \binom{n+1}{0} a^{n+1-k} + \binom{n+1}{1} a^n b + \cdots$$

$$+ \binom{n+1}{k} a^{n+1} b^k + \cdots + \binom{n+1}{n+1} b^{n+1}.$$

11. Use the binomial theorem to write each of the following in expanded form. (a) $(c + d)^5$, (b) $(x - y)^6$, (c) $(2 + x)^7$, (d) $(1 + y)^8$, (e) $(a - b)^{10}$.

12. Partially verify that your results in Problem 11 are correct by evaluating both sides of the equality (i) in part (a) with $c = 1$ and $d = 1$, (ii) in part (b) with $x = 2$ and $y = 1$, (iii) in part (c) with $x = -1$.

13. Use the binomial theorem to prove Theorem 6.23.6. [HINT: Let $a = b = 1$.]

7

The system of integers

7.1 The integers

People all over the world make many, many uses of the system of natural numbers. Whenever we are concerned with sets of objects, we may need to know how many elements are in these sets or in sets we can construct out of them by taking unions or Cartesian products. Natural numbers are the indispensable tools to use on such problems.

But for some very practical problems the natural numbers are inadequate. Consider, for example, the problem of measuring temperature where the height of a column of mercury serves as our measure. It is customary when using the so-called Fahrenheit scale to say that the temperature is 32 degrees at the temperature at which water freezes, and 212 degrees at the temperature at which water boils. The scale on the thermometer is laid out as we laid out points for the natural numbers on a number line, with the points for 32 and 212 determined by the above decision. But anyone who has lived in a cold climate discovers that the mercury sometimes falls to points below the point labeled 0. And if we have only natural numbers, none of the points below the point 0 are labeled at all.

More generally, thinking of the usual geometric line which runs off in both directions, it is striking that no points to the left of the origin are labeled with numbers when we construct a number line as we did before for the natural numbers.

If we interpret natural numbers as vectors, as we did in Section 6.10, we may think of the arrow which represents, say, 3, as a description of a trip from one point to another. But when we do, and have only the natural numbers, we can only describe trips from left to right. And yet it is perfectly clear that it makes sense to talk about trips (or arrows) from right to left.

When we think of the process of addition of natural numbers by a nomograph such as that in Figure 6.15, it is very natural to ask whether we can't extend the scales somehow down below zero and get a device that is helpful. Such a question is especially pertinent if we use the nomograph of Figure 6.15 for subtraction. (See Problem 1, Section 6.13.) For example, if we wish to find $7 - 9$ using the nomograph we should evidently draw a line through the point labeled 7 on the left axis and the point labeled 9 on the middle axis. Our answer should be found where the line crosses the right axis. But our line crosses this axis below the origin and there are no labeled points there.

Of course, this result is tied in with a basic weakness of the system of natural numbers—it is not closed under subtraction.

Putting this another way, it is a fact about the natural number system that sometimes a simple equation such as $x + a = b$, where a and b are in N, has no solution in N. If we had a system which was closed under subtraction, we could always find a solution for such an equation, namely, $x = b - a$.

Still another way to look at this difficulty is to observe that although 0 is the identity element for addition in N since $0 + n = n + 0 = n$ for all n, there is usually no inverse element with respect to addition for a given natural number. In fact, the only element in N that has an inverse with respect to addition is 0 itself.

So let us start on a program which we hope will remedy these weaknesses. Evidently what we should do is to invent some new kind of numbers which will help. But we don't want to throw away the natural numbers, of course. It will be a good thing if, in our new system, we can do every kind of applied problem we could do with natural numbers alone, as well as some problems that we can't work with the natural numbers. We shall accomplish this if some of our newly invented numbers behave just as the natural numbers do. In other words, we should try to invent a new system which is an *extension* of the natural number system, in the sense that certain of our new numbers behave just like natural numbers.

In our new system we would like $x + a = b$ to have a solution whenever a and b are natural numbers (or new numbers that behave like natural numbers). We may then think of the new number which is intended to be the solution for $x + a = b$ as being determined by the pair (a,b) where a and b are natural numbers. But we expect that there will be many different equations which all have the same solution. For example, $x + 5 = 7$, $x + 6 = 8$, and $x + 10 = 12$ all have the solution $x = 2$ in N. When will $x + a = b$ and $x + c = d$ have the same solution in the system of natural numbers? When $b - a = d - c$. Suppose $x = b - a$ and $x = d - c$. Then $b = x + a = (d - c) + a = a + (d - c) = (a + d) - c = (d + a) - c$. But this implies $b + c = d + a$. Consequently, $b - a = d - c$ implies $b + c = d + a$ in N.

Now, whenever a, b, c, and d are natural numbers either $b + c = d + a$ or not. Both $b + c$ and $a + d$ are always well defined natural numbers since N is closed under addition.

If we think of representing the new numbers we are inventing by means of pairs of integers, we should evidently think of certain pairs as corresponding to the same new number—the *same* solution of two different equations. The above condition for two equations to have the same solution suggests that we should consider $N \times N$, the class of all ordered pairs of natural numbers and define a relation E in $N \times N$ by saying $(a,b)E(c,d)$ if and only if $b + c = d + a$. Then, for example, $(0,4)E(3,7)$ since $4 + 3 = 7 + 0$, $(7,2)E(15,10)$ since $2 + 15 = 10 + 7$, and $(5,5)E(17,17)$ since $5 + 17 = 17 + 5$.

It is not at all difficult to prove that this relation is an equivalence relation in $N \times N$. Some of its equivalence classes are:

$$\{(0,0),(1,1),(2,2),(3,3),\ldots\}$$

$$\{(1,0),(2,1),(3,2),(4,3),\ldots\}$$

$$\{(0,1),(1,2),(2,3),(3,4),\ldots\}$$

$$\{(2,0),(3,1),(4,2),(5,3),\ldots\}$$

$$\{(0,2),(1,3),(2,4),(3,5),\ldots\}$$

$$\{(3,0),(4,1),(5,2),(6,3),\ldots\}$$

$$\{(0,3),(1,4),(2,5),(3,6),\ldots\}.$$

We may take each one of these classes as representing the solution to a collection of equations. For example, the first class represents the solution to the equations $x + 0 = 0$, $x + 1 = 1$, $x + 2 = 2$, $x + 3 = 3, \ldots$. The fifth class represents the solution to the equations $x + 0 = 2$, $x + 1 = 3$, $x + 2 = 4$, $x + 3 = 5, \ldots$.

We shall take our new numbers to actually be these classes. Each equivalence class for E in $N \times N$ will be called an **integer.** This is the basic definition of this chapter.

Let us recall that we may write $[(a,b)]$ to stand for the equivalence class to which the pair (a,b) belongs. For example, the seven classes above could be identified as $[(0,0)]$, $[(1,0)]$, $[(0,1)]$, $[(3,1)]$, $[(2,4)]$, $[(5,2)]$, and $[(1,4)]$, respectively. With this device we can assign convenient names to the integers. If n is a natural number and $n \neq 0$, we write "^+n" for the integer $[(0,n)]$ and we write "^-n" for the integer $[(n,0)]$. We read "^+n" as "positive n" and "^-n" as "negative n." We shall write "$\bar{0}$" for the integer $[(0,0)]$. We read "$\bar{0}$" as "zero bar."

Thus, $^-3 = \{(3,0),(4,1),(5,2),(6,3),\ldots\}$ and we read "$^-3$" as "negative three." Also $^+49 = [(0,49)] = \{(0,49),(1,50),(2,51),(3,52),\ldots\}$ and we read "$^+49$" as "positive forty-nine."

If we use a certain type of naming system for the natural numbers along with this procedure for naming the integers, we would say that our naming system for the integers is of the same type. For example, "$^{+}576$" and "$^{-}462$" are evidently decimal names for certain integers and "$^{+}10110_2$" and "$^{-}1010111_2$" are evidently **binary** names for certain integers.

Let n be a nonzero natural number. If $x = [(0,n)] = {}^{+}n$, then we say that x is **positive.** If $x = [(n,0)] = {}^{-}n$, then we say that x is **negative.** Every integer is either 0 or positive or negative. (A proof of this will be given at the end of this section).

A positive integer is also said to have a **positive sign** or **plus sign.** A negative integer is said to have a **negative sign** or a **minus sign.** This terminology can be a bit misleading because when we think of the symbols "+" and "−" as signs, then it is names for integers rather than integers themselves which have signs. "$^{+}3$" has the sign "+" as part of it, but the integer $^{+}3$ does not have "+" as part of it. Thus, it could be thought that we are semantically ignorant when we say $^{+}3$ has a plus sign. But we need not interpret the "has" here in the sense of "has as a part of itself." All we mean when we say that $^{+}3$ has a plus (or positive) sign is that $^{+}3$ is positive.

If we take this approach, we may properly speak of the integers as **signed numbers** although it is actually the numerals (names) for integers which have signs as part of themselves.

Of course, $\bar{0}$ is neither positive nor negative and $\bar{0}$ does not have a sign. Neither does "$\bar{0}$" have (contain) a sign.

The integer $^{+}5$ is by definition the set $\{(0,5),(1,6),(2,7),(3,8),\ldots\}$ and we suggested above that this new number was supposed to correspond to the solution of the equations $x + 0 = 5$, $x + 1 = 6$, $x + 2 = 7$, $x + 3 = 8$, ... But, in the natural number system, the natural number 5 is the solution to these equations. Hence, in some sense, $^{+}5$ must behave as 5 does. More generally, we expect the collection of nonnegative integers $\{\bar{0},{}^{+}1,{}^{+}2,{}^{+}3,\ldots\}$ to behave very much like N.

That this is indeed so we shall see in detail later. Before we look at these details we must say what it means to add and multiply integers. And before we do that we should look at various ways to interpret these new beasts called integers. We should try to develop some feel for them before we start performing operations on them.

Definition 7.1.1 We define the relation E in $N \times N$ by saying that $(a,b)E(c,d)$ if and only if $b + c = d + a$.

Theorem 7.1.1. E is an equivalence relation on $N \times N$.

PROOF: E is clearly reflexive. When $(a,b)E(c,d)$ we have $b + c = d + a$ which implies $d + a = b + c$. Then $(c,d)E(a,b)$ and we see that E is symmetric. If $(a,b)E(c,d)$ and $(c,d)E(e,f)$, then $b + c = d + a$ and $d + e = f + c$. Hence $(b + c) + (d + e) = (d + a) + (f + c)$ and it follows that $(b + e) + (c + d) = (f + a) + (c + d)$. But this implies (Theorem 6.4.6) that $b + e = f + a$. Hence $(a,b)E(e,f)$. Therefore, E is transitive and the proof is complete.

Definition 7.1.2. An **integer** is an equivalence class in $N \times N$ with respect to E.

Theorem 7.1.2. If x is an integer then either $x = [(0,0)]$ or there is a nonzero natural number n such that $x = [(n,0)]$ or $x = [(0,n)]$.

PROOF: Let x be an integer. Since x is an equivalence class in $N \times N$ there are natural numbers a and b such that $x = [(a,b)]$. Either $a = b$ or $a < b$ or $a > b$. (Theorem 6.9.7) If $a = b$, then $x = [(a,a)] = [(0,0)]$ since $(a,a)E(0,0)$. If $a < b$, then there is a nonzero natural number n such that $a + n = b$. Then $x = [(a,a + n)] = [(0,n)]$ since $(a,a + n)E(n,0)$. If $a > b$, then there is a nonzero natural number n such that $a = b + n$. In this case, $x = [(b + n,b)] = [(n,0)]$ since $(b + n,b)E(n,0)$.

Definition 7.1.3. If n is a nonzero natural number we write "^+n" for the integer $[(0,n)]$ and "^-n" for the integer $[(n,0)]$. We let $\bar{0} = [(0,0)]$. We read "^+n" as "positive n" and "^-n" as "negative n." (Note that by Theorem 7.1.2 each integer gets a name by this procedure.)

Theorem 7.1.3. If n and m are nonzero natural numbers, then (a) $^+m = {}^+n$ implies $m = n$, and (b) $^-m = {}^-n$ implies $m = n$.

PROOF: See Problem 6.

Definition 7.1.4. Suppose x is an integer. If $x = {}^+n$ for some nonzero natural number n, then x is said to be **positive** or to **have a positive sign.** If $x = {}^-n$ for some nonzero natural number n, then x is said to be **negative** or to **have a negative sign.** The integer $\bar{0}$ is neither positive nor negative and has no sign.

Theorem 7.1.4. If x is any integer, then either x is positive or x is negative or $x = \bar{0}$, and at most one of these can be true.

PROOF: The first part follows immediately from Theorem 7.1.2 and Definition 7.1.4. To prove the last part, note that when n and m are in N we have $(0,n)E(m,0)$ implying $n + m = 0 + 0 = 0$. But this is true when and only when $n = m = 0$.

Problems

1. Which of the following are true? (E is the relation defined in this section.)

   (a) $(7,10)E(2,5)$
   (b) $(53,75)E(52,75)$
   (c) $(7,40)E(40,7)$
   (d) $(7,40)E(100,133)$
   (e) $(40,7)E(133,100)$

(f) $(10^6 + 6, 10^6)E(10^{12} + 6, 10^{12})$
(g) $(47,47)E(1006,1006)$
(h) $(57 + 6,43)E(43 + 6,29)$.

2. Display seven members of each of the following equivalence classes:

(a) [(0,6)]
(b) [(6,0)]
(c) [(44,66)]
(d) [(1437,496)]
(e) [(77,76)]
(f) [(197,198)]
(g) [(59,59)]
(h) [(3478,10,000)]
(i) [(693,342)].

3. Find the decimal name for each integer (equivalence class) in Problem 2.
4. Display several of the pairs which belong to each of the following classes (integers):

(a) $^{+}33$ (b) $^{-}33$ (c) $^{-}5674$ (d) $^{+}1000$.

5. Group together those of the following pairs that belong to the same equivalence class with respect to E.

(a) (5,7)	(b) (9,14)	(c) (39,41)
(d) (53,4)	(e) (9,79)	(f) (1000,1002)
(g) (70,140)	(h) (215,264)	(i) (0,5)
(j) (90,95)	(k) (56,58)	(l) (49,0).

6. Prove Theorem 7.1.3.

7.2 Interpretations of integers

About all we have done so far is tell what an integer is and show how to name it. We must, of course, proceed to describe certain operations on integers—operations which are somehow natural extensions of the operations of addition, multiplication, subtraction, and division on the natural numbers. But before we do this to get the *system* of integers, let us look at other interpretations of the integers.

We would hope that among the integers we shall be able to find one that somehow represents, say, $7 - 10$. The integer that does this, it turns out, is $^{-}3$ or [(10,7)]. More generally, we may think of the integer $[(a,b)]$ as the new number we have invented to be the result $b - a$ where a and b are natural numbers. Thus, $^{-}14$ is the number we invented to be the result $0 - 14$ or $1 - 15$ or $14 - 28$. $^{+}21$ is the number we invented to be the result $21 - 0$ or $22 - 1$ or $123 - 102$. *Since in the natural number system* we already have $21 - 0 = 22 -$

$1 = 123 - 102 = 21$, we see again that $^{+}21$ had best end up acting like the natural number 21.

A geometric way to think of the integers is as follows. We choose a point on a horizontal line and label it "$\overline{0}$." We call this point the **origin.** We then label the point n units to the right of the origin as "^{+}n." We label the point n units to the left of the origin as "^{-}n." We thus end up with a number line as shown in Figure 7.1.

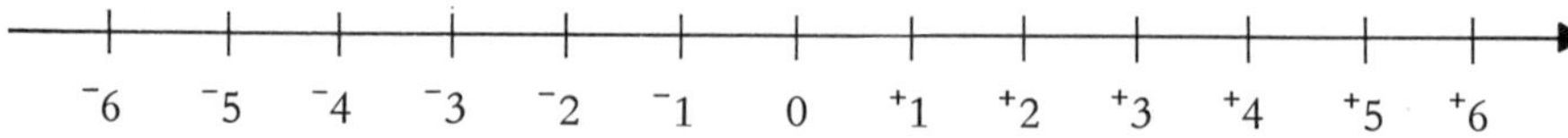

FIGURE 7.1

Each integer is assigned a point by this procedure. The labeled points to the right of the origin correspond to positive integers. The labeled points to the left of the origin correspond to the negative integers.

The number line constructed with integral points on it has a satisfying symmetry that is lacking when we construct a number line with points for only the natural numbers.

But let us, for a moment, return to this more restricted number line where only natural numbers are assigned points. Consider arrows that may be drawn on this line between two points corresponding to natural numbers. We now allow arrows that run from right to left as well as from left to right.

As we did once before, let us use "(n,m)" to stand for the arrow from the point n to the point m. In Section 6.10, where we considered only arrows running towards the right, we needed $m \geq n$. But here, we are relaxing that restriction and we shall consider A, the collection of all arrows (n,m) where n and m may be any natural numbers. Three members of A are shown in Figure 7.2. They are (0,3), (6,4), and (9,5).

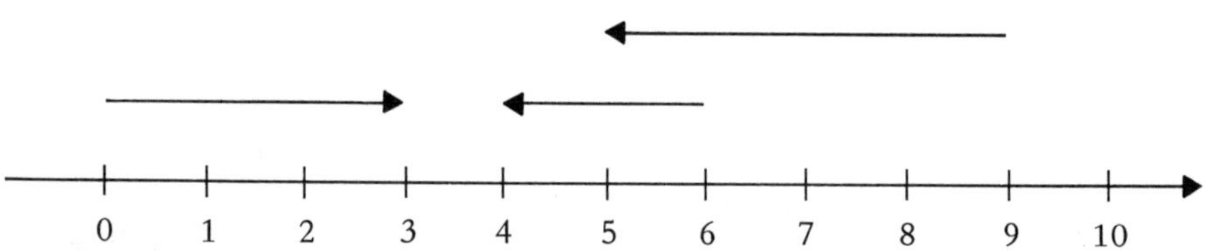

FIGURE 7.2

But notice that we may think of A as $N \times N = \{(n,m) \mid n \in N \text{ and } m \in N\}$. And such a collection of pairs as $\{(0,3), (1,4), (2,5), (3,6), \ldots\}$ can be thought of as a collection of arrows—these arrows running towards the right and having length 3. Then $^{+}3 = \{(0,3), (1,4), (2,5), (3,6), \ldots\}$ may be interpreted as a collection of arrows—the collection of all those arrows in A which run from left to right and have length 3. We would call this the **vector interpretation** of $^{+}3$ or the interpretation of $^{+}3$ as a **vector.** As in Section 6.10, a **vector** is a class of arrows all having the same length and direction.

$^-5 = \{(5,0), (6,1), (7,2), \ldots\}$ may be interpreted as a class of arrows (i.e. as a vector)—the class of all arrows in A running from right to left and having length 5.

There is another way to interpret these integers we have invented. We may think of ^-n as a new kind of number we have invented to act as an inverse element with respect to addition for the natural number n. And ^+n is simply the number we invented in our new system to act like our old friend, the natural number n.

To summarize the above ways of thinking about integers, let us put down some ways to think of the integer $^-2$.

By definition, $^-2 = \{(2,0), (3,1), (4,2), (5,3), \ldots\} = [(2,0)]$, where the equivalence class is with respect to the relation E of the last section.

We may think of $^-2$ as the number we invented to be the solution of the equations $x + 2 = 0$, $x + 3 = 1$, $x + 4 = 2$, $x + 5 = 3, \ldots$.

We may think of $^-2$ as the number corresponding to the point 2 units to the left of the origin on a number line.

We may think of $^-2$ as the vector which is the collection of all arrows (on the number line of Figure 7.2) which have length 2 and run from right to left.

We may think of $^-2$ as the number we invented to be the additive inverse for the natural number 2 (or of something that behaves like the natural number 2).

Let us henceforth use "I" as a name for the collection of all integers. That is, let $I = \{\ldots {}^-3, {}^-2, {}^-1, \bar{0}, {}^+1, {}^+2, {}^+3, \ldots\}$.

Problems

1. In a sketch, represent each of the following integers by an arrow running between two points for natural numbers on a number line. (You will no doubt want to use different scales in different sketches.)

(a) $^+4$ (b) $^-4$ (c) $^+9$
(d) $^-9$ (e) $^-7{,}500{,}000$ (f) $^+100{,}000$.

2. Explain how one of our geometric interpretations of integers solves (approximately) the problem of measuring temperatures colder than 0 degrees Fahrenheit.

3. Reconstruct the nomograph of Figure 6.15 and, by using integers, extend all three scales somewhat below 0. Following the procedure of Problem 1, Section 6.13, conjecture what should be the value of the following:

(a) $^+7 - {}^+10$ (b) $^+7 - {}^-5$ (c) $^-7 - {}^-7$
(d) $^-5 - {}^+3$ (e) $^-7 - {}^-10$.

4. Describe six different ways to think of the following:

(a) $^-6$ (b) $^+6$ (c) $\bar{0}$
(d) $^-1000$.

5. Find the decimal names for these:

(a) $^{-}11011_2$ (b) $^{-}212_3$ (c) $^{-}ete9_{12}$.

7.3 Addition

We discovered in an earlier section that when a, b, c, and d are natural numbers and $a \leq b$ and $c \leq d$, then $(b + d) - (a + c) = (b - a) + (d - c)$. (See Theorem 6.14.5.) Since we may think of the integers $[(a,b)]$ and $[(c,d)]$ as the numbers we invented to be the results $b - a$ and $d - c$ when we drop the restrictions that $a \leq b$ and $c \leq d$, we make the following definition which applies to any two integers:

$$[(a,b)] +_I [(c,d)] = [(a + c, b + d)].$$

This operation $+_I$ in I is called **addition.** When x and y are in I, $x +_I y$ is called the **sum** of x and y. Applying the definition to some special cases, we get:

$$^{+}3 +_I {}^{+}2 = [(0,3)] +_I [(0,2)] = [(0 + 0, 3 + 2)] = [(0,5)] = {}^{+}5$$

$$^{-}4 +_I {}^{-}2 = [(4,0)] +_I [(2,0)] = [(4 + 2, 0 + 0)] = [(6,0)] = {}^{-}6$$

$$^{+}7 +_I {}^{-}5 = [(0,7)] +_I [(5,0)] = [(5,7)] = [(0,2)] = {}^{+}2$$

$$^{-}6 +_I {}^{+}2 = [(6,0)] +_I [(0,2)] = [(6,2)] = [(4,0)] = {}^{-}4.$$

It is a straightforward matter to show that $+_I$ is indeed an operation and that I is closed under $+_I$. It is also easy to make use of the commutativity and associativity of addition of natural numbers to establish the commutativity and associativity of addition in I. For example, the equations $[(a,b)] +_I [(c,d)] = [(a + c, b + d)] = [(c + a, d + b)] = [(c,d)] +_I [(a,b)]$ which use the commutativity of addition in N to justify the middle equality, prove that $+_I$ is commutative in I.

For any $x = [(a,b)]$ in I we have $\bar{0} +_I x = x +_I \bar{0} = [(a,b)] +_I [(0,0)] = [(a + 0, b + 0)] = [(a,b)] = x$. Hence, $\bar{0}$ is the additive identity in I. Also, $^{+}n +_I {}^{-}n = {}^{-}n +_I {}^{+}n = [(n,0)] +_I [(0,n)] = [(n + 0, 0 + n)] = [(n,n)] = [(0,0)] = \bar{0}$. Hence, ^{-}n is the additive inverse for ^{+}n and ^{+}n is the additive inverse for ^{-}n. Clearly $\bar{0}$ is the additive inverse for $\bar{0}$. Since every integer is either $\bar{0}$ or ^{-}n or ^{+}n for some natural number n, we see that every integer has an additive inverse in I. We have reached one of the goals we set for ourselves.

The way in which we have defined addition of integers agrees with our old notion of vector addition when we interpret integers as vectors. The vector diagrams in Figure 7.3 illustrate the sums which we worked out earlier.

Or we may interpret such sums in terms of trips. A trip of 3 units to the east followed by a trip of 2 units to the east is equivalent to a single trip of 5 units to the east. And $^{+}3 +_I {}^{+}2 = {}^{+}5$. A trip of 7 units to the east followed by a trip of

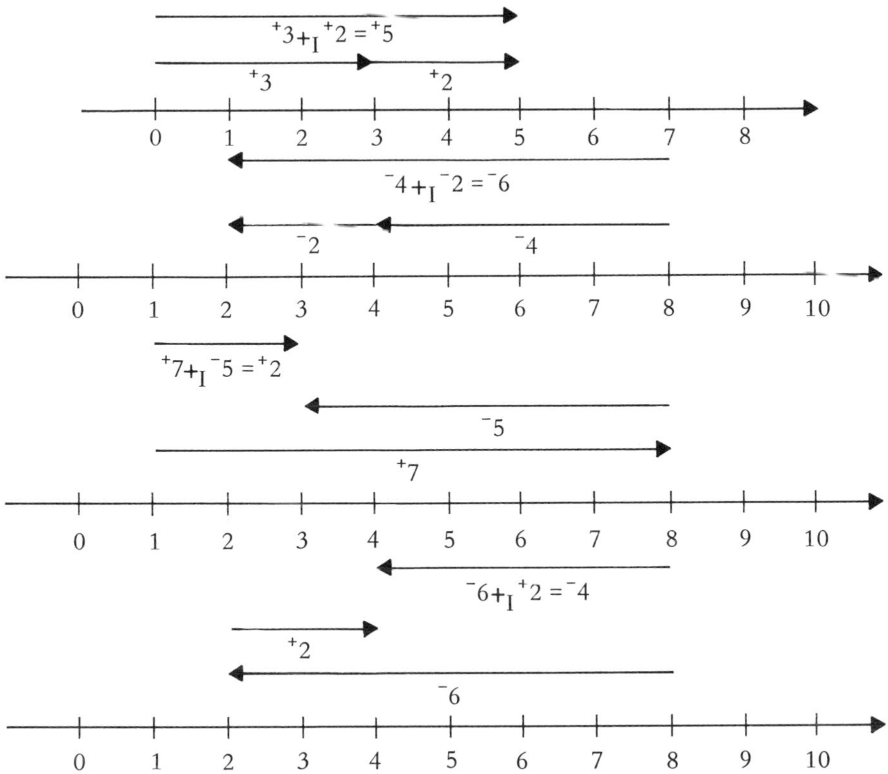

FIGURE 7.3

5 units to the west is equivalent to a single trip of 2 units to the east. And $^+7 +_I {}^-5 = {}^+2$.

We may restate the definition of addition of integers in terms of a flow chart as in Figure 7.4.

We should, however, also look in detail at the process of addition when more conventional names are used for the integers. Consider the following example: $^+2 +_I {}^-5 = (0,2) +_I (5,0) = (5,2) = (3,0) = {}^-3$. Calculations of a similar kind prove:

$$^-2 +_I {}^+5 = {}^+3,\ {}^-2 +_I {}^-5 = {}^-7, \text{ and } {}^+2 +_I {}^+5 = {}^+7.$$

In order to describe an effective process for getting such results as the above it is helpful to have the notion of the **absolute value** of an integer. If n is a nonzero natural number and $x = {}^+n$ or $x = {}^-n$, then we say that the **absolute value** of x is n. The absolute value of $\overline{0}$ we take to be 0. We shall use "$|x|$" to stand for the absolute value of x. Notice that as we have defined it, the absolute value of an integer is always a natural number. From our previous work, we

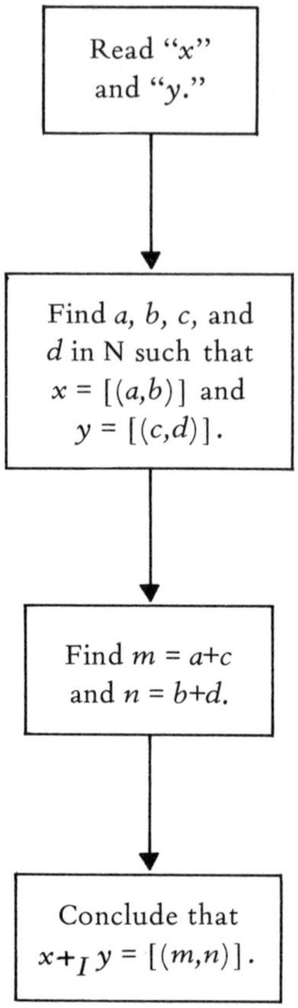

FIGURE 7.4 A flow diagram for the definition of addition in *I*.

have algorithms for adding natural numbers and for subtracting natural numbers when it is possible. We make use of this fact and the general results suggested by the above examples to build the machine in Figure 7.5 for adding integers. We assume that x and y are given in terms of names for natural numbers with a "+" or a "−" affixed.

This procedure is not as complicated as it appears to be in Figure 7.5. The process may be summarized by saying that when x and y have the same sign, we get $x + y$ by finding $|x| + |y|$ and affixing the common sign, and when x and y have different signs, we find $|x|$ and $|y|$, subtract the smaller of these from the larger, and then affix the sign of the number with the larger absolute value.

Naturally enough, we call the inverse operation for $+_I$ **subtraction** and denote it by "$-_I$". Since every integer has an additive inverse in I we see that for any x and y in I, we may find $x -_I y$ by adding to x the inverse element for y with respect to $+_I$. In particular, we shall have for any x in I and any natural

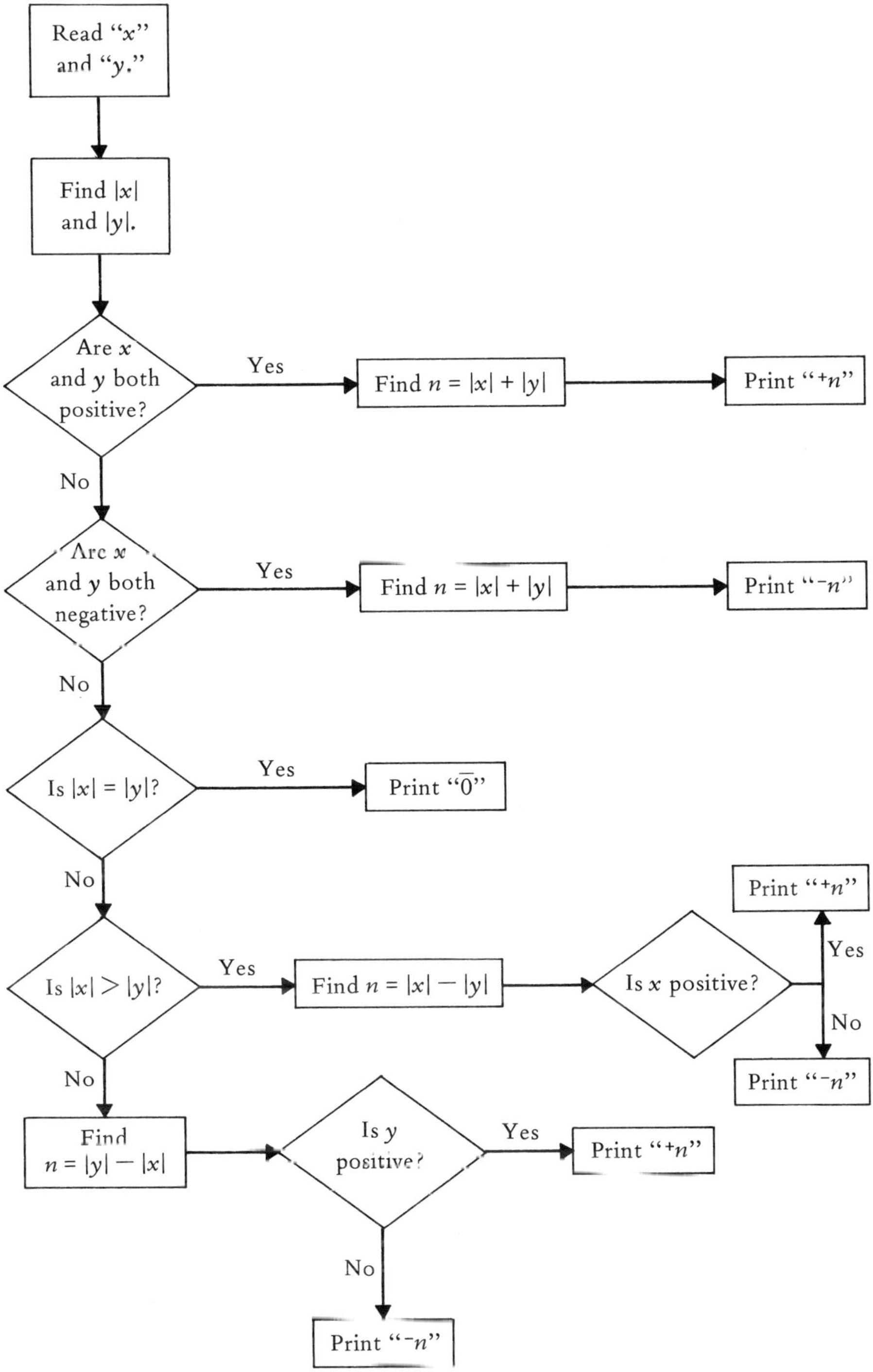

FIGURE 7.5 Flow chart for addition in *I*.

number n, $x -_I \overline{0} = x +_I \overline{0} = x$ and $x -_I {}^+n = x +_I {}^-n$ and $x -_I {}^-n = x +_I {}^+n$. I is closed under subtraction as we had hoped.

It will be helpful if we call the additive inverse of an integer the **opposite** of that integer. Then the opposite of ${}^+n$ is ${}^-n$, the opposite of ${}^-n$ is ${}^+n$, and the opposite of $\overline{0}$ is $\overline{0}$.

What we discovered about subtraction in I can be put as: to subtract y from x, find the sum of x and the opposite of y. Hence, we may always perform a subtraction in I by taking an opposite and performing an addition.

Definition 7.3.1. $[(a,b)] +_I [(c,d)] = [(a + c, b + d)]$.

Theorem 7.3.1. If x, y, z, and w are in I and $x = y$ and $z = w$, then $x + z = y + w$.

PROOF: Let $x = [(a,b)]$, $y = [(c,d)]$, $z = [(e,f)]$, and $w = [(g,h)]$ where $a, b, \ldots, h$ are in N. Suppose $x = y$ and $z = w$. Then $a + d = c + b$ and $e + h = g + f$ and $(a + d) + (e + h) = (c + b) + (g + f)$. We have $x +_I z = [(a + e, b + f)]$ and $y +_I w = [(c + g, d + h)]$. These two integers are the same if $(a + e) + (d + h) = (c + g) + (b + f)$. But this last equality clearly follows from the fact that $(a + d) + (e + h) = (c + b) + (g + f)$.

Theorem 7.3.2. Addition in I is commutative and associative and $\overline{0}$ is the identity element for addition in I.

PROOF: See Problem 5.

Theorem 7.3.3. If n is a nonzero natural number, ${}^-n$ is the additive inverse for ${}^+n$ and vice versa.

Theorem 7.3.4. Every integer has an additive inverse in I.

Definition 7.3.2. If n is a natural number and $x = {}^+n$ or $x = {}^-n$, we say that the absolute value of x is n. Notation: $|x| = n$. We take $|\overline{0}|$ to be 0.

Theorem 7.3.5. If x and y are in I we may find $x +_I y$ by the following procedure:

(a) If x and y have a common sign, find $|x| + |y|$ and affix this sign.
(b) If x and y differ in sign subtract the smaller of the numbers $|x|$ and $|y|$ from the larger and affix the sign that goes with the larger.

Definition 7.3.3. **Subtraction** is the inverse operation for addition in I.

Theorem 7.3.6. I is closed under subtraction.

Definition 7.3.4. If x is in I, the *opposite* of x is the additive inverse of x.

Theorem 7.3.7. To find $x -_I y$ we may add x and the opposite of y.

PROBLEMS

1. Identify each of the following integers as an equivalence class for a certain pair by direct application of the definition of addition:

(a) $[(0,5)] +_I [(0,50)]$ (b) $[(7,13)] +_I [(13,7)]$
(c) $[(53,794)] +_I [(7,8)]$ (d) $[(43,2)] +_I [(1,40)]$
(e) $[(3974,974)] +_I [(4,3)]$ (f) $[(32,5)] +_I [(72,40)]$.

2. Convert the name for each of the integers in Problem 1 to decimal form and find the sum by the method of Figure 7.5.

3. Find each of the following sums:

(a) $^{+}75 +_I {}^{-}93 +_I {}^{+}149$ (b) $^{+}360 +_I {}^{-}492 +_I {}^{-}75$
(c) $^{-}3974 +_I {}^{-}795 +_I {}^{+}432$.

4. Find each of the following differences:

(a) $^{+}795 -_I {}^{+}432$ (b) $^{+}795 -_I {}^{+}1743$
(c) $^{-}432 -_I {}^{+}347$ (d) $^{-}976 -_I {}^{-}1064$
(e) $^{+}753 -_I {}^{-}795$.

5. Prove that addition is associative in I.

6. Prove that the procedure of Figure 7.5 (i.e. Theorem 7.3.5) is valid by considering separately each of the cases

(a) x and y both positive
(b) x and y both negative
(c) x positive and y negative and $|x| > |y|$
(d) x positive and y negative and $|x| < |y|$
(e) x positive and y negative and $|x| = |y|$
(f) x negative and y positive and $|x| > |y|$
(g) x negative and y positive and $|x| < |y|$
(h) x negative and y positive and $|x| = |y|$
(i) $x = \overline{0}$
(j) $y = \overline{0}$.

7. Find decimal names for

(a) $^{+}6 +_I {}^{+}6 +_I {}^{+}6 +_I {}^{+}6 +_I {}^{+}6$
(b) $^{-}7 +_I {}^{-}7 +_I {}^{-}7 +_I {}^{-}7$
(c)† $\overline{0} -_I {}^{+}4 -_I {}^{+}4 -_I {}^{+}4$
(d) $\overline{0} -_I {}^{-}5 -_I {}^{-}5 -_I {}^{-}5$.

† As we did in I, we let $x -_I y -_I z$ stand for $(x -_I y) -_I z$ and use a similar device for longer strings of subtractions.

8. Find the decimal name for

(a) $|^{-}74|$ (b) $|^{+}7 +_I {}^{+}5|$ (c) $|^{+}7 -_I {}^{+}5|$
(d) $|^{+}7 +_I {}^{-}5|$ (e) $|^{-}7 +_I {}^{-}5|$ (f) $|^{-}73 +_I {}^{+}64|$.

9. Find x if

(a) $x +_I {}^{+}6 = {}^{+}7$ (b) $x +_I {}^{-}6 = {}^{+}7$
(c) $x +_I {}^{-}6 = {}^{-}7$ (d) $x -_I {}^{+}13 = {}^{+}20$
(e) $x +_I {}^{+}13 = {}^{+}20$ (f) $x -_I x = x$
(g) $x +_I {}^{-}3974 = {}^{+}796$ (h) $x -_I {}^{+}479 = {}^{-}39{,}000$.

[HINT: when $x +_I y = z$ then we may add the opposite of y to both sides of this equation to find x.]

10. Argue that the hint given in Problem 9 does yield the unique solution to the given equation.

11. Construct a partial table for addition in I. Have the table contain a row and a column for each of the integers $^{-}6, {}^{-}5, \ldots, \bar{0}, {}^{+}1, {}^{+}2, \ldots, {}^{+}6$.

12. Construct a partial table for subtraction in I. Have rows and columns as in Problem 11.

7.4 Multiplication

We discovered in Section 6.14 (Theorem 6.14.7) that when a, b, c, and d are natural numbers and $a \leq b$ and $c \leq d$, then $(b - a) \cdot (d - c) = (bd + ac) - (bc + ad)$. Since we may think of the integers $[(a,b)]$ and $[(c,d)]$ as the numbers we invented to be the results of the subtractions $b - a$ and $d - c$ when there are no restrictions on the relative sizes of the given natural numbers, we are led to give the following definition for **multiplication** of integers:

$$[(a,b)] \cdot_I [(c,d)] = [(bc + ad, bd + ac)].$$

We, of course, call $[(a,b)] \cdot_I [(c,d)]$ the *product* of $[(a,b)]$ and $[(c,d)]$, or else $[(a,b)]$ *times* $[(c,d)]$.

If we were checking all logical details we would first want to prove that the above definition does indeed give us an operation. That is, we would want to show that if x, y, z, and w are in I and $x = y$ and $z = w$, then $x \cdot_I z = y \cdot_I w$. This will be done in the appendix to this section.

Let us look at some examples of the use of the definition.

$$\bar{0} \cdot_I {}^{+}2 = [(0,0)] \cdot_I [(0,2)] = [(0 \cdot 2 + 0 \cdot 0, 0 \cdot 2 + 0 \cdot 0)] = [(0,0)] = \bar{0}$$

$$^{+}2 \cdot_I {}^{+}3 = [(0,2)] \cdot_I [(0,3)] = [(0 \cdot 3 + 2 \cdot 0, 2 \cdot 3 + 0 \cdot 0)] = [(0,6)] = {}^{+}6$$

$$^{-}2 \cdot_I {}^{+}5 = [(2,0)] \cdot_I [(0,5)] = [(2 \cdot 5 + 0 \cdot 0, 2 \cdot 0 + 0 \cdot 5)] = [(10,0)] = {}^{-}10$$

$$^{-}4 \cdot_I {}^{-}4 = [(4,0)] \cdot_I [(4,0)] = [(4 \cdot 0 + 0 \cdot 4, 0 \cdot 0 + 4 \cdot 4)] = [(0,16)] = {}^{+}16.$$

It is quite easy to prove that multiplication is commutative and associative in I and that I is closed under multiplication. Also $^{+}1$ is the identity element for multiplication in I as we see from the following:

$$[(0,1)] \cdot_I [(a,b)] = [(a,b)] \cdot_I [(0,1)] = [(a \cdot 1 + b \cdot 0, b \cdot 1 + a \cdot 0)] = [(a,b)].$$

The integer $^{-}1$ has the interesting and easily established property that when n is a natural number $^{-}1 \cdot_I {}^{+}n = {}^{-}n$ and $^{-}1 \cdot_I {}^{-}n = {}^{+}n$. Hence, whenever x is an integer, $^{-}1 \cdot_I x$ is the additive inverse or opposite for x. Multiplication by $^{-}1$ in I changes the sign on the integer.

Multiplication of any integer by $\bar{0}$ yields $\bar{0}$. The proof of this is easy and will be left as a problem.

If we wish to find the product of ^{-}m and ^{+}n we may write

$$\begin{aligned} {}^{-}m \cdot_I {}^{+}n &= [(m,0)] \cdot_I [(0,n)] = [(0 \cdot 0 + m \cdot n, 0 \cdot n + m \cdot 0)] \\ &= [(m \cdot n, 0)] = {}^{-}(mn). \end{aligned}$$

Similar consideration of the other possible cases yields the conclusion that for any nonzero natural numbers m and n:

$$^{+}m \cdot_I {}^{+}n = {}^{+}(mn)$$

$$^{-}m \cdot_I {}^{+}n = {}^{-}(mn)$$

$$^{+}m \cdot_I {}^{-}n = {}^{-}(mn)$$

$$^{-}m \cdot_I {}^{-}n = {}^{+}(mn).$$

The last three are sometimes summarized by saying that a positive times a negative gives a negative and a negative times a negative gives a positive.

We may use these facts and an algorithm for multiplication of natural numbers to construct a simple procedure for multiplying integers as in Figure 7.6.

In this procedure, we simply take absolute values of the integers involved, multiply the resulting natural numbers, and then affix the proper sign to the name of the product.

One fact that follows easily from the procedure of Figure 7.6 is that $|x \cdot_I y| = |x| \cdot |y|$ whenever x and y are integers.

As we might hope, multiplication distributes across addition in I. That is, if x, y, and z are integers, then $x \cdot_I (y +_I z) = (x \cdot_I y) +_I (x \cdot_I z)$. And, of course, since multiplication is commutative in I, we also have $(x +_I y) \cdot_I z = (x \cdot_I z) +_I (y \cdot_I z)$.

In teaching multiplication of integers to young children we could probably not use the basic definition as given in this section. We could explain the process of multiplication in terms of Figure 7.6. But we should be able to give motivation to the way the signs are handled there. If we assume that we want the distributive law to hold, then one way to supply such motivation is as follows:

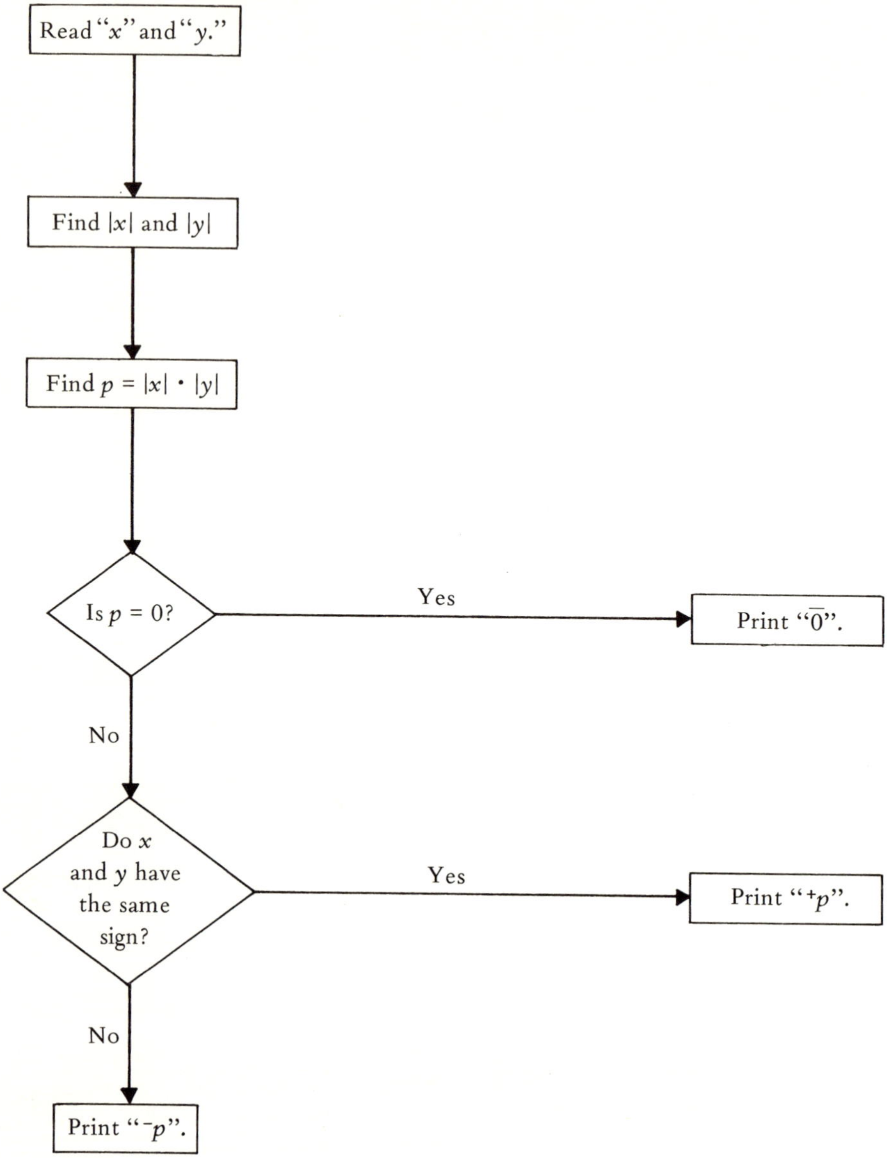

FIGURE 7.6 Flow chart for multiplication of integers.

When n is a natural number we know that $^+n +_I {}^-n = \overline{0}$. If m is also a natural number we should have $^+m \cdot_I (^+n +_I {}^-n) = \overline{0}$. But if the distributive law holds we should have $^+m \cdot_I (^+n +_I {}^-n) = (^+m \cdot_I {}^+n) +_I (^+m \cdot_I {}^-n)$. And if this is to be $\overline{0}$, we must have $^+m \cdot_I {}^-n$ be the opposite of $^+m \cdot_I {}^+n$.

Also $^-m \cdot_I ({}^+n +_I {}^-n) = ({}^-m \cdot_I {}^+n) +_I ({}^-m \cdot_I {}^-n) = \bar{0}$ implies we should have $^-m \cdot_I {}^-n$ be the opposite of $^-m \cdot_I {}^+n$.

One fact that is often useful and is a direct extension of a result in N is the **cancellation law:**

If x, y, and z are in I and $z \neq \bar{0}$, then $x \cdot_I z = y \cdot_I z$ implies $x = y$.

Definition 7.4.1. $[(a,b)] \cdot_I [(c,d)] = [(bc + ad, bd + ac)]$.

Theorem 7.4.1. If x, y, z, and w are in I and $x = y$ and $z = w$, then $x \cdot_I z = y \cdot_I w$.

PROOF: Let $x = [(a,b)]$, $y = [(c,d)]$, $z = [(e,f)]$, and $w = [(g,h)]$. Suppose $x = y$ and $z = w$. Then (i) $b + c = d + a$ and (ii) $f + g = h + e$. By Definition 7.4.1, $x \cdot_I z = [(be + af, bf + ae)]$ and $y \cdot_I w = [(dg + ch, dh + cg)]$. So $x \cdot_I z = y \cdot_I w$ if $bf + ae + dg + ch = dh + cg + be + af$. Multiplying both sides of equation (i) by f, of equation (i) by e, of equation (ii) by d, and of equation (ii) by c yields the equations

$$bf + cf = df + af$$

$$ae + de = be + ce$$

$$dg + df = dh + de$$

$$ch + ce = cf + cg.$$

Adding together all the left sides of these and all the right sides yields $(bf + ae + dg + ch) + (cf + de + df + ce) = (dh + cg + be + af) + (cf + de + df + ce)$. By Theorem 6.4.6, this implies $bf + ae + dg + ch = dh + cg + be + af$. Hence, $x \cdot_I z = y \cdot_I w$.

Theorem 7.4.2. Multiplication in I is commutative and associative and $^+1$ is the identity element for multiplication.

PROOF: See Problem 3.

Theorem 7.4.3. If x is in I, then $x \cdot_I \bar{0} = \bar{0} \cdot_I x = \bar{0}$.

Theorem 7.4.4. If n and m are nonzero natural numbers and $nm = p$, then $^+n \cdot_I {}^+m = {}^+p$, $^-n \cdot_I {}^+m = {}^-p$, $^+n \cdot_I {}^-m = {}^-p$, $^-n \cdot_I {}^-m = {}^+p$.

PROOF: See Problem 3.

Theorem 7.4.5. If n is a natural number, then $^-1 \cdot_I {}^+n = {}^-n$ and $^-1 \cdot_I {}^-n = {}^+n$.

PROOF: This follows immediately from Theorem 7.4.4 since $1 \cdot n = n$.

Theorem 7.4.6. If x and y are integers, then $|x \cdot_I y| = |x| \cdot |y|$.

PROOF: The proof is easily constructed by considering the different possible cases and using Theorem 7.4.3 and Theorem 7.4.4.

Theorem 7.4.7. If x, y, and z are in I, then $x \cdot_I (y +_I z) = (x \cdot_I y) +_I (x \cdot_I z)$.

PROOF: Let $x = [(a,b)]$, $y = [(c,d)]$, and $z = [(e,f)]$. We have $x \cdot_I (y +_I z) = [(a,b)] \cdot_I [(c + e,d + f)] = [(b(c + e) + a(d + f),b(d + f) + a(c + e))] = [(bc + be + ad + af,bd + bf + ac + ae)]$. Also $x \cdot_I z = [(be + af,bf + ae)]$. Hence, $(x \cdot_I y) +_I (x \cdot_I z) = [(bc + ad + be + af,bd + ac + bf + ae)] = [(bc + be + ad + af,bd + bf + ac + ae)]$. Hence, $x \cdot_I (y +_I z) = (x \cdot_I y) +_I (x \cdot_I z)$.

Theorem 7.4.8. If x, y, and z are in I and $z \neq \bar{0}$ and $x \cdot_I z = y \cdot_I z$, then $x = y$.

PROOF: Let $x = [(a,b)]$, $y = [(c,d)]$, and $z = [(e,f)]$. If $z \neq \bar{0}$, then $e \neq f$. Then either $f > e$ or $e > f$. We shall take care of the case where $f > e$. The other possibility may be handled similarly. Suppose that $x \cdot_I z = y \cdot_I z$. Then $[(be + af,bf + ae)] = [(de + cf,df + ce)]$. This implies $be + af + df + ce = bf + ae + de + cf$. Hence, $f(b + c) + e(a + d) = f(a + d) + e(b + c)$. But when $f > e$ this implies $f(a + d) - e(a + d) = f(b + c) - e(b + c)$ or $(f - e)(a + d) = (f - e)(b + c)$. Since $f - e \neq \bar{0}$, this implies $a + d = b + c$. That is, $x = y$.

PROBLEMS

1. Directly from the definition for multiplication in I, find

(a) $[(1,3)] \cdot_I [(2,2)]$ (b) $[(0,6)] \cdot_I [(1,0)]$
(c) $[(0,5)] \cdot_I [(0,6)]$ (d) $[(3,0)] \cdot_I [(5,0)]$
(e) $[(1,0)] \cdot_I [(1,0)]$ (f) $[(1,53)] \cdot_I [(7,20)]$
(g) $[(49,7)] \cdot_I [(31,9)]$.

2. Write each integer involved in Problem 1 in decimal form and check the results of your work by carrying through your multiplications by means of the process of Figure 7.6.

3. The basic properties of multiplication in I follow in a straightforward manner from our definition and the basic properties of addition and multiplication in N. The only difficult part is avoiding mechanical mistakes in manipulating the rather long expressions involved. Try your hand at proving

(a) that multiplication is commutative in I
(b) that multiplication is associative in I
(c) Theorem 7.4.3
(d) Theorem 7.4.4.

4. Find simple names for

(a) ${}^{+}11010_2 \cdot_I {}^{-}10_2$ (b) ${}^{-}210_3 \cdot_I {}^{-}121_3$
(c) ${}^{+}et_{12} \cdot_I {}^{-}98_{12}$ (d) ${}^{-}1001_2 \cdot_I {}^{-}1001_2$.

5. Construct a partial table for multiplication in I. Have a row and a column for each of the numbers $^-6$, $^-5$, $^-4$, . . . $^+4$, $^+5$, $^+6$.

7.5 The nonnegative integers

We now have the operations of addition and multiplication defined on I and have, thus, built a system $(I, +_I, \cdot_I)$. We have suggested in several places that this system is intended to be an extension of the system of natural numbers. That is, we want this new system to contain, in some sense, the old one $(N, +, \cdot)$. We are now in a position to verify that we have accomplished this.

Let P be the collection of nonnegative integers. $P = \{\bar{0}, {}^+1, {}^+2, {}^+3, \ldots\}$. We claim that the system $\{P, +_I, \cdot_I\}$ is isomorphic to the system $\{N, +, \cdot\}$. The isomorphism that establishes this fact is the function f defined on P by the equations $f(\bar{0}) = 0$, $f({}^+n) = n$. Note that f could just as well be defined by: $f(x) = |x|$ for x in P. f is clearly a one-to-one correspondence between P and N. (See Theorem 7.1.3.) To establish that f is an isomorphism between the two systems, suppose that x and y are in P and different from $\bar{0}$. Then there are nonzero natural numbers n and m such that $x = {}^+n$ and $y = {}^+m$. Let $s = n + m$ and $p = nm$. We have $f(x +_I y) = f({}^+n +_I {}^+m) = f({}^+s) = s = n + m = f({}^+n) + f({}^+m) = f(x) + f(y)$. And $f(x \cdot_I y) = f({}^+n \cdot_I {}^+m) = f({}^+p) = p = n \cdot m = f({}^+n) \cdot f({}^+m) = f(x) \cdot f(y)$. The cases where $x = \bar{0}$ and $y = \bar{0}$ are similar. Therefore, f is an isomorphism between $(P, +_I, \cdot_I)$ and $(N, +, \cdot)$.

Nonnegative integers behave exactly like natural numbers as far as the operations of addition and multiplication are concerned. Any true statement about addition and multiplication in N can be translated into a true statement about addition and multiplication of nonnegative integers.

In view of this result, we shall henceforth feel free to consider "$\bar{0}$" as a name for either the natural number 0 or the integer $\bar{0}$. And when n is a nonzero natural number we may follow the commonly accepted procedure and consider "n" as a name for either the natural number n or the positive integer ${}^+n$—whichever seems convenient or appropriate at the time. We shall use this device especially often in later chapters. However, in the rest of this chapter we shall usually write "${}^+n$" as a name for the integer.

From this point on, we shall drop the subscript "I" on the operation symbols for I. This will cause no confusion since when it is possible to interpret the results of the operations as being in either I or N, we shall get corresponding results in either system.

We may think of "$5 + 7 = 12$" as saying the same as "${}^+5 +_I {}^+7 = {}^+12$" or vice versa. We may think of "$4 \cdot 5 = 20$" and "${}^+4 \cdot_I {}^+5 = {}^+20$" as saying the same thing.

Any applied problem that we can solve using the system $(N, +, \cdot)$ we can also solve using nonnegative integers. But the entire system of integers is a richer system than $(N, +, \cdot)$ since we can do some things in I which we cannot do in the natural number system. So in the future we shall act as if the system of integers had absorbed the natural number system. The natural numbers are still there but they are disguised as nonnegative integers. While natural numbers and nonnegative integers are conceptually different beasts, we shall often feel free to ignore the distinctions between them.

We are, then, in a position to verify that one of our basic interpretations of integers is correct. We have said that when a and b are natural numbers we may think of the integer $[(a,b)]$ as the result of subtracting a from b. With the conventions we discussed above, this is indeed the case. We have that when a and b are any natural numbers different from 0:

$$b - a = {}^{+}b - {}^{+}a = {}^{+}b + {}^{-}a = [(0,b)] + [(a,0)] = [(0 + a,b + 0)] = [(a,b)].$$

A similar argument goes through when a or b is 0.

A result such as the above is tied in with another matter of notation that is in common use. If x is an integer, it is standard practice to write "$-x$" to mean $({}^{-}1)x$. That is, "$-x$" is a name for the opposite of x. We shall read "$-x$" as "minus x." With this notation we may restate two of our results about subtraction in I as:

$$x - (-y) = x + y \quad \text{and} \quad x + (-y) = x - y.$$

Note that in both of these equations, the symbol "$-$" is being used in two different senses. The very first occurrence of "$-$" in the first equation is a case where it is meant to name the operation of subtraction. The "$-$" that occurs on the right side of the second equation is also to be interpreted in this fashion. But the other occurrences of "$-$" in the two equations are to be taken differently. They are to be interpreted as indicating the opposite of an integer. We may avoid confusion due to this ambiguous use of "$-$" by taking "$-$" to mean the operation of subtraction whenever it is possible to do so and, otherwise, taking it to be attached to the name of an integer to indicate that we are to take the opposite of that integer.

We actually now have four different uses for the symbol "$-$." It denotes the operation of subtraction in N. It occurs as a super-prefix on a name for a nonzero natural number to give a name for a negative integer. It denotes the operation of subtraction in I. It may be used as a prefix on a name for an integer to indicate the opposite of that integer.

With this notation we may restate the definition of the absolute value of an integer as follows:

$$|x| = x \quad \text{if } x \text{ is nonnegative}$$

$$|x| = -x \quad \text{if } x \text{ is negative.}$$

We said earlier that the absolute value of an integer is always a natural number. But we are now identifying our nonnegative integers with the natural numbers. So we may say that the absolute value of any integer is a nonnegative integer. Note that $-x$ is positive when x is a negative integer.

We must resist the temptation to claim that $|-x| = x$. If we interpret the assertation that $|-x| = x$ in the usual way, it will be true only if the equality holds no matter what integer is substituted for "x." And $|-(^-2)| = |^+2| = {}^+2 \neq {}^-2$. In fact, whenever x is negative $|-x| \neq x$.

Since $^+3$ is the opposite of $^-3$ in I and we have agreed that we may write simply "3" for $^+3$ if we wish, we now see that we have $-(^+3) = -3 = {}^-3$. So we may drop the super-prefix "$-$" down if we wish and cause no confusion. It is common practice in mathematical literature to do this.

Since $(P, +_I, \cdot_I)$ is isomorphic to $(N, +, \cdot)$ we see that P is closed under addition and multiplication because N is.

It is also helpful to note that the collection of all *positive* integers is closed under addition and multiplication. That is, the sum or product of any two positive integers is positive. That this is so follows from the above isomorphism and the facts that in N, $n + m = 0$ implies $n = m = 0$, and $nm = 0$ implies $n = 0$ or $m = 0$.

It will be convenient to use "I^+" to denote the collection of all positive integers, and "I^-" to denote the collection of all negative integers. Then $I = I^- \cup \{0\} \cup I^+$.

Theorem 7.5.1. Let $P = \{\bar{0}, {}^+1, {}^+2, {}^+3, \ldots\}$. The system $\{P, +_I, \cdot_I\}$ is isomorphic to the system $\{N, +, \cdot\}$.

Definition 7.5.1. We shall feel free to write "$+$" for "$+_I$," "$\cdot$" for "$\cdot_I$," "0" for "$\bar{0}$," and "n" for "^+n" when n is a nonzero natural number. [NOTE: In view of Theorem 7.5.1 this will usually cause no confusion.]

Theorem 7.5.2. If a and b are natural numbers then $b - a = [(a,b)]$.

Definition 7.5.2. We shall write "$-x$" for the opposite of x. We read "$-x$" as "minus x."

Theorem 7.5.3. $|x| = x$ if x is nonnegative. $|x| = -x$ if x is negative.

Definition 7.5.3. I^+ is the collection of all positive integers. I^- is the collection of all negative integers.

Theorem 7.5.4. The set P of Theorem 7.5.1 is closed under addition and multiplication. I^+ is also closed under these operations.

PROBLEMS

1. Let $Q = \{\bar{0}, {}^-1, {}^-2, {}^-3, \ldots\}$ be the collection of all nonnegative integers.

   (a) Is the system $(Q, +_I)$ isomorphic to the system $(N, +)$? Explain.
   (b) Is the system $(Q, +_I, \cdot_I)$ isomorphic to the system $(N, +, \cdot)$? Explain.

2. Let $A = \{\bar{0}, {}^+2, {}^+4, {}^+6, \ldots\}$ be the system of even nonnegative integers.

   (a) Is the system $(A, +_I)$ isomorphic to the system $(N, +)$? Explain.

(b) Is the system $(A, \cdot_I)$ isomorphic to the system $(N, \cdot)$? Explain.
(c) Is the system $(A, +_I, \cdot_I)$ isomorphic to the system $(N, +, \cdot)$? Explain.

3. For x in I, let $f(x) = x + 1$.

(a) Is the function f a one-to-one correspondence between the system I and itself?
(b) Is f an isomorphism between $(I, +)$ and itself?

4. Do Problem 3 if we take it that f is defined by $f(x) = x + k$ where k is a fixed integer.
5. Is I^- closed under addition? Under multiplication? Explain.
6. In the part of this section where we were showing that the function f was an isomorphism between $(P, +_I, \cdot_I)$ and $(N, +, \cdot)$ we did not carry through the details when $x = \bar{0}$ or $y = \bar{0}$. Do so.
7. In our demonstration that $[(a,b)]$ is indeed $b - a$ when a and b are natural numbers, we did not carry through the details for the cases where $a = \bar{0}$ or $b = \bar{0}$. Do so.
8. Let us define a relation S in I by saying that xSy when and only when x and y are both 0, or x and y have the same sign. Show that S is an equivalence relation and find its equivalence classes.
9. Show that $(I, +_I, \cdot_I)$ is **not** isomorphic to $(N, +, \cdot)$.
10. Disprove each of the following:

(a) If x and y are in I, then $|x + y| = |x| + |y|$.
(b) If x and y are in I, then $|x - y| = |x| - |y|$.

7.6 Interpretations of addition and multiplication

If n is a natural number different from 0, we may think of the integer ^+n as representing a trip of n units to the right along the number line. And ^-n represents a trip of n units to the left. If x and y are integers, $x + y$ represents a trip which has the same effect as a trip x followed by a trip y.

Since $^+1 \cdot x = x$ for all integers x and since multiplication distributes across addition in I we have such a result as $^+3 \cdot x = (^+1 + {}^+1 + {}^+1)x = x + x + x$. Hence, multiplication by $^+3$ corresponds to building up a trip out of 3 consecutive trips x. For example, $^+3 \cdot {}^-2 = {}^-2 + {}^-2 + {}^-2 = {}^-6$. And 3 trips of 2 units to the left is equivalent to a single trip of 6 units to the left.

More generally, multiplication by a positive integer corresponds to repeated addition in the sense that when n is a natural number and x is any integer, we have $^+n \cdot x = x + x + x + \cdots + x$ where there are n terms on the right.

We know that $^-1 \cdot x$ is the opposite of x. If we think of x as a certain trip then $^-1 \cdot x$ is a trip of the same distance as x but in the opposite direction. Then we may think of such a product as $^-3 \cdot {}^+4 = (^-1 \cdot {}^+3) \cdot {}^+4 = {}^-1 \cdot (^+3 \cdot {}^+4)$ as corresponding to the opposite of a trip consisting of 3 consecutive trips of 4 units to the right.

$^-2 \cdot (^-5) = (^-1 \cdot {}^+2) \cdot {}^-5 = {}^-1 \cdot ({}^+2 \cdot {}^-5)$ so that $^-2 \cdot {}^-5$ corresponds to the opposite of a trip consisting of 2 trips of 5 units to the left.

In general, when n is a nonzero natural number, $^-n \cdot x = {}^-1 \cdot ({}^+n \cdot x)$. So $^-n \cdot x$ corresponds to the opposite of a trip consisting of n consecutive trips x.

We may also look at the matter from a somewhat different point of view. We have that $x + (^-n \cdot y) = x + (^-1 + {}^-1 + \cdots + {}^-1)y = x + {}^-1 \cdot y + {}^-1 \cdot y + {}^-1 \cdot y + \cdots + {}^-1 \cdot y = x - y - y - y - \cdots - y$ where there are n terms of the same form in each place where "$\cdots$" appears. This says that when it comes to adding ^-ny to an integer, we may accomplish the same result by subtracting y exactly n times. In this sense, multiplication by a negative integer may be thought of as repeated subtraction.

Suppose two people called A and B are playing a game for money. If A wins n dollars from B in one play of the game, let us say that the outcome of the game is ^+n. If B wins n dollars from A, let us say that the outcome is ^-n. Then the outcome of two plays of the game is $x + y$ if x is the outcome on the first play and y is the outcome on the second. For example, if the outcome of the first play is $^+3$ (i. e. A wins 3 dollars on the first play) and the outcome on the second play is $^-5$ (i. e. B wins 5 dollars), then the outcome of the two plays together is $^+3 + {}^-5 = {}^-2$ (i. e. B wins 2 dollars). Then we may think of $^+n \cdot x$ as being the outcome of n successive plays of the game when each individual play has the outcome x. We may think of $^-n \cdot x$ as the opposite (i.e. the outcome if the players were interchanged) of the outcome of n successive plays of the game where each individual play has outcome x.

Suppose a particle or a block rests on a surface. We may apply force to the block by pushing on it with various objects. Let us use ^+n to designate a force of n pounds acting towards the right and ^-n to designate a force of n pounds acting towards the left. Then $x + y$ is the **resultant** of the two forces x and y in the sense that there will be the same effect† on the block from the single force $x + y$ as there is from the two forces x and y acting together. For example, $^+3 + {}^-5 = {}^-2$ tells us that a single force of 2 pounds acting towards the left is equivalent to a force of 3 pounds towards the right and a force of 5 pounds to the left acting simultaneously. Here $^+n \cdot x$ corresponds to the resultant of n forces x acting simultaneously. And $^-n \cdot x$ corresponds to the opposite (i. e. in the opposite direction) of the resultant of n forces x acting simultaneously. The opposite of any force may be thought of as the force required to exactly cancel out the effect of the given force.

Suppose we have a very large bank account. Let us use ^+n to designate a deposit of n dollars in this account and ^-n to designate a withdrawal of n dollars. Then $^-100 + {}^+50 = {}^-50$ tells us we have the same result if we withdraw 50

† We are concerned here with the effect on the block as far as its possible motion is concerned. In actual practice, two very large forces acting on the block in opposite directions might crush the block. But this is not the sort of thing we are interested in. If the two large opposite forces exactly balance each other, the block will not move.

dollars as we do when we withdraw 100 dollars and then deposit 50 dollars. We may then think of $^+n \cdot x$ as the result of n successive transactions of x. We may think of $^-n \cdot x$ as the opposite of n successive transactions of x. The opposite of a given transaction may be thought of as the transaction we should perform to exactly cancel out the effect of the given transaction.

Problems

1. What one single transaction might we perform at a bank to have the same effect as

(a) 5 withdrawals of 50 dollars
(b) 5 withdrawals of 50 dollars and 3 deposits of 75 dollars
(c) 10 deposits of 200 dollars and 7 withdrawals of $270
(d) the opposite of 8 deposits of $40
(e) the opposite of 9 withdrawals of $14
(f) 12 deposits of 140 dollars followed by the opposite of 12 deposits of 140 dollars
(g) 12 deposits of $140 followed by the opposite of 12 withdrawals of $140.

2. If necessary, introduce symbols that stand for the quantities involved (being careful to indicate your choice of directions or choice of signs), and then find an expression in terms of these symbols that would give us

(a) the resultant force acting on a body of weight W pounds falling through the atmosphere at such a speed that the air resistance is R pounds.
(b) the outcome of purchasing n lottery tickets at a cost of C dollars each and then winning nothing.
(c) the outcome of purchasing n lottery tickets at a cost of C dollars each and then winning a prize of P dollars.
(d) the outcome of purchasing n lottery tickets at a cost of C dollars each, selling K of these for S dollars each, and then winning a prize of P dollars on one of the tickets you still hold.
(e) the outcome of manufacturing several items at a certain cost per item and then selling some of them at a certain price per item.
(f) the outcome of buying some shares of a certain stock and then later selling those shares.
(g) the result to you when, near the end of the income tax form, you have finally computed the total tax due and the amount already withheld from your salary.

3. On a certain television program, a flying submarine appears. Instruct the engineer who designs the machine as to how he may combine its altimeter and depth meter so as to use only one dial on the instrument panel.

4. Discuss some of the properties of the operation M defined in I by $^+nM^+m =$

$^+(nm)$, $^-nM^+m = {}^+nM^-m = {}^-(nm)$, $^-nM^-m = {}^-(nm)$, $\overline{0}Mx = xM\overline{0} = \overline{0}$. Here n and m are nonzero natural numbers and x is any integer.

5. Discuss the properties of the operation M^* defined by: $^+nM^{*+}m = {}^+(nm)$, $^+nM^{*-}m = {}^+(nm)$, $^-nM^{*+}m = {}^-(nm)$, $^-nM^{*-}m = {}^-(nm)$, $\overline{0}M^*x = xM^*\overline{0} = \overline{0}$. Here n and m are nonzero natural numbers and x is any integer.

7.7 Division

We define division in I to be the inverse operation for multiplication in I. We use "$\div$" to denote division in I as we did in N. Since division is defined in terms of multiplication and the system of nonnegative integers with addition and multiplication is isomorphic to the system of natural numbers with addition and multiplication, we see that $^+m \div {}^+n = {}^+q$ in I if and only if $m \div n = q$ in N.

We carry over the usual terminology for N to I and say that w is divisible by z if $w \div z$ is an integer. We also say that z is a **divisor** of w when w is divisible by z.

By the cancellation law mentioned at the end of Section 7.4, we know that if x, y, z, and w are integers such that $xz = w$ and $yz = w$ and $z \neq 0$, then $x = y$. Hence, if $z \neq 0$ and there is an integer x such that $xz = w$, then there is exactly one such x. It follows that any multiplication fact $xz = w$ with $z \neq 0$ automatically yields a division fact $x = w \div z$.

If we make use of this and our results about multiplying integers we may conclude that if m and n are natural numbers and $m \div n = q$, then

$$^+m \div {}^+n = {}^+q$$

$$^+m \div {}^-n = {}^-q$$

$$^-m \div {}^+n = {}^-q$$

$$^-m \div {}^-n = {}^+q.$$

And, further, if x and y are integers and $|x|$ is not divisible by $|y|$ in N, then x is not divisible by y in I.

This yields a simple direct process for division of integers which may be summarized by the flow chart in Figure 7.7.

Notice that the procedure tells us that $|x \div y| = |x| \div |y|$ whenever either side of this equality makes sense.

Some cases of special interest are the following which hold for any integer x:

$$x \div {}^-1 = {}^-x$$

$$x \div {}^+1 = x$$

$$x \div x = {}^+1 \quad \text{when} \quad x \neq 0.$$

Of course, I is not closed under division. In particular, division by 0 is impossible in I. Division is not commutative or associative in I. But we would

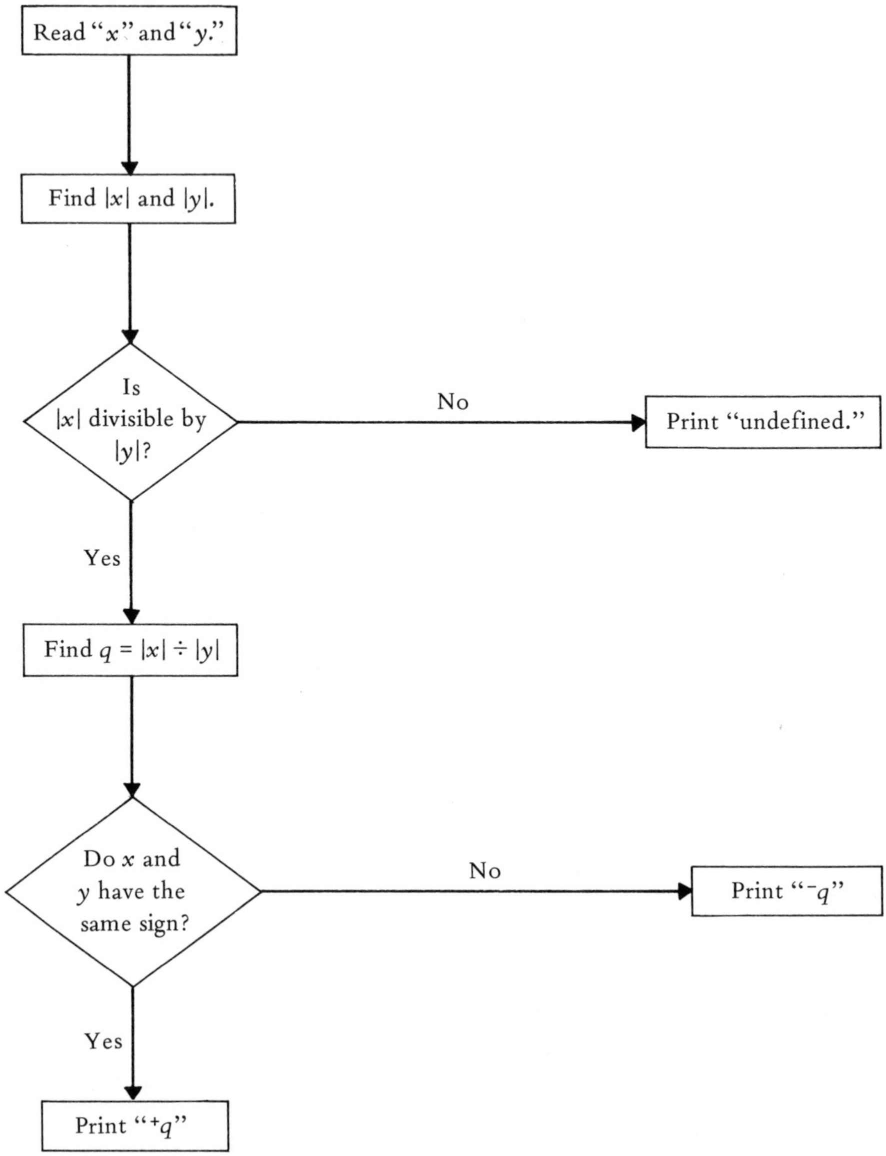

FIGURE 7.7 A flow chart for division of integers. The machine computes $x \div y$.

expect several results similar to those that hold for division in N. And these expectations are not in vain. The following hold whenever x, y, z, and w are in I, $y \neq 0$, $w \neq 0$, and the left sides of the equations are defined:

(i) $x \div y = z \div w$ implies $xw = zy$

(ii) $(x \div y) + (z \div y) = (x + z) \div y$
(iii) $(x \div y) + (z \div w) = (xw + zy) \div (wy)$
(iv) $(x \div y) \cdot (z \div w) = (xz) \div (yw)$.

As illustrations of (i) we have

$$^{+}15 \div {}^{-}3 = {}^{-}25 \div {}^{+}5 \quad \text{while} \quad {}^{+}15 \cdot {}^{+}5 = {}^{-}25 \cdot {}^{-}3 = {}^{+}75$$

$$^{-}36 \div {}^{-}12 = {}^{+}27 \div {}^{+}9 \quad \text{while} \quad {}^{-}36 \cdot {}^{+}9 = {}^{+}27 \cdot {}^{-}12 = {}^{-}324.$$

As examples of (ii) we have:

$$({}^{-}15 \div {}^{+}3) + ({}^{-}39 \div {}^{+}3) = {}^{-}5 + {}^{-}13 = {}^{-}18$$
$$\text{while} \quad ({}^{-}15 + {}^{-}39) \div {}^{+}3 = {}^{-}54 \div {}^{+}3 = {}^{-}18$$

$$({}^{+}144 \div {}^{-}12) + ({}^{+}36 \div {}^{-}12) = {}^{-}12 + {}^{-}3 = {}^{-}15$$
$$\text{while} \quad ({}^{+}144 + {}^{+}36) \div {}^{-}12 = {}^{+}180 \div {}^{-}12 = {}^{-}15$$

$$({}^{+}75 \div {}^{-}15) + ({}^{-}45 \div {}^{-}15) = {}^{-}5 + {}^{+}3 = {}^{-}2$$
$$\text{while} \quad ({}^{+}75 + {}^{-}45) \div {}^{-}15 = {}^{+}30 \div {}^{-}15 = {}^{-}2$$

To illustrate (iii) we have:

$$({}^{-}4 \div {}^{+}2) + ({}^{-}18 \div {}^{-}3) = {}^{-}2 + {}^{+}6 = {}^{+}4$$
$$\text{while} \quad ({}^{-}4 \cdot {}^{-}3 + {}^{-}18 \cdot {}^{+}2) \div ({}^{-}3 \cdot {}^{+}2)$$
$$= ({}^{+}12 + {}^{-}36) \div {}^{-}6 = {}^{-}24 \div {}^{-}6 = {}^{+}4.$$

$$({}^{+}49 \div {}^{-}7) + ({}^{-}16 \div {}^{-}4) = {}^{-}7 + {}^{+}4 = {}^{-}3$$
$$\text{while} \quad ({}^{+}49 \cdot {}^{-}4 + {}^{-}16 \cdot {}^{-}7) \div ({}^{-}7 \cdot {}^{-}4)$$
$$= ({}^{-}196 + {}^{+}112) \div {}^{+}28 = {}^{-}84 \div {}^{+}28 = {}^{-}3.$$

As examples of (iv) we have:

$$({}^{+}24 \div {}^{-}6) \cdot ({}^{-}15 \div {}^{+}3) = {}^{-}4 \cdot {}^{-}5 = {}^{+}20$$
$$\text{while} \quad ({}^{+}24 \cdot {}^{-}15) \div ({}^{-}6 \cdot {}^{+}3) = {}^{-}360 \div {}^{-}18 = {}^{+}20$$

$$({}^{-}32 \div {}^{-}8) \cdot ({}^{+}14 \div {}^{-}7) = {}^{+}4 \cdot {}^{-}2 = {}^{-}8$$
$$\text{while} \quad ({}^{-}32 \cdot {}^{+}14) \div ({}^{-}8 \cdot {}^{-}7) = {}^{-}448 \div {}^{+}56 = {}^{-}8.$$

Definition 7.7.1. **Division** in I is denoted by "$\div$" and is the inverse of multiplication in I. When $x \div y$ is an integer we say that x is **divisible** by y and y is a **divisor** of x.

Theorem 7.7.1. If x, y, and z are in I and $y \neq 0$ and $x = zy$, then $x \div y = z$.

Theorem 7.7.2. If m, n, and q are natural numbers and $m \div n = q$, then ${}^{+}m \div {}^{+}n = {}^{+}q$, ${}^{+}m \div {}^{-}n = {}^{-}q$, ${}^{-}m \div {}^{+}n = {}^{-}q$, and ${}^{-}m \div {}^{-}n = {}^{+}q$.

PROOF: This follows immediately from Theorem 7.7.1 and Theorem 7.4.3.

Theorem 7.7.3. If x and y are integers, $x \div y$ is defined in I if and only if $|x| \div |y|$ is defined in N.

PROOF: If $y = 0$ then $x \div y$ is undefined since when $x \neq 0$ there is no integer z such that $x = 0 \cdot z$ and when $x = 0$ we have $x = 0 \cdot z$ for any z. Thus, division by 0 is impossible in I as it is in N. If $y \neq 0$, we know that there is a nonzero natural number n such that $y = {}^{+}n$ or $y = {}^{-}n$. Hence, when $|x| \div |y|$ is defined in N we may apply Theorem 7.7.2 and find $x \div y$ defined in I. If $x \div y = z$ in I we must have $x = z \cdot y$ and $y \neq 0$. Then $|x| = |z| \cdot |y|$ where $|y| \neq 0$ and $|x|$ is divisible by $|y|$ in I.

Theorem 7.7.4. If x, y, z, and w are in I, and y divides x, and w divides z, then $x \div y = z \div w$ implies $xw = zy$.

PROOF: See Problem 2.

Theorem 7.7.5. If x, y, and z are in I and y divides x and y divides z, then $(x \div y) + (z \div y) = (x + z) \div y$.

PROOF: Suppose $x \div y = n$ and $z \div y = m$. Then $x = ny$ and $z = my$. Note that $y \neq 0$. Then $x + z = ny + my = (n + m) \cdot y$. Hence, $(x + z) \div y = n + m$.

Theorem 7.7.6. If x, y, z, and w are in I and y divides x and w divides z, then $(x \div y) + (z \div w) = (xw + zy) \div yw$.

PROOF: See Problem 2.

Theorem 7.7.7. If x, y, z, and w are in I, and y divides x, and w divides z, then $(x \div y) \cdot (z \div w) = (xz) \div (yw)$.

PROOF: Suppose $x \div y = n$ and $z \div w = m$. Then $y \neq 0$ and $w \neq 0$, so $yw \neq 0$. Also, $x = ny$ and $z = mw$. Hence, $xz = (ny) \cdot (mw) = (nm) \cdot (yw)$. It follows that $(xz) \div (yw) = nm$.

PROBLEMS

1. Use four different numerical examples to verify the assertion

(a) (i) (b) (ii) (c) (iii) (d) (iv).

2. Since division in I is much like division in N, many of the proofs of properties of division in N can be carried over bodily to give proofs for I. Prove

(a) Theorem 7.7.4. (b) Theorem 7.7.6.

3. Find decimal names for

(a) ${}^{+}144 \div {}^{-}36$ (b) ${}^{-}144 \div {}^{+}36$
(c) ${}^{-}144 \div {}^{-}36$ (d) ${}^{-}1554 \div {}^{+}21$
(e) ${}^{-}1554 \div {}^{-}21$ (f) ${}^{-}52{,}358 \div {}^{+}94$.

4. (a) What integer multiplied by $^{+}6$ gives $^{-}96$?
   (b) What integer multiplied by $^{-}6$ gives $^{-}96$?
   (c) What integer multiplied by $^{-}6$ gives $^{+}96$?
   (d) What integer multiplied by 0 gives $^{-}7$?

5. (a) If your checking account contains \$684, how many \$57 checks can you write without being overdrawn?
   (b) How many deposits of \$35 each should we make to build up a bank account from \$3925 to \$4310?
   (c) From \$3925 to \$3540?
   (d) From A dollars to B dollars?

6. Does the process of division by factoring work in I? Illustrate your answer with several examples.

7. Construct a partial table for division in I. Have the table show a row and a column for each of the numbers $^{-}12, ^{-}11, \ldots, ^{+}11, ^{+}12$.

8. An integer is said to be *even* if it is divisible by 2.
   (a) Show that if an integer x is not even, then there is an integer k such that $x = 2k + 1$.
   (b) Show that if x is not even, then x^2 is not even.

7.8 Inequalities

The point $^{+}8$ is to the right of the point $^{+}4$ on the number line. The point $^{+}6$ is to the right of the point $^{-}4$. The point $^{+}1$ is to the right of the point $^{-}1000$. The point $^{-}50$ is to the right of the point $^{-}52$.

Guided by the idea that "$\geq$" in I should, as it does in N, mean the same as "at least as far to the right on the number line," we state the following definition:

> If x and y are integers, we say that $x \geq y$ if there is a nonnegative integer k such that $x = y + k$.

For example, $^{+}8 \geq {}^{+}4$ since $^{+}4$ is nonnegative and $^{+}8 = {}^{+}4 + {}^{+}4$, $^{+}6 \geq {}^{-}4$ since $^{+}6 = -4 + {}^{+}10$, $^{+}1 \geq {}^{-}1000$ since $^{+}1 = {}^{-}1000 + {}^{+}1001$, and $^{-}50 \geq {}^{-}52$ since $^{-}50 = {}^{-}52 + {}^{+}2$.

Of course, this relation $\geq$ is a different relation from the relation we denoted by "$\geq$" in N. Strictly speaking, we should probably use something like "$>_I$" to name this new relation. It is a relation in I rather than in N. However, using the same symbol to refer to the two different relations will cause no confusion. The reason for this is that the relation $\geq$ as defined above in the collection of **nonnegative** integers is isomorphic to the old relation $\geq$ in N. So when we see "$7 \geq 4$" we may think of it as an assertion about integers or about natural numbers as we please. That $\geq$ in $P = \{\overline{0}, {}^{+}1, {}^{+}2, \ldots\}$ is isomorphic to $\geq$ in N follows from the facts that the system $(P, +_I, \cdot_I)$ is isomorphic to the system $(N, +, \cdot)$ and that $>$ is defined in the same way in terms of addition in both systems.

We build on the above definition of $\geq$ in I to define $>$, $\leq$, and $<$ in I in the obvious fashion. We also carry over to I the meanings of "maximum of a set," "minimum of a set," and "between."

We read "$x \geq y$" as "x is greater than or equal to y" just as we did in N. Similar vocalizations are used for "$x > y$," "$x \leq y$," and "$x < y$."

In N we had $x \geq 0$ for every x. This is, of course, no longer true in I. In fact, we have, when x is an integer

(a) $x > 0$ if and only if x is positive, and
(b) $x < 0$ if and only if x is negative.

Also, when x is an integer

(a) x is nonnegative if and only if $x \geq 0$, and
(b) x is nonpositive if and only if $x \leq 0$.

These results certainly agree with our geometric interpretations since all the points corresponding to positive integers are to the right of the origin and all the points corresponding to negative integers are to the left of the origin.

Since we know that I^+ is closed under addition and multiplication and we now see that $x > 0$ if and only if $x \in I^+$, we have:

If $x > 0$ and $y > 0$, then $x + y > 0$ and $xy > 0$.

Since the set P of nonnegative integers is also closed under addition and multiplication, and we now know that $x \geq 0$ if and only if $x \in P$, we have:

If $x \geq 0$ and $y \geq 0$, then $x + y \geq 0$ and $xy \geq 0$.

Since I itself is closed under subtraction we see that $x = y + z$ if and only if $x - y = z$. It follows that when x and y are integers,

$x \geq y$ if and only if $x - y$ is nonnegative.

And, consequently,

$x > y$ if and only if $x - y$ is positive.

Many of the basic theorems about inequalities in I are simple restatements of theorems about natural numbers. All of the assertions numbered from 1 through 14 on page 156 *do* hold if we interpret x, y, and z as integers there. With relatively modest modifications, the proofs given for these assertions about natural numbers become valid for the corresponding assertions about integers.

In I, the assertion that $x \geq y$ if and only if $x + z \geq y + z$ is especially useful since each integer has an additive inverse in I.

Suppose we are told that x is a number such that $x + {}^{+}3 < {}^{-}7$. We get a clearer notion of the condition x satisfies when we claim that the given inequality is logically equivalent to the inequality $(x + {}^{+}3) + {}^{-}3 \leq {}^{-}7 + {}^{-}3$ or $x \leq {}^{-}10$. In other words, $x + {}^{+}3 \leq {}^{-}7$ if and only if $x \leq {}^{-}10$. From another point of view we have shown that $\{x \mid x + {}^{+}3 \leq {}^{-}7\} = \{x \mid x \leq {}^{-}10\} = \{\cdots, {}^{-}13, {}^{-}12, {}^{-}11, {}^{-}10\}$. We see immediately from this that $^{-}10$ is the maximum for $\{x \mid x + {}^{+}3 \leq {}^{-}7\}$. It is also clear that this set has no minimum.

As another example we have $\{x \mid x - {}^{-}5 \geq {}^{-}17\} = \{x \mid (x - {}^{-}5) + {}^{-}5 \geq {}^{-}17 + {}^{-}5\} = \{x \mid x \geq {}^{-}22\} = \{{}^{-}22, {}^{-}21, {}^{-}20, \ldots\}$. The minimum of this set is $^{-}22$ and there is no maximum for the set.

Given the names of two integers x and y can we easily tell whether $x > y$ or $x < y$? Assuming that the names are constructed by one of our standard procedures we can do so, because we have a simple procedure for determining the relative sizes of two natural numbers.

We first note that with any of our usual systems it is easy to tell whether a number is positive or negative. If a number is negative it is less than 0. If it is positive, it is greater than 0. Hence, if either $x = 0$ or $y = 0$ we can tell immediately whether $x > y$ or $x < y$.

In the case where x and y are different from 0 and have different signs, our decision is easy since any negative integer is less than any positive integer.

If x and y are both positive, we compare the natural numbers $|x|$ and $|y|$ and conclude $x > y$ when $|x| > |y|$, and $x < y$ when $|x| < |y|$.

If x and y are both negative, we compare the natural numbers $|x|$ and $|y|$ and conclude $x > y$ when $|x| < |y|$, and $x < y$ when $|x| > |y|$. Note well the change in order here.

The entire procedure is summarized by the flow chart in Figure 7.8.

Definition 7.8.1. If $x \in I$ and $y \in I$ we say that $x \geq y$ if there is a nonnegative integer k such that $x = y + k$. We carry over the definitions of $>$, $\leq$, $<$, "between," "maximum," and "minimum" from N.

Theorem 7.8.1. The relation $\geq$ in the nonnegative integers is isomorphic to the relation $\geq$ in N.

Theorem 7.8.2. Suppose $x \in I$. (a) $x > 0$ if and only if x is positive. (b) $x < 0$ if and only if x is negative.

Theorem 7.8.3. Suppose $x \in I$. (a) $x \geq 0$ if and only if x is nonnegative. (b) $x \leq 0$ if and only if x is nonpositive.

Theorem 7.8.4. Suppose $x \in I$ and $y \in I$. (a) If $x > 0$ and $y > 0$, then $x + y > 0$ and $xy > 0$. (b) If $x \geq 0$ and $y \geq 0$, then $x + y \geq 0$ and $xy \geq 0$.

Theorem 7.8.5. Suppose $x \in I$ and $y \in I$. (a) $x \geq y$ if and only if $x - y$ is nonnegative. (b) $x > y$ if and only if $x - y$ is positive.

Theorem 7.8.6. Suppose x, y, z, and w are integers.

(a) $x \geq x$.
(b) $x \geq y$ and $y \geq z$ implies $x \geq z$.

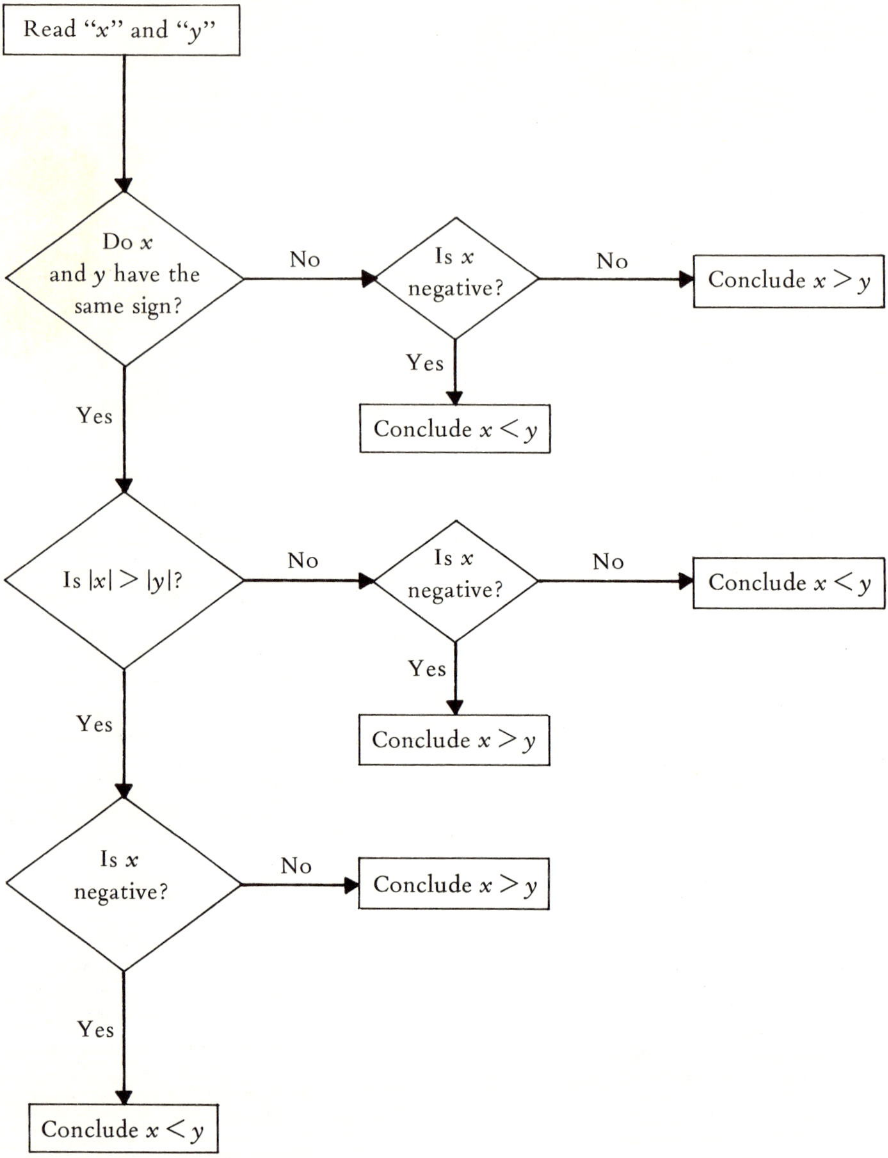

FIGURE 7.8 Flow chart for deciding whether $x < y$ or $x > y$ when x and y are integers and $x \neq 0$ and $y \neq 0$ and $x \neq y$.

(c) $x > y$ and $y > z$ implies $x > z$.
(d) $x \geq y$ or $y \geq x$.
(e) $x \geq y$ and $y \geq x$ implies $x = y$.
(f) $x > y$ or $x = y$ or $x < y$ and at most one of these can hold.

(g) $x + 1 > x$.
(h) $x \geq y$ if and only if $x + z \geq y + z$.
(i) $x > y$ if and only if $x + z > y + z$.
(j) $x \geq y$ and $z \geq w$ implies $x + z \geq y + w$.
(k) $x \geq y$ and $z > w$ implies $x + z > y + w$.
(l) $x < y$ implies $x + 1 \leq y$.
(m) There is no integer x such that $0 < x < 1$.

PROOF: Many of these results may be established by proofs which are simple restatements of proofs of corresponding assertions for N. We shall prove only parts (d), (k), and (l) here as illustrations.

Let x and y be given integers. Then, since I is closed under subtraction, $x - y$ is an integer. $x - y$ must be either nonnegative or nonpositive. In the former case, $x \geq y$. In the latter case, $-(x - y) = y - x$ must be nonnegative, and, hence, $y \geq x$ by Theorem 7.8.5. So we must have $x \geq y$ or $y \geq x$ and this proves (d).

Suppose that $x \geq y$ and $z > w$. Then there is a nonnegative integer k and a positive integer m such that $x = y + k$ and $z = w + m$. Then $x + z = (y + k) + (w + m) = (y + w) + (k + m)$. By Theorem 7.8.4, $k + m$ is nonnegative. And, indeed, $k + m$ must be positive since $k + m = 0$ implies $k = m = 0$ (Theorem 6.2.3). We may conclude that $x + z > y + w$ and we have proved part (k).

Suppose that $x < y$. Then there is a positive integer k such that $y = x + k$. But if k is a positive integer (i.e. a nonzero natural number) there is a nonnegative integer (i.e. a natural number) m such that $k = m + 1$ (Theorem 6.4.7). Hence, $y = x + (m + 1) = (x + 1) + m$ where m is a nonnegative integer. Therefore $y \geq x + 1$ and $x + 1 \leq y$. This proves part (l).

Theorem 7.8.7. If x is negative and y is positive, then $x < y$.

PROOF: Apply Theorem 7.8.2 and Theorem 7.8.6.

Theorem 7.8.8. Suppose x and y are positive. Then $x > y$ if and only if $|x| > |y|$.

Theorem 7.8.9. Suppose x and y are negative. Then $x > y$ if and only if $|x| < |y|$.

PROOF: If x and y are negative, then there are natural numbers n and m such that $x = {}^-n$ and $y = {}^-m$. We have $|x| = n$ and $|y| = m$.

If $x > y$ then there is a positive integer k such that $x = y + k$ or ${}^-n = {}^-m + k$. Hence, ${}^-n + (m + n) = ({}^-m + k) + (m + n)$ or $m = n + k$. Therefore, $n < m$ or $|x| < |y|$.

If $|x| < |y|$, then $n < m$ and there is a positive integer k such that $m = n + k$. Hence, $m + ({}^-m + {}^-n) = (n + k) + ({}^-m + {}^-n)$. This implies ${}^-n = {}^-m + k$. Therefore, ${}^-n > {}^-m$ or $x > y$.

PROBLEMS

1. Which of the following are true?

(a) $^+75 \geq {}^+35$ (b) $^+5 \geq 0$
(c) $^+5 \geq {}^-2$ (d) $^+5 \geq {}^-1000$
(e) $^-50 < 0$ (f) $^-50 < {}^-37$
(g) $^-50 > {}^-6$ (h) $^-574 > {}^+3$
(i) $^-4 < {}^+1$ (j) $^-75 < {}^-156$.

2. Describe each of the following sets in roster notation and find the maximum and minimum of the set if they exist.

(a) $\{x \mid {}^+8 \leq x \leq {}^+13\}$ (b) $\{x \mid {}^+8 < x < {}^+13\}$
(c) $\{x \mid 0 \leq x \leq {}^+5\}$ (d) $\{x \mid 0 < x < {}^+5\}$
(e) $\{x \mid {}^-6 \leq x \leq {}^+2\}$ (f) $\{x \mid {}^-6 < x < {}^+2\}$
(g) $\{x \mid x \geq {}^+3\}$ (h) $\{x \mid x \leq {}^-4\}$
(i) $\{x \mid x + {}^+3 \geq {}^-15\}$ (j) $\{x \mid x + {}^-8 \geq {}^+16\}$
(k) $\{x \mid x + 1 \geq x\}$ (l) $\{x \mid x - {}^-7 \leq {}^-4\}$
(m) $\{x \mid {}^+7 - x \geq {}^-4\}$.

3. Prove Theorem 7.8.2 (b).
4. Prove parts (a) through (k) of Theorem 7.8.6.
5. Prove Theorem 7.8.7.
6. For each of the following assertions, decide whether it is true or false and explain carefully how you know. x, y, z, and w are integers.

(a) If x is between y and z, then $x + w$ is between $y + w$ and $z + w$.
(b) $x \leq y$ if and only if $x - z \leq y - z$.
(c) $x < 2x$.
(d) If $x + y > 0$ then $x > 0$ and $y > 0$.
(e) If $x + y > 0$ then $x > 0$ or $y > 0$.
(f) $x < y$ and $y < z$ implies $x < z$.
(g) x is between y and z if and only if $x - y$ and $x - z$ differ in sign.

7. Describe the following sets in roster notation:

(a) $\{x \mid x - {}^+3 \geq {}^-7 \text{ and } x + {}^-5 \leq {}^+6\}$
(b) $\{x \mid x - {}^+3 \geq {}^-7 \text{ or } x + {}^-5 \leq {}^+6\}$
(c) $\{x \mid {}^-7 \leq x \leq {}^+10 \text{ and } {}^+1 \leq x \leq {}^+12\}$
(d) $\{x \mid {}^-7 \leq x \leq {}^+10 \text{ or } {}^+1 \leq x \leq {}^+12\}$
(e) $\{x \mid {}^-5 \leq x < {}^-1 \text{ and } {}^-1 < x < {}^+7\}$
(f) $\{x \mid {}^-5 \leq x < {}^-1 \text{ or } {}^-1 < x < {}^+7\}$.

7.9 More about inequalities

$^+5 < ^+8$ in I and yet $(^-1)(^+5) > (^-1)(^+8)$ in I. Things are more complicated in I than in N when we get involved with multiplication and equalities. Consider the following examples:

$$
\begin{array}{lll}
^+5 < ^+12 & \text{and} & ^+5\cdot{}^+2 < {}^+12\cdot{}^+2 \\
^+5 < ^+12 & \text{but} & ^+5\cdot{}^-2 > {}^+12\cdot{}^-2 \\
^-6 < ^+2 & \text{and} & ^-6\cdot{}^+3 < {}^+2\cdot{}^+3 \\
^-6 < ^+2 & \text{but} & ^-6\cdot{}^-2 > {}^+2\cdot{}^-2 \\
^-10 < ^-4 & \text{but} & ^-10\cdot 0 = {}^-4\cdot 0 \\
^-10 < ^-4 & \text{but} & ^-10\cdot{}^-5 > {}^-4\cdot{}^-5.
\end{array}
$$

The general result that explains such cases is as follows:

Let x, y, and z be integers. (a) If $z > 0$, then $x < y$ when and only when $xz < yz$. (b) If $z < 0$, then $x < y$ when and only when $xz > yz$. (Note the reversal of order in part (b)).

Similar statements carry over to the cases where $x \leq y$, $x > y$, or $x \geq y$. In other words, multiplication or division (when it is possible) of both sides of an inequality by a positive number preserves the inequality, but multiplication or division (when it is possible) of both sides of an inequality by a **negative** number *reverses* the inequality.

Suppose as an application of the above, that we wish to consider the collection of integers $A = \{x \mid {}^-2x + {}^+7 \leq {}^+3\}$. We see that $A = \{x \mid {}^-2x \leq {}^+3 + {}^-7\} = \{x \mid 2x \leq {}^-4\} = \{x \mid {}^-2x \leq {}^-2\cdot{}^+2\} = \{x \mid x \geq {}^+2\}$. Then $\min(A) = {}^+2$ and $\max(A)$ does not exist. It is much easier to see the nature of A from the description "$\{x \mid x \geq {}^+2\}$" than it is from the description "$\{x \mid {}^-2x + {}^+7 \leq {}^+3\}$."

From our intuitive grasp of the number line the following basic properties of I are clear:

If x is an integer, then there is no integer y such that $x < y < x + 1$.

There do exist nonempty sets of integers which have no minimum elements. (An example would be I itself.)

Any nonempty **bounded** set of integers has both a maximum and a minimum.

Any finite set of integers is bounded, and conversely.

Consider the set $B = \{x \mid |x| \leq 4\}$. We note that when x is any integer, $|x| \geq 0$. And if we look at a number line with integral points we see that we may think of $|x|$ geometrically in terms of distance. In fact, $|x|$ is the distance between the point x and the origin. Thus, we may think of the set B above as

$\{x|x$ is no more than 4 units away from the origin$\}$. This interpretation helps us to see that $B = \{x|{}^-4 \leq x \leq {}^+4\} = \{{}^-4, {}^-3, {}^-2, {}^-1, 0, {}^+1, {}^+2, {}^+3, {}^+4\}$. B is bounded and finite, and $\max(B) = {}^+4$, and $\min(B) = {}^-4$.

In more general terms, we shall have for any natural number n, $\{x||x| \leq n\} = \{x|$the point x is no more than n units away from the origin$\} = \{x|{}^-n \leq x \leq {}^+n\}$.

When x and y are integers we may think of $|x - y|$ as the distance between the point x and the point y on a number line. We may think of this as a generalization of the above result since when $y = 0$ we have $|x - y| = |x - 0| = |x|$.

As examples for the general case we have

$$|{}^+7 - {}^+3| = |{}^+4| = 4,$$

$$|{}^+3 - {}^+7| = |{}^-4| = 4,$$

$$|{}^-7 - {}^+3| = |{}^-7 + {}^-3| = |{}^-10| = 10,$$

$$|{}^+3 - {}^-7| = |{}^+3 + {}^+7| = |{}^+10| = 10,$$

$$|{}^-15 - {}^-4| = |{}^-15 + {}^+4| = |{}^-9| = 9, \qquad \text{and}$$

$$|{}^-4 - {}^-15| = |{}^-4 + {}^+15| = |{}^+9| = 9$$

while the distance between the points ${}^+7$ and ${}^+3$ is 4 units, the distance between the points ${}^-7$ and ${}^+3$ is 10 units, and the distance between the points ${}^-15$ and ${}^-4$ is 9 units.

This geometric interpretation is helpful in such simplifications as the following: $\{x||x - {}^+5| \leq 6\} = \{x|$the point x is no more than 6 units from the point ${}^+5\} = \{{}^-1, 0, {}^+1, {}^+2, {}^+3, {}^+4, {}^+5, {}^+6, {}^+7, {}^+8, {}^+9, {}^+10, {}^+11\}$. See Figure 7.9.

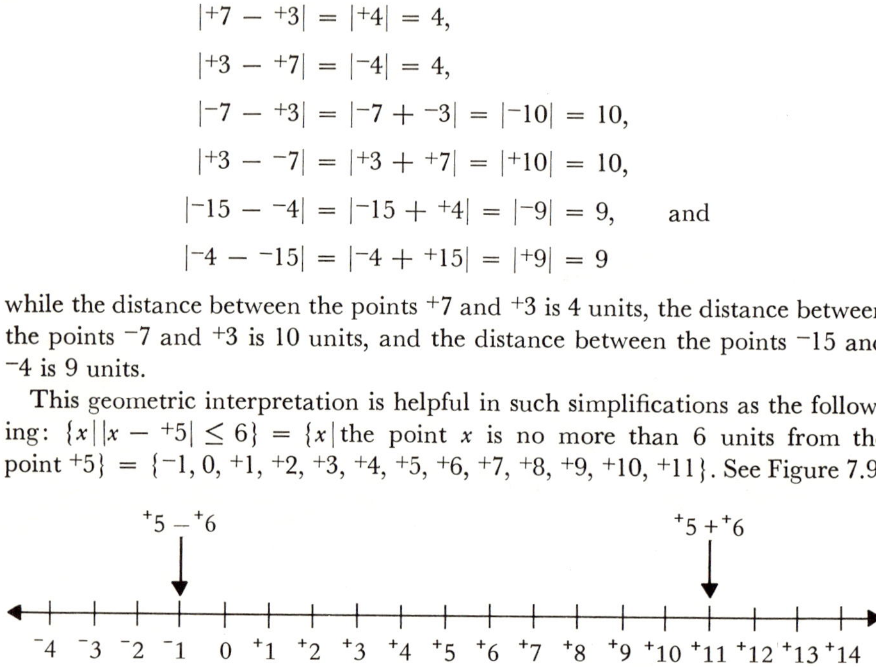

FIGURE 7.9

In general, we have that when b is a nonnegative integer, $\{x||x - a| \leq b\} = \{x|a - b \leq x \leq a + b\}$.

There is one other important fact about absolute values and inequalities. Notice that

$$|{}^+7 + {}^+5| = 12 \quad \text{while} \quad |{}^+7| + |{}^+5| = 12,$$

$$|{}^-7 + {}^+5| = 2 \quad \text{while} \quad |{}^-7| + |{}^+5| = 12,$$

$$|{}^+7 + {}^-5| = 2 \quad \text{while} \quad |{}^+7| + |{}^-5| = 12, \qquad \text{and}$$

$$|{}^-7 + {}^-5| = 12 \quad \text{while} \quad |{}^-7| + |{}^-5| = 12.$$

We might be lead to conjecture that in every case $|x + y| \leq |x| + |y|$. And this is indeed true for any integers x and y. The equality will hold whenever x and y have the same sign, and the strong inequality will hold whenever x and y have opposite signs.

Theorem 7.9.1. Suppose x, y, and z are integers.

(a) If $z > 0$, then $x < y$ if and only if $xz < yz$.
(b) If $z < 0$, then $x < y$ if and only if $xz > yz$.

PROOF: We shall prove part (b). The proof for (a) is similar. Suppose $z < 0$. Then z is negative.

If $x < y$, then $y > x$ and by Theorem 7.8.5(b), $y - x$ is positive. Hence, $z(y - x)$ is negative. Therefore, by Theorem 7.9.2, $z(y - x) < 0$. Hence, $zy - zx < 0$. Then, by Theorem 7.8.5, $zx > zy$.

If $zx > zy$, then $zx - zy$ is positive. Hence, $z(x - y)$ is positive. Then $x - y \neq 0$. But if $x > y$ then $x - y$ is positive and $z(x - y)$ would be negative. Since this is not the case we must have $x < y$ by Theorem 7.8.6(f).

Theorem 7.9.2. If x is an integer, there is no integer y such that $x < y < x + 1$.

PROOF: Suppose $x < y$. Then, by Theorem 7.8.6(l), $x + 1 \leq y$. Hence, we cannot have $y < x + 1$ by Theorem 7.8.6(f).

Theorem 7.9.3. Any nonempty bounded set of integers is finite and has both a maximum and a minimum.

PROOF: Let A be a nonempty bounded set of integers. Let $B = \{y \mid y = |x|$ for some x in $A\}$. B must be bounded for, if not, A will be unbounded. Then B is finite. But it follows that A must be finite since there are at most two different integers for each absolute value in A. But when A is finite we may, by comparing all possible pairs of elements in A, find $\min(A)$ and $\max(A)$.

Theorem 7.9.4. Any finite nonempty set of integers is bounded.

PROOF: Suppose A is finite and nonempty. Then $\max(A)$ and $\min(A)$ exist by the argument in the proof of Theorem 7.9.3. Then all the points in A fall between $\min(A) - 1$ and $\max(A) + 1$.

Theorem 7.9.5. If x is an integer and n is a natural number, $|x| \leq n$ if and only if $^-n \leq x \leq {}^+n$.

PROOF: We consider two cases.

Suppose $x < 0$. Then $|x| = -x$. Hence, $|x| \leq n$ if and only if $-x \leq n$. But by Theorem 7.9.1(b), $-x \leq n$ if and only if $(^-1)(-x) \geq (^-1)(n)$ which is equivalent to $x \geq {}^-n$. Hence, when $x < 0$, $^-n \leq x < 0$ if and only if $|x| \leq n$.

Suppose $x \geq 0$. Then $|x| = x$ and $|x| \leq n$ if and only if $x \leq n$. Hence, when $x \geq 0$, $|x| < n$ if and only if $0 \leq x \leq {}^+n$.

These two results together with Theorem 7.8.6(f) yield the theorem.

Theorem 7.9.6. If x, a, and b are integers and b is nonnegative, then $\{x \mid |x - a| \leq b\} = \{x \mid a - b \leq x \leq a + b\}$.

PROOF: By Theorem 7.9.5, $|x - a| \leq b$ if and only if ${}^-b \leq x - a \leq b$. By Theorem 7.8.6(h), this last holds if and only if $a - b \leq x \leq a + b$.

Theorem 7.9.7. If x and y are integers, $|x + y| \leq |x| + |y|$.

PROBLEMS

1. If necessary, introduce symbols to stand for the given quantities involved and state the given condition in terms of an inequality or inequalities.

(a) We wish to sell a certain number of identical items at a fixed price and gross at least \$1000. [SOLUTION: Let n be the number of items to be sold. Let p be the selling price in dollars for each item. We want $np \geq 1000$.]

(b) We wish to work a certain number of hours at a fixed hourly wage and earn at least \$2000.

(c) We own some shares of a certain stock and wish to sell these shares when we can gross at least \$500 from the sale.

(d) We have a fixed monthly income from an annuity and another fixed monthly income from a retirement fund. We wish to work a certain number of hours during the month at a fixed hourly wage so that we get a total income of at least \$1000, but not more than \$1500 next month.

(e) An association has two types of members who pay different amounts of dues. The association wants to collect at least \$50,000 in dues from its members.

(f) A manufacturer has a chance to stock up at special prices on some parts that he uses. There are three kinds of parts at three different unit prices. He wishes to pay no more than a total of \$4700 for the parts.

(g) The same situation as in part (f), but there are also limitations on the amount of storage the manufacturer has available, and also in the quantity of any particular item available from the seller.

(h) We wish to drive from town A to town B, from B to C, and then return to A. Unrealistically, but in view of known road conditions, we plan to travel at different constant speeds on each of the trips between towns. We wish our total driving time to be less than 15 hours.

(i) A manufacturer makes four different kinds of small items. He can make no more than 100 of any one kind each month. And in any month he cannot make a total of more than 220 items. Each item is sold at a certain unit price. He wants the total selling price of his month's production to be at least \$2000.

(j) Take the set up as in part (i), but assume further that the manufacturer knows the unit cost of production for each kind of item and wants his total monthly production costs to be no more than \$1500.

2. Each time a certain game for two people is played, the winner gets \$10 from the loser. Two people, I and II, play the game several times with the final outcome being that I wins a total of A dollars (A could be negative). What, if anything, can be said about the number of times the game was played if it is known that

(a) $|A| \geq 50$
(b) $|A| = 120$ and one of the players won every play of the game
(c) $|A| \leq 120$ and one of the players won every play of the game
(d) $A = 0$
(e) neither player won more than 4 plays of the game and $|A| = 40$
(f) neither player won more than 4 plays of the game and $|A| = 30$
(g) neither player won more than 4 plays of the game and $A = 0$
(h) neither player won more than 4 plays of the game and $|A| = 50$?

3. Describe each of the following sets in roster notation and find the maximum and minimum of the set if they exist:

(a) $\{x \mid {}^{+}2x \leq {}^{+}4\}$
(b) $\{x \mid {}^{-}2x \geq {}^{-}8\}$
(c) $\{x \mid {}^{-}2x \geq {}^{+}8\}$
(d) $\{x \mid {}^{+}5x - {}^{+}7 \leq {}^{+}3\}$
(e) $\{x \mid {}^{-}2x + {}^{-}47 \leq {}^{-}1579\}$
(f) $\{x \mid {}^{-}4x + {}^{-}8 \leq {}^{+}12\}$
(g) $\{x \mid {}^{+}2x \leq {}^{+}3\}$
(h) $\{x \mid {}^{+}2x \geq x\}$
(i) $\{x \mid {}^{-}2x + {}^{+}1 \leq {}^{-}2x\}$.

4. Do what was requested in Problem 3 for the following sets:

(a) $\{x \mid |x| \leq {}^{+}1\}$
(b) $\{x \mid |x| \leq 0\}$
(c) $\{x \mid |x - {}^{+}3| \leq {}^{+}2\}$
(d) $\{x \mid |x - {}^{-}3| \leq {}^{+}2\}$
(e) $\{x \mid |x - {}^{-}3| \leq {}^{-}2\}$
(f) $\{x \mid {}^{+}2|x| + {}^{-}7 \leq {}^{+}5\}$
(g) $\{x \mid |x| \geq 2\}$
(h) $\{x \mid {}^{+}3|x - {}^{+}5| + {}^{+}6 \leq {}^{+}27\}$
(i) $\{x \mid {}^{-}7|x - {}^{-}6| - {}^{+}15 \geq {}^{+}8\}$.

5. Which of the sets in Problem 3 are (i) finite, (ii) bounded?
6. Which of the sets in Problem 4 are (i) finite, (ii) bounded?
7. Prove Theorem 7.9.1(a).
8. For each of the following assertions, decide whether it is true or false and explain carefully how you know. x, y, z, and w are integers here:

(a) $x \neq 0$ if and only if $|x| \not< 0$.
(b) If x is between y and z, then xw is between yw and zw.

(c) If x is between y and z and $w \neq 0$, then xw is between yw and zw.
(d) If z is a divisor of both x and y and $x \leq y$, then $x \div z \leq y \div z$.
(e) If $xy > 0$ then $x > 0$ or $y > 0$.
(f) If $xy < 0$, then $x < 0$ or $y < 0$.
(g) Given any x and y, there is a z such that $xz > y$.
(h) Given any x and y, if $x \neq 0$, then there is a z such that $xz > y$.
(i) $x^2 \geq 0$.
(j) If $x \neq 0$, then $x^2 > 0$.
(k) If $x^2 \leq 9$, then $x \leq 3$.
(l) If $x \leq 3$, then $x^2 \leq 9$.

9. Prove Theorem 7.9.7. [HINT: Consider the various possible cases suggested by the examples given immediately before the first statement of the theorem.]

7.10 Some graphical devices

We may extend the nomograph given in Figure 6.15 on page 166 so that it may be used for adding integers. The result is shown in Figure 7.10.

Since subtraction is the inverse operation for addition, this nomograph can also be used for subtractions. To find $z - x$, draw a line through the point z on the middle axis and the point x on the left axis. This line will hit the right axis in the point $y = z - x$.

We may also use integral points on number lines to construct a slide rule for addition. We lay out such a line along the bottom edge of a piece of cardboard or wood, and then again lay out such a line along the top edge of another. To find $z = x + y$, we put the two edges together and slide the top one along until the origin on the top scale is over the point x on the bottom scale. Directly under the point y on the top scale we shall find the point z which is the sum. See Figure 7.11 where the computation of $^-5 + {}^+7$ is illustrated. The slide rule can also be used to carry out subtractions.

The graphical device for the addition of natural numbers as shown on page 167 can also be extended so as to be useful with integers. The result is shown in Figure 7.12.

We see that in this figure it is useful, as it was once before, to construct two perpendicular number lines which intersect at a point which is the origin on each line.

These same perpendicular axes occur when we extend to I the graphical device for multiplication as given on page 168. The result is given in Figure 7.13, which enables us to carry out multiplications for some integers with small absolute values.

Figure 7.13 may also be used in a limited way to carry out divisions since division is the inverse operation for multiplication.

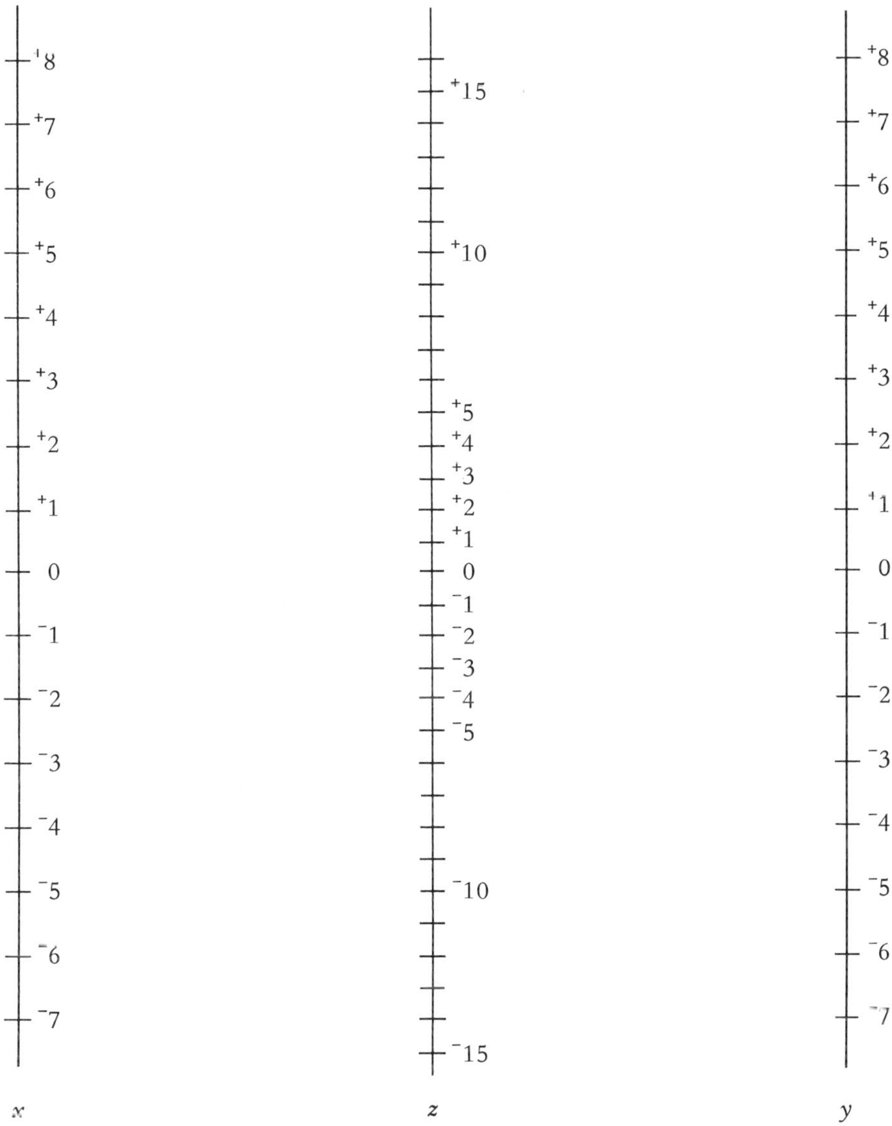

FIGURE 7.10 Nomograph for the addition of integers. To find $z = x + y$, draw the line from x on the left axis to y on the right axis. This line crosses the middle axis at z.

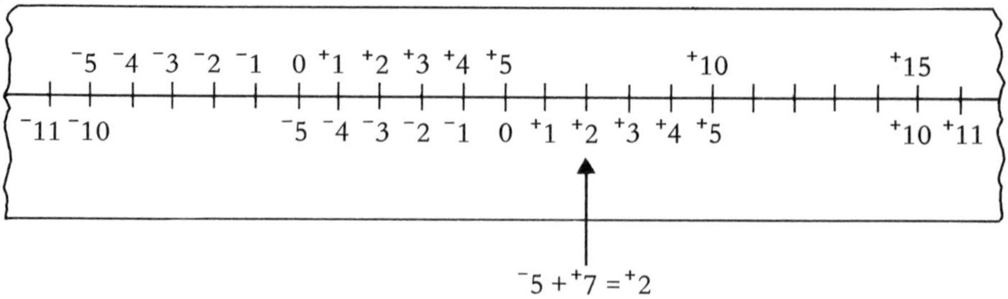

FIGURE 7.11 A slide rule for addition of integers.

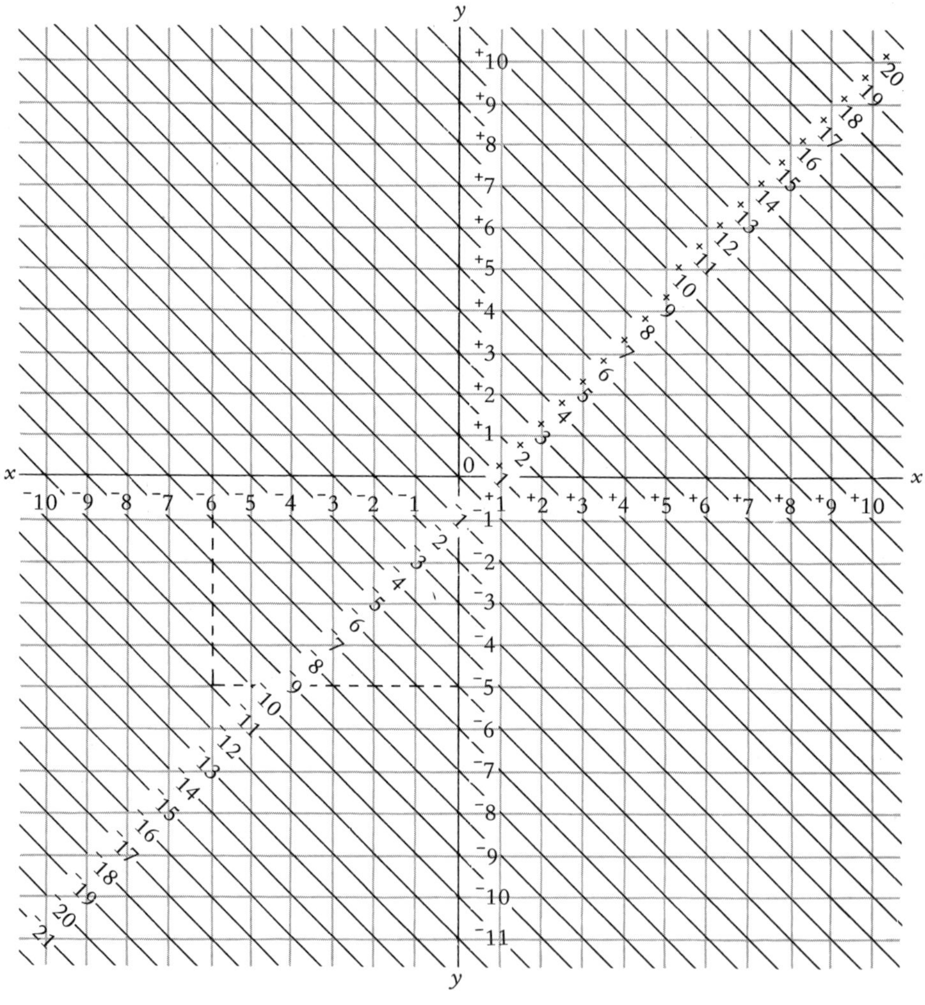

FIGURE 7.12 To find the sum of x and y, draw a vertical through x on the horizontal axis and a horizontal through y on the vertical axis. The lines will intersect on the line labeled $z = x + y$. The computation of $^-6 + {}^-5 = {}^-11$ is shown.

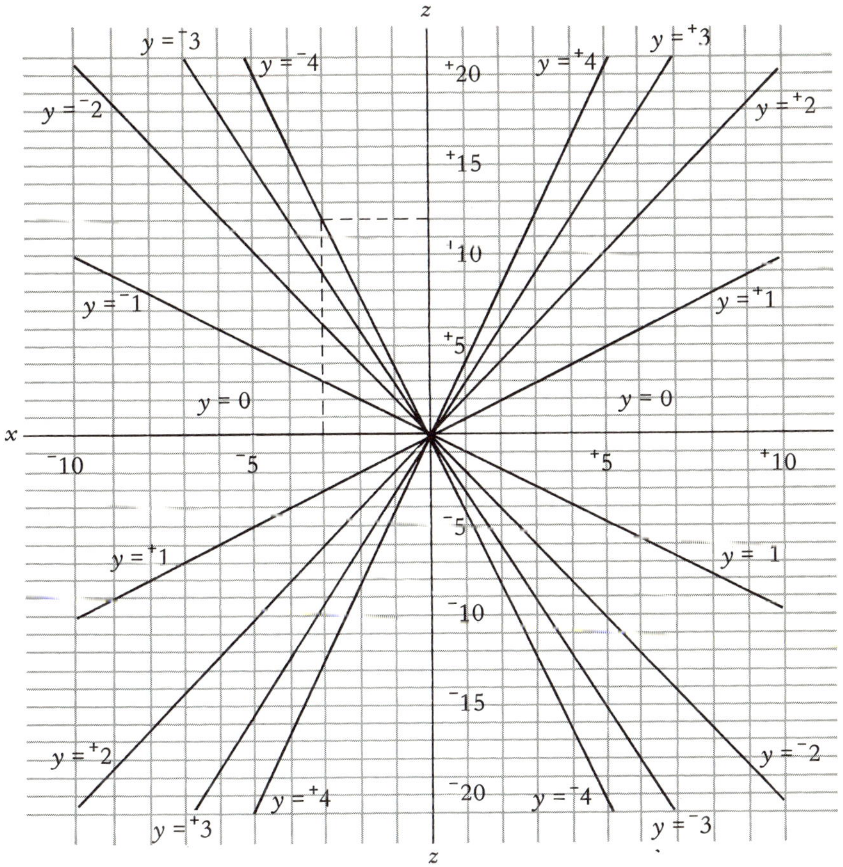

FIGURE 7.13 To find $z = xy$, draw the vertical at x to the line labeled y. From there draw the horizontal to z on the vertical axis. The computation of $^-3 \cdot {}^-4 = {}^+12$ is shown.

PROBLEMS

1. Verify, by means of several examples, that the device actually does what is claimed for it in

(a) Figure 7.10 (b) Figure 7.12 (c) Figure 7.13.

2. Show, in a sketch, what a slide rule for addition would look like when it is used to calculate

(a) $^+5 + {}^+7$ (b) $^-6 + {}^-4$ (c) $^-10 + {}^+6$.

3. Describe carefully how to use Figure 7.12 to perform subtractions with integers. Illustrate your method with a particular example.

4. Describe carefully how to use Figure 7.13 to perform divisions with integers. Illustrate your method with a particular example.

5. How could Figure 7.13 be extended so that it could be used to evaluate some products xy where $y = {}^+5, {}^+6, {}^-5$, or ${}^-6$?

6. Describe in detail how to use an addition slide rule for performing subtractions of integers. Illustrate your method with an example and a sketch.

*7.11 Modular arithmetic

The even integers are those that are multiples of 2. The numbers . . . , ${}^-6$, ${}^-4$, ${}^-2$, 0, ${}^+2$, ${}^+4$, ${}^+6$, . . . are even. Let us introduce a relation M in I by saying that $x \, M \, y$ if and only if $x - y$ is even. For example, ${}^+6 \, M \, {}^+10$ since ${}^+6 - {}^+10 = {}^-4$ and ${}^-4$ is even, ${}^+7 \, M \, {}^+15$ since ${}^+7 - {}^+15 = {}^-8$ and ${}^-8$ is even, and ${}^-10{,}751 \, M \, {}^+1$ since ${}^-10{,}751 - {}^+1 = {}^-10{,}752$ and ${}^-10{,}752$ is even.

It is not difficult to show that M is an equivalence relation in I. Its equivalence classes are $\{\ldots, {}^-6, {}^-4, {}^-2, 0, {}^+2, {}^+4, {}^+6, \ldots\}$ and $\{\ldots, {}^-5, {}^-3, {}^-1, {}^+1, {}^+3, {}^+5, \ldots\}$. The first of these sets is the collection of all even integers. The second is the set of all odd integers. But we could also use our standard notation for equivalence classes and identify these classes as $[0]$ and $[{}^+1]$, respectively. Two numbers that belong to the same equivalence class for M are sometimes said to have the same **parity.**

There is a standard notation for the relation M. Where we have written "$x \, M \, y$" the more usual terminology is "$x \equiv y \pmod 2$." This is read as "x is congruent to y modulo 2" or "x is congruent to y mod 2." We shall now adopt this more common notation. $x \equiv y \equiv z \pmod 2$ means that $x \equiv y \pmod 2$ and $y \equiv z \pmod 2$.

We see that $x \equiv 0 \pmod 2$ if and only if x is even. And $x \equiv {}^+1 \pmod 2$ if and only if x is odd.

It is quite easy to show that if $x \equiv 0 \pmod 2$ and $y \equiv 0 \pmod 2$ then $x + y \equiv 0 \pmod 2$. In other words, the sum of two even integers is even. For, suppose x and y are even. Then there are integers n and m such that $x = 2n$ and $y = 2m$. Then $x + y = 2n + 2m = 2(n + m)$ so that $x + y$ must be even.

More generally, suppose $x \equiv y \pmod 2$ and $z \equiv w \pmod 2$. We claim that we must then have $x + z \equiv y + w \pmod 2$. The proof of this is straightforward. $x \equiv y \pmod 2$ implies that $x - y$ is even. $z \equiv w \pmod 2$ implies that $z - w$ is even. But then $(x - y) + (z - w)$ is even. And $(x - y) + (z - w) = (x + z) - (y + w)$. Hence, $x + z \equiv y + w \pmod 2$.

As an illustration of how we might use this result, suppose that we consider the numbers $x = {}^+53{,}471$ and $y = {}^-7{,}546{,}170$. It is easy to see that $x \equiv 1 \pmod 2$ and $y \equiv 0 \pmod 2$. Then $x + y \equiv 1 + 0 \equiv 1 \pmod 2$ and we get this result without actually finding the sum of x and y at all.

It is quite easy to prove that if $x \equiv 0 \pmod 2$ and y is any integer, then

$xy \equiv 0 \pmod{2}$. That is, the product of an even integer with any integer is even. We leave the proof as a problem.

More generally, if $x \equiv y \pmod{2}$ and $z \equiv w \pmod{2}$ then $xz \equiv yw \pmod{2}$. This may be proved as follows. $x \equiv y \pmod{2}$ implies that $x - y$ is even. $z \equiv w \pmod{2}$ implies that $z - w$ is even. Then $xz - yw = xz - yz + yz - yw = z(x - y) + y(z - w)$. But $z(x - y)$ is even since $x - y$ is. And $y(z - w)$ is even since $z - w$ is. Hence, their sum is even, and we conclude that $xz \equiv yw \pmod{2}$.

The results above may be interpreted as suggesting that we may build in a very natural way an algebraic system on the collection of equivalence classes $A = \{[0], [^+1]\}$. We see that if x and y are in the same one of these classes, and z and w are in the same class, then $x + z$ and $y + w$ are in the same class.

Let us define the operations of **addition modulo 2** and **multiplication modulo 2** in A by saying that $[x] +_2 [y] = [x + y]$ and $[x] \cdot_2 [y] = [xy]$.

For example, $[^+1] +_2 [0] = [^+1 + 0] = [^+1]$ and $[^+1] \cdot_2 [^+1] = [^+1 \cdot {}^+1] = [^+1]$. The results of these and similar calculations may be summarized in the following tables:

$+_2$	[0]	[+1]
[0]	[0]	[+1]
[+1]	[+1]	[0]

$\cdot_2$	[0]	[+1]
[0]	[0]	[0]
[+1]	[0]	[+1]

If we drop the brackets, we see that we have here the clock arithmetic for a two-hour clock as discussed on page 96.

And from another point of view, since [0] is the set of even integers and $[^+1]$ is the set of odd integers, the above tables can be thought of as asserting the same as the following tables:

+	even	odd
even	even	odd
odd	odd	even

·	even	odd
even	even	even
odd	even	odd

The last two tables simply summarize several assertions such as "the sum of an odd and an even integer is odd" and "the product of two odd integers is odd."

The systems $(\{[0], [^+1]\}, +_2, \cdot_2)$, $(\{0,1\}, \underset{+}{\sim}, \underset{\cdot}{\sim})$ (see page 90), and ({even, odd}, +,) are all isomorphic.

The nice thing about putting things in terms of equivalence classes is that it is easy to generalize the situation from this point of view. Let k be a positive integer. We shall say that $x \equiv y \pmod{k}$ if and only if $x - y$ is a multiple of k. This relation is an equivalence relation.

If, for instance, $k = 4$, then the equivalence classes are

$$[0] = \{\ldots, {}^-8, {}^-4, 0, {}^+4, {}^+8, \ldots\}$$

$$[^+1] = \{\ldots, {}^-11, {}^-7, {}^-3, {}^+1, {}^+5, {}^+9, \ldots\}$$

$$[^+2] = \{\ldots, ^-10, ^-6, ^-2, ^+2, ^+6, ^+10, \ldots\}$$

$$[^+3] = \{\ldots, ^-9, ^-5, ^-1, ^+3, ^+7, ^+11, \ldots\}.$$

For any integer k, we may define the operations of addition modulo k and multiplication modulo k on the collection of equivalence classes by the equations:

$$[x] +_k [y] = [x + y], \qquad \text{and}$$

$$[x] \cdot_k [y] = [xy].$$

In Figure 7.14 we have given tables for the operations that result when $k = 3$, 4, 5, and 6. In these tables we have, for simplicity, omitted the square brackets around the numerals. In the form in which the tables are given, they may be thought of as giving the operations for a clock arithmetic for a k hour clock.

It is quite easy to show for any integer k that the collection of equivalence classes of integers modulo k is closed under $+_k$ and $\cdot_k$, that $+_k$ and $\cdot_k$ are commutative and associative, that $\cdot_k$ distributes across $+_k$, and that [0] and $[^+1]$ are the identity elements for $+_k$ and $\cdot_k$, respectively.

Modular addition and multiplication are very useful tools and occur in many places in mathematics. And the congruence symbol "$\equiv$" is most often used in these places. Note that in our terminology we may write either "$4 + 3 \equiv 1$ (mod 6)" or "$[4] +_6 [3] = [1]$" and mean essentially the same thing in each case. In fact, if we take $+_6$ to be defined by the table in Figure 7.14, we may even write "$4 +_6 3 = 1$." But it would probably be confusing to write "$4 + 3 = 1$" since the symbol "$+$" is usually taken to mean ordinary addition in one of the more common number systems.

We may think of the table for $+_6$ in Figure 7.14 as either an abbreviation for the table for addition modulo 6, which is actually an operation on equivalence classes, or as a table which gives us a system $(\{0,1,2,3,4,5\},+_6)$ which is isomorphic to the system $(\{[0],[1],[2],[3],[4],[5]\},+_6)$ of addition of integers mod 6. Similar remarks apply to the other tables in Figure 7.14.

Definition 7.11.1. $x \equiv y$ (mod k) if and only if $x - y$ is a multiple of k. We read "$x \equiv y$ (mod k)" as "x is congruent to y modulo(or mod) k."

Theorem 7.11.1. For any positive integer k, the relation of congruence modulo k is an equivalence relation.

PROOF: Let k be a positive integer. Since $0 = 0 \cdot k$, 0 is a multiple of k. Hence, $x \equiv x$ (mod k). If $x \equiv y$ (mod k), then $x - y$ is a multiple of k. And when $x - y$ is a multiple of k, so is $y - x$. Hence, $x \equiv y$ (mod k) implies $y \equiv x$ (mod k). Suppose $x \equiv y$ (mod k) and $y \equiv z$ (mod k). Then $x - y$ and $y - z$ are multiples of k. That is, there are integers n and m such that $x - y = nk$ and $y - z = mk$. Therefore, $x - z = (x - y) + (y - z) = nk + mk = (n + m)k$ and $x - z$ is seen to be a multiple of k.

That is, $x \equiv y \pmod{k}$ and $y \equiv z \pmod{k}$ implies $x \equiv z \pmod{k}$. This completes the proof.

$k = 3$

$+_3$	0	1	2
0	0	1	2
1	1	2	0
2	2	0	1

$\cdot_3$	0	1	2
0	0	0	0
1	0	1	2
2	0	2	1

$k = 4$

$+_4$	0	1	2	3
0	0	1	2	3
1	1	2	3	0
2	2	3	0	1
3	3	0	1	2

$\cdot_4$	0	1	2	3
0	0	0	0	0
1	0	1	2	3
2	0	2	0	2
3	0	3	2	1

$k = 5$

$+_5$	0	1	2	3	4
0	0	1	2	3	4
1	1	2	3	4	0
2	2	3	4	0	1
3	3	4	0	1	2
4	4	0	1	2	3

$\cdot_5$	0	1	2	3	4
0	0	0	0	0	0
1	0	1	2	3	4
2	0	2	4	1	3
3	0	3	1	4	2
4	0	4	3	2	1

$k = 6$

$+_6$	0	1	2	3	4	5
0	0	1	2	3	4	5
1	1	2	3	4	5	0
2	2	3	4	5	0	1
3	3	4	5	0	1	2
4	4	5	0	1	2	3
5	5	0	1	2	3	4

$\cdot_6$	0	1	2	3	4	5
0	0	0	0	0	0	0
1	0	1	2	3	4	5
2	0	2	4	0	2	4
3	0	3	0	3	0	3
4	0	4	2	0	4	2
5	0	5	4	3	2	1

FIGURE 7.14

Theorem 7.11.2. If $x \equiv y \pmod{k}$ and $z \equiv w \pmod{k}$ then $x + z \equiv y + w \pmod{k}$ and $xz \equiv yw \pmod{k}$.

PROOF: $(x + z) - (y + w) = (x - y) + (z - w)$ and this is a multiple of k if $x - y$ and $z - w$ are. Also $xz - yw = xz - zy + zy - yw = z(x - y) + y(z - w)$ and this is a multiple of k if $x - y$ and $z - w$ are.

Definition 7.11.2. In the collection of equivalence classes of integers modulo k we define **modular addition** and **multiplication** by the equations

$$[x] +_k [y] = [x + y]$$

$$[x] \cdot_k [y] = [xy].$$

Theorem 7.11.3. The collection of equivalence classes of integers modulo k is closed under modular addition and multiplication. And these operations are commutative and associative, and the latter distributes across the former. $[0]$ and $[1]$ are the identity elements for $+_k$ and $\cdot_k$, respectively.

PROOF: We shall prove only part of this theorem, leaving the rest of the proof as a problem. To establish that modular multiplication distributes across modular addition we need only note that $[x] \cdot_k ([y] +_k [z]) = [x] \cdot_k [y + z] = [x(y + z)] = [xy + xz] = [xy] +_k [xz] = ([x] \cdot_k [y]) +_k ([x] \cdot_k [z])$ where the third equality is justified by the distributive law in I.

PROBLEMS

1. Which of the following are true?

(a) $30 \equiv 0 \pmod 2$ (b) $57 \equiv 0 \pmod 2$
(c) $30 \equiv 5 \pmod 5$ (d) $57 \equiv 43 \pmod 3$
(e) $5764 \equiv 64 \pmod{100}$

2. Display several members of each equivalence class for the relation of congruence modulo k when k is

(a) 3 (b) 5 (c) 6 (d) 7 (e) 8.

3. Prove that the product of any integer with an even integer is even.

4. Construct tables for the operations of addition and multiplication modulo k when k is

(a) 7 (b) 8.

5. By checking a few cases partially verify that the tables for $+_4$ and $\cdot_4$ in Figure 7.14 are correct when we interpret the addition in terms of a 4-hour clock and interpret multiplication as repeated addition.

6. If possible, find an x such that

(a) $1 + x \equiv 0 \pmod 2$ (b) $4 + x \equiv 2 \pmod 5$
(c) $3x \equiv 1 \pmod 5$ (d) $4x \equiv 3 \pmod 5$
(e) $4x \equiv 1 \pmod 6$ (f) $2x \equiv 1 \pmod 3$

(g) $2x \equiv 1 \pmod 4$ (h) $3x \equiv 4 \pmod 6$
(i) $2x \equiv 4 \pmod 5$.

7. Find, if possible, the inverse element with respect to $+_k$ and the inverse element with respect to $\cdot_k$ for each equivalence class when k is

(a) 3 (b) 4 (c) 5 (d) 6 (e) 7 (f) 8.

8. Explain the details of the argument in the proof of Theorem 7.11.2.
9. Complete the proof of Theorem 7.11.3.

*7.12 Groups

When we consider the system of integers with the operation of addition we find that the following are some basic properties of the system.

(i) I is closed under $+$.
(ii) $+$ is associative in I.
(iii) There is an identity element for $+$ in I.
(iv) Each element in I has an inverse with respect to $+$.

Many systems with a wide variety of uses have similar properties. Such systems are called **groups.** More precisely, if G is a set and $\oplus$ is an operation on G, then the system $(G, \oplus)$ is said to be a **group** if

(i) G is closed under $\oplus$.
(ii) $\oplus$ is associative in G.
(iii) There is an identity element for $\oplus$ in G.
(iv) Each element in G has an inverse element with respect to $\oplus$.

If the system $(G, \oplus)$ is a group, then we say that the set G is a **group with respect to** $\oplus$. We saw above that I is a group with respect to addition. On the other hand, I is not a group with respect to multiplication because it is not true that each integer has an inverse element with respect to multiplication. In particular, $^+2$ has no multiplicative inverse in I.

If a system $(G, \oplus)$ is a group and, furthermore, $\oplus$ is commutative in G, then $(G, \oplus)$ is said to be a **commutative group** or an **abelian group.**

I is a commutative group with respect to addition.

N, the set of natural numbers, is not a group with respect to addition since most elements in N do not have additive inverses in N.

The system of integers modulo 3 is a commutative group with respect to addition. The additive inverses of 0, 1, and 2 are 0, 2, and 1 respectively.

The nonzero elements in $\{0,1,2,3,4\}$ form a commutative group with respect to multiplication as defined by the table for $\cdot_5$ in Figure 7.14. 1 is the identity and the multiplicative inverses for 1, 2, 3, and 4 are 1, 3, 2, and 4, respectively.

PROBLEMS

1. Which of the following form a group?

   (a) the even integers with respect to addition
   (b) the odd integers with respect to addition
   (c) the even integers with respect to multiplication
   (d) the odd integers with respect to multiplication
   (e) the set $\{0\}$ with respect to the operation $\oplus$ defined by $0 \oplus 0 = 0$.

2. Taking each of the tables in Figure 7.14 as defining an operation on a certain set, and, thus, giving us a system, which of these systems are groups? [HINT: See Theorem 7.11.3.]

3. Is the system $(\{1,2,3, \ldots, k-1\}, \cdot_k)$ a group when k is

   (a) 2 (b) 3 (c) 4 (d) 5 (e) 6 (f) 7 (g) 8?

[NOTE: 0 is not in the set. Refer to Figure 7.14 and your work in Problem 7.11.4. Also note Theorem 7.11.3.]

4. In view of your results in Problem 3, what condition do you think k must satisfy in order that the system $(\{1,2, \ldots, k-1\}, \cdot_k)$ be a group? It may help to check the case $(\{1,2, \ldots, 8\}, \cdot_9)$ first.

5. Which of the following systems are groups?

   (a) $(\{\phi, \{1\}, \{2\}, \{1,2\}\}, \cup)$
   (b) $(\{1,2,3\}, W)$ as defined on page 100
   (c) $(\{a,b,c\}, K)$ as defined in Problem 8, Section 5.2
   (d) $(\{a,b,c,d\}, L)$ as defined in Problem 8, Section 5.2
   (e) $(\{a,b,c,d\}, N)$ as defined in Problem 8, Section 5.2.

6. Which of the systems of Problem 5 are commutative groups?

7. Let A be a set and let F be the collection of all one-to-one correspondences which map A **onto** A. Is F a group with respect to composition? Explain. Is F a commutative group with respect to composition?

8. For each integer k, define a function f_k on I by taking $f_k(x) = x + k$. Let H be the collection of all such functions. Show that H is a group with respect to composition. Is it isomorphic to some familiar group?

9. Figure 7.15 shows an equilateral triangle ABC. If we rotate this triangle counterclockwise through 120° about its center, it falls back on itself. Thus, such a rotation determines a mapping v_1 of the triangle onto itself. Similarly, a rotation through 240° determines a mapping v_2 of the triangle onto itself. Let v_0 be the mapping which maps each point of the triangle on itself. In other words, let v_0 be the mapping determined by a rotation of 0°. Consider the set v_0, v_1, v_2 under the operation of composition. Show that this system is a group. It is called the rotation group for the triangle. Is it isomorphic to some familiar group?

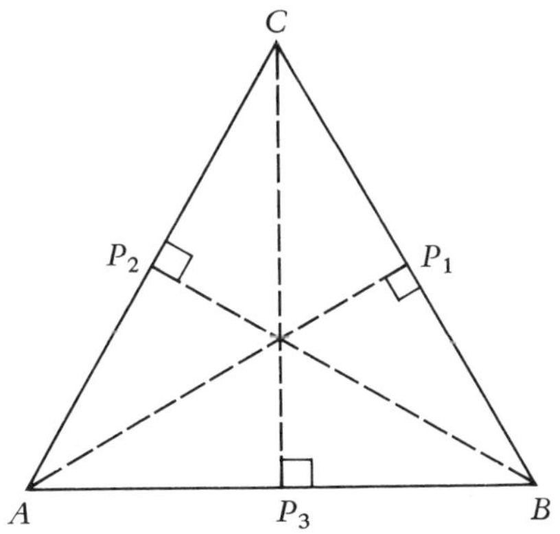

FIGURE 7.15

10. Referring again to Figure 7.15, let m_1 be the mapping determined by a rotation through 180° about the line AP_1, m_2 be that determined by a rotation through 180° about BP_2, and m_3 that determined by a rotation through 180° about CP_3. Construct an operation table for composition on the set of mappings $\{v_0,v_1,v_2,m_1,m_2,m_3\}$. Show that the resulting system is a group. It is called the **group of symmetries** of the triangle.

11. Using rotations about the center through multiples of 90° and also rotations through 180° about the dotted lines shown in Figure 7.16, construct the operation tables for the group of symmetries of the square. [SUGGESTION: Use d_1 and d_2 for the mappings arising from rotation about the diagonals, and h and v for those from rotations about the horizontal and vertical.]

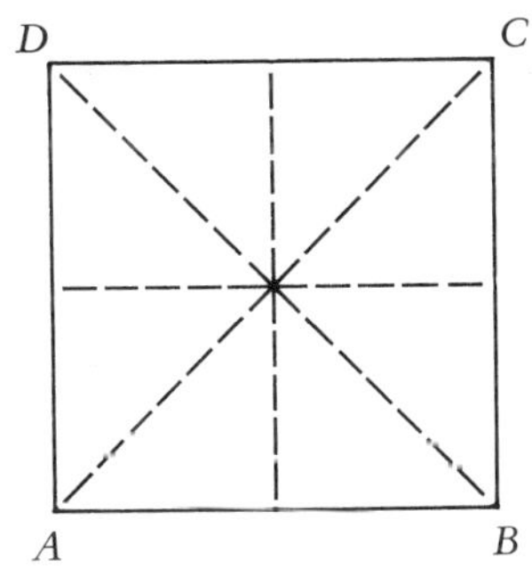

FIGURE 7.16

12. In how many different ways may we define an operation on $\{1,2\}$ so that it becomes a group? (See Problem 9, Section 5.2.)

*7.13 The theory of groups

If $(G, \oplus)$ is a group we know that

(i) G is closed under $\oplus$,
(ii) $\oplus$ is associative in G,
(iii) there is an identity element for $\oplus$ in G, and
(iv) each element in G has an inverse element with respect to $\oplus$.

From these four properties for the system $(G, \oplus)$ we can deduce some other properties that the system must have. The body of theorems that assert these properties make up what would be called the **theory of groups.** Any assertion that we deduce from the four given properties must hold when applied to any particular group. Since there are many, many different interesting and useful systems that are groups, the development of a portion of the theory of groups is a very fruitful procedure. We get results we can use in many different places.

This points up one important aspect of the abstract axiomatic approach that is so typical of mathematics. Using this approach, we first carefully define an abstract concept such as that of a group. We then deduce some of the logical consequences of the properties or axioms assumed (for example, the properties (i), (ii), (iii), and (iv) above.) And, if we are fortunate as we are with the notion of a group, the results we obtain may be used in a great many concrete situations. Actually, a so-called pure mathematician might be interested in developing the logical consequences of the axioms whether they are seen to be useful or not. Many of the results discovered by pure mathematicians in the pursuit of the **interesting** rather than of the **useful** have, as a matter of fact, subsequently turned out to be most helpful in very practical situations.

Let us see some of the consequences of the axioms for a group. We first establish two facts about groups which were actually discussed before. (See page 101.)

Theorem 7.13.1. There is exactly one identity element in a group.

Theorem 7.13.2. Each element in a group has exactly one inverse element.

PROOFS: Let $(G, \oplus)$ be a group. To prove the first assertion we note that by property (iii) there is an identity element. It remains to prove that there could not be two identity elements. But suppose e_1 and e_2 are both identity elements. Then $e_1 \oplus e_2 = e_1$ since e_2 is an identity. Also $e_1 \oplus e_2 = e_2$ since e_1 is an identity. Since $\oplus$ is an operation on G, $e_1 \oplus e_2$ must be a unique element of G and, therefore, $e_1 = e_2$. This completes the proof.

To prove the second assertion, we first note that if x is in G, then x has an inverse. It remains to prove that there could not be two different inverses for x. Suppose z and w are both inverses for x and e is the identity element for the group. Then $x \oplus z = z \oplus x = e$ and $x \oplus w = w \oplus x = e$.

From the equality $x \oplus z = e$ and the closure property (i) we get $w \oplus (x \oplus z) = w \oplus e$. By property (ii) this implies $(w \oplus x) \oplus z = w \oplus e$. By the definition of the identity element e and by the fact that w is an inverse for x we get from this last equality that $e \oplus z = w$. But, since e is the identity, $e \oplus z = z$. Hence, $z = w$ and the proof is complete.

Theorem 7.13.3. If x and y are elements in a group and ^-x and ^-y are their inverses, then the inverse of $x \oplus y$ is $(^-y) \oplus (^-x)$. (Note the order.)

PROOF: The proof of this is straightforward. We have by property (ii) that $(x \oplus y) \oplus ((^-y) \oplus (^-x)) = [(x \oplus y) \oplus (^-y)] \oplus (^-x) = [x \oplus (y \oplus (^-y))] \oplus (^-x) = [x \oplus e] \oplus (^-x) = x \oplus (^-x) = e$. In a similar fashion, we may show that $((^-y) \oplus (^-x)) \oplus (x \oplus y) = e$. Hence, by Theorem 7.13.2 above, $(^-y) \oplus (^-x)$ is *the* inverse for $x \oplus y$.

Theorem 7.13.4. If a and b are given elements in a group, then the equation $x \oplus a = b$ has the unique solution $x = b \oplus (^-a)$ where ^-a is the inverse for a.

PROOF: We shall leave the proof of this as an exercise.

PROBLEMS

1. Prove Theorem 7.13.4. [HINT: First prove that if $x = b \oplus (^-a)$, then $x \oplus a = b$. Then show that if $x \oplus a = b$, then $x = b \oplus (^-a)$.]

2. Make use of Theorem 7.13.4 in the appropriate group (see Figure 7.14) to solve the equations

(a) $x + {}^+3 = {}^+6$ (b) $x +_2 1 = 0$
(c) $x \cdot_3 2 = 1$ (d) $x +_4 2 = 3$
(e) $x +_4 3 = 1$ (f) $x +_5 2 = 3$
(g) $x +_5 3 = 2$ (h) $x +_5 4 = 1$
(i) $x +_5 1 = 0$ (j) $x +_6 3 = 1$.

3. Discuss the material about inverse operations on page 108 using the language of groups.

4. Any group that is isomorphic to the group of integers mod k under addition is said to be a **cyclic group of order k.** Show that if a is a nonzero integer and we define an operation M by the table

M	1	a	a^2	a^3
1	1	a	a^2	a^3
a	a	a^2	a^3	1
a^2	a^2	a^3	1	a
a^3	a^3	1	a	a^2

then $(\{1, a, a^2, a^3\}, M)$ is a cyclic group.

5. An element g of G is said to be a **generator** for the group $(G, \oplus)$ if every element of G can be written in the form $g \oplus g \oplus \cdots \oplus g$ for some number of terms. Find a generator if there is one for

(i) the group of Problem 4
(ii) the integers mod 4 under addition
(iii) the integers mod 5 under addition
(iv) the integers which are not multiples of 3 under mod 3 multiplication
(v) the integers which are not multiples of 5 under mod 5 multiplication
(vi) $(I, +)$.

6. Verify Theorem 7.13.3 for all possible x and y when we have the group of integers mod 3 under addition.

7. If $(G, \oplus)$ is a group and $H \subseteq G$, then $(H, \oplus)$ may be a group when we restrict the operation $\oplus$ to act only on elements of H. If so, H is said to be a **subgroup** of G (with respect to $\oplus$). For example, $(\{0,2\}, +_4)$ is a subgroup of $(\{0,1,2,3\}, +_4)$.

(a) Find three different subgroups of $(\{0,1,2,3,4,5\}, +_6)$.
(b) Are there any subgroups of the additive group of integers mod 5 except for the trivial ones with the sets $\{0\}$ and $\{0,1,2,3,4,5\}$?
(c) Find some nontrivial subgroups of the group of symmetries of the triangle (see Problem 11, Section 7.12).
(d) Find some nontrivial subgroups of the group of symmetries of the square (see Problem 12, Section 7.12).

8

The rational number system

8.1 The rational numbers

If we wish to cut a 3 foot board into 2 equal pieces it is difficult to say how to do it if we only have integers to work with. There is no solution in I for the simple equation $2x = 3$.

If we set up a number line to use as a scale for a thermometer and, at a certain time, the top of the mercury column falls between two integral points on the scale, there is no very helpful way to report this fact if we have only integers to work with. More generally, when we have only integers, it is somehow unfortunate that there are so many points on a number line that are not assigned to any numbers.

We now wish to build a new number system that will, at least in part, take care of such difficulties. We want our new system to be an extension of the system of integers.

It is a basic weakness of I that we can so seldom carry out the process of division. Or, putting this another way, when we are given integers a and b we can so seldom find an x such that $bx = a$. We hope that the new numbers we invent will enable us to solve such an equation.

But if we simply make up a new number which will be a solution of say $2x = 3$, we need to note that this same new number probably should be the solution to a batch of similar equations. In I itself, what is the situation when x is a number which satisfies both $bx = a$ and $dx = c$ when b and d are different from 0? In such a case, we must have $x = a \div b$ and $x = c \div d$. But we have seen that in I, $a \div b = c \div d$ implies that $ad = cb$. Now when a, b, c, and d are integers, ad and cb are always well defined integers and, hence, we can always tell whether $ad = cb$ is true or not. We shall follow this lead.

We evidently want to make our new numbers correspond to pairs of integers. The new number x for which we hope to have $2x = 3$ will correspond to the pair (3,2). But this number should also correspond to other pairs for other equations that should have this same solution.

Let D be the collection of all nonzero integers. We define a relation E in $I \times D$ by saying that:

$$(a,b)\ E\ (c,d) \text{ if and only if } ad = cb.$$

For example,

$$(6,2)\ E\ (9,3) \text{ since } 6\cdot 3 = 9\cdot 2,$$

$$(6,2)\ E\ (15,5) \text{ since } 6\cdot 5 = 15\cdot 2, \qquad \text{and}$$

$$(3,1)\ E\ (6,2) \text{ since } 3\cdot 2 = 6\cdot 1.$$

Note that $2x = 6$, $3x = 9$, $5x = 15$, and $1\cdot x = 3$ all have the same solution in I. Namely, $x = 3$.

We have $(^-5,1)\ E\ (20,^-4)$ since $^-5\cdot{}^-4 = 20\cdot 1$. Also $(^-5,1)\ E\ (^-25,5)$ since $^-5\cdot 5 = {}^-25\cdot 1$. And $(10,^-2)\ E\ (^-25,5)$ since $10\cdot 5 = {}^-25\cdot{}^-2$.

Note that the equations $1\cdot x = {}^-5$, $^-4x = 20$, $^-5x = {}^-25$, and $^-2x = 10$ all have the same solution in I. Namely, $x = {}^-5$.

It can be shown that, as we might expect, E is an equivalence relation in $I \times D$. Some of its equivalence classes are:

$\{(0,1), (0,^-1), (0,2), (0,^-2), (0,3), (0,^-3), \ldots\}$,
$\{(1,1), (^-1,^-1), (2,2), (^-2,^-2), (3,3), (^-3,^-3), \ldots\}$,
$\{(1,2), (^-1,^-2), (2,4), (^-2,^-4), (3,6), (^-3,^-6), \ldots\}$,
$\{(^-1,2), (1,^-2), (^-2,4), (2,^-4), (^-3,6), (3,^-6), \ldots\}$,
$\{(2,1), (^-2,^-1), (4,2), (^-4,^-2), (6,3), (^-6,^-3), \ldots\}$, and
$\{(13,17), (^-13,^-17), (26,34), (^-26,^-34), (39,51), (^-39,^-51), \ldots\}$.

Each such equivalence class we call a **rational number.** That is, we have as the basic definition of this chapter:

> A **rational number** is an equivalence class in $I \times D$ for the relation E defined above.

We may use our usual method for identifying equivalence classes and claim that the above sets (or rational numbers) are, respectively, $[(0,1)]$, $[(1,1)]$, $[(^-1,^-2)]$, $[(^-3,6)]$, $[(2,1)]$, and $[(^-13,^-17)]$. But there is a standard terminology in use around the world which replaces this rather cumbersome notation. For the equivalence class $[(a,b)]$ we shall write simply "a/b" or "$\frac{a}{b}$". We call the expression "a/b" or "$\frac{a}{b}$" a **fraction.** If "a" names an integer and "b" names a nonzero integer, then "a/b" is a fraction and, hence, names a rational number.

In the fraction "a/b" we call a the **numerator** and b the **denominator.** We read "a/b" as "a over b" or "a bth's." For example, "7/9" may be read as "seven over nine" or "seven ninths," "49/123" may be read as "forty-nine over one hundred twenty-three" or "forty-nine one hundred and twenty-thirds."

From our definition of E we see that

$$[(a,b)] = [(c,d)] \text{ if and only if } ad = cb.$$

Hence, when we put down names for these classes as fractions we have:

$$a/b = c/d \text{ if and only if } ad = cb.$$

Note that b and d must be different from 0 here. If "a" names an integer, the expression "$a/0$" is simply meaningless.

On the other hand, note that when a is any nonzero integer we have $0/a = 0/1$ since $0 \cdot 1 = 0 \cdot a = 0$.

We may give a flow chart for deciding whether or not two fractions name the same rational number. See Figure 8.1.

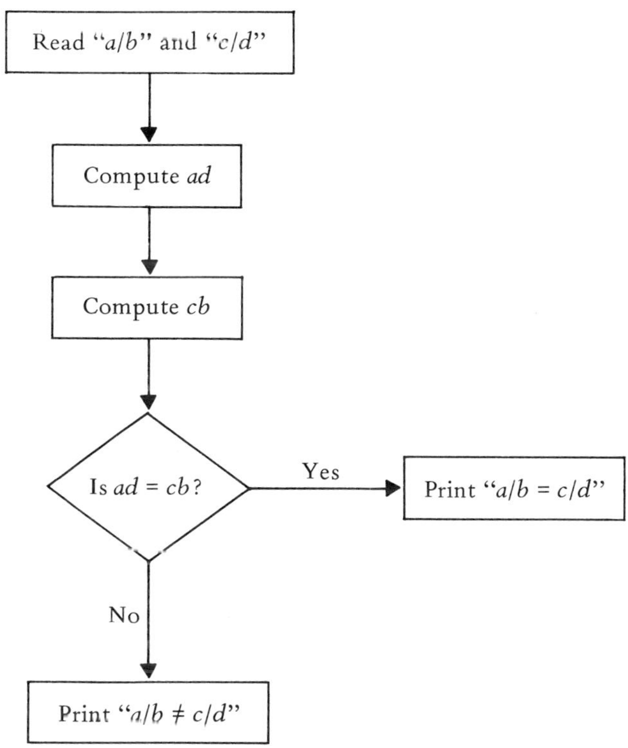

FIGURE 8.1 Flow chart for testing for equality with fractions. b and d are assumed to be different from 0.

Since we have introduced the rational number 3/1 as a solution to the equation $1 \cdot x = 3$ and we know that $x = 3$ is a solution to this equation in I, we would expect 3/1 to act in some sense like the integer 3. Similarly, we expect $n/1$ to somehow act like the integer n. We shall return to the details of this matter later.

One basic fact about rational numbers follows almost immediately from our definition:

If a, b, and c are integers and $b \neq 0$ and $c \neq 0$, then $(ac)/(bc) = a/b$.

This **cancellation law for fractions** holds because the denominator $bc \neq 0$ when $b \neq 0$ and $c \neq 0$, and $(ac)/(bc) = a/b$ if $(ac)b = a(bc)$. But this last equality follows from the associativity and commutativity of multiplication in I.

When we make use of the above result in such a case as

$$\frac{111}{153} = \frac{37 \cdot 3}{51 \cdot 3} = \frac{37}{51},$$

we would say that we have simplified the fraction "111/153" by cancelling out the common factor of 3 from the numerator and denominator. Similar terminology is used in other such cases.

Definition 8.1.1. Let D be the collection of nonzero integers. Let E be the relation defined on $I \times D$ by $(a,b)\ E\ (c,d)$ if and only if $ad = cb$.

Theorem 8.1.1. E is an equivalence relation in $I \times D$.

PROOF: Suppose (a,b), (c,d), and (e,f) are in $I \times D$. $(a,b)\ E\ (a,b)$ since $ab = ab$.

If $(a,b)\ E\ (c,d)$ then $ad = cb$. Then, clearly, $cb = ad$ and this implies $(c,d)\ E\ (a,b)$.

Suppose that $(a,b)\ E\ (c,d)$ and $(c,d)\ E\ (e,f)$. Then $ad = cb$ and $cf = ed$. Therefore, $(ad)(cf) = (cb)(ed)$. This implies that $(afc)d = (ebc)d$. But $d \neq 0$, so this yields $afc = ebc$. If $c \neq 0$, then it follows that $af = eb$ and, hence $(a,b)\ E\ (e,f)$. If $c = 0$, then a must also be 0 since $ad = cb = 0$ and $d \neq 0$. Also if $c = 0$, then $e = 0$ since $ed = cf = 0$ and $d \neq 0$. But clearly $(0,b)\ E\ (0,f)$ since $0 \cdot f = b \cdot 0$.

Thus, we have shown that E is reflexive, symmetric, and transitive, and E is, therefore, an equivalence relation.

Definition 8.1.2. A **rational number** is an equivalence class in $I \times D$ with respect to the relation E.

Definition 8.1.3. We use "a/b" or "$\frac{a}{b}$" to denote the rational number $[(a,b)]$.

We say that "a/b" is a **fraction** and call a the **numerator** and b the **denominator** of the fraction.

Theorem 8.1.2. If a, b, c, and d are integers and b and d are not 0, then $a/b = c/d$ if and only if $ad = cb$.

Theorem 8.1.3. If a, b, and c are integers and b and c are not 0, then $(ac)/(bc) = a/b$.

PROBLEMS

1. Let E be the relation defined in this section. Which of the following are true?

(a) $(5,1)\ E\ (10,2)$
(b) $(^-5,^-2)\ E\ (5,2)$
(c) $(^-6,5)\ E\ (6,^-5)$
(d) $(1549,3)\ E\ (25{,}533,\ 51)$
(e) $(23,^-621)\ E\ (^-2,54)$
(f) $(0,17)\ E\ (0,29)$
(g) $(^-11{,}106,\ 19)\ E\ (6170,^-5)$.

2. Find six other members of the equivalence class in $I \times D$ with respect to E which contains

(a) $(7,1)$
(b) $(^-7,1)$
(c) $(36,4)$
(d) $(^-5,7)$
(e) $(0,9)$
(f) $(14,2)$
(g) $(2,^-11)$
(h) $(6170,^-5)$
(i) $(^-16,^-9)$.

3. Which of the following are true?

(a) $6/2 = 3/1$
(b) $37/2 = 18/1$
(c) $^-6/2 = 6/^-2$
(d) $23{,}829/13 = 7943/1$
(e) $^-2555/73 = 140/^-4$
(f) $0/72 = 0/^-4$
(g) $5771/^-16 = 22{,}084/^-64$
(h) $3/51 = {^-2}/34$
(i) $24{,}168/23{,}256 = 689/663$
(j) $^-100{,}000/1{,}000{,}000 = 1{,}000{,}000/^-10{,}000{,}000$
(k) $^-5{,}793{,}406/^-11{,}586{,}812 = 1/2$
(l) $397/4 = 794/7$.

4. Show that if we take E as defined above to be a relation in $I \times I$, then E is **not** an equivalence relation in $I \times I$. [HINT: Show that E is not transitive in $I \times I$.]

5. We may use other names aside from decimal names for the integers that occur in a fraction. Which of the following are true?

(a) $111111_T/111_T = 1111_T/11_T$
(b) $10{,}000_2/1000_2 = 10_2/1$
(c) $^-1010_2/101_2 = 110100_2/^-110_2$
(d) $1110_2/11_2 = 101010_2/1001_2$

(e) $2210_3/221_3 = 121211_3/1210_3$
(f) $^{-}10{,}212_3/^{-}11_3 = 222_3/1$
(g) $100{,}000_{12}/1000_{12} = 10{,}000{,}000_{12}/100{,}000_{12}$
(h) $^{-}1000_3/100_3 = {}^{-}100{,}000_3/10{,}000_3$.

6. "a/b" is sometimes read as "the ratio of a to b." Some other terminology is also often used. For instance, "$a/b = c/d$" is sometimes read as "a is to b as c is to d." And the expression "$a:b :: c:d$" is sometimes used to mean $a/b = c/d$. Although this notation is becoming less and less common it does sometimes crop up in the literature. Show that

(a) 5 is to 9 as 70 is to 126
(b) $^{-}6$ is to 4 as 36 is to $^{-}24$
(c) $144:12 :: {}^{-}48:{}^{-}4$
(d) ab is to ac as b is to c
(e) if a is to b as c is to d, and c is to d as e is to f, then a is to b as e is to f.

8.2 Interpretations of rational numbers

We have said that when a and b are integers and b is not 0, we may think of a/b as the number we invented to be the solution to the equation $bx = a$.

Intimately tied up with this interpretation is the one where we think of a/b as the number we invented to be the result $a \div b$.

A somewhat different way to look at it would be to think of the number $1/b$ as the number we invented to be the multiplicative inverse for the integer b. And then a/b may be thought of as a times $1/b$.

But we may also interpret rational numbers in a geometric rather than an algebraic fashion. Suppose we consider a number line which has points on it for the integers. It is quite easy geometrically to find the point on the line that cuts the segment from the point 0 to the point 1 into two pieces of equal length. We call this point the point 1/2. Notice that if we double the distance from the origin to the point 1/2 we get a distance of one unit.

Similarly, the point 1/3 on the number line is obtained when we divide the segment from 0 to 1 into 3 pieces of equal length and pick out the first division point moving from the left.

More generally, we find the point $1/n$ by taking the first division point at the left when the segment from 0 to 1 is split up into n pieces all of the same length.

If p is an integer, the point p/n is obtained by finding the point $1/n$ and then determining the point that is a distance from the origin of $|p|$ times the distance from the origin to $1/n$ in the appropriate direction—to the right if p is positive and to the left if p is negative.

Several points obtained in this fashion are shown in Figure 8.2.

Points that are assigned to rational numbers in this manner are called **rational points** on the line.

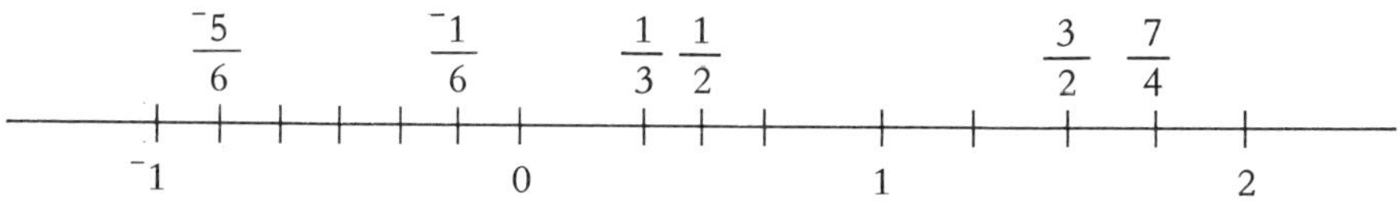

FIGURE 8.2

Every rational number is assigned to a point on the line by the above procedure. To make this absolutely clear in view of what we did above we need to note that, given any rational number, there is a fraction that names it which has a positive denominator. But this is quite easy to establish. For suppose $x = r/s$ where s is a negative integer. Then we have $x = ((^-1)r)/((^-1)s)$. But $(^-1)s$ is a positive integer.

A little thought convinces us that, in sharp contrast to the line with only integral points associated with numbers, there are many, many rational points on any segment of the line. For example, the distinct points 1/2, 1/3, 1/4, 1/5, . . . are all points on the segment between 0 and 1. We shall pursue this matter further in a later section. But it is tied in with a very important use of the rational numbers which we shall now discuss.

With many physical devices for making measurements, the operator observes the position of a hairline or a pointer on a scale. This scale may be around a circular dial as on a clock, or around a cylinder as on a micrometer, or along a straight line as on a ruler. When the operator "takes a reading" from the device he reports a number that indicates the position of the hairline or pointer. These reported numbers are always (or almost always) rational numbers. Usually the scale has quite a few rational points marked on it. But there will be only a finite number of these **marked** points on any actual scale. The "reading" that the operator takes is either a name for the number assigned to the marked point closest to the pointer or it may be the number obtained by imagining equally spaced points (perhaps ten of them) between two adjacent marked points and taking the rational number corresponding to the closest one of these. Thus, most measurements in the physical and biological sciences are given in terms of rational numbers.

Someone, evidently in a very materialistic mood, once said something to the effect that if you can't measure it you shouldn't talk about it. Although this is a grossly oversimplified and somewhat misleading admonition to give to even a budding scientist, there is a grain of truth hidden in it. Certainly science could not be science as we know it today without a great deal of emphasis on measurements. And since rational numbers are the primary mathematical tools used for measuring, they are of extreme importance.

As we might expect from our work with other kinds of numbers, rational numbers may be interpreted as vectors. On the number line, as before, let us write "(x,y)" to designate the arrow running from the initial point x to the terminal point y. We now allow x and y to be any rational points. Let A be the collection of all such arrows. Let us say that two arrows in A are in the relation M

to one another if they have the same length and the same direction. M is an equivalence relation in A. The equivalence classes in A with respect to M are called **vectors.** Any arrow in a given equivalence class (or vector) will be said to *represent* that vector.

As we might expect, the rational number $1/2$ behaves like the vector represented by the arrow $(0,1/2)$. In other words, $1/2$ behaves like the vector pointing to the right and having length $1/2$. $^-3/2$ behaves like the vector pointing to the left and having length $3/2$. In Figure 8.3 we have represented several vectors that correspond to rational numbers.

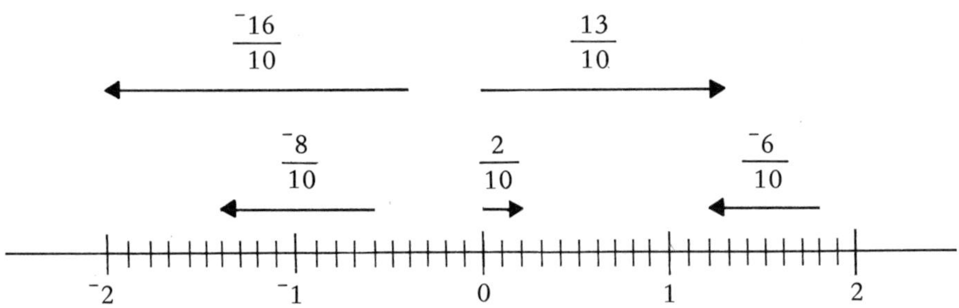

FIGURE 8.3

Another geometric representation of rational numbers may be obtained as follows. Suppose we construct two perpendicular number lines as we have done before. Let p and q be integers and q be positive. We may easily construct a line segment with length p/q in our figure by the following procedure. Find the point q on the horizontal axis. Draw the line L perpendicular to the horizontal axis at that point. Starting from the horizontal axis, move out along L $|p|$ units—moving up if p is positive, and down if p is negative. From the point thus obtained draw a straight line L^* to the origin. If we now draw the line which is perpendicular to the horizontal axis at the point 1, and draw a horizontal from the point where this line hits L^*, we shall find the point p/q on the vertical axis. In Figure 8.4 we have shown such constructions of $7/5$ and $^-10/3$.

Another geometric technique to find the point p/q on a number line is the following. Here p and q may be either positive or negative—but of course, q cannot be 0. On two perpendicular number lines as constructed before, find the point p on the vertical axis and the point q on the horizontal axis. Draw the line L between these two points. Through the point 1 on the horizontal axis draw the line parallel to L. This line will hit the vertical axis at the point p/q. We have shown the construction of $3/5$ and $^-7/4$ by this procedure in Figure 8.5.

We shall postpone the justifications for these last two graphical procedures until a later chapter.

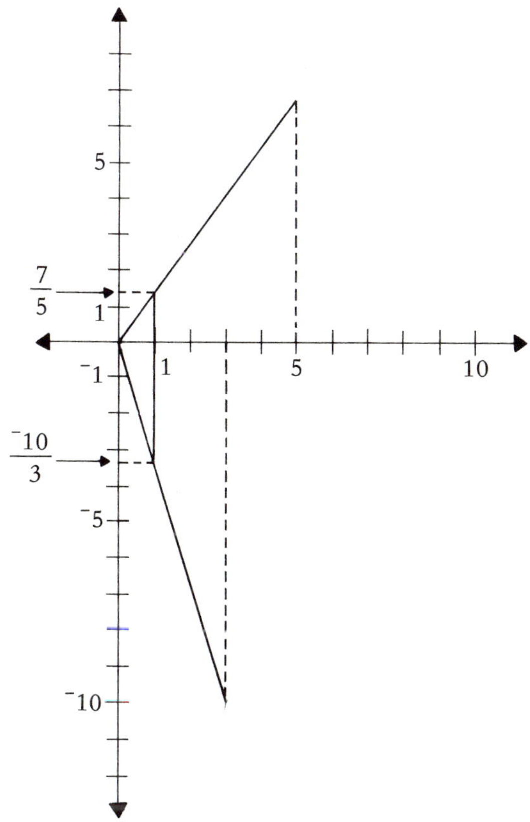

FIGURE 8.4

To summarize the interpretations that we have considered for rational numbers, we shall put down several ways to think of the rational number 1/2.

1/2 is, by definition, the equivalence class {(1,2), (⁻1,⁻2), (2,4), (⁻2,⁻4), (3,6), (⁻3,⁻6), . . .}.

1/2 is the number we invented to be the solution to each of the equations $2x = 1$, $^{-}2x = {}^{-}1$, $4x = 2$, $^{-}4x = {}^{-}2$, $6x = 3$, $^{-}6x = {}^{-}3$, ...

1/2 is the number we invented to be the result of the divisions $1 \div 2$, $^{-}1 \div {}^{-}2$, $2 \div 4$, $^{-}2 \div {}^{-}4$, $3 \div 6$, $^{-}3 \div {}^{-}6$, ...

1/2 is the number corresponding to the point which divides the segment from the point 0 to the point 1 into two pieces of equal length.

1/2 is the number corresponding to the point constructed by the method of Figure 8.4 (or Figure 8.5) when $p = 1$ and $q = 2$.

We shall, henceforth, use "F" as a name for the collection of all rational numbers.

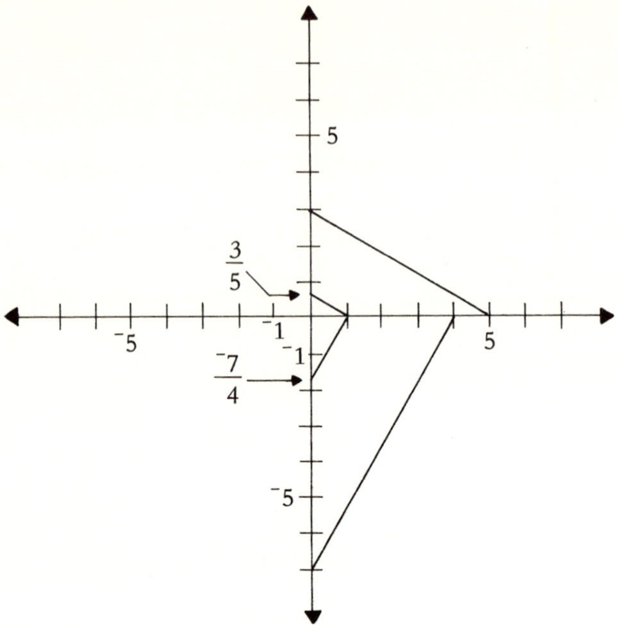

FIGURE 8.5

PROBLEMS

1. State at least four different ways to interpret the rational number

(a) ⁻1/2 (b) 3/1 (c) ⁻4/5 (d) 73/15
(e) 37/2 (f) ⁻7/8 (g) ⁻10/2.

2. On a number line, draw an arrow to represent the vector corresponding to each of the rational numbers in Problem 1.

3. For each rational number in Problem 1 construct the corresponding point on a number line by the method of

(a) Figure 8.4 (b) Figure 8.5.

4. For each of the following fractions give six other fractions that name the same rational number.

(a) "7/1" (b) "⁻43/2" (c) "729/10,935"
(d) "⁻3/⁻10" (e) "⁻14/2."

5. Verify geometrically that 7/5 = 14/10 by finding both the points 7/5 and 14/10 by the method of

(a) Figure 8.4 (b) Figure 8.5.

6. A geometric procedure for dividing a segment into 5 pieces of equal length is as follows. Suppose the segment we wish to divide up is the segment between points A and B (see Figure 8.6). At point A, construct the perpendicular L at A. Choosing some convenient unit, start at A and lay off 5 consecutive segments of unit length along L. Let P_1, P_2, P_3, P_4, and P_5 be the points we determine in this fashion. Draw the line L_5 through P_5 and B. Then draw the line L_i through P_i parallel to L_5 ($i = 1, 2, 3, 4$). Let L_i hit our given segment in the point Q_i. The points Q_1, Q_2, Q_3, and Q_4 split up the segment from A to B into 5 pieces of equal length. Describe and illustrate a similar procedure for splitting a segment into n pieces of equal length when n is

(a) 4 (b) 6.

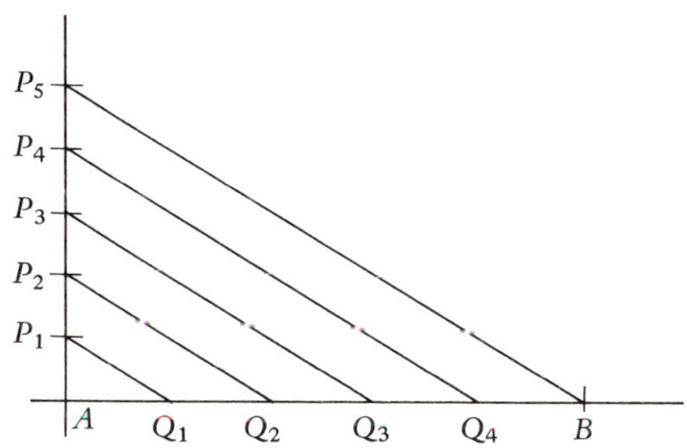

FIGURE 8.6

8.3 Addition

In Section 7.5 we discovered that when a, b, c, and d are integers and b is a divisor of a and d is a divisor of c, then $(a \div b) + (c \div d) = (ad + cb) \div (bd)$. In view of our interpretation of a rational number as the result of a division of one integer by another, we should base our definition of addition of rationals on this. In F, we define the operation of **addition** by:

$$a/b +_F c/d = (ad + cb)/(bd).$$

Here, a, b, c, and d can be any integers except that we must have $b \neq 0$ and $d \neq 0$. Note that the "$+$" that occurs on the right side of the equality in our definition refers to addition of integers, whereas, we have used "$+_F$" to denote the operation of addition in F.

We can be led to the same definition by a different route. When n is a positive integer we may think of $1/n$ as the length of one of the segments obtained when we divide the segment from 0 to 1 into n pieces of equal length. p segments of length $1/n$ laid end to end give a segment of length p/n. $p/n +_F q/n$ should be the length of $p + q$ segments of length $1/n$. That is, we should have $p/n +_F$

$q/n = (p + q)/n$. Now we know from our definition of rational numbers that when $d \neq 0$ we have $a/b = (ad)/(bd)$. Also when $b \neq 0$, we have $c/d = (bc)/(bd)$. Hence, we should have

$$a/b +_F c/d = (ad)/(bd) +_F (cb)/(bd) = (ad + cb)/(bd).$$

Either interpretation of rational numbers leads to this same formalization of what it means to add two rational numbers.

As some numerical illustrations of the definition we have:

$$6/1 +_F 5/1 = (6\cdot 1 + 5\cdot 1)/(1\cdot 1) = 11/1$$

$$7/2 +_F 4/2 = (7\cdot 2 + 4\cdot 2)/(2\cdot 2) = 22/4$$

$$^-6/7 +_F {}^-5/^-2 = (^-6\cdot{}^-2 + {}^-5\cdot 7)/(7\cdot{}^-2) = {}^-23/^-14$$

$$5/^-3 +_F {}^-3/5 = (5\cdot 5 + {}^-3\cdot{}^-3)/(^-3\cdot 5) = 34/^-15.$$

A schematic representation of the calculation of the sum of two rational numbers is shown in Figure 8.7.

But there is no need to memorize the somewhat hearty procedure of Figure 8.7. It is much closer to our intuitive notions of rational numbers if we first recall that it is easy to add two rationals when they are expressed as fractions that both have the same denominator. We simply have $p/n +_F q/n = (p + q)/n$. And then we should recall the basic fact that multiplying the numerator and denominator of a fraction by a nonzero integer gives us another fraction that names the same rational number. Then we have $a/b +_F c/d = (ad)/(bd) +_F (bc)/(bd) = (ad + bc)/(bd) = (ad + cb)/(bd)$.

Basic properties of addition in F are established by proofs that are quite good examples of simply grinding away.

A proof that addition in F is commutative is:

> $a/b +_F c/d = (ad + cb)/(bd) = (cb + ad)/(bd) = (cb + ad)/(db) = c/d +_F a/b$. Here the first and last equalities are justified by reference to the definition of $+_F$. The second equality is justified by the commutativity of addition in I, and the third by the commutativity of multiplication in I.

A proof that addition in F is associative is obtained when we note that

$$\begin{aligned}(a/b +_F c/d) +_F e/f &= (ad + cb)/(bd) +_F e/f \\ &= ((ad + cb)f + e(bd))/((bd)f) = (adf + cbf + ebd)/(bdf)\end{aligned}$$

and

$$\begin{aligned}a/b +_F (cf + ed)/(df) &= (a(df) + (cf + ed)b)/(b(df)) \\ &= (adf + cfb + edb)/(bdf) = (adf + cbf + ebd)/(bdf).\end{aligned}$$

A proof that $0/1$ is the identity element for addition in F is:

$$a/b +_F 0/1 = (a\cdot 1 + 0\cdot b)/(b\cdot 1) = a/b.$$

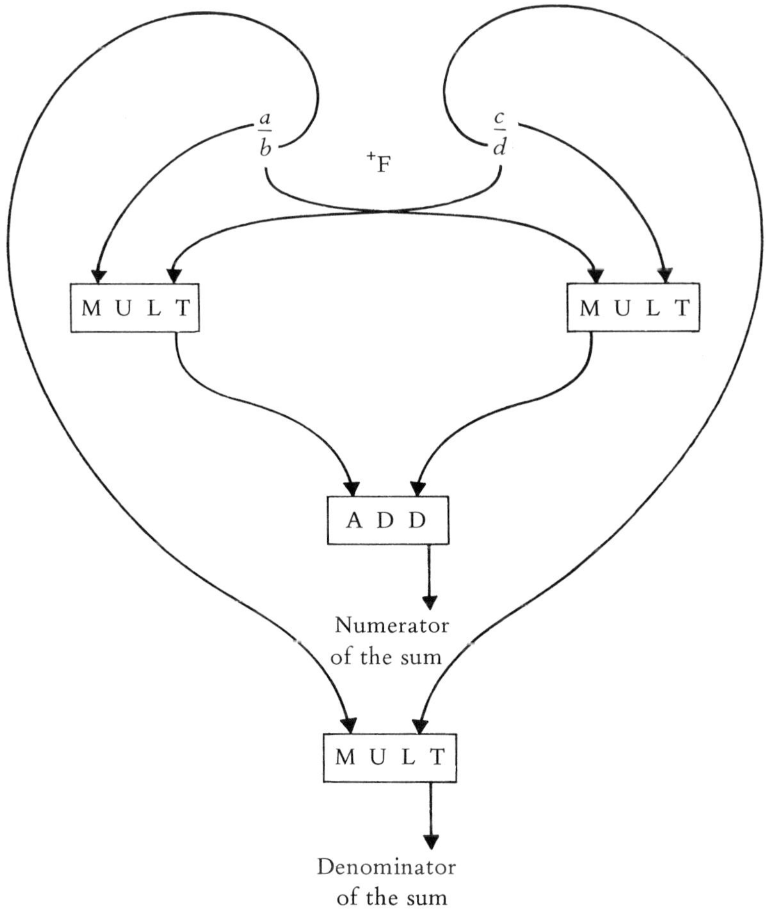

FIGURE 8.7

We know that $(^{-}1)a$ is the opposite or additive inverse for a in I. We have in F

$$\begin{aligned} a/b +_F ((^{-}1)a)/b &= (ab + (^{-}1)ab)/(bb) \\ &= ((a + (^{-}1)a)b)/(b^2) = (0 \cdot b)/(b^2) = (0)/(b^2) = 0/1. \end{aligned}$$

Hence, $((^{-}1)a)/b$ is the additive inverse for a/b in F. We shall also call $((^{-}1)a)/b$ the **opposite** of a/b.

Subtraction in F is, of course, defined to be the inverse operation for addition in F. Since each rational number has a rational additive inverse, we see that F is closed under subtraction. And we may perform a subtraction in F by adding on the additive inverse. That is,

$$a/b -_F c/d = a/b +_F ((^{-}1)c)/d.$$

As we did in *I*, we shall sometimes write "$-x$" for the opposite of the rational number x.

When working with rational numbers, questions of efficiency and convenience arise as they always do in any practical endeavor. For example, it might be logically correct to give the result of an addition as 5,794,317/17,382,981. But it would certainly be more helpful to give the same result as "1/3" since 5,794,317/17,382,981 = 1/3. Among all the fractions that might be used to name a given rational number, it is natural to have a preference for those fractions whose numerators and denominators have smaller absolute values. There is a very practical reason for this preference aside from laziness and an easier intuitive grasp of the meaning of shorter names. If we are going to carry out computations with the rational numbers, we cut down our chances for mechanical errors if we use integers with shorter names. And integers with smaller absolute values *do* have shorter names.

A fraction that will be as simple as possible as a name for a given rational number is a fraction **in lowest terms.** A fraction "a/b" is said to be **in lowest terms** if a and b have no common divisors except 1 and $^-1$. In other words, "a/b" is in lowest terms if $cd = a$ and $ed = b$ implies $d = 1$ or $d = {}^-1$.

"51/3" is not in lowest terms since $17 \cdot 3 = 51$ and $1 \cdot 3 = 3$. However, we have $51/3 = 17/1$ and "17/1" is in lowest terms.

Note that it is **fractions** (i.e. *names* for rational numbers) that are said to be in lowest terms—not the numbers themselves. Lack of recognition of this semantic distinction can lead to some confusion. For example, consider the following 3-step argument:

(a) 51/3 is not in lowest terms.
(b) $51/3 = 17/1$
(c) 17/1 is not in lowest terms.

Assertion (c) is obtained from (a) and (b) by substitution of equals for equals. Something is wrong! It is statement (a) itself. It is the fraction (name) "51/3" that is not in lowest terms. It does not make sense to say that (the number) 51/3 is in lowest terms. We cannot apply the definition of "being in lowest terms" to the *number* 51/3 because 51/3 does not even have a numerator or a denominator. Of course, the fraction (name) "51/3" has both a numerator and a denominator. And that is exactly the point. It is a property of a name to be in lowest terms—not a property of a number.

A fraction is in lowest terms when we have simplified it as much as possible by cancelling out common factors from the numerator and denominator.

We discovered earlier that any given rational number has at least one name in the form of a fraction with a positive denominator. It is helpful to adopt the practice of usually using fractions that have positive denominators. But we shall occasionally want to do otherwise.

Combining the last comments we may state, as a general rule, that among all the possible fractions that name a rational number, we have a slight preference for the one that is in lowest terms and has a positive denominator. In any particular case, however, it may be a task of no small magnitude to put a result in this preferred form. We shall discuss this matter further in a later section. At this stage, we should evidently simply try to find, by trial and error, a common divisor for the numerator and denominator of a fraction that we wish to simplify.

Definition 8.3.1. We define **addition** in F by $a/b +_F c/d = (ad + cb)/(bd)$.

Theorem 8.3.1. If x, y, z, and w are in F and $x = y$ and $z = w$, then $x +_F z = y +_F w$.

PROOF: Suppose $x = a/b$, $y = c/d$, $z = e/f$, $w = g/h$, and $x = y$ and $z = w$. Here $a, b, \ldots, h$ are all integers and b, d, f, and h are not 0. Since $x = y$ and $z = w$ we have $ad = cb$ and $eh = fg$. Also $x +_F z = a/b +_F e/f = (af + eb)/(bf)$ and $y +_F w = c/d +_F g/h = (ch + gd)/(dh)$. Now $(af + eb)/(bf) = (ch + gd)/(dh)$ if $(af + eb)dh = (ch + gd)bf$. But this last equality holds if and only if $afdh + ebdh = chbf + gdbf$. From $ad = cb$ we deduce $adfh = cbfh$. From $eh = fg$ we deduce $ehbd = fgbd$. From $adfh = cbfh$ and $ehbd = fgbd$ we deduce that $adfh + ehbd = cbfh + fgbd$. But this last implies $afdh + ebdh = chbf + gdbf$. And we saw above that this equality implies $x +_F z = y +_F w$.

Theorem 8.3.2. Addition in F is commutative and associative and $0/1$ is the identity element for addition in F.

Theorem 8.3.3. $a/b +_F c/b = (a + c)/b$.

PROOF: See Problem 2.

Theorem 8.3.4. If a/b is a rational number, then $((^-1)a)/b$ is its additive inverse.

Definition 8.3.2. **Subtraction** in F (denoted by "$-_F$") is the inverse operation for addition in F.

Theorem 8.3.5. F is closed under subtraction and $a/b -_F c/d = a/b +_F ((^-1)c)/d$.

Theorem 8.3.6. Any rational number has a name in the form of a fraction in lowest terms.

PROOF: Let x be a rational number. Then there are integers a and b with $b \neq 0$ such that $x = a/b$. If $a = 0$, then $x = 0/1$ and "0/1" is in lowest terms. Suppose that $a \neq 0$. Either a and b have a common divisor different from 1 and $^-1$, or not. If not, then we have that which we seek. If so, then for some integers c, d, and e we have $a = cd$, $b = ed$, and $1 < |d|$. Then $x = cd/ed = c/e$. Note that from $1 < |d|$ it follows that $|c| = |1| \cdot |c| <$

$|d|\cdot|c| = |dc| = |a|$ and $|e| = |1|\cdot|e| < |d|\cdot|e| = |de| = |b|$. Now we go through the same argument with "c/e". If c and e have no common divisors aside from 1 and $^-1$ we have completed the proof. If they do, then we may simplify the fraction further. But such simplification reduces the absolute value of both the numerator and the denominator. Thus, the process must eventually stop† and when it does we have found a name for x in the desired form.

PROBLEMS

1. Find a simple fraction which names each of the following:

(a) $7/3 +_F 5/3$	(b) $73/3 +_F {}^-75/3$
(c) $73/3 -_F 75/3$	(d) $13/17 +_F 4/17$
(e) $13/17 +_F {}^-13/17$	(f) ${}^-75/17 +_F 63/17$
(g) $5/{}^-2 +_F 7/{}^-2$	(h) $333/3 +_F 111/{}^-1$
(i) $1/2 +_F 1/3$	(j) $1/2 -_F 3/4$
(k) $33/91 +_F 71/12$	(l) $32/3 +_F {}^-17/6$
(m) $10072/51 +_F 17/3$	(n) $699/3 -_F 47/5$
(o) $7/35 +_F 19/70$	(p) $396/39 -_F 753/72$.

2. Directly from our original definition of addition in F prove that $a/b +_F c/b = (a + c)/b$.

3. Express each of the following rational numbers by means of a fraction which is in lowest terms and has a positive denominator.

(a) $2/{}^-4$	(b) $225/5$
(c) $72/{}^-3$	(d) ${}^-12/144$
(e) $27/{}^-243$	(f) $10{,}000/{}^-100$
(g) ${}^-25/625$	(h) $221/{}^-51$

(i) $851/1517$.

4. Is F closed under addition? Explain how you know.

---

5. Is $(F, +_F)$ a group? Explain.

6. Let $T = \{x \mid x \in F \text{ and } x = a/1 \text{ for some } a \in I\}$. Is $(T, +_F)$ a group? Explain.

7. Let $S = \{x \mid x \in F \text{ and } x = a/2 \text{ for some } a \text{ in } I\}$. Is $(S, +_F)$ a group? Explain.

8. Generalize your conclusions of Problems 6 and 7.

† To be specific, there could not be more than min $\{|a|, |b|\}$ steps required.

8.4 Multiplication

We discovered in Section 7.5 that when a, b, c, and d are integers, and b is a divisor of a, and d is a divisor of c, then $(a \div b) \cdot (c \div d) = (ac) \div (bd)$ in I. Thinking of the rational number a/b as representing $a \div b$, we are led to define **multiplication** in F by

$$(a/b) \cdot_F (c/d) = (ac)/(bd).$$

Note that if a/b and c/d are rational numbers then $b \neq 0$ and $d \neq 0$. Hence, $bd \neq 0$ and $(ac)/(bd)$ is a rational number.

As some numerical illustrations of the definition consider:

$$(5/1) \cdot_F (6/1) = (5 \cdot 6)/(1 \cdot 1) = 30/1$$

$$(^-6/1) \cdot_F (2/1) = (^-6 \cdot 2)/(1 \cdot 1) = {}^-12/1$$

$$(1/2) \cdot_F (3/4) = (1 \cdot 3)/(2 \cdot 4) = 3/8$$

$$(1/17) \cdot_F (17/1) = (1 \cdot 17)/(17 \cdot 1) = 1/1$$

$$(^-1/5) \cdot_F (5/^-1) = (^-1 \cdot 5)/(5 \cdot {}^-1) = 1/1.$$

The commutativity and associativity of multiplication in I immediately yield that multiplication is commutative and associative in F.

$1/1$ is the identity element for multiplication in F since $(a/b) \cdot_F (1/1) = (a \cdot 1)/(b \cdot 1) = a/b$.

Multiplication distributes across addition in F.

In striking contrast to the situation in I, most elements in F have inverses with respect to multiplication! If $a \neq 0$ and $b \neq 0$, then

$$(a/b) \cdot_F (b/a) = (ab)/(ba) = (ab)/(ab) = 1/1.$$

We need both a and b different from 0 here so that a/b and b/a are both rational numbers.

The above result has the consequence that:

> If x is a rational number and $x \neq 0/1$, then x has a rational multiplicative inverse x^{-1} such that $x^{-1} \cdot_F x = x \cdot_F x^{-1} = 1/1$.

We call the multiplicative inverse of x the **reciprocal** of x. The reciprocal of a/b is b/a when $a \neq 0$. $0/b$ does not have a reciprocal. We shall sometimes use "x^{-1}" to name the reciprocal of x.

We define **division** in F to be the inverse operation for multiplication in F. Then $x \div_F y = z$ for x, y, and z in F if and only if z is the unique rational number such that $x = y \cdot_F z$. But suppose $x = a/b$ and $y = c/d$ where $c \neq 0$. We claim that $x \div_F y = (a/b) \div_F (c/d) = (a/b) \cdot_F (d/c) = x \cdot_F y^{-1}$. To prove this we first note that when $z = x \cdot_F y^{-1}$ we do have $x = y \cdot_F z$. For we have

$y \cdot_F z = (c/d) \cdot_F ((a/b) \cdot_F (d/c)) = (c/d) \cdot_F (d/c) \cdot_F (a/b) = ((c/d) \cdot_F (d/c)) \cdot_F (a/b) = (1/1) \cdot_F (a/b) = a/b$.

We next show that if $x = y \cdot_F w$, then $w = x \cdot_F y^{-1} = z$. This will show that z is the *unique* rational number such that $x = y \cdot_F z$. $x = y \cdot_F w$ implies that $y^{-1} \cdot_F x = y^{-1} \cdot_F (y \cdot_F w) = (y^{-1} \cdot_F y) \cdot_F w = (1/1) \cdot_F w = w$. But $y^{-1} \cdot_F x = x \cdot_F y^{-1}$.

We conclude that, in F, when we wish to divide by a rational number x which is different from $0/1$, we may instead multiply by the reciprocal of x.

As long as $x \neq 0/1$ and x is a rational number it has a reciprocal and, hence, division of any rational by that x yields a result. This is a vast extension over the situation in I.

When we wish to multiply or divide two fractions we would prefer to do so efficiently and to end up with a simple name for the result if we can. One device that is quite often efficient is to make use of cancellation as much as possible. For example, we could write

$$\text{“}52/4 \cdot_F 4/52 = \overset{1}{\cancel{52}} \cdot \overset{1}{\cancel{4}} / \overset{1}{\cancel{4}} \cdot \overset{1}{\cancel{52}} = 1/1.\text{”}$$

Similarly,

$$\overset{1}{\cancel{25}}/\overset{1}{\cancel{3}} \cdot_F \overset{9}{\cancel{27}}/\overset{4}{\cancel{100}} = 9/4.$$

And

$$144/10 \div_F 3/50 = \overset{48}{\cancel{144}}/10 \cdot_F \overset{5}{\cancel{50}}/\cancel{3} = 240/1.$$

A basic fact about multiplication in F is that in order for a product to be $0/1$ one of the factors must be $0/1$. That is, $x \cdot_F y = 0/1$ implies $x = 0/1$ or $y = 0/1$.

Definition 8.4.1. We define **multiplication** in F by $(a/b) \cdot_F (c/d) = (ac)/(bd)$.

Theorem 8.4.1. If x, y, z, and w are in F and $x = y$ and $z = w$, then $x \cdot_F z = y \cdot_F w$.

PROOF: Suppose $x = a/b$, $y = c/d$, $z = e/f$, and $w = g/h$ and $x = y$ and $z = w$. Here $a, b, \ldots, h$ are integers and b, d, f, and h are different from 0. $x = y$ implies $ad = cb$. $z = w$ implies $eh = fg$. $x \cdot_F z = (a/b) \cdot_F (e/f) = (ae)/(bf)$. Also $y \cdot_F w = (c/d) \cdot_F (g/h) = (cg)/(dh)$. Now $(ae)/(bf) = (cg)/(dh)$ if $(ae)(dh) = (cg)(bf)$. But this last equality holds if $(ad)(eh) = (cb)(fg)$ and this equation follows from $ad = cb$ and $eh = fg$.

Theorem 8.4.2. Multiplication in F is commutative and associative and $1/1$ is the identity element for multiplication in F.

Theorem 8.4.3. If a and b are nonzero integers then $(a/b) \cdot_F (b/a) = 1/1$.

Theorem 8.4.4. Multiplication distributes across addition in F.

PROOF: Let $x = a/b$, $y = c/d$, and $z = e/f$ where $a, b, \ldots, f$ are integers and $b \neq 0$, $d \neq 0$, and $f \neq 0$. $a/b \cdot_F (c/d +_F e/f) = a/b \cdot_F ((cf + ed)/df) = (a(cf + ed))/(b(df))$. Also $(a/b \cdot_F c/d) +_F (a/b \cdot_F e/f) = ((ac)/(bd)) +_F ((ae)/(bf)) = ((ac)(bf) + (ae)(bd))/((bd)(bf)) = (a(cf + ed))/(b(df))$. Hence, $x \cdot_F (y +_F z) = (x \cdot_F y) +_F (x \cdot_F z)$.

Theorem 8.4.5. If x is a rational number and $x \neq 0/1$ then x has a rational inverse with respect to multiplication.

Definition 8.4.2. If $a \neq 0$, the rational number b/a is called the **reciprocal** of a/b.

Definition 8.4.3. **Division** in F is the inverse operation for multiplication in F.

Theorem 8.4.6. If $c \neq 0$, then $(a/b) \div_F (c/d) = (a/b) \cdot_F (d/c)$.

Theorem 8.4.7. If $x \cdot_F y = 0/1$, then $x = 0/1$ or $y = 0/1$.

PROOF: Let $x = a/b$ and $y = c/d$ and suppose $x \cdot_F y = 0/1$. Then, by Definition 8.4.1, $(ac)/(bd) = 0/1$ and, hence, $ac = 0$. But, by Theorem 7.4.9, this implies $a = 0$ or $c = 0$. Hence, $x = 0/b$ or $y = 0/d$. But $0/b = 0/1$ and $0/d = 0/1$.

PROBLEMS

1. Find simple fractions which name the following:

   (a) $1/3 \cdot_F 7/2$
   (b) $^-2/3 \cdot_F 7/4$
   (c) $^-1/2 \cdot_F 8/3$
   (d) $^-3/4 \cdot_F {}^-16/9$
   (e) $10^4/36 \cdot_F 30/10^3$
   (f) $144/51 \cdot_F 17/36$
   (g) $5376/4965 \cdot_F 0/76$
   (h) $^-2/5 \cdot_F {}^-5/2$
   (i) $76/3 \cdot_F {}^-3/{}^-76$
   (j) $49/36 \cdot_F 6/7$
   (k) $512/2 \cdot_F 3/32$.

2. Prove that division in F is not

   (a) commutative
   (b) associative.

3. Is F closed under multiplication? Explain how you know.
4. Prove that addition does not distribute across multiplication in F.
5. Does division distribute across addition in F? Explain how you know.

---

6. Is $(F, \cdot_F)$ a group? Explain.
7. Let $Q = \{x \mid x \in F \text{ and } x \neq 0/1\}$.
   (a) Is $(Q, \cdot_F)$ a group? Explain.
   (b) Is $(Q, \div_F)$ a group? Explain.

8.5 Integral rational numbers

We invented the rational number $a/1$ to be a solution of the equation $1 \cdot x = a$ when a is an integer. But in the system of integers, a is already a solution for $1 \cdot x = a$. Thus, we would expect the rational number $a/1$ to behave like the integer a. We may now show that this is so.

Any rational number $a/1$ where a is an integer is called an **integral rational number.** Let T be the collection of all integral rational numbers. We claim that the system $(T, +_F, \cdot_F)$ is isomorphic to the system $(I, +, \cdot)$. To establish this fact we simply note that the function f defined by $f(a/1) = a$ has the properties that it is a one-to-one-correspondence between T and I, and $f(a/1 +_F b/1) = f((a + b)/1) = a + b = f(a/1) + f(b/1)$, and $f(a/1) \cdot_F f(b/1) = f((ab)/1) = ab = f(a/1) \cdot f(b/1)$. Hence, f is an isomorphism between the two systems.

Thus, we see that the rational number system is, as we had hoped, an extension of the system of integers. As far as addition and multiplication go the integral rational numbers and the integers behave just the same.

So, in the future, we may in many instances write either "a" or "$a/1$" for the rational number $a/1$ and cause no confusion. And we shall now drop the subscript "F" on the operation symbols for F and feel free to interpret $x + y$ and $x \cdot y$ (or xy) as results either in I or in F as seems appropriate.

In particular, we now take "0" as a name for the rational number $0/1$, or for the integer 0 (or $\overline{0}$), or even for the natural number 0, whichever is convenient.

Let us call a rational number **positive** if it can be named by a fraction "a/b" where a and b are both positive integers. A rational number that is not 0 and is not positive is said to be **negative.** The set of nonnegative integral rational numbers is $\{0/1, 1/1, 2/1, 3/1, \cdots\}$.

We note that the system of nonnegative integral rational numbers with addition and multiplication is isomorphic to the system of nonnegative integers with these operations. But since this latter system is isomorphic to $(N, +, \cdot)$, so is the former system.

Thus, we may think of the expression "16" as a name for the rational number $16/1$, or for the integer $^+16$, or for the natural number 16, whichever is convenient.

We have said that when a and b are integers we may think of a/b as the number we invented to be the result of the division $a \div b$ and we may now verify that this is so. Using the convention above we have that when a and b are integers and $b \neq 0$

$$a \div b = (a/1) \div (b/1) = (a/1) \cdot (1/b) = (a \cdot 1)/(1 \cdot b) = a/b.$$

And it follows that, in F, the number a/b *is* the solution for the equation $bx = a$ when $b \neq 0$. For we have that when $x = a/b$, $bx = (b/1) \cdot (a/b) = (b \cdot a)/(1 \cdot b) = a/1 = a$. And it can be shown that $x = a/b$ is the only solution to the equation.

In view of the fact that when a and b are integers we have $a/b = a \div b$ when a and b are thought of as rational numbers, we now extend the meaning of the fraction line and may take "x/y" or "$\frac{x}{y}$" to mean $x \div y$ when x and y are any rational numbers whatsoever.†

For example, $(5/2)/(3/10) = 5/2 \div 3/10 = 5/2 \cdot 10/3 = 50/6 = 25/3$, and $(3/4 - 7/3)/(4/5) = (3/4 - 7/3) \div (4/5) = ((9 - 28)/12) \div (4/5) = (^-19/12) \cdot (5/4) = {}^-95/48$.

The rules for adding, subtracting, multiplying, and dividing fractions carry over to the case where the numerators and denominators are rational numbers rather than integers.

Suppose a and b are positive integers and b goes into a q times with a remainder of r. We may make use of rational numbers to express this as:

$$a/b = q + r/b.$$

That this is so follows from the nature of our algorithm for division. For we have $a - qb = r$ and, hence, $(a - qb)/b = r/b$. This implies $a/b - (qb)/b = r/b$ or $a/b = q + rb$.

It is true that in F we can solve many equations and carry out many division problems that we could not attack before. But we must realize one important fact. F is still *not closed under division*. Division by 0 is still impossible in F. The reasons for this are essentially the same as in I. We have $0 \cdot x = 0$ for all x in F. Hence, when $y \neq 0$ there is no rational number x such that $0 \cdot x = y$. Therefore, we must leave $y \div 0$ undefined when $y \neq 0$. On the other hand, since $0 \cdot x = 0$ for all x, there is not a unique value x such that $0 \cdot x = 0$ and, therefore, we must leave $0 \div 0$ undefined.

Definition 8.5.1. If x is a rational number and $x = a/1$ for some integer a, then x is said to be an **integral** rational number. Let T be the collection of all integral rational numbers.

Theorem 8.5.1. The system $(T, +_F, \cdot_F)$ is isomorphic to the system $(I, +, \cdot)$.

Definition 8.5.2. If "n" names an integer we shall also consider "n" as a name for the rational number $n/1$. We shall use "$+$" and "$\cdot$" for "$+_F$" and "$\cdot_F$," respectively.

Definition 8.5.3. A rational number is said to be **positive** if it can be written as a fraction whose numerator and denominator are positive integers. A rational number that is not positive and is not 0 is said to be **negative.**

---

† We let the fraction sign, as well as its counterpart the division sign, take precedence over "$+$" and "$-$." For example, $5 + 3/7 = 5 + (3/7)$ and this is different from $(5 + 3)/7$. We also let division take precedence over multiplication. For example $3/7 \cdot 7/4 = (3/7) \cdot (7/4) = 3/4$ and this is different from $(3/(7 \cdot 7))/4$ and from $3/((7 \cdot 7)/4)$.

Theorem 8.5.2. If Q is the set of nonnegative integral rational numbers, then $(Q, +, \cdot)$ is isomorphic to $(N, +, \cdot)$.

Theorem 8.5.3. If a and b are integers (or integral rational numbers), then $a/b = a \div b$.

Definition 8.5.4. If x and y are rational numbers and $y \neq 0$, then we let $x/y = x \div y$.

Theorem 8.5.4. If x, y, z, and w are rational numbers with $y \neq 0$ and $w \neq 0$, then

(a) $(x/y) + (z/w) = (xw + zy)/yw$
(b) $(x/y) - (z/w) = (x/y) + (-z/w)$
(c) $(x/y) \cdot (z/w) = (xz)/(yw)$
(d) $(x/y)/(z/w) = (xw)/(yz)$ provided $z \neq 0$.

PROOF: We shall prove only part (a) since the proofs of the other parts are similar. Let $x = a/b$, $y = c/d$, $z = e/f$, and $w = g/h$ where $a, b, \ldots, h$ are integers. Then $(x/y) + (z/w) = ((a/b)/(c/d)) + ((e/f)/(g/h)) = ((a/b) \div (c/d)) + ((e/f) \div (g/h)) = (a/b)(d/c) + (e/f)(h/g) = (ad/bc) + (eh/fg) = (adfg + ehbc)/bcfg$ where we have used Definition 8.5.4, Theorem 8.4.6, Definition 8.4.1, and Definition 8.3.1. We also have $(xw + zy)/yw = ((a/b)(g/h) + (e/f)(c/d)) \div ((c/d)(g/h)) = ((ag/bh) + (ec/fd)) \div (cg/dh) = ((agfd + ecbh)/bhfd) \div (cg/dh) = ((agfd + ecbh)/bhfd)(dh/cg) = (adfg + ehbc)/bcfg$ where we have used Definition 8.5.4, Definition 8.4.1, Definition 8.3.1, and Theorem 8.4.6. Since the end results of these two strings of equalities are the same, the proof is complete.

Theorem 8.5.5. Let a, b, q, and r be positive integers with $r < b$. b goes into a q times with a remainder of r if and only if $a/b = q + (r/b)$.

PROBLEMS

1. Let T be the collection of all integral rational numbers. Define the mapping h on T by letting $h(a/1) = -a$. Is h an isomorphism between the systems

(a) $(T, +_F)$ and $(I, +)$
(b) $(T, +_F, \cdot_F)$ and $(I, +, \cdot)$?

2. With T as in Problem 1 define the mapping k on T by letting $k(a/1) = |a|$. Is k an isomorphism between the systems $(T, +_F)$ and $(N, +)$?

3. Explain exactly what "5" means in terms of our definitions when we interpret it as referring to

(a) a natural number
(b) an integer
(c) a rational number.

4. Is the set T of Problem 1 closed under

(a) addition (b) multiplication
(c) subtraction (d) division?

Explain how you know.

5. Let G be the function defined by letting $G(a/b) = b/a$ when a and b are integers. Is G an isomorphism between the system $(F, \cdot_F)$ and itself? Explain.

6. Let Q be the collection of all nonzero rationals. Is the function G of Problem 5 an isomorphism between the system $(Q, \cdot_F)$ and itself? Explain.

7. Find simple names for each of the following:

(a) $(1/2 + 7/3)/(3/10)$
(b) $(4/17)/(9/34)$
(c) $(4 + 7/2)/(6 - 4/13)$
(d) $(51/3 - 37/2)/(1/2 - 1/3)$
(e) $(((1/2 + 1/3)/(1/4 + 1/5))/(1/6 + 1/7))/3$
(f) $((1/2 + 1/3)/((1/4 + 1/5)/(1/6 + 1/7)))/3$.

---

8. Let $P = \{x \mid x \in F \text{ and } x \text{ is positive}\}$. Is $(P, \cdot_F)$ a group? Explain.

8.6 Positive rational numbers

We have said that a rational number x is **positive** if there exist positive integers a and b such that $x = a/b$. And if a rational number x is not 0 and is not positive, then it is said to be **negative.** Let us use "F^+" as a name for the collection of all positive rational numbers and "F^-" for the collection of all negative rational numbers. F^+ is the set of numbers corresponding to those rational points on the number line which lie to the right of the origin.

As we did in I, we shall in F say that a positive number *has a positive sign* and that a negative number *has a negative sign*. But to say that a rational number *has* a positive sign is not to say that the symbol "+" must somehow be found in any name for the number. For example, $^-3/^-2$ is positive since $^-3/^-2 = 3/2$.

Since 0 is not a positive rational number it follows that given any rational number x we must have that x is negative or $x = 0$ or x is positive, and only one of these can be the case.

Suppose $x \in F^+$. Then there are positive integers a and b such that $x = a/b$. But $a/b = ((^-1)a)/((^-1)b)$ and $(^-1)a$ and $(^-1)b$ are negative. Hence, x can also be written as a fraction where the numerator and denominator are both negative.

Similarly, if $x = c/d$ where c and d are both negative integers then $x = ((^-1)c)/((^-1)d)$ so that x can be written as a fraction where both the numerator and denominator are positive. And x must be positive. We thus have:

> A rational number x is positive if and only if it can be written as a fraction whose numerator and denominator have the same sign.

Since the integer 0 has no sign we may also conclude from the above result that:

> A rational number is negative if and only if it can be written as a fraction whose numerator and denominator differ in sign.

If we use a positive denominator we may restate the above as:

> Suppose $x = a/b$ where b is positive. Then x is positive if and only if a is positive and x is negative if and only if a is negative.

And further:

> Suppose $x = a/b$ where b is positive. Then x is nonnegative if and only if a is nonnegative, and x is nonpositive if and only if a is nonpositive.

We extend the notion of the **absolute value** of a number to F by stating the definition:

If a and b are integers and $x = a/b$ then $|x| = |a|/|b|$.

It immediately follows that $|0| = 0$ and that if $x \neq 0$, then $|x|$ is positive.

In terms of the number line we may think of $|x|$ as the distance between the point x and the origin.

Making use of the results of Section 7.4, it is quite easy to show that when x and y are rational numbers

$$|xy| = |x||y| \qquad \text{and}$$

$$|x \div y| = |x| \div |y| \qquad \text{when} \quad y \neq 0.$$

However, as we would expect from our earlier work, the absolute value of a sum is not, in general, equal to the sum of the absolute values. For example,

$$|1/2 + {}^{-}1/2| = |0| = 0 \qquad \text{and yet} \qquad |1/2| + |{}^{-}1/2| = 1/2 + 1/2 = 1.$$

It is not at all difficult to prove that F^{+} is closed under addition and multiplication. That is, the sum of any two positive rational numbers is a positive rational. And the product of any two positive rational numbers is a positive rational.

The rational number system can be built up by a different route than the one we have followed. We could have first defined positive rational numbers in terms of equivalence classes on $N \times I^{+}$ and then extended this system to the entire system F by forcibly introducing additive inverses for the positive rationals. This alternate route would, perhaps, be somewhat closer to the usual sequence in teaching arithmetic where fractions are introduced before negative numbers are mentioned. We chose the pattern of development that we **did** use because this pattern is somewhat closer to the other patterns we have followed and will follow later for extending one number system to another.

It is hoped that our route is more typical and, hence, more illuminating of the grand strategy for building such extensions.

Definition 8.6.1. A rational number x is said to be **positive** (or to **have a positive sign**) if there exist positive integers a and b such that $x = a/b$.

Theorem 8.6.1. 0 is not positive.

PROOF: If a and b are integers such that $0 = 0/1 = a/b$, then by the definition of rational numbers $0 \cdot b = a \cdot 1$. Hence, $a = 0$ and cannot be a positive integer.

Definition 8.6.2. A rational number that is not positive and is different from 0 is said to be **negative** (or to **have a negative sign**).

Theorem 8.6.2. If x is a rational number then exactly one of the following must hold:

(i) x is negative
(ii) $x = 0$
(iii) x is positive.

Theorem 8.6.3. A rational number is positive if and only if it can be written as a fraction whose numerator and denominator have the same sign.

Theorem 8.6.4. A rational number is negative if and only if it can be written as a fraction whose numerator and denominator have opposite signs.

Theorem 8.6.5. If $b > 0$ and $x = a/b$, then x has the same sign (in F) as a does (in I).

Definition 8.6.3. If a and b are integers and $b \neq 0$ and $x = a/b$ we define the **absolute value** of x to be $|x| = |a|/|b|$.

Theorem 8.6.6. If x is in F, then $|x|$ is nonnegative. And $|x| = 0$ if and only if $x = 0$.

Theorem 8.6.7. If x and y are in F then

(a) $|xy| = |x| \cdot |y|$ and
(b) $|x \div y| = |x| \div |y|$ when $y \neq 0$.

PROOF: Suppose $x = a/b$ and $y = c/d$ where a, b, c, and d are integers. Then $xy = ac/bd$ and $|xy| = |ac|/|bd| = (|a||c|)/(|b||d|) = (|a|/|b|)(|c|/|d|) = |x| \cdot |y|$ by Theorem 7.4.6. Also, using the same theorem, we have when $c \neq 0$, $|x \div y| = |(a/b)(d/c)| = |ad/bc| = |ad|/|bc| = (|a||d|)/(|b||c|) = (|a|/|b|) \cdot (|d|/|c|) = (|a|/|b|) \div (|c|/|d|) = |x| \div |y|$.

Definition 8.6.4. Let $F^+ = \{x \mid x \text{ is a positive rational number}\}$ and $F^- = \{x \mid x \text{ is a negative rational number}\}$.

Theorem 8.6.8. F^+ is closed under addition and multiplication.

PROOF: This follows easily from the fact that I^+ is closed under addition and multiplication. We leave the details to be done as an exercise.

PROBLEMS

1. In F, let us define the relation R by saying $x \, R \, y$ is and only if $x = y = 0$ or if x and y have the same sign. Is R an equivalence relation in F? If so, what are its equivalence classes?

2. Is F^+ closed under

(a) subtraction (b) division?

Explain how you know.

3. Is F^- closed under

(a) addition (b) subtraction
(c) multiplication (d) division?

Explain how you know.

4. Prove Theorem 8.6.8.

5. Let $Q = F^+ \cup \{0\}$ and $S = F^- \cup \{0\}$. Is the system $(Q, +)$ isomorphic to the system $(S, +)$? Explain.

6. If x, y, and z are positive rational numbers, we say that x is y **per cent** (or $y\%$) of z if and only if $x/z = y/100$. In other words, x is y per cent of z when $x = z(y/100)$. To find y per cent of z, we multiply z by $y/100$. For example, 5 is 1 **per cent** of 500 since $5 = 500(1/100)$. What number is

(a) 3 per cent of 960
(b) 7 per cent of 14,000
(c) 6 per cent of 5400
(d) 1/2 per cent of 17,500?

7. (a) 3200 is 6 per cent of what number?
(b) 560 is 8 per cent of what number?
(c) 43,000 is 3 per cent of what number?

8. (a) 75 is what per cent of 75,000?
(b) 3600 is what per cent of 18,000?
(c) 596 is what per cent of 298?

8.7 Inequalities

One-third of a pie is larger than one-fourth of it. We would get less money if we inherited one-fifth of our rich uncle's estate than if we inherited two-ninths of it. If seven-eighths of the people in a town support a certain candidate for mayor, he has more support than if only three-quarters of them favor him.

Clearly, there are many, many practical situations where we need to compare the relative sizes of two rational numbers.

As a direct analogue of the corresponding definition in I we have:

> When x and y are rational numbers we say that $x \geq y$ if there is a nonnegative rational number z such that $x = y + z$.

This is the third time we have used "$\geq$" to name a relation. We might have labeled this new relation by means of a symbol such as "$\geq_F$." But there is no need to do this. Since the system of integral rational numbers with addition and multiplication is isomorphic to the system $(I, +, \cdot)$ it follows that the relation $\geq$ as defined above in the set of integral rational numbers is isomorphic to the relation $\geq$ we had in I. Thus, we are free to interpret such a statement as $6 \geq {}^-5$ as either an assertion about integers or an assertion about the rational numbers $6/1$ and ${}^-5/1$.

We define the relations $>$, $\leq$, and $<$ in F in terms of $\geq$ as we have done before. We also carry over to F our definitions for "between," "maximum," and "minimum."

In terms of the number line, $x \geq y$ in F if and only if the rational point x is at least as far to the right as the rational point y. Since F is closed under subtraction we have the result similar to one in I that when x and y are rational numbers

> $x \geq y$ if and only if $x - y$ is nonnegative.

We may build on this result to get a rather simple procedure for comparing the sizes of two rational numbers. Suppose a, b, c, and d are integers and b and d are positive. We know that $a/b \geq c/d$ if and only if $a/b - c/d$ is nonnegative. But $a/b - c/d = (ad)/(bd) - (cb)/(bd) = (ad - cb)/bd$. When b and d are positive integers, so is bd and, hence, this last fraction names a nonnegative rational number if and only if $ad - cb$ is nonnegative (in I). And $ad - cb$ is nonnegative if and only if $ad \geq cb$. We have thus shown that when a, b, c, and d are integers:

> If b and d are positive, then $a/b \geq c/d$ if and only if $ad \geq cb$.

In a similar fashion we may prove that:

> If b and d are positive, then $a/b > c/d$ if and only if $ad > cb$.

We now have a very convenient method for comparing two rational numbers when we write them as fractions with positive denominators. The procedure is summarized by the flow chart of Figure 8.8.

As examples of the use of the above we see that

$$1/3 > 1/4 \qquad \text{since} \qquad 4 \cdot 1 = 4 > 1 \cdot 3 = 3,$$

$$1/5 < 2/9 \qquad \text{since} \qquad 1 \cdot 9 = 9 < 2 \cdot 5 = 10,$$

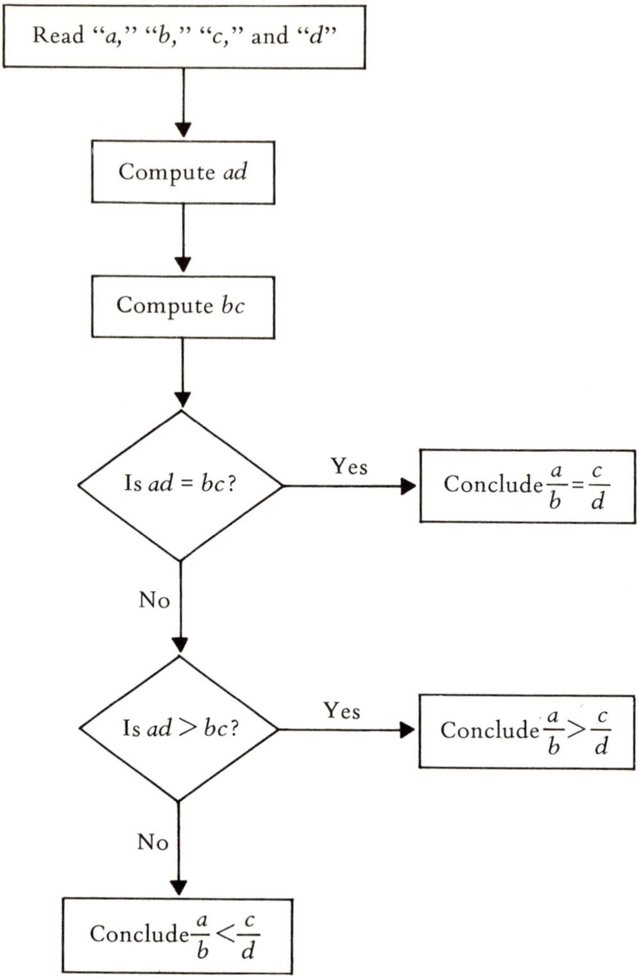

FIGURE 8.8 Flow chart for comparing a/b and c/d when b and d are positive.

$$7/8 > 3/4 \quad \text{since} \quad 7 \cdot 4 = 28 > 3 \cdot 8 = 24,$$

$$^-7/5 < 3/5 \quad \text{since} \quad ^-7 \cdot 5 = {}^-35 < 3 \cdot 5 = 15,$$

$$^-5/9 > {}^-17/3 \quad \text{since} \quad ^-5 \cdot 3 = {}^-15 > {}^-17 \cdot 9 = {}^-153,$$

and

$$^-1073/1000 < {}^-107/100 \quad \text{since} \quad ^-107{,}300 < {}^-107{,}000.$$

One immediate result of the above method is:

If a, b, and c are integers and c is positive, then $a/c \geq b/c$ if and only if $a \geq b$.

Also

> If a, b, and c are integers and c is positive, then $a/c > b/c$ if and only if $a > b$.

These results tie in nicely with our geometric notions about rational numbers. When c is a positive integer, $1/c$ is the length of any one of the segments obtained when we break up the segment from 0 to 1 into c pieces of equal length. And $a/c = a(1/c)$ represents the point reached from the origin by laying out $|a|$ consecutive segments of length $1/c$ in the proper direction—to the right if a is positive and to the left if a is negative. b/c may be interpreted similarly. We certainly should have the point a/c at least as far to the right as the point b/c when and only when $a \geq b$.

Many of the properties that we discovered for $\geq$ in N carry over to F. In fact, the assertions numbered from 2 to 12 on page 156 are all true if we take it that x, y, z, and w are rational numbers.

And some additional properties of $\geq$ in I carry over to F. In particular, we have that when x, y, and z are rational numbers

(a) $x > 0$ if and only if x is positive
(b) $x < 0$ if and only if x is negative
(c) If $z > 0$, then $x \geq y$ if and only if $xz \geq yz$
(d) If $z < 0$, then $x \geq y$ if and only if $xz \leq yz$
(e) If $z \geq 0$, then $|x - y| \leq z$ if and only if $y - z \leq x \leq y + z$
(f) $|x + y| \leq |x| + |y|$.

Since all nonzero elements in F have reciprocals, results (c) and (d) above are especially useful. If, for instance, we wish to find the maximum and minimum for the set $\{x|^-6 \leq 7 - 3x \leq 4\}$, we may do so as follows.

$$\begin{aligned}\{x|^-6 \leq 7 - 3x \leq 4\} &= \{x|^-6 - 7 \leq {}^-3x \leq 4 - 7\}\\ &= \{x|^-13 \leq {}^-3x \leq {}^-3\} = \{x|(^-1/3)(^-13) \geq (^-1/3)(^-3x)\\ &\qquad \geq (^-1/3)(^-3)\} = \{x|13/3 \geq x \geq 1\} = \{x|1 \leq x \leq 13/3\}.\end{aligned}$$

We see immediately that the maximum of this set is 13/3 and the minimum is 1. Note well the reversal of the inequalities at the third step above.

As another example, we have $\{x||2x - 1| \leq 6\} = \{x|1 - 6 \leq 2x \leq 1 + 6\} = \{x|(1/2)(^-5) \leq (1/2)(2x) \leq (1/2)(7)\} = \{x|^-5/2 \leq x \leq 7/2\}$ so that this set has a maximum of 7/2 and a minimum of $^-5/2$.

Despite the fact that some of the basic facts about inequalities in I carry over to F, the order situation in F is strikingly different from that in I. It is false in F that when $x < y$ we must have $x + 1 \leq y$. For example, when $x = 1/2$ and $y = 1$, we have $x < y$ and yet $x + 1 = 3/2 > 1$.

It is also false in F that there is no number between 0 and 1. For example, we have $0 < 1/2 < 1$.

In fact, we have the following theorem which points up a very important difference between I and F:

> If x and y are rational numbers and $x < y$, then there is a rational number z such that $x < z < y$.

This is easy to prove because we can show that $(x + y)/2$ is between x and y. This may be done as follows. From $x < y$ it follows that $x + x < y + x$ and also $x + y < y + y$. Then $2x < x + y < 2y$. And multiplying both sides of these two inequalities by the positive number $1/2$ we obtain $x < (x + y)/2 < y$.

Whenever we have two distinct rational numbers x and y we must have $x < y$ or $y < x$. So it follows from what we just proved that given any two distinct rational numbers there is a rational number that is between them. In terms of the number line this tells us that between any two rational points there is another rational point.†

A truly remarkable fact follows from these considerations. Between any two distinct rational numbers there are *infinitely many* rational numbers.

For suppose that x and y are in F and $x \neq y$. There is no loss in generality if we suppose $x < y$. We know there is an x_1 such that $x < x_1 < y$. But then there must be an x_2 such that $x < x_2 < x_1$. Then there must be an x_3 such that $x < x_3 < x_2$. And so forth without end. The distinct numbers $x_1, x_2, x_3, \ldots$ are all between x and y.

In N and I we found that bounded sets were always finite. We see that this is not true in F. We shall dig more deeply into this result and related matters in the next section.

Definition 8.7.1. When x and y are in F, we say that $x \geq y$ if there is a non-negative rational number z such that $x = y + z$. We define $>$, $\leq$, $<$, "between," $\max(A)$, and $\min(A)$ as before.

Theorem 8.7.1. Let Q be the collection of integral rational numbers. The relation $\geq$ of Definition 8.7.1 in Q is isomorphic to the relation $\geq$ in I.

Theorem 8.7.2. Suppose $x \in F$ and $y \in F$. $x \geq y$ if and only if $x - y$ is non-negative.

PROOF: See the proof of Theorem 7.8.4.

Theorem 8.7.3. If a, b, c, and d are integers and b and d are positive, then $a/b \geq c/d$ when and only when $ad \geq bc$.

Theorem 8.7.4. When a, b, and c are integers and c is positive, $a/c \geq b/c$ if and only if $a \geq b$.

† Because F has this property, the rationals are said to be *dense in themselves.*

Theorem 8.7.5. Suppose x, y, z, and w are in F. Then

(a) $x \geq x$
(b) $x \geq y$ and $y \geq z$ implies $x \geq z$
(c) $x > y$ and $y > z$ implies $x > z$
(d) $x \geq y$ or $y \geq x$
(e) $x \geq y$ and $y \geq x$ implies $x = y$
(f) $x > y$ or $y > x$ or $x = y$ and only one of these may hold
(g) $x + 1 > x$
(h) $x \geq y$ if and only if $x + z \geq y + z$
(i) $x > y$ if and only if $x + z > y + z$
(j) $x \geq y$ and $z \geq w$ implies $x + z \geq y + w$
(k) $x \geq y$ and $z > w$ implies $x + z > y + w$.

PROOF: The proofs are similar to those for similar statements in I. They will be left as exercises.

Theorem 8.7.6. Suppose x, y, and z are in F. Then

(a) $x > 0$ if and only if x is positive
(b) $x < 0$ if and only if x is negative
(c) If $z > 0$, then $x \geq y$ if and only if $xz \geq yz$
(d) If $z < 0$, then $x \geq y$ if and only if $xz \leq yz$
(e) If $z \geq 0$, then $|x - y| \leq z$ if and only if $y - z \leq x \leq y + z$
(f) $|x + y| \leq |x| + |y|$.

Theorem 8.7.7. Between any two distinct rational numbers there is another rational number.

Theorem 8.7.8. Between any two distinct rational numbers there are infinitely many distinct rational numbers.

PROBLEMS

1. Which of the following are true?

(a) $37/19 \geq 14/19$
(b) $^-37/19 \geq {}^-14/19$
(c) $^-5/4 \geq {}^-2/3$
(d) $^-59/3 \geq {}^-75/3$
(e) $5/17 \geq {}^-73/4$
(f) $5/^-7 \geq {}^-2/3$
(g) $100/57 \geq 99/56$
(h) $^-75/81 \geq {}^-25/27$
(i) $^-43/6 \geq 42/^-5$
(j) $\dfrac{1021}{1000} \geq \dfrac{102}{100}$
(k) $\dfrac{^-2574}{1000} \geq \dfrac{^-19{,}650}{10{,}000}$.

2. Find the maximum and minimum (if any) for each of the following sets:

(a) $\{x | 2x \leq 1\}$
(b) $\{x | {}^-3/4 \leq x \leq 7/5\}$
(c) $\{x | {}^-2x \leq 1\}$
(d) $\{x | {}^-3/4 \leq x + 1 \leq 0\}$
(e) $\{x | |x| \leq 1/17\}$
(f) $\{x | {}^-2 \leq 3 + 4x \leq 7\}$
(g) $\{x | {}^-4 \leq 5 - 6x \leq 3\}$.

3. (a) We have 520 items we wish to sell to gross at least $1780. What selling prices are feasible ones for us to set?
   (b) We purchase 150 shares of a certain stock at $57 per share. The market price of the stock has fallen very low. We wish to sell all the shares when the price gets back up to where we shall lose no more than a total of $1500 on these shares. Which market prices will allow us to do this?
   (c) In a certain town of 7500 people, it is known that at least 1/3 are below voting age. How many people in the town could be of voting age?
   (d) Would you rather get as a prize 13/49 of $10,000 or 1/4 of $9,750?

4. Suppose a, b, and c are positive integers. Prove that $a/b \geq a/c$ if and only if $c \geq b$.

---

5. Prove Theorem 8.7.5.
6. Prove Theorem 8.7.6.
7. What is the isomorphism that establishes Theorem 8.7.1?

8.8 Least upper bounds

Let A be the set of rational numbers between 0 and 1. That is, let $A = \{x | x \in F \text{ and } 0 < x < 1\}$. 1/2, 1/3, 1/4, 1/5, and 1/6 are all in A. In fact if n is any positive integer than $1/n$ is in A. 2/3, 3/4, 4/5, 5/6, and 6/7 are all in A. In fact, $n/(n + 1)$ is in A whenever n is a positive integer.

The rational points on the number line which are in A all fall between the point 0 and the point 1. A is certainly a bounded set. And yet A is infinite.

We shall now put this notion of a set being bounded a bit more precisely and less geometrically than we have done before. We shall say that a rational number a is an **upper bound** for a nonempty set S of rationals if $x \leq a$ for all x in S. A set which has an upper bound is said to be **bounded above.** We shall say that a number b is a **lower bound** for S if $b \leq x$ for all x in S. A set which has a lower bound is said to be **bounded below.** If a set is bounded above and also bounded below then we say that it is simply **bounded.**

For the set $A = \{x | 0 < x < 1\}$, 1, 5, 7, 9, and 10^6 are all upper bounds. $^-1000$, $^-4$, $^-2$, and 0 are all lower bounds for A. A is a bounded set.

The set $B = \{x | x = 1/n \text{ where } n \in I^+\}$ is bounded above by 2 and below by $^-2$. Hence, B is a bounded set.

The set $C = \{x | x = n/(n + 1) \text{ where } n \in I^+\}$ has 10 as an upper bound and $^-50$ as a lower bound. Hence, C is bounded.

Notice that when we once find an upper bound for a set, then any larger number will also be an upper bound for the same set. Also if we find any one lower bound for a set, any smaller number will also be a lower bound.

What is the maximum of the above set $B = \{1, 1/2, 1/3, 1/4, \ldots\}$? Evidently, 1 is the largest element in B so that $\max(B) = 1$. What is the minimum of B? There is none! We see that $1/(n + 1) < 1/n$ for any positive integer n. And any element in B must be in the form $1/n$ for some positive integer n. So, given any element in B, there is another element in B which is smaller. Hence, B can have no minimum.

This situation is in marked contrast to that in N where every nonempty set of natural numbers has a minimum. And it is also quite different from what we know about I. Any nonempty bounded set of integers has both a maximum and a minimum. Yet B is a bounded set of rationals with no minimum.

The bounded set $C = \{1/2, 2/3, 3/4, 4/5, \ldots\}$, as defined above has a minimum of $1/2$ and yet has no maximum.

The bounded set $A = \{x \mid 0 < x < 1\}$ has neither a maximum nor a minimum. For let x be any element of A. Then $0 < x$ and there is a rational number y such that $0 < y < x$. Then $y \in A$. Since $y \in A$, x cannot be the minimum of A. Thus, A does not have a minimum. A similar argument may be used to show that A has no maximum.

Notice that the argument showing that A has no minimum establishes that there is no such thing as the smallest positive rational number. Given any positive rational number, we can find a smaller one.

In other words, it does not make sense to speak of the *next* rational point to the right of the origin on the number line. There is no such animal. Given *any* rational point to the right of the point 0, there is a rational point that falls between the given point and 0. So the given point cannot be the *next* point to the right of 0.

Putting this another way, we claim that there is no such thing as the rational point which is closest to the origin, and yet to the right of the origin. Given any rational point to the right of the origin, we can find another point to the right of 0 that is closer to 0.

Although the set $A = \{x \mid 0 < x < 1\}$ has no maximum, there is some special connection between the number 1 and this set. 1 is an upper bound for A. But, more than this, no number less than 1 can be an upper bound for A. That is, 1 is the **least upper bound** for A. Or in other words, $1 = \min(\{x \mid x \text{ is an upper bound for } A\})$. This notion is useful more generally.

If S is a set of rational numbers we say that a number a is the **least upper bound** for S if a is the minimum of the collection of all upper bounds for S.

We state more briefly that a is the least upper bound for S by writing "$a = \text{lub}(S)$."

For the set A above we have $\text{lub}(A) = 1$.

For the set B above we have $\text{lub}(B) = 1$. Notice that when a set S has a maximum element, then $\max(S) = \text{lub}(S)$.

For the set C above we have $\text{lub}(C) = 1$.

How do we tell if a number is a lub† for a set? First of all, a set must be bounded above if it is to have a lub. For if a set is not bounded above it has no upper bounds at all, let alone a **least** upper bound.

Suppose a *is* an upper bound for a set S. How can we show that it is the **least** upper bound? We can prove that it is by establishing any one of the following:

(i) whenever b is an upper bound for S we have $a \leq b$
(ii) if $b < a$, then b is not an upper bound for S
(iii) if $b < a$, then there is some element x is S such that $b < x \leq a$
(iv) given any positive rational number d, no matter how small, there is a member x of S such that $a - d < x \leq a$.

For example, we may use the form (iv) to show that $\text{lub}(C) = 1$ for the set C above. But, before we do, we must mention a basic property of the rationals called the **Archimedean property:**

If x and y are positive rationals, then there is a positive integer n such that $nx > y$.

In other words, the positive integral multiples of a fixed positive rational get arbitrarily large. This ties in with the geometric idea that if we have a segment with nonzero rational length, we can build up a segment that is as long as we please by laying enough copies of our given segment end to end.

Accepting this fact, we shall show that the lub of $C = \{1/2, 2/3, 3/4, 4/5, \ldots\}$ is 1. We first see that when n is any positive integer we have $n < n + 1$ so that $n/(n + 1) < 1$. Hence 1 is an upper bound for C. Let d be any positive rational. By the **Archimedean property** of F, there is a positive integer n such that $nd > 1$. Then $d > 1/n$. Hence $(^-1)d < (^-1)(1/n)$ and $1 - d < 1 - (1/n) = n/n - 1/n = (n - 1)/n < 1$. If $n = 1$ then $d > 1$ and $1 - d < 0$ so that every member of C is between $1 - d$ and 1. If $n \neq 1$, then $(n - 1)/n$ is a member of S which falls between $1 - d$ and 1. We conclude by (iv) that $\text{lub}(C) = 1$.

If we look at things from the left instead of from the right on the number line, we are led to make the following definitions for a set S of rational numbers. A number b is said to be the **greatest lower bound** (glb) for S if b is the maximum of the collection of all lower bounds for S.

For the sets A, B, and C above we have $\text{glb}(A) = 0$, $\text{glb}(B) = 0$, and glb $\text{glb}(C) = 1/2$. Note that when S has a minimum we have $\min(S) = \text{glb}(S)$.

Suppose b is a lower bound for S. Then $b = \text{glb}(S)$ if any one of the following holds:

† We shall henceforth use "lub" as an abbreviation for "least upper bound."

(i) if c is a lower bound for S, then $c \leq b$
(ii) if $c > b$, then c is not a lower bound for S
(iii) if $c > b$, then there is a member x of S such that $b \leq x < c$
(iv) given any positive rational d, no matter how small, there is a member x of S such that $b \leq x < b + d$.

It is quite easy to use the Archimedean property and condition (iv) to prove that lub$(B) = 0$ for the set B above.

A very deep question can now be asked about F. Is it the case that every nonempty set that has an upper bound in F also has a least upper bound in F? It turns out the answer is negative. Of this, more later.

Definition 8.8.1. A number a is said to be an **upper bound** for a nonempty set S of rationals if $x \leq a$ for all x in S. b is said to be a **lower bound** for S if $b \leq x$ for all x in S.

Definition 8.8.2. If a nonempty set S has an upper bound it is said to be **bounded above.** If S has a lower bound it is said to be **bounded below.** If S is bounded above and bounded below, it is said to be **bounded.**

Definition 8.8.3. lub(S) is the minimum of the collection of all upper bounds for S. glb(S) is the maximum of the set of all lower bounds for S.

Theorem 8.8.1. (a) If a set S has a maximum, then lub(S) = max(S).
(b) If a set S has a minimum, then glb(S) = min(S).

Theorem 8.8.2. (The Archimedean property) If x and y are positive rationals then there is an integer n such that $nx > y$.

PROOF: Let $x = a/b$ and $y = c/d$ where a, b, c, and d are positive integers. If n is a positive integer we have $nx = n(a/b) = (na)/b$. Now $(na)/b > c/d$ if and only if $nad > bc$. But by Theorem 6.16.2 we know that whatever the values of a, b, c, and d, there does exist an n which makes this last equality hold.

Theorem 8.8.3. Suppose that a is an upper bound for a nonempty set S. Then a = lub(S) if and only if one of the following holds:

(i) whenever b is an upper bound for S, we have $a \leq b$
(ii) if $b < a$, then b is not an upper bound for S
(iii) if $b < a$, then there is an x in S such that $b < x \leq a$
(iv) if d is any positive number, then there is an x in S such that $a - d < x \leq a$.

PROOF: Part (i) is simply a restatement of Definition 8.8.3. (ii) is simply a restatement of the meaning of (i). (iii) is a restatement of (ii). (iv) is a restatement of (iii) since $b < a$ if and only if there is a positive d such that $b = a - d$.

Theorem 8.8.4. Suppose b is a lower bound for a nonempty set S. Then $b = \text{glb}(S)$ if and only if one of the following holds:

(i) whenever a is a lower bound for S we have $b \geq a$
(ii) if $a > b$, then a is not a lower bound for S
(iii) if $a > b$, then there is an x in S such that $b \leq x < a$
(iv) if d is positive, then there is an x in S such that $b \leq x < b + d$.

Problems

1. For each of the following sets find the maximum, the minimum, the lub, and the glb if they exist.

(a) F
(b) ϕ
(c) $\{1\}$
(d) $\{^-77/4\}$
(e) $\{^-1, ^+1\}$
(f) $\{x \mid {}^-1 \leq x \leq 2\}$
(g) $\{x \mid {}^-1 < x \leq 2\}$
(h) $\{x \mid {}^-1 \leq x < 2\}$
(i) $\{x \mid {}^-1 < x < 2\}$
(j) $\{x \mid x \leq {}^-5\}$
(k) $\{x \mid x < {}^-5\}$
(l) $\{x \mid x > {}^-17\}$
(m) $\{x \mid x \geq {}^-17\}$
(n) $\{x \mid x < {}^-2 \text{ or } x > 5\}$
(o) $\{x \mid 5x - 7 = 6\}$
(p) $\{x \mid x = k^2 \text{ for some } k \text{ in } I\}$
(q) $\{x \mid |x - 3| \leq 2\}$
(r) $\{x \mid |x - 3| < 2\}$
(s) $\{x \mid x = 1/10^n \text{ for some } n \text{ in } I^+\}$
(t) $\{x \mid x = 10^n \text{ for some } n \text{ in } I^+\}$
(u) $\{x \mid x^2 \neq 0\}$
(v) $\{x \mid x = (2n)/(n + 1) \text{ for some } n \text{ in } I^+\}$.

2. Find an integer n such that $nx > y$ when

(a) $x = 1/1000$, $y = 1/10$
(b) $x = 1/97{,}453$, $y = 679$
(c) $x = 1/5{,}000{,}000$, $y = 6{,}000{,}000$
(d) $x = 1/5000$, $y = {}^-1$.

3. Prove Theorem 8.8.1.

4. Prove that if glb(S) and lub(S) exist, then glb(S) $\leq$ lub(S). Under what conditions does the equality hold?

5. Carefully justify your conclusions in Problem 1 for parts (i), (l), (s), and (v).

6. If a set S is unbounded, does it follow that neither glb(S) nor lub(S) exists? Explain.

7. Which of the sets in Problem 1 are

(i) bounded (ii) finite?

8. Suppose k is a fixed positive integer. Let $S = \{x | x = k/10^n$ for some n in $I^+\} = \{k/10, k/10^2, k/10^3, \ldots\}$. What is glb($S$)? Explain how you know.

9. Consider the set $D = \{x | x = k/10^n$ for some k in I and some n in $I^+\}$. What can be said about glb(D) and lub(D)?

8.9 Proper fractions

Whenever the denominator of a fraction has an absolute value which is larger than the absolute value of the numerator, the fraction names a rational number whose absolute value is less than 1. Such a fraction is called a **proper fraction.** For example, "5/7," "$^-$9/11," "8/$^-$14," and "$^-$13/$^-$27" are all proper fractions. "7549/86" is not a proper fraction, and, hence, is called an **improper fraction.**

On the number line, the rational point corresponding to any proper fraction must fall between the point $^-1$ and the point 1. Where does the point corresponding to the fraction "7594/86" fall on the line? Since $88 \cdot 86 = 7568$ and $7594 - 7568 = 26$, we see that $7594/86 = (88 \cdot 86 + 26)/86 = (88 \cdot 86)/86 + 26/86 = 88 + 26/86 = 88 + 13/43$. This tells us that the point 7594/86 falls between the point 88 and the point 89.

It is common practice to write "$88\frac{13}{43}$" to name the number $88 + 13/43$. Such a convention amounts to suppressing the "+" sign and closing up the resulting expression. We read "$88\frac{13}{43}$" as "eighty-eight and thirteen forty-thirds." Note that the "+" sign comes back in as an "and" when we verbalize the expression. It is common practice to refer to $88\frac{13}{43}$ as a **mixed number.** However, this terminology seems a bit unfortunate. What is actually involved is a sort of **mixed numeration.** In the expression "$88\frac{13}{43}$" we have used two different notations—the usual notation for an integer and also that for a (proper) fraction where the numerator and denominator are written in decimal notation. In our work here, we shall say that we have expressed a rational number as a **mixed numeral** if we have written a name for the number in the form

$$\text{“}n\,\frac{a}{b}\text{”} \qquad \text{or} \qquad \text{“}^-n\,\frac{a}{b}\text{”}$$

where "n" names a nonzero natural number and "a/b" names a proper fraction.

We take it that

$$\text{“}^-n\,\frac{a}{b}\text{”}$$

is a name for the opposite of

$$n\,\frac{a}{b}\,.$$

And

$$\text{“}n\,\frac{a}{b}\text{”}$$

means $n + a/b$. We have then that $^-7\frac{1}{9} = {}^-(7 + 1/9) = {}^-7 - 1/9$. We read "$-7\frac{1}{9}$" as "negative seven and one-ninth." Notice that this actually means the same as "negative(seven and one-ninth)" and has a different meaning from "(negative seven) and one-ninth."

Every rational number can be expressed as either a proper fraction or a mixed numeral.

In fact, suppose that a and b are positive integers. Then by the algorithm of Section 6.20 for deciding whether or not b is a divisor of a, we can find non-negative integers (natural numbers) q and r such that $a = qb + r$ and $r < b$. Hence,

$$a/b = (qb + r)/b = (qb)/b + r/b = q + r/b = q\frac{r}{b}\cdot$$

And "r/b" is a proper fraction since $r < b$. If $q = 0$ then "a/b" is a proper fraction. If $q \neq 0$, then

$$\text{“}q\frac{r}{b}\text{”}$$

is a mixed numeral for a/b.

If a/b is negative and not a proper fraction, then we may use the above procedure to express $|a/b|$ as a mixed numeral and then add a "$-$" sign to get the mixed numeral for a/b.

Notice that if we follow the above procedure very precisely we would express 48/12 in mixed numerals as "$4\frac{0}{12}$." But it is customary to drop the "0/12" and simply write "4" when we are asked to put 48/12 in mixed numerals. In other words, we drop the proper fraction when it names 0.

The process of converting an improper fraction "a/b" into a mixed numeral is essentially the process of division in Section 6.20. When we are given positive integers a and b and use the division process to find a quotient q and a remainder r, we then have that $a/b = q + r/b$.

$$\text{“}q\frac{r}{b}\text{”}$$

is a mixed numeral for a/b.

Having rational numbers written as mixed numerals often makes it easy to compare their sizes. For instance, it is very easy to see that $57\frac{13}{19} > 56\frac{17}{39}$, and not so easy to see that $1083/19 > 2184/39$. And yet, these two inequalities actually say precisely the same thing.

It is an easy matter to convert a mixed numeral name to an improper fraction. We simply express the integral part as a fraction, reinsert the "+" sign, and carry out the indicated addition. For example, $8\frac{13}{17} = 8/1 + 13/17 = (8\cdot 17 + 13\cdot 1)/1\cdot 17 = 149/17$. Also $^-14\frac{3}{10} = -(14 + 3/10) = -(140 + 3)/10 = {}^-143/10$.

There is a possibility for confusion with the use of **mixed numerals** that should be mentioned. In all of the systems N, I, and F we have used simple juxtaposition to indicate multiplication. We have taken "xy" to mean $x\cdot y$.

Hence, we must be careful to avoid interpreting "$4\frac{98}{100}$" as a name for $4 \cdot (98/100)$. When we use mixed numerals, it is almost as if we had switched over to using simple juxtaposition to indicate addition rather than multiplication. However, mixed numerals always have the special form of a name for a natural number and then a proper fraction—with perhaps a "$-$" sign as a prefix. So when we see this form we should expect that a mixed numeral is intended and not the name of a product of two rationals.

Partly because of this chance for confusion and for other reasons, most people who work a great deal with numbers tend to avoid mixed numerals except in the special form which we shall discuss in the next section.

Mixed numerals are usually very cumbersome when multiplications and divisions are involved. When we wish to multiply or divide two rational numbers written as mixed numerals it is usually best to first convert these names to improper fractions. An exception to this rule is the case where the mixed numerals are in the special form discussed in the next section.

PROBLEMS

1. Write the following numbers as mixed numerals:

(a) $47/10$ (b) $^-74/3$
(c) $4753/71$ (d) $3005/100$
(e) $^-471/101$ (f) $(^-471)/(^-32)$
(g) $943/(^-47)$.

2. Write the following as improper fractions:

(a) $6\frac{4}{7}$ (b) $-7\frac{1}{9}$
(c) $43\frac{57}{100}$ (d) $30\frac{5}{1000}$
(e) $374\frac{940}{1000}$ (f) $-43\frac{1}{43}$
(g) $17\frac{3}{16}$.

3. Simplify the following, putting your answer as a fraction or a mixed numeral—whichever seems most convenient.

(a) $4\frac{2}{7} + 9\frac{3}{7}$ (b) $5\frac{2}{9} - 4\frac{1}{9}$
(c) $-3\frac{1}{17} + 2\frac{1}{6}$ (d) $43\frac{2}{9} - 64\frac{5}{9}$
(e) $37\frac{2}{3} - 5\frac{3}{17}$ (f) $(5\frac{2}{3}) \cdot (\frac{5}{7})$
(g) $(3\frac{1}{3})(-7\frac{3}{5})$ (h) $5\frac{3}{4} \div 6\frac{1}{3}$
(i) $4\frac{2}{17} \div 8\frac{1}{2}$ (j) $(4\frac{2}{5} - 7\frac{3}{5}) \div (10\frac{1}{3} + \frac{7}{9})$
(k) $\dfrac{\frac{3}{17} + \frac{4}{3}}{\frac{1}{9} - \frac{2}{7}}$ (l) $\dfrac{\frac{3}{10}(\frac{9}{100} + \frac{4}{10})}{\frac{7}{100}(\frac{2}{5} + \frac{7}{100})}$.

4. We have said that when a rectangle has one side b units long and the other side h units long, then it has an area of bh square units. We carry over this

definition to the case where b and h are any positive rational numbers (and may fail to be integers). Find the area of each figure in Figure 8.9 where the unit distance is one inch.

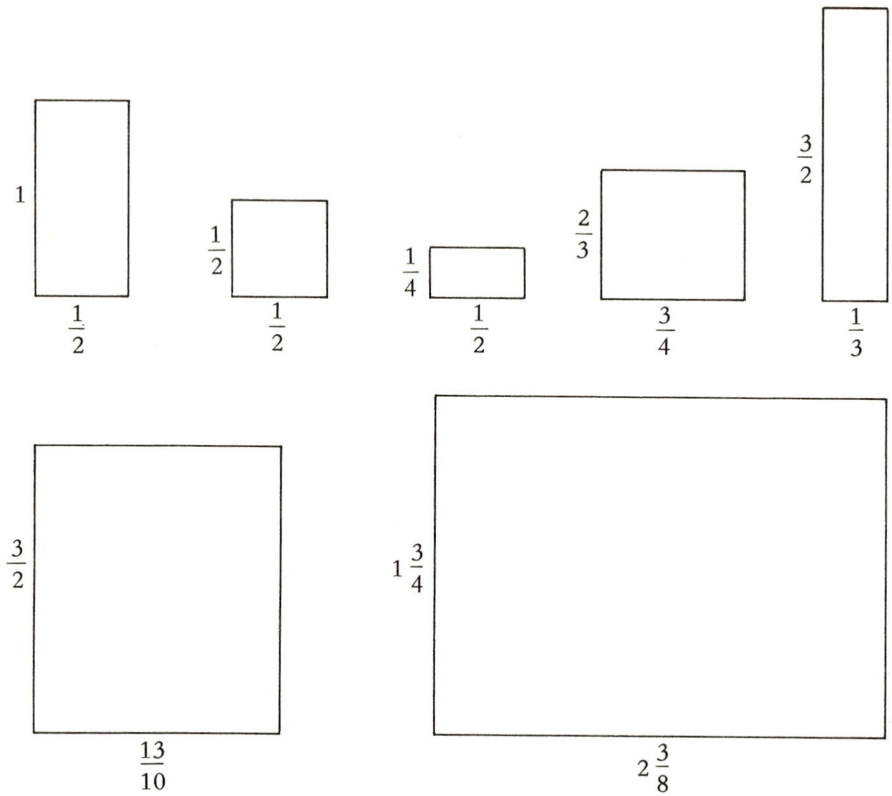

FIGURE 8.9

5. Consider again the first three rectangles in Figure 8.9. Use copies of the given rectangle in each case to construct a rectangle with sides of integral length. And thus argue that the area found for each in Problem 4 is reasonable.

6. A right triangle is a triangle which has one right angle. Such a triangle may be thought of as one-half of a rectangle as in Figure 8.10. If b and h are the lengths of the legs (the two shorter sides), then we see that the area of the triangle should be $(1/2)bh$ or $bh/2$. Find the area of each figure in Figure 8.11 where the unit distance is one inch.

7. A parallelogram is a four-sided figure in which the opposite sides are parallel. We may cut up a parallelogram and put it together again to form a rectangle. This is illustrated in Figure 8.12 where in each case the triangle cut off on the right end is moved to the left and added on to form a rectangle. It

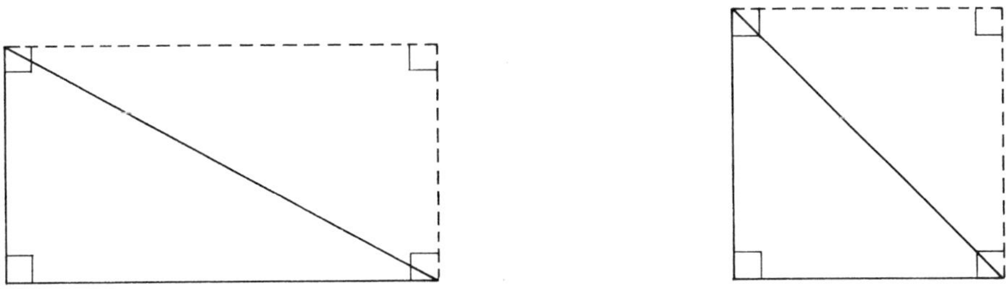

FIGURE 8.10

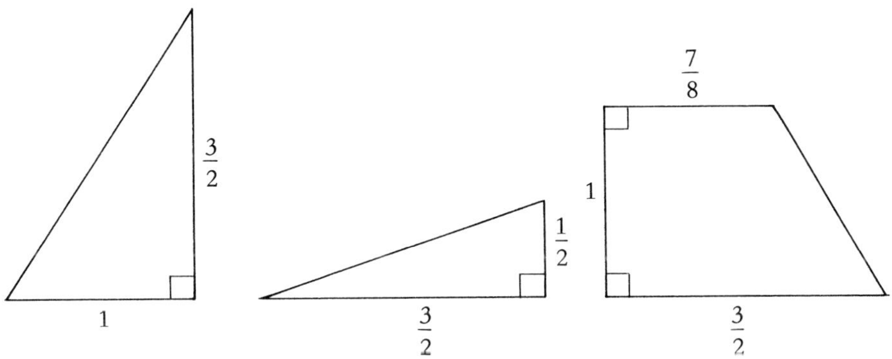

FIGURE 8.11

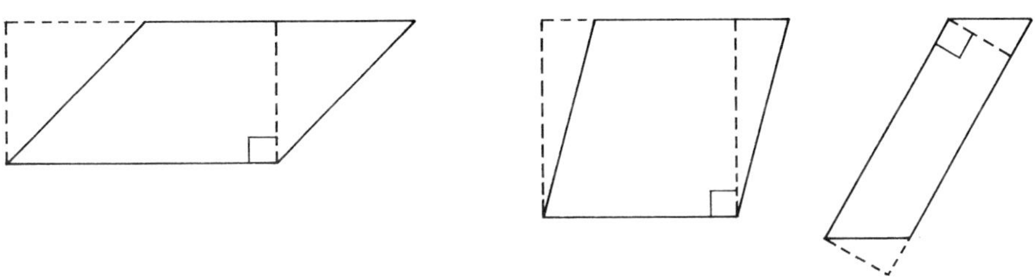

FIGURE 8.12

follows that if b is the length of one side of a parallelogram and h is the distance along the perpendicular to the other side, then the area of the parallelogram is bh. We would usually refer to h as an altitude of the parallelogram. Find the area of each figure in Figure 8.13 where the unit distance is one inch.

8. Given any triangle we may think of it as one-half of a parallelogram. This fact is illustrated in Figure 8.14 where in each case the dotted line is parallel to the opposite side. Thus we may get the area of a triangle by finding $(1/2)bh$ where b is the length of one side and h (the altitude) is the length of the perpen-

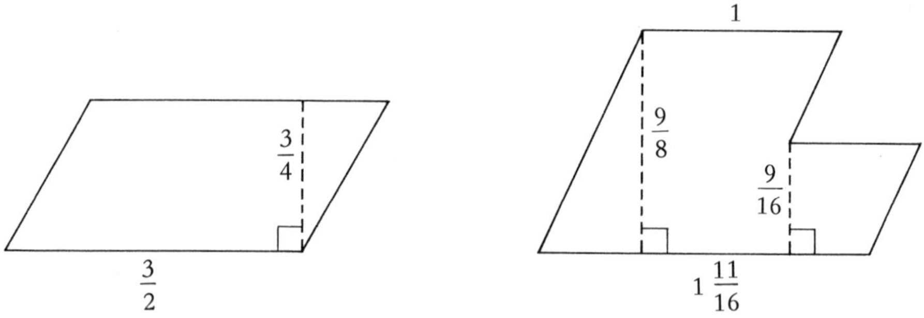

FIGURE 8.13

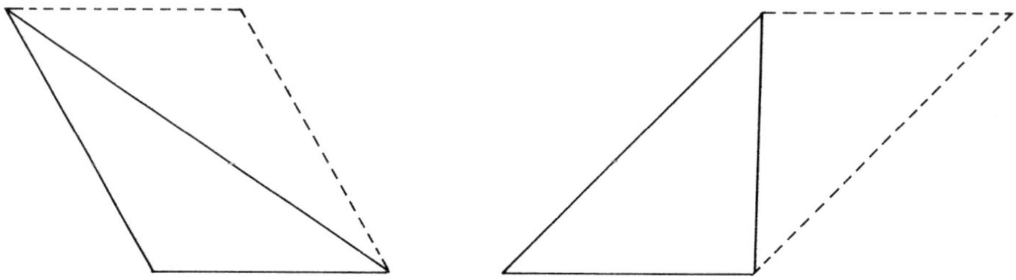

FIGURE 8.14

dicular from the opposite vertex to this side. Find the area of each figure in Figure 8.15 where the unit of distance is one inch.

8.10 Decimal fractions

The decimal name for the natural number 10^n is very simple—a "1" followed by n "0" 's. There are some simple standard names for those rational numbers that may be written as fractions whose denominators are powers of ten. Some of these are displayed in the following examples.

$$.2 = 2/10$$

$$.25 = 25/100$$

$$.637 = 637/1000$$

$$.0042 = \frac{0042}{10{,}000} = 42/10{,}000.$$

In general, "$.a_1a_2 \cdots a_n$" where "a_1," "a_2," . . . and "a_n" are decimal digits, is taken as a name for the rational number $a_1a_2 \cdots a_n/10^n$. Here, as shown by the fourth example above, we violate our usual convention and allow initial zeroes in the decimal name for a rational number.

We take "$-.a_1a_2 \cdots a_n$" to mean $-a_1a_2 \cdots a_n/10^n$.

"$.a_1a_2 \cdots a_n$" or "$-.a_1a_2 \cdots a_n$" is said to be a **proper decimal fraction.** The "." in these expressions is called the **decimal point.**

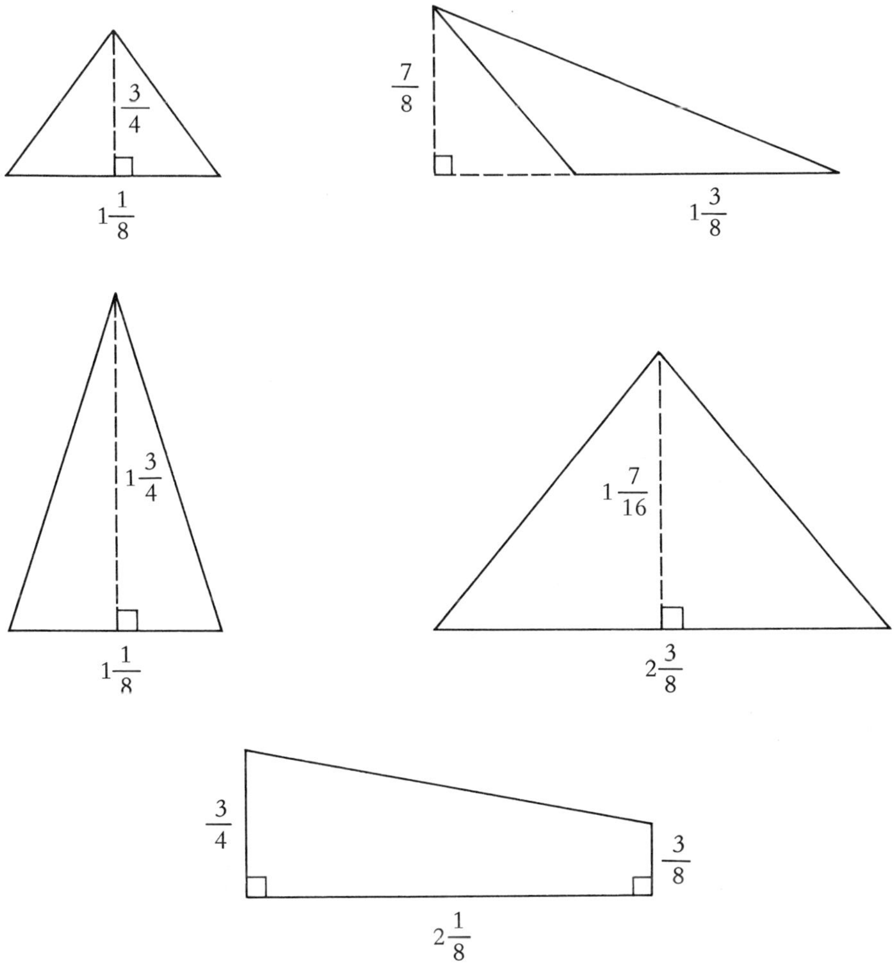

FIGURE 8.15

If "b" is the decimal name for a positive integer we write "$b.a_1a_2 \cdots a_n$" as a name for $b + (a_1a_2 \cdots a_n/10^n)$. We call "$b.a_1a_2 \cdots a_n$" a **decimal fraction.** We see then that a decimal fraction is a special form of a mixed numeral. We write "$-b.a_1a_2 \cdots a_n$" for the opposite of $b.a_1a_2 \cdots a_n$. As some examples, we have:

$$4.98 = 4 + 98/100 = 498/100$$

$$3.1416 = 3 + 1416/10{,}000 = 31{,}416/10{,}000$$

$$-5.74 = -(5 + 74/100) = -\tfrac{574}{100}$$

$$-3794.2 = -(3794 + 2/10) = -\tfrac{37942}{10}.$$

We read "75.431" as "seventy-five and four hundred and thirty-one thousandths." "-4.02" is read as "negative four and two hundredths." Note that this last does not mean the same as "(negative four) and two hundredths." The latter is $^-4 + .02 = -3.98$.

We have that $576.64 = 57{,}664/100$, $3.5126 = 35{,}126/10{,}000$, and $532.006 = 532{,}006/1000$. More generally, if the "c"'s and "a"'s are decimal digits,

$$c_1c_2 \cdots c_k \odot a_1a_2 \cdots a_n = c_1c_2 \cdots c_ka_1a_2 \cdots a_n/10^n.$$

In a decimal fraction the decimal point is printed on a lower level than would be a point indicating the operation of multiplication. "5.4" means five and four tenths, whereas "$5 \cdot 4$" means 20.

Using decimal fractions, it is very easy to compare the sizes of two rational numbers. We may carry over, almost without change, our method for deciding the relative sizes of natural numbers expressed as decimals. For instance, the following are seen, almost immediately, to be true:

$$.099 < .100$$

$$.099 < .1$$

$$.0043 > .0005$$

$$7.5432901 > 7.5432897$$

$$-5.47 < -5.40$$

$$-.00021 < -.00007$$

$$-43.00496 > -43.00502.$$

Notice that when the numbers involved are both negative we may compare their absolute values and then reverse the conclusion obtained. We should remember that $x < y$ if and only if $-x > -y$.

Another advantage of the decimal fraction notation is that the algorithms which we developed for addition, subtraction, multiplication, and division of natural numbers written in decimal notation can be modified quite easily so that they may be used equally well with decimal fractions. We shall discuss this matter in the next section.

But before digging into algorithms which work with decimal fractions we should consider the basic question of whether or not each rational number can be expressed as a decimal fraction. The answer is that some rational numbers cannot be so expressed.

Many rational numbers *do* have simple decimal fraction forms. For instance, $1/2 = .5$, $1/4 = .25$, $1/8 = .125$, $1/10 = .1$, $1/20 = .05$, and $1/40 = .025$. But even the simple number $1/3$ cannot be expressed as a decimal fraction. To prove this, first note that when we multiply 1, 2, 3, 4, 5, 6, 7, 8, or 9 by 3 we get a product whose decimal name does not have a "0" in the units place. Now suppose that $1/3$ *can* be written as a decimal fraction. Then there are positive integers k and n such that $1/3 = k/10^n$. In fact, we may choose the k and n here so that n is as small as possible. Then the decimal name for k cannot have a "0" in the units place of its decimal name. For if there is such a "0" we could cancel 10 from the numerator and denominator of the fraction "$k/10^n$" and get a smaller value of n. But if the decimal name for k does not have a "0" in its units place, then neither does that for $3k$. This contradicts the equality $10^n = 3k$ which follows from $1/3 = k/10^n$. We conclude that $1/3$ cannot be expressed as a decimal fraction.

But we should not give up on decimal fractions too soon. We can find decimal fractions which are very close to $1/3$. For example, $1/3 - .333333 = 1/3 - 333{,}333/1{,}000{,}000 = 1{,}000{,}000/3{,}000{,}000 - 999{,}999/3{,}000{,}000 = 1/3{,}000{,}000$ so that .333333 differs from $1/3$ by only one three-millionth.

Since .333333 is so close to $1/3$ we would expect that we may carry out certain calculations with .333333 in place of $1/3$ and make only very little difference. In fact, in an applied problem where some of the numbers involved are measurements obtained as scale readings, the numbers involved are already, in some sense, only approximations to true values and it would seem that, in many such cases, we would be able to treat $1/3$ as if it were the same as .333333 and make no essential difference. Such considerations open a wide range of questions about approximations and the analysis of errors. We shall dig a little more into these matters later in Section 10.9.

Was it simply a lucky choice when we considered $1/3$ above? Or can any rational number be approximated closely by a decimal fraction? The affirmative answer that we find to this question is closely tied in with the fact that the rational numbers expressible as decimal fractions are **dense** in the rational numbers. That is, given any two distinct rational numbers there is a rational expressible as a decimal fraction which falls between them. We may establish this as follows. First suppose that a/b and c/d are distinct rational numbers. There is no loss in generality if we assume that $a/b < c/d$. Then $c/d - a/b$ is positive. Let $c/d - a/b = p/q$ where p and q are positive integers. Then $1/q \leq p/q$. Let us find an n such that $10^n > q$. That such an n exists is clear since we may, for example, take n to be any natural number greater than the number of digits in the decimal name for q. We have $0 < 1/10^n < 1/q \leq p/q$. Consider the numbers $0, 1/10^n, 2/10^n, 3/10^n, \ldots$. By the Archemedean property of the rationals, some of these must eventually be at least as large as c/d. Let $(k + 1)/10^n$ be the first such. Then $k/10^n < c/d$. We claim that we must also have $a/b < k/10^n$. For if $a/b \geq k/10^n$ then we have $^{-}a/b \leq {}^{-}k/10^n$ and this combined with $c/d \leq (k + 1)/10^n$ yields $c/d - a/b \leq (k + 1)/10^n - k/10^n - 1/10^n$ and this is impossible because $1/10^n < c/d - a/b$. Since $k/10^n$ can be expressed as a decimal fraction and $a/b < k/10^n < c/d$, we have shown that the numbers expressible as decimal fractions are dense in the positive rationals. The proof for the case where a/b and c/d differ in sign is trivial, and that where a/b and c/d are negative is taken care of by considering their opposites.

The above proof may be thought of geometrically as follows. Suppose a/b and c/d are as given in Figure 8.16. We take the segment from 101 to 102 and divide it into 10 equal parts, then into 100 equal parts, then into 1000 equal parts, etc. Eventually, one of our division points must fall between the point

a/b and the point c/d. Such a division point corresponds to a number expressible as a decimal fraction.

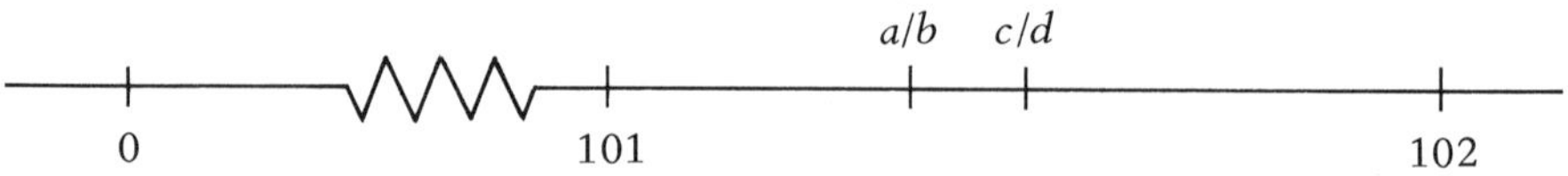

FIGURE 8.16

PROBLEMS

1. Express the following numbers as decimal fractions:

(a) $4/10$ (b) $426/1000$ (c) $527/100$
(d) $375/250$ [HINT: First multiply the numerator and denominator by 4.]
(e) $35/5$ (f) $^-75/10^6$ (g) $(^-364)/(^-10^7)$
(h) $5{,}376{,}700/10^5$ (i) $436/10^7$ (j) $31{,}000{,}001/10^7$.

2. Express the following numbers as fractions in lowest terms:

(a) .4 (b) .75 (c) .33 (d) 1.44
(e) -8.1 (f) .00111 (g) .00555
(h) 3.95 [HINT: Note that a natural number k and 10^n have a common factor only if 2 or 5 is a common factor. This follows from the facts that $10^n = (2\cdot 5)^n = 2^n\cdot 5^n$ and 2 and 5 are divisible only by themselves and 1 in N.]

3. Find the decimal names for the following results by converting to ordinary fractions, carrying out the operation, and then converting back to decimal fractions:

(a) $.3 + .5$ (b) $.6 + .7$
(c) $3.6 + .004$ (d) $(3)\cdot(3.3)$
(e) $(7.5)\cdot(.021)$ (f) $(5300.1)\cdot(1.011)$
(g) $(4.42) \div (221)$ (h) $(4.42) \div (2.21)$
(i) $(4.42) \div (.00221)$ (j) $(325) \div (.001)$
(k) $(.035) \div (7)$ (l) $(.035) \div (700)$
(m) $(.035) \div (.0007)$ (n) $1000 \div (.000001)$
(o) $53{,}476\cdot(.00001)$.

4. Find a decimal fraction that names a number that is between the numbers

(a) $2/3$ and $3/4$ (b) .5 and .6
(c) $71/51$ and $71/50$ (d) $^-1/573$ and $2/7596$
(e) $1/10^8$ and $1/10^9$.

5. Prove that $1/7$ cannot be written as a decimal fraction.

6. By how much does $1/7$ differ from (a) .14287, (b) .1428714287?

7. Prove that when n is any natural number the following can be written as decimal fractions:

(a) $1/2^n$ (b) $1/5^n$.

8.11 Operations with decimal fractions

If we wish a simple name for the sum of .67025 and .43 we may write

$$\begin{aligned} .67025 + .43 &= 67{,}025/100{,}000 + 43/100 \\ &= 67{,}025/100{,}000 + 43{,}000/100{,}000 \\ &= (67{,}025 + 43{,}000)/100{,}000 = 110{,}025/100{,}000 \\ &= 100{,}000/100{,}000 + 10{,}025/100{,}000 = 1.10025. \end{aligned}$$

More generally, suppose that "$.a_1a_2 \cdots a_n$" and "$.b_1b_2 \cdots b_k$" are decimal fractions and $n > k$. Then

$$\begin{aligned} .a_1a_2 \cdots a_n + .b_1b_2 \cdots b_k &= a_1a_2 \cdots a_n/10^n + b_1b_2 \cdots b_k/10^k \\ &= a_1a_2 \cdots a_n/10^n + b_1b_2 \cdots b_k00 \cdots 0/10^n \end{aligned}$$

where there are $n - k$ "0" 's after "b_k" in the last numerator. We now have the two fractions with a common denominator and may simply add the two natural numbers in the numerator and use the sum as the numerator of our result. The denominator of the sum will be 10^n.

This entire procedure can be thought of as simply a modification of our previous method for adding natural numbers in decimal form. All that we need to do is keep the decimal points straight and treat missing digits as if they were "0" 's. The conventional, compact way to display the above calculation is

$$\begin{array}{r} .67025 \\ +\ .43 \\ \hline 1.10025. \end{array}$$

This can be considered to be an abbreviation of

$$\begin{array}{r} 67{,}025/100{,}000 \\ +\ 43{,}000/100{,}000 \\ \hline 110{,}025/100{,}000. \end{array}$$

Since we already have an algorithm for addition of integers the above technique yields an algorithm for adding any two decimal fractions. Since a subtraction may be thought of as an addition of an opposite, we also have an efficient mechanical process for subtracting when the numbers are expressed as decimal fractions.

When $x = c_1c_2 \cdots c_p.a_1a_2 \cdots a_n$ and $y = d_1d_2 \cdots d_q.b_1b_2 \cdots b_k$ where the a's, b's, c's, and d's are decimal digits we have

$$xy = \frac{c_1c_2 \cdots c_pa_1a_2 \cdots a_n}{10^n} \cdot \frac{d_1d_2 \cdots d_qb_1b_2 \cdots b_k}{10^k}.$$

Hence, to get xy we need only multiply the natural numbers named in the numerators and then divide by 10^{n+k}. That is, to multiply two decimal fractions, act as if the decimal point were not there and multiply, then put in a decimal point in the result $n + k$ places from the right. Since we have an algorithm for multiplying natural numbers and know how to take care of the signs when we multiply integers, the above technique gives us an efficient mechanical process for multiplying any two numbers which can be expressed as decimal fractions.

With x and y as given above we make use of the fact that we may divide by multiplying by the reciprocal and have

$$\begin{aligned} x \div y &= \frac{c_1c_2 \cdots c_pa_1a_2 \cdots a_n}{10^n} \cdot \frac{10^k}{d_1d_2 \cdots d_qb_1b_2 \cdots b_k} \\ &= \frac{10^k}{10^n} \cdot \frac{c_1c_2 \cdots c_pa_1a_2 \cdots a_n}{d_1d_2 \cdots d_qb_1b_2 \cdots b_k}. \end{aligned}$$

The last fraction in this product may not be expressible as a decimal fraction. But if it is and we can find such a decimal fraction then the decimal fraction for $x \div y$ is easily obtained since multiplying by $10^k/10^n$ simply amounts to moving the decimal point around.

For example, $10.83 \div .057 = (1083/100) \div (57/1000) = 10 \cdot (1083/57)$. And using the algorithm of Section 6.20 we have

```
     19
57)1083
    57
   ---
    513
    513 .
    ---
```

Hence, $10.83 \div .057 = 10 \cdot 19 = 190$. But note that since dividing by .057 is equivalent to multiplying by 1000/57, and since we may multiply by 1000 by moving the decimal point appropriately we may combine all the above into a single array as:

```
          190.
.051)10.830
         5 7
        ----
         5 13
         5 13 .
        ----
```

If e and f are positive integers, how can we tell when e/f can be expressed as a decimal fraction? And if it can be so expressed, how can we find the appropriate decimal? The answers to both these questions are given by a slight modification of the standard division algorithm for natural numbers.

Suppose e/f can be expressed as a decimal fraction. Then there are positive integers k and n such that $e/f = k/10^n$. Then $(e \cdot 10^n)/f = k$. Also if this last equality holds for some positive integers n and k then e/f is expressible as a decimal fraction. In other words e/f can be written as a decimal fraction if and only if f is a divisor of some one of the numbers e, $10e$, $100e$, $1000e$, ...

Suppose we wish to express 117/225 as a decimal fraction. Clearly 225 is not a divisor of 117. Is 225 a divisor of 1170? We find out by using the division algorithm for natural numbers.

$$\begin{array}{r} 5 \\ 225\overline{)1170} \\ 1125 \\ \hline 45. \end{array}$$

Since we do not get a remainder of 0, 225 is **not** a divisor of 1170. Is 225 a divisor of 11700? We have

$$\begin{array}{r} 52 \\ 225\overline{)11700} \\ 1125 \\ \hline 450 \\ 450 \\ \hline 0. \end{array}$$

Hence, $11700 \div 225 = 52$. And this yields $(117 \cdot 100) \div 225 = 52$ and, hence, $117/225 = 52/100 = .52$. This may also be written as $((117 \cdot 100)/225)/100 = .52$. That is, to get our decimal fraction we first move the decimal point in the numerator two places to the right (by multiplying by 100) to get a multiple of 225. But then we move the decimal point to the left two places by dividing by 100. The net result, so to speak, is to leave the decimal point where it was in the first place, supplying "0"'s as needed and carrying through the division algorithm as if there were no decimal points present. The conventional way to write up the above problem is:

$$\begin{array}{r} .52 \\ 225\overline{)117.00} \\ 112\ 5 \\ \hline 4\ 50 \\ 4\ 50\ . \\ \hline \end{array}$$

When e/f can be expressed as a decimal fraction, we may find it by carrying out the division algorithm and handling the decimal point as indicated above. That is, we tack on "0"'s to the numeral for e and continue the division algorithm until we get a remainder of 0. We then divide the quotient obtained by 10^n where n is the number of "0"'s we tacked on. But this amounts to keeping the digits in the proper columns and lining up the decimal points in the numeral for e and in the quotient.

If, on the other hand, e/f cannot be written as a decimal fraction, then f divides none of the numbers e, $10e$, $100e$, $1000e$, ... This means that the algorithm described above never terminates. We never get a remainder of 0 at any stage.

In actual practice, can we tell when this last is the case? Suppose we have carried through the division algorithm repeatedly and discovered that f is not a divisor of, say, e, $10e$, $100e$, . . . , $10^{20}e$. Can we stop? Or must we keep on going, hoping that we will soon come across a larger number in the form $10^n e$ which is a multiple of f? The answer, of course, depends on the relative sizes of e and f, but there is a way to tell if we have gone far enough. Let us first consider some numerical examples.

Suppose we wish to express 8/11 as a decimal fraction. We have

$$
\begin{array}{r}
.7272\cdots \\
11\overline{)8.0000} \\
\underline{7\,7} \\
30 \\
\underline{22} \\
80 \\
\underline{77} \\
30 \\
\underline{22} \\
8 \\
\cdot \\
\cdot \\
\cdot\,\cdot
\end{array}
$$

Here it is clear after a few steps that the pattern "72" will repeat indefinitely and, hence, 11 divides no natural number in the form $8 \cdot 10^n$. That is, 8/11 *cannot* be expressed as a decimal fraction.

Consider the rational number 5/6. We have:

$$
\begin{array}{r}
.8333\cdots \\
6\overline{)5.00000} \\
\underline{4\,8} \\
20 \\
\underline{18} \\
20 \\
\underline{18} \\
20 \\
\underline{18} \\
2 \\
\cdot \\
\cdot \\
\cdot\,\cdot
\end{array}
$$

And again it is easy to see that we shall never get a remainder of 0. Hence, 5/6 is not expressible as a decimal fraction.

For the rational number 1/7 we have:

```
      .1428571428
   7)1.0000000000
      7
     ---
      30
      28
      --
       20
       14
       --
        60
        56
        --
         40
         35
         --
          50
          49
          --
           10
            7
           --
            30
            28
            --
             20
             14
             --
              60
              56
              --
               4
               .
                .
                 . .
```

And here again we see that a pattern is emerging which will continue on and on without ever giving us a remainder of 0. 1/7 is not expressible as a decimal fraction.

In general, when e and f are positive integers, we can be certain that e/f cannot be expressed as a decimal fraction if we have not obtained a 0 remainder in the division algorithm after carrying out $f - 1$ steps in the process past the stage where we begin "bringing down" only "0"'s. This is because there are only $f - 1$ different possible nonzero subremainders if we use the efficient division algorithm of Section 6.20. And, therefore, if we do not find a 0 remainder in the number of steps indicated, then we must find ourselves, at some stage, in exactly the same situation as we were earlier. And then the pattern will begin to repeat. The calculation that goes with 1/7 illustrates this nicely. There we ran through all 6 possible subremainders in the order 3, 2, 6, 4, 5, 1 and then we were again at the stage where we were asking how many times 7 goes into 10—just as we were at the very beginning of the process.

The numerical examples for 8/11, 5/6, and 1/7 above are all, in a sense, illustrations of a faulty algorithm. The processes never stop to give us final results. And yet the expressions being generated by the algorithms are useful. If we simply quit working at any step, we may think of the string of digits we have generated up to that point as being the name of an approximation to the given rational number. And we would hope that the approximations generated in this way would get better and better as we go through more and more steps. This is indeed true. We shall return to some details of this matter in the next chapter.

It is sometimes convenient to summarize such results as that in the example for 5/6 above by asserting that $5/6 = .8333\frac{2}{6}$. We mean by this simply that $5/6 = (8333\frac{1}{3}) \div 10^4$. Similar interpretations are put on other such *mixed decimals*. For computational purposes, such mixed decimals are usually avoided.

PROBLEMS

1. Find the decimal fraction name for each of the following (i) by converting to ordinary fractions, performing the indicated operation on these fractions, and then converting back to a decimal fraction; (ii) by using the more usual algorithm for decimal fractions.

(a) $6.43 + 2.59$ (b) $736.5 + 6.794$
(c) $37.1 + .0000552$ (d) $396 - .0052$
(e) $52.79 - 379.61$ (f) $-5.606 + 7.5$
(g) $-3.007 + 9.6$ (h) $-7.51 - 9.69$
(i) $(.7) \cdot (.6)$ (j) $(7.5) \cdot (6.9)$
(k) $(7.5) \cdot (-6.9)$ (l) $(-7.5) \cdot (-6.9)$
(m) $(75{,}000)(.005)$ (n) $(.00052)(.071)$
(o) $(33{,}000)(1.01)$ (p) $(21.84) \div 5.6$
(q) $(2184) \div .0056$ (r) $(.02184) \div (560)$
(s) $(72.4) \div (.0001)$ (t) $(.0059) \div (.000002)$
(u) $(15{,}000) \div (.03)$.

2. Which of the following can be expressed as decimal fractions? Justify your conclusions.

(a) 3/25 (b) 1/625 (c) 2/15
(d) 51/3 (e) 3/8 (f) 989/2300
(g) 4/9.

3. Prove or disprove:

(a) $2/3 = .66666\frac{2}{3}$ (b) $1/7 = .142857\frac{1}{7}$
(c) $1/9 = .11111\frac{1}{9}$ (d) $3/31 = .0906\frac{21}{31}$.

4. Note that $4/25 = .16$ and $4/16 = .25$. Also $13/50 = .26$ and $13/26 = .50$. Conjecture and prove a generalization suggested by these results.

5. 10^n may be written as a product of primes as $2^n 5^n$. If e and f are positive integers show that if "e/f" is a fraction in lowest terms, then e/f can be expressed as a decimal fraction if and only if the only prime factors of f are 2 and 5.

8.12 Integral exponents

In the realm of finance, **interest** is money that is paid for the use of money. If we deposit 1000 dollars in a savings account that pays an annual interest rate of 5%, then this money will "earn" 1000(.05) or 50 dollars interest in one year. (See problem 8.6.6 for a discussion of percentages.) In general, a principle of P dollars invested for n years at an annual simple interest rate of i will earn niP dollars in interest. The amount we would have available at the end of the n years would be $P + niP = P(1 + ni)$.

But if the bank compounds the interest annually, then the interest earned in any one year itself earns interest during the following year. At the end of the first year we would have $P(1 + i)$ dollars in our account. At the end of the second the amount would be $P(1 + i) + iP(1 + i) = P(1 + i) \cdot 1 + P(1 + i) \cdot i = P(1 + i)(1 + i)$. The amount available after n years with the interest compounded annually would be $P(1 + i)(1 + i) \cdots (1 + i)$ where there are n factors of $1 + i$.

Clearly, it would be helpful in such problems if we carried over notions about exponents to the rational number system.

When x is a rational number and n is a positive integer, it is easy to generalize, in a very natural way, our previous definition for x^n. We take it that $x^1 = x$ and $x^{n+1} = x \cdot x^n$. Since F is closed under multiplication we see that x^n is a well-defined rational number whenever $x \in F$ and $n \in I^+$. As we did in N, we take it that $x^0 = 1$ for all $x \neq 0$ but we leave 0^0 undefined.

With this terminology we may say that the amount at the end of n years for an investment of P dollars at an interest rate i compounded annually is $P(1 + i)^n$. Note that this formula gives a reasonable result even when $n = 0$ since P dollars invested for no time (0 years) will amount to $P(1 + i)^0 = P$ dollars.

The basic properties that we discovered before for natural number exponents carry over to F. When x and y are rational numbers and n and m are positive integers we have:

(i) $(xy)^n = x^n y^n$
(ii) $(x/y)^n = x^n/y^n$
(iii) $x^n \cdot x^m = x^{n+m}$
(iv) $x^n/x^m = x^{n-m}$ when $n \geq m$
(v) $(x^n)^m = x^{nm}$.

The proofs of these are quite easy in terms of simple inductions. In fact, the proofs given for the same equations in Chapter 6 may be carried over bodily.

To remind us of the nature of proofs by induction, we shall prove the first equality. We leave the others as exercises. Let x and y be rational numbers. Let $A = \{k \mid k \in I^+ \text{ and } (xy)^k = x^k y^k\}$. We first note that $1 \in A$. For we have $(xy)^1 = xy = x^1 y^1$. Next suppose that $n \in A$. Then, using our basic definition for exponents we have $(xy)^{n+1} = (xy)(xy)^n = (xy)(x^n y^n) = (xx^n)(yy^n) = (x^{n+1})(y^{n+1})$. Hence, $n \in A$ implies $n + 1 \in A$. Since $1 \in A$, we see that $A = I^+$ and this completes the proof.

Note that if $n = 0$ and x and y are not 0 then $(xy)^n = 1$ and also $x^n y^n = 1 \cdot 1 = 1$ so that $(xy)^n = x^n y^n$ holds in this case also.

Suppose we know that $x^n = y^n$ where x and y are rational and n is a positive integer. Can we conclude that $x = y$? Clearly we cannot. For example, we have $2^2 = (^-2)^2$ and yet $^-2 \neq 2$. However, we can show that if $x^n = y^n$ then we must have $x = y$ or $x = -y$. For suppose $x = a/b$ and $y = c/d$ and $x^n = y^n$ where a, b, c, and d are integers. Then we have $(a/b)^n = (c/d)^n = a^n/b^n = c^n/d^n$, $a^n d^n = c^n b^n$, and $(ad)^n = (cb)^n$ where we have used properties (ii) and (i) above. But it is easy to see from the multiplication algorithm for integers (Figure 7.6) that $(ad)^n = (cb)^n$ implies $(|ad|)^n = (|cb|)^n$. Now if one of the natural numbers $|ad|$ and $|cb|$ is larger than the other, their nth powers cannot be equal (Theorem 6.9.20). Hence, we must have $|ad| = |cb|$. This implies that $ad = cb$ or $ad = -cb$. Then $a/b = c/d$ or $a/b = -c/d$. That is, $x = y$ or $x = -y$.

We would now like to attach a meaning to x^{-n} where n is a positive integer. If we wish property (iv) above to hold, we should have $x^{-n} = x^{0-n} = x^0/x^n = 1/x^n$. And this is, indeed, the meaning we take for x^{-n}. That is, when x is any rational number different from 0 and n is a positive integer, then we define x^{-n} to be $1/x^n$.

For example, $2^{^-2} = 1/2^2 = 1/4$, $(1/2)^{^-3} = 1/(1/2)^3 = 1/(1/8) = 8$, $10^{^-4} = 1/10^4 = 1/10{,}000 = .0001$, $(.01)^{^-2} = 1/(.01)^2 = 1/.0001 = 10{,}000$, and $(^-3/2)^{^-3} = 1/(^-3/2)^3 = 1/(^-27/8) = {}^-8/27$.

We now have it that x^y makes sense whenever x is a rational number and y is any integer—except that x^y is undefined when $x = 0$ and $y \leq 0$. We would hope that the properties (i), (ii), (iii), (iv), and (v) above all hold when n and m are any integers. It is easy to prove that they do when they make sense.

For instance, to show that $(x/y)^n = x^n/y^n$ when n is any integer we need only check the cases where $n = 0$ and $n = -k$ for some positive integer k. We have $(x/y)^0 = 1 = 1/1 = x^0/y^0$ as long as x and y are not zero. And $(x/y)^{-k} = 1/(x/y)^k = 1/(x^k/y^k) = (1/x^k)/(1/y^k) = x^{-k}/y^{-k}$. Similar proofs go through for the other four basic properties.

Negative exponents are useful in many practical problems. If, for example, we wish to know what principle P to invest at an interest rate of i compounded annually for n years in order to accumulate A dollars, we have that $A = P(1 + i)^n$ and, therefore, $P = A/(1 + i)^n$ or $P = A(1 + i)^{-n}$. This last result is one of the standard formulas of the mathematics of finance.

Definition 8.12.1. If x is a rational number, we let $x^1 = x$ and $x^{n+1} = x \cdot x^n$ when n is a positive integer. If $x \neq 0$ we let $x^0 = 1$.

Theorem 8.12.1. If $x \in F$, $y \in F$, $n \in I^+$, and $m \in I^+$ then

(a) $(xy)^n = x^n y^n$
(b) $(x/y)^n = x^n/y^n$ when $y \neq 0$
(c) $x^n \cdot x^m = x^{n+m}$
(d) $x^n/x^m = x^{n-m}$ when $n \geq m$
(e) $(x^n)^m = x^{nm}$.

Theorem 8.12.2. If $x \in F$, $y \in F$, $n \in I^+$, and $x^n = y^n$, then $x = y$ or $x = -y$.

Definition 8.12.2. If $x \in F$ and $n \in I^+$, we let $x^{-n} = 1/x^n$.

Theorem 8.12.3. The conclusions of Theorem 8.12.1 hold when n and m are any integers provided that both sides of the equations are defined.

Problems

1. Make a table showing the values of x^y for $x = {}^-2, {}^-1, {}^-1/2, {}^-1/10, 0, 1/10, 2/10, 1/2, 1, 2$, and $y = {}^-3, {}^-2, {}^-1, 0, 1, 2, 3$. Use the y values as column headings.

2. Find decimal fractions naming 10^n when $n = 0, {}^-1, {}^-2, {}^-3, {}^-4, {}^-5, {}^-6, {}^-7$.

3. What is the value of

(a) $({}^-1)^3$ (b) $({}^-1)^4$
(c) $({}^-1)^{1000}$ (d) $({}^-1)^{5743}$
(e) $({}^-1)^{379,432}$ (f) $({}^-1)^{2n}$
(g) $({}^-1)^{2n+1}$?

(In parts f, and g, n is a positive integer.)

4. If $x = y$, then, of course, $x^y = y^x$. Are there cases where $x^y = y^x$ and yet $x \neq y$?

5. Show that when n is a positive integer and x is rational, we have $x^n = 0$ if and only if $x = 0$. [Hint: See Theorem 8.4.7.]

6. Simplify the following:

(a) $3^{-4}/3^{-3}$ (b) $(.4)^2/(.2)^4$
(c) $(.43)^7/(.43)^5$ (d) $(1/2)^4 \cdot 2^5$
(e) $(1/2)^{-4} \cdot 2^{-5}$ (f) $((.3)^4(.3)^7)/(.3)^{12}$
(g) $(.1)^{-2}$ (h) $(.3)^4/(.09)^5$ [Hint: $(.3)^2 = .09$]
(i) $(.36)^{-5}/(.6)^{-9}$ (j) $(1.01)^{17} \cdot (1.01)^{-18} + (1.01)^{18} \cdot (1.01)^{-17}$
(k) $(3^7 + 3^6)/3^6$ (l) $(4^{-3} + 4^{-2})/(4^{-2} + 4^{-1})$.

7. A principle of \$100 is invested for 5 years at 5% interest. Find the total amount available at the end of the period if the interest is (a) simple interest, (b) compounded annually.

8. Do Problem 7 for a ten year period. [HINT: $(1 + .05)^{10} = ((1 + .05)^5)^2$.]

9. How much money must we invest at 5% interest compounded annually for 5 years in order to have an amount of $50,000 at the end of the period?

10. Do Problem 9 for a 10 year period.

11. A ball falls from a table top 3 feet above the floor. Each time it hits the floor it bounces up to 3/4 of the height from which it fell. How far has the ball travelled by the nth time it hits the floor when n is

(a) 2 (b) 3
(c) 4 (d) 5
(e) 6?

12. Suppose the ball in Problem 11 bounces only to 1/10 of the height of its fall. Answer the questions of Problem 11 in this case.

13. If possible, find a rational number x such that

(a) $x^2 = 1/4$ (b) $x^2 = 1/9$
(c) $x^3 = .001$ (d) $x^{-3} = 27/8$
(e) $x^4 = 1$ (f) $x^{-6} = 10^6$
(g) $x^3 = x^{12}$ (h) $(2x)^5 = 1$
(i) $2x^{-5} = 64$ (j) $x^5 + x^7 = 2$
(k) $x^4 = x^{-5}$ (l) $x^6 = x^6 + 1$.

14. When x is a rational number and n is a positive integer, let us define the operation ★ of **repeated exponentiation** by taking $x \star 1 = x^x$ and $x \star (n + 1) = x^{x \star n}$. For example, $2 \star 1 = 2^2 = 4$, $2 \star 2 = 2^{2 \star 1} = 2^4 = 16$, $2 \star 3 = 2^{2 \star 2} = 2^{16} = 155{,}536$, and $2 \star 4 = 2^{2 \star 3} = 2^{155{,}536}$. Give some other examples and investigate the properties of this operation. Is $10 \star 10$ a large number? How many "0"'s are there in its decimal name?

15. Show that when n and m are positive integers and x and y are rational numbers, then

(a) $(x/y)^n = x^n/y^n$ (b) $x^n \cdot x^m = x^{n+m}$
(c) $x^n/x^m = x^{n-m}$ when $n \geq m$ (d) $(x^n)^m = x^{nm}$.

16. Show that when n and m are positive integers and x and y are rational numbers, then

(a) $(xy)^{-n} = x^{-n} \cdot y^{-n}$ (b) $x^{-n} \cdot x^{-m} = x^{-n-m}$.

17. Show that when n is a positive integer and m is a negative integer and x is rational, then

(a) $x^n \cdot x^m = x^{n+m}$ (b) $(x^n)^m = x^{nm}$.

18. Show that when n and m are negative integers and x is rational

$$(x^n)^m = x^{nm}.$$

8.13 Fractional exponents

If we wish to construct a square field that contains one million square feet, we need the length x of the side of the square to be such that $x^2 = 1{,}000{,}000$. That is, x must be a number whose square is 1,000,000.

If we have a cubical box which contains 1728 cubic inches and we wish to find the length x of one side of the cube we are looking for a number x such that $x^3 = 1728$. That is, x is a number whose cube is 1728.

In general, let us call a number x an nth **root** of a number y if $x^n = y$. Thus, 3 is a second root (or, as it is more usually put, a **square root**) of 9 since $3^2 = 9$. 1000 is a square root of 1,000,000 since $1000^2 = 1{,}000{,}000$. Note that $^-1000$ is also a square root of 1,000,000 since $(^-1000)^2 = 1{,}000{,}000$. $1/4$ and $^-1/4$ are both square roots of $1/16$ since $(1/4)^2 = (^-1/4)^2 = 1/16$. $1/3$ is a third root (or, as it is more usually put, a **cube root**) of $1/27$ since $(1/3)^2 = 1/27$. 12 is a cube root of 1728 since $12^3 = 1728$. .1 and $-.1$ are fourth roots of .0001 since $(.1)^4 = (-.1)^4 = .0001$.

On the other hand, there can be no rational square root of $^-1$ since the square of any rational number is nonnegative. In general, there can be no rational nth root of a negative number when n is even.

It is customary to introduce a special symbolism for roots of numbers. If n is odd, we shall write "$\sqrt[n]{a}$" for the nth root of a. That is, when n is odd, $x = \sqrt[n]{a}$ if and only if $x^n = a$. When n is even "$\sqrt[n]{a}$" stands for the *positive* nth root of a. When n is even, $x = \sqrt[n]{a}$ if and only if $x^n = a$ and $x \geq 0$.

We call the symbol "$\sqrt{}$" a *radical*. We usually delete the "2" in "$\sqrt[2]{}$" and write simply "$\sqrt{}$."

We have $\sqrt{1{,}000{,}000} = 1000$ since $1000^2 = 1{,}000{,}000$ and $1000 \geq 0$. Similarly, $\sqrt{1/16} = 1/4$, $\sqrt[3]{1/27} = 1/3$, $\sqrt[3]{1728} = 12$, and $\sqrt[4]{.0001} = .1$. On the other hand, $\sqrt{^-1}$ is undefined in F.

If we think of a^n as the result of an operation P on the numbers a and n so that $a^n = a\ P\ n$, then the "root operation" is the closest thing we can get to an inverse operation for P. If we take the operation R to be defined by $a\ R\ n = \sqrt[n]{a}$, then we will have $(a\ P\ n)\ R\ n = a$ whenever a is a positive rational number and n is a positive integer. For example,

$$(3\ P\ 5)\ R\ 5 = (3^5)\ R\ 5 = \sqrt[5]{3^5} = 3,$$

$$((1/2)\ P\ 3)\ R\ 3 = (1/2)^3\ R\ 3 = \sqrt[3]{(1/2)^3} = 1/2,$$

$$(4\ P\ 2)\ R\ 2 = (4^2)\ R\ 2 = \sqrt{4^2} = 4.$$

In the case where n is even, we put the restriction that $x \geq 0$ when $x = \sqrt[n]{a}$ so that $a\ R\ n = \sqrt[n]{a}$ *is* an operation and yields a unique result when it yields any. If we simply defined $x = \sqrt{a}$, for instance, to mean that x is a square root of a, then we would not know whether to take $x = 4 \; R \; 2 = \sqrt{4}$ to be 2 or to be $^-2$. Both 2 and $^-2$ are square roots of 4. To remove this ambiguity we insist that $\sqrt[n]{a}$ is the **positive** nth root of a whenever n is even.

Suppose that $x = \sqrt[n]{a}$. We might imagine that x is some power of a so that $x = a^y$. But $x^n = a$ and if we want the usual properties for exponents to hold we should have $a = x^n = (a^y)^n = a^{yn}$. Hence, it appears that we should have $yn = 1$ or $y = 1/n$. This sort of reasoning leads us to make the following definition:

> If n is a positive integer we define $a^{1/n}$ to be $\sqrt[n]{a}$.

We have for examples, $9^{1/2} = 3$, $1{,}000{,}000^{1/2} = 1000$, $(1/16)^{1/2} = 1/4$, $1728^{1/3} = 12$, $(1/27)^{1/3} = 1/3$, and $(.0001)^{1/4} = .1$. On the other hand, $(^-1)^{1/2}$ is undefined.

We extend this definition in a natural way by saying:

> When n and m are both integers and n is positive, we define $a^{m/n}$ to be $(a^m)^{1/n}$. That is $a^{m/n} = \sqrt[n]{a^m}$.

For instance, $8^{2/3} = (8^2)^{1/3} = 64^{1/2} = 4$, $(1/9)^{3/2} = ((1/9)^3)^{1/2} = (1/729)^{1/2} = 1/27$, $(1/32)^{^-2/5} = ((1/32)^{^-2})^{1/5} = (1/(1/32)^2)^{1/5} = (1024)^{1/5} = 4$.

The manner in which the definition of $a^{m/n}$ is stated is not the most convenient one for purposes of computation. But the choice was a deliberate one because of a peculiar dependence on the order in which things are done. We would certainly like $(^-2)^{4/2}$ to be the same as $(^-2)^2$ since $4/2 = 2$. But note that $((^-2)^{1/2})^4$ is undefined since $(^-2)^{1/2}$ is undefined. If we had defined $a^{m/n}$ to be $(a^{1/n})^m$ then $(^-2)^{4/2}$ would fail to exist. Nevertheless, when $(a^{1/n})^m$ is a well-defined rational number it **will** be the same as $a^{m/n} = (a^m)^{1/n}$. And the actual computation of $a^{m/n}$ is usually easier to do in the form $(a^{1/n})^m$ than in the form $(a^m)^{1/n}$. For instance, the above examples may be done more simply as follows:

$$8^{2/3} = (8^{1/3})^2 = 2^2 = 4$$

$$(1/9)^{3/2} = ((1/9)^{1/2})^3 = (1/3)^3 = 1/27$$

$$(1/32)^{^-2/5} = ((1/32)^{1/5})^{^-2} = (1/2)^{^-2} = 1/(1/2)^2 = 4.$$

Since any rational number may be written as a ratio of integers with the integer in the denominator positive, we now know what it means to assert that $a^y = x$ when a, y, and x are any rational numbers.

We would hope that the basic properties of exponents would carry over to the case where the exponents are not integers. In large measure, they do.

However, we must qualify them somewhat since x^y is sometimes undefined in F. If x, y, and z are rational numbers and all the expressions involved define rational numbers, then

$$\text{(i)}\ \ x^z \cdot y^z = (xy)^z$$
$$\text{(ii)}\ \ x^z/y^z = (x/y)^z$$

$$\text{(iii)}\ x^y \cdot x^z = x^{y+z}$$
$$\text{(iv)}\ x^y/x^z = x^{y-z}$$
$$\text{(v)}\ (x^y)^z = x^{yz}.$$

A somewhat stronger statement can be made. Whenever the left side of any one of the above equations is defined, then so is the right side and the asserted equality holds.

However, there are cases for every one of the above equalities where the left side fails to be defined while the right side makes sense. If, for example, we let $x = {}^-1$ and $y = z = 1/2$ in (iv) we see that $({}^-1)^{1/2-1/2} = ({}^-1)^0 = 1$ and yet $({}^-1)^{1/2}$ and $({}^-1)^{1/2}$ are undefined. To get similar examples for the others, try the following instances. In (i) and (ii) let $x = y = {}^-1$ and $z = 1/2$. In (iii), let $x = {}^-1$, $y = 1/2$ and $z = {}^-1/2$. In (v), let $x = {}^-1$, $y = 1/2$, and $z = 2$.

If n is even we may think of $\sqrt[n]{x}$ (or $x^{1/n}$) as the positive number which yields x when it is raised to the nth power. When n is odd we have the same interpretation except that we should omit the word "positive." Hence, we may think of radicals as tools introduced to help us solve certain equations.

Faced with the problem of solving the equation $x^2 = 4$ we would claim that one solution is $x = \sqrt{4} = 2$. But we must be careful to realize that there is another solution to this equation. Namely, $x = {}^-2$. That is, the solution set for the equation $x^2 = 4$ is $\{\sqrt{4}, -\sqrt{4}\}$ or $\{2, {}^-2\}$.

In view of the meaning of the radicals involved it is not too difficult to establish (see Theorem 8.12.2) that:

when n is even:

(a) if $a < 0$, then the equation $x^n = a$ has no rational solution

(b) when the equation $x^n = a$ does have a rational solution, then its solution set is $\{\sqrt[n]{a}, -\sqrt[n]{a}\}$.

When n is odd and the equation $x^n = a$ has a rational solution, then the equation has the solution set $\{\sqrt[n]{a}\}$.

Taking our universal set to be the set of rational numbers, we may state these results in somewhat different terms as follows:

If n is even and $a < 0$, then $\{x \mid x^n = a\} = \phi$.

If n is even and $\{x \mid x^n = a\} \neq \phi$, then $\{x \mid x^n = a\} = \{\sqrt[n]{a}, -\sqrt[n]{a}\}$.

If n is odd and $\{x \mid x^n = a\} \neq \phi$, then $\{x \mid x^n = a\} = \{\sqrt[n]{a}\}$.

There is a striking weakness in the above conclusions in that, except in the case where n is even and $a < 0$, they give no procedure for deciding whether or not $\{x \mid x^n = a\}$ is empty. We shall look at this matter again later in Section 10.1.

There are algorithms available for finding roots of rational numbers when they exist. But they are of little practical value and we shall not discuss them here.

Definition 8.13.1. If $x^n = a$, then we say that x is an nth **root** of a. $x = \sqrt[n]{a}$ if and only if $x^n = a$ and, in case n is even, $x \geq 0$.

Definition 8.13.2. If n is a positive integer and m is an integer, $a^{m/n} = \sqrt[n]{a^m} = (a^m)^{1/n}$.

Theorem 8.13.1. If x, y, and z are rational numbers, then each of the following equations holds whenever the left side is defined:

(i) $x^z y^z = (xy)^z$
(ii) $x^z/y^z = (x/y)^z$
(iii) $x^y \cdot x^z = x^{y+z}$
(iv) $x^y/x^z = x^{y-z}$
(v) $(x^y)^z = x^{yz}$.

PROOF: Suppose that $x^z = w$ and $y^z = v$. Let $z = a/b$ where $b > 0$. Then $x^{a/b} = \sqrt[b]{x^a} = w$ and $y^{a/b} = \sqrt[b]{y^a} = v$. Hence, $x^a = w^b$ and $y^a = v^b$. Then, by Theorem 8.12.3, $x^a \cdot y^a = (xy)^a = w^b v^b = (wv)^b$. If b is even then w and v are positive and, therefore, wv is also. It follows that $(xy)^{a/b} = wv$. That is, $(xy)^z = x^z y^z$. Also $x^a/y^a = (x/y)^a = w^b/v^b = (w/v)^b$, by Theorem 8.12.3, and therefore, $(x/y)^{a/b} = w/v$. Hence, $(x/y)^z = x^z/y^z$.

If $x^z = w$ and $x^y = u$ where $z = a/b$ and $y = c/d$ and $b > 0$ and $d > 0$, then $x^{a/b} = w$ and $y^{c/d} = u$. Hence, $x^a = w^b$ and $x^c = u^d$. Then $(x^a)^d = (w^b)^d$ and $(x^c)^b = (u^d)^b$. Then we may use Theorem 8.12.3 and get $x^{ad} = w^{bd}$ and $x^{cb} = u^{bd}$. Consequently, $(x^{ad})(x^{cb}) = (w^{bd})(u^{bd})$ and, by Theorem 8.12.3 again, $x^{ad+cb} = (wu)^{bd}$. If b is even, then w is positive, and, more than that, x must be positive since $x^{a/b}$ is defined. Similar comments apply if d is even. If b or d is even, then bd is even and, since x must be positive, w and u are both positive. If bd is even, then b is even or d is even. Hence, if bd is even, wu is positive. It follows that $x^{(ad+cb)/bd} = wu$. That is, $x^{y+z} = x^y \cdot x^z$.

We shall leave the proofs of parts (iv) and (v) as exercises.

Theorem 8.13.2. Suppose $a \in F$ and $n \in I^+$. Let $S = \{x \mid x^n = a \text{ and } x \in F\}$. (a) If n is even and $a < 0$, then $S = \phi$. (b) If n is even and $S \neq \phi$, then $S = \{\sqrt[n]{a}, -\sqrt[n]{a}\}$. (c) If n is odd and $S \neq \phi$, then $S = \{\sqrt[n]{a}\}$.

Problems

1. Find

(a) $\sqrt{25}$	(b) $\sqrt[3]{1}$
(c) $\sqrt[3]{-1}$	(d) $\sqrt[3]{8}$
(e) $\sqrt[3]{-8}$	(f) $\sqrt[5]{32}$
(g) $\sqrt[6]{64}$	(h) $\sqrt[7]{128}$

(i) $\sqrt[4]{10{,}000}$ (j) $\sqrt[4]{100{,}000{,}000}$
(k) $\sqrt[8]{100{,}000{,}000}$ (l) $\sqrt{144}$
(m) $\sqrt[4]{81}$ (n) $\sqrt[4]{.0001}$
(o) $\sqrt{.81}$ (p) $\sqrt[4]{81/16}$.

2. Restate your conclusions of Problem 1 by using fractional exponents.
3. Evaluate

(a) $0^{3/4}$ (b) $4^{3/2}$
(c) $49^{3/2}$ (d) $32^{2/5}$
(e) $10{,}000^{3/4}$ (f) $1^{7/5}$
(g) $1{,}000{,}000^{5/6}$ (h) $(1/27)^{1/3}$
(i) $(1/27)^{-1/3}$ (j) $(8/27)^{-2/3}$
(k) $(.0001)^{-3/4}$ (l) $(-1/32)^{-3/5}$
(m) $(-1/16)^{-3/4}$ (n) $(8/1728)^{-2/3}$
(o) $(-.027)^{-4/3}$.

4. Find the solution set for each of the following equations:

(a) $x^2 = 1$ (b) $x^3 = 1$
(c) $x^3 = {}^-1$ (d) $x^4 = 1$
(e) $x^4 = {}^-1$ (f) $x^5 = 1$
(g) $x^5 = {}^-1$ (h) $x^7 = 0$
(i) $x^5 = 32$ (j) $x^5 = {}^-32$
(k) $x^6 = 1{,}000{,}000$ (l) $x^6 = {}^-1{,}000{,}000$
(m) $(2x)^3 = 1$ (n) $(2x)^4 = 16$
(o) $(3x)^2 = 36$
(p) $(x + 1)^2 = 4$ [HINT: $(x + 1)^2 = 4$ if and only if $x + 1 = 2$ or $x + 1 = {}^-2$.]
(q) $(2x + 1)^2 = 4$ (r) $(7 - 4x)^3 = 27$
(s) $(7 - 4x)^2 = 9$ (t) $(x - 1)^4 = 1$
(u) $3x^3 + 81 = 0$ (v) $3x^4 + 81 = {}^-162$.

5. Verify that there are instances for parts (i), (ii), (iii), and (v) of Theorem 8.13.1 where the left side fails to define a rational number and yet the right side does.
6. Prove Theorem 8.13.1, part (iv).
7. Prove Theorem 8.13.1, part (v).
8. Disprove each of the following. x, y, and z are rational numbers.

(a) If $x \leq y$, then $x^z \leq y^z$.
(b) If $x \leq y$, then $z^x \leq z^y$.
(c) If $0 < x < y$, then $x^z < y^z$.
(d) If $0 < x < y$ and $z \geq 0$, then $x^z < y^z$.
(e) $\sqrt{x^2} = x$.

(f) $(x + y)^z = x^z + y^z$.
(g) $\sqrt{x + y} = \sqrt{x} + \sqrt{y}$.

9. Suppose we define the operation E in F by stating that $x \; E \; y = x^y$. Prove that

(a) F is not closed under E
(b) E is not commutative
(c) E is not associative
(d) There is no identity element for E in F
(e) E does not distribute across addition in F.

10. Prove Theorem 8.13.2.

8.14 Other names for fractions

We have said that when "a_1," "a_2," . . . , and "a_n" are decimal digits then $.a_1a_2 \cdots a_n = a_1a_2 \ldots a_n/10^n$. In view of what we now know about exponents we may now claim that $.a_1a_2 \cdots a_n = (a_1a_2 \cdots a_n)(10^{-n})$. But if we remember here the basic meaning of "$a_1a_2 \cdots a_n$" as a name for a natural number we have

$$\begin{aligned} .a_1a_2 \cdots a_n &= (a_1 \cdot 10^{n-1} + a_2 \cdot 10^{n-2} + \cdots + a_{n-1} \cdot 10 + a_n)(10^{-n}) \\ &= a_1(10^{-1}) + a_2(10^{-2}) + \cdots + a_{n-1}(10^{n-1}) + a_n(10^{-n}). \end{aligned}$$

That is, $.a_1a_2 \cdots a_n$ is a_1 tenths and a_2 hundredths and . . . and $a_n 10^n$ths. For example, .65304 is 6 tenths, 5 hundredths, 3 thousandths, and 4 hundred-thousandths.

We see how this all ties together nicely when we look at the meaning of such a numeral as "534.705." It is given by

$$(5)10^2 + (3)10^1 + (4)10^0 + (7)10^{-1} + (0)10^{-2} + (5)10^{-3}.$$

More generally, if the a's and b's are all decimal digits, we have

$$\begin{aligned} b_kb_{k-1} \cdots b_1b_0.a_1a_2 \cdots a_n = (b_k)10^k + (b_{k-1})10^{k-1} + \cdots + (b_1)10^1 + (b_0)10^0 \\ + (a_1)10^{-1} + (a_2)10^{-2} + \cdots + (a_n)10^{-n}. \end{aligned}$$

But when we realize this meaning for decimal fractions, it is easy to adopt a similar device if we wish to use a base other than 10 for our system of numerals.

In the base 2, if the c's and d's are all binary digits we should have

$$\begin{aligned} (c_kc_{k-1} \cdots c_1c_0.d_1d_2 \cdots d_n)_2 = (c_k)2^k + (c_{k-1})2^{k-1} + \cdots + (c_1)2^1 \\ + (c_0)2^0 + (d_1)2^{-1} + (d_2)2^{-2} + \cdots \\ + (d_n)2^{-n}. \end{aligned}$$

The expression on the left here would be called a **binary fraction.**

Thus, $10011.110001_2 = 2^2 + 2 + 1 + 1/2 + 1/4 + 1/64$. Of course, it is also true that

$$.110001_2 = \frac{110001_2}{1000000_2} = 110001_2/2^6.$$

In a binary fraction we should probably refer to the "." as the **binary point.**

The algorithms for carrying out operations with rational numbers written in decimal notation carry over as we would expect to other numeration systems. For example, in the base 2, omitting the subscript "2" we have:

```
  110001.101001
+   1010.11001
---------------
  111100.001011,
```

and

```
   1101.011
   ×11.01
   --------
   1101 011
  110101 1
 1101011
 ----------
 101011.01111,
```

and

```
          101.0001
     ------------
 101 ) 11001.0101
       101
       ---
         101
         101
         ---
          00101
            101
            ---
              0.
```

While in the decimal numeration system we have $1/5 = .2$, in base two notation $5 = 101_2$ and carrying out the division algorithm with 1 and 101_2 we have

```
        .00110011 . . .
     ------------------
 101 ) 1.000000
         101
        ----
          110
          101
          ---
           1000
            101
           ----
             110
             101
             ---
               1
               .
               .
               . .
```

Evidently, $1/101_2$ cannot be expressed as a binary fraction. But we would expect $.00110011_2$ to be a good approximation to $1/101_2 = 1/5$. In fact, $.00110011_2 = 1/8 + 1/16 + 1/128 + 1/256 = (32 + 16 + 2 + 1)/256 = 51/256$ and $1/5 - 51/256 = (256 - 255)/(5 \cdot 256) = 1/1280$.

In the decimal system we find that 1/2, 1/4, 1/5, 1/8, and, of course, 1/10 can be expressed as decimal fractions. 1/3, 1/6, 1/7, and 1/9 cannot be so expressed.

In the base 12 system of numeration, however, we find that 1/2, 1/3, 1/4, 1/6, 1/8, 1/9, and 1/12 can be expressed as base 12 fractions. 1/5, 1/7, 1/10, and 1/11 cannot be so expressed. The fact that so many simple fractions can be expressed as base 12 fractions is one of the main reasons for the support given by some people for the adoption of a base 12 numeration system.

In the base 12 system we have $1/2 = .6_{12}$, $1/3 = .4_{12}$, $1/6 = .2_{12}$, $1/8 = .16_{12}$, $1/9 = .14_{12}$, and, of course, $1/12 = .1_{12}$.

On the other hand, if we attempt the division algorithm for 1/5 with base 12 numerals we have

$$\begin{array}{r} .244\cdots \\ 5\overline{)1.000\cdots} \\ t \\ \hline 20 \\ 18 \\ \hline 2 \\ \cdot \\ \cdot \\ \cdot\,\cdot \end{array}$$

Clearly, the process never terminates to give a base 12 fraction for 1/5.

In a similar fashion for 1/10 with base 12 numerals we have $1/10 = 1/t$ and we get

$$\begin{array}{r} .122\ldots \\ t\overline{)1.000\ldots} \\ t \\ \hline 20 \\ 18 \\ \hline 20 \\ 18 \\ \hline 2 \\ \cdot \\ \cdot \\ \cdot\,\cdot \end{array}$$

And again the process does not stop.

We may use mixed numerals and write

$$1/5 = \left(.244\frac{2}{5}\right)_{12} \quad \text{and} \quad 1/10 = \left(.122\frac{2}{t}\right)_{12}.$$

But 1/5 and 1/10 cannot be expressed *exactly* as (unmixed) base 12 fractions.

Problems

1. Express each of the following as ratios of integers which are written in the decimal notation.

(a) $.101_2$ (b) $.0011_2$
(c) 1010.10101_2 (d) $.01211_3$
(e) 221.0021_3 (f) $.ee_{12}$
(g) $.89te_{12}$ (h) $t9.02_{12}$.

2. Express each of the following as binary fractions or decide that they cannot be so expressed.

(a) 1/16 (b) 5/24 (c) 3/14 (d) 127/128 (e) 47/8.

3. Express each of the following as base 3 fractions or decide that they cannot be so expressed.

(a) 1/9 (b) 26/27 (c) 111/81 (d) 150/5 (e) 5/6.

4. Express each of the following as base 12 fractions or decide that they cannot be so expressed.

(a) 1/144 (b) 23/24 (c) 17/192 (d) 75/36 (e) 10/81.

5. As a different approach to Problem 4, note that $10/81 = 10(1/3)^4 = 10(4/12)^4 = t_{12}(.4_{12})^4$ so that we may easily find the base 12 fraction for 10/81 by multiplying out the above, rather than using the division algorithm. Use this method to check your results in parts *b*, *c*, and *d* of Problem 4.

6. Show that $1/11 = 1/e_{12}$ cannot be expressed as a base 12 fraction.

8.15 Some graphical devices

Scales on measuring devices are usually labeled by decimal fractions. Rulers are often an exception in this country where eighths, sixteenths, and thirty-seconds of an inch are common scale units. We may, with rational numbers, use some of the graphical devices that we have previously used for performing operations in N or I. However, a new problem arises because the rational numbers are dense on the line. When we use a geometric construction to find a point on a rational number line, the best we can hope to do by eye in determining the rational number corresponding to that point is to give a number which is close to the exact one.

In Figure 8.17 we have constructed a nomograph similar to that of Figure 6.15 for carrying out the operation of addition. We have marked points on the outer scales for the decimal fractions -5, -4.9, $-4.8, \ldots,$ 0, .1, . . . , 4.9, 5.0. The nomograph may be used to give the precise answer to a problem of adding two of these particular numbers. For other rational numbers we must estimate their sum. We may use this same figure for subtraction.

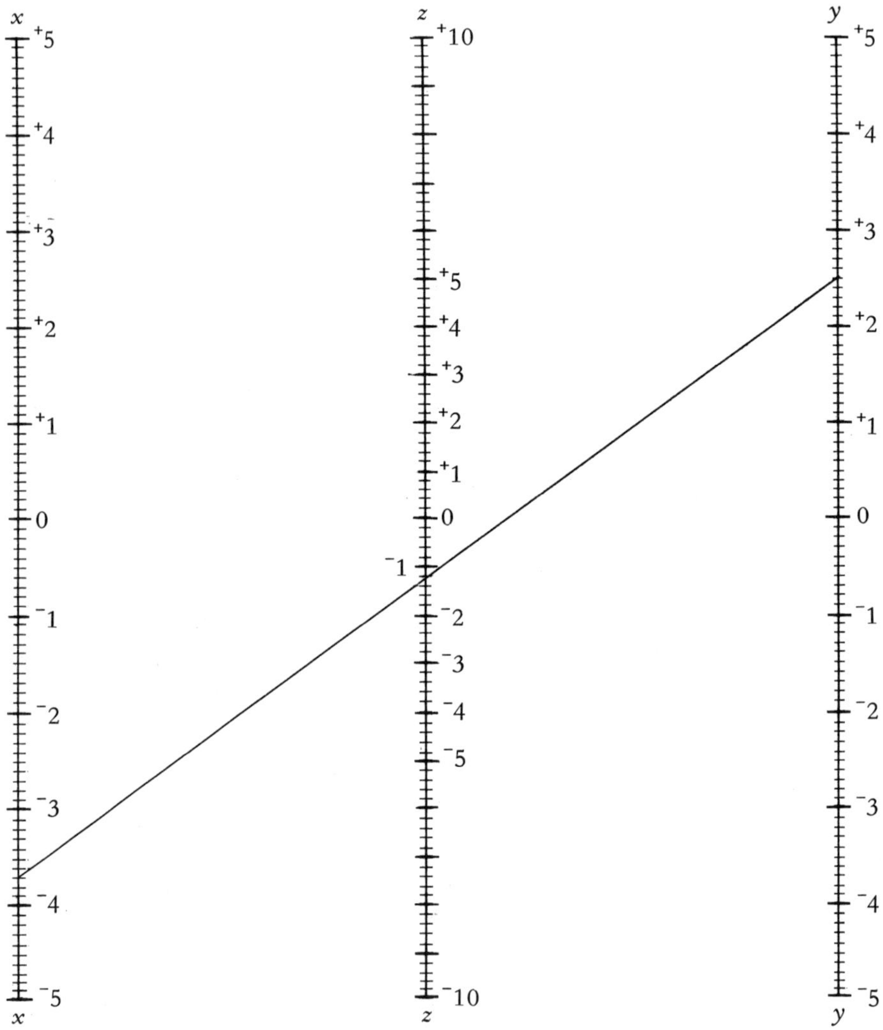

FIGURE 8.17 Nomograph for addition of rationals. To find $z = x + y$, draw the line from x on the left axis to y on the right axis. This line will cross the middle axis at z. The computation of $-3.7 + 2.5 = -1.2$ is shown.

The device in Figure 6.16 for adding natural numbers does not carry over well to the rationals. But the method of construction behind Figure 6.16 may be used to give us a graphical method for adding rationals. It is as follows. Construct horizontal and vertical number lines intersecting at their origins. To find the sum of the rational numbers x and y, construct a vertical line through the point x on the horizontal axis. Construct a horizontal line through the point y on the vertical axis. Let P be the point of intersection of these two constructed lines. Draw the line L which passes through the point 1 on the horizontal axis and the point 1 on the vertical axis. Through P, draw a line L' parallel to L. The line L' will intersect the horizontal axis in the point $x + y$. See Figure 8.18 where the construction of $-1/3 + 1/2$ is shown.

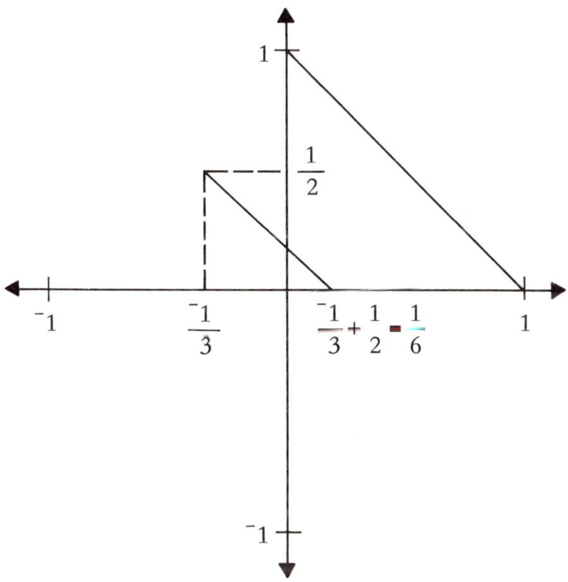

FIGURE 8.18

We may use a similar method for subtractions. With the axes and the line L constructed as before, we may find $z - y$ by the following technique. Draw a horizontal line through the point y on the vertical axis. Draw a line L'' through the point z on the horizontal axis and parallel to L. Let P' be the point of intersection of these two constructed lines. The vertical line drawn through P' will hit the horizontal axis at the point $x = z - y$. The construction of $-3/2 - 1/3$ is shown in Figure 8.19.

The slide rule for addition that we described in Section 7.10 will also work for addition of rationals if we take the number lines to have rational points on them. The same device may also be used for subtraction of rationals. See Figure 8.20.

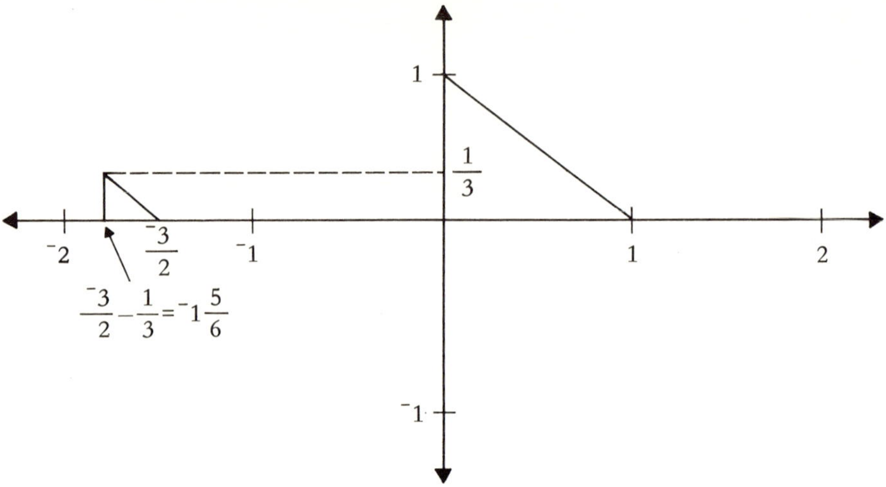

FIGURE 8.19

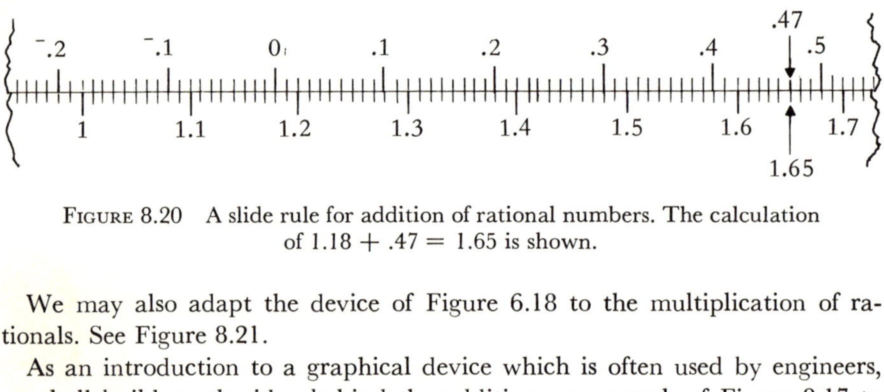

FIGURE 8.20 A slide rule for addition of rational numbers. The calculation of $1.18 + .47 = 1.65$ is shown.

We may also adapt the device of Figure 6.18 to the multiplication of rationals. See Figure 8.21.

As an introduction to a graphical device which is often used by engineers, we shall build on the idea behind the addition nomograph of Figure 8.17 to get a nomograph for multiplication of rational numbers from a certain restricted class.

Let T be the collection of all integral powers of 2. That is, let $T = \{\ldots, 2^{\bar{}3}, 2^{\bar{}2}, 2^{\bar{}1}, 2^0, 2^1, 2^2, 2^3, \ldots\} = \{\ldots, 1/8, 1/4, 1/2, 1, 2, 4, 8, \ldots\}$. T is closed under multiplication since we have $2^x \cdot 2^y = 2^{x+y}$ and I is closed under addition. If we wish to multiply two members of T, we may do so by finding the exponents that correspond to them and then adding these exponents to get the exponent of the product. In a certain sense, we may multiply in T by adding. But we know how to construct a nomograph for adding integers. If we construct such a nomograph and put the "2^x" label on the point we would ordinarily call "x," we shall get a nomograph for multiplying in T. The result is shown in Figure 8.22.

Notice that the axes in Figure 8.22 are certainly not number lines in the usual sense. On the usual number line, when $x - y = z - w$ then the distance be-

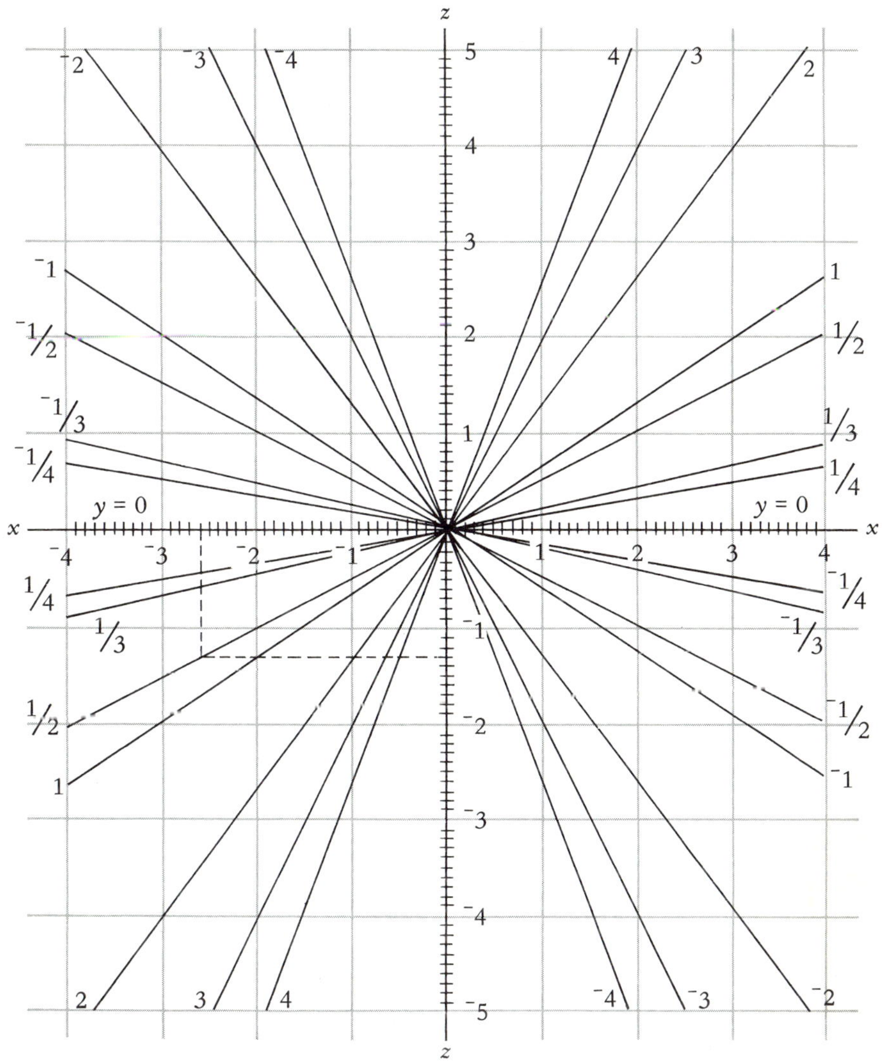

FIGURE 8.21 Graphical device for finding the product xy for $y = \pm 1, \pm 2, \pm 3, \pm 4, \pm 1/2, \pm 1/3, \pm 1/4$. To find $z = xy$, draw the vertical at x to the line for y. The horizontal from there will hit the vertical axis at z. The calculation of $(-2.6)\ (1/2) = -1.3$ is shown.

tween the point x and the point y is the same as the distance between the point z and the point w. This is not true on the axes in the nomograph of Figure 8.22.

If we put the x scale from Figure 8.22 on the top edge of a piece of cardboard and the y scale on the bottom edge of another, we may construct a slide rule for multiplying in T. We would use the same procedure as that used with the slide rule for addition in Figure 8.20. Such a slide rule for multiplication would

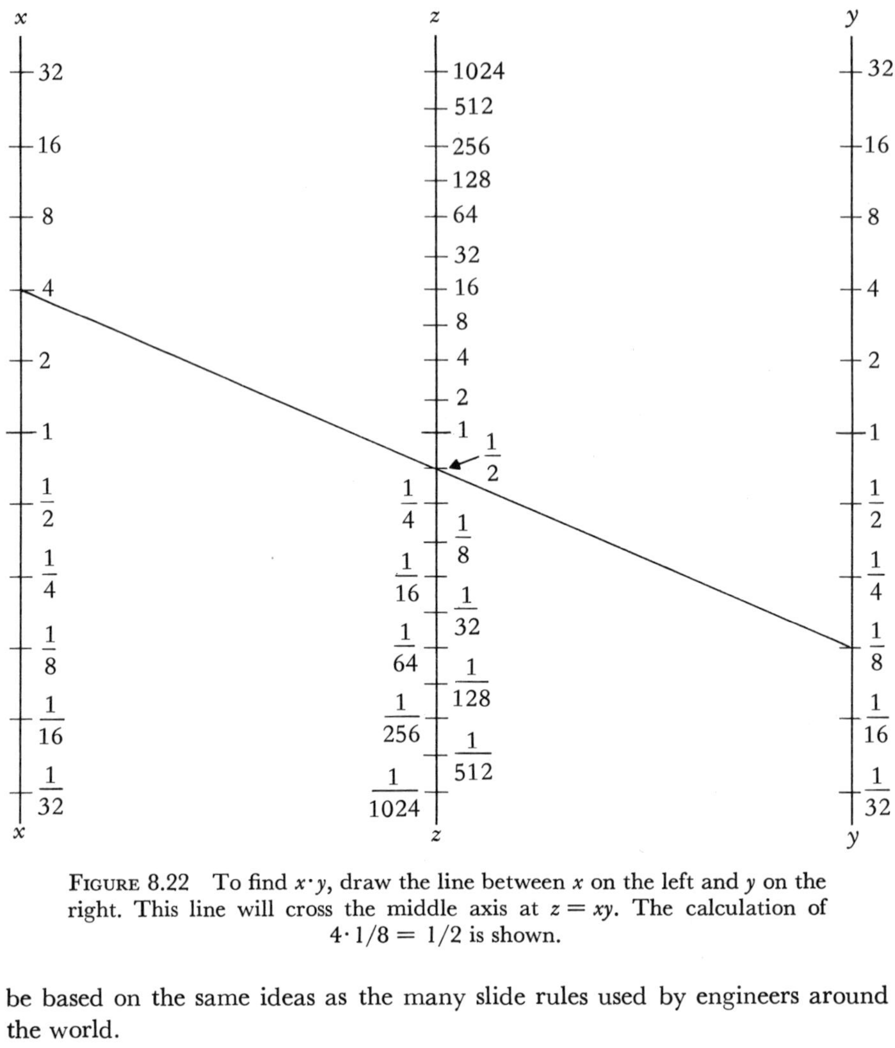

FIGURE 8.22 To find $x \cdot y$, draw the line between x on the left and y on the right. This line will cross the middle axis at $z = xy$. The calculation of $4 \cdot 1/8 = 1/2$ is shown.

be based on the same ideas as the many slide rules used by engineers around the world.

PROBLEMS

1. Use Figure 8.17 to find

(a) $3.2 + 1.6$ (b) $3.2 + (-2.7)$
(c) $3.2 + (-4.9)$ (d) $-3.4 + 3.7$
(e) $-3.4 - (3.7)$ (f) $-4.1 - (-2.5)$.

2. Construct a nomograph for addition (or subtraction) similar to that of Figure 8.17, but with the left and right scales having points marked on them for $-1.00, -.99, -.98, \ldots, 0, 0.1, \ldots, .99, 1.00$ (graph paper will be helpful here). Illustrate the use of your nomograph with some examples.

3. Use the construction of Figure 8.18 or of Figure 8.19 to evaluate the sums or differences of Problem 1.

4. Show in a sketch how a slide rule for addition would look when used to calculatc

(a) $2.14 + 1.97$ (b) $2.14 - 3.53$.

5. Use Figure 8.21 to evaluate

(a) $(3)(1)$ (b) $(^{-}3)(^{-}1)$
(c) $(^{-}3)(1/2)$ (d) $(^{-}3)(^{-}1/2)$
(e) $(3.6)(1/4)$ (f) $(3.6)(^{-}1/4)$
(g) $(^{-}3.6)(1/4)$ (h) $(^{-}3.6)(^{-}1/4)$.

6. Describe how to use Figure 8.21 for performing divisions. Illustrate your process with several examples.

7. Construct a nomograph for multiplication in the set $\{\ldots, 3^{-3}, 3^{-2}, 3^{-1}, 3^0, 3^1, 3^2, 3^3, \ldots\} = \{\ldots, 1/27, 1/9, 1/3, 1, 3, 9, 27, \ldots\}$. Illustrate its use with several examples.

8. Construct a slide rule for multiplying in

(a) the set T defined in this section
(b) the set definced in Problem 7.

---

9. Is the set of Problem 7 a group with respect to multiplication? Explain.

10. Show that the group of Problem 9 is isomorphic to the group $(I, +)$.

11. Suppose b is a nonzero rational number and we let $B = \{x \mid x = b^n$ for some n in $I\}$.

(a) Show that $(B, \cdot)$ is a group.
(b) Is this group isomorphic to $(I, +)$? (Be careful).

8.16 Some uses of prime numbers in F

In Section 8.2 we discussed the problem of reducing a fraction to lowest terms. Writing natural numbers as products of primes can be very helpful in this regard. If we are given a fraction "p/q" and can write $|p|$ and $|q|$ as products of primes, then it is quite easy to reduce the fraction to lowest terms. For example, we have $527/155 = (17\cdot31)/(5\cdot31) = 17/5$ and the last fraction here is clearly in lowest terms since 17 and 5 are prime.

One practical approach for reducing "p/q" to lowest terms is to use the sequence $p_1, p_2, p_3, \ldots$ of successively larger primes in the procedure of Figure 8.23.

This same procedure would work if we simply used successive elements of the sequence 2, 3, 4, 5, 6, 7, . . . of natural numbers as our trial common divisors. But it would be much less efficient with this sequence.

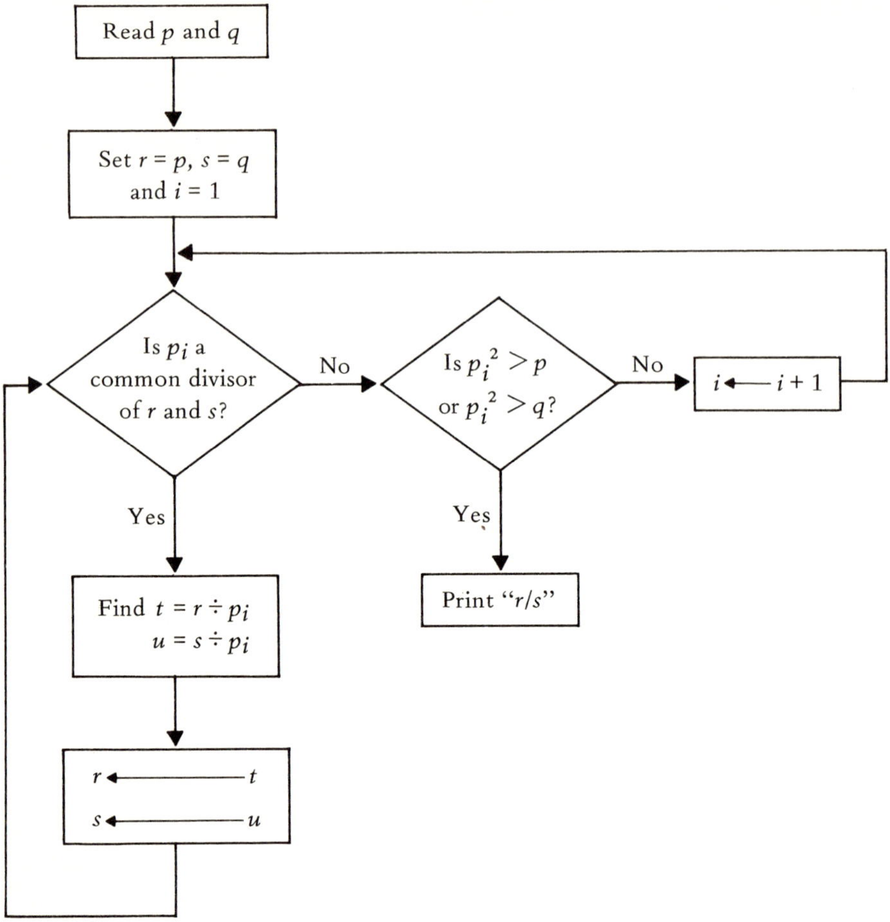

FIGURE 8.23 Flow chart for reducing "p/q" to lowest terms. p_i is the ith prime.

The product of all the prime numbers that are "cancelled out" by the procedure of Figure 8.23 would be the number that is called the **greatest common divisor** of p and q. When the greatest common divisor of two integers is 1, then the integers are said to be **relatively prime.** The problem of expressing "p/q" in lowest terms is the problem of finding another fraction "r/s" where r and s are relatively prime and $p/q = r/s$.

We defined addition of rational numbers by saying that

$$a/b + c/d = (ad + bc)/bd.$$

But when we are faced with an actual problem of such an addition we may be able to save ourselves some labor by using another procedure rather than by applying the definition directly. For example, we may write $5/36 + 1/48 =$

$5/(3 \cdot 12) + 1/(4 \cdot 12) = (4 \cdot 5)/(3 \cdot 4 \cdot 12) + (3 \cdot 1)/(3 \cdot 4 \cdot 12) = 20/144 + 3/144 = 23/144$. This is simpler than writing $5/36 + 1/48 = (48 \cdot 5 + 36 \cdot 1)/(36 \cdot 48) = (240 + 36)/1728 = 276/1728$ and then recognizing that $276/1728 = 23/144$.

We call 144 the **least common multiple** of 36 and 48. It is the smallest number which is a multiple of both 36 and 48.

When faced with the problem of finding a simple fraction to express the sum $a/b + c/d$, we shall be following a quite efficient procedure if we find the least common multiple of b and d, convert "a/b" and "c/d" to equivalent fractions with this denominator, and then perform our addition.

If we can express b and d as products of primes, then the problem of finding their least common multiple is quite easy to solve. For example, since $636 = 2 \cdot 318 = 2^2 \cdot 159 = 2^2 \cdot 3 \cdot 53$ and $360 = 2 \cdot 180 = 2^2 \cdot 90 = 2^3 \cdot 45 = 2^3 \cdot 3 \cdot 15 = 2^3 \cdot 3^2 \cdot 5$, it is easy to see that the least common multiple of 636 and 360 is $2^3 \cdot 3^2 \cdot 5 \cdot 53 = 19{,}080$. In general, we include as a factor in the least common multiple the highest power of each prime that appears as a factor in either of the given numbers.

Using such a method in an addition of rationals we would have $5/636 + 1/360 = 5/(2^2 \cdot 3 \cdot 53) + 1/(2^3 \cdot 3^2 \cdot 5) = (2 \cdot 3 \cdot 5 \cdot 5)/(2^3 \cdot 3^2 \cdot 5 \cdot 53) + (53 \cdot 1)/(2^3 \cdot 3^2 \cdot 5 \cdot 53) = (150 + 53)/(2^3 \cdot 3^2 \cdot 5 \cdot 53) = 203/19{,}080$. This last fraction happens to be in lowest terms since $203 = 7 \cdot 29$ and neither of these last prime factors appears as a factor in the denominator. Of course, the procedure does not always yield a fraction which is in lowest terms.

Expressing natural numbers as products of primes can also be helpful in finding the results of multiplying or dividing fractions. We cancel out common factors in all possible ways and end up with a fraction that is automatically in lowest terms. For example,

$$(144/636) \cdot (53/360) = ((2^4 \cdot 3^2)/(2^2 \cdot 3 \cdot 53)) \cdot (53/(2^3 \cdot 3^2 \cdot 5)) = 1/(2 \cdot 3 \cdot 5) = 1/30 \text{ and}$$

$$(527/1000) \div (155/51) = (527/1000) \cdot (51/155) = ((17 \cdot 31)/(2^3 \cdot 5^3)) \cdot ((17 \cdot 3)/(5 \cdot 31)) = (3 \cdot 17^2)/(2^3 \cdot 5^4) = 867/5000.$$

PROBLEMS

1. Express each of the following as a product of primes:

(a) 48	(b) 1000	(c) 221
(d) 3948	(e) 2401	(f) 833
(g) 1,000,000	(h) 144	(i) 1728
(j) 216	(k) 101	(l) 11,211
	(m) 11,111.	

2. Find the greatest common divisor of each of the following pairs of numbers:

(a) 48, 1000 (b) 1000, 3948
(c) 221, 2401 (d) 3948, 2401
(e) 2401, 833 (f) 144, 1728
(g) 101, 11211 (h) 216, 1728.

3. Reduce each of the following fractions to lowest terms:

(a) 48/1000 (b) 1000/3948
(c) 3948/1000 (d) 2401/3948
(e) 833/2401 (f) 144/1000
(g) 144/1728 (h) 11211/101
(i) 51/1173 (j) 333/11211
(k) 57772/9999.

4. Find the least common multiple of each pair of numbers given in Problem 2.
5. Express each of the following as a fraction in lowest terms:

(a) $3/7 + 4/49$ (b) $2/3 + 1/51$
(c) $3/100 + 7/25$ (d) $5/360 + 7/144$
(e) $5/36 + 14/144$ (f) $72/51 + 13/54$
(g) $537/13 + 1/169$ (h) $(36/7)\cdot(196/144)$
(i) $(101/48)\cdot(221/11211)$ (j) $(1000/48) \div (125/36)$
(k) $(625/1728) \div (4000/196)$.

6. Express each of the following as a fraction in lowest terms:

(a) $1 + 1/2 + 1/3 + 1/4$
(b) $1 + 1/2 + 1/3 + 1/4 + 1/5$
(c) $1/2 + 1/4$
(d) $1/2 + 1/4 + 1/8$
(e) $1/2 + 1/4 + 1/8 + 1/16$
(f) $1/2 + 1/4 + 1/8 + 1/16 + 1/32$
(g) $1 + 1/4 + 1/16$
(h) $1 + 1/4 + 1/16 + 1/64$
(i) $2/3 + (2/3)^2 + (2/3)^3 + (2/3)^4 + (2/3)^5$.

7. Let p and q be relatively prime. Find a necessary and sufficient condition in terms of the prime factors of q that p/q can be expressed as a terminating decimal. [HINT: p/q can be so expressed if and only if there is an integer k and a natural number n such that $p(10)^n = kq$. What does 10^n look like as a product of primes?]

8. Generalize the result in Problem 7 to establish a necessary and sufficient condition on q that p/q can be expressed as a base b fraction when p and q are relatively prime. In particular what is the situation when the base of the numeration system is prime?

*8.17 Fields

We have now studied three different number systems—the natural numbers, the integers, and the rational numbers. In each of these systems, we have basic operations of addition and multiplication. Some fundamental properties of the three systems are summarized in Figure 8.24. We may think of our development of *I*, and then, of *F* as successful attempts to remove some "No"'s from spots in this table.

	ADDITION				
	Closure	Commutative	Associative	Identity element	Each element has an inverse
N	Yes	Yes	Yes	Yes	No
I	Yes	Yes	Yes	Yes	Yes
F	Yes	Yes	Yes	Yes	Yes

	MULTIPLICATION					
	Closure	Commutative	Associative	Identity element	Each non-zero element has an inverse	Multiplication distributes across addition
N	Yes	Yes	Yes	Yes	No	Yes
I	Yes	Yes	Yes	Yes	No	Yes
F	Yes	Yes	Yes	Yes	Yes	Yes

FIGURE 8.24

In view of Figure 8.24, it is natural to ask if there are any other systems that have the same basic properties as our system of rational numbers. It is fruitful to make the following definition:

If S is a set and $\boxplus$ and $\boxdot$ are operations in S, we say that the system $(S, \boxplus, \boxdot)$ is a *field* if (i) S is closed under $\boxplus$ and $\boxdot$, (ii) $\boxplus$ and $\boxdot$ are commutative and associative, (iii) there is an identity element for $\boxplus$ in S (let us call it 0) and there is an identity element (let us call it 1) for $\boxdot$ in S, (iv) each element in S has an inverse element with respect to $\boxplus$ and each element different from 0 in S has an inverse element with respect to $\boxdot$, and (v) $\boxdot$ distributes across $\boxplus$ in S. The properties (i), (ii), (iii), (iv), and (v) are called **field axioms**.

Consider the system $(A, \underset{+}{\sim}, \underset{\cdot}{\sim})$ where $A = \{0, 1\}$ and $\underset{+}{\sim}$ and $\underset{\cdot}{\sim}$ are defined by the tables on page 90. That is, $\underset{+}{\sim}$ and $\underset{\cdot}{\sim}$ are defined by

$\underset{+}{\sim}$	0	1
0	0	1
1	1	0

$\underset{\cdot}{\sim}$	0	1
0	0	0
1	0	1

It is a straightforward matter to check that (i) A is closed under $\underset{+}{\sim}$ and $\underset{\cdot}{\sim}$, (ii) $\underset{+}{\sim}$ and $\underset{\cdot}{\sim}$ are commutative and associative in A, (iii) 0 is the identity element for $\underset{+}{\sim}$ and 1 is the identity element for $\underset{\cdot}{\sim}$, (iv) each element in A has an inverse element with respect to $\underset{+}{\sim}$ (in fact, $^-0 = 0$ and $^-1 = 1$) and each nonzero element in A has an inverse element with respect to $\underset{\cdot}{\sim}$ $(1^{-1} = 1)$, and finally, (v) $\underset{\cdot}{\sim}$ distributes across $\underset{+}{\sim}$ in A. Hence, $(A, \underset{+}{\sim}, \underset{\cdot}{\sim})$ is a field.

On the other hand, $(N, +, \cdot)$ and $(I, +, \cdot)$ are not fields because in neither do we, in general, have multiplicative inverses. Of course, $(N, +, \cdot)$ also fails to be a field because it has, in general, no additive inverses.

It is not difficult to check that the system $(\{0, 1, 2\}, \oplus, \odot)$, where $\oplus$ and $\odot$ are defined by the tables on page 90, is a field.

But the system $(\{0, 1, 2, 3\}, \overset{+}{\sim}, \overset{\cdot}{\sim})$, where $\overset{+}{\sim}$ and $\overset{\cdot}{\sim}$ are defined by the tables in Problem 2, Section 5.1, fails to be a field. In this system, 2 has no multiplicative inverse.

The three systems with finite sets that we considered as examples above are isomorphic to certain systems of modular arithmetic as discussed in Section 7.11. A general result about such systems is that the arithmetic of integers modulo n is a field if and only if n is a prime. We shall not prove this theorem, but we note that it is consistent with our discoveries about the above examples.

Why study such an abstract idea as that of a field? One of the main reasons is one of efficiency. If we can discover some general theorems that follow from the field axioms, we know that we can use these theorems in any particular case of a field. Since there are many useful systems that *are* fields, we find that theorems about fields apply in many different cases and we need not start from scratch each time we deal with a system that happens to be a field.

Let us look at some theorems that follow from the field axioms. Suppose $(S, \boxplus, \boxdot)$ is a field. Let $-x$ be the additive inverse for x and let x^{-1} be the multiplicative inverse for x.

(a) If x and y are in S and $x \boxplus y = x$, then $y = 0$. We may prove this as follows. $x \boxplus y = x$ implies $-x \boxplus (x \boxplus y) = -x \boxplus x$ since x has an additive inverse and S is closed under addition. But then we have $(-x \boxplus x) \boxplus y = 0$, and $0 \boxplus y = 0$, and $y = 0$ by the associativity of addition, the definition of $-x$, and the fact that 0 is the additive identity.

(b) $x \boxdot 0 = 0$ for all x in S. We know that $x \boxplus 0 = x$. Since S is closed under $\boxdot$ and $\boxdot$ distributes across $\boxplus$, this yields $x \boxdot (x \boxplus 0) = x \boxdot x$ and $(x \boxdot x) \boxplus (x \boxdot 0) = x \boxdot x$. By (a), this gives $x \boxdot 0 = 0$.

(c) If a and b are in S and $a \neq 0$, then the equation $(a \boxdot x) \boxplus b = 0$ has the unique solution $x = -b \boxdot a^{-1}$ in S.

We can see that $-b \boxdot a^{-1}$ does indeed satisfy the equation since $[a \boxdot (-b \boxdot a^{-1})] \boxplus b = [(-b \boxdot a^{-1}) \boxdot a] \boxplus b = [-b \boxdot (a^{-1} \boxdot a)] \boxplus b = [-b \boxdot 1] \boxplus b = -b \boxplus b = 0$.
Also, when $a \neq 0$, we see that $(a \boxdot x) \boxplus b = 0$ implies $[(a \boxdot x) \boxplus b] \boxplus (-b) = 0 \boxplus -b$. And this yields $(a \boxdot x) \boxplus (b \boxplus (-b)) = -b$ and then $(a \boxdot x) \boxplus 0 = -b$ and, hence, $a \boxdot x = -b$. Therefore, $(a \boxdot x) \boxdot a^{-1} = -b \boxdot a^{-1}$. This gives $(x \boxdot a) \boxdot a^{-1} = -b \boxdot a^{-1}$, and $x \boxdot (a \boxdot a^{-1}) = -b \boxdot a^{-1}$, and, finally $x \boxdot 1 = x = -b \boxdot a^{-1}$. This shows that $-b \boxdot a^{-1}$ is the unique solution to the equation.

PROBLEMS

1. Which of the following systems are fields?

   (a) $(\{\phi, \{1\}, \{2\}, \{1,2\}\}, \cup, \cap)$
   (b) $(\{[0], [1], [2], [3], [4]\}, +_5, \cdot_5)$ (see Section 7.11)
   (c) $(\{a, b, c, d\}, M, L)$ as defined by the tables in Problem 8, Section 5.2
   (d) $(\{\text{odd}, \text{even}\}, \boxplus, \boxdot)$ as defined by the tables in Problem 8, Section 5.2.

2. Prove that in any field $(S, \boxplus, \boxdot)$

   (a) $-x = (^-1) \boxdot x$
   (b) $-(x \boxplus y) = (-x) \boxplus (-y)$
   (c) $-(-x) = x$
   (d) $(x \boxdot y)^{-1} = x^{-1} \boxdot y^{-1}$ if $x \neq 0$ and $y \neq 0$
   (e) $(x^{-1})^{-1} = x$ if $x \neq 0$
   (f) $x \boxdot y = 0$ implies $x = 0$ or $y = 0$. [HINT: If $y \neq 0$, then y^{-1} is in S.]

3. If $A = \{0\}$ and $0 + 0 = 0$ and $0 \cdot 0 = 0$, is $(A, +, \cdot)$ a field?
4. Show that if $(S, \boxplus, \boxdot)$ is a field and S contains at least two elements then $0 \neq 1$.
5. Let $(S, \boxplus, \boxdot)$ be a system. Show that this system is a field if and only if

   (a) S is a commutative group with respect to $\boxplus$
   (b) the elements different from 0 in S form a commutative group with respect to $\boxdot$, and

(c) $\boxed{\cdot}$ distributes across $\boxed{+}$ in S. (Here 0 is the identity element of the group $(S, \boxed{+})$.)

6. Show that if $k \neq 1$ and the system $(\{[0], [1], [2], \ldots, [k-1],\} +_k, \cdot_k)$ of the arithmetic of integers mod k is a field, then k is a prime. [HINT: Use the result in Problem 2(f).]

7. Suppose that i, j, and k are positive integers less than p where p is a prime. Prove that:

(a) If $ij \equiv ik \pmod{p}$, then $j = k$.
(b) The equivalence classes $[i], [2i], [3i], \ldots, [(p-1)i]$ are all distinct and different from $[0]$. (See Section 7.11.)
(c) Each equivalence class has a multiplicative inverse with respect to $\cdot_p$.

8. Use Problem 7 to show that when k is prime, then $(\{[0], [1], \ldots, [k-1]\},$ $+_k, \cdot_k)$ is a field.

9

Sequences

9.1 Null sequences

About 2500 years ago a Greek by the name of Zeno put forth a series of arguments that seem to show that, contrary to our everyday experience, motion is actually impossible. These arguments have come to be known as Zeno's paradoxes. One of the simplest of these arguments can be put as follows. Suppose a frog is at one end of a log and wishes to move to the other. Whether the frog moves in a smooth crawl or in more natural hops, it is clear that before the frog can get from one end of the log to the other he must pass over the midpoint of the log. And if he reaches this midpoint he must, before he reaches the end he is heading towards, pass over the point halfway between the midpoint and that end. But if he gets to that point he must still pass over the point halfway between there and the end. And so forth without end. Clearly, no mere frog can do all of these infinitely many things, and, hence, he can never get to the other end of the log. It follows that all motion must be some sort of illusion.

If we imagine that the log is one unit long on a number line, we can represent the frog's situation by the picture in Figure 9.1 where we have shown some of the halfway points.

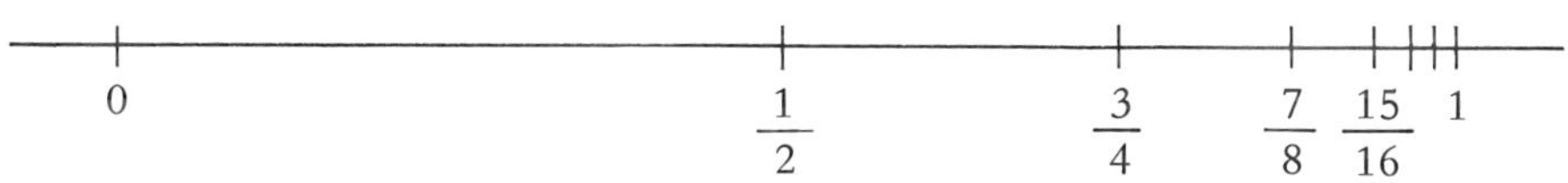

FIGURE 9.1

The coordinate of the first halfway point is 1/2. That of the second is 3/4. That of the third is 7/8. Et cetera. We have here what is called a **sequence.**† We evidently have a halfway point corresponding to each positive integer. We have the *first* halfway point, 1/2, then the *second* halfway point, 3/4, then the *third*, 7/8, etc. It is this sort of thing we mean when we say we have a **sequence.** More precisely, we have the definition:

A **sequence** is a function whose domain is I^+, the set of positive integers.

Sequences arise in many different connections aside from Zeno's paradox. We now shall elaborate on the idea of a sequence and return to Zeno's argument in the next section.

If f is a sequence we shall call $f(1)$ the *first term* in the sequence, $f(2)$ the *second term* in the sequence, $f(3)$ the *third term* in the sequence, etc. It is customary to use subscripts instead of functional notation for the terms in a sequence. Thus, we might let $f(n) = a_n$ and have a_n for the nth term in the sequence for $n = 1, 2, 3, \ldots$. It is also customary to display a sequence by giving the first few terms of it in the proper order. For example, the sequence of coordinates of halfway points in Figure 9.1 might be displayed as

$$1/2, 3/4, 7/8, 15/16, \ldots.$$

When a sequence is displayed this way it is assumed that the pattern is so obvious that we can figure out what the successive terms in the sequence are. It would be more precise to give, along with the first few terms in the sequence, a general formula that may be used to compute the nth term of the sequence for any positive integer n. In the following examples of sequences, we have not given a general formula for the nth term since we assume that in each case the procedure for generating successive terms is obvious.

(a) .3, .33, .333, .3333, . . .
(b) 11, 11, 11, 11, 11, . . .
(c) $1, -1, 1, -1, 1, -1, \ldots$
(d) .1, .01, .001, .0001, . . .
(e) 1, 1/2, 1/3, 1/4, 1/5, . . .
(f) 1, 1/2, 1/4, 1/8, 1/16, . . .
(g) 3, 5, 7, 9, 11, . . .
(h) 4, 14, 24, 34, 44, . . .
(i) 45, 40, 35, 30, . . .
(j) 2, 4, 8, 16, 32, . . .
(k) 1/2, 2/3, 3/4, 4/5, 5/6, . . .
(l) $1, -1/3, 1/9, -1/27, 1/81, -1/243, \ldots$
(m) 60, 60/4, 60/16, 60/64, . . .

† Or an **infinite sequence.** But we shall usually omit the adjective "infinite."

(n) 1, 1/4, 1/9, 1/16, 1/25, . . .
(o) 10, 10^2, 10^3, 10^4, . . .
(p) 1/3 − .3, 1/3 − .33, 1/3 − .333, 1/3 − .3333, . . .
(q) 1/4 − .24, 1/4 − .249, 1/4 − .2499, 1/4 − .24999, . . .
(r) .9, .99, .999, .9999, . . .
(s) .1, −.1, .01, −.01, .001, −.001, .0001, −.0001, . . .
(t) $10^6/1$, $10^6/2$, $10^6/3$, $10^6/4$, . . .
(u) −1/3, −1/6, −1/9, −1/12, −1/15, . . .
(v) .1, .11, .111, .1111, . . .
(w) $(497^{674})/2$, $(497^{674})/4$, $(497^{674})/8$, $(497^{674})/16$, . . .
(x) 1/2, 1/2 + 1/4, 1/2 + 1/4 + 1/8, 1/2 + 1/4 + 1/8 + 1/16, . . .

Giving sequences in this manner is equivalent to giving the first few entries in a table displaying the values of the function (i.e., terms of the sequence). For example, we may think of (a) above as a condensation of the table

x	1	2	3	4	. . .
$f(x)$	.3	.33	.333	.3333	. . .

.

We shall usually write "$(y_1, y_2, y_3, \ldots)$" to designate the sequence whose terms are y_1, y_2, y_3, etc.

It will be convenient if we think of the sequence labeled (a) above as having the terms $a_1, a_2, a_3, a_4, \ldots$. Similarly, the terms in (b) we shall call $b_1, b_2, b_3, \ldots$. Et cetera.

It would be natural to call the sequence (b) a **constant** sequence since evidently each term in (11, 11, 11, . . .) is equal to the same number.

There is a common feature shared by the sequences (d), (e), and (f) above. This common aspect has to do with the long-run behaviour of the sequences. What happens in (.1, .01, .001, . . .), (1, 1/2, 1/3, . . .), and (1, 1/2, 1/4, 1/8, . . .) as we go further and further out? Evidently, in each case the terms get smaller and smaller and, in fact, get very close to 0. Such sequences are said to be **null** sequences. Somewhat more precisely we have:

> A **null** sequence is one in which the terms approach 0 as we go further out in the sequence.

Evidently (n) is also a null sequence. However, (a), (b), and (c) are clearly not null sequences.

(s) is a null sequence. While it cannot be said of (s) that the terms get smaller each time you move to the next term, it can be said that the terms approach 0 as you go further out in the sequence.

Evidently what is meant by saying that the terms in a sequence approach 0 is that the absolute values of the terms get small as you move out in the sequence. But we really mean more than that they simply get small. We do not want the constant sequence $(10^{-27}, 10^{-27}, 10^{-27}, \ldots)$ to be a null sequence even though the terms in the sequence get (are) small. When we say the terms approach 0 we mean that the absolute values of the terms get *arbitrarily* small far out in the sequence. But to say that these absolute values get arbitrarily small is to say that by going out far enough in the sequence we can make the absolute values of the terms smaller than any preassigned positive number. It is such an analysis of what it means for the terms to approach 0 that leads us to the precise definition of what it means for a sequence of rational numbers to be a null sequence:

> If $(z_1, z_2, z_3, \ldots)$ is a sequence of rational numbers, we say that it is a **null** sequence if, given any positive d, there is an N such that $|z_n| < d$ whenever $n > N$.

This may be restated as follows: $(z_1, z_2, z_3, \ldots)$ is a null sequence if $d > 0$ implies that there is an N such that $n > N$ implies $|z_n| < d$.

In other words, $(z_1, z_2, z_3, \ldots)$ is a null sequence when, no matter how small a positive number d we are given, we can find an N such that each term z_n with $n > N$ has absolute value less than d.

The fact that this definition is quite complicated reflects the fact that we are dealing here with an infinite set (the set of all the terms in the sequence) and so we must be very careful to be sure we know what we are talking about. Zeno's paradox itself points to the fact that great care must be exercised if we are involved with arguments about infinite sets.

To say that $(z_1, z_2, z_3, \ldots)$ is a null sequence is to say that there is a machine (or process) G that behaves as follows: Given *any* positive number d, no matter how small, the machine G (see Figure 9.2) will give us a number N which has the property that $n > N$ implies $|z_n| < d$. If we change d, the value of N we get from G will usually change. In general, we would expect that as we make d smaller, the value of N that G gives us will increase. We expect that we shall need to go further out in the sequence if we wish to insure that $|z_n|$ is smaller.

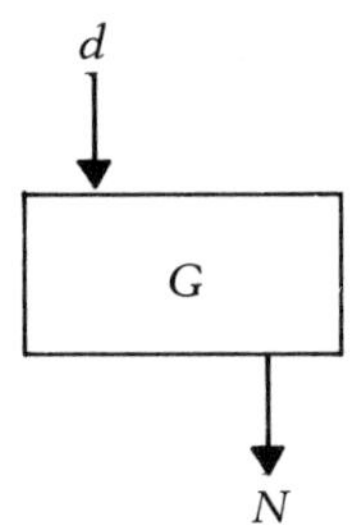

FIGURE 9.2

To prove that $(z_1, z_2, z_3, \ldots)$ is a null sequence we must in effect tell how to construct a machine G which will do the above job for that sequence. This point of view again indicates why the study of infinite sequences is somewhat subtle and complicated. To assert that a sequence is null is to assert the existence of a process or function or machine that has certain properties.

Let us illustrate the use of the precise definition of a null sequence by proving the intuitively obvious fact that $(e_1, e_2, e_3, e_4, \ldots) = (1, 1/2, 1/3, 1/4, \ldots)$ is a null sequence. A proof can be given as follows. Let d be any positive number. We wish to show that by taking n large enough we can make $|e_n| = |1/n| < d$. But $|1/n| = 1/n$ and since n and d are both positive we also know that $1/n < d$ if and only if $n > 1/d$. Let us, then, take $N = 1/d$. Then we have that $n > N$ implies $n > 1/d$. This, in turn, implies $1/n < d$ or $|1/n| = |e_n| < d$. We conclude that $(1, 1/2, 1/3, 1/4, \ldots)$ is a null sequence since we have shown that given any positive d, $n > N$ where $N = 1/d$ implies $|1/n| < d$.

Putting this another way, we have shown that the machine G in Figure 9.3 has the property that when it is given any positive d, it produces an N with the property that $n > N$ implies $|e_n| < d$.

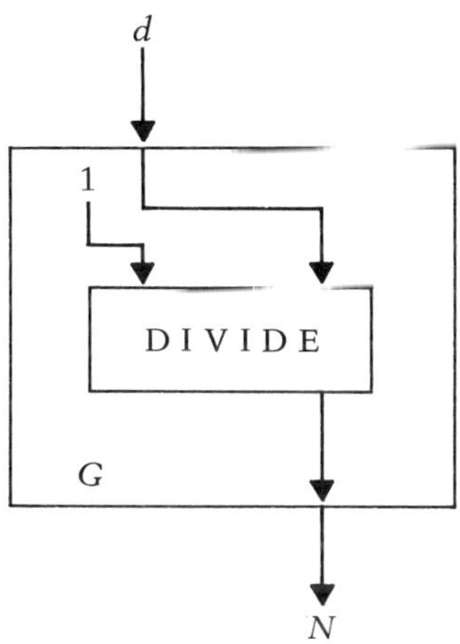

FIGURE 9.3

It is easy to use the precise definition of a null sequence and the fact that $|z_n| = |-z_n| = ||z_n||$ to show that if $(z_1, z_2, z_3, \ldots)$ is a null sequence then $(-z_1, -z_2, -z_3, \ldots)$ and $(|z_1|, |z_2|, |z_3|, \ldots)$ are also null sequences.

In fact, suppose that a machine G as in Figure 9.2 has the property that, given any positive d, it puts out an N such that $n > N$ implies $|z_n| < d$. This same machine will serve for the other two sequences since when $n > N$, we have $|-z_n| = ||z_n|| = |z_n| < d$.

Multiplying each term of a null sequence by a constant k yields a sequence that must be null. For surely if z_n approaches 0 as n gets large so does kz_n. (A more detailed argument is given at the end of this section.)

Suppose we are given the two sequences $(z_1, z_2, z_3, \ldots)$ and $(w_1, w_2, w_3, \ldots)$.

There are several sequences we can build in a natural way out of them. We might consider, for example, the sequences $(z_1 + w_1, z_2 + w_2, z_3 + w_3, \ldots)$, $(z_1 - w_1, z_2 - w_2, z_3 - w_3, \ldots)$, and $(z_1w_1, z_2w_2, z_3w_3, \ldots)$. It is not at all difficult to believe that these three sequences must be null sequences if $(z_1, z_2, z_3, \ldots)$ and $(w_1, w_2, w_3, \ldots)$ are null sequences. And this fact may be proved rigorously from the precise definition.

Suppose B is a number and $(z_1, z_2, z_3, \ldots)$ has the property that $|z_n| < B$ for all n. Then we say that $(z_1, z_2, z_3, \ldots)$ is a **bounded** sequence.

If $(z_1, z_2, z_3, \ldots)$ is a bounded sequence and $(w_1, w_2, w_3, \ldots)$ is a null sequence, then surely the sequence $(z_1w_1, z_2w_2, z_3w_3, \ldots)$ must be a null sequence. For the terms in $(w_1, w_2, w_3, \ldots)$ approach 0 and no $|z_n|$ is larger than B.

Definition 9.1.1. A **sequence** is a function whose domain is I^+. If f is a sequence we call $f(n)$ the nth **term** of the sequence. We shall usually use subscripts to indicate terms in sequences. If $z_n = f(n)$ for n in I^+, then we may denote the sequence f by $(z_1, z_2, z_3, \ldots)$ or $(z_1, z_2, \ldots, z_n, \ldots)$.

Definition 9.1.2. $(z_1, z_2, z_3, \ldots)$ is a **constant sequence** if for some number b, $z_n = b$ for all n in I^+.

Definition 9.1.3. $(z_1, z_2, z_3, \ldots)$ is a **null sequence** if, given any positive d, there is an N such that $n > N$ implies $|z_n| < d$.

Theorem 9.1.1. $(1, 1/2, 1/3, 1/4, \ldots)$ is a null sequence.

Theorem 9.1.2. If $(z_1, z_2, z_3, \ldots)$ is a null sequence, then $(-z_1, -z_2, -z_3, \ldots)$ and $(|z_1|, |z_2|, |z_3|, \ldots)$ are null sequences.

Theorem 9.1.3. If $(z_1, z_2, z_3, \ldots)$ is a null sequence, so is $(kz_1, kz_2, kz_3, \ldots)$ where k is any constant.

PROOF: If $k = 0$, then the result is obvious. Suppose $k \neq 0$. Let d be a given positive number. If $(z_1, z_2, z_3, \ldots)$ is a null sequence, there is an N such that $n > N$ implies $|z_n| < d/|k|$. Then $n > N$ implies $|k||z_n| = |kz_n| < d$. Hence, $(kz_1, kz_2, kz_3, \ldots)$ is a null sequence.

We may restate our argument for the case $k \neq 0$ as follows: Suppose G is a machine which, given any positive d, puts out an N such that $n > N$ implies $|z_n| < d$. We construct a machine H as in Figure 9.4. We claim that given any positive d, H puts out an N such that $n > N$ implies $|kz_n| < d$. This follows from the fact that $n > N$ implies $|z_n| < d/|k|$ as can be seen from the way H is designed.

Theorem 9.1.4. If $(z_1, z_2, z_3, \ldots)$ and $(w_1, w_2, w_3, \ldots)$ are null sequences, so are $(z_1 + w_1, z_2 + w_2, z_3 + w_3, \ldots)$, $(z_1 - w_1, z_2 - w_2, z_3 - w_3, \ldots)$, and $(z_1w_1, z_2w_2, z_3w_3, \ldots)$.

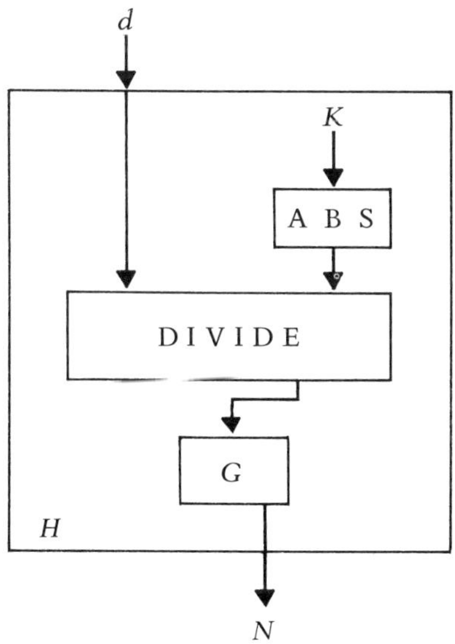

FIGURE 9.4

PROOF: Let a positive d be given. Find N_1 and N_2 such that $n > N_1$ implies $|z_n| < d/2$ and $n > N_2$ implies $|w_n| < d/2$. Let $N_3 = \max\{N_1, N_2\}$. Then $n > N_3$ implies $|z_n + w_n| \leq |z_n| + |w_n| < d/2 + d/2 = d$. Hence, $(z_1 + w_1, z_2 + w_2, z_3 + w_3, \ldots)$ is a null sequence. Also $n > N_3$ implies $|z_n - w_n| = |z_n + (-w_n)| \leq |z_n| + |-w_n| = |z_n| + |w_n| < d/2 + d/2 = d$. Hence, $(z_1 - w_1, z_2 - w_2, z_3 - w_3, \ldots)$ is a null sequence. Find N_4 such that $n > N_4$ implies $|z_n| < 1$. Let $N_5 = \max\{N_2, N_4\}$. Then $n > N_5$ implies $|z_n w_n| = |z_n||w_n| < |w_n| < d/2 < d$. Hence, $(z_1w_1, z_2w_2, z_3w_3, \ldots)$ is a null sequence.

Definition 9.1.4. $(z_1, z_2, z_3, \ldots)$ is said to be a **bounded** sequence if there is a number B such that $|z_n| < B$ for all n.

Theorem 9.1.5. Suppose $(z_1, z_2, z_3, \ldots)$ is a bounded sequence and $(w_1, w_2, w_3, \ldots)$ is a null sequence. Then $(z_1w_1, z_2w_2, z_3w_3, \ldots)$ is a null sequence.

PROOF: Suppose $|z_n| < B$ for all n and $(w_1, w_2, w_3, \ldots)$ is a null sequence. Let d be a positive number. Find an N such that $n > N$ implies $|w_n| < d/B$. Then $n > N$ implies $|z_n w_n| = |z_n||w_n| < B|w_n| < B(d/B) = d$ and, hence, $(z_1w_1, z_2w_2, z_3w_3, \ldots)$ is a null sequence.

PROBLEMS

1. Find the next two terms for each of sequences (a) through (x).
2. Describe or find an expression for the nth term for each sequence (a) through (x).

3. Which sequences from (a) through (x) are null sequences? (No proofs required.)

4. Suppose $(z_1, z_2, z_3, \ldots)$ is a sequence with the property that for some number b, $z_{n+1} = z_n + b$ for all n in I^+. Then the sequence is said to be an **arithmetic progression.** b is said to be the **common difference** for the sequence. Which of the sequences (a) through (x) are arithmetic progressions?

5. Use the precise definition of a null sequence to prove that $(0, 0, 0, \ldots)$ is a null sequence.

6. Which of the following conditions insure that $(z_1, z_2, z_3, \ldots)$ is a null sequence? Which insure that $(z_1, z_2, z_3, \ldots)$ is *not* a null sequence?

(a) $|z_n| < 2$ for all n.
(b) $|z_n| > 2$ for all n.
(c) $z_{10,000} = 0$.
(d) $|z_n| < .0000001$ for $n > 5$.
(e) $|z_n| > .0000001$ for $n > 5$.
(f) $|z_n| < 1/n$ for all n.
(g) $z_{5n} > 1$ for all n.
(h) $|z_{2n}| < 1/n$ for all n.
(i) $z_{2n} = 0$ for all n.
(j) $z_n = 0$ for all n such that $n > 10^6$.
(k) For some positive d and every N there is an n such that $n > N$ and $z_n > d$.
(l) $(z_1, z_2, z_3, \ldots)$ is an arithmetic progression.

7. Find sequences $(z_1, z_2, z_3, \ldots)$ and $(w_1, w_2, w_3, \ldots)$ which are not null sequences and yet (a) $(z_1 + w_1, z_2 + w_2, z_3 + w_3, \ldots)$ is a null sequence, (b) $(z_1w_1, z_2w_2, z_3w_3, \ldots)$ is a null sequence.

8. Find sequences $(z_1, z_2, z_3, \ldots)$ and $(w_1, w_2, w_3, \ldots)$ which are both null sequences and $(z_1/w_1, z_2/w_2, z_3/w_3, \ldots)$ is (a) a null sequence, (b) a sequence which is not null.

---

9. Discuss the proof of Theorem 9.1.4 in terms of machines which for the various sequences tell us how far out we must go to insure that the terms are sufficiently small. (See Figure 9.3 and Figure 9.4.)

10. Prove that any null sequence is bounded.

9.2 Convergent sequences

In Zeno's paradox as discussed in the previous section, the sequence whose nth term was the distance covered by the frog when he had reached the nth halfway point was discovered to be $(1/2, 3/4, 7/8, 15/16, \ldots)$. This sequence is not a null sequence. And yet the terms do approach something as we go further and further out in the sequence. In fact, if we look at how much each of these terms differs from 1 we get the sequence $(1/2 - 1, 3/4 - 1, 7/8 - 1, 15/16 - 1, \ldots) = (-1/2, -1/4, -1/8, -1/16, \ldots)$. Indeed, this last sequence has as its nth term the negative of the distance the frog must travel from the nth halfway point to reach the end of the log. This sequence *is* a null sequence. Since the difference between 1 and the nth term in the first sequence goes towards 0, it would be natural to say that the terms of the first sequence approach 1 as we go further out in the sequence. (See Figure 9.1.) We shall say that the first sequence **converges** to 1. More generally, we have:

> The sequence $(z_1, z_2, z_3, \ldots)$ is said to **converge** to L if $(z_1 - L, z_2 - L, z_3 - L, \ldots)$ is a null sequence. If $(z_1, z_2, z_3, \ldots)$ converges to L then we say that the sequence is **convergent**, call L the **limit** of the sequence, and write $\lim_{n\to\infty} z_n = L$. We read "$\lim_{n\to\infty} z_n$" as "the limit of z_n as n goes to infinity." If a sequence is not convergent then we say that it is **divergent** and that it has no limit.

For example, $(11, 11, 11, 11, \ldots)$ converges and has the limit 11. $(1, 1/2, 1/3, 1/4, \ldots)$ converges and has the limit 0. $(1/2, 3/4, 7/8, 15/16, \ldots)$ converges and has the limit 1. On the other hand, $(1, 2, 3, 4, \ldots)$ and $(1, -1, 1, -1, 1, -1, \ldots)$ both diverge and have no limit.

We may put these examples in the language of limits as follows: $\lim_{n\to\infty} 11 = 11$, $\lim_{n\to\infty} 1/n = 0$, $\lim_{n\to\infty} (1 - 1/2^n) = 1$, $\lim_{n\to\infty} n$ does not exist, and $\lim_{n\to\infty} (-1)^{n+1}$ does not exist.

The idea here is that $(z_1, z_2, z_3, \ldots)$ converges to L if the terms in the sequence approach L as we go further out. If the sequence diverges, it never settles down in such a way that the terms approach any fixed number.

Evidently a sequence is a null sequence if and only if it converges to 0.

Notice that the symbol "∞" which we read as "infinity" in reading "$\lim_{n\to\infty} z_n$" has, as yet, no meaning by itself. "∞" only makes sense in the context of describing the limit of a sequence. "∞" should not, of course, be treated as if it were somehow the name of a number. It is not.

We may now take care of one aspect of Zeno's paradox. What is the apparent force of Zeno's argument? It seems to have something to do with an assumption that it must take an infinite amount of time to cover infinitely many pieces of the log which have nonzero length. But does such an assumption hold up? Suppose the frog naively travels at the rate of one yard per minute and the log is one yard long. The amount of time he needs to reach the first halfway point is 1/2 minute. To reach the second halfway point it takes him 3/4 of a minute. . . . To reach the nth halfway point it takes him $1 - 1/2^n$ minutes. It is clear that he can reach any particular one of the halfway points in less than a minute. And it takes him a total of $\lim_{n\to\infty} (1 - 1/2^n) = 1$ minute to traverse the entire log. We shall turn to another way of looking at this in Section 9.5.

If Zeno's paradox is meant to be an argument that the frog will take an infinite amount of time to move from one end of the log to the other (as implied by asserting that he will *never* reach the other end), then it does not seem to have much force.

Suppose $(z_1, z_2, z_3, \ldots)$ and $(w_1, w_2, w_3, \ldots)$ are given sequences. As we commented once before, there are several sequences we can construct out of these two. $(z_1 + w_1, z_2 + w_2, z_3 + w_3, \ldots)$, $(z_1 - w_1, z_2 - w_2, z_3 - w_3, \ldots)$, $(kz_1, kz_2, kz_3, \ldots)$, $(z_1w_1, z_2w_2, z_3w_3, \ldots)$, and $(z_1/w_1, z_2/w_2, z_3/w_3, \ldots)$ are such. It is natural to ask what the convergence of $(z_1, z_2, \ldots)$ and $(w_1, w_2, \ldots)$ will imply about the convergence of the others. If we think of $\lim_{n\to\infty} z_n = L$ and $\lim_{n\to\infty} w_n = M$ as asserting that z_n gets close to L far out in the sequence and w_n gets close to M far out in the sequence, our intuition leads us to the correct conjecture about the limits of the other sequences we constructed. We shall have $\lim_{n\to\infty} (z_n + w_n) = L + M$, $\lim_{n\to\infty} (z_n - w_n) = L - M$, $\lim_{n\to\infty} (kz_n) = kL$, $\lim_{n\to\infty} (z_nw_n) = LM$, and $\lim_{n\to\infty} (z_n/w_n) = L/M$ where in the last case we assume w_n is never 0 and $M \neq 0$. These results are sometimes put briefly (and somewhat imprecisely) as:

(i) The limit of a sum is the sum of the limits.
(ii) The limit of a difference is the difference of the limits.
(iii) A constant may be taken outside the limit sign.
(iv) The limit of a product is the product of the limits.
(v) The limit of a quotient is the quotient of the limits provided the limit in the denominator is different from 0.

Definition 9.2.1. The sequence $(z_1, z_2, z_3, \ldots)$ is said to **converge** to L if $(z_1 - L, z_2 - L, z_3 - L, \ldots)$ is a null sequence. If $(z_1, z_2, z_3, \ldots)$ converges to L we say that it is **convergent** and has the **limit** L, and we write $\lim_{n\to\infty} z_n = L$. If a sequence is not convergent, then we say it is **divergent** and has no limit.

Theorem 9.2.1. The limit of a constant sequence is the constant.

Theorem 9.2.2. A sequence is a null sequence if and only if it converges to zero.

PROOF: $z_n - 0 = z_n$ for all n.

Theorem 9.2.3. If $\lim_{n\to\infty} z_n = L$ and $\lim_{n\to\infty} w_n = M$ then

(a) $\lim_{n\to\infty} (z_n + w_n) = L + M$

(b) $\lim_{n\to\infty} (z_n - w_n) = L - M$

(c) $\lim_{n\to\infty} kz_n = kL$

(d) $\lim_{n\to\infty} z_n w_n = LM$

(e) $\lim_{n\to\infty} z_n/w_n = L/M$ if w_n is never 0 and $M \neq 0$.

PROOF: By hypothesis $(z_1 - L, z_2 - L, z_3 - L, \ldots)$ and $(w_1 - M, w_2 - M, w_3 - M, \ldots)$ are null sequences. Hence, by Theorem 9.1.4, $(z_1 + w_1 - (L + M), z_2 + w_2 - (L + M), z_3 + w_3 - (L + M), \ldots)$ and $(z_1 - w_1 - (L - M), z_2 - w_2 - (L - M), z_3 - w_3 - (L - M), \ldots)$ are null sequences. This proves (a) and (b).

By Theorem 9.1.3, $(kz_1 - kL, kz_2 - kL, kz_3 - kL, \ldots)$ is a null sequence. This proves (c).

Let $y_n = (z_n - L)(w_n - M)$. $(y_1, y_2, y_3, \ldots)$ is a null sequence by Theorem 9.1.4. We have $y_n = z_n w_n - Lw_n - Mz_n + LM$. Using the rather standard trick of adding and subtracting LM, we then get $y_n = z_n w_n - LM + LM - Lw_n - Mz_n + LM = z_n w_n - LM + L(M - w_n) - M(z_n - L)$. Therefore, $z_n w_n - LM = y_n + L(w_n - M) + M(z_n - L)$. By Theorem 9.1.3, $(L(w_1 - M), L(w_2 - M), L(w_3 - M), \ldots)$ and $(M(z_1 - L), M(z_2 - L), M(z_3 - L), \ldots)$ are null sequences. Hence, $(z_1 w_1 - LM, z_2 w_2 - LM, z_3 w_3 - LM, \ldots)$ is the term by term sum of three null sequences and must itself be a null sequence by Theorem 9.1.4. This completes the proof of (d).

Suppose $M \neq 0$ and w_n is never 0 and we let $v_n = z_n/w_n - L/M$. We have $v_n = (z_n M - w_n L)/w_n M = (z_n M - LM + LM - w_n L)/w_n M = (M(z_n - L) + L(M - w_n)/w_n M$. Hence, $v_n w_n = (z_n - L) + (L/M)(M - w_n)$. Therefore by Theorem 9.1.2, Theorem 9.1.3, and Theorem 9.1.4, $(v_1 w_1, v_2 w_2, \ldots)$ is a null sequence. Let a positive number d be given. Find N_1 such that $n > N_1$ implies $|w_n - M| < |M|/2$. Note that when $n > N_1$ we have w_n within a distance $|M|/2$ of M on the number line. With $M \neq 0$, this implies $|w_n| > |M|/2$ when $n > N_1$. Find an N_2 such that $n > N_2$ implies $|v_n w_n| < |M|d/2$. Let $N_3 = \max\{N_1, N_2\}$. Then $n > N_3$ implies $(|M|/2)d > |v_n w_n| = |v_n||w_n| > (|M|/2)|v_n|$. Hence, $n > N_3$ implies $|v_n| < d$. Therefore, $(v_1, v_2, v_3, \ldots)$ is a null sequence and the proof of (e) is complete.

PROBLEMS

1. Find the limit (if any) for each sequence (a) through (x) from the last section. (No proofs required. [HINT FOR (v): Consider $.1 - 1/9$, $.11 - 1/9$, $.111 - 1/9, \ldots$.])

2. Find sequences $(z_1, z_2, z_3, \ldots)$ and $(w_1, w_2, w_3, \ldots)$ which are both divergent and yet the following sequence converges:

(a) $(z_1 + w_1, z_2 + w_2, z_3 + w_3, \ldots)$
(b) $(z_1 - w_1, z_2 - w_2, z_3 - w_3, \ldots)$
(c) $(z_1 w_1, z_2 w_2, z_3 w_3, \ldots)$
(d) $(z_1/w_1, z_2/w_2, z_3/w_3, \ldots)$.

3. Which of the following conditions insure that the sequence $(z_1, z_2, z_3, \ldots)$ converges? Which of them insure that $(z_1, z_2, z_3, \ldots)$ diverges?

(a) $z_n = 6$ whenever $n > 5050$.
(b) $z_n < 0$ whenever $n > 36$.
(c) $-10 - 1/n < z_n < -10 + 1/n$ for all n.
(d) $|z_n - 2| < 1/n$ for all n.
(e) $(z_1 - 10, z_2 - 10, z_3 - 10, \ldots)$ is a null sequence.
(f) For every number L there is a positive d such that for every N there is an n such that $n > N$ and $|z_n - L| > d$.
(g) $(z_1 - 4, z_2 - 4, z_3 - 4, \ldots)$ is not a null sequence.
(h) $(z_1 - w_1, z_2 - w_2, z_3 - w_3, \ldots)$ is a null sequence and $(w_1, w_2, w_3, \ldots)$ converges.
(i) $(z_1 - 7, z_2 - 7, z_3 - 7, \ldots)$ is a convergent sequence.
(j) $(z_2, z_4, z_6, z_8, \ldots)$ is a convergent sequence.
(k) $z_n \geq n$ whenever $n > 76$.
(l) $2^n z_n \leq 1$ for all n.
(m) $|2^n z_n| \leq 1$ for all n.
(n) There is a number k such that $(kz_1, kz_2, kz_3, \ldots)$ converges.
(o) There is a number k such that $(z_1 + k, z_2 + k, z_3 + k, \ldots)$ converges.
(p) $(2z_1, 2z_2, 2z_3, \ldots)$ diverges.
(q) $(z_1/1, z_2/2, z_3/3, z_4/4, \ldots)$ converges.

4. Let us say that $(z_1, z_2, z_3, \ldots)\ E\ (w_1, w_2, w_3, \ldots)$ if $\lim_{n\to\infty} z_n = \lim_{n\to\infty} w_n$. (a) Is E an equivalence relation on the set of all sequences of rationals? (b) Is E an equivalence relation on the set of all convergent sequences of rationals?

5. Is it possible to have $(z_1, z_2, z_3, \ldots)$ convergent and $(w_1, w_2, w_3, \ldots)$ divergent while the following is convergent?

(a) $(z_1 + w_1, z_2 + w_2, z_3 + w_3, \ldots)$
(b) $(z_1w_1, z_2w_2, z_3w_3, \ldots)$
(c) $(w_1/z_1, w_2/z_2, w_3/z_3, \ldots)$.

Explain.

9.3 Monotone sequences

A sequence such as (.3, .33, .333, . . .) has the property that the terms get larger and larger as we move out in the sequence. Such sequences are important and a special terminology is used in connection with them.

If $z_{n+1} \geq z_n$ for all n, then $(z_1, z_2, z_3, \ldots)$ is said to be a **monotone increasing** sequence.

In a similar vein we have: If $z_{n+1} \leq z_n$ for all n then $(z_1, z_2, z_3, \ldots)$ is said to be **monotone decreasing** sequence.

For example, (1, 2, 3, . . .), (.1, .11, .111, . . .) and (1/2, 3/4, 7/8, 15/16, . . .)

are all monotone increasing. $(-1, -2, -3, \ldots)$, $(2, 3/2, 4/3, 5/4, 6/5, \ldots)$, and $(1.1, 1.01, 1.001, 1.0001, \ldots)$ are all monotone decreasing.

If a sequence is monotone increasing and we represent the terms in the sequence by points on a number line, then as we move further out in the sequence we never move to the left on the line. A similar remark applies to a monotone decreasing sequence.

A sequence is said to be a **monotone** sequence if it is monotone increasing or monotone decreasing.

As we see from the above examples, some monotone sequences are convergent and some are divergent.

If a sequence is both monotone increasing and convergent, can anything helpful be said about the relationship between the limit and the terms of the sequence? If L is the limit of the sequence then the terms of the sequence approach L as we move out in the sequence. But if the sequence is monotone increasing the terms can never decrease as we move further out in the sequence. Hence, the terms in the sequence must be approaching the limit L from *below*. If we think of the terms of the sequence as coordinates of points on a number line we must always be moving to the right (if we move at all) as we go out in the sequence. The points must be getting closer and closer to the limit point L. And, hence, the points in the sequence must be sneaking up on L *from the left*. We may summarizc this result as the theorem:

> If $(z_1, z_2, z_3, \ldots)$ is monotone increasing and $\lim_{n\to\infty} z_n = L$, then $z_n \leq L$ for all n.

Does the converse of this theorem hold? Evidently not. For example, we have $1 - 1/2^n \leq 2$ for all n and yet $\lim_{n\to\infty} (1 - 1/2^n) = 1 \neq 2$.

If the sequence $(z_1, z_2, z_3, \ldots)$ has the property that $z_{n+1} > z_n$ for all n, then the sequence is said to be a **strictly increasing** sequence. For example, $(1, 2, 3, 4, \ldots)$ is a strictly increasing sequence. **Strictly decreasing** is defined similarly.

Definition 9.3.1. If $z_{n+1} \geq z_n$ for all n, then $(z_1, z_2, \ldots)$ is said to be a **monotone increasing** sequence. If $z_{n+1} \leq z_n$ for all n, then $(z_1, z_2, \ldots)$ is said to be a **monotone decreasing** sequence. A sequence is said to be **monotone** if it is monotone increasing or monotone decreasing.

Theorem 9.3.1. If $(z_1, z_2, \ldots)$ is monotone increasing then $z_{n+k} \geq z_n$ for all n in I^+ and all natural numbers k.

PROOF: See Problem 3.

Theorem 9.3.2. If $(z_1, z_2, \ldots)$ is monotone increasing and $\lim_{n\to\infty} z_n = L$, then $z_n \leq L$ for all n.

PROOF: Suppose $\lim_{n\to\infty} z_n = L$ and p is a positive integer such that $z_p > L$. If $(z_1, z_2, \ldots)$ is monotone increasing, then $z_{p+k} \geq z_p > L$ for all positive integers k. If $z_p - L = d$ then $d > 0$ and $|z_{p+k} - L| \geq d$ for all positive k. But then $(z_1 - L, z_2 - L, z_3 - L, \ldots)$ cannot be a null sequence and we cannot have $\lim_{n\to\infty} z_n = L$. Hence, there must not be a p such that $z_p > L$. Therefore, $z_n \leq L$ for all n.

Definition 9.3.2. If $z_{n+1} > z_n$ for all n then $(z_1, z_2, \ldots)$ is said to be a **strictly increasing** sequence. If $z_{n+1} < z_n$ for all n, then $(z_1, z_2, \ldots)$ is said to be a **strictly decreasing** sequence.

PROBLEMS

1. Which of the sequences (a) through (x) in Section 9.1 is (i) monotone increasing, (ii) monotone decreasing, (iii) strictly increasing, (iv) strictly decreasing?
2. What can be said about a sequence that is both monotone increasing and monotone decreasing?
3. Use induction on k to prove Theorem 9.3.1.
4. Give an example of a sequence $(z_1, z_2, z_3, \ldots)$ that has the following property or else argue that it is impossible to do so.

(a) $z_n \leq 1$ for all n and $\lim_{n\to\infty} z_n = 2$.
(b) $z_n < 1$ for all n and $\lim_{n\to\infty} z_n = 1$.
(c) The sequence is strictly decreasing and divergent.
(d) The sequence is strictly decreasing and convergent.
(e) The sequence is strictly increasing and $\lim_{n\to\infty} z_n = 5$ and $z_4 = 5$.
(f) The sequence is monotone increasing and $\lim_{n\to\infty} z_n = 5$ and $z_4 = 5$.
(g) The sequence is strictly increasing and convergent, and every term in the sequence is a positive integer.
(h) The sequence is monotone increasing and convergent, and every term in the sequence is a positive integer.
(i) The sequence is monotone and yet $(-z_1, -z_2, -z_3, \ldots)$ is not monotone.
(j) The sequence is monotone and yet $(|z_1|, |z_2|, |z_3|, \ldots)$ is not monotone.

9.4 Subsequences

The sequences $(1, 1, 2, 1/2, 3, 1/3, 4, 1/4, 5, 1/5, \ldots)$ and $(2, .1, 4, .01, 8, .001, 16, .0001, \ldots)$ have the peculiar property that they do not converge and yet if we pick out only the even numbered terms in either one of the sequences we get a null sequence. Sequences which we construct by picking out only certain terms of a given sequence are called **subsequences** of the original. More precisely we have:

> If $(i_1, i_2, i_3, \ldots)$ is a strictly increasing sequence of positive integers and $(z_1, z_2, z_3, \ldots)$ is a given sequence, then $(z_{i_1}, z_{i_2}, z_{i_3}, \ldots)$ is said to be a **subsequence** of $(z_1, z_2, z_3, \ldots)$. We shall call $(i_1, i_2, i_3, \ldots)$ the **indexing sequence** for the subsequence.

For example, the indexing sequence $(2, 4, 6, 8, \ldots)$ yields the subsequence $(1, 1/2, 1/3, 1/4, \ldots)$ from the first sequence given in this section and the subsequence $(.1, .01, .001, .0001, \ldots)$ from the second. The indexing sequence $(1, 3, 5, 7, \ldots)$ yields the subsequence $(1, 2, 3, 4, \ldots)$ from the first sequence.

If $(z_1, z_2, z_3, \ldots)$ converges to L we would expect any subsequence of $(z_1, z_2, z_3, \ldots)$ also to converge to L. This is because terms far out in $(z_1, z_2, z_3, \ldots)$ must be close to L, and far out in a subsequence we find only terms which originally were far out in $(z_1, z_2, z_3, \ldots)$.

This result may be put another way which will be helpful to us in the future. Suppose we know that a sequence $(z_1, z_2, z_3, \ldots)$ is convergent but we don't know what its limit is. We may be able to pick out a certain subsequence and find *its* limit. But then we know that this limit must indeed be the limit of the original sequence. In other words, if $(z_1, z_2, z_3, \ldots)$ is convergent and a subsequence of this sequence has limit L, then $\lim_{n\to\infty} z_n = L$.

On the other hand, it is perfectly possible to have a subsequence convergent while the original sequence is divergent. The sequences given at the beginning of this section are examples of such a situation.

However, the sequences given at the beginning of this section do a lot of bouncing around and are certainly not monotone. What if a monotone sequence has a convergent subsequence? Must we then have the original sequence convergent? We must.

We may argue for this result as follows. Suppose $(z_1, z_2, z_3, \ldots)$ is a monotone increasing sequence and $(z_{i_1}, z_{i_2}, z_{i_3}, \ldots)$ is a subsequence such that $\lim_{k\to\infty} z_{i_k} = L$. Evidently the subsequence is monotone increasing since the original is. Then by a result of the preceding section $z_{i_k} \leq L$ for all k. Could there be a z_p such that $z_p > L$? We claim not. We do this because, given any p, there is certainly a k such that $i_k > p$ and if $z_p > L$ we would have $z_{i_k} \geq z_p > L$ which is impossible. Hence, $z_n \leq L$ for all n. We know that z_{i_k} gets close to L if we take k large enough. But we now have that when $n > i_k$, $z_{i_k} \leq z_n \leq L$. So z_n must get close to L when n is large. That is, $(z_1, z_2, z_3, \ldots)$ converges to L. A similar argument takes care of the casc where $(z_1, z_2, z_3, \ldots)$ is monotone decreasing. We may state what we have established as the theorem:

> If a subsequence of a monotone sequence converges to L, then the sequence converges to L.

Definition 9.4.1. If $(i_1, i_2, i_3, \ldots)$ is a strictly increasing sequence of positive integers, then $(z_{i_1}, z_{i_2}, z_{i_3}, \ldots)$ is said to be a **subsequence** of the sequence $(z_1, z_2, z_3, \ldots)$. The sequence $(i_1, i_2, i_3, \ldots)$ is said to be the **indexing sequence** for the subsequence.

Theorem 9.4.1. If $\lim_{n\to\infty} z_n = L$, then every subsequence of $(z_1, z_2, z_3, \ldots)$ converges to L.

PROOF: Suppose $\lim_{n\to\infty} z_n = L$ and $(z_{i_1}, z_{i_2}, z_{i_3}, \ldots)$ is a subsequence of $(z_1, z_2, z_3, \ldots)$. Let d be a given positive number. Find N such that $n > N$ implies $|z_n - L| < d$. Since $(i_1, i_2, i_3, \ldots)$ is a strictly increasing sequence of positive integers there is a K such that $i_K > N$. Then whenever $k > K$ we have $i_k > N$ and hence $|z_{i_k} - L| < d$. Therefore, $\lim_{k\to\infty} z_{i_k} = L$.

Theorem 9.4.2. Any subsequence of a monotone sequence is monotone.

PROOF: From the nature of an indexing sequence $i_{k+1} > i_k$. Hence, $z_{i_{k+1}} \geq z_{i_k}$ if $(z_1, z_2, \ldots)$ is monotone increasing, and $z_{i_{k+1}} \leq z_{i_k}$ if $(z_1, z_2, \ldots)$ is monotone decreasing.

Theorem 9.4.3. If a subsequence of a monotone sequence converges to L, then the sequence converges to L.

PROOF: Suppose $(z_{i_1}, z_{i_2}, z_{i_3}, \ldots)$ is a subsequence of the monotone increasing sequence $(z_1, z_2, z_3, \ldots)$ and $\lim_{k\to\infty} z_{i_k} = L$. We first note that we must have $z_n \leq L$ for all n. (The proof of this was given previously.) Let d be a positive number. Find a K such that $k > K$ implies $|z_{i_k} - L| < d$. There is no loss in generality if we assume K is a positive integer. Note that since $z_{i_k} \leq L$, $|z_{i_k} - L| < d$ is equivalent to $L - d < z_{i_k} \leq L$. Let $N = i_{K+1}$. Since $(z_1, z_2, \ldots)$ is monotone increasing, $n > N$ implies $z_n \geq z_N = z_{i_{K+1}}$. Hence, $n > N$ implies $L - d < z_{i_{K+1}} \leq z_n \leq L$. That is, $n > N$ implies $|z_n - L| < d$. Therefore, $\lim_{n\to\infty} z_n = L$.

PROBLEMS

1. Evidently a sequence must be divergent if it is possible to find two subsequences of it which converge to different limits. Use this result to show that each of the following sequences is divergent.

(a) $(0, 1, 0, 1, 0, 1, \ldots)$
(b) $(1, 1/2, 1/2, 2/3, 1/3, 3/4, 1/4, 4/5, 1/5, 5/6, 1/6, \ldots)$
(c) $(2, 1/2, 2, 3, 1/3, 3/2, 4, 1/4, 4/3, 5, 1/5, 5/4, 6, 1/6, 6/5, \ldots)$
(d) $(1, 1/2, 1, 1, 1/3, 1, 1, 1, 1/4, 1, 1, 1, 1, 1/5, \ldots)$
(e) $(1/2, -2/3, 3/4, -4/5, 5/6, -6/7, 7/8, -8/9, \ldots)$

2. Suppose $(z_1, z_2, z_3, \ldots)$ is a sequence with the property that $z_n = 4$ for

infinitely many values of n, and also $z_n = 4.0001$ for infinitely many values of n. What may we conclude about the convergence of the sequence? Explain.

3. Construct a sequence which has subsequences which converge to 1, 2, and 3.

4. Consider the sequence (1, 1/2, 2/2, 1/3, 2/3, 3/3, 1/4, 2/4, 3/4, 4/4, 1/5, 2/5, 3/5, 4/5, 5/5, 1/6, 2/6, 3/6, 4/6, 5/6, 6/6, 1/7, . . .). Find (if possible) a subsequence of this sequence which converges to (a) 0, (b) 1, (c) 1/2, (d) 1/4, (e) 1/72, (f) 47/6941. (g) Comment on this wonderful sequence. (h) In particular, if we consider the points on a number line which correspond to points in the sequence, what does the collection of all of them look like?

5. If we return to our basic notion that a sequence is a function defined on I^+, we may interpret the process of building a subsequence in terms of taking the composition of certain kinds of functions. Do so. (You may wish to consider the functions involved as machines and draw some pictures. Let $f(n) = z_n$ and $g(k) = i_k$ in our definition of subsequence.)

6. Construct a sequence $(z_1, z_2, z_3, \ldots)$ which has the property that $z_{2n+2} > z_{2n}$ for all n, $z_{2n+3} < z_{2n+1}$ for all n, and (a) the sequence converges, (b) the sequence diverges.

7. Every sequence is a subsequence of itself. Show that it is possible, however, to have two different sequences each of which is a subsequence of the other.

9.5 Geometric progressions

Let us return for another look at Zeno's poor, old, potentially immobile frog on the log. Suppose, as we did before, that the log is one yard long and the frog moves at a rate of one yard per minute. It will take him 1/2 minute to reach the midpoint of the log from where he started. It will take him $1/2 + 1/4$ minutes to reach the next halfway point. And $1/2 + 1/4 + 1/8$ minutes to reach the next one. Et cetera. It would be natural to say that the total amount of time required to traverse the log is

$$1/2 + 1/4 + 1/8 + 1/16 + \cdots$$

where the sum goes on without end. And it is in this form that Zeno's argument about the frog's being unable to reach the other end of the log may seem to have some force. How can a sum of infinitely many non-zero positive numbers be anything but infinite? How indeed?

In order to answer the question we must first realize that the notion of "a sum of infinitely many numbers" must be clarified. We have carefully defined what it means to add two rational numbers and using the associative law we see how the sum of n rational numbers is well defined when n is any positive integer. But we have not yet said in any way what it might possibly mean to add infinitely many numbers together. There seems to be only one reasonable way to explain what such a sum could mean.

From the expression $1/2 + 1/4 + 1/8 + \cdots$ we can construct the so-called **partial sums** $1/2$, $1/2 + 1/4$, $1/2 + 1/4 + 1/8$, $1/2 + 1/4 + 1/8 + 1/16, \ldots$.

The sequence formed by these partial sums is (1/2, 3/4, 7/8, 15/16, . . .) and we know that this converges to 1. This limit is what we then take to be the sum of *all* the numbers 1/2, 1/4, 1/8, 1/16, . . . and we write

$$1 = 1/2 + 1/4 + 1/8 + \cdots .$$

In other words, we **define** the sum of infinitely many numbers to be the limit, if there is one, for the sequence of partial sums. In more detail we have:

> Given the sequence $(a_1, a_2, a_3, \ldots)$, the **sum** of all the terms in the sequence, which we denote by "$a_1 + a_2 + a_3 + \cdots$," is defined to be the limit, if there is one, for the sequence $(a_1, a_1 + a_2, a_1 + a_2 + a_3, \ldots)$. We call $a_1 + a_2 + a_3 + \cdots$ an **infinite series.** a_n is called the nth **term of the series.** $a_1 + a_2 + a_3 + \cdots + a_n$ is called the nth **partial sum** for the series. If $a_1 + a_2 + a_3 + \cdots = L$ then we say that the series **converges** to L. If the sequence of partial sums diverges we say that the series **diverges.**

Putting the above in a somewhat different way, we may say that we make the definition

$$a_1 + a_2 + a_3 + \cdots = \lim_{n \to \infty} (a_1 + a_2 + \cdots + a_n).$$

Notice first of all that some sequences of terms yield series which have no sum. For example, $1 + 1 + 1 + 1 + \cdots$, $1 - 1 + 1 - 1 + 1 - 1 + \cdots$, and $1 + 2 + 3 + 4 + \cdots$ all diverge and have no sum. The corresponding sequences of partial sums are (1, 2, 3, . . .), (1, 0, 1, 0, 1, 0, . . .), and (1, 3, 6, 10, . . .).

It is a very deep and subtle question as to exactly which sequences $(a_1, a_2, a_3, \ldots)$ yield series $a_1 + a_2 + a_3 + \cdots$ which converge. We shall not dig deeply into this matter, but shall take care of one useful special case.

In the series $1/2 + 1/4 + 1/8 + \cdots$ the terms make up the sequence (1/2, 1/4, 1/8, . . .). This sequence has the property that to get the $(n + 1)$st term we multiply the nth term by one-half. Sequences with a similar property arise in many connections and there is a special terminology used for them.

A sequence in the form $(a, ar, ar^2, ar^3, \ldots)$ is called a **geometric progression.** r is called the **common ratio** for the progression.

(1/2, 1/4, 1/8, . . .) is a geometric progression with $a = 1/2$ and $r = 1/2$.
$(1, -1, 1, -1, 1, -1, \ldots)$ is a geometric progression with $a = 1$ and $r = -1$.
(1, .1, .01, .001, .0001, . . .) is a geometric progression with $a = 1$ and $r = .1$.
(5, 50, 500, 5000, . . .) is a geometric progression with $a = 5$ and $r = 10$.

Certain geometric progressions are null sequences. Whever $a = 0$ then certainly $(a, ar, ar^2, ar^3, \ldots)$ is a null sequence. But, also, if $|r| < 1$, then r^n goes to 0 as n gets large. We may conclude that $(a, ar, ar^2, ar^3, \ldots)$ is a null sequence whenever $|r| < 1$.

Suppose we consider a series whose terms form a geometric progression:

$$a + ar + ar^2 + ar^3 + \cdots .$$

Such a series is called a **geometric series.**

Under what conditions does such a series have a sum? Let S_n be the nth partial sum for the series. Then $S_1 = a$, $S_2 = a + ar$, $S_3 = a + ar + ar^2$, etc. Notice that $S_{n+1} = S_n + ar^n$.

If $a = 0$, then $S_n = 0$ for all n and the series has the sum 0. Suppose $a \neq 0$. If $r = 1$, then $S_n = na$ and in this case the series clearly has no sum.

If $r \neq 1$, we claim that $S_n = a((1 - r^n)/(1 - r))$ for all n. We may prove this by induction on n. Let

$$A = \left\{k \,\middle|\, S_k = a\frac{1 - r^k}{1 - r}\right\}.$$

Since

$$S_1 = a\frac{1 - r^1}{1 - r} = a,$$

we see that $1 \in A$. Suppose $k \in A$. Then

$$S_k = a\frac{1 - r^k}{1 - r}$$

and

$$S_{k+1} = a\frac{1 - r^k}{1 - r} + ar^k = a\frac{1 - r^k}{1 - r} + ar^k\frac{1 - r}{1 - r}$$

$$= a\frac{1 - r^k + r^k - r^{k+1}}{1 - r} = a\frac{1 - r^{k+1}}{1 - r}.$$

Hence, $k + 1 \in A$. We conclude that $A = I^+$. That is,

$$S_n = a\frac{1 - r^n}{1 - r}$$

for all n in I^+.

When $|r| < 1$, r^n approaches 0 when n gets large. Therefore, we have:

$$\text{If} \quad |r| < 1, \quad \text{then} \quad a + ar + ar^2 + ar^3 + \cdots = \frac{a}{1 - r}.$$

It is not difficult to show that when $a \neq 0$, the geometric series diverges for $|r| \geq 1$.

The above results are extremely useful. For example,

$$.1 + .01 + .001 + .0001 + \cdots = \frac{.1}{1 - .1} = \frac{1}{9}$$

$$.3 + .03 + .003 + .0003 + \cdots = \frac{.3}{1 - .1} = \frac{1}{3}$$

$$.45 + .0045 + .000045 + .00000045 + \cdots = \frac{.45}{1 - .01} = \frac{5}{11}$$

$$1 + \frac{4}{5} + \frac{16}{25} + \frac{4^3}{5^3} + \cdots = \frac{1}{1 - 4/5} = 5$$

$$.1 - .01 + .001 - .0001 + .00001 - \cdots = \frac{.1}{1 + .1} = \frac{1}{11}.$$

In each case, we simply recognize that the series is geometric, determine the first term a and the common ratio r, note that $|r| < 1$, and then calculate $a/(1 - r)$ to find the sum. If the series is geometric and $|r| \geq 1$, then the series diverges and has no sum unless we have the trivial case where $a = 0$.

Definition 9.5.1. The sequence $(a, ar, ar^2, ar^3, \ldots)$ is said to be a **geometric progression** with **common ratio** r.

Theorem 9.5.1. If $a = 0$ or $|r| < 1$ then the geometric progression $(a, ar, ar^2, ar^3, \ldots)$ is a null sequence.

PROOF: If $a = 0$ or $r = 0$ the result is obvious. Suppose $a \neq 0$ and $|r| = p/q$ where $p \in I^+$ and $q \in I^+$. If $|r| < 1$, then $p < q$ and $q/p > 1$. Let $q/p = 1 + s$. Note that $s > 0$. Then, by the binomial theorem, (Problem 10, Section 6.23) we may write $(q/p)^n = (1 + s)^n$ as $1^n + n \cdot 1^{n-1} \cdot s + \cdots = 1 + ns + \cdots$ where the omitted terms are all positive. Then $(q/p)^n \geq 1 + ns$. Since s is fixed here we can make $1 + ns$ and, hence, $(q/p)^n$ as large as we please by taking n large enough. But this means that we can make $(p/q)^n = 1/(q/p)^n$ as small as we please by taking n large enough. It follows that $\lim_{n \to \infty} |r|^n = 0$ and, therefore, $\lim_{n \to \infty} ar^n = 0$.

Definition 9.5.2. If $(a_1, a_2, a_3, \ldots)$ is a sequence of numbers, the **sum** of all of them is the limit, if there is one, of the sequence $(a_1, a_1 + a_2, a_1 + a_2 + a_3, a_1 + a_2 + a_3 + a_4, \ldots)$. We write "$a_1 + a_2 + a_3 + \cdots$" for this sum and call $a_1 + a_2 + a_3 + \cdots$ an **infinite series** whose nth **term** is a_n. A series with sum L is said to **converge** to L. A series which has no sum is said to **diverge.**

Theorem 9.5.2. If $a \neq 0$, $a + ar + ar^2 + \cdots$ converges if and only if $|r| < 1$. When it converges its sum is $a/(1 - r)$.

[NOTE: There are, of course, many infinite series which are not geometric series. So the above theorem simply scratches the surface of the study of the convergence of infinite series.]

PROBLEMS

1. For each of the following infinite series find the first few terms of the sequence of partial sums, and then find the sum of the series if it has one.

(a) $1 + 0 + 1 + 0 + 1 + 0 + 1 + 0 + \cdots$
(b) $-1 + 2 - 1 + 2 - 1 + 2 - 1 + 2 - \cdots$
(c) $1 + .2 + .04 + .008 + .0016 + \cdots$
(d) $.6 + .06 + .006 + .0006 + \cdots$
(e) $.31 + .0031 + .000031 + .00000031 + \cdots$
(f) $5 - 5 + 5 - 5 + 5 - 5 + \cdots$
(g) $.9 + (.9)^2 + (.9)^3 + (.9)^4 + \cdots$
(h) $3 + .9 + (.9)^2 + (.9)^3 + (.9)^4 + \cdots$
(i) $.357 + .000357 + .000000357 + \cdots$
(j) $.42 + .0021 + .000021 + .00000021 + \cdots$
(k) $.572 + .0001 + .00001 + .000001 + .0000001 + \cdots$
(l) $.101 + .000101 + .000000101 + .000000000101 + \cdots$
(m) $.07 + .0007 + .000007 + .00000007 + \cdots$.

2. If S_n is the nth partial sum for an infinite series which has a sum S, then $\lim_{n\to\infty} S_n = S$. Use this fact to prove that if a_n is the nth term of a convergent series then $\lim_{n\to\infty} a_n = 0$. [HINT: $\lim_{n\to\infty} a_n = \lim_{n\to\infty} (S_n - S_{n-1})$ where we take $S_0 = 0$.]

3. Show that if every term in a series is positive then the sequence of partial sums for the series is strictly increasing.

4. Show that if $a_1 + a_2 + a_3 + \cdots$ and $b_1 + b_2 + b_3 + \cdots$ are convergent series then so are (i) $(a_1 + b_1) + (a_2 + b_2) + (a_3 + b_3) + \cdots$, (ii) $(a_1 - b_1) + (a_2 - b_2) + (a_3 - b_3) + \cdots$, (iii) $ka_1 + ka_2 + ka_3 + \cdots$ where k is a constant. [HINT: Let $(S_1, S_2, S_3, \ldots)$ be the sequence of partial sums for the first series and $(T_1, T_2, T_3, \ldots)$ be the sequence of partial sums for the second. Then use some of our basic theorems about limits.]

5. The **sigma notation** for sums is often used in connection with infinite series. If $(a_1, a_2, a_3, \ldots)$ is a sequence of numbers we define $\sum_{k=1}^{1} a_k = a_1$ and $\sum_{k=1}^{n+1} a_k = \sum_{k=1}^{n} a_k + a_{n+1}$. We read "$\sum_{k=1}^{n} a_k$" as "the sum of a_k as k runs from 1 to n" or as "sigma as k goes from 1 to n of a_k." We also may write "$a_1 + a_2 + a_3 + \cdots$" as "$\sum_{k=1}^{\infty} a_k$" which we read as "the sum as k goes from 1 to infinity of a_k." Use the sigma notation to restate the results in Problem 4 in more detail.

6. If $(a_1, a_2, a_3, \ldots)$ is an arithmetic progression (see Problem 4, Section 9.1), under exactly what conditions will $a_1 + a_2 + a_3 + \cdots$ converge?

7. Obtain a desk calculator that multiplies automatically and convince yourself that it is reasonable that r^n goes to 0 as n gets large when $|r| < 1$ by

finding the approximate value of r^n for $n = 5, 10, 50, 100$ when r is (a) .9, (b) .99, (c) .9999. [HINT: $r^{100} = (r^{50})^2$.]

8. Show that a geometric series converges if and only if its terms form a null sequence.

---

9. Suppose $a_n > 0$ for all n. Let $b_n = (-1)^{n+1}a_n$. Then $b_1 + b_2 + b_3 + b_4 + b_5 + \cdots = a_1 - a_2 + a_3 - a_4 + a_5 - \cdots$. Such a series is called an alternating series. Let S_n be the nth partial sum for this series. Show that if $a_{n+1} \leq a_n$ for all n then (a) $S_{2n+2} \geq S_{2n}$ for all n and (b) $S_{2n+1} \leq S_{2n-1}$ for all n.

10. Let $a_1 - a_2 + a_3 - a_4 + a_5 - \cdots$ be an alternating series as defined in Problem 9. Suppose that the series converges to a number S and that $a_{n+1} \leq a_n$ for all n. Prove that $|S_n - S| \leq a_n$ for all n. [HINT: Note that $(S_2, S_4, S_6, \ldots)$ is a monotone increasing sequence and $(S_1, S_3, S_5, \ldots)$ is a monotone decreasing sequence. Then we must have $S_{2n} \leq S \leq S_{2n+1}$ for all n.]

11. Suppose $a_1 + a_2 + a_3 + \cdots = S$ and $b_1 + b_2 + b_3 + \cdots = T$ where $a_n \geq 0$ and $b_n \geq 0$ for all n. Show that if $a_n \leq b_n$ for all n then $S \leq T$.

12. Another of Zeno's paradoxes runs as follows. Achilles offers to race with a tortoise and, being big-hearted, he gives the tortoise a head start. But then, to his surprise, Achilles finds he can never catch the tortoise, let alone pass him. For when Achilles gets to the point where the tortoise started he finds that the tortoise has moved ahead a bit to a new point. And when Achilles reaches this new point, the tortoise has moved on to a third point. And so on, without end. Discuss the situation in terms of sequences assuming a head start of 50 yards, a speed of 500 yd/min for Achilles, and a speed of 5 yd/min for the tortoise. In particular, determine, as the sum of an infinite series, how long it will take Achilles to catch the tortoise.

9.6 Infinite decimals

Let us look more closely than we have to date at the sequence (.3, .33, .333, . . .). Here the nth term is named by the decimal fraction which has n "3"'s after the decimal point. That is, the nth term is $a_n = .333\ldots3 = .999\ldots9/3 = 999\ldots9/(3 \cdot 10^n) = (10^n - 1)/(3 \cdot 10^n) = 1/3 - 1/(3 \cdot 10^n)$ where in each case where ". . ." occurs, there are n digits in the string. It is clear that $\lim_{n\to\infty} (1/3 - 1/(3 \cdot 10^n)) = 1/3$. That is, (.3, .33, .333, . . .) converges to 1/3. Because this is so we sometimes write "1/3 = .333 . . . ." The last expression on the right is called an **infinite decimal fraction.** An **infinite** decimal fraction is strikingly different from an ordinary (finite) decimal fraction. The ordinary decimal fraction we introduced in Section 8.10 always has a final digit. Infinite decimals are quite another cup of tea. We need to make clear what these new beasts are to name, if anything.

Suppose that ("d_1," "d_2," "d_3," . . .) is an infinite sequence of decimal digits. Then we shall call the expression "$.d_1d_2d_3\ldots$" an **infinite decimal fraction.**

Following the lead we discovered for the sequence (.3, .33, .333, . . .) we shall say that the infinite decimal fraction "$.d_1d_2d_3 \ldots$" is a name for the rational number (if there is one) that is the limit of the sequence ($.d_1$, $.d_1d_2$, $.d_1d_2d_3$, $.d_1d_2d_3d_4$, . . .).

Thus, ".1111 . . ." is a name for the limit (if there is one) of (.1, .11, .111, . . .). There is a rational limit for this last sequence. We note that $.111 \ldots 1 = .999 \ldots 9/9 = 999 \ldots 9/(9 \cdot 10^n) = (10^n - 1)/(9 \cdot 10^n) = 1/9 - 1/(9 \cdot 10^n)$ where n is the number of digits in each string containing a ". . . ." Then the limit of (.1, .11, .111, . . .) is 1/9. In other words, we may assert that $.111 \ldots = 1/9$.

".999 . . ." is a name for the limit (if there is one) of (.9, .99, .999, . . .). But $.999 \ldots 9 = 999 \ldots 9/10^n = (10^n - 1)/10^n = 1 - 1/10^n$ where n is the number of digits in each string containing a ". . . ." Since $\lim_{n \to \infty} (1 - 1/10^n) = 1$ we see that $.999 \ldots = 1$. Thus, we have it that the use of infinite decimals has introduced a rather strange name for an old friend.

It seems to bother some people that we have such a result as $.999 \ldots = 1$. They evidently have the feeling that the number named on the left side of this equality is somehow always a bit less than 1 and could not actually be equal to 1. It is possible that part of the confusion here has to do with semantics. The left side of the equality is ".999 . . . ." And in order to know whether the equality is true or not we must ask what ".999 . . . " means. Now ".999 . . . " is not an ordinary decimal fraction and our definition of the meaning of decimal fractions as given in Section 8.10 simply does not apply to it. When we ask what ".999 . . . " means we must refer to some definition which gives meaning to infinite decimal fractions. But the only reasonable such definition leads us to the conclusion that ".999 . . . " is a name for the limit of (.9, .99, .999, . . .). And, as we saw above, this limit is 1. Hence, we must have $.999 \ldots = 1$.

It may also be that part of the difficulty lies in the fact that we might expect that when we have two infinite decimals $.d_1d_2d_3 \ldots$ and $.e_1e_2e_3 \ldots$ and have $.d_1d_2d_3 \ldots = .e_1e_2e_3 \ldots$, that then it must follow that $d_1 = e_1$, $d_2 = e_2$, $d_3 = e_3$, etc. But *this is not so*. We have seen that $1 = 1.0000 \ldots = .9999 \ldots$. We also have, for example, $.50000 \ldots = .49999 \ldots$ and yet some (in fact, all) of the digits in the two strings are different.

Let us allow a decimal name for an integer to be written in front of the decimal point in an infinite decimal fraction and interpret the result in a natural way. Then we have $7.333 \ldots = 7 + .333 \ldots = 7 + \frac{1}{3} = 7\frac{1}{3}$, $347.111 \ldots = 347 + .111 \ldots = 347 + \frac{1}{9} = 347\frac{1}{9}$, $3.999 \ldots = 3 + .999 \ldots = 3 + 1 = 4$, $-7.333 \ldots = -(7 + .333 \ldots) = -7\frac{1}{3}$, and $-4.999 \ldots = -(4 + .999 \ldots) = -(4 + 1) = -5$. A decimal name for an integer followed by an infinite decimal fraction we shall call simply an **infinite decimal.**

Consider the infinite decimal ".454545 . . . ." If this expression names anything, then it names the limit of the sequence (.4, .45, .454, .4545, . . .). Note that this sequence is monotone increasing. It is easier to deal with the subsequence (.45, .4545, .454545, . . .) than with the original. And we know from our work in Section 9.4 that if this subsequence converges to, say, x, then the original monotone sequence converges to x. In the subsequence (.45, .4545, .454545, . . .), the nth term is $b_n = .4545 \ldots 45$ where there are $2n$ digits. But notice that we may write

$$b_n = .45 + .0045 + .000045 + .00 \ldots 045$$

where there are $2n - 2$ "0's" before the "45" in the last term. Thus, b_n is the sum of the first n terms in the geometric progression (.45, .0045, .000045, . . .) where the first term is .45 and the common ratio is .01. Hence,

$$b_n = .45\left(\frac{1 - (.01)^n}{1 - .01}\right) = \frac{45}{99}(1 - (.01)^n).$$

Clearly $\lim_{n \to \infty} b_n = 45/99$. Hence, we have shown that

$$.454545 \ldots = \frac{45}{99} = \frac{5}{11}.$$

It turns out that something similar to the above procedure may be used with any infinite decimal where there is a certain block of digits that repeats over and over again to generate the infinite decimal fraction. Such infinite decimals are called **repeating** decimals. It is customary to use a bar over a block of digits to indicate that the block is to be repeated endlessly. This notation is illustrated in the following:

$$.\overline{3} = .3333 \ldots$$

$$.\overline{1} = .1111 \ldots.$$

$$.\overline{9} = .9999 \ldots.$$

$$.\overline{45} = .454545 \ldots$$

$$36.37\overline{6} = 36.376666 \ldots$$

$$5.479\overline{14} = 5.479141414 \ldots$$

$$-2.4\overline{9} = -2.49999 \ldots$$

$$.1\overline{876} = .1876876876 \ldots.$$

Applying the method we used with $.\overline{45}$ to a more general case, we have:

If "d_1", "d_2", . . . , and "d_k" are decimal digits, then

$$.\overline{d_1d_2 \ldots d_k} = \frac{d_1d_2 \ldots d_k}{10^k - 1}.$$

For example, $.\overline{743} = 743/999$, $.\overline{0042} = 42/9999$, and $7.\overline{456} = 7 + .\overline{456} = 7\frac{456}{999}$.

Even more generally, we have that when "b_1," "b_2," . . . , "b_n," "d_1," "d_2," . . . , and "d_k" are all decimal digits, then

$$.b_1b_2 \ldots b_n\overline{d_1d_2 \ldots d_k} = .b_1b_2 \ldots b_n + 10^{-n}\left(\frac{d_1d_2 \ldots d_k}{10^k - 1}\right).$$

For example, $.74\overline{36} = .74 + 10^{-2}(.\overline{36}) = .74 + 10^{-2}(36/99) = 74/100 + 36/9900 = 7362/9900$, $.040\overline{6} = .040 + 10^{-3}(.\overline{6}) = .040 + 10^{-3}(6/9) = 40/1000 + 2/3000 = 122/3000$, and $54.7\overline{7} = 54.7 + 10^{-1}(.\overline{7}) = 54.7 + 10^{-1}(7/9) = 547/10 + 7/90 = 4930/90$.

It would appear that any repeating decimal can be handled in this fashion and, hence, names a rational number. This is indeed true.

Note that any ordinary (finite) decimal fraction can be written as an infinite repeating decimal by using an infinite string of "0's." For example, $7.56 = 7.56\overline{0} = 7.56000 \ldots$ and $.041 = .041\overline{0} = .041000 \ldots$.

The even more remarkable fact is that **any** rational number can be expressed as an infinite fraction. We shall prove this in the appendix to this section. The proof amounts to showing that when we are given p/q where p and q are integers, the division algorithm for $p \div q$ yields an infinite decimal that is repeating. And, furthermore, this infinite decimal is actually a name for p/q.

If we combine the two main results above, we find that an infinite decimal names a rational number if and only if it is a repeating decimal. This is a very important fact.

It follows almost immediately from this† that infinite decimals that are not repeating decimals cannot be names for rational numbers. It is easy to find infinite decimals that are not repeating. Some examples are:

$$.101001000100001 \ldots$$

$$.1234567891011121314 \ldots$$

$$.43433433343333 \ldots, \quad \text{and}$$

$$.1231223122231222223 \ldots.$$

These expressions cannot be names for rational numbers.

Our next major chore will be to invent some things for such nonrepeating decimals to name. It will turn out that when we have done so, we shall have taken care of many of the shortcomings of the rational number system.

Definition 9.6.1. If ("d_1," "d_2," "d_3," . . .) is a sequence of decimal digits then the expression "$.d_1d_2d_3 \ldots$" is said to be an **infinite decimal fraction.**

Definition 9.6.2. "$.d_1d_2d_3 \ldots$" is a name for the limit (if there is one) of the sequence $(.d_1, .d_1d_2, .d_1d_2d_3, \ldots)$. If "$n$" is the decimal name for a natural

† For the details see the appendix to this section.

number, we write "$n.d_1d_2d_3 \ldots$" for $n + .d_1d_2d_3 \ldots$ and "$- n.d_1d_2d_3 \ldots$" for the opposite of $n + .d_1d_2d_3 \ldots$.

Theorem 9.6.1. (a) $.333\ldots = 1/3$, (b) $.111\ldots = 1/9$, (c) $.999\ldots = 1$.

Definition 9.6.3. If "b_1," "b_2," ..., "b_k," "d_1," "d_2," "d_3," ..., "d_n" are all decimal digits then "$.b_1b_2 \ldots b_kd_1d_2 \ldots d_nd_1d_2 \ldots d_nd_1d_2 \ldots d_n \ldots$" is said to be an (infinite) **repeating decimal fraction.** We write "$.b_1b_2 \ldots b_k\overline{d_1d_2 \ldots d_n}$" for the above infinite decimal.

Theorem 9.6.2. $.b_1b_2 \ldots b_n\overline{d_1d_2 \ldots d_k} = .b_1b_2 \ldots b_n + 10^{-n}\left(\dfrac{d_1d_2 \ldots d_k}{10^k - 1}\right).$

PROOF: $(.b_1, .b_1b_2, \ldots, .b_1b_2 \ldots b_n, .b_1b_2 \ldots b_nd_1, .b_1b_2 \ldots b_nd_1d_2, \ldots, .b_1b_2 \ldots b_nd_1d_2 \ldots d_n, .b_1b_2 \ldots b_nd_1d_2 \ldots d_nd_1, \ldots)$ is a monotonic increasing sequence of rational numbers. Hence, by Theorem 9.4.1 and Theorem 9.4.3, this sequence converges to a rational number x if and only if the subsequence $(.b_1b_2 \ldots b_nd_1d_2 \ldots d_k, .b_1b_2 \ldots b_nd_1d_2 \ldots d_kd_1d_2 \ldots d_k, \ldots)$ converges to x. If we let x_p be the pth term of this subsequence, we have $x_p = .b_1b_2 \ldots b_n + .00 \ldots 0d_1d_2 \ldots d_k + .00 \ldots 0d_1d_2 \ldots d_k + \cdots + .00 \ldots 0d_1d_2 \ldots d_k$ where there are n "0's" in the first string of "0's," $n + k$ in the second, $n + 2k$ in the third, ..., and $n + (p - 1)k$ in the last. We see then that x_p is $.b_1b_2 \ldots b_n$ plus the first p terms in a geometric progression which has a first term of $(.d_1d_2 \ldots d_k) \cdot 10^{-n}$ and a common ratio of 10^{-k}. Hence,

$$\lim_{p\to\infty} x_p = .b_1b_2 \ldots b_n + \frac{(.d_1d_2 \ldots d_k)10^{-n}}{1 - 10^{-k}}$$

$$= .b_1b_2 \ldots b_n + 10^{-n}\left(\frac{d_1d_2 \ldots d_k}{10^k - 1}\right).$$

This completes the proof.

Theorem 9.6.3. If a and b are positive integers and $a < 10^n$ and $b < 10^k$ and $x = (a/10^n) + b/(10^n(10^k - 1))$, then x can be expressed as an infinite decimal.

PROOF: a has a decimal name in the form "$b_1b_2 \ldots b_n$" where the "b_i's" are decimal digits. b has a decimal name in the form $d_1d_2 \ldots d_k$ where the "d_i's" are decimal digits. By Theorem 9.6.2, $x = .b_1b_2 \ldots b_n\overline{d_1d_2 \ldots d_k}$.

Theorem 9.6.4. If x is a positive rational number and $x < 1$, then x can be expressed as a repeating decimal fraction.

PROOF: Let $x = p/q$ where p and q are positive integers and $p < q$. By repeated use of the division algorithm we may find nonnegative integers $s_1, s_2, s_3, \ldots$ and $r_1, r_2, r_3, \ldots$ such that $(10^ip)/q = s_i + (r_i/q)$ and $r_i < q$ for $i = 1,2,3, \ldots$. Since the $r_1,r_2,r_3, \ldots$ are all less than q there must be an r_j and an r_n such that $r_j = r_n$ and $j > n$. Then $(10^jp)/q - s_j = r_j/q = r_n/q = (10^np)/q - s_n$. Hence, $(10^j - 10^n)(p/q) = s_j - s_n$ and $p/q =$

$(s_j - s_n)/(10^j - 10^n) = (s_j - s_n)/(10^n(10^{j-n} - 1))$. Let $j - n = k$. Since $p < q$ we must have $s_j - s_n < 10^n(10^k - 1)$. By the division algorithm there are nonnegative integers a and b such that $(s_j - s_n)/(10^k - 1) = a + (b/(10^k - 1))$ where $b < 10^k - 1$. Since $(s_j - s_n)/(10^n(10^k - 1)) < 1$ we see that $a < 10^n$. We now see that $p/q = a/10^n + b/(10^n(10^k - 1))$ where $a < 10^n$ and $b < 10^k - 1$. By Theorem 9.6.3, it follows that p/q can be expressed as an infinite decimal.

Theorem 9.6.5. Any rational number can be expressed as a repeating decimal.

PROOF: Any rational number x can be written as an integer plus a proper fraction, or else as the opposite of such a number. Hence, by Definition 9.6.2 and Theorem 9.6.4, x can be expressed as a repeating infinite decimal.

Theorem 9.6.6. If "d_1," "d_2," . . . , and "e_1," "e_2," . . . are decimal digits and $.d_1d_2d_3 \ldots = .e_1e_2e_3 \ldots$ then (a) $d_n = e_n$ for all n or (b) there is a k such that $d_i = e_i$ for $i < k$, $d_k = e_k + 1$, and $d_i = 0$ and $e_i = 9$ for $i > k$, or (c) there is a k such that $d_i = e_i$ for $i < k$, $d_k + 1 = e_k$, and $d_i = 9$ and $e_i = 0$ for $i > k$.

PROOF: Suppose "d_1," "d_2," . . . and "e_1," "e_2," . . . are decimal digits such that none of the conditions (a), (b) or (c) holds. Let j be the smallest positive integer such that $d_j \neq e_j$. There is no loss in generality if we assume that $d_j > e_j$. If $d_j > e_j + 1$, then for all $k > j$, we have $.d_1d_2 \ldots d_k - .e_1e_2 \ldots e_k \geq 10^{-j}$. Hence, we cannot have $.d_1d_2d_3 \ldots = .e_1e_2e_3 \ldots$ since $(.d_1, .d_1d_2, .d_1d_2d_3, \ldots)$ and $(.e_1, .e_1e_2, .e_1e_2e_3, \ldots)$ cannot converge to the same limit. If $d_j = e_j + 1$ then there is a smallest positive m such that $m > j$ and $d_m \neq 0$ and $e_m \neq 9$. Then for $k > m$ we must have $.d_1d_2 \ldots d_k - .e_1e_2 \ldots e_k > 10^{-m}$ and again we cannot have $.d_1d_2d_3 \ldots = .e_1e_2e_3 \ldots$. This completes the proof.

Theorem 9.6.7. A nonrepeating infinite decimal cannot be a name for a rational number.

PROOF: If x is rational, then it can be expressed as a repeating decimal (Theorem 9.6.5). Then by Theorem 9.6.6, x cannot be expressed as a nonrepeating decimal.

PROBLEMS

1. Express each of the following as a ratio of integers. (a) .56565656 . . . , (b) 7.13131313 . . . , (c) $.\overline{6}$, (d) $.\overline{67}$, (e) $5.37\overline{621}$, (f) $-3.9\overline{39}$, (g) $-573.\overline{375}$.

2. Express each of the following as infinite decimals. (a) 6, (b) 37/4, (c) 3/8, (d) 5/6, (e) 3/7, (f) 13/14, (g) 523/9, (h) 1/17.

3. Let G be the collection of all infinite repeating decimals. If $x \in G$, let $f(x)$ be the rational number named by x. Is f a one-to-one correspondence between G and F, the set of all rational numbers? Explain carefully.

4. Suppose we let H be the collection of all infinite repeating decimals that do not end in an infinite string of "0's." Is the function f of Problem 2 a one-to-one correspondence between H and F? Explain carefully.

5. Let us assume that, as is the case with ordinary decimals, infinite decimals may be multiplied by powers of 10 by moving the decimal point around. Let us also assume that infinite decimals may be subtracted in a natural fashion. Then the problem of expressing an infinite repeating decimal as a ratio of integers may be solved somewhat differently. Suppose, for example, that $x = .1\overline{876}$. Then $10x = 1.\overline{876}$ and $10{,}000x = 1{,}876.\overline{876}$. Hence, $10{,}000x - 10x = 9990x = 1{,}876.876876876\ldots - 1.876876876\ldots = 1875$. And, finally, $x = 1875/9990$. Use this method to work Problem 1.

10

The system of real numbers

10.1 The real numbers

The system of rational numbers is a much richer system than the system of integers. Many equations that have no solution in I *do* have solutions in F. For example, the equations $2x = 1$ and $7x = 5$ have no solutions in I, but they may be solved in F.

If, however, we consider equations in the form $x^n = k$ where n and k are given positive integers, we discover a surprising fact. We can solve no more equations in this form in F than we could in I itself! Let us see why this is so.

Suppose that p, q, r, n, and k are all integers and that q and n are positive. If "p/q" is in lowest terms and $p/q = r$ we claim that we must have $q = 1$. For if $p/q = r$, then $p = rq$ and $p/q = (rq)/q$ and, therefore, if $q \neq 1$, "p/q" is not in lowest terms.

Now suppose that "p/q" is in lowest terms and that $(p/q)^n = k$. This implies that $p^n/q^n = k$. But when "p/q" is in lowest terms, "p^n/q^n" must be also. (See Problem 7.) Therefore, by what we proved above we must have $q^n = 1$. But this implies that $q = 1$. (See Problem 6.) Therefore, $p^n = k$. Since any rational number may be expressed as a fraction in lowest terms, this tells us that if x is a rational number such that $x^n = k$, then x is actually an integer.

In other words, when n and k are integers and n is positive, the equation $x^n = k$ has a solution in F only when it already had a solution in I. In moving from I to F we have gained nothing in terms of solving equations in the form $x^n = k$.

Looked at from another point of view, the above result says that if the equation $x^n = k$ does not have an integral solution, then it has *no solution at all* in F.

If we consider the special case of the equation $x^2 = 2$, we can easily check that it has no integral solution. But then it follows that it has no solution whatsoever in F. In other words, "$\sqrt{2}$" does not name a rational number.

This particular result may be put more geometrically. Suppose we construct a square with a side one unit long on a number line as in Figure 10.1.

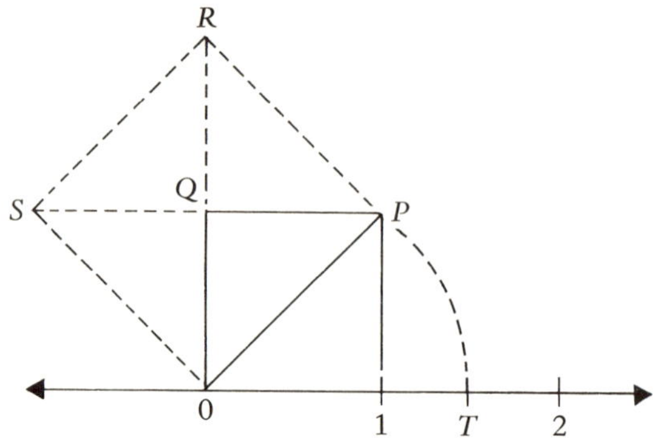

FIGURE 10.1

Let us draw in the diagonal OP. If we construct three copies of the triangle OQP and arrange them as shown we get the square $OPRS$. Since our original square contains one square unit of area, the triangle OPQ has an area of $1/2$ square unit. But then the square $OPRS$ has $4 \cdot 1/2 = 2$ square units of area. Hence, the length of OP should be a number whose square is 2. But we know that there is no rational number whose square is 2. If we construct a circle which passes through P and has its center at the origin, the circle will cross the number line at the point T. Since the length of OT is the same as that of OP, this point T does not have a rational coordinate. Although the rational points are dense on the line in the sense that between any two distinct rational points there is another rational point, we now see that there are some points on the line that are not rational. We have *not* taken care of *all* the points on the line when we assigned rational coordinates to some of them.

The task we now set ourselves is the building of a number system which is an extension of the rational number system and which *does* enable us to assign a number to each point on the line.

Following our geometric intuition, we might try to identify any particular point P on the line by giving a sequence of rational points that sneak up on P from the left. When the terms in two different sequences both approach the point P from the left we would expect these terms to get close together as you go far out in the sequences. We shall follow this lead.

Let S be the collection of all bounded, monotone increasing sequences of rational numbers. We shall define a relation E on S by saying that

$$(x_1, x_2, x_3, \ldots)\ E\ (y_1, y_2, y_3, \ldots)$$

when $(x_1 - y_1, x_2 - y_2, x_3 - y_3, \ldots)$ is a null sequence.

It is quite easy to see that E is an equivalence relation. $(x_1, x_2, x_3, \ldots)\ E\ (x_1, x_2, x_3, \ldots)$ since $(0, 0, 0, \ldots)$ is certainly a null sequence. $(x_1, x_2, x_3, \ldots)\ E$

$(y_1, y_2, y_3, \ldots)$ implies $(y_1, y_2, y_3, \ldots)\ E\ (x_1, x_2, x_3, \ldots)$ since $(y_1 - x_1, y_2 - x_2, y_3 - x_3, \ldots)$ is a null sequence when $(x_1 - y_1, x_2 - y_2, x_3 - y_3, \ldots)$ is. (See Theorem 9.1.2.) If $(x_1, x_2, x_3, \ldots)\ E\ (y_1, y_2, y_3, \ldots)$ and $(y_1, y_2, y_3, \ldots)\ E\ (z_1, z_2, z_3, \ldots)$ then $(x_1 - y_1, x_2 - y_2, x_3 - y_3, \ldots)$ and $(y_1 - z_1, y_2 - z_2, y_3 - z_3, \ldots)$ are null sequences. Hence, $((x_1 - y_1) + (y_1 - z_1), (x_2 - y_2) + (y_2 - z_2), (x_3 - y_3) + (y_3 - z_3), \ldots) = (x_1 - z_1, x_2 - z_2, x_3 - z_3, \ldots)$ is a null sequence. (See Theorem 9.1.4.) So we have established that E is reflexive, symmetric, and transitive.

As might be expected in view of our earlier work, we have as the basic definition of this chapter:

> An equivalence class in S for the relation E defined above is called a **real number.**

Since $(0, 0, 0, \ldots)$ is a bounded monotone increasing sequence, the equivalence class $[(0, 0, 0, \ldots)]$ is a real number. This number (or class) consists of *all* the bounded monotone increasing sequences $(x_1, x_2, x_3, \ldots)$ for which $(x_1 - 0, x_2 - 0, x_3 - 0, \ldots)$ is a null sequence. In other words, $[(0, 0, 0, \ldots)]$ is the collection of all monotone increasing null sequences.

Since $(.3, .33, .333, .3333, \ldots)$ is a bounded monotone increasing sequence, the class $[(.3, .33, .333, .3333, \ldots)]$ is a real number. This class contains all sequences $(x_1, x_2, x_3, \ldots)$ such that $(x_1 - .3, x_2 - .33, x_3 - .333, \ldots)$ is a null sequence. For example, this class contains the sequence $(1/3, 1/3, 1/3, \ldots)$ since $(1/3 - .3, 1/3 - .33, 1/3 - .333, \ldots) = (1/30, 1/300, 1/3000, \ldots)$ is a null sequence.

Since $(.1, .101, .101001, .1010010001, .101001000100001, \ldots)$ is a bounded monotone increasing sequence of rationals, the equivalence class determined by it is a real number.

We may construct the real number that will give us the length of the segment OT in Figure 10.1 as follows. Consider the sequence $(a_1, a_2, a_3, \ldots)$ where, for each n, $a_n = k_n/10^n$ and k_n is the largest integer such that $(k_n/10^n)^2 < 2$. Thus, $a_1 = 14/10$ since $(14/10)^2 = 196/100 < 2$ and $(15/10)^2 = 225/100 > 2$. $a_2 = 141/100$ since $(141/100)^2 = 19{,}881/10{,}000 < 2$ and $(142/100)^2 = 20{,}164/10{,}000 > 2$. $a_3 = 1414/1000$ since $(1414/1000)^2 = 1{,}999{,}396/1{,}000{,}000 < 2$ and $(1415/1000)^2 = 2{,}002{,}225/1{,}000{,}000 > 2$. And so forth. Recall that when $0 < x < y$ in F, then $x^2 < y^2$, so that when we discover, for example, that $(142/100)^2 > 2$, then certainly $(k/100)^2 > 2$ for any integer k such that $k > 142$. Note that we could describe a_n here by saying that it is the largest integral multiple of 10^{-n} whose square is less than 2. Calculations similar to those above show that $(a_1, a_2, a_3, a_4, a_5, a_6, \ldots) = (1.4, 1.41, 1.414, 1.4142, 1.41421, 1.414214, \ldots)$. Evidently this sequence is a bounded monotone increasing sequence.† So $a = [(a_1, a_2, a_3, \ldots)]$ is a real number. Since $a_n{}^2$ is very close to 2

† Note that despite initial indications we do not get a repeated pattern of "14" here.

when n is large we might expect this real number a to be one whose square is two. This expectation will not be in vain. The real number a will be the exact length of segment OT in Figure 10.1.

We may look at the construction of the sequence $(a_1, a_2, a_3, \ldots)$ geometrically as if we were using stronger and stronger magnifying glasses on the number line. See Figure 10.2.

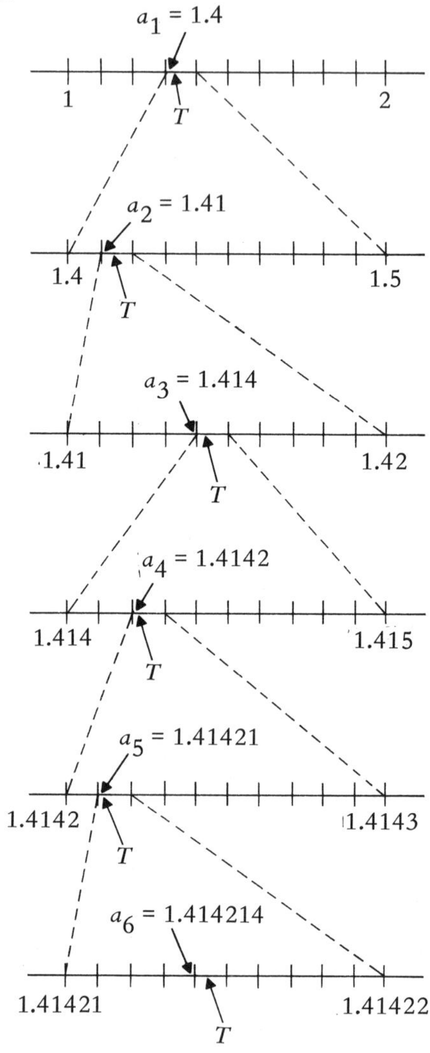

FIGURE 10.2

The special form of the sequence $(a_1, a_2, a_3, \ldots)$ suggests a general method for constructing names for real numbers.

To discuss this method we first need to distinguish those real numbers which are said to be positive. We shall say that a real number $[(x_1, x_2, x_3, \ldots)]$ is positive if there is at least one x_n in the sequence which is a positive rational. Of course, $(x_1, x_2, x_3, \ldots)$ is monotone increasing and, if it has one positive rational term, then it has infinitely many.

If a real number is not equal to $[(0, 0, 0, \ldots)]$ and is not positive, we shall say that it is **negative.** A positive real number will be said to *have a positive sign.* A negative real number will be said to *have a negative sign.*

Suppose "b" is the decimal name for a positive integer and "d_1," "d_2," "d_3," ... are decimal digits. Then $(b.d_1, b.d_1d_2, b.d_1d_2d_3, \ldots)$ is a bounded monotone increasing sequence of rational numbers. Hence, $[(b.d_1, b.d_1d_2, b.d_1d_2d_3, \ldots)]$ is a real number. We shall write the infinite decimal "$b.d_1d_2d_3 \ldots$" as a name for this real number. We see then that any infinite decimal names a real number. It can also be shown that any positive real number has an infinite decimal that names it.

This naming procedure introduces some ambiguities since we saw earlier that certain infinite decimals are also names for rational numbers. But these ambiguities will cause no difficulties. We shall verify this later when we consider carefully the sense in which the real number system is an extension of F.

Theorem 10.1.1. If $x \in F$, $n \in I^+$, $k \in I$, and $x^n = k$, then $x \in I$.

Definition 10.1.1. Let S be the collection of all bounded monotone increasing sequences of rational numbers. Define the relation E on S by saying that $(x_1, x_2, x_3, \ldots)\ E\ (y_1, y_2, y_3, \ldots)$ if $(x_1 - y_1, x_2 - y_2, x_3 - y_3, \ldots)$ is a null sequence.

Theorem 10.1.2. E is an equivalence relation in S.

Definition 10.1.2. An equivalence class in S with respect to E is said to be a **real number.**

Definition 10.1.3. A real number $[(x_1, x_2, x_3, \ldots)]$ is said to be **positive** if there is at least one x_n which is a positive rational.

Definition 10.1.4. If "b" is a decimal name for a natural number and "d_1," "d_2," "d_3," ... are decimal digits then we write "$b.d_1d_2d_3 \ldots$" to name the real number $[(b.d_1, b.d_1d_2, b.d_1d_2d_3, \ldots)]$.

Theorem 10.1.3. Any positive real number has an infinite decimal name.

PROOF: Let $x = [(x_1, x_2, x_3, \ldots)]$ be a positive real number. Since $(x_1, x_2, x_3, \ldots)$ is bounded, there is a smallest natural number k such that $x_n < k$ for all n. Since at least one x_n is positive, k cannot be zero and $k - 1$ is a natural number. Let "b" be the decimal name for $k - 1$. Consider the numbers b, $b + 1/10$, $b + 2/10, \ldots$, and $b + 9/10$. There must be a largest one of these numbers, say $b + d_1/10$, which is not an upper bound for $\{x_1, x_2, x_3, \ldots\}$. This is because b itself is not such an upper bound. Consider the numbers

$b + d_1/10$, $b + d_1/10 + 1/100$, $b + d_1/10 + 2/100, \ldots$, and $b + d_1/10 + 9/100$. There must be a largest one of these, say $b + d_1/10 + d_2/100$, which fails to be an upper bound for the terms in the sequence. We next consider the numbers $b + d_1/10 + d_2/100$, $b + d_1/10 + d_2/100 + 1/1000, \ldots$, and $b + d_1/10 + d_2/100 + 9/1000$ and proceed as before. In this manner, we generate an infinite decimal "$b.d_1d_2d_3 \ldots$." It remains to show that $x = [(b.d_1, b.d_1d_2, b.d_1d_2d_3, \ldots)]$. Note that for each n there must be a rational number x_{k_n} from the sequence $(x_1, x_2, x_3, \ldots)$ such that $b.d_1d_2 \ldots d_n < x_{k_n} \leq b.d_1d_2 \ldots d_n + 10^{-n}$. For, otherwise, we selected d_n incorrectly. Moreover, $b.d_1d_2 \ldots d_n + 10^{-n}$ is an upper bound for $\{x_1, x_2, x_3, \ldots\}$ and also an upper bound for $\{b.d_1, b.d_1d_2, b.d_1d_2d_3, \ldots\}$. Let e be a given positive number. Select an n such that $10^{-n} < e$. Find k_n as given above. Let $N = \max\{n, k_n\}$. When $j > N$ we have $b.d_1d_2 \ldots d_j \geq b.d_1d_2 \ldots d_n$ and $x_j \geq x_{k_n}$. Hence, when $j > N$ we have both $b.d_1d_2 \ldots d_n \leq b.d_1d_2 \ldots d_j \leq b.d_1d_2 \ldots d_n + 10^{-n}$ and $b.d_1d_2 \ldots d_n < x_j \leq b.d_1d_2 \ldots d_n + 10^{-n}$. Therefore, $|x_j - b.d_1d_2 \ldots d_j| < 10^{-n} = e$. It follows that $(x_1 - b.d_1, x_2 - b.d_1d_2, x_3 - b.d_1d_2d_3, \ldots)$ is a null sequence and this completes the proof.

Problems

1. Which of the following sequences are in S? Of those that are in S, are there any that are in the relation E to one another? (S and E are the collection and relation defined in this section.)

(a) (.1, .01, .001, .0001, . . .)
(b) (1, 2, 3, 4, . . .)
(c) (.1, .12, .123, .1234, . . . ,
.123456789, .12345678910, .1234567891011, . . .)
(d) (.1, .1001, .1001001, .1001001001, . . .)
(e) (.79, .7979, .797979, . . .)
(f) (5, 5, 5, 5, . . .)
(g) (.1001001, .1001001001001, .1001001001001001001, . . .)
(h) (.4, .49, .499, .4999, . . .)
(i) (5.1, 5.01, 5.001, 5.0001, . . .)
(j) (−1, −.1, −.01, −.001, −.0001, . . .)
(k) (24/5, 124/25, 624/125, . . .)
(l) (−2, −.2, −.02, −.002, −.0002, . . .)
(m) (−1, −1/2, −1/3, −1/4, −1/5, . . .)
(n) (.1, .121, .121221, .1212212221, .121221222122221, . . .).

2. Find an infinite decimal that names the real number identified by each sequence in S given in Problem 1. (An infinite string of "0's" is, of course, allowable in an infinite decimal.)

3. Find the first four terms in the sequence $(b_1, b_2, b_3, \ldots)$ where, for each n, $b_n = k_n/10^n$ and k_n is the largest integer such that $(k_n/10^n)^2 < 3$. What

algebraic property do you expect the real number $[(b_1, b_2, b_3, \ldots)]$ to possess?

4. Find the first three terms in the sequence $(c_1, c_2, c_3, \ldots)$ where, for each n, $c_n = k_n/10^n$ and k_n is the largest integer such that $(k_n/10^n)^3 < 2$. What algebraic property do you expect the real number $[(c_1, c_2, c_3, \ldots)]$ to possess?

5. Which of the following are rational numbers? (a) $\sqrt{36}$, (b) $\sqrt{35}$, (c) $\sqrt{100}$, (d) $\sqrt{10}$, (e) $\sqrt{1/9}$, (f) $\sqrt{-4}$, (g) $\sqrt{289}$.

6. Show that if q is a positive integer and $q^n = 1$, then $q = 1$. [HINT: If q is a positive integer, then $q \geq 1$. Use induction to show that if $q \neq 1$, then $q^n > 1$ and, hence, $q^n \neq 1$.]

7. Show that if "p/q" is a fraction in lowest terms then so is "p^n/q^n." [HINT: Use the fundamental theorem of arithmetic to write p and q as products of primes. Then show that if p and q have no prime factor in common, neither do p^n and q^n.]

8. Show that if $x = [(x_1, x_2, x_3, \ldots)]$ and x is negative, then $x_n < 0$ for every n.

9. Show that we may have x nonnegative and yet $x = [(x_1, x_2, x_3, \ldots)]$ with $x_n < 0$ for every n.

10. Show that if x and y are real numbers and $x = y$ and x is positive, then y is positive.

11. Show that if $(x_1, x_2, \ldots) \in (0, 0, 0, \ldots)$, then $[(x_1, x_2, \ldots)]$ is not positive.

10.2 Interpretations of real numbers

Let us consider again the sequence (.1, .101, .101001, .1010010001, . . .). This sequence is bounded and monotone increasing. The real number $t =$ [(.1, .101, .101001, .1010010001, . . .)] is, by definition, the collection of all bounded monotone increasing sequences $(x_1, x_2, x_3, \ldots)$ such that $(x_1 - .1, x_2 - .101, x_3 - .101001, \ldots)$ is a null sequence. But there are several other ways to think of the real number t.

The sequence (.1, .101, .101001, . . .) has no limit in F. That is, there is no rational number to which this sequence converges. We may think of t as the number we invented to be the limit of this sequence. t is the number which is approximated very closely by the terms far out in the sequence.

Putting this more geometrically, we may think of t as the coordinate of the point P on a number line which is approached from the left by the rational points .1, .101, .101001, .1010010001, ... Or, in other words, t is the distance between the origin and this point P.

While the set {.1, .101, .101001, . . .} is bounded above, it does not have a least upper bound in F. That is, there is no rational number which is lub ({.1, .101, .101001, . . .}). We may think of t as the number we invented to be this least upper bound.

Finally, we may also think of t as the number named by the infinite decimal ".1010010001 . . . ."

Let $a = [(a_1, a_2, a_3, \ldots)] = [1.4, 1.41, 1.414, 1.4142, 1.41421, 1.414214, \ldots)]$ where, as in the last section, a_n is the largest integral multiple of 10^{-n} such that $a_n^2 < 2$. Then, by definition, a is the collection of all bounded monotone increasing sequences $(x_1, x_2, x_3, \ldots)$ such that $(x_1 - a_1, x_2 - a_2, x_3 - a_3, \ldots)$ is a null sequence. But we may also think of a as the number we invented to be the limit of $(a_1, a_2, a_3, \ldots)$. Or we may think of a as the number we invented to be lub $(\{a_1, a_2, a_3, \ldots\})$. Or we may think of a as the number named by the infinite decimal "1.414214 . . . ." And, in view of the special way we constructed a, we know that a_n^2 approaches 2 as n gets large. So we may also think of a as the number we invented to be $\sqrt{2}$. And, thus, we may also think of a as the length of the segment OT in Figure 10.1.

In general, whenever $(x_1, x_2, x_3, \ldots)$ is a bounded monotone increasing sequence of rationals we may, aside from the basic definition, think of the real number $x = [(x_1, x_2, x_3, \ldots)]$ as the number we invented to be the limit of $(x_1, x_2, x_3, \ldots)$, or as the number we invented to be lub $(\{x_1, x_2, x_3, \ldots\})$. And, depending on the special nature of the sequence $(x_1, x_2, x_3, \ldots)$ there may, of course, be other interpretations for x.

From this point onward, let us use "R" to name the collection of all real numbers.

Problems

1. Describe at least four ways to interpret each of the following real numbers.

   (a) $[(.01, .01001, .010010001, .01001000100001, \ldots)]$
   (b) the real number defined in Problem 3, Section 10.1
   (c) the real number defined in Problem 4, Section 10.1
   (d) $[(.9, .99, .999, .9999, \ldots)]$,
   (e) $[(.6, .60, .600, .6000, .60000, \ldots)]$
   (f) $[(-4, -3.7, -3.64, -3.637, -3.6364, -3.63637, -3.636364, \ldots)]$.

2. Let $(a_1, a_2, a_3, \ldots)$ be the sequence defined above in this section. Find the first five terms in the sequence $(2 - a_1^2, 2 - a_2^2, 2 - a_3^2, \ldots)$.

3. Construct an illustration similar to that in Figure 10.2 for the sequence in Problem 3, Section 10.1.

4. Argue informally in terms of the least upper bound interpretation to show that when $(x_1, x_2, x_3, \ldots)$ is a bounded monotone increasing sequence of rationals then $[(x_1, x_2, x_3, \ldots)] = [(x_{k+1}, x_{k+2}, x_{k+3}, \ldots)]$ for any positive integer k.

5. What must be the sum of the series $10^{-1} + 10^{-3} + 10^{-6} + \cdots + 10^{-(1+2+3+\cdots+n)} + \cdots$?

---

6. Let T be the collection of all monotone increasing (but not necessarily bounded) sequences of rational numbers. Define the relation E on T by saying that $(x_1, x_2, x_3, \ldots)\ E\ (y_1, y_2, y_3, \ldots)$ if $(x_1 - y_1, x_2 - y_2, x_3 - y_3, \ldots)$ is a null

sequence. Is E an equivalence relation on T? If so, discuss the nature of its equivalence classes.

10.3 Addition

We define the *sum* of two real numbers by means of the equation

$$[(x_1, x_2, x_3, \ldots)] +_R [(y_1, y_2, y_3, \ldots)] = [(x_1 + y_1, x_2 + y_2, x_3 + y_3, \ldots)].$$

That is, to find the sum of two real numbers we take monotonic increasing sequences of rationals that identify them, construct a new sequence by adding the two given sequences term by term, and then use this sequence to identify the real number that is the sum.

For example, if $x = [(.1, .11, .111, .1111, \ldots)]$, $y = [(.1, .10, .101, .1010, .10100, .101001, .1010010, .10100100, .101001000, .1010010001, \ldots)]$, $z = [(.8, .88, .888, \ldots)]$, and $w = [(.5, .54, .545, .5454, .54544, .545445, .5454454, .54544544, .545445444, .5454454445, \ldots)]$, then $x +_R y = [(.2, .21, .212, .2121, .21211, .212112, \ldots)]$, $z +_R w = [(1.3, 1.42, 1.433, 1.4342, 1.43432, 1.43433432, \ldots)]$, and $y +_R w = [(.6, .64, .646, .6464, .64644, .646446, \ldots)]$.

It is quite easy to check that this definition of addition makes sense in R. We first note that if $(x_1, x_2, x_3, \ldots)$ and $(y_1, y_2, y_3, \ldots)$ are both bounded monotonic increasing sequences in F then $(x_1 + y_1, x_2 + y_2, x_3 + y_3, \ldots)$ is also such a sequence in F. For suppose $x_n \leq A$, $x_{n+1} \geq x_n$, $y_n \leq B$, and $y_{n+1} \geq y_n$ for all n. Then $x_n + y_n \leq A + B$ and $x_{n+1} + y_{n+1} \geq x_n + y_n$ for all n.

Also note that if $x = [(x_1, x_2, x_3, \ldots)]$, $y = [(y_1, y_2, y_3, \ldots)]$, $z = [(z_1, z_2, z_3, \ldots)]$, and $w = [(w_1, w_2, w_3, \ldots)]$ where $x = z$ and $y = w$, then $x +_R z = y +_R w$. For if $(x_1 - z_1, x_2 - z_2, x_3 - z_3, \ldots)$ and $(y_1 - w_1, y_2 - w_2, y_3 - w_3, \ldots)$ are both null sequences, then $(x_1 + y_1 - (z_1 + w_1), x_2 + y_2 - (z_2 + w_2), x_3 + y_3 - (z_3 + w_3), \ldots) = ((x_1 - z_1) + (y_1 - w_1), (x_2 - z_2) + (y_2 - w_2), (x_3 - z_3) + (y_3 - w_3), \ldots)$ is also a null sequence. (Theorem 9.1.4.)

So addition is an operation under which R is closed. Furthermore, it is quite easy to see that addition is commutative and associative in R. The identity element for addition in R is the real number $[(0, 0, 0, \ldots)]$.

When x and y are real numbers and $x = [(x_1, x_2, x_3, \ldots)]$ and $y = [(y_1, y_2, y_3, \ldots)]$, then we may think of x_n as a rational approximation to x and y_n as a rational approximation to y. And $x_n + y_n$ is then a rational approximation to $x +_R y$. In many applications, it suffices to deal with rational approximations to the real numbers involved, rather than with the exact values of the real numbers. Indeed, the standard practice in most applications is to use terminating decimals as approximations to the real numbers and carry out the addition on the approximations. For this reason and some other reasons, we shall give no general algorithm for the addition of real numbers.

Let us consider the problem of finding the additive inverse for a real number. Suppose x is a positive real number. Then x may be named by an infinite decimal "$b.d_1d_2d_3 \ldots$" where "b" is a decimal name for a natural number and the "d's" are decimal digits. Let $x_n = b.d_1d_2 \ldots d_n$. Then, by the manner in which we assigned a meaning to the infinite decimal we have $x = [(x_1, x_2, x_3, \ldots)]$. If we wish to construct an additive inverse for x we might consider the sequence $(-x_1, -x_2, -x_3, \ldots)$, but this sequence is not necessarily a monotone increasing sequence. In fact, unless $(x_1, x_2, x_3, \ldots)$ is a constant sequence, $(-x_1, -x_2, -x_3, \ldots)$ must *fail* to be monotone increasing. Hence, $(-x_1, -x_2, -x_3, \ldots)$ will not, in general, be in S, the collection of all bounded monotone increasing sequences of rationals, and "$[(-x_1, -x_2, -x_3, \ldots)]$" does not make sense. We may get out of this difficulty quite easily.

Suppose we let $y_n = -x_n - 10^{-n}$ so that $y_1 = -x_1 - .1$, $y_2 = -x_2 - .01$, $y_3 = -x_3 - .001, \ldots$. For example, if $(x_1, x_2, x_3, x_4, \ldots) = (.7, .77, .777, .7777, \ldots)$, then $(y_1, y_2, y_3, y_4, \ldots) = (-.8, -.78, -.778, -.7778, \ldots)$. We claim that in the general case $(y_1, y_2, y_3, \ldots)$ is a bounded monotone increasing sequence. For we know that $x_n \geq 0$ for all n so that $-x_n \leq 0$ and, therefore, $y_n = -x_n - 10^{-n} \leq 0$ for all n. Hence, $(y_1, y_2, y_3, \ldots)$ is bounded. And, furthermore, since the decimal names for x_n and x_{n+1} are identical except for the digit "d_{n+1}" and $d_{n+1} \leq 9$, we have $x_{n+1} + 10^{-(n+1)} = x_n + d_{n+1} \cdot 10^{-(n+1)} + 10^{-(n+1)} = x_n + (d_{n+1} + 1) \cdot 10^{-(n+1)} \leq x_n + 10^{-n}$. Hence, $y_{n+1} = -x_{n+1} - 10^{-(n+1)} \geq -x_n - 10^{-n} = y_n$. Therefore, $(y_1, y_2, y_3, \ldots)$ is monotone increasing.

Let $y = [(y_1, y_2, y_3, \ldots)]$. Then we have $x +_R y = [(-.1, -.01, -.001, -.0001, \ldots)] = [(0, 0, 0, 0, \ldots)]$. Therefore, y is the additive inverse for x. It is natural to call y the *opposite* for (or of) x and to use "$-b.d_1d_2d_3 \ldots$" as a name for it. We now see that every infinite decimal names some real number. We have accomplished the task we set for ourselves at the end of Chapter 9.

By a method similar to that used to show that each positive real number has a decimal name, it can be shown that each negative real number is the opposite of some positive real. We shall not go into the details of this matter. But we may conclude that *every* real number has an additive inverse or an opposite. And it also follows that each real number has an infinite decimal name. Of course, $[(0, 0, 0, \ldots)] = 0.00000 \ldots$.

We define *subtraction* in R to be the inverse operation for addition. Since each real number has an additive inverse, R is closed under subtraction. In R, as in F and I, we use the minus sign as a symbol for the operation of subtraction and also as a prefix on the name of a number to indicate the opposite of that number. Then the rules for handling signs in sums and differences in F carry over to R.

Definition 10.3.1. We define addition in R by saying that $[(x_1, x_2, x_3, \ldots)] +_R [(y_1, y_2, y_3, \ldots)] = [(x_1 + y_1, x_2 + y_2, x_3 + y_3, \ldots)]$.

Theorem 10.3.1. If x, y, z, and w are in R and $x = z$ and $y = w$, then $x +_R y = z +_R w$.

Theorem 10.3.2. Addition is commutative and associative in R and [(0, 0, 0, . . .)] is the identity element for addition in R.

Theorem 10.3.3. If x is a positive real number then x has an additive inverse in R.

Theorem 10.3.4. Each negative real number is the additive inverse for a unique positive real number.

Definition 10.3.2. If x is the additive inverse (or the **opposite**) of the positive real number $b.d_1d_2d_3 \ldots$, then we say that $x = -b.d_1d_2d_3 \ldots$. (Note that as a consequence of Theorem 10.3.4 and the fact that [(0, 0, 0, . . .)] is its own opposite, we now have it that each real number has an infinite decimal name and each infinite decimal names a real number.) In general, $-x$ will be the opposite for x.

Theorem 10.3.5. The opposite of a negative real number is positive and *vice versa*.

Definition 10.3.3. **Subtraction** in R is the inverse operation for addition in R. Notation: $-_R$.

Theorem 10.3.6. R is closed under subtraction and $x -_R y = x +_R (-y)$.

PROBLEMS

1. With x, y, z, and w as given on page 411 find (a) $y +_R z$, (b) $x +_R w$, (c) $x +_R z$.
2. Prove that addition in R is commutative and associative.
3. Prove that [(0, 0, 0, . . .)] is the identity element for addition in R.
4. Show that [(−.4, −.34, −.334, −.3334, . . .)] is the additive inverse for [(.3, .33, .333, .3333, . . .)].
5. Interpreting the infinite decimals as names of real numbers prove that (a) 1.0000 . . . = .9999 . . . , (b) −2.9999 . . . = −3.0000 . . . .
6. Prove that the set of *positive* real numbers is closed under addition.
7. Show that when x is a positive real number then $-x$ is a negative real number.

---

8. Prove Theorem 10.3.4.
9. Prove Theorem 10.3.5.

10.4 Multiplication

Suppose a and b are the real numbers given by a = [(.7, .727, .727227, .7272272227, . . .)] and b = [(.5, .55, .555, .5555, . . .)]. What should we take to be the product of a and b? It is very natural to take $a \cdot_R b$ to be the real number determined by the sequence obtained by multiplying term by term the sequences which determine a and b. That is, we take it that $a \cdot_R b = ((.7)(.5),$ $(.727)(.55),$ $(.727227)(.555),$ $(.7272272227)(.5555),$ $\ldots) = [(.35,$ $.39985,$ $.403610985,$ $.40397472220985,$ $\ldots)]$. When we try to generalize this approach to define multiplication in R, we find that we must first restrict our attention to nonnegative real numbers and then take care of other cases by another technique.

In general, when x and y are nonnegative real numbers we can find monotone increasing sequences of rationals which represent them and have all their terms nonnegative. In fact, we may simply use the sequences that go with their infinite decimal names to accomplish this.

Suppose $x = [(x_1, x_2, x_3, \ldots)]$ and $y = [(y_1, y_2, y_3, \ldots)]$ where $x_n \geq 0$ and $y_n \geq 0$ for all n. Then we define the **product** of x and y to be

$$x \cdot_R y = [(x_1y_1, x_2y_2, x_3y_3, \ldots)].$$

$(x_1y_1, x_2y_2, x_3y_3, \ldots)$ is a bounded monotone increasing sequence of rationals. For when $0 \leq x_n \leq A$, $0 \leq y_n \leq B$, $x_{n+1} \geq x_n$, and $y_{n+1} \geq y_n$ for all n, then we have $x_ny_n \leq AB$ and $x_{n+1}y_{n+1} \geq x_ny_n$ for all n. Hence, $[(x_1y_1, x_2y_2, x_3y_3, \ldots)]$ is a real number.

To confirm that we have indeed defined an operation on the positive reals, suppose that x, y, z, and w are positive real numbers and that $x = z$ and $y = w$. Let $x = [(x_1, x_2, x_3, \ldots)]$, $y = [(y_1, y_2, y_3, \ldots)]$, $z = [(z_1, z_2, z_3, \ldots)]$, and $w = [(w_1, w_2, w_3, \ldots)]$ where all the terms in these sequences are nonnegative. We know that all four of the given sequences are bounded and also that $(x_1 - z_1,$ $x_2 - z_2, x_3 - z_3, \ldots)$ and $(y_1 - w_1, y_2 - w_2, y_3 - w_3, \ldots)$ are null sequences. To show that $x \cdot_R y = z \cdot_R w$, we need only show that $(x_1y_1 - z_1w_1, x_2y_2 - z_2w_2,$ $x_3y_3 - z_3w_3, \ldots)$ is a null sequence. We have for any n that $x_ny_n - z_nw_n =$ $x_ny_n - y_nz_n + y_nz_n - z_nw_n = y_n(x_n - z_n) + z_n(y_n - w_n)$. Since $(y_1, y_2, y_3, \ldots)$ is bounded and $(x_1 - z_1, x_2 - z_2, x_3 - z_3, \ldots)$ is a null sequence, $(y_1(x_1 - z_1),$ $y_2(x_2 - z_2), \ldots)$ is a null sequence. (Theorem 9.1.5.) Similarly, $(z_1(y_1 - w_1),$ $z_2(y_2 - w_2), \ldots)$ is a null sequence. Since the sum term by term of two null sequences yields a null sequence, we conclude that $x \cdot_R y = z \cdot_R w$.

To extend multiplication to all of R, we first introduce the notion of the **absolute value** of a real number. When x is real we let

$$|x| = x \quad \text{if } x \text{ is nonnegative and}$$

$$|x| = -x \quad \text{if } x \text{ is negative.}$$

We then say that $x \cdot_R y = -(|x| \cdot_R |y|)$ when x and y differ in sign and $x \cdot_R y =$ $|x| \cdot_R |y|$ when x and y are both negative. These definitions make sense since we have already described how to find the product of two nonnegative reals, and $|x|$ and $|y|$ are always nonnegative.

With multiplication defined on all of R in this fashion, it follows that R is closed under the operation and that the operation is commutative and associative. Also multiplication distributes across addition in R. The proofs of these facts are straightforward and somewhat tedious. We shall omit them.

It is easy to see that $[(1, 1, 1, \ldots)]$ is the identity for multiplication in R.

And it is easy to believe that every real number different from $[(0, 0, 0, \ldots)]$ has a multiplicative inverse or reciprocal. But it is not so easy to prove this as we might hope. When $x = [(x_1, x_2, x_3, \ldots)]$, we might expect that $(1/x_1, 1/x_2, 1/x_3, \ldots)$ would be a sequence which would determine the reciprocal of x. But, unfortunately, $(1/x_1, 1/x_2, 1/x_3, \ldots)$ will not, in general, be a monotone increasing sequence. One device to get around this when x is positive is the following. Suppose $x = [(x_1, x_2, x_3, \ldots)]$ where $x_n > 0$ for all n. Consider the sequence $(-1/x_1, -1/x_2, -1/x_3, \ldots)$. Since $x_{n+1} \geq x_n > 0$ we have $1/x_n \geq 1/x_{n+1}$ and hence, $-1/x_n \leq -1/x_{n+1}$ for all n. Let $y_n = -1/x_n$. Then $y = [(y_1, y_2, y_3, \ldots)]$ is a real number. Also $x \cdot_R y = [(-1, -1, -1, \ldots)]$ so that $-y$ is the multiplicative inverse for x. When x is negative a similar device may be used.

We define **division** to be the inverse operation for multiplication in R. Since every real number except $[(0, 0, 0, \ldots)]$ has a reciprocal, division, except by $[(0, 0, 0, \ldots)]$, is always possible in R.

We shall give no algorithms for multiplication and division in R. In most applications, when we wish to multiply or divide real numbers, we instead perform the indicated operations on rational approximations to the given real numbers. When x_n and y_n are good rational approximations to the real numbers x and y, respectively, we expect $x_n y_n$ and x_n/y_n to be good approximations to $x \cdot_R y$ and $x \div_R y$. When we are working with desk calculators or electronic computers where all numbers are treated as terminating decimals, it is always the practice to deal only with rational approximations. We shall return to this topic in Section 10.9.

Definition 10.4.1. Suppose x and y are nonnegative real numbers. Let $x = [(x_1, x_2, x_3, \ldots)]$ and $y = [(y_1, y_2, y_3, \ldots)]$ where $x_n \geq 0$ and $y_n \geq 0$ for all n. We define the **product** of x and y to be $x \cdot_R y = [(x_1y_1, x_2y_2, x_3y_3, \ldots)]$.

Definition 10.4.2. We define the absolute value of a real number x by taking

$$|x| = x \quad \text{if } x \text{ is nonnegative and}$$

$$|x| = -x \quad \text{if } x \text{ is negative.}$$

Definition 10.4.3. $x \cdot_R y = -(|x| \cdot_R |y|)$ when x and y differ in sign. $x \cdot_R y = |x| \cdot_R |y|$ when x and y are negative.

Theorem 10.4.1. If $x = y$ and $z = w$, then $x \cdot_R z = y \cdot_R w$.

Theorem 10.4.2. R is closed under multiplication. Multiplication is commutative and associative and distributes across addition in R. $[(1, 1, 1, \ldots)]$ is the identity element for multiplication in R.

Theorem 10.4.3. Every real number different from $[(0, 0, 0, \ldots)]$ has a multiplicative inverse (or **reciprocal**) in R.

Definition 10.4.4. **Division** is the inverse operation for multiplication in R. Notation: $\div_R$.

Theorem 10.4.4. If $y \neq [(0, 0, 0, \ldots)]$, then $x \div_R y$ is a well-defined real number. In fact, $x \div_R y = x \cdot_R y^{-1}$ where y^{-1} is the reciprocal for y.

PROBLEMS

1. Find each of the following products by displaying the first few terms in a sequence identifying the product.

(a) $[(.5, .5, .5, .5, \ldots)] \cdot_R [(.3, .3, .3, .3, \ldots)]$
(b) $[(.3, .33, .333, .3333, \ldots)] \cdot_R [(3, 3, 3, 3, \ldots)]$
(c) $[(.1, .101, .101001, .1010010001, \ldots)] \cdot_R [(3, 3, 3, 3, \ldots)]$
(d) $[(.1, .101, .101001, .1010010001, \ldots)] \cdot_R [(.79, .7979, .797979, \ldots)]$
(e) $[(.67, .6767, .676767, \ldots)] \cdot_R [(.79, .7979, .797979, \ldots)]$
(f) $(.676767\ldots) \cdot_R (.797979\ldots)$
(g) $(-.676767\ldots) \cdot_R (.797979\ldots)$
(h) $(-.676767\ldots) \cdot_R (-.797979\ldots)$
(i) $(.181881888188881\ldots) \cdot_R (.16161616\ldots)$
(j) $[(.9, .99, .999, \ldots)] \cdot_R [(.9, .99, .999, \ldots)]$

2. Prove that when x, y, and z are nonnegative real numbers then (a) $x \cdot_R y = y \cdot_R x$, (b) $(x \cdot_R y) \cdot_R z = x \cdot_R (y \cdot_R z)$, (c) $x \cdot_R (y +_R z) = (x \cdot_R y) +_R (x \cdot_R z)$.

3. Prove that $[(1, 1, 1, \ldots)]$ is the identity element for multiplication in R.

4. Prove that the set of positive real numbers is closed under multiplication.

5. Referring to Section 9.1 and Problem 3, Section 9.1, find a good rational approximation to $\sqrt{2} \cdot_R \sqrt{3}$.

10.5 Rational real numbers

We have said that the real number $x = [(x_1, x_2, x_3, \ldots)]$ may be thought of as the number we invented to be the limit of the bounded monotone increasing sequence of rationals $(x_1, x_2, x_3, \ldots)$.

But in the rational number system itself, any constant sequence is a bounded monotone increasing sequence which already has a limit in F. Hence, if we expect R to be an extension of F it should turn out to be the case that the real number $[(a, a, a, \ldots)]$ will behave exactly like the rational number a. And this is indeed the situation.

We shall say that a real number x is a **rational real number** if there is a rational number a such that $x = [(a, a, a, \ldots)]$. A real number that is not rational is said to be **irrational.**

Let R^* be the collection of all rational real numbers. We claim that the system $(R^*, +_R, \cdot_R)$ is isomorphic to the system $(F, +, \cdot)$. In fact, for x in R^* and $x = [(a, a, a, \ldots)]$ the function f defined by $f(x) = a$ is an isomorphism between the two systems.

The real number $x = [(a, a, a, \ldots)]$ is quite a different beast from the rational number a. x is the collection of all bounded monotone increasing sequences of rationals $(x_1, x_2, x_3, \ldots)$ such that $(x_1 - a, x_2 - a, x_3 - a, \ldots)$ is a null sequence. Yet we now see that in view of the way we defined addition and multiplication in R, such an x behaves exactly like the rational number a as far as these operations go.

So we shall do as we have done in similar situations in the past and consider an expression such as "1/2" to be a name either for the rational number 1/2 or for the real number $[(1/2, 1/2, 1/2, \ldots)]$. And this deliberate ambiguity will cause no confusion as long as we are dealing with the operations of addition and multiplication and concepts that can be defined in terms of them. We may in most instances treat any rational real number as if it were identical with its corresponding rational number.

We may now also drop the subscript "R" on the symbols for the operations of addition, multiplication, subtraction, and division in R.

In view of the conventions justified by the above isomorphism we may now say that a real number is rational if and only if it can be written as a ratio of integers. To show that a real number x is rational we need only find integers p and q such that $x = p/q$.

We have seen that any real number may be named by an infinite decimal. But infinite *repeating* decimals we have seen to be names for rational numbers. In view of the above isomorphism, we now see that a real number is rational if and only if it has an infinite decimal name which is a repeating decimal.

Of course, if the digits in an infinite decimal are, from a certain point on, all "0's," we shall feel free to omit them and use the resulting terminating decimal to name the real number.

Any infinite decimal that is not repeating is the name of an irrational real number. For example, .01001000100001 . . . is an irrational number.

In view of the route by which we extended each of the number systems we have considered, we may now consider N, I, and F all to be subsets of R. And, indeed, we have $N \subset I \subset F \subset R$.

When we wish to perform calculations with rational real numbers, we may use any devices we developed for such purposes in F. And, as we have indicated before, when dealing with *any* real numbers we must often carry out operations on rational approximations rather than on the real numbers themselves.

It is a strange fact that, from a practical point of view, the system of real numbers is probably the most useful number system there is, and yet when it comes to doing actual arithmetic operations we *use* real numbers hardly at all—we fall back on rational approximations instead. This, no doubt, has something to do with the inherent complexity of irrational numbers. An integer can be identified by giving a single ordered pair of natural numbers. And a rational number can, in turn, be identified by giving a single pair of integers. But in order to identify an irrational real number we must evidently get involved with some sort of infinite process—either an infinite sequence of decimal digits, or an infinite sequence of rational numbers which determines the irrational.

And yet without these complicated beasts called irrational numbers we would have points on the number line that could not be assigned coordinates. A line should, we presume, have no breaks or gaps in it. Our geometric intuition forces these irrational numbers on us.

It is possible to pursue these ideas more deeply by making other distinctions between various kinds of real numbers. We shall return to this matter in Section 10.8.

PROBLEMS

1. The symbol "3" may now be thought of as a name for a natural number, for an integer, for a rational number, or for a real number. Explain exactly what "3" means under each of these four interpretations.

2. As a partial verification that the function f defined in this section has the properties claimed for it, show that it is an isomorphism between the systems $(R^{*+}, +_R, \cdot_R)$ and $(F^+, +, \cdot)$ where R^{*+} is the set of positive rational real numbers and F^+ is the set of positive rational numbers.

3. Are the systems $(R^*, +_R, \cdot_R, -_R, \div_R)$ and $(F, +, \cdot, -, \div)$ isomorphic? Explain. (Here R^* is the collection of all rational real numbers.)

4. Is the set of irrational real numbers closed under (a) addition, (b) multiplication?

5. Could the sum of a rational number and an irrational number be rational? Explain.

6. Could the product of a rational number and an irrational number be rational? Explain. What if the rational number is different from 0?

---

7. To verify that the function f defined in this section is indeed a one-to-one correspondence, show that (i) if a and b are rational numbers and $a \neq b$, then $[(a, a, a, \ldots)] \neq [(b, b, b, \ldots)]$, (ii) if $[(a, a, a, \ldots)] \neq [(b, b, b, \ldots)]$ then $a \neq b$.

8. By considering the various possible signs on the real numbers involved, complete the proof, started in Problem 2, that shows that the function f defined in this section is an isomorphism between $(R^*, +_R, \cdot_R)$ and $(F, +, \cdot)$.

10.6 Inequalities

We may carry over to R much of our work with inequalities in F. We first say what it means for one real number to be at least as large as another by stating the definition:

> If x and y are in R, we say that $x \geq y$ if there is a nonnegative real number z such that $x = y + z$.

We then define $>$, $\leq$, $<$, "maximum," "minimum," "between," "upper bound," "lower bound," "bounded," "lub," and "glb" just as we did in F.

And many of the theorems about inequalities in F carry over to R. In particular, if x, y, and z are real numbers then (a) $x \geq y$ and $y \geq z$ implies $x \geq z$, (b) $x \geq y$ and $y \geq x$ implies $x = y$, (c) $x > y$ or $x = y$ or $x < y$ and at most one of these can hold, (d) $x \geq y$ if and only if $x + z \geq y + z$, (e) $x \geq y$ and $z > 0$ implies $xz \geq yz$, (f) $x \geq y$ and $z < 0$ implies $xz \leq yz$, and (g) $x > 0$ implies that there is an integer k such that $kx > y$. We shall not dig into the proofs of these and similar theorems which are analogues of theorems in F. Instead, let us see how F and R differ in regard to $\geq$.

There *is* a basic difference between the order situation in F and that in R. In F, it is possible to have a set of numbers which is bounded above and yet has no lub. An example of such a set is $\{x \mid x \in F \text{ and } x^2 < 2\}$. There is no lub in F for this set because $\sqrt{2}$ is not in F. But this sort of thing *cannot* happen in R. That is:

> If A is a nonempty set of real numbers which is bounded above, then there is a real number u such that lub $(A) = u$.

More briefly, we may say that any nonempty set of reals that is bounded above has a real lub.

Because this is true, the real number system is said to be *complete*. We shall prove the completeness of R in the appendix to this section. Let us here look more closely at the significance of this property.

Consider the set $A = \{.1, .101, .101001, .1010010001, \ldots\}$. Each element of A is a rational number. That is, $A \subseteq F$. Also A is a bounded set. In particular, if $x \in A$ then $x < .2$. Also, if $x \in A$, then $x < .102$. And if $x \in A$, then $x < .101002$. If $x \in A$, then $x < .1010010002$. But when we try to find the *smallest* number that is an upper bound for A, we find that we must move outside of F. The least upper bound of A is the real number $.1010010001 \ldots$. And this number is irrational.

The completeness of R assures us that, in general, when a set of numbers has at least one upper bound, then it must have a *smallest* upper bound also. In other words, if the collection of upper bounds for a set is not empty, then that collection actually has a minimum. This may sound innocuous enough. But recall that it is very easy to construct a set of real numbers (or of rational numbers) that does *not* have a minimum. For example, $\{x \mid x \in F \text{ and } x > 0\}$ has no minimum. The completeness of R tells us that if C is any nonempty set of real numbers, then the collection $\{x \mid x \in R \text{ and } x \text{ is an upper bound for } C\}$ is either empty or it has a minimum element. This is a very powerful conclusion about a very broad class of collections.

Without going into much detail let us mention some of the immediate consequences of the completeness of R.

If $(a_1, a_2, a_3, \ldots)$ is a bounded monotone increasing sequence of real numbers,† then the sequence converges in R. A similar assertion does not hold in F. For example, the sequence (.1, .101, .101001, .1010010001, . . .) is a bounded monotone increasing sequence in F, but the sequence does not converge in F. The limit of the sequence is the number .1010010001 . . . which is irrational and is not in F. Until we expand the system F to get the system R, the number .1010010001 . . . is not there for the sequence to converge to.

If $b_1 + b_2 + b_3 + \cdots$ is a series whose terms are all positive real numbers and whose partial sums are bounded, then the series converges in R. A similar statement does not hold in F. For example, the series $.1 + .001 + .000001 + .0000000001 + \cdots + 10^{-(1+2+3+\cdots+n)} + \cdots$ has all positive rational terms and it can be seen that all its partial sums are less than .2. Yet this series does not converge in F. The sum of the series is the number .1010010001 . . . which is not in F.

Any bounded sequence of real numbers has a subsequence which converges in R. A similar statement does not hold in F. For example, (.1, .101, .101001, .1010010001, . . .) is a bounded sequence in F. Yet every subsequence of this sequence fails to converge in F. This is because each subsequence has the limit .1010010001 . . . which is not in F.

When we leave the completeness property and consider how rational and irrational points are distributed on the line, we discover that any segment of nonzero length must contain both rational and irrational points. For it can be shown that between any two distinct real numbers there is a rational number and also an irrational number. Thus, it is sometimes said that the irrational points, as well as the rational points, are *dense* on the line. A general proof of this result may be built along the lines of the following example. Let $x =$ .101001000100001 . . . and $y =$ .101001001001001 . . . . Then .1010010005 is a rational number that is between x and y, and .1010010005050050005000005 . . . is an irrational number that is between x and y.

When two real numbers are written as infinite decimals we may usually determine which is larger by comparing the names a digit at a time as we did in Section 8.10. But some difficulties do arise with this procedure. For example, if we carry over blindly the digit by digit comparison procedure, we would evidently conclude by it that $.999\ldots < 1.000\ldots$. But this is false since $.999\ldots = 1$. However, we may get out of this difficulty by agreeing not to use a decimal name that has an infinite string of "9's" at the end. For example, instead of "47.9543999 . . ." we should write "47.9544" when we wish to compare the size of this number with another.

Even with this modification some difficulties remain. Suppose we have

† We carry over to R all the basic definitions for sequences and series as given in Chapter 9.

checked the first 1000 digits in the names for x and y and found them to be the same. May we conclude that $x = y$? Evidently not, since the names for x and y may differ in the next decimal place. Thus, there may be very real practical and theoretical difficulties in deciding the relative sizes of two given real numbers.

The following is an illustration of the kind of difficulty that might arise. Let $x = .d_1d_2d_3 \ldots$ where $d_n = 1$ if the nth digit in the decimal name for $\sqrt{2}$ starts a string of seven "5's," and $d_n = 0$ otherwise. Is $x = 0$? To prove that x does equal 0 we would evidently need to prove that no string of seven "5's" ever occurs in the decimal name for $\sqrt{2}$. And we could never hope to prove that x does equal 0 by simply grinding out successive decimal digits for $\sqrt{2}$. But we could prove that $x \neq 0$ and, hence, that $x > 0$ by finding a string of seven "5"'s in the decimal expansion of $\sqrt{2}$.

Of course, in most applications when two numerals agree out to quite a few digits past the decimal point, they may be considered, for all practical purposes, to be names of the same number.

Definition 10.6.1. If x and y are real numbers we say that $x \geq y$ if there is a nonnegative real number z such that $x = y + z$. We define $>$, $\leq$, $<$, "maximum," "minimum," "bounded," "lub," and "glb" as before.

Theorem 10.6.1. Suppose x, y, and z are in R.

(a) $x \geq y$ and $y \geq z$ implies $x \geq z$.
(b) $x \geq y$ and $y \geq x$ implies $x = y$.
(c) $x > y$ or $x = y$ or $x < y$ and at most one of these can hold.
(d) $x \geq y$ if and only if $x + z \geq y + z$.
(e) $x \geq y$ and $z > 0$ implies $xz \geq yz$.
(f) $x \geq y$ and $z < 0$ implies $xz \leq yz$.
(g) $x > 0$ implies that there is an integer k such that $kx > y$.

Theorem 10.6.2. The system of real numbers is *complete* in the sense that any nonempty set of real numbers that has an upper bound has a least upper bound.

PROOF: Let A be a nonempty set of real numbers and suppose that b is a real upper bound for A. Among the numbers $\ldots, -.2, -.1, 0, .1, .2, .3, \ldots$ there must be a largest one which is not an upper bound for A. This is because some integral multiple of .1 must be larger than b and, hence, be an upper bound for A. Also, A is not empty and must contain some real number, say x. Some integral multiple of .1 must be smaller than x and, hence, fail to be an upper bound for A. Let u_1 be the largest integral multiple of .1 which is not an upper bound for A.

More generally, let u_n be the largest integral multiple of 10^{-n} which is not an upper bound for A. The sequence $(u_1, u_2, u_3, \ldots)$ is a bounded monotone increasing sequence of rational numbers. Let u be the real number determined by this sequence so that $u = [(u_1, u_2, u_3, \ldots)]$. We claim that $u = \text{lub } (A)$.

First let us verify that u is an upper bound for A. Suppose it is not. Then there is a real number x in A such that $u < x$. Let $d = x - u$. Then $d > 0$ and there exists an n such that $10^{-n} < d$. Now $u_n \leq u$ and, hence, $u_n + 10^{-n} < u + d = x$. Therefore, $u_n + 10^{-n}$ is not an upper bound for A. But this is a contradiction since u_n was chosen to be the largest integral multiple of 10^{-n} that was not an upper bound for A. We conclude that u must be an upper bound for A.

To show that u is the *least* upper bound for A, we shall show that no real number less than u can be an upper bound for A. Suppose $t < u$. Let $u - t = d^*$. Then $d^* > 0$. Hence, there exists a k such that $2 \cdot 10^{-k} < d^*$. Then there must be at least one integral multiple of 10^{-k} which is between t and u. The largest of these must be u_k. But, since u_k is not an upper bound for A, there is a number x in A such that $u_k < x$. Since $t < u_k$, this yields $t < x$. This shows that t is not an upper bound for A and finishes the proof of the theorem.

Theorem 10.6.3. Between any two distinct real numbers there is a rational number and also an irrational number.

Problems

1. Find the real number (if any) that is the least upper bound of each of the following sets of real numbers. (a) $\{.1, -.1, .101, -.101, .101001, -.101001, .1010010001, -.1010010001, \ldots\}$, (b) $\{x \mid x < 1\}$, (c) $\{x \mid x \leq 1\}$, (d) $\{\sqrt{2}\}$, (e) $\{x \mid x^2 < 3\}$, (f) $\{x \mid x^4 < 6\}$, (g) $\{x \mid x^2 > 2\}$, (h) $\{.2, .232, .232332, .2323323332, \ldots\}$.

2. Is the system of natural numbers complete in the sense that every nonempty set of natural numbers which is bounded above has a lub which is a natural number? Explain.

3. Is the system of integers complete? (See Problem 2.) Explain.

4. Show that any nonempty set of real numbers which is bounded below has a real glb. [Hint: If A is the given set let $B = \{y \mid y = -x \text{ for some } x \text{ in } A\}$. What can be said about lub (B)?]

5. Find, if possible, a rational number and an irrational number between each of the two given numbers. (a) 1, 2; (b) .10564, .10563; (c) 1, .9999 . . . ; (d) .56565656 . . . , .5656656665 . . . ; (e) $\sqrt{2}$, $\sqrt{3}$; (f) 10^{-50}, 10^{-51}; (g) −.4769534, −.4769533; (h) 3×10^{-7}, 4×10^{-7}; (i) $\sqrt{36}$, $\sqrt{37}$.

6. Prove Theorem 10.6.3.

7. Show that if $(x_1, x_2, x_3, \ldots)$ is a bounded monotone increasing sequence of real numbers, then there is a real number to which $(x_1, x_2, x_3, \ldots)$ converges. [HINT: What can be said about the set $\{x_1, x_2, x_3, \ldots\}$?]

8. Suppose that $a_n \geq 0$ for all n and that all the partial sums of the series $a_1 + a_2 + a_3 + \ldots$ are less than a number B. Show that the series converges to a real number. [HINT: See Problem 7.)

9. Prove that any bounded sequence of real numbers has a subsequence which is convergent in R.

10.7 Exponents

We carry over from F to R the basic definitions for exponents. If a is real and $a \neq 0$, we let $a^0 = 1$. When n is a natural number we let $a^{n+1} = a \cdot a^n$ and $a^{-n} = 1/a^n$. When p is a positive integer we say that $y = a^{1/p} = \sqrt[p]{a}$ if $y^p = a$ and $y \geq 0$ when p is even. When p and q are integers and p is positive we let $a^{q/p} = \sqrt[p]{a^q}$.

When a is real and n is a natural number, a^n is a well-defined real number as long as a and n are not both 0. This is because R is closed under multiplication. Also, since division, except by 0, is always possible in R, a^{-n} is a real number when $a \neq 0$ and n is a natural number.

When p is a positive integer, $\sqrt[p]{a}$ may or may not be a real number. If $a < 0$ and $p = 2k$ so that p is even, then $\sqrt[p]{a}$ is not a real number. This follows from the fact that if b is any real number then $b^2 \geq 0$ so that $b^p = b^{2k} = (b^2)^k \geq 0$. So we could not have $b^p = a$ when $a < 0$.

On the other hand, if p is odd, then $\sqrt[p]{a}$ is a real number for any real a. And when $a \geq 0$, $\sqrt[p]{a}$ is a real number for any positive integer p.

Putting these results another way, we may say that the equation $x^p = a$ has a real solution whenever $a \geq 0$ or whenever p is odd. But $x^p = a$ has no real solution when p is even and $a < 0$.

To suggest why the above general assertions hold, let us consider a particular case. Suppose we wish to show that the equation $x^4 = 6$ has a solution in R. We shall accomplish this if we show that $\sqrt[4]{6}$ is a real number. Note that $1^4 = 1$ and $1 < 6$. Also $2^4 = 16$ and $16 > 6$. We next find that $(1.5)^4 = 5.0725$ and $(1.6)^4 = 6.6536$. Let $x_1 = 1.5$. We next carry out computations to discover that $(1.56)^4 < 6$ and $(1.57)^4 > 6$. So we let $x_2 = 1.56$. We then find that $(1.561)^4 < 6$ and $(1.565)^4 > 6$. So we let $x_3 = 1.564$. And we continue in this fashion to generate a sequence $(1.5, 1.56, 1.564, \ldots)$. x_n, the nth term of this sequence, is the largest integral multiple of 10^{-n} whose fourth power is less than 6. We claim that $x = [(1.5, 1.56, 1.564, \ldots)] = \sqrt[4]{6}$. Since $(1.5, 1.56, 1.564, \ldots)$ is clearly a bounded monotone increasing sequence of rationals and $x > 0$ and $x^4 = [(x_1^4, x_2^4, x_3^4, \ldots)]$, all we need to do to establish this is to show that

$(6 - x_1^4, 6 - x_2^4, 6 - x_3^4, \ldots)$ is a null sequence. But because of the manner in which x_n is defined we have for each n, $x_n^4 < 6 \leq (x_n + 10^{-n})^4$. Therefore,

$$\begin{aligned} 0 < 6 - x_n^4 &\leq (x_n + 10^{-n})^4 - x_n^4 \\ &= [(x_n + 10^{-n})^2 + x_n^2]\cdot[(x_n + 10^{-n})^2 - x_n^2] \\ &= [(x_n + 10^{-n})^2 + x_n^2][(x_n + 10^{-n}) + x_n][(x_n + 10^{-n}) - x_n]. \end{aligned}$$

Here we have twice used the fact that $z^2 - w^2 = (z + w)(z - w)$. But clearly we have $x_n < 2$ and $x_n + 10^{-n} < 2$ for every n. So $6 - x_n^4$ must be less than $(2^2 + 2^2)(2 + 2)(10^{-n}) = 32 \cdot 10^{-n}$. And this number can be made as small as we please by taking n large enough. Hence, $x = \sqrt[4]{6}$.

We see then that the real number x constructed above is a solution of the equation $x^4 = 6$. But we should note that there is also another solution to this equation. Namely, $-\sqrt[4]{6}$.

In general, when p is even and a is positive, the equation $x^p = a$ has the two real solutions $\sqrt[p]{a}$ and $-\sqrt[p]{a}$.

When p is odd, the equation $x^p = a$ has exactly one real solution. It is $\sqrt[p]{a}$.

Tables which give rational approximations for roots of numbers are available in various books. And several small electronic desk calculators are now available that quickly compute approximate roots of numbers when they exist.

A flow chart for finding the approximate value of $\sqrt[p]{a}$ when a is positive may be given as in Figure 10.3. The process diagrammed yields x_n, the largest multiple of 10^{-n} which is less than $\sqrt[p]{a}$. The diagrammed process is very inefficient since it simply computes $(10^{-n})^p$, $(2\cdot 10^{-n})^p$, $(3\cdot 10^{-n})^p$, ... until it finds a number at least as great as a, and then takes the proper choice for x_n.

We now see that when a is positive $a^{q/p} = \sqrt[p]{a^q}$ is a real number whenever p and q are integers and p is positive. And we evidently have a method for finding rational approximations to $a^{q/p}$ with any desired degree of accuracy.

But we have still not said what "a^x" might mean if x happens to be an irrational number. When x is irrational we cannot express it as the ratio of two integers and, hence, cannot treat a^x as if it were the integral power of a certain root.

But there is a very natural way to define a^x when a is positive and x is *any* real number. Suppose $x = [(x_1, x_2, x_3, \ldots)]$ where $(x_1, x_2, x_3, \ldots)$ is a bounded monotone increasing sequence of rationals. Then $(a^{x_1}, a^{x_2}, a^{x_3}, \ldots)$ is a sequence of real numbers. We take the limit of this sequence to be a^x. Hence, we may find the approximate value of a^x by taking x_n to be a rational number that approximates x and computing a^{x_n}.

As a result of the above definition, we find that a^x is a well-defined real number whenever a is positive and x is real. We shall not go into the details of the verification of this. The rules for handling exponents as developed in Chapter 8 carry over to R. Also let us note one important fact that the intuition easily accepts. When $a > 1$ and $x < y$, then $a^x < a^y$.

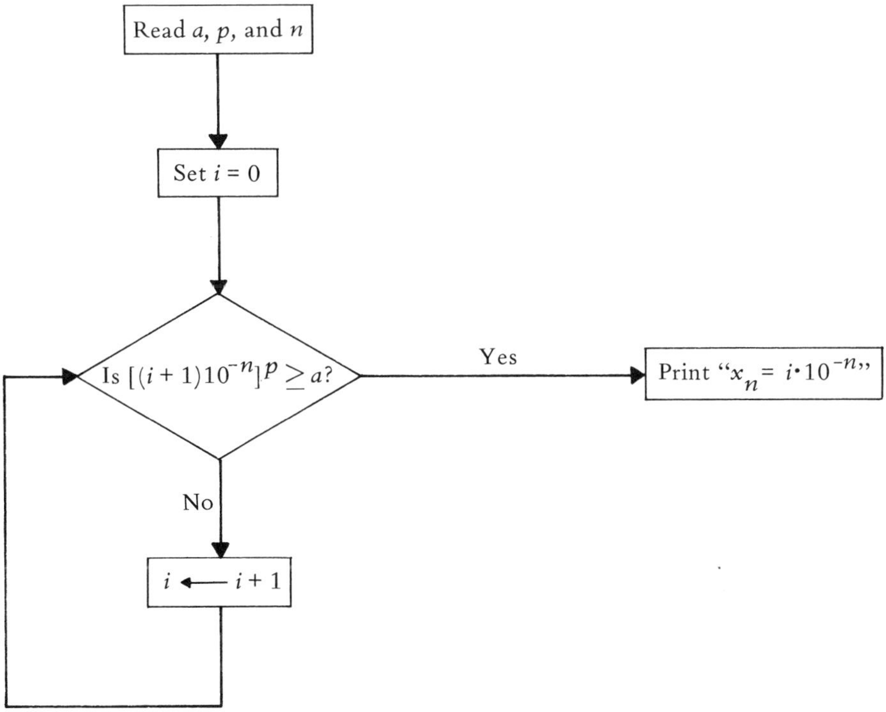

FIGURE 10.3

We may now give a rather convincing reason, promised long ago, for leaving 0^0 undefined. We know that for each positive integer n we have $0^{1/n} = \sqrt[n]{0} = 0$. Therefore, the sequence $(0^1, 0^{1/2}, 0^{1/3}, 0^{1/4}, \ldots)$ converges to 0. Since $\lim_{n\to\infty} 1/n = 0$, it would seem we should have $0^0 = 0$. On the other hand, $(1/n)^0 = 1$ for each positive integer n, and the sequence $(1^0, (1/2)^0, (1/3)^0, (1/4)^0, \ldots)$ converges to 1. So it seems we should have $0^0 = 1$. If we were to define 0^0 to be some number k, we would want $\lim_{n\to\infty} a_n^{b_n} = k$ whenever $\lim_{n\to\infty} a_n = 0$ and $\lim_{n\to\infty} b_n = 0$. But we see that we could not hope to have this. Because with the proper choices of a_n and b_n we can make $a_n^{b_n}$ approach 0 or we can make it approach 1. And, in fact, we can even make $a_n^{b_n}$ converge to other values. For example, with $a_n = 1/(2^n n)$ and $b_n = 1/n$, it can be shown that $\lim_{n\to\infty} a_n^{b_n} = 1/2$.

PROBLEMS

1. Use the method applied in this section to $\sqrt[4]{6}$ to find the first three digits in the decimal name for (a) $\sqrt{5}$, (b) $\sqrt[4]{2}$, (c) $\sqrt[3]{7}$, (d) $\sqrt[3]{-5}$, (e) $\sqrt[3]{5}$, (f) $\sqrt[3]{27}$, (g) $\sqrt{.15}$, (h) $\sqrt[3]{.000001}$, (i) $\sqrt[4]{.0001}$, (j) $\sqrt[3]{.14}$.

2. Consulting a table of square roots, find the approximate value for each of the following. [HINT: $\sqrt[4]{a} = \sqrt{\sqrt{a}}$.] (a) $\sqrt{17}$, (b) $\sqrt{270}$, (c) $\sqrt{27000}$, (d) $\sqrt[4]{17}$, (e) $\sqrt{10}$, (f) $\sqrt{.11} = \sqrt{11/100} = \sqrt{11}/10$, (g) $\sqrt{.007} = \sqrt{70/10{,}000} = \sqrt{70}/100$, (h) $\sqrt{.003}$, (i) $\sqrt{.0034}$, (j) $\sqrt{.00062}$.

3. Describe, but do not carry out, the calculations we could use to find the approximate value of (a) $2^{\sqrt{2}}$, (b) $(\sqrt{2})^{\sqrt{2}}$.

4. Using tables of roots or any other computational aids that are available, find the approximate values for the first few terms in the sequence $\sqrt{2}$, $\sqrt[3]{3}$, $\sqrt[4]{4}$, $\sqrt[5]{5}$, ... Conjecture the limit of this sequence.

5. Let us define the sequence $(c_1, c_2, c_3, \ldots)$ by taking $c_1 = k$ and $c_{n+1} = \sqrt{c_n}$. Find the first few terms in the sequence when (a) $k = 1$, (b) $k = 2$, (c) $k = 10{,}000$, (d) $k = 0$, (e) $k = .01$, (f) $k = .0001$. Conjecture the limit of the sequence for various values of k.

6. Suppose a is a real number greater than 1, and n and m are integers. Prove that if $n < m$, then $a^n < a^m$ and $a^{1/n} > a^{1/m}$.

7. Given that $\lim_{n\to\infty} \sqrt[n]{k} = 1$ for $k > 0$ and $\lim_{n\to\infty} \sqrt[n]{n} = 1$, find $\lim_{n\to\infty} c_n$ if c_n is given by (i) $\sqrt[n]{2n}$, (ii) $\sqrt[n]{n^2}$, (iii) $\sqrt[n]{3^n n}$.

8. It can be shown that the function f defined by $f(x) = 10^x$ is a one-to-one correspondence between R and R^+, the set of positive reals. Show further that this f is an isomorphism between the systems $(R, +)$ and $(R^+, \cdot)$.

10.8 Algebraic real numbers

If $a_0, a_1, a_2, \ldots,$ and a_n are real numbers and n is a natural number, an expression in the form

$$a_nx^n + a_{n-1}x^{n-1} + \cdots + a_1x + a_0$$

is said to be a (real) **polynomial.** The numbers $a_n, a_{n-1}, \ldots, a_1, a_0$ are said to be the **coefficients** of the polynomial. The number n is said to be the **degree** of the polynomial if $a_n \neq 0$. An equation that asserts that a polynomial is equal to 0 is said to be a **polynomial equation.** A function f is said to be a **polynomial function** if there is a fixed polynomial whose value at every x is $f(x)$.

As illustrations we have:

$5x - 7$ is a polynomial of degree 1.

$5x - 7 = 0$ is a first degree polynomial equation.

If $f(x) = 5x - 7$ for all x then f is said to be a first degree polynomial function.

$x^2 - 2$ is a polynomial of degree 2.

$x^2 - 2 = 0$ is a second degree polynomial equation.

If $f(x) = x^2 - 2$ for all x, then f is a second degree polynomial function.

First degree polynomial equations or functions are also said to be **linear.** Second degree polynomial equations or functions are also said to be **quadratic.** Similarly, **"cubic"** is used in place of "third degree polynomial," **"quartic"** in place of "fourth degree polynomial," and **"quintic"** in place of "fifth degree polynomial."

Any polynomial equation in the form $x - a = 0$ where a is a natural number has a solution in N, namely $x = a$.

We may think of our earlier work of developing the system of integers as a successful attempt to get a number system in which the polynomial equations $x - a = 0$ and $x + a = 0$ both have solutions for any natural number a.

We may think of our work of developing the rational number system as a successful attempt to get a system in which the linear equation $a_1x + a_0 = 0$ has a solution whenever a_1 and a_0 are integers and $a_1 \neq 0$.

We may think of our work in developing the real number system as a partially successful attempt to get a system in which quadratic equations with integral coefficients have a solution. The success was only partial since an equation such as $x^2 + 1 = 0$ has no solution in R. However, we did find that $x^2 - 2 = 0$ does have a solution (in fact, two solutions) in R although it has no solutions in F.

A real number that is a solution of a polynomial equation is said to be a **zero** of the polynomial involved. Thus $\sqrt{2}$ and $-\sqrt{2}$ are both zeroes of the polynomial $x^2 - 2$.

Perhaps any real number can be thought of as a zero of a properly chosen polynomial with integral coefficients. That is, it may be that any real number x is **algebraic** in the sense that there are integers $a_n, a_{n-1}, \ldots, a_1, a_0$ such that x is a solution for the equation $a_nx^n + a_{n-1}x^{n-1} + \cdots + a_1x + a_0 = 0$.

But it turns out that this is not so. There are real numbers which are not solutions to any polynomial equation with integral coefficients. Such real numbers are said to be **transcendental.**

It is no small chore to prove that transcendental real numbers actually exist. But we have already seen examples of such numbers. It can be proved, for instance, that the number 0.101001000100001 . . . is transcendental.

But there are two other transcendental numbers of a different kind that crop up repeatedly in pure and applied mathematics. The first of these is π, which is defined to be the ratio of the circumference to the diameter in a circle. The value of π has been worked out to many decimal places. One rational approximation to π that is not in error by more than 10^{-10} is 3.1415926536. Such a value may be computed by making use of the fact (which we accept without proof) that $\pi/4 = 1 - 1/3 + 1/5 - 1/7 + 1/9 - \cdots$ where the expression on the right is an infinite series as discussed in Section 9.5.

A second transcendental number that is useful in many ways is the number usually called e and defined by

$$e = \lim_{n \to \infty} (1 + 1/n)^n.$$

A rational approximation that is within 10^{-15} of e is 2.718281828459045. We shall not give any formal proof for the existence of the limit used to define e. However, the following table which gives the approximate value for $(1 + 1/n)^n$ for various values of n strongly suggests to the intuition that the limit does exist and has a value in the proper neighborhood.

n	1	2	4	8	16	32
$(1 + 1/n)^n$	2	2.25	2.4414	2.5658	2.6379	2.6770

n	64	128	256	512	1024	2048
$(1 + 1/n)^n$	2.6973	2.7077	2.7130	2.7156	2.7170	2.7176

The number π arises very naturally as soon as we begin to deal with such simple geometric configurations as circles. And yet the number π is *not* algebraic, so that we find again that our geometric ideas force us to deal with quite complicated numbers. And the complication is not simply that quite a bit of labor is required to find a good rational approximation to the number. It is somehow a more fundamental complication than that. We can never hope to express π as the solution of a polynomial equation with integral coefficients.

We shall not attempt to give any sort of proof of the deep fact that π *is* transcendental. All of the proofs available require the use of a great deal of machinery from the branch of mathematics called the calculus.

The transcendental number e can be thought of as arising out of a very natural question in the mathematics of finance. If A dollars are invested for n years at an annual interest rate of i for n years and the interest is compounded annually, we shall have $A(1 + i)^n$ in our account at the end of n years. If the compounding is quarterly we shall have $A(1 + i/4)^{4n}$. If the interest is compounded k times a year, we shall have $A(1 + i/k)^{kn}$. Suppose we let $m = k/i$. Then $k = mi$, and the amount in our account when the interest is compounded k times a year is $A(1 + 1/m)^{min} = A[(1 + 1/m)^m]^{in}$. But when k is large so that m is large this is approximately equal to Ae^{in}. And we may think of this last quantity as the amount we would have at the end of the period if the interest is compounded *continuously* over the entire n years.

Thus, we find again that a quite straightforward applied problem leads us to a transcendental number. We shall not, of course, attempt to give a proof here that e *is* transcendental.

Both algebraic and transcendental real numbers are dense on the line. That is, between any two distinct real numbers there is an algebraic as well as a transcendental real number.

PROBLEMS

1. Find a polynomial with integral coefficients for which the given number is a zero. (a) $\sqrt{5 - \sqrt{7}}$ [SOLUTION: Let $x = \sqrt{5 - \sqrt{7}}$. Then $x^2 = 5 - \sqrt{7}$ and $-\sqrt{7} = x^2 - 5$. Hence, $7 = (x^2 - 5)^2 = (x^2)^2 - 10x^2 + 25$. Therefore, $x^4 - 10x^2 + 18 = 0$. $\sqrt{5 - \sqrt{7}}$ is a zero of the polynomial $x^4 - 10x^2 + 18$.], (b) $2/3$, (c) $-7/43$, (d) $\sqrt{77}$, (e) $-\sqrt{77}$, (f) $\sqrt[6]{4}$, (g) $\sqrt{7 + \sqrt{5}}$, (h) $\sqrt[3]{2 + \sqrt{6}}$, (i) $\sqrt[3]{6} - 2$, (j) $\sqrt{2} + \sqrt{3}$.

2. Assuming that it is known that e is transcendental, show that e^n must be transcendental for each positive integer n.

3. Show that every rational number is an algebraic number.

4. Show that the set of transcendental numbers is a proper subset of the set of irrational numbers.

5. Show that if x is an algebraic number and $x \neq 0$, then $-x$ and $1/x$ are algebraic numbers.

6. If x is a transcendental number does it follow that $-x$ and $1/x$ are transcendental? Explain.

10.9 Computations with approximations

We have said several times that when we are faced with performing arithmetic operations on real numbers, we often find approximate results by performing the given operations on terminating decimals that are approximations to the given real numbers. We shall now consider some of the techniques and problems that arise in doing this.

Let us first note that any positive real number can be written as a power of ten or else as a product of a number between 1 and 10 and a power of ten. For example,

$$15{,}432 = 1.5432 \times 10^4$$

$$.00015 = 1.5 \times 10^{-4}$$

$$.0000047 = 4.7 \times 10^{-6}$$

$$1{,}500{,}000{,}000 = 1.5 \times 10^9$$

$$.1010010001\ldots = (1.010010001\ldots) \times 10^{-1}.$$

When a number is expressed this way it is said to be written in **scientific notation** (or, sometimes, **standard notation**). Such notation is very convenient when we are dealing with very large or very small numbers. For example, the speed of light is usually given as 3×10^{10} cm/sec rather than as 30,000,000,000 cm/sec. The former expression is easier to comprehend and easier to deal with in calculations.

Electronic computers that use so-called **floating-point** arithmetic express numbers in a fashion very similar to scientific notation. In computers, it is sometimes the case that the number is written as a product of a power of ten and a number between .1 and 1. Of course, all the machine needs to record, in regard to the power of ten, is the exponent required. Thus, for computational purposes a machine might use the expression ".57600432 −3" to identify the number $.57600432 \times 10^{-3}$. The machine may well use another base besides the base 10 for its numerals. The bases 2, 8, 10, and 16 are all in common use in computers.

Scientific notation is useful for indicating the accuracy with which measurements are made. Suppose, for example, that we are interested in what the temperature is on a certain summer morning. If we are inside an air-conditioned building, we might ask someone who has been outside what the temperature is. He might reply, "About 70 degrees." If we assume that our observer is estimating the temperature from the way it feels to him it might be reasonable to suppose that he is correct to the nearest 10 degrees. That is, we would probably be safe in supposing that the temperature is closer to 70° than it is to 60° or 80°. In other words, the temperature is between 65° and 75°. We may indicate this by reporting the temperature as 7×10 degrees. If, on the other hand, we know that our observer has looked at a thermometer outside, it might be natural to suppose that the correct temperature is 70° to the nearest degree. That is, that the true temperature is between 69.5 and 70.5. We may indicate this by reporting the temperature to be 7.0×10 degrees. When the numeral "0" is written after the decimal point here, it is clear that it means something about the accuracy of the measurement. Although we have $7 = 7.0$, we take it that a reported measurement of 7.0×10 is different from 7×10.

We would say that "7.0×10" has two **significant digits.** "7×10" has only one **significant digit.** The number of significant digits in "70" is unclear. A reported measurement of 70 could be correct to the nearest 10 or to the nearest unit.

If we measure the speed of light and report our result as "3.0×10^{10}" cm/sec, then we are claiming that our result is correct to the nearest 10^9 cm/sec. That is, we claim as a result of our experiment that the true speed of light is between 2.95×10^{10} cm/sec and 3.05×10^{10} cm/sec. If we were to report our conclusion as "30,000,000,000 cm/sec," it would not be clear whether we were claiming the speed to be between 2.5×10^{10} and 3.5×10^{10} or between 29,999,999,999.5 and 30,000,000,000.5. When we use scientific notation and "0's" occur at the end of the first numeral, we know that they are meant to be there to indicate something about the accuracy of our measurement.

The digits that occur in the first numeral when the result of a measurement is written in scientific notation are all *significant* in the sense that they tell you something about the accuracy of the measurement.

If we refer to the table on page 64 we find that the population of the U.S.A. in 1965 is given as 195 million. In scientific notation, we would write this population as 1.95×10^8. This is a better form than 195,000,000 since we might interpret this last figure as if it were exact to the nearest person (or the nearest 10 people, or the nearest 100 people, or the nearest 1000 people, etc.) We would say that 1.95×10^8 has 3 significant digits. The number of significant digits in "195,000,000" is not clear since we do not know, without some extra information, whether the "0's" are there just as "place holders" or whether they mean more than that.

When performing arithmetic operations with numbers that are only approximations to true values we should do so sensibly and make an effort not to be led astray by our algorithms. Suppose, for example, that we are working as a surveyor and we find ourselves in the following situation. We measure two sides of a triangle by means of a surveyor's chain (i.e., a steel measuring tape) and find them to be 102.51 feet and 98.43 feet, respectively. The third side of the triangle is inaccessible and we are forced to estimate its length by eye. We estimate it to be 1.5×10^2 feet. To find the perimeter of the triangle we might write: $102.51 + 98.43 + 150 = 350.94$. But if we do so, we are surely misleading ourselves or someone who uses our work. We know one side of the triangle only to the nearest 10 feet. And yet we seem mysteriously to find the perimeter correct to the nearest .01 feet! We should instead recognize the limitations of our measurements and perform our computation of the perimeter as follows:

$$\begin{array}{ll} 1.0251 & \times 10^2 \\ .9843 & \times 10^2 \\ 1.5 & \times 10^2 \\ \hline 3.\cancel{5094} & \times 10^2. \end{array}$$

Our conclusion is that the perimeter is 3.5×10^2 feet. And, sensibly, this implies that we know the perimeter only to the nearest 10 feet.

The general rule for handling addition of approximations is to

(a) write all the numbers as multiples of the same power of 10, so that only significant digits appear to the left of the decimal points,
(b) add the numbers in the usual fashion, and then
(c) round off the result so that it contains the same number of significant digits to the right of the decimal point as in the number with the smallest number of significant digits to the right of the decimal point.

For example, if we wish to add the measurements 4.71×10^4, 7.5×10^5, 7.37×10^2, and 9.64×10^{-1}, we would write:

$$\begin{array}{ll} .471 & \times 10^5 \\ 7.5 & \times 10^5 \\ .00737 & \times 10^5 \\ .00000964 & \times 10^5 \\ \hline 7.97837964 & \times 10^5. \end{array}$$

We would then round this off and claim that the sum is 8.0×10^5. This process essentially disregards the last two numbers in the sum. But this is appropriate since 7.5×10^5 is correct only to the nearest 10^4 and, hence, the potential error in this number swamps the other two (relatively) small numbers.

To *round off* a number to the nearest 10^n is to find an integral multiple of 10^n which is as close as possible to the number. For instance, if we round off $\sqrt{2}$ to the nearest tenth we get 1.4. Rounding off 75,430 to the nearest 1000 yields 75,000. Rounding off .00745 to the nearest hundredth yields .01. And rounding off .000059 to the nearest thousandth yields 0.

To round off 750 to the nearest hundred, we may take either 700 or 800 since both 700 and 800 differ from 750 by 50. Similarly, rounding off .095 to the nearest hundredth yields either .09 or .10.

In carrying out calculations where many numbers may be rounded off, it is customary, in the cases where there are two possible choices, to choose the larger one when the preceding digit is odd and the smaller when it is even. Then when we are rounding to the nearest ten-thousandth we would get .0064 from .00645, and .0068 from .00675. The reason behind such a procedure is that we expect that the errors introduced will tend to balance themselves out.

Let us now turn to a case where we must deal with products of approximate numbers. Suppose we have a field in the form of a right triangle. We measure one leg of the triangle carefully and find it to be $156.73 = 1.5673 \times 10^2$ ft. We find that we cannot accurately measure the other leg of the triangle, but we estimate it to be 150 ft to the nearest 10 feet. That is, the other leg has a length of 1.5×10^2 ft. What is the area of the triangle? We might claim that the area, in square feet, is given by

$$1/2(156.73)(150) = 11{,}754.75.$$

But surely this would be misleading since a report of an area of 11,754.75 will be taken to mean that the true area is between 11,754.745 and 11,754.755.

Actually, we know only that the first leg has a length between 156.725 and 156.735, while the other leg has a length between 145 and 155. Hence, the true area must be between

$$1/2(156.725)(145) = 11{,}362.5625$$

and

$$1/2(156.735)(155) = 12{,}146.9625.$$

But these values differ by almost 1000. So to claim that we know the area to the nearest 10^{-2} is certainly unwarranted.

The standard procedure for taking care of such a situation is to note that the measurement of 1.5×10^2 has only two significant digits and then round off our product so that it has only two significant digits. That is, we would conclude from our first calculation that the area is 1.2×10^4 sq ft.

In general, when multiplying or dividing two numbers that are consequences of measurements, we round off the result so that it contains the same number of significant digits as the factor with the smallest number of significant digits.

Note that in our calculation of the area of the triangle, we recognized that the "1" and the "2" in the fraction "1/2" were not the results of measurements. We did not need to consider the significant digits in them. If you like, we may think of these as 1.000 . . . and 2.000 . . . .

When many calculations are carried out with many different numbers, the problem of accumulation of errors due to rounding off becomes important and quite complex. People who make use of large electronic calculators certainly must be concerned with such matters. We shall not delve more deeply into the subject here.

PROBLEMS

1. Write each of the following in scientific notation. (a) 5,002, (b) .0653, (c) 7,500,000, (d) 6000.4, (e) 695, (f) .00000117, (g) 637.692, (h) 356×10^4, (i) 7976×10^{-10}.

2. Where possible, determine the number of significant digits in each of the numerals given in Problem 1.

3. Assuming that *all* of the numbers involved are the results of measurements, find the correct results of the following. (a) 5,002 + 473.51, (b) 3,004 + 697.3 + 47.92, (c) $(5.76 \times 10^7) + (6.32 \times 10^6) + (7.3 \times 10^7)$, (d) $(4.00 \times 10^3) + (5.6 \times 10^3) + (7.36 \times 10^4)$, (e) $(4.302 \times 10^{-4}) + (3.7 \times 10^{-5}) + (6.529 \times 10^{-3})$, (f) .00153 + .075 + .0006, (g) $(4.72 \times 10) + (4.98 \times 10^{-1}) + (3 \times 10^{-1})$, (h) $(7.5 \times 10^2)\cdot(3.751 \times 10^3)$, (i) $(7.516 \times 10^6)\cdot(6.93 \times 10^{-3})$, (j) $(3 \times 10^{-4})\cdot(3.697 \times 10^{-6})\cdot(8.6 \times 10^5)$, (k) $(.00153 + .097)(3.76 \times 10^{-3})$, (l) $(3.6793 - 3.654)\cdot(7.521 \times 10^3)$, (m) $(.000534 - .000539)\cdot(5.782 \times 10^6)$.

4. Suppose the two sides of a rectangle are measured as 5.0×10^{-1} and 6.43 miles. Between what two values must the true area of the rectangle fall?

5. Suppose that when a certain computer does its calculations it identifies numbers by "writing" first an algebraic sign (+ or −), then 8 decimal digits, and then an algebraic sign (+ or −) and two digits to indicate an integral exponent between −91 and +91. Then the computer would, for example, express 602 (or $.602 \times 10^3$) as "+60200000, +3" and it would express −.043 (or $-.43 \times 10^{-1}$) as "−43000000, −1." Exactly how many different numbers can the machine name? Are these numbers evenly spaced on a portion of the number line?

*10.10 Logarithms

Let us consider the function f defined on R by the equation $f(x) = 10^x$. Some typical values of this function are shown in the following table:

x	−4	−3	−2	−1	0	1	2	3
$f(x)$	.0001	.001	.01	.1	1	10	100	1000

As this table suggests, and as we asserted in a preceding section, it can be shown that when $x < y$, then $10^x < 10^y$. That is, $x < y$ implies $10^x < 10^y$. One conclusion that we may draw from this is that f never maps two different real numbers on the same real number.

What is the range of f? It is R^+, the collection of all positive reals. Accepting this fact without proof, we may make use of the conclusion immediately above to see that f is actually a one-to-one correspondence between R and R^+. For if R^+ is the range of f, then 10^x is a positive real number for each real x. And, moreover, each positive real number is a value of f at some x. Also we know that $f(x) = f(y)$ implies $x = y$.

Since f is a one-to-one correspondence between R and R^+, it has an inverse function f^{-1} which maps R^+ onto R. Some typical values of the inverse function are given in the following table:

y	.0001	.001	.01	.1	1	10	100	1000
$f^{-1}(y)$	-4	-3	-2	-1	0	1	2	3

If we wish to describe the function f^{-1} we could say that $f^{-1}(y)$ is the power to which we must raise 10 in order to get y. For instance, to show how we may obtain a value not given in the above table, let us find $f^{-1}(\sqrt{10})$. $f^{-1}(\sqrt{10})$ is the power to which we must raise 10 in order to get $\sqrt{10}$. But $\sqrt{10} = 10^{1/2}$. So evidently $f^{-1}(\sqrt{10}) = 1/2$. Similarly, $f^{-1}(\sqrt[3]{10}) = 1/3$.

It is common practice in mathematics to call $f^{-1}(y)$ the *logarithm to the base* 10 of y. For example, we would say that the logarithm to the base 10 of $\sqrt{10}$ is 1/2. This is written more briefly as $\log_{10} \sqrt{10} = 1/2$.

The two basic properties of the logarithm function which make it extremely useful are the following which hold for any real z and any positive x and y.

$$\log_{10} (xy) = \log_{10} x + \log_{10} y$$

$$\log_{10} (x^z) = z \log_{10} x.$$

These are easily proved. Suppose that $a = \log_{10} x$ and $b = \log_{10} y$. Then $10^a = x$ and $10^b = y$. Hence, $xy = 10^a \cdot 10^b = 10^{a+b}$. Therefore, $\log_{10} (xy) = a + b = \log_{10} x + \log_{10} y$. Also $x^z = (10^a)^z = 10^{az} = 10^{z \log_{10} x}$. Hence, $\log_{10} (x^z) = z \log_{10} x$.

To illustrate how these basic properties may be used, let us accept the facts that, correct to 5 decimal places, we have $\log_{10} 2 = .30103$, $\log_{10} 3 = .47712$, $\log_{10} 7 = .84510$, and $\log_{10} 11 = 1.04139$. Then we may compute $\log_{10} 6 = \log_{10} 2 \cdot 3 = \log_{10} 2 + \log_{10} 3 = .30103 + .47712 = .77815$. Also $\log_{10} 10 = 1 = \log_{10} 2 \cdot 5 = \log_{10} 2 + \log_{10} 5$. Hence, $\log_{10} 5 = 1 - \log_{10} 2 = 1 - .30103 = .69897$. $\log_{10} 9 = \log_{10} 3^2 = 2 \log_{10} 3 = 2(.47712) = .95424$. $\log_{10} \sqrt{7} = \log_{10} 7^{1/2} = (1/2) \log_{10} 7 = (1/2)(.84510) = .42255$. $\log_{10} 20 = \log_{10} (2 \cdot 10) =$

1.30103. $\log_{10} 200 = \log_{10} (2 \cdot 10^2) = 2.30103$. $\log_{10} (.0002) = \log_{10} (2 \cdot 10^{-4}) = .30103 - 4 = -3.69897$.

If $y = \log_{10} x$, then we say that x is the anti-log to the base 10 of y. That is, the anti-log to the base 10 of y is 10^y.

Tables of logarithms to the base ten are widely available. They usually give the (approximate values of) logarithms of numbers between 1 and 10. With such a table, it is easy to find the (approximate value of the) logarithm of any real number. We simply write our number in scientific notation and then make use of the additive property of logarithms. This is what was done in the above calculation of the logarithms of 20, 200, and .0002.

With tables of logarithms available we may also find the (approximate value of the) antilog for a given number. We then have a tremendously useful computational aid. For example, if we wish to find the approximate value of $\sqrt{7}$, we may find that $\log_{10} \sqrt{7} = .42255$ and then use our tables to find that the antilog of .42255 is 2.6458. Hence, $\sqrt{7}$ is approximately 2.6458. To find the approximate value of $\sqrt{2}^{\sqrt{2}}$ we find $\log_{10} \sqrt{2} = 1/2\ (.30103) = .15052$. Then $\log_{10} \sqrt{2}^{\sqrt{2}} = \sqrt{2}\ \log_{10} \sqrt{2} = \sqrt{2}\ (.15052) = (1.4142)(.15052) = .21287$. We then find the antilog of .21287 to be 1.6326. Hence, $\sqrt{2}^{\sqrt{2}} = 1.6326$ where, of course, we expect the last equality to be only an approximation.

Notice that, since the tabulated values for the logarithms are only approximations to the true values, we must use the methods of the last section in carrying out the calculations and rounding off properly.

PROBLEMS

1. Find the logarithm to the base 10 of each of the following: (a) 10^4, (b) 1,000,000, (c) 1,000,000,000, (d) .01, (e) .000001, (f) $\sqrt[5]{10}$, (g) $\sqrt[5]{100}$, (h) $\sqrt[3]{.0001}$.

2. Use the logarithms of 2, 3, 7, and 11 as given in this section to compute (a) $\log_{10} 21$, (b) $\log_{10} 49$, (c) $\log_{10} 77$, (d) $\log_{10} \sqrt[3]{11}$, (e) $\log_{10} 42$, (f) $\log_{10} 32$, (g) $\log_{10} 11{,}000$, (h) $\log_{10} \sqrt{11{,}000}$, (i) $\log_{10} \sqrt[5]{11}$, (j) $\log_{10} \sqrt[3]{539}$, (k) $\log_{10} .0003$, (l) $\log_{10} \sqrt{.0003}$.

3. Show that if we know the logarithms of 2, 3, 7, and 11, then we may easily find the logarithms of all the natural numbers 2, 3, 4, 5, 6, . . . , 12. [HINT: The fundamental theorem of arithmetic.]

4. Find in the library a 5-place table of common logarithms (i.e. logarithms to the base 10) and use it to find the logarithm to the base 10 of (a) 1.5, (b) 1.52, (c) 152, (d) 152,000, (e) .00152, (f) $\sqrt{.00152}$, (g) 7374, (h) .07374, (i) $(.07374)\cdot(.00152)$, (j) 5370, (k) 4.57×10^4, (l) 4.57×10^{-3}.

5. Find the anti-log to the base 10 of (a) 1, (b) 2, (c) -4, (d) .30103, (e) $\log_{10} 5.6$.

6. Use a 5-place table of logarithms to find the anti-log to the base 10 of

(a) .30103, (b) .99078, (c) .75051, (d) .75542, (e) .44560, (f) .46015, (g) .76020, (h) .66766, (i) .43062, (j) 3.75051, (k) .75051 − 2, (l) 6.66766, (m) −4.56038.

7. Use a 5-place table of logarithms to find the approximate decimal value of (a) $\sqrt[7]{7}$, (b) $\sqrt[3]{121}$, (c) $3^{4/3}$, (d) $\sqrt{1543}$, (e) $\sqrt[6]{1{,}540{,}000}$, (f) $\sqrt{3}^{\sqrt{3}}$.

8. In the branch of mathematics called the calculus, it is very helpful to use logarithms to the base e rather than logarithms to the base 10. (See Section 10.8 for the definition of the number e.) We say that y is the logarithm to the base e of x (or else y is the natural logarithm of x) and write $y = \log_e x$ if $e^y = x$. The notation "$y = \ln x$" is sometimes used in place of "$y = \log_e x$." Find (a) $\log_e e$, (b) $\log_e e^2$, (c) $\log_e \sqrt{e}$, (d) $\ln (1/e)$, (e) $\ln \sqrt[5]{e^2}$.

*10.11 Ordered fields

If we refer to the definition of a field given in Section 8.17, we see that the system $(R, +, \cdot)$ is a field. Are the systems $(R, +, \cdot)$ and $(F, +, \cdot)$ isomorphic? Evidently they are not, since the system of real numbers is complete while the system of rational numbers is not. The notion of completeness is defined in terms of the relation $\geq$. And in each of the sets of numbers I, F, and R we defined the relation $\geq$ in terms of the operation of addition and the collection of numbers which we called positive. In each of the sets I, F, and R we found that the multiplicative identity, 1, was positive and that the additive identity, 0, was not positive. And we found that the set of positive numbers was closed under addition and under multiplication. Further, we found for each of the sets I, F, and R, that if x is a number in the set, then either $x = 0$ or x is positive or $-x$ is positive and at most one of these can hold.

If, in any given field, we can find a set of elements which has properties similar to the set of positive numbers in F and R, then perhaps we can define a relation similar to $\geq$ in the field, and, thus, *order* the field. We are consequently led to make the following definition which applies to any field $(A, \oplus, \odot)$:

> We shall say that $(A, \oplus, \odot)$ is an **ordered field** if there is a subset P of A such that (i) the multiplicative identity of the field is in P, (ii) 0, the additive identity of the field, is not in P, (iii) P is closed under addition and multiplication, and (iv) if $x \in A$, then either $x = 0$ or $x \in P$ or $-x \in P$ and at most one of these can hold.

Then we see that the systems $(F, +, \cdot)$ and $(R, +, \cdot)$ are both ordered fields since we may take P, in each case, to be the collection of positive numbers.

On the other hand, the system $(\{0,1\}, +_2, \cdot_2)$ of integers mod 2 is a field, but not an ordered field. In view of conditions (i) and (ii) the only possible candidate for P is the subset $\{1\}$. But we have both $1 \in \{1\}$ and $-1 \in \{1\}$ since $-1 \equiv 1 \pmod 2$, so that condition (iv) is violated.

Similarly, it may be shown that the field $(\{0,1,2\}, +_3, \cdot_3)$ of integers mod 3 is not an ordered field.

In any ordered field it is possible to define a relation $\geq$ by stating that $x \geq y$ means that there is an element z in the field such that $x = y + z$ and $z = 0$ or $z \in P$. We may then show that for all elements x, y, and z in the field (a) $x \geq x$, (b) $x \geq y$ or $y \geq x$, (c) $x \geq y$ and $y \geq z$ implies $x \geq z$, (d) $x \geq y$ if and only if $x + z \geq y + z$, and (e) $x \geq y$ and $z \in P$ implies $xz \geq yz$.

It is then possible to introduce the ideas of upper and lower bounds for sets in A and also the notion of the lub of a set. But then it makes sense, for any ordered field, to ask if it is complete or not.

A result related to this is one that throws a great deal of light on the nature of the real number system. It can be shown that *any* ordered field which *is* complete must be isomorphic to the system $(R, +, \cdot)$. Thus, in a certain sense, there is *only one* complete ordered field—the immensely useful one that was constructed in this chapter.

Problems

1. Prove that $(\{0,1,2\}, +_3, \cdot_3)$ is not an ordered field.
2. Prove that if 1 is the multiplicative identity in an ordered field, then $1 + 1$ and $1 \cdot 1$ are positive (i.e., are in the set P with which we may order the field).
3. Prove that if 0 is the additive identity and 1 is the multiplicative identity of an ordered field, then $0 \leq 1$.
4. Prove that in any ordered field (a) $x \geq x$, (b) $x \geq y$ or $y \geq x$, (c) $x \geq y$ and $y \geq z$ implies $x \geq z$, (d) $x \geq y$ if and only if $x + z \geq y + z$, (e) $x \geq y$ and $z \in P$ implies $xz \geq yz$.

11

Some uses of the real numbers

11.1 Similar triangles

We may use real numbers to measure the lengths of segments of straight lines. We are, as a consequence, able to deal with the basic geometric notion of similarity.

Suppose that we are given the two triangles ABC and $A^*B^*C^*$ in Figure 11.1, and suppose that the size of the angle† at A is the same as that at A^*. Let us write "$\measuredangle A = \measuredangle A^*$" to assert this fact.‡ Suppose further that $\measuredangle B = \measuredangle B^*$ and§ $\measuredangle C = \measuredangle C^*$. Then we say that the two triangles are *similar*.

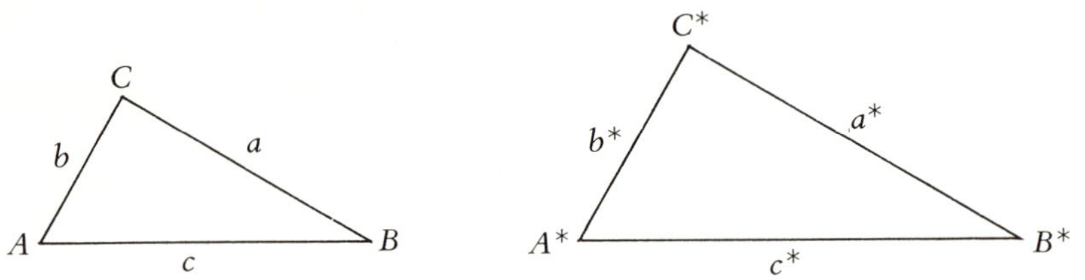

FIGURE 11.1

In general, we say that any two triangles are *similar* if it is possible to pair up the angles in the two triangles in such a way that the two angles in each pair have the same size.

---

† In this informal treatment of some geometric notions, we assume that the reader is familiar with the idea of an angle as something that may be measured with a protractor.

‡ This is a slight violation of more precise terminology. Usually "$\measuredangle A = \measuredangle A^*$" is taken to mean that the angle A and the angle A^* are the same angle. We are taking it to mean that A and A^* are the same size or *have the same measure*.

§ Actually, once we know that $\measuredangle A = \measuredangle A^*$ and $\measuredangle B = \measuredangle B^*$, then it must follow that $\measuredangle C = \measuredangle C^*$. This is due to the fact that in any triangle the measures of the three angles must add up to 180°.

One of the most useful facts in mathematics is that in similar triangles the lengths of corresponding sides are proportional. In Figure 11.1, where the lower case letters designate the lengths of the sides, we have $a/a^* = b/b^* = c/c^*$. This is what is meant by saying that the lengths of corresponding sides are proportional.

Suppose DEF and $D^*E^*F^*$ are triangles and $\measuredangle D = \measuredangle D^*$, $\measuredangle E = \measuredangle E^*$, and $\measuredangle F = \measuredangle F^*$. Let d be the length of the side opposite angle D. Let e be the length of the side opposite angle E. Et cetera. Suppose that $d = 3$, $e = 4$, $f = 5$, and $d^* = 7$. Then, since the triangles are similar, we can find the lengths e^* and f^*. For we must have $d/d^* = 3/7 = e/e^* = 4/e^*$ and, hence, $e^* = 4(7/3) = 28/3$. And we must have $d/d^* = 3/7 = f/f^* = 5/f^*$ and, hence, $f^* = 5(7/3) = 35/3$.

We may make use of similar triangles to justify several of the graphical devices we have used at various places in this text.

Let us first consider the device in Figure 6.18 for multiplying natural numbers. The construction and use of this device may be described as follows. Suppose we are given two natural numbers x and y and we wish to find the product xy. We construct the usual two perpendicular axes meeting at their origins. At the point 1 on the horizontal axis we construct a vertical line and find the point P on it y units above the horizontal axis. We then draw the line L through P and the origin. Next we draw a vertical line at the point x on the horizontal axis and note the point Q where this line hits the line L. See Figure 11.2. Clearly the two triangles we have constructed are similar. Hence, $z/y = x/1$ and, therefore, $z = xy$. This shows that the procedure given for the use of Figure 6.18 is valid.

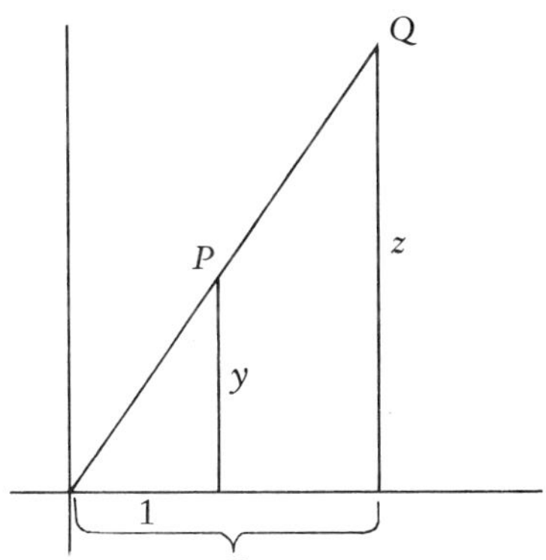

FIGURE 11.2

Notice that this justification for a method applies equally well if x and y are positive integers, positive rationals, or positive reals. To check on the somewhat more general device of Figure 7.13, it is only necessary to consider the various possible combinations of signs for x and y.

In Figure 8.4, we gave a procedure for constructing a segment with length p/q where p and q are integers. We may now justify that procedure as follows. Suppose p and q are given positive integers. We first construct the usual perpendicular axes. At the point q on the horizontal axis, we construct a vertical line and find the point P on it which is p units above the axis. We draw the line L through the origin and P. At the point 1 on the horizontal axis, construct a vertical line which hits L at the point Q. See Figure 11.3. Clearly the constructed triangles are similar. Hence $z/p = 1/q$ and, therefore, $z = p/q$ as was the claimed result for the procedure. We note that this procedure is essentially the reverse of the one described just above. And this is appropriate since division is the inverse operation for multiplication and p/q may be thought of as $p \div q$.

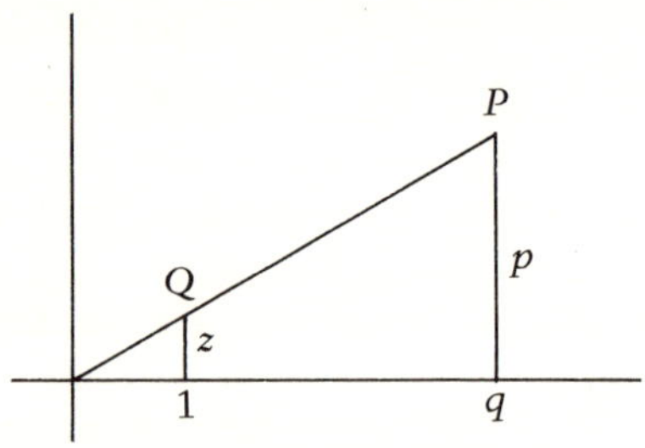

FIGURE 11.3

We may use similar triangles to show why the nomograph for addition as given in Figure 6.15 works. Suppose we constructed three equally spaced vertical lines as in Figure 11.4. We cut them with a single horizontal line to get three

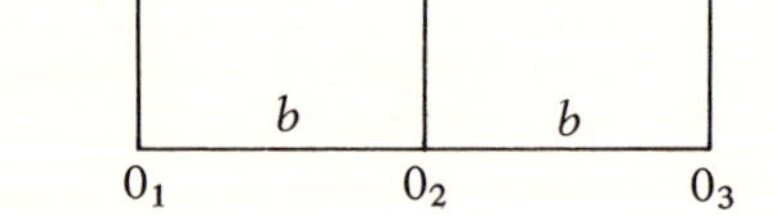

FIGURE 11.4

points we shall use as origins. Taking the left and right lines as number lines with the same unit distance on both, draw the line connecting the point x on the left line to the point y on the right. Let this line hit the middle vertical in the point Q. In the diagram we suppose that $x > y$. Drawing a horizontal line through the point y on the right axis, we get two similar triangles. If b, h, and r

are the lengths shown in the figure, we then have $r/h = b/2b = 1/2$. Then $r = h/2$ and we have the length from 0_2 to Q, in terms of the unit distance on the left and right axes, given by $y + h/2 = y + (x - y)/2 = (x + y)/2$. It follows that if we mark a scale on the middle vertical using a unit distance that is $1/2$ of the unit distance on the other verticals, the point Q will have a coordinate which is the desired sum. A similar argument applies if $x \leq y$.

Note that, again, this justification applies equally well when x and y are rational or real as long as they are both positive.

Let us now make use of similar triangles to prove one of the most famous and useful theorems in geometry. Suppose we are given a triangle ABC where angle C is a right angle. On the side AB let us construct a square of side c as shown in Figure 11.5. Suppose further that we draw the line from C which is perpendicular to AB. Let this line intersect AB in the point D, and let d and e be the lengths indicated in the figure.

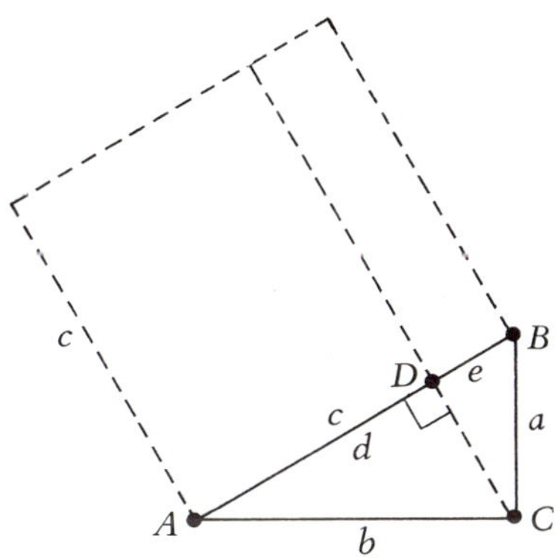

FIGURE 11.5

It is easy to see that triangle ABC is similar to triangle ACD. Hence, $c/b = b/d$ and, therefore, $b^2 = cd$. Note that cd is the area of the larger rectangle inside the square we constructed.

It is easy to see that triangle ABC is similar to triangle CBD. Hence, $a/c = e/a$ and, therefore, $a^2 = ec$. Note that ec is the area of the smaller rectangle inside our square.

But the total area of the square we constructed is c^2. Therefore, we have $a^2 + b^2 = ec + dc = c^2$. This result is referred to as the **Pythagorean Thoerem.** It may be put in other words as: In a right triangle, the square of the length of the hypotenuse† is equal to the sum of the squares of the lengths of the other two sides.

† In a right triangle, the side opposite the right angle is called the **hypotenuse.** It is always the longest side in the triangle.

PROBLEMS

1. Suppose triangle DEF is similar to triangle $D^*E^*F^*$ and $\measuredangle D = \measuredangle D^*$, $\measuredangle E = \measuredangle E^*$, and $\measuredangle F = \measuredangle F^*$. Find the lengths e^* and f^* (when the usual labeling procedure is used) if (i) $d = e = f = d^* = 10$, (ii) $d = e = f = 10$ and $d^* = 50$, (iii) $d = e = 20$, $f = 10$, and $d^* = 50$, (iv) $d = 5$, $e = 12$, $f = 13$, and $d^* = 10$, (v) $d = 5$, $e = 12$, $f = 13$, and $d^* = 1$, (vi) $d = \sqrt{3}$, $e = \sqrt{5}$, $f = 2\sqrt{2}$, and $d^* = 3$, (vii) $d = 4$, $e = \sqrt{7}$, $f = \sqrt{9}$, and $d^* = 16$, (viii) $d = 10$, $e = 5$, $f = 4$, and $d^* = 17$. (Stay alert on the last part here.)

2. Write up a justification for the procedure of Figure 8.5 when p and q are positive integers.

3. Explain in detail the similarities claimed for the triangles in the proof given for the Pythagorean theorem.

4. Suppose that in triangle ABC, angle C is a right angle. Find the missing length of side if (i) $a = 3$, $b = 4$, (ii) $a = 1$, $b = 1$, (iii) $a = 1$, $b = \sqrt{3}$, (iv) $a = 5$, $c = 5\sqrt{2}$, (v) $a = 12$, $c = 13$, (vi) $a = 5\sqrt{3}$, $b = 5$, (vii) $a = 10$, $c = 9$ (Careful!).

5. Suppose a, b, and c are the lengths of the sides in triangle ABC. It can be shown that if $a^2 + b^2 = c^2$, then the triangle is a right triangle. This is the converse of the Pythagorean theorem. Which of the triangles DEF given in Problem 1 are right triangles?

6. Let T be the set of all triangles. Is the similarity relation an equivalence relation on T? If so, describe the equivalence classes for the relation.

7. Find the area of each right triangle DEF or $D^*E^*F^*$ as given in Problem 1.

8. On the basis of your results in Problem 7, conjecture a relationship between the areas of two similar triangles in terms of the ratio of the lengths of corresponding sides.

9. A **quadrilateral** is a figure with four (straight) sides. Suppose that for the quadrilaterals $ABCD$ and $A^*B^*C^*D^*$ we have $\measuredangle A = \measuredangle A^*$, $\measuredangle B = \measuredangle B^*$, $\measuredangle C = \measuredangle C^*$, and $\measuredangle D = \measuredangle D^*$. Does it follow that the lengths of corresponding sides of the quadrilaterals are proportional? Explain.

10. Justify the nomograph of Figure 7.10 where negative numbers are allowed.

11.2 Analytic geometry

With the real number system available for use, each point on a number line corresponds to a number and each real number corresponds to a point on the line. But then we may translate statements about the geometry of the number line into statements about numbers and *vice versa*.

When we want to do **plane** geometry it turns out that we may use a horizontal and vertical axis in the plane in such a fashion that each point in the plane corresponds to an ordered pair of real numbers. We may then translate statements of plane geometry into statements about real numbers and *vice versa.*

Many of the most powerful techniques in pure and applied mathematics are built on this interplay between geometry and the real number systems.

Suppose we have a horizontal and vertical number line in a plane as in Figure 11.6. Let us, for convenience, call the horizontal axis the x-axis and the vertical axis the y-axis. (Any other two variables would actually serve as well as "x" and "y" here.) Given a point P in the plane we may draw the perpendicular from P to the x-axis. This perpendicular will hit the axis in a point corresponding to a real number. We call this real number the **x-coordinate** of P. Similarly, when we draw the perpendicular from P to the y-axis, the real number corresponding to the point of intersection is called the **y-coordinate** of P. If x_0 is the x-coordinate for P and y_0 is the y-coordinate for P, then we say that P *is* the point (x_0,y_0). It is legitimate to do this since the pair (x_0,y_0) is uniquely determined by the point P and, furthermore, when we are given the pair (x_0,y_0) we may find the point P. Thus, we have a one-to-one correspondence between the collection of all points in the plane and $R \times R$, the collection of all ordered pairs of real numbers.

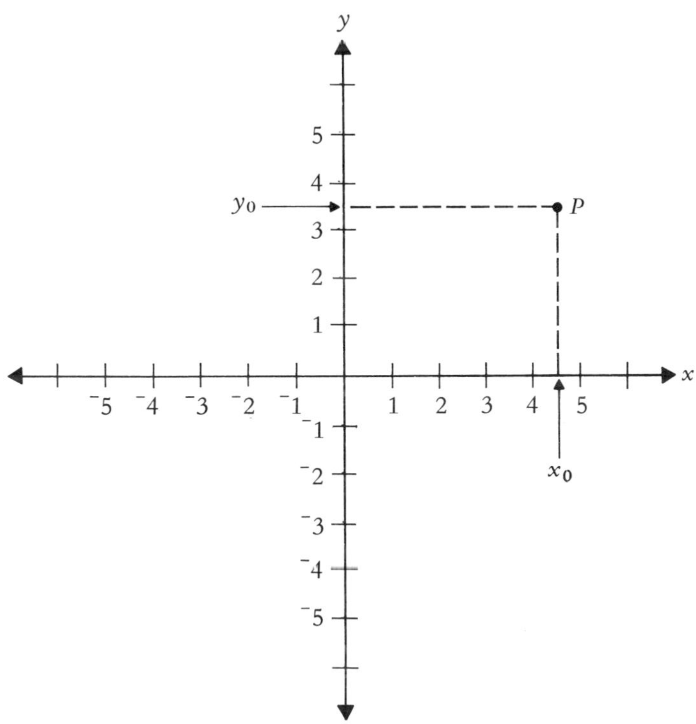

FIGURE 11.6

In Figure 11.7, we have shown the points (3,4), $(-2,5)$, (0,1), (2,0), $(-1,-4)$, $(3,-2)$, and $(0,-5)$. Notice that when $P = (x_0,y_0)$ we may think of x_0 as the directed distance from P to the y-axis, and of y_0 as the directed distance from P to the x-axis.

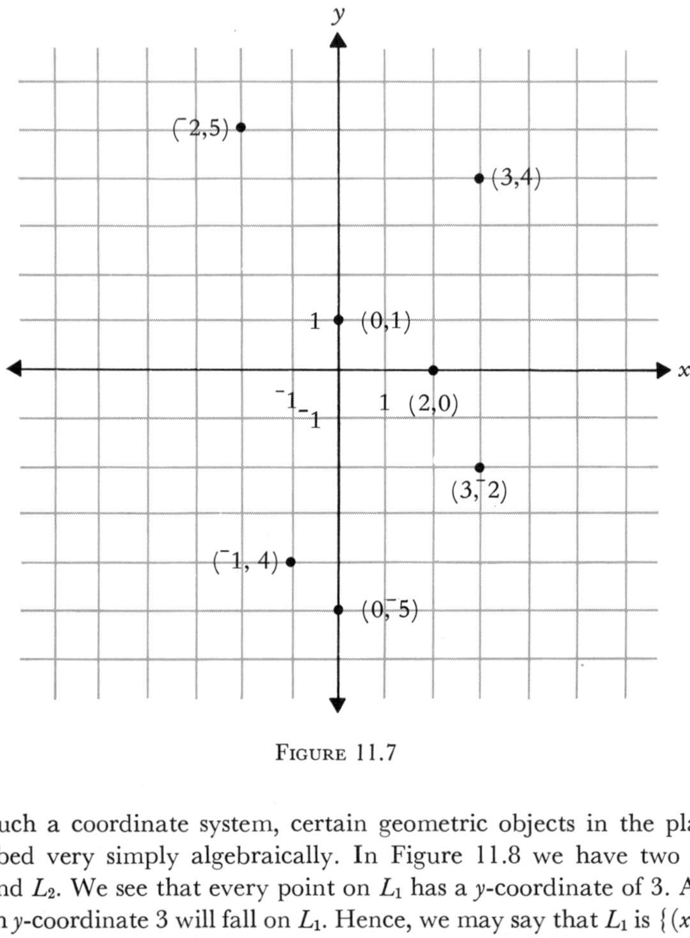

FIGURE 11.7

With such a coordinate system, certain geometric objects in the plane can be described very simply algebraically. In Figure 11.8 we have two straight lines L_1 and L_2. We see that every point on L_1 has a y-coordinate of 3. Also any point with y-coordinate 3 will fall on L_1. Hence, we may say that L_1 is $\{(x,y)\,|\,y = 3\}$ where we are, of course, thinking of this set as a collection of **points**—the points identified by the ordered pairs of real numbers satisfying the stated condition. Even more briefly, it is customary to say that L_1 *is* the graph of the equation $y = 3$. This terminology is an alternative way to put the assertion that L_1 is $\{(x,y)\,|\,y = 3\}$.

The line L_2 in Figure 11.8 is $\{(x,y)\,|\,x = -5\}$. Or we would say that L_2 is the graph of $x = -5$.

Given any equation (or inequality) involving x and y, the collection of points whose x- and y-coordinates satisfy the equation (or inequality) is said to be the **graph** of the equation (or inequality).

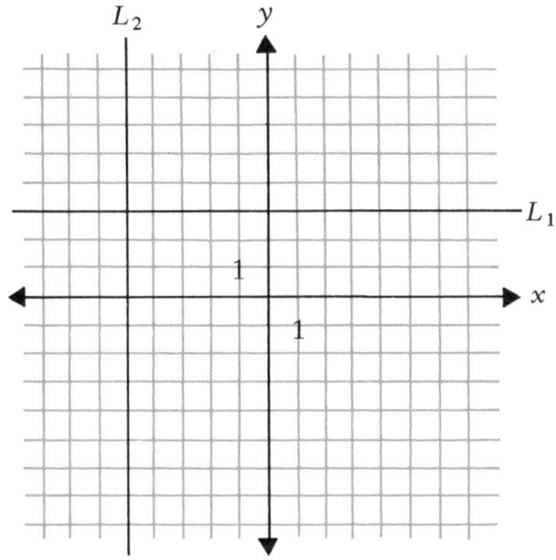

FIGURE 11.8

In Figure 11.9, we have displayed the graphs of several equations and inequalities.

If f is a function whose domain and range are subsets of R, then by a **graph** of f we mean a graph of the equation $y = f(x)$. It is customary to have the vertical axis used for coordinates corresponding to **values** of the function.

In Figure 11.10 we have displayed the graphs of several functions.

Note that in view of the nature of a function, any vertical line in the plane must intersect the graph of a function at most once.

In many applications of mathematics to the real world there are two properties being considered whose measured values can best be thought of as depending on the other, so that when we consider their corresponding measured values we get a function. The graphs of such functions are helpful in understanding the phenomena being studied.

In considering the motion of a ball thrown directly upwards we might measure s, the distance above the earth in yards, at the time t in seconds and be led to a graph such as that in Figure 11.11(a).

In this same motion problem we might measure the velocity v of the ball (in yards per second) at time t (in seconds) and get a graph such as that in Figure 11.11(b). When the ball is moving up we take the velocity to be positive and when down, negative.

In considering the growth of a human being, we might measure his height h in inches at the age t in years and get a graph such as that in Figure 11.11(c).

In considering the flight of a rocket, we might measure F the force in thousands of pounds on the rocket due to the propulsion system when the rocket is x feet above the earth and get a graph such as that in Figure 11.11(d).

In a weather study we might measure the barometric pressure P in inches of mercury at the time t in hours and get a graph such as that in Figure 11.11(e).

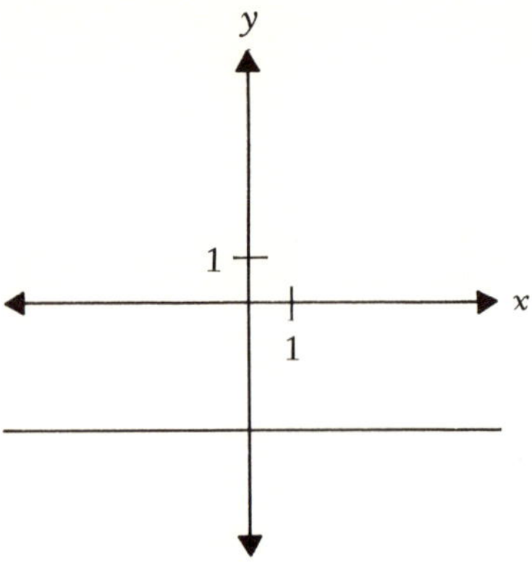

The graph of $y = {}^{-}3$

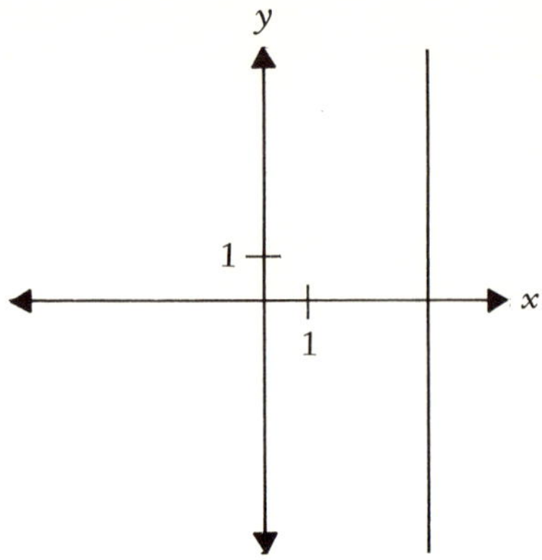

The graph of $x = 4$

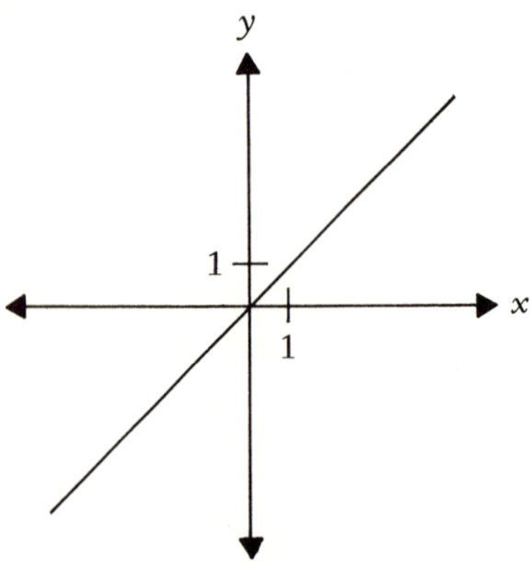

The graph of $y = x$

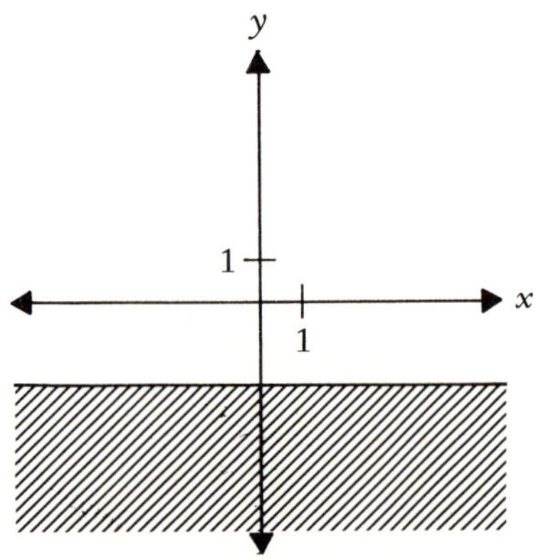

The graph of $y \le {}^{-}2$

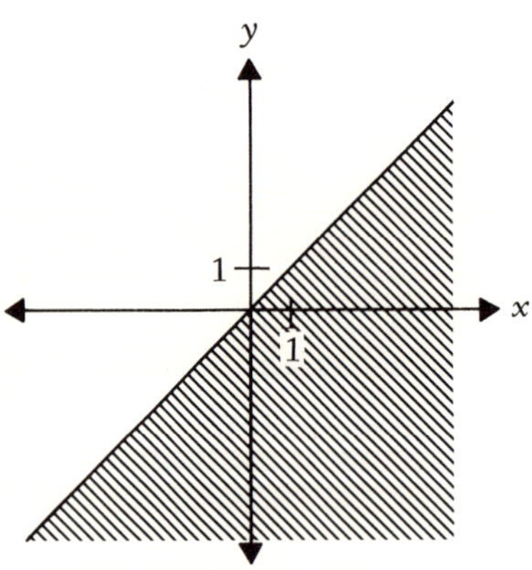

The graph of $y \le x$

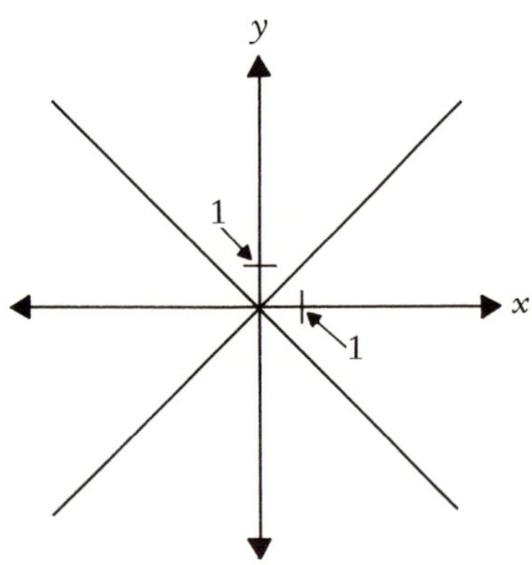

The graph of $|x| = |y|$

FIGURE 11.9

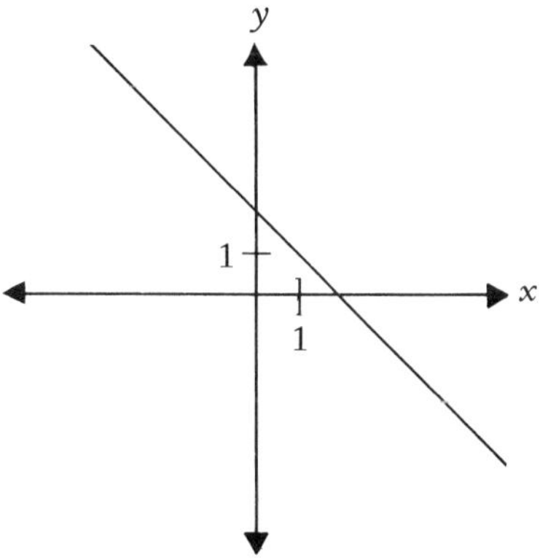

The graph of $x + y = 2$

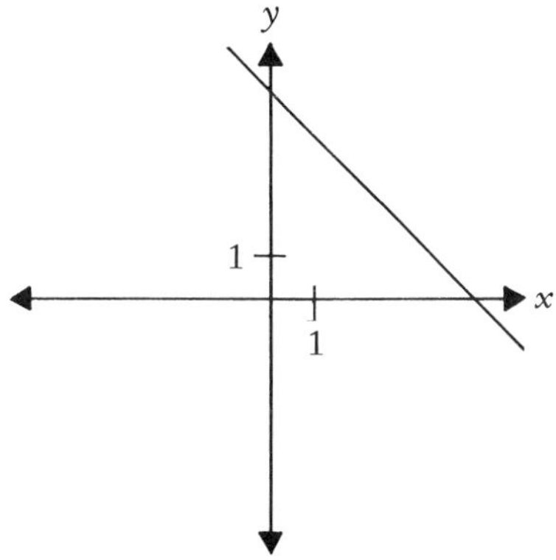

The graph of $x + y = 5$

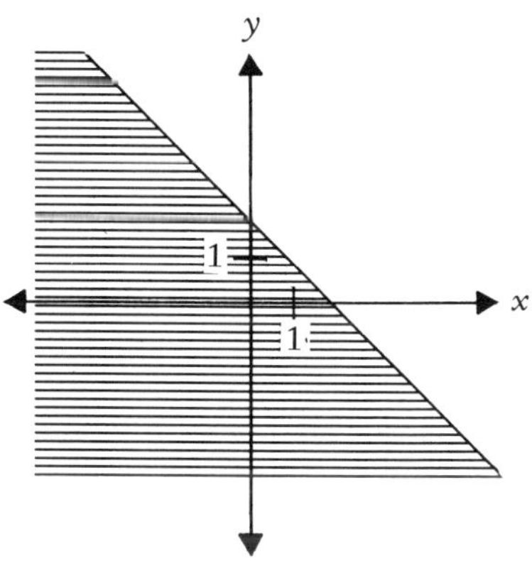

The graph of $x + y \leq 2$

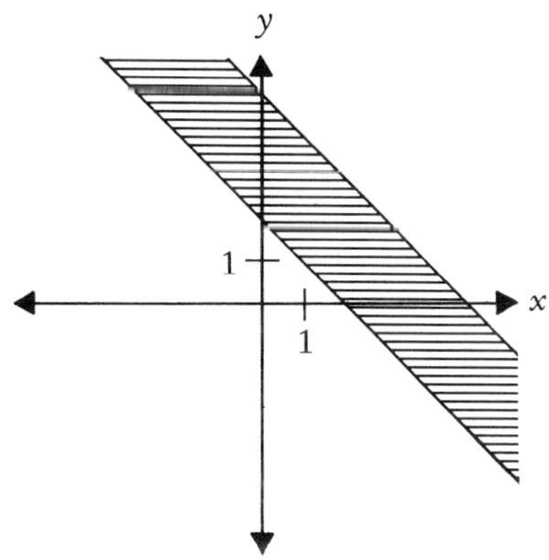

The graph of $2 \leq x + y \leq 5$

FIGURE 11.9 (continued)

In another consideration of the weather, we might measure the temperature T in degrees Fahrenheit at the time t in hours and get a graph such as that in Figure 11.11(f).

The usefulness of graphs in such situations as the above and many, many similar practical endeavors can hardly be overemphasized.

PROBLEMS

1. Construct a pair of coordinate axes as in Figure 11.7 and plot the points (7,0), (0,7), $(-7,0)$, $(0,-7)$, (3,3), (5,2), $(-5,2)$, $(-2,-5)$, $(3,-6)$, $(-5,-5)$, $(-7,-7)$, (6,1), (0,0). Label each point with its ordered pair of coordinates.

2. Graph each of the following: (a) $y = 2x$, (b) $y \leq 2x$, (c) $y \geq 2x$, (d) $y \leq 4$, (e) $y \geq -4$, (f) $|y| \leq 4$, (g) $|y| \geq 4$.

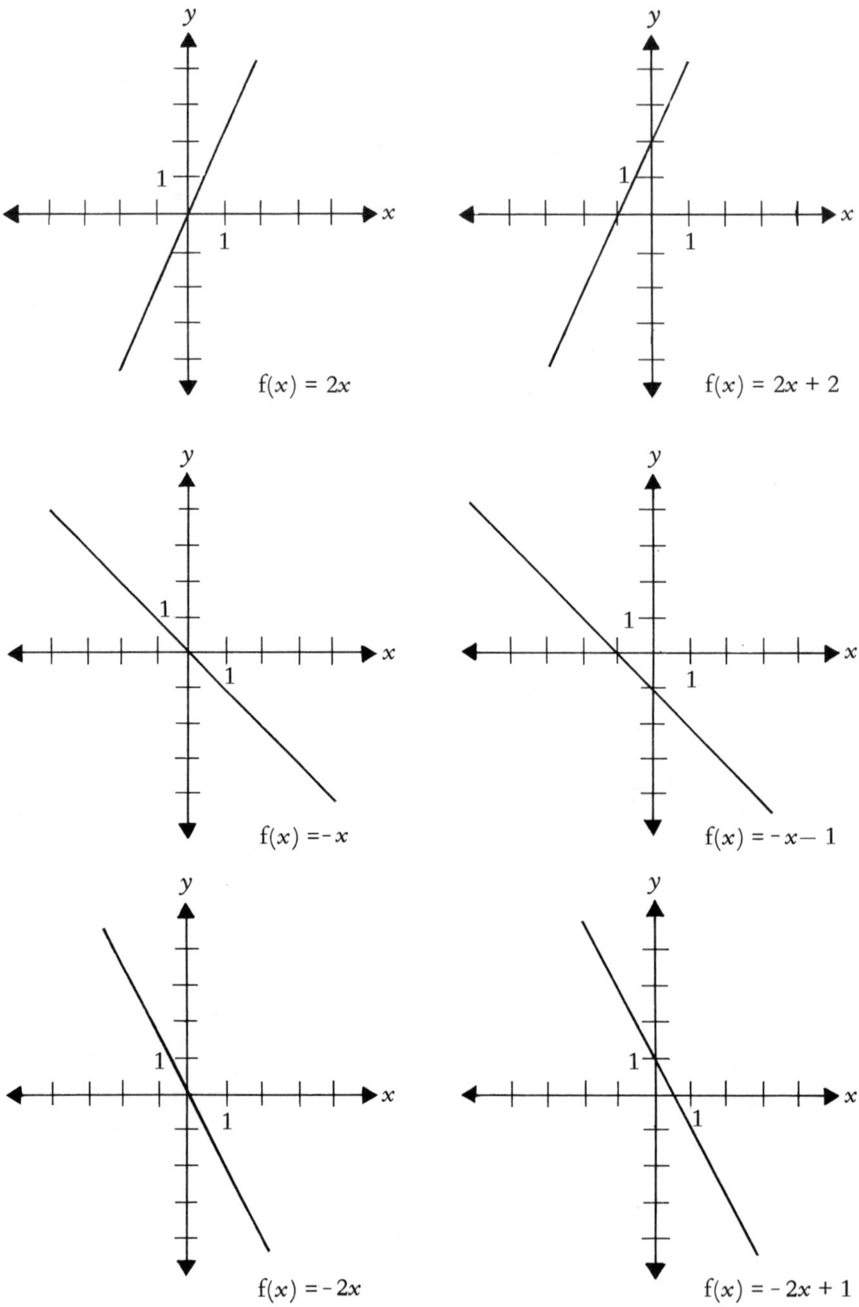

FIGURE 11.10 The graphs of several functions.

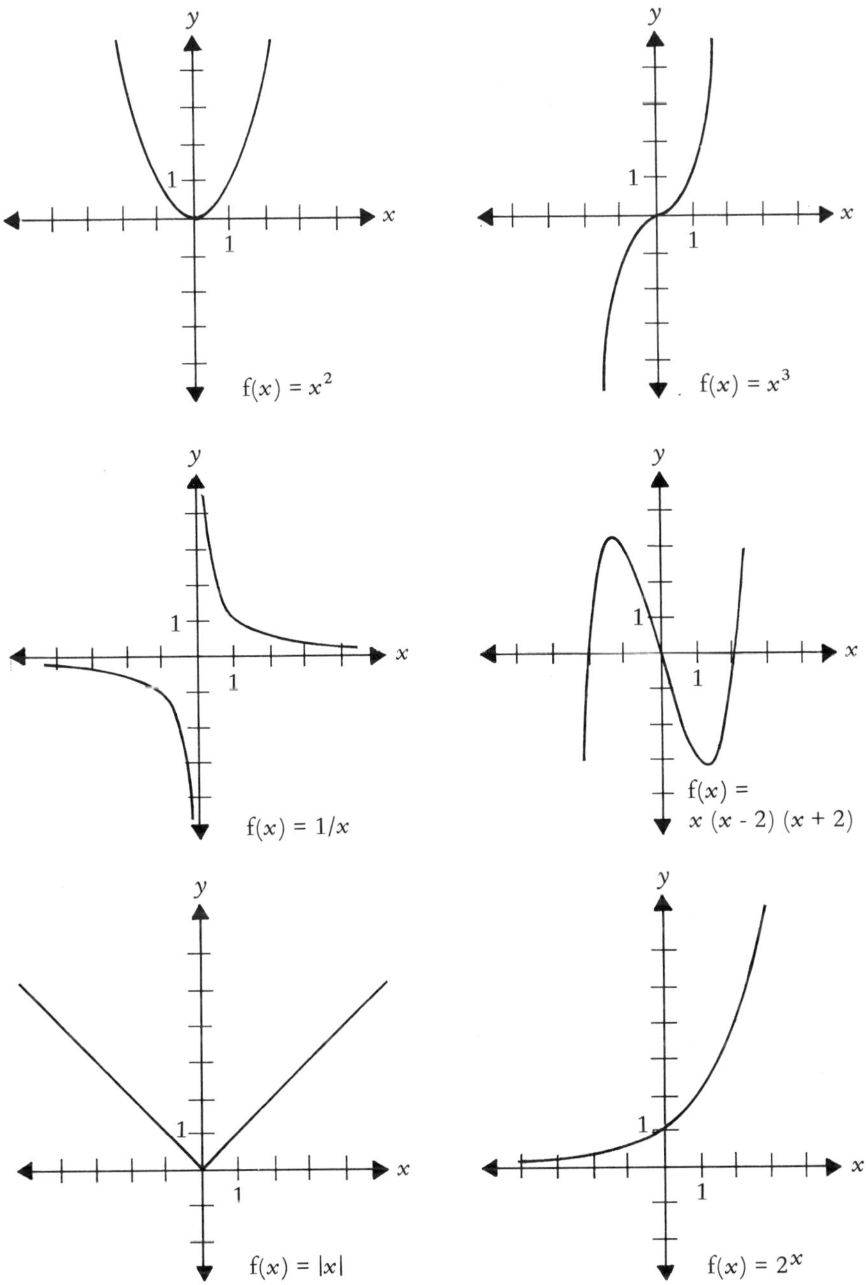

FIGURE 11.10 (continued)

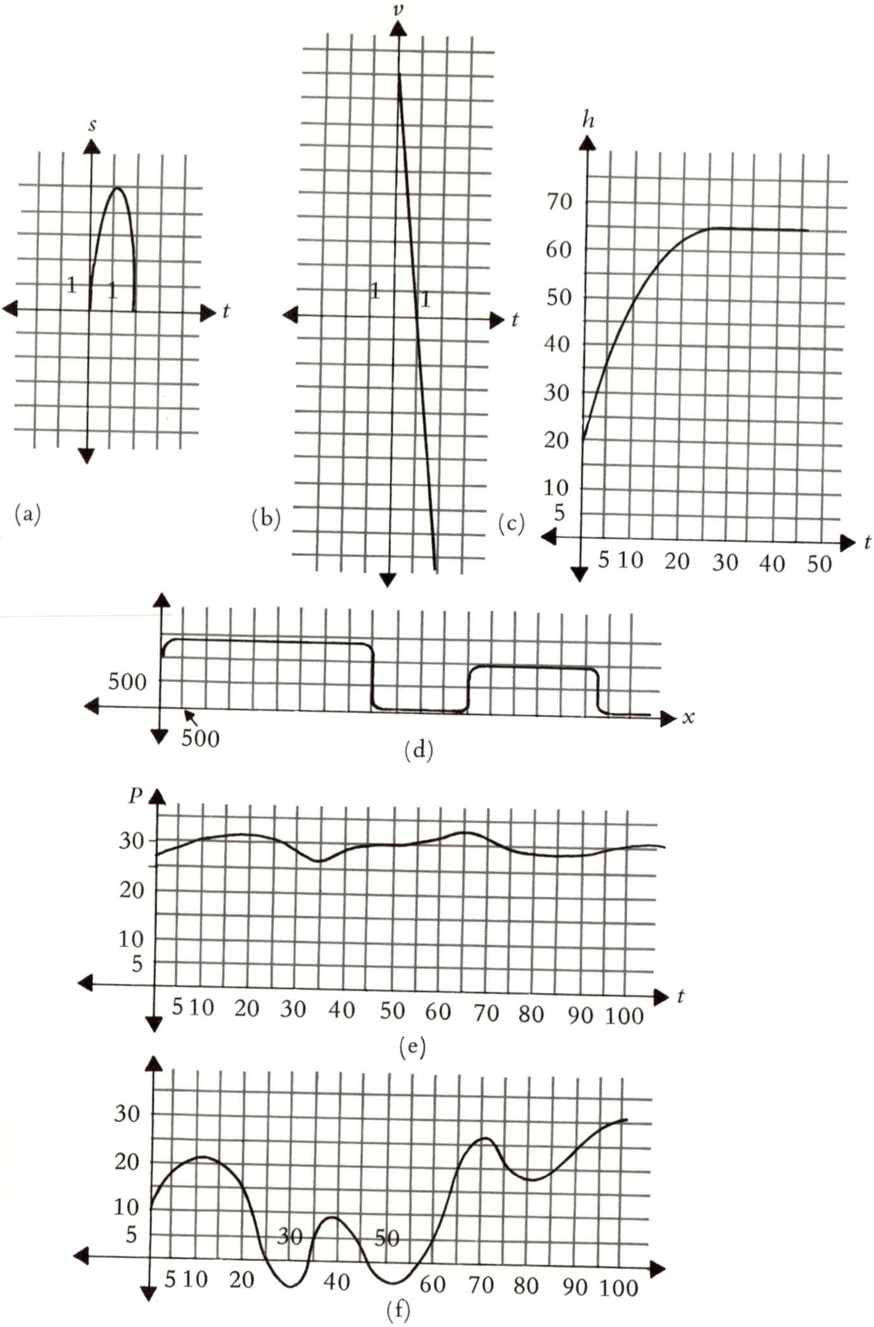
s
1
1
t
(a)
v
1
1
t
(b)
h
70
60
50
40
30
20
10
5
5 10 20 30 40 50
t
(c)
500
500
x
(d)
P
30
20
10
5
5 10 20 30 40 50 60 70 80 90 100
t
(e)
30
20
10
5
30
50
5 10 20
40
60 70 80 90 100
(f)

FIGURE 11.11

3. Sketch a graph of the function f if $y = f(x)$ and $f(x)$ is (a) $2x$, (b) $3x$, (c) $x - 1$, (d) $-2x$.

4. From Figure 11.11(a) determine (a) how high the ball is when $t = 1/2$, (b) the greatest height the ball attains, (c) the total time that the ball is in flight.

5. From Figure 11.11(b) determine (a) for what times the ball is moving upwards, (b) for what times the ball is moving downward, (c) at what time the ball attains its maximum height.

6. From Figure 11.11(c) determine (a) at what age the person attains half his maximum height, (b) the age at which the person seems to be growing fastest.

7. From Figure 11.11(e) determine the times at which it would be appropriate for the TV weather man to say that the barometer is (a) rising, (b) falling, (c) steady.

8. From Figure 11.11(f) determine (a) the maximum temperature for the time period considered, (b) the minimum temperature, (c) the times when it would be appropriate to say that it is warming up.

9. In a coordinate system plot the points (1,3) and (2,5). Draw a straight line through these two points. Find the coordinates of seven other points on this line and conjecture a condition that x and y must satisfy if (x,y) is a point on the line.

10. Do Problem 9 starting with the points (a) (2,7) and (4,−1), (b) (0,−3) and (1,−1), (c) (0,−3) and (−1,−5).

11. Sketch a rough graph which, with your age in years on the horizontal axis, shows your (a) weight in pounds, (b) height in inches, (c) distance from Washington, D.C., in miles, (e) number of living siblings, (f) number of living grandparents.

11.3 Proportionality and linearity

In many applied problems it is the case that one measured quantity y is **proportional** to another measured quantity x. What this means is that there is a real number k such that $y = kx$ for all x values. For example, the area of a rectangle with a fixed width is proportional to its height, the output of a radio amplifier we expect to be (approximately) proportional to its input, and the gross income from the sale of items at a fixed price is proportional to the number of items sold.

A particular instance of proportionality arises in connection with changing scales of measurement. If y is the length of an object in inches and x is its length in feet, then $y = 12x$. If y is the weight of an object in ounces and x is its weight in pounds then $y = 16x$. There are many situations where it is necessary to do such things as switch from meters to inches, pounds to grams, seconds to minutes, days to years, miles to light-years, etc.

The graph of $y = kx$ is especially simple geometrically. Figure 11.12 shows the graph of $y = kx$ for various values of k. (See also Figure 7.13.) The graph of $y = kx$ is always a straight line through the origin (0,0). We call the number k the **slope** of such a straight line. If we are at one point on the line $y = kx$ and move to the right along the line so that the x-coordinate increases by 1, then the y-coordinate increases by k. Thus, k may be thought of as the change in y per unit increase in x.

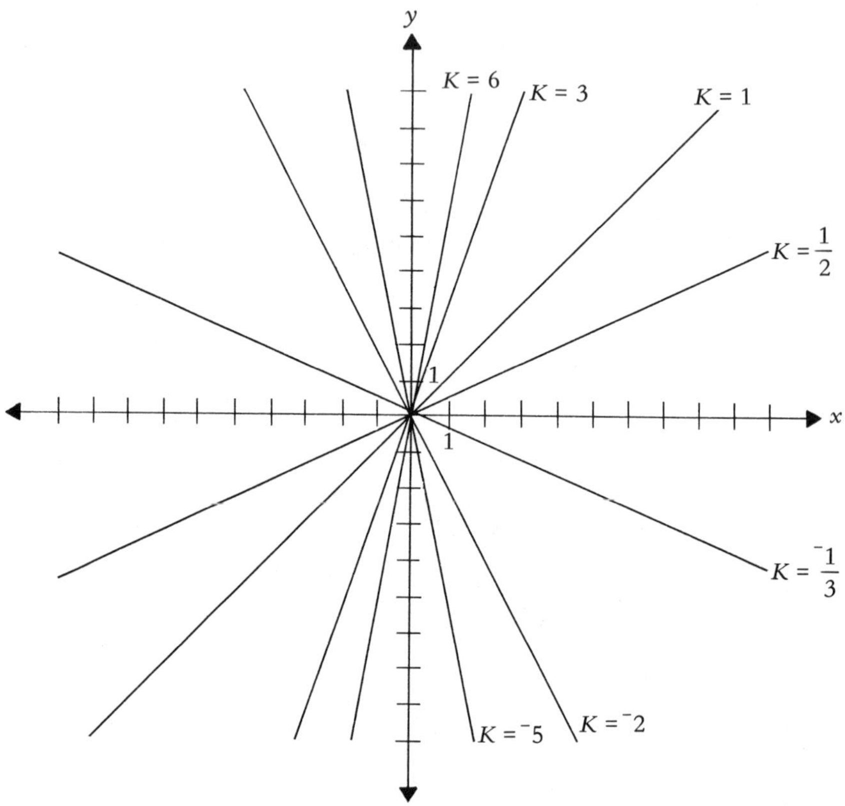

FIGURE 11.12

The graph of $y = kx + b$ is obtained from the graph of $y = kx$ by simply sliding the straight line $y = kx$ up a distance of b when $b > 0$ and down a distance of $|b|$ when $b < 0$. Figure 11.13 shows the graph of $y = kx + 1$ for various values of k. Figure 11.14 shows the graph of $y = kx - 2$ for various values of k.

The line $y = kx + b$ is said to have the **slope** k and y-**intercept** b. The slope k may be interpreted as it was before for the special case $b = 0$.

In view of the above results it is natural to call a function f a linear function

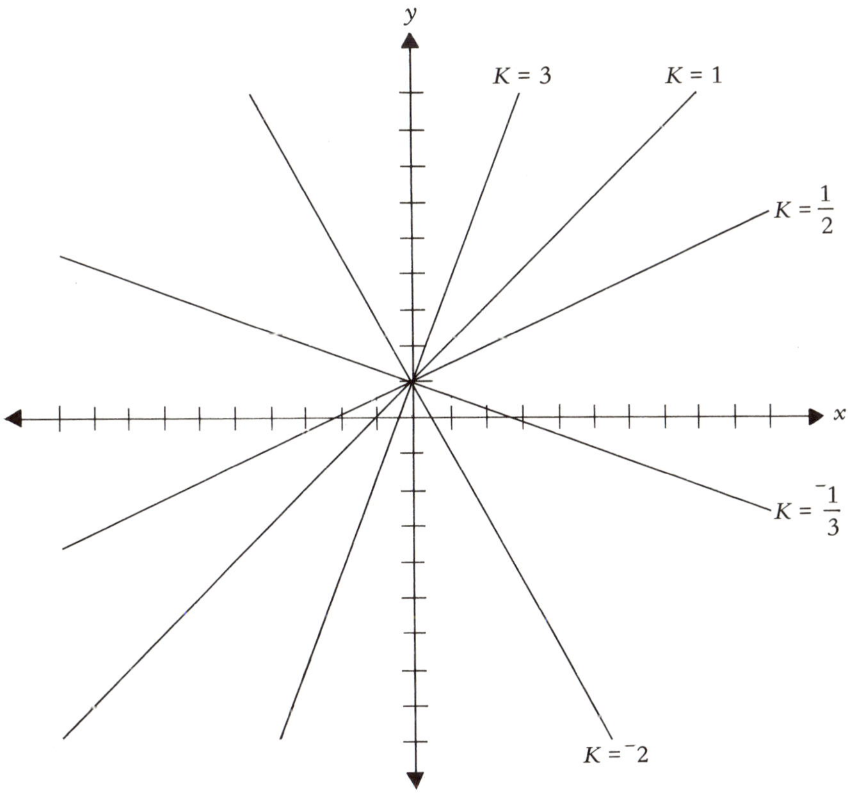

FIGURE 11.13

if there are constants k and b such that $f(x) = kx + b$ for all x. The graph of a linear function is always a straight line. And any nonvertical straight line is the graph of some linear function. A vertical straight line could not be the graph of a function.

With this result in hand, it is very easy to graph such a function as the one defined by $f(x) = -2x + 7$. We know its graph is a straight line. And we know that two points determine a straight line. So we may simply calculate that $(0,f(0)) = (0,7)$ and $(1,f(1)) = (1,5)$, plot these points, and draw the straight line through them.

Almost all of the geometry of straight lines in the plane can now be translated into assertions about linear functions. For example, it can be shown that two nonvertical straight lines are parallel if and only if they have the same slope, and they are perpendicular if and only if the slope of one is the negative reciprocal of the slope of the other. We shall not pursue these matters of **analytic geometry** further except to comment that the notion of *distance* in the plane can be expressed in terms of coordinates.

The distance between the points (x_1,y_1) and (x_2,y_2) is $\sqrt{(x_2 - x_1)^2 + (y_2 - y_1)^2}$.

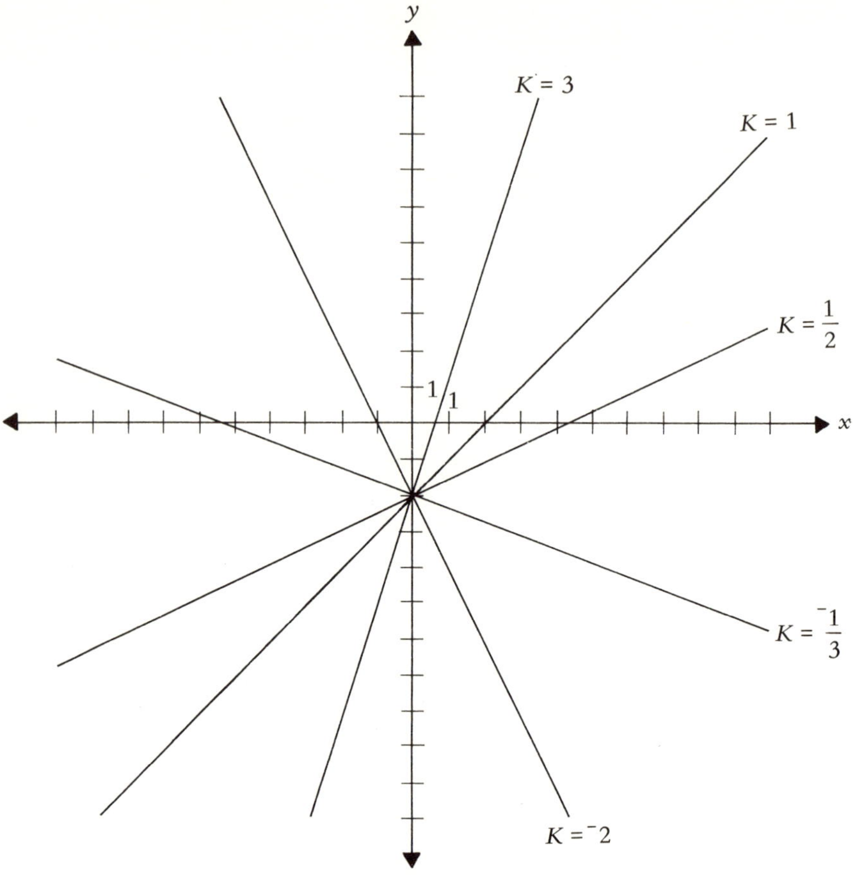

FIGURE 11.14

This result is an immediate consequence of the Pythagorean theorem which we established earlier. See Figure 11.15.

PROBLEMS

1. Graph $y = kx$ for k equal to (a) -3, (b) $1/4$, (c) 4, (d) 0, (e) $-1/10$, (f) $17/4$.

2. Graph $y = kx + 5$ for the values of k given in Problem 1.

3. Graph $y = kx - 4$ for the values of k given in Problem 1.

4. Graph $x + y = b$ for b equal to $-5, -4, -3, -2, -1, 0, 1, 2, 3, 4$, and 5. (Then see Figure 7.12.)

5. Find the distance between the two given points: (a) (0,0) and (3,4), (b) (6,3) and (9,7), (c) $(-2,1)$ and (2,4), (d) $(-2,4)$ and (2,4).

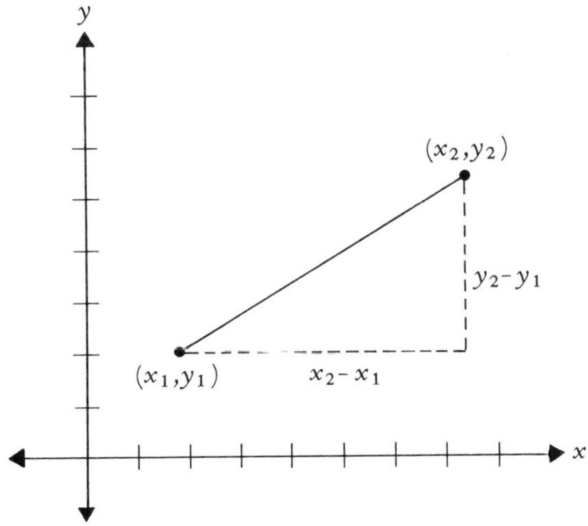

FIGURE 11.15

6. Notice that the distance from (0,0) to (x,y) is $\sqrt{x^2+y^2}$. Graph the equation (a) $x^2+y^2=1$, (b) $x^2+y^2=4$.

7. Notice that the distance from (a,b) to (x,y) is $\sqrt{(x-a)^2+(y-b)^2}$. Graph the equation (a) $(x-1)^2+(y-1)^2=1$, (b) $(x-2)^2+y^2=4$, (c) $(x-3)^2+(y-1)^2=9$.

8. Suppose y and x are measured quantities connected by a function f so that $y=f(x)$. Find the function f if

(a) y is proportional to x and $y=5$ when $x=1$.
(b) y is proportional to x and $y=7$ when $x=3$.
(c) y is proportional to z, z is proportional to x, $y=15$ when $z=3$, and $z=4$ when $x=2$.
(d) $y+3$ is proportional to x and $y=0$ when $x=1$.
(e) $y-b$ is proportional to x, $y=32$ when $x=0$, and $y=212$ when $x=100$.

9. Which linear functions are one-to-one correspondences between R and R?

10. Show that if there is a k such that $f(x)=kx$ for all real x then $f(ax_1+bx_2)=af(x_1)+bf(x_2)$.

11. Let f be the function defined by $f(x)=2x$. Diagram f as a mapping by drawing two parallel (real) number lines and sketching in the arrows showing (a) what the mapping does to $x=0, \pm 1, \pm 2, \pm 3, \pm 4$; (b) what the mapping does to the interval $0 \leq x \leq 1$, (c) what the mapping does to the interval $2 \leq x \leq 4$. (d) Explain why we might call the slope 2 here a *magnification factor*.

12. Sketch a graph of the line and find an equation for it if the line (a) passes

through (0,0) and (1,6), (b) passes through (0,2) and has slope -2, (c) passes through (1,4) and has slope 1, (d) passes through $(1,-6)$ and has slope -4.

13. Find an equation for the vertical line which passes through the point $(a,0)$. What is the slope of a line which is perpendicular to such a line?

11.4 Rates of change

In weather forecasting it is probably more important to know whether the barometer reading is rising or falling than to know what the reading is. In following the path of a space vehicle it is important to know at any given time not only where the vehicle is, but also its velocity. In considering difficulties that may be caused by overpopulation in the future, it is important to know something not only about current population figures but also something about the rates of change of various populations.

We have seen that it is often the case that one measured quantity may be thought of as a function of another. The idea of the rate of change of a function underlies a great variety of applications and, indeed, lies behind an entire branch of mathematics known as **the calculus.**

Suppose f is a function whose domain and range are subsets of R. The *average rate of change* of f as x changes from a to b is defined to be

$$\frac{f(b) - f(a)}{b - a}.$$

If we have a particle moving along a line so that its coordinate at time x is $f(x)$, then $(f(b) - f(a))/(b - a)$ is the average velocity of the particle between the times $x = a$ and $x = b$.

If $f(x)$ is a population at time x, in years, then $(f(b) - f(a))/(b - a)$ is the average rate of change of the population over the time span from a to b.

If $f(x)$ is the barometric pressure at time x, then $(f(b) - f(a))/(b - a)$ is the average rate of change in this pressure over the time interval from a to b.

It is natural to consider what happens to the average rate of change of a function as we take shorter and shorter intervals. In particular, suppose h_1, h_2, h_3, . . . is a null sequence with no term equal to 0. We might consider the sequence r_1, r_2, r_3, . . . where

$$r_n = \frac{f(a + h_n) - f(a)}{(a - h_n) - a} = \frac{f(a + h_n) - f(a)}{h_n}.$$

If this sequence of average rates of change converges to a certain number R no matter what null sequence h_1, h_2, h_3, . . . we use, then we shall call R the *instantaneous rate of change* of f at a. In fact, since R depends on a we may consider it to be the value of a function f' at a and write

$$f'(a) = \lim_{n \to \infty} \frac{f(a + h_n) - f(a)}{h_n}.$$

We call this function f' the **derivative** of f. The process of finding f', the derivative, for a given function f is called **differentiation.** A function that has a derivative at a is said to be **differentiable** at a.

Let us solve one easy problem of differentiation. Suppose $L(x) = kx + b$ and $h_1, h_2, h_3, \ldots$ is a null sequence with no zero terms. We have

$$\begin{aligned} r_n &= \frac{[k(a + h_n) + b] - (ka + b)}{h_n} \\ &= \frac{ka + kh_n + b - ka - b}{h_n} \\ &= k. \end{aligned}$$

Then certainly $\lim_{n \to \infty} r_n = k$ and we have $R = L'(a) = k$ for any a. In other words, if $L(x) = kx + b$, then L has a derivative L' and $L'(x) = k$ for all x.

As another example, let $s(x) = x^2$ and $h_1, h_2, h_3, \ldots$ be a null sequence with no zero terms. Then the nth term in the sequence of average rates of change for s at a is

$$\begin{aligned} r_n &= \frac{(a + h_n)^2 - a^2}{h_n} \\ &= \frac{a^2 + 2ah_n + h_n^2 - a^2}{h_n} \\ &= 2a + h_n. \end{aligned}$$

Clearly, $\lim_{n \to \infty} r_n = 2a$ and therefore:

If $s(x) = x^2$, then $s'(x) = 2x$.

There are several interpretations that may be given to the value of a derivative.

$f'(a)$ may be thought of as the instantaneous rate of change of f with respect to a unit change in x at $x = a$. Or if $y = f(x)$, then $f'(a)$ is the instantaneous rate of change of y with respect to x at $x = a$.

$f'(a)$ may be thought of as the number that has the property that $f(x) - f(a)$ is approximately equal to $f'(a)(x - a)$ when x is close to a. Or, putting this somewhat differently, when x is close to a, the function $f(x)$ acts very much like the linear function $T(x) = f'(a)(x - a) + f(a)$.

$f'(a)$ may be thought of as the slope of the line tangent to the curve $y = f(x)$ at $x = a$. The **tangent line** is the straight line that fits the curve most closely at the point. In fact, the tangent line at $x = a$ has the equation $y = f'(a)(x - a) + f(a)$. And $f'(a)$ is the slope of this line. See Figure 11.16.

When $y = f(x)$ and y and x represent certain measured quantities, the value of f' may represent certain useful quantities. If we are considering a particle

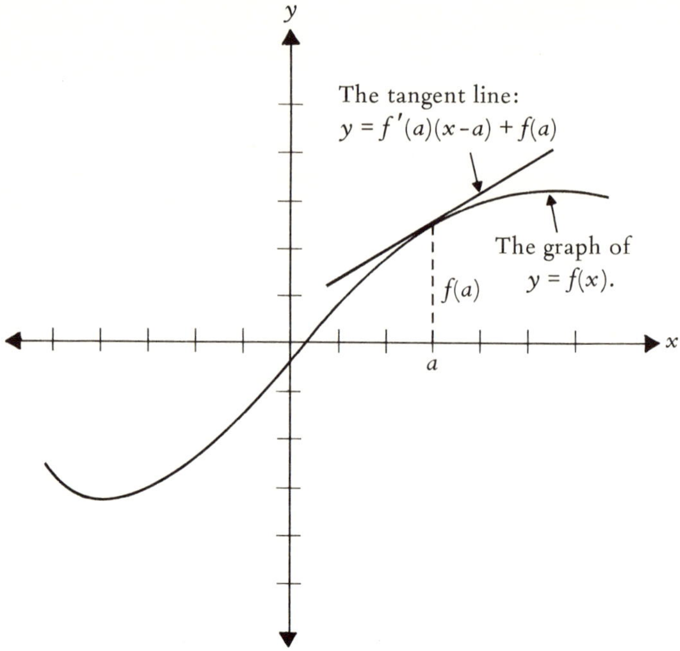

FIGURE 11.16

moving on a line which has coordinate y at time x, then $f'(a)$ is the **velocity** of the particle at time a. If y is the velocity of the particle at time x, then $f'(a)$ is the **acceleration** of the particle.

Many of the basic facts about derivatives are quite easy to establish from the basic definition. We list some of these in a table below:

If $f(x)$ is given by	then $f'(x)$ is given by
k	0
x^n	nx^{n-1} $(n \neq 0)$
$kg(x)$	$kg'(x)$
$g(x) + h(x)$	$g'(x) + h'(x)$
$g(x)h(x)$	$g(x)h'(x) + g'(x)h(x)$
$g(x)/h(x)$	$(h(x)g'(x) - g(x)h'(x))/(h(x))^2$
$g \circ h(x)$	$(g' \circ h(x))h'(x)$

In most courses in calculus considerable time is spent in discussing the use of the results in the above table. But it is the *idea* of a derivative that is important. Results such as those tabulated above are, in large part, simply tools to help solve technical problems of differentiation. But, of course, the two things are inseparable. If you wish to *use* the idea of a derivative you immediately run into technical problems of differentiation of particular functions.

PROBLEMS

1. Prove that a constant function has a derivative function whose value is always 0.

2. Sketch a careful graph of $y = x^2$ on the interval $0 \leq x \leq 2$. At the point on the curve where $x = 1$, draw the straight line (by eye) that seems to fit the curve best. Find the slope of this line and compare it with the derivative of $s(x) = x^2$ at $x = 1$.

3. By using the graphical technique of Problem 2 at various points, sketch a rough graph of the derivative of the function graphed in (a) Figure 11.11(a), (b) Figure 11.11(b), (c) Figure 11.11(c), (d) Figure 11.11(f).

4. Suppose f is differentiable at a and that $f'(a) > 0$. Prove that f must be an increasing function at a, in the sense that $f(a + h) > f(a)$ when h is close to 0 and positive and $f(a + h) < f(a)$ when h is close to 0 and negative.

5. Translate the weather report's usual statements about the barometer rising, falling, or being steady into statements about the derivative of a certain function. (See Problem 4.)

6. If a ball is thrown straight up into the air and y is the distance above the ground at time t, suppose $y = f(t)$. What can be said about f' at the time when the ball reaches its maximum height?

7. The **second derivative** of a function is the derivative (if there is one) of the derivative of the function. Let f'' be the second derivative of f and find f'' if $f(x)$ is given by: (a) 6, (b) $7x$, (c) x^2, (d) x^3, (e) x^7.

8. Using the fact that even when n is not an integer $f(x) = x^n$ implies $f'(x) = nx^{n-1}$, find $f'(x)$ if $f(x)$ is given by (a) $\sqrt{x}$, (b) $\sqrt[3]{x}$, (c) $\sqrt[10]{x}$, (d) $1/\sqrt{x}$, (e) $1/\sqrt[4]{x}$.

11.5 Areas under curves

If you travel on your bicycle at a velocity of 10 mph between 1 P.M. and 4 P.M., you will go 30 miles. If we think of your velocity as a (constant) function of time, we may construct the graph shown in Figure 11.17. But when we do so, we note that the distance of 30 miles that was traveled may be thought of geometrically as the area of the rectangle that lies above the horizontal axis, below the graph of the velocity function, and between the vertical lines at $t = 1$ and at $t = 4$. Or, as we shall say, the distance is given by the area that lies under the graph of the velocity function and over the interval from $t = 1$ to $t = 4$.

If you pedal your bike so that you go at 10 mph between 1 P.M. and 3 P.M., and then go at 15 mph between 3 P.M. and 4 P.M., you will travel $2 \cdot 10 +$

$1 \cdot 15 = 35$ miles. In Figure 11.18 we have again graphed the velocity as a function of time. And again we note that the distance traveled may be thought of as an area—that of the union of the two rectangles shown. That is, the distance is again the area that lies under the graph of the velocity function and over the interval from 1 to 4.

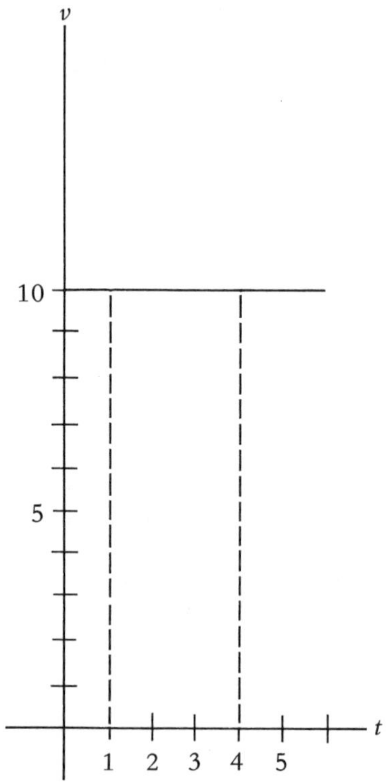

FIGURE 11.17

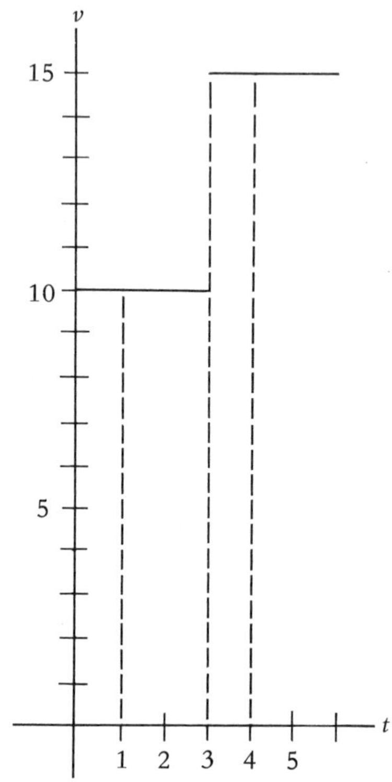

FIGURE 11.18

More generally, suppose we have the velocity v in mph given as a function of the time t in hours by $v = f(t)$. And suppose the graph of f is as shown in Figure 11.19. Such a curve suggests a somewhat more realistic velocity situation than the two earlier graphs. In riding a bicycle it would be quite unusual to hold exactly the same velocity for any extended length of time.

How can we use the graph of Figure 11.19 to determine the distance traveled between $t = 1$ and $t = 4$? It is natural (and correct) to expect this distance to be given by the area that lies under the graph of the velocity function and over the interval from 1 to 4—that is, the area under the curve which falls between the dotted lines in Figure 11.19.

Now it turns out that many, many other useful quantities aside from distances can be thought of in terms of areas under certain curves. So a general notation

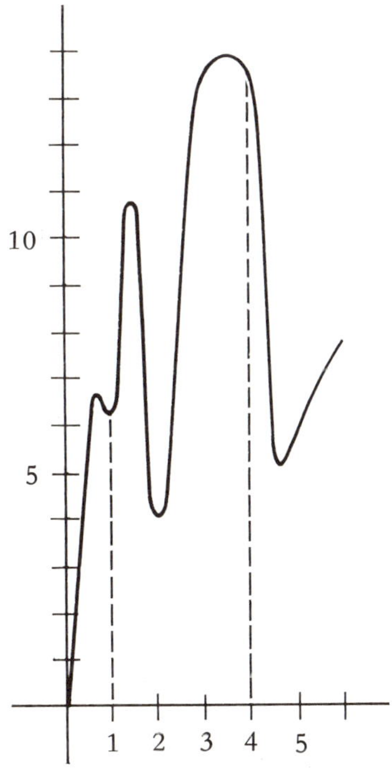

FIGURE 11.19

is introduced for such situations. Suppose g is a function which maps a subset of R into R^+. We shall write

$$\int_a^b g(x)dx$$

for the area that lies under the graph of $y = g(x)$ and over the interval from $x = a$ to $x = b$. "$\int_a^b g(x)dx$" is also read as "the **definite integral** from a to b of $g(x)dx$." The symbol "$\int$" is called an **integral sign** and a and b are called the **lower** and **upper limits** for the integral, respectively.

For example, we discovered in Figure 11.17 that if $g(x) = 10$ for all x then $\int_1^4 g(x)dx = 30$. In Figure 11.18 we discovered that if $g(x) = 10$ for all $x \leq 3$ and $g(x) = 15$ for $x > 3$, then $\int_1^4 g(x)dx = 35$. We claimed that the distance traveled between $t = 1$ and $t = 4$ by the bike with the velocity function f graphed in Figure 11.19 is given by $\int_1^4 f(t)dt$.

The idea of the definite integral of a function is one of the fundamental notions of the branch of mathematics called calculus. In some books you will find simply "$\int_a^b g$" used in place of "$\int_a^b g(x)dx$." And this less conventional notation has something to be said for it logically. The value of $\int_a^b g(x)dx$ depends on the numbers a and b and on the function g. It does not depend on what symbol we use to label the horizontal axis when we graph g.

In Figure 11.20, we have graphed the function h defined by the equation $y = h(x) = 2x$. $\int_0^3 h(x)dx$ is the area that lies under the graph of $y = h(x)$ and over the interval from 0 to 3. That is, $\int_0^3 h(x)dx$ is the area of the triangle shown in the figure. We see that the triangle has an altitude of 6 and a base of 3 so that $\int_0^3 h(x)dx = 1/2 \cdot 6 \cdot 3 = 9$. $\int_3^5 h(x)dx$ is the area of the trapezoid shown in the figure. If we think of this trapezoid as made up of a triangle and a rectangle we get $\int_3^5 h(x)dx = 2 \cdot 6 + 1/2 \cdot 4 \cdot 2 = 16$.

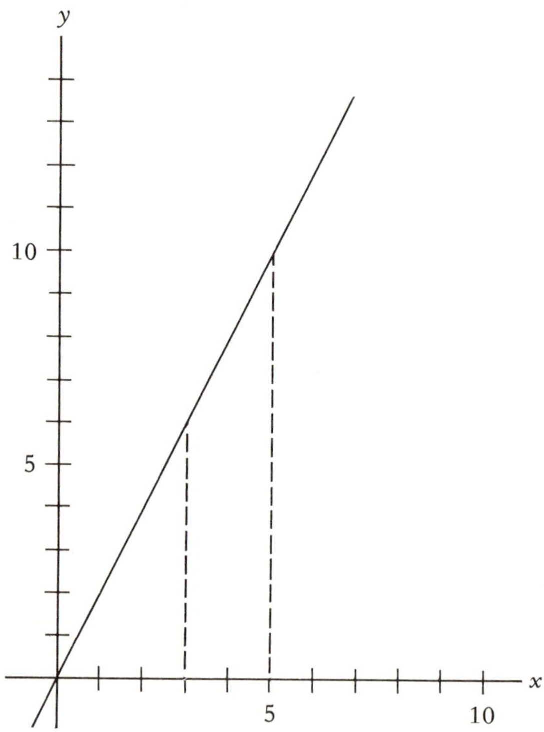

FIGURE 11.20

We have said that the idea of the derivative of a function and the notion of the definite integral of a function are both fundamental to the calculus. A basic result in this field which states a relationship between these two ideas is called **the fundamental theorem of calculus.** One form of this theorem may be put somewhat imprecisely as follows: if g is a well-behaved function and we define the function G by taking $G(x) = \int_a^x g(t)dt$, then the derivative G' of G is given $G'(x) = g(x)$.

Let us try to get some feeling for why this theorem is true. Suppose g has the graph shown in Figure 11.21. Let $G(x) = \int_a^x g(t)dt$ for all x. Then $G(x)$ is the area that lies under the graph $y = g(t)$ and over the interval from a to x. This area is shown in the figure. To find $G'(x)$ we suppose that $h_1, h_2, h_3, \ldots$ is a null sequence with nonzero terms and try to find

$$\lim_{n \to \infty} \frac{G(x + h_n) - G(x)}{h_n}.$$

But $G(x + h_n)$ is the area under the graph of $y = g(t)$ and over the interval from a to $x + h_n$. Therefore, $G(x + h_n) - G(x)$ is the area that lies under the graph of $y = g(t)$ and over the interval from x to $x + h_n$. But when h_n is small this area is very close to the area of the small rectangle shown in Figure 11.21. The area of this rectangle is $g(x) \cdot h_n$. Hence, when h_n is small we have $(G(x + h_n) - G(x))/h_n$ approximately the same as $g(x) \cdot h_n/h_n = g(x)$. So we would expect that $G'(x) = g(x)$.

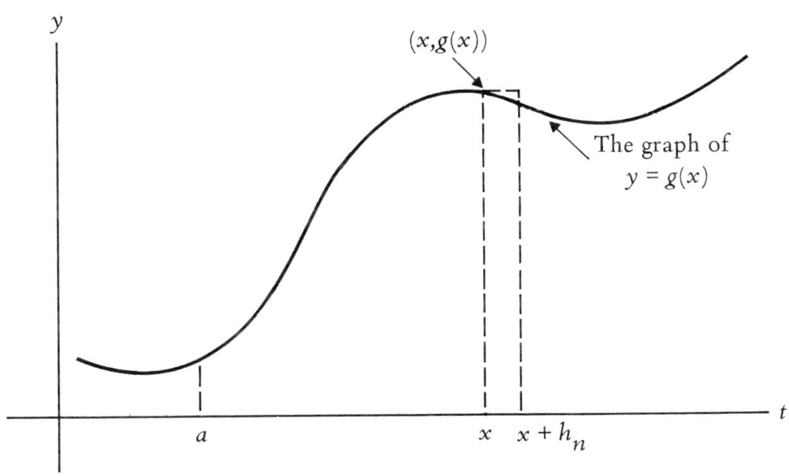

FIGURE 11.21

It develops that this fundamental theorem is very helpful in yielding practical procedures for finding the values of various definite integrals. But we shall not pursue this matter further here. The interested reader may consult any standard calculus text.

PROBLEMS

1. With h as given in Figure 11.20 find (a) $\int_0^5 h(x)dx$, (b) $\int_1^6 h(x)dx$, (c) $\int_6^6 h(x)dx$, (d) $\int_0^b h(x)dx$.

2. Suppose $g(x)$ is the greatest integer in x. That is, $g(x)$ is the largest integer which does not exceed x. Graph the function g. Find (a) $\int_0^2 g(x)dx$, (b) $\int_0^4 g(x)dx$, (c) $\int_3^6 g(x)dx$, (d) $\int_2^\pi g(x)dx$, (e) $\int_{\sqrt{2}}^\pi g(x)dx$.

3. Argue that if c is a positive real number and f is a function such that $f(x) = c$ for all x, then $\int_a^b f(x)dx = c(b - a)$ whenever $b \geq a$.

4. Using results we discovered in the last section, verify the special cases of the fundamental theorem of calculus where $a = 0$, x is positive and g is defined by (a) $g(t) = 2$, (b) $g(t) = k$ where $k > 0$, (c) $g(t) = t$, (d) $g(t) = 2t$, for all t.

5. A stone falls from rest so that its velocity t seconds after release is $32t$ ft per sec. How far will the stone fall (a) in the first 3 seconds, (b) in the first 5 seconds, (c) in the time interval between $t = 3$ and $t = 5$, (d) in the time interval between $t = 5$ and $t = 6$? (We have taken a downward velocity as positive here.)

6. Show that if y is proportional to t then the area which lies under the graph of y as a function of t and over the interval from 0 to t is proportional to t^2.

7. Argue for the result that

$$\int_a^b g(x)dx + \int_b^c g(x)dx = \int_a^c g(x)dx$$

where $a < b < c$.

8. Verify that $\int_a^b (f(x) + g(x))dx = \int_a^b f(x)dx + \int_a^b g(x)dx$ for the special case where $a = 1$, $b = 5$, and f and g are defined by (a) $f(x) = 1$, $g(x) = 3$, (b) $f(x) = x$, $g(x) = 2$, (c) $f(x) = x$, $g(x) = x$.

9. Do you think it is true that when $0 \leq f(x) \leq g(x)$ for $a \leq x \leq b$, we must have $\int_a^b f(x)dx \leq \int_a^b g(x)dx$? Explain.

10. We have in this section assumed that we were dealing with "well-behaved" functions. A function that is not well-behaved is the function k defined by $k(x) = 0$ if x is irrational, $k(x) = 1$ if x is rational. Discuss the graph of $y = k(x)$. What do you think should be the value (if any) of $\int_0^1 k(x)dx$?

11.6 Vector spaces

When we have a rectangular coordinate system in a plane we may think of each point as being identified by a pair of real numbers—the coordinates of the point. Alternatively, we may think of any point in such a plane as identifying the arrow which runs from the origin of the coordinate system to the point. This arrow may, in turn, be thought of as representing the **vector** which is the equivalence class of all arrows having the same length and direction as the given arrow.

So when x_1 and x_2 are real numbers we shall feel free to speak of the ordered pair (x_1,x_2) as the **point** (x_1,x_2) or as the **vector** (x_1,x_2).

A point in ordinary space may be identified by giving three real numbers. For example, we may locate any point P in a room by giving, in order, its distance from the front wall, its distance from the right wall, and its distance from the floor. (See Figure 11.22.) So when x_1, x_2, and x_3 are real we may think of the ordered triple (x_1,x_2,x_3) as representing a point P in space. But we may also think of (x_1,x_2,x_3) as the vector given by the arrow running from the origin to the point P.

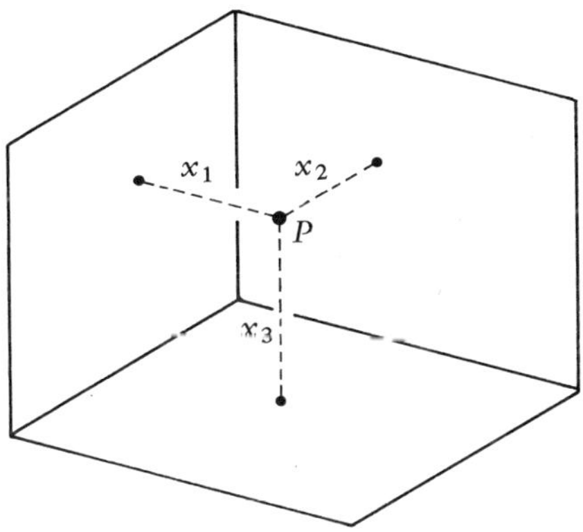

FIGURE 11.22

Such considerations lead us to the following generalization:

> When $x_1, x_2, \ldots, x_n$ are real numbers we shall refer to the ordered n-tuple $(x_1,x_2,x_3, \ldots ,x_n)$ as a **point** or a **vector** in ***n*-dimensional space.** The number x_i is said to be the ith **coordinate** of the point $(x_1,x_2, \ldots ,x_n)$ and to be the ith **component** of the vector $(x_1,x_2, \ldots ,x_n)$.

We shall let R^n be the collection of all n-dimensional points (or vectors).

Ordered n-tuples and this geometric terminology turn out to be very useful in many different connections.

For example, if we wish to consider the motion of a particle in ordinary 3-dimensional space we would be interested (at any given instant) in its three space coordinates and *also* in three numbers giving its velocity in each of the 3 directions. Thus, the situation of the particle is described by a 6-tuple (x,y,z,v_x,v_y,v_z) and the collection of all such possible 6-tuples could be called the **state-space.** Each point in this 6-dimensional space represents a possible state of the particle. Although it may be a bit difficult to visualize a 6-dimensional space, the geometric language we have used here tends to be quite helpful.

If we consider the experiment of throwing a die 10 times and let X_i be the number obtained on the ith toss, then the result of the 10 tosses may be represented as $(X_1, X_2, \ldots, X_{10})$. It is convenient to speak of a particular such outcome as a **sample point** and the collection of all possible sample points as the **sample space** for the experiment.

Suppose a person has four different bank accounts. If, at a given time, there are x_1 dollars in the first account, x_2 in the second, x_3 in the third, and x_4 in the fourth, we might refer to (x_1,x_2,x_3,x_4) as his **account vector.**

Suppose a store sells n different items and during a certain month sells x_i of the ith item. Then the n-tuple $(x_1,x_2, \ldots ,x_n)$ represents the situation in regard to sales for the store during that month. We would call $(x_1,x_2, \ldots ,x_n)$ the **sales vector** for the month.

Suppose that we make a list of the usual foods eaten in a certain locality and number them from 1 to n. If, in a given week, a person eats x_i ounces of the ith food for $i = 1,2, \ldots ,n$, it would be natural to refer to $(x_1,x_2, \ldots ,x_n)$ as that person's **diet vector** for the week.

Suppose that in a classroom we find out that there are f_1 students with no living grandparents, f_2 students with exactly one living grandparent, f_3 with exactly 2, f_4 with 3, and f_5 with 4. Then we could call (f_1,f_2,f_3,f_4,f_5) a **frequency vector** for these observations.

If one teacher observes in his classroom a frequency vector of (f_1,f_2,f_3,f_4,f_5) for the number of living grandparents, and another teacher observes a frequency vector of (g_1,g_2,g_3,g_4,g_5) for the same thing in his classroom, then $(f_1 + g_1,\ f_2 + g_2,\ f_3 + g_3,\ f_4 + g_4,\ f_5 + g_5)$ is the frequency vector for the group obtained by putting the two classes together.

If, in one week, a person has a diet vector of $(x_1,x_2, \ldots ,x_n)$ and in the next week has a diet vector of $(y_1,y_2, \ldots ,y_n)$, then his diet vector for the two-week period is $(x_1 + y_1,\ x_2 + y_2, \ldots,\ x_n + y_n)$.

If a person has an account vector of (x_1,x_2,x_3,x_4) and makes a transaction of y_1 with the first account (y_1 being positive for a deposit and negative for a withdrawal), y_2 with the second account, y_3 with the third account, and y_4 with the fourth, then his account vector becomes $(x_1 + y_1,\ x_2 + y_2,\ x_3 + y_3,\ x_4 + y_4)$.

Such examples lead us to define the addition of two vectors in R^n by saying

$$(x_1,x_2, \ldots ,x_n) + (y_1,y_2, \ldots ,y_n) = (x_1 + y_1,\ x_2 + y_2, \ldots,\ x_n + y_n).$$

Clearly such addition is commutative and associative and $(0,0, \ldots ,0)$ is the identity for addition in R^n. R^n is closed under this addition and $(-x_1,\ -x_2, \ldots,\ -x_n)$ is the additive inverse for $(x_1,x_2, \ldots ,x_n)$.

Notice, however, that for the sum of two vectors to make sense, the two vectors must have the same dimension. When it is understood that the dimension of the space is fixed, we may save ourselves some writing and use "$\vec{x}$" to denote $(x_1,x_2, \ldots ,x_n)$ and "$\vec{y}$" to denote $(y_1,y_2, \ldots ,y_n)$.

Notice, incidentally, that this vector addition we have introduced here agrees with that used in a more limited context in Chapters 6 and 7.

PROBLEMS

1. Suppose a company owns three different plants all of which manufacture the same item. Explain how it might be useful for the directors of the company to refer to such a thing as a monthly (a) production vector, (b) labor cost vector, (c) material cost vector.

2. Would vector addition be helpful to directors of the company of Problem 1 in preparing their annual report? Explain.

3. Evaluate: (a) $(5,4) + (-5,3)$, (b) $(6) + (7)$, (c) $(5,2) + (6)$, (d) $(7,4,-9) + (6,9,-4)$, (e) $(5,6,7,8,9,10) + (10,9,8,7,6,5)$, (f) $(5,7,8,4,3) + (-5,7,-8,-4,3)$.

4. In a rectangular coordinate system in the plane, plot each of the following vectors and their sum (a) $(1,1)$, $(1,2)$, (b) $(1,1)$, $(1,-1)$, (c) $(2,1)$, $(-2,-1)$, (d) $(-1,5)$, $(4,-1)$, (e) $(4,10)$, $(-3,-7)$, (f) $(5,-4)$, $(-6,3)$, (g) $(-5,5)$, $(-4,4)$.

5. Prove that vector addition in R^n is (a) commutative, (b) associative.

6. If we define subtraction in R^n to be the inverse operation for addition in R^n, is R^n closed under subtraction? Explain how you know.

---

7. Argue that the notation "R^n" for the collection of all n-dimensional vectors agrees with the notation of Section 6.22 if we think of the "n" as standing for the set $\{1,2, \ldots ,n\}$.

8. Is R^n a group under vector addition?

11.7 Matrices

Suppose that a store sells n different items and that $(x_1,x_2, \ldots ,x_n)$ is its sales vector for a certain month. That is, x_i is the number of the ith item that were sold during the month. The management of the store decides on the price at which each item is to be sold. Suppose it was the case that the ith item was sold at a price of a_i dollars per item. We would refer to $(a_1,a_2, \ldots ,a_n)$ as the *price vector* or *pricing policy* for the month. The *gross sales* for the month would be $s = a_1x_1 + a_2x_2 + \cdots + a_nx_n$. This is because the store would take in a_ix_i dollars during the month from sales of the ith item. There is a helpful form which expresses the computation of the gross sales as a sort of multiplication.

$$(a_1,a_2, \ldots ,a_n) \cdot \begin{pmatrix} x_1 \\ x_2 \\ \vdots \\ x_n \end{pmatrix} = (a_1x_1 + a_2x_2 + \cdots + a_nx_n).$$

Here we have chosen to write the sales vector as a column rather than a row. The reason for this will appear shortly. We can describe the operation somewhat imprecisely as follows: we lift up the column on the right and lay it along the left row, then we multiply corresponding terms and add the resulting products.

If, for example, we sell 15 of the first item and 37 of the second at respective prices of \$50 and \$13 per item, we have a gross sales of $(50, 13)\begin{pmatrix}15\\37\end{pmatrix} = (50 \cdot 15 + 13 \cdot 37) = (1231)$.

Now suppose we have two different stores selling the same two items. Let x_{11} be the amount of the first item sold in the first store. Let x_{21} be the amount of the second item sold in the first store. Let x_{12} be the amount of the first item sold in the second store. Let x_{22} be the amount of the second item sold in the second store. Suppose both stores use the same price vector (a_1,a_2). Then the gross sales in the first store is $S_1 = a_1x_{11} + a_2x_{21}$ and that in the second is $S_2 = a_1x_{12} + a_2x_{22}$. These results may be summarized by the array

$$\begin{pmatrix}S_1\\S_2\end{pmatrix} = (a_1,a_2) \cdot \begin{pmatrix}x_{11} & x_{12}\\x_{21} & x_{22}\end{pmatrix} = \begin{pmatrix}a_1x_{11} + a_2x_{21}\\a_1x_{12} + a_2x_{22}\end{pmatrix}.$$

Note that a pattern similar to that used earlier occurs again. To find S_1 we (so to speak) lift up the first column in the array

$$\begin{pmatrix}x_{11} & x_{12}\\x_{21} & x_{22}\end{pmatrix},$$

lay it along the row (a_1,a_2), multiply corresponding terms, and then add. To get S_2 we do the same thing with the second column of the array

$$\begin{pmatrix}x_{11} & x_{12}\\x_{21} & x_{22}\end{pmatrix}.$$

Now suppose we have two stores as before but we wish to consider the effects of two different pricing policies. Let (a_{11},a_{12}) be the first policy we wish to consider and (a_{21},a_{22}) the second. That is, if we use the first policy we shall charge a_{11} dollars for the first item and a_{12} dollars for the second. With the second policy we shall charge a_{21} dollars for the first item and a_{22} for the second. Let S_{ij} be the gross sales when the ith policy is used at the jth store. Then the four possible gross sales are given by

$$\begin{pmatrix}S_{11} & S_{12}\\S_{21} & S_{22}\end{pmatrix} = \begin{pmatrix}a_{11} & a_{12}\\a_{21} & a_{22}\end{pmatrix} \cdot \begin{pmatrix}x_{11} & x_{12}\\x_{21} & x_{22}\end{pmatrix}$$

$$= \begin{pmatrix}a_{11}x_{11} + a_{12}x_{21} & a_{11}x_{12} + a_{12}x_{22}\\a_{21}x_{11} + a_{22}x_{21} & a_{21}x_{12} + a_{22}x_{22}\end{pmatrix}.$$

And the same pattern is evident here that we had before. For example to get S_{21} we take the first column from

$$\begin{pmatrix}x_{11} & x_{12}\\x_{21} & x_{22}\end{pmatrix}$$

and lay it along the second row in

$$\begin{pmatrix} a_{11} & a_{12} \\ a_{21} & a_{22} \end{pmatrix},$$

multiply corresponding terms, and then add. As a numerical example we have

$$\begin{pmatrix} 10 & 15 \\ 13 & 12 \end{pmatrix} \cdot \begin{pmatrix} 10 & 20 \\ 30 & 5 \end{pmatrix} = \begin{pmatrix} 10 \cdot 10 + 15 \cdot 30 & 10 \cdot 20 + 15 \cdot 5 \\ 13 \cdot 10 + 12 \cdot 30 & 13 \cdot 20 + 12 \cdot 5 \end{pmatrix}$$
$$= \begin{pmatrix} 550 & 275 \\ 490 & 320 \end{pmatrix}.$$

For instance, if the second price vector is used by the first store, the gross sales there will be \$490.

This type of "multiplication" turns out to be useful in many different cases. We shall call any rectangular array of real numbers a **matrix.** If the array has n rows and m columns of numbers we shall say the matrix is $n \times m$, which we read as "n by m." The numbers that appear in the array are called **elements** in the matrix.† If A is a matrix we shall use "a_{ij}" to name the element in the ith row and the jth column in A. Note that when we use these double subscripts, the first subscript always indicates the row and the second the column. This is a standard convention followed in most applications of matrices.

If A is an $n \times m$ matrix and B is an $m \times p$ matrix we follow the lead of the above examples and define the **matrix product** of A and B to be the matrix C whose element in the ith row and the jth column is given by

$$c_{ij} = a_{i1}b_{1j} + a_{i2}b_{2j} + \cdots + a_{im}b_{mj}.$$

This is an algebraic formulation of the pattern we discovered in the above example. To find the element in the ith row and jth column of $A \cdot B$ we take the jth column in B, lay it along the ith row in A, multiply corresponding terms, and then add the products. Note that for the operation to make sense, A must have the same number of columns as B has rows.

Generalizing our notion of addition of vectors, we may also define **matrix addition** for two matrices of the same size. We simply take $A + B$ to be equal to the matrix D where $d_{ij} = a_{ij} + b_{ij}$.

The system of 1×1 matrices with the operations of addition and multiplication is isomorphic to the system $(R, +, \cdot)$. And therefore, the entire system of matrices with addition and multiplication as defined above is, in a certain sense, an extension of the real number system. But in most cases it is not customary to refer to a matrix that is not 1×1 as a number. And, of course, the collection of *all* matrices is not closed under either addition or multiplication.

† From another point of view we could define an $n \times m$ matrix to be a mapping from $\{1, 2, \ldots, n\} \times \{1,2, \ldots, m\}$ into R. And the values of the mapping would be the elements in the matrix.

There is, however, a certain class of matrices that it *is* very helpful to interpret as numbers of a certain kind. We shall discuss this case in Section 11.8.

Let us illustrate the processes of matrix multiplication and addition.

$$(2\ \ -1)\cdot\begin{pmatrix}1\\-2\end{pmatrix} = (2\cdot 1 + (-1)(-2)) = (4)$$

$$(1,1)\cdot\begin{pmatrix}1\\-1\end{pmatrix} = (1\cdot 1 + 1\cdot(-1)) = (0)$$

$$\begin{pmatrix}3\\6\\1\end{pmatrix} + \begin{pmatrix}1\\-2\\4\end{pmatrix} = \begin{pmatrix}4\\4\\5\end{pmatrix}$$

$$(3\ \ 6\ \ 1)\cdot\begin{pmatrix}1\\-2\\4\end{pmatrix} = (3 - 12 + 4) = (-5)$$

$$\begin{pmatrix}1\\-2\\4\end{pmatrix}\cdot(3\ \ 6\ \ 1) = \begin{pmatrix}3 & 6 & 1\\-6 & -12 & -2\\12 & 24 & 4\end{pmatrix}$$

$$\begin{pmatrix}1\\0\\2\end{pmatrix}\cdot(0\ \ 1\ \ 4) = \begin{pmatrix}0 & 1 & 4\\0 & 0 & 0\\0 & 2 & 4\end{pmatrix}$$

$$\begin{pmatrix}1 & 1\\0 & 2\end{pmatrix}\cdot\begin{pmatrix}1 & 2\\4 & 0\end{pmatrix} = \begin{pmatrix}5 & 2\\8 & 0\end{pmatrix}$$

$$\begin{pmatrix}7 & 16\\0 & -5\end{pmatrix}\cdot\begin{pmatrix}1 & 0\\0 & 1\end{pmatrix} = \begin{pmatrix}7 & 16\\0 & -5\end{pmatrix}$$

$$\begin{pmatrix}1 & 0\\0 & 1\end{pmatrix}\cdot\begin{pmatrix}7 & 6\\4 & -3\end{pmatrix} = \begin{pmatrix}7 & 6\\4 & -3\end{pmatrix}$$

$$\begin{pmatrix}1 & 0 & 0\\3 & 6 & 2\\0 & 1 & 0\end{pmatrix}\cdot\begin{pmatrix}3\\1\\0\end{pmatrix} = \begin{pmatrix}3\\15\\1\end{pmatrix}$$

$$\begin{pmatrix}2 & 0 & 0\\0 & 5 & 0\\0 & 0 & 6\end{pmatrix}\cdot\begin{pmatrix}1 & 0 & 0\\0 & -1 & 0\\0 & 0 & 2\end{pmatrix} = \begin{pmatrix}2 & 0 & 0\\0 & -5 & 0\\0 & 0 & 12\end{pmatrix}.$$

We see that matrix multiplication is not commutative. It is, however, associative in the sense that $(A\cdot B)\cdot C = A\cdot(B\cdot C)$ whenever the indicated multiplications are possible. There is no identity element for multiplication in the collection of all matrices.

However, if we fix n and consider only $n \times n$ (so called *square*) matrices then the matrix A which has $a_{ii} = 1$ for $1,2,\ldots,n$ and $a_{ij} = 0$ for $i \neq j$ is the identity for multiplication. For example, it is easy to check that

$$\begin{pmatrix}1 & 0\\0 & 1\end{pmatrix}\begin{pmatrix}2 & 3\\4 & 5\end{pmatrix} = \begin{pmatrix}2 & 3\\4 & 5\end{pmatrix}\begin{pmatrix}1 & 0\\0 & 1\end{pmatrix} = \begin{pmatrix}2 & 3\\4 & 5\end{pmatrix}$$

and

$$\begin{pmatrix}1 & 0 & 0\\0 & 1 & 0\\0 & 0 & 1\end{pmatrix}\begin{pmatrix}-1 & 0 & 4\\1 & 5 & 6\\-10 & 0 & 2\end{pmatrix} = \begin{pmatrix}-1 & 0 & 4\\1 & 5 & 6\\-10 & 0 & 2\end{pmatrix}\begin{pmatrix}1 & 0 & 0\\0 & 1 & 0\\0 & 0 & 1\end{pmatrix}$$
$$= \begin{pmatrix}-1 & 0 & 4\\1 & 5 & 6\\-10 & 0 & 2\end{pmatrix}.$$

Furthermore, the following are easy to verify.

$$\begin{pmatrix}1 & 0\\0 & 1\end{pmatrix}\begin{pmatrix}5\\-3\end{pmatrix} = \begin{pmatrix}5\\-3\end{pmatrix}, \quad \begin{pmatrix}1 & 0 & 0\\0 & 1 & 0\\0 & 0 & 1\end{pmatrix}\begin{pmatrix}6\\0\\-4\end{pmatrix} = \begin{pmatrix}6\\0\\-4\end{pmatrix}.$$

And, in general, we shall have

$$\begin{pmatrix}1 & 0 & 0 \ldots 0\\0 & 1 & 0 \ldots 0\\0 & 0 & 1 \ldots 0\\\vdots & \vdots & \vdots \quad \vdots\\0 & 0 & 0 \ldots 1\end{pmatrix}\begin{pmatrix}x_1\\x_2\\\vdots\\x_n\end{pmatrix} = \begin{pmatrix}x_1\\x_2\\\vdots\\x_n\end{pmatrix}$$

where the matrix on the left is the multiplicative identity for $n \times n$ matrices. The fact that such an identity matrix exists suggests that, at least in certain cases, we should be able to find matrices which are multiplicative inverses of certain other matrices. This does indeed occur and the idea of an inverse matrix (with respect to multiplication) for a given matrix turns out to be extremely useful. We shall pursue this matter a bit further in the next section.

Problems

1. Suppose that we are planning a long trip where we expect to travel by car on 5 successive days. Let x_i be the number of hours that we plan to drive on the ith day for $i = 1,2,3,4,5$. Then we may think of (x_1,x_2,x_3,x_4,x_5) as a description of our entire journey and even, if we wish, refer to (x_1,x_2,x_3,x_4,x_5) as a *time vector* for the trip. Suppose that with some knowledge of the predicted weather conditions we plan to travel at an average speed of a_i miles per hour on the ith day. We might call (a_1,a_2,a_3,a_4,a_5) the speed policy for our trip. What will be the total distance we will drive in the 5 days? Represent this result as a matrix product.

2. Suppose that in the situation in Problem 1 we wish also to consider two different speed policies for the given time vector. Label the components of these two policies appropriately and show how the distances covered using the two policies may be worked out in a matrix product.

3. Discuss the situation in Problem 2 if we wish to consider 4 different speed policies.

4. Discuss the situation in Problem 1 if we wish to compare the results of two different speed policies *and* two different time vectors.

5. Suppose we consider n different kinds of food and that the ith one of these has c_i calories per ounce for $i = 1,2, \ldots, n$. Suppose that, in a week, a person consumes x_i ounces of the ith food for $i = 1,2, \ldots, n$. Then $(x_1, x_2, \ldots, x_n)$ is the person's *diet vector* for the week. (a) Represent the total calorie intake of the person for the week in terms of a matrix product. (b) Consider the situation where we want to compare the calorie intake for 4 different diet vectors.

6. Find (if possible) the following products and sums.

$$\text{(a)}\ (2\quad 5)\cdot\begin{pmatrix}5\\2\end{pmatrix},\quad \text{(b)}\ (2\quad 5)\cdot\begin{pmatrix}5\\2\\3\end{pmatrix},\quad \text{(c)}\ \begin{pmatrix}5\\2\end{pmatrix}\cdot(2\quad 5),$$

$$\text{(d)}\ (0\quad 1\quad 2)\cdot\begin{pmatrix}5\\6\\0\end{pmatrix},\quad \text{(e)}\ \begin{pmatrix}3&0\\0&4\end{pmatrix}\cdot\begin{pmatrix}1&0\\0&-1\end{pmatrix},\quad \text{(f)}\ \begin{pmatrix}0&3\\0&4\end{pmatrix}\cdot\begin{pmatrix}1&0\\5&2\end{pmatrix},$$

$$\text{(g)}\ (6)\cdot(3\quad -1\quad 5),\quad \text{(h)}\ (3\quad 5\quad 9)+(-3\quad -5\quad 9),$$

$$\text{(i)}\ \begin{pmatrix}7&0\\5&2\end{pmatrix}+\begin{pmatrix}7&0\\-1&6\end{pmatrix},\quad \text{(j)}\ \begin{pmatrix}5&0\\1&4\end{pmatrix}+\begin{pmatrix}7&0\\-1&6\end{pmatrix},$$

$$\text{(k)}\ \begin{pmatrix}3&-1\\4&2\end{pmatrix}\cdot\begin{pmatrix}6&2\\-6&4\end{pmatrix},\quad \text{(l)}\ \begin{pmatrix}0&1\\-1&0\end{pmatrix}\cdot\begin{pmatrix}0&1\\-1&0\end{pmatrix},$$

$$\text{(m)}\ \begin{pmatrix}1&1\\-1&1\end{pmatrix}\cdot\begin{pmatrix}1&1\\-1&1\end{pmatrix},\quad \text{(n)}\ \begin{pmatrix}1&0&0\\0&1&0\\0&0&1\end{pmatrix}\cdot\begin{pmatrix}3&2&1\\0&1&6\\4&0&0\end{pmatrix},$$

$$\text{(o)}\ \begin{pmatrix}3&0&5\\-1&2&2\\6&-4&0\end{pmatrix}\cdot\begin{pmatrix}-1&1&2\\0&1&4\\1&1&6\end{pmatrix},\quad \text{(p)}\ \begin{pmatrix}1&0&0\\0&0&1\\0&1&0\end{pmatrix}\cdot\begin{pmatrix}0&1&0\\1&0&0\\0&0&1\end{pmatrix},$$

$$\text{(q)}\ \begin{pmatrix}1&0&0\\0&1&0\\0&0&1\end{pmatrix}\begin{pmatrix}0&1&0\\0&2&1\\1&2&0\end{pmatrix},\quad \text{(r)}\ \begin{pmatrix}7&4&3\\-1&0&0\\1&5&1\end{pmatrix}\begin{pmatrix}1&0&0\\0&1&0\\0&0&1\end{pmatrix},$$

$$\text{(s)}\ \begin{pmatrix}1&0\\0&1\end{pmatrix}\begin{pmatrix}0\\5\end{pmatrix},\quad \text{(t)}\ \begin{pmatrix}1&0&0\\0&1&0\\0&0&1\end{pmatrix}\begin{pmatrix}6\\10\\17\end{pmatrix},\quad \text{(u)}\ \begin{pmatrix}1&1\\0&1\end{pmatrix}\begin{pmatrix}1&-1\\0&1\end{pmatrix},$$

$$\text{(v)}\ \begin{pmatrix}2&5\\-3&6\end{pmatrix}\begin{pmatrix}2/9&-5/27\\1/9&2/27\end{pmatrix}.$$

7. Let M_{nm} be the collection of all $n \times m$ matrices. (a) Is M_{nm} closed under addition? (b) Is addition commutative and associative in M_{nm}? (c) Is there an identity element for addition in M_{nm}? (d) Does each element in M_{nm} have an additive inverse? (e) Is M_{nm} closed under subtraction?

8. The special case of multiplying together a 1×1 matrix and a $1 \times n$ matrix is called a **scalar** multiplication. We have $(c)(x_1x_2 \ldots x_n) = (cx_1cx_2 \ldots cx_n)$. And this is usually written simply as

$$c(x_1x_2 \ldots x_n) = (cx_1cx_2 \ldots cx_n).$$

If $\vec{x}$ and $\vec{y}$ are $1 \times n$ matrices (vectors) show that $c\vec{x} + c\vec{y} = c(\vec{x} + \vec{y})$ and $(c + d)\vec{x} = c\vec{x} + d\vec{x}$.

9. When $\vec{x} = (x_1, x_2, \ldots, x_n)$ and $\vec{y} = (y_1, y_2, \ldots, y_n)$ are n-dimensional vectors then the single element in the matrix product

$$(x_1x_2 \ldots x_n) \cdot \begin{pmatrix} y_1 \\ y_2 \\ \vdots \\ y_n \end{pmatrix} = (x_1y_1 + x_2y_2 + \cdots + x_ny_n)$$

is called the **dot product** or **inner product** of $\vec{x}$ and $\vec{y}$ and is usually written as $\vec{x} \cdot \vec{y}$. Investigate the properties of this operation and give some numerical examples.

11.8 Systems of linear equations

Lean beef has about 170 calories and about 20 grams of protein per 100 grams. Cooked potatoes have about 92 calories and about 3 grams of protein per 100 grams. Is it possible to eat a meal of steak and potatoes and nothing else and get 200 calories and 15 grams of protein?

Suppose it can be done by eating x_1 grams of steak and x_2 grams of potatoes. Then we must have

$$170 \cdot \frac{x_1}{100} + 92 \cdot \frac{x_2}{100} = 200$$

and

$$20 \cdot \frac{x_1}{100} + 3 \cdot \frac{x_2}{100} = 15.$$

These equations can be simplified to yield

$$1.7x_1 + .92x_2 = 200$$

$$.2x_1 + .03x_2 = 15.$$

We could answer our question in the affirmative if we could find positive numbers x_1 and x_2 which make both these equations true. So it would be helpful to *solve* the *system of equations* in the sense that we would like to find all pairs (x_1,x_2) which make both the equations true.

This is a special case of a large class of problems that arise in applications. If $a_{11},a_{12},\ldots,a_{1n}$, $a_{21},a_{22},\ldots,a_{2n},\ldots,a_{n1},a_{n2},\ldots,$ and a_{nn} are real numbers, and $b_1,b_2,\ldots,$ and b_n are real numbers, then

$$\begin{aligned} a_{11}x_1 + a_{12}x_2 + \cdots + a_{1n}x_n &= b_1 \\ a_{21}x_1 + a_{22}x_2 + \cdots + a_{2n}x_n &= b_2 \\ \vdots \qquad\qquad\qquad\qquad & \quad \vdots \\ a_{n1}x_1 + a_{n2}x_2 + \cdots + a_{nn}x_n &= b_n \end{aligned}$$

is said to be a *system of n linear equations in n unknowns.* A *solution* of this system will be any n-tuple $(x_1,x_2,\ldots,x_n)$ which satisfies all the equations simultaneously.

In the above example, we had $n = 2$ and $a_{11} = 1.7$, $a_{12} = .92$, $a_{21} = .2$, $a_{22} = .03$, $b_1 = 200$, and $b_2 = 15$.

One of the very nice uses of matrices is discovered when we note that the system of the example or the general system above can be written as a single matrix equation. The first system is

$$\begin{pmatrix} 1.7 & .92 \\ .2 & .03 \end{pmatrix}\begin{pmatrix} x_1 \\ x_2 \end{pmatrix} = \begin{pmatrix} 200 \\ 15 \end{pmatrix}.$$

And if we let

$$A = \begin{pmatrix} a_{11} & a_{12} & \cdots & a_{1n} \\ a_{21} & a_{22} & \cdots & a_{2n} \\ \vdots & \vdots & & \vdots \\ a_{n1} & a_{n2} & \cdots & a_{nn} \end{pmatrix}$$

and†

$$\vec{x} = \begin{pmatrix} x_1 \\ x_2 \\ \vdots \\ x_n \end{pmatrix} \quad \text{and} \quad \vec{b} = \begin{pmatrix} b_1 \\ b_2 \\ \vdots \\ b_n \end{pmatrix},$$

then the general system is simply

$$A\vec{x} = \vec{b}.$$

When the system of equations is written in the simple form $A\vec{x} = \vec{b}$, the temptation is strong to solve the equation (system) by multiplying both sides on the left by the multiplicative inverse for the matrix A. And, indeed, this works when A has such an inverse. For we shall then have that $A \cdot \vec{x} = \vec{b}$ if and only if $A^{-1}(A\vec{x}) = A^{-1}\vec{b}$. But this holds if and only if $(A^{-1} \cdot A)\vec{x} = \vec{x} = A^{-1}\vec{b}$.

† Here, as a convenience, we use our vector notation for an $n \times 1$ matrix.

So if we had a method for finding A^{-1} when it exists, we would be able to solve the equation (system) in a very straightforward manner. There are mechanical routines (and computer programs) for finding the inverse of a square matrix when it has one. We shall not discuss these routines here. But we shall show how the method works when we have A^{-1} given to us.

Suppose

$$A = \begin{pmatrix} 2 & 0 \\ 0 & 3 \end{pmatrix}.$$

Then

$$A^{-1} = \begin{pmatrix} 1/2 & 0 \\ 0 & 1/3 \end{pmatrix}$$

since

$$\begin{pmatrix} 1/2 & 0 \\ 0 & 1/3 \end{pmatrix}\begin{pmatrix} 2 & 0 \\ 0 & 3 \end{pmatrix} = \begin{pmatrix} 1 & 0 \\ 0 & 1 \end{pmatrix}.$$

Then we may solve the equation $A\vec{x} = \vec{b}$ by writing

$$\begin{pmatrix} x_1 \\ x_2 \end{pmatrix} = \vec{x} = A^{-1}\vec{b} = \begin{pmatrix} 1/2 & 0 \\ 0 & 1/3 \end{pmatrix}\begin{pmatrix} b_1 \\ b_2 \end{pmatrix} = \begin{pmatrix} b_1/2 \\ b_2/3 \end{pmatrix}.$$

This checks with the immediate solution we find for the system

$$2x_1 + 0x_2 = b_1$$

$$0x_1 + 3x_2 = b_2.$$

If

$$A = \begin{pmatrix} 1 & 1 \\ 0 & 1 \end{pmatrix},$$

then

$$A^{-1} = \begin{pmatrix} 1 & -1 \\ 0 & 1 \end{pmatrix}$$

since

$$A^{-1}A = \begin{pmatrix} 1 & -1 \\ 0 & 1 \end{pmatrix}\begin{pmatrix} 1 & 1 \\ 0 & 1 \end{pmatrix} = \begin{pmatrix} 1 & 0 \\ 0 & 1 \end{pmatrix}.$$

Therefore,

$$A\vec{x} = \begin{pmatrix} 2 \\ 3 \end{pmatrix} \text{ yields } \vec{x} = \begin{pmatrix} 1 & -1 \\ 0 & 1 \end{pmatrix}\begin{pmatrix} 2 \\ 3 \end{pmatrix} = \begin{pmatrix} -1 \\ 3 \end{pmatrix}.$$

And if we have the system

$$x_1 + x_2 = 2$$

$$x_2 = 3,$$

then clearly $x_2 = 3$ and, hence, $x_1 = -1$.

If

$$A = \begin{pmatrix} 2 & 5 \\ -3 & 6 \end{pmatrix} \quad \text{and} \quad \vec{b} = \begin{pmatrix} 5 \\ 0 \end{pmatrix}$$

then the equation $A\vec{x} = \vec{b}$ has solution

$$\begin{pmatrix} x_1 \\ x_2 \end{pmatrix} = A^{-1}\vec{b} = \begin{pmatrix} 2/9 & -5/27 \\ 1/9 & 2/27 \end{pmatrix} \begin{pmatrix} 5 \\ 0 \end{pmatrix} = \begin{pmatrix} 10/9 \\ 5/9 \end{pmatrix},$$

since

$$A^{-1} = \begin{pmatrix} 2/9 & -5/27 \\ 1/9 & 2/27 \end{pmatrix}.$$

This last may be checked by noting that

$$\begin{pmatrix} 2/9 & -5/27 \\ 1/9 & 2/27 \end{pmatrix} \begin{pmatrix} 2 & 5 \\ -3 & 6 \end{pmatrix} = \begin{pmatrix} 1 & 0 \\ 0 & 1 \end{pmatrix}.$$

To return to the problem with which we started this section, it can be shown that the inverse of

$$\begin{pmatrix} 1.7 & .92 \\ .2 & .03 \end{pmatrix}$$

is the matrix

$$\begin{pmatrix} -30/133 & 920/133 \\ 200/133 & -1700/133 \end{pmatrix}.$$

And therefore, the system of equations

$$1.7x_1 + .92x_2 = 200$$

$$.2x_1 + .03x_2 = 15$$

has the solution

$$\begin{pmatrix} x_1 \\ x_2 \end{pmatrix} = \begin{pmatrix} -30/133 & 920/133 \\ 200/133 & -1700/133 \end{pmatrix} \begin{pmatrix} 200 \\ 15 \end{pmatrix} = \begin{pmatrix} 7800/133 \\ 14500/133 \end{pmatrix}.$$

Hence, if we eat 7800/133 grams of steak and 14500/133 grams of potatoes we shall get the desired distribution of calories and protein.

If we are given the system

$$x_1 + x_2 = 1$$

$$x_1 + x_2 = 2,$$

then clearly there could not be a solution since we cannot have both

$$x_1 + x_2 = 1 \quad \text{and} \quad x_1 + x_2 = 2.$$

Hence, the matrix $\begin{pmatrix} 1 & 1 \\ 1 & 1 \end{pmatrix}$ must not have an inverse. There are, of course, many other matrices that have no inverses.

Problems

1. Verify, in each case, that the second matrix in the pair is the inverse of the first.

(a) $\begin{pmatrix} 6 & 0 \\ 0 & -10 \end{pmatrix}\begin{pmatrix} 1/6 & 0 \\ 0 & -1/10 \end{pmatrix}$, (b) $\begin{pmatrix} 0 & 1 \\ 1 & 0 \end{pmatrix}\begin{pmatrix} 0 & 1 \\ 1 & 0 \end{pmatrix}$,

(c) $\begin{pmatrix} 7 & 4 \\ 3 & 0 \end{pmatrix}\begin{pmatrix} 0 & 1/3 \\ 1/4 & -7/12 \end{pmatrix}$,

(d) $\begin{pmatrix} 5 & -2 \\ 2 & -1 \end{pmatrix}\begin{pmatrix} 1 & -2 \\ 2 & -5 \end{pmatrix}$, (e) $\begin{pmatrix} -7 & 10 \\ 3 & -4 \end{pmatrix}\begin{pmatrix} 2 & 5 \\ 3/2 & 7/2 \end{pmatrix}$,

(f) $\begin{pmatrix} 1 & 0 & 0 \\ 0 & 0 & 1 \\ 0 & 1 & 0 \end{pmatrix}\begin{pmatrix} 1 & 0 & 0 \\ 0 & 0 & 1 \\ 0 & 1 & 0 \end{pmatrix}$, (g) $\begin{pmatrix} 0 & 1 & 0 \\ 1 & 0 & 0 \\ 0 & 0 & 1 \end{pmatrix}\begin{pmatrix} 0 & 1 & 0 \\ 1 & 0 & 0 \\ 0 & 0 & 1 \end{pmatrix}$,

(h) $\begin{pmatrix} 1 & 0 & 2 \\ 2 & -1 & 3 \\ 4 & 1 & 8 \end{pmatrix}\begin{pmatrix} -11 & 2 & 2 \\ -4 & 0 & 1 \\ 6 & -1 & -1 \end{pmatrix}$, (i) $\begin{pmatrix} 4 & 0 & 5 \\ 0 & 1 & -6 \\ 3 & 0 & 4 \end{pmatrix}\begin{pmatrix} 4 & 0 & -5 \\ -18 & 1 & 24 \\ -3 & 0 & 4 \end{pmatrix}$,

(j) $\begin{pmatrix} 2 & 3 & 4 \\ 5 & 6 & 7 \\ 8 & 9 & 0 \end{pmatrix}\begin{pmatrix} -21/10 & 6/5 & -1/10 \\ 28/15 & -16/15 & 1/5 \\ -1/10 & 1/5 & -1/10 \end{pmatrix}$,

(k) $\begin{pmatrix} 0 & 1 & 0 & 0 \\ 0 & 0 & 1 & 0 \\ 0 & 0 & 0 & 1 \\ 1 & 0 & 0 & 0 \end{pmatrix}\begin{pmatrix} 0 & 0 & 0 & 1 \\ 1 & 0 & 0 & 0 \\ 0 & 1 & 0 & 0 \\ 0 & 0 & 1 & 0 \end{pmatrix}$, (l) $\begin{pmatrix} 1 & 2 & -1 \\ 0 & 1 & 3 \\ 0 & 0 & 1 \end{pmatrix}\begin{pmatrix} 1 & -2 & 7 \\ 0 & 1 & -3 \\ 0 & 0 & 1 \end{pmatrix}$.

2. Let

$$A = \begin{pmatrix} a & b \\ c & d \end{pmatrix} \quad \text{and} \quad e = ad - cb.$$

Suppose that $e \neq 0$ and let

$$B = \begin{pmatrix} d/e & -b/e \\ -c/e & a/e \end{pmatrix}.$$

Show that

$$AB = BA = \begin{pmatrix} 1 & 0 \\ 0 & 1 \end{pmatrix}.$$

3. Solve each of the following systems of equations. (See Problem 1.)

(a) $\begin{aligned} 6x_1 &= 30 \\ -10x_2 &= 30 \end{aligned}$ (b) $\begin{aligned} 7x_1 + 4x_2 &= 7 \\ 3x_1 \qquad &= 12 \end{aligned}$

(c) $\begin{aligned} 5x_1 - 2x_2 &= 0 \\ 2x_1 - x_2 &= 0 \end{aligned}$ (d) $\begin{aligned} 5x_1 - 2x_2 &= 6 \\ 2x_1 - x_2 &= 14 \end{aligned}$

(e) $\begin{aligned} -7x_1 + 10x_2 &= 3 \\ 3x_1 - 4x_2 &= 17 \end{aligned}$ (f) $\begin{aligned} x_1 \qquad + 2x_3 &= 4 \\ 2x_1 - x_2 + 3x_3 &= 0 \\ 4x_1 + x_2 + 8x_3 &= 10 \end{aligned}$

(g) $\begin{aligned} -11x_1 + 2x_2 + 2x_3 &= 1 \\ -4x_1 \qquad + x_3 &= 2 \\ 6x_1 - x_2 - x_3 &= 3 \end{aligned}$ (h) $\begin{aligned} 4x_1 \qquad + 5x_3 &= 4 \\ x_2 - 6x_3 &= 5 \\ 3x_1 \qquad + 4x_3 &= 4 \end{aligned}$

(i) $$\begin{aligned} 2x_1 + 3x_2 + 4x_3 &= 10 \\ 5x_1 + 6x_2 + 7x_3 &= 11 \\ 8x_1 + 9x_2 &= 4 \end{aligned}$$

(j) $$\begin{aligned} x_1 + 2x_2 - x_3 &= 143 \\ x_2 + 3x_3 &= 0 \\ x_3 &= 1000 \end{aligned}$$

(k) $$\begin{aligned} x_1 + x_2 + x_3 &= 1 \\ 5x_2 + 7x_3 &= -4 \\ 2x_1 + 2x_2 + 2x_3 &= -3 \end{aligned}$$

(l) $$\begin{aligned} x_1 - 2x_2 + 7x_3 &= 0 \\ x_2 + x_3 &= 0 \\ -2x_1 + 4x_2 + 14x_3 &= 0 \end{aligned}$$

4. A **permutation matrix** is one whose elements are all either 0 or 1, and such that each row and each column contains exactly one 1. List all $n \times n$ permutation matrices for (a) $n = 2$, (b) $n = 3$. By trial and error find the inverse for each of these matrices.

5. If A is an $n \times n$ matrix,

$$\vec{x} = \begin{pmatrix} x_1 \\ x_2 \\ \cdot \\ \cdot \\ \cdot \\ x_n \end{pmatrix} \quad \text{and} \quad \vec{b} = \begin{pmatrix} 0 \\ 0 \\ 0 \\ \cdot \\ \cdot \\ \cdot \\ 0 \end{pmatrix}$$

where $\vec{b}$ is $n \times 1$, then $A\vec{x} = \vec{b}$ is said to be a **homogeneous** system of equations. (a) Show that if $\vec{x}$ and $\vec{y}$ are solutions of such a system, then so are $c\vec{x}$ and $\vec{x} + \vec{y}$. (b) Show that if A has an inverse matrix, then the system has the unique (trivial) solution $\vec{b}$.

6. Suppose A is an $n \times n$ matrix and I is the $n \times n$ identity matrix. If k is a real number such that the matrix $A - kI$ has no inverse, then k is said to be a **characteristic value** or an **eigenvalue** for the matrix A. (Here, kI is the matrix with k's along the main diagonal running from upper left to lower right, and 0's elsewhere.) Show that the given matrix has the indicated characteristic value

(a) $\begin{pmatrix} 10 & 4 \\ 3 & 11 \end{pmatrix}$, $k = 7$, (b) $\begin{pmatrix} 4 & 8 \\ 3 & 2 \end{pmatrix}$, $k = -2$, (c) $\begin{pmatrix} 1 & 1 \\ 0 & 1 \end{pmatrix}$, $k = 1$.

11.9 The complex number system

The system of all one by one matrices with matrix addition and matrix multiplication is isomorphic to the system of real numbers $(R, +, \cdot)$. Hence, the system of all matrices is in some sense an extension of $(R, +, \cdot)$.

There are certain 2 by 2 matrices that *are* called numbers and are extremely useful. Consider the set C of all 2×2 matrices in the form $\begin{pmatrix} a & b \\ -b & a \end{pmatrix}$ where a and b are real numbers. We claim that C is closed under matrix addition and multiplication. For we have

$$\begin{pmatrix} a & b \\ -b & a \end{pmatrix} + \begin{pmatrix} c & d \\ -d & c \end{pmatrix} = \begin{pmatrix} a + c & b + d \\ -(b + d) & a + c \end{pmatrix}$$

and

$$\begin{pmatrix} a & b \\ -b & a \end{pmatrix} \cdot \begin{pmatrix} c & d \\ -d & c \end{pmatrix} = \begin{pmatrix} ac - bd & ad + bc \\ -bc - ad & -bd + ac \end{pmatrix} = \begin{pmatrix} ac - bd & ad + bc \\ -(ad + bc) & ac - bd \end{pmatrix}.$$

And, furthermore, in C, multiplication is commutative—in contrast to the general situation for matrix multiplication. We have

$$\begin{pmatrix} c & d \\ -d & c \end{pmatrix} \cdot \begin{pmatrix} a & b \\ -b & a \end{pmatrix} = \begin{pmatrix} ca - db & cb + da \\ -da - cb & -db + ca \end{pmatrix} = \begin{pmatrix} ac - bd & ad + bc \\ -(ad + bc) & ac - bd \end{pmatrix}.$$

This result with the preceding equations establishes the claimed commutativity.

Let C^* be the subset of C consisting of all matrices in the form $\begin{pmatrix} a & 0 \\ 0 & a \end{pmatrix}$ where a is real. It is easy to see that $(C^*, +, \cdot)$ is isomorphic to $(R, +, \cdot)$.

So, in the system $(C, +, \cdot)$, we may think of $\begin{pmatrix} a & 0 \\ 0 & a \end{pmatrix}$ as behaving exactly like the real number a. In particular, $\begin{pmatrix} 1 & 0 \\ 0 & 1 \end{pmatrix}$ behaves like the real number 1 and is the identity for multiplication in C. $\begin{pmatrix} -1 & 0 \\ 0 & -1 \end{pmatrix}$ behaves like -1 and is the additive inverse for $\begin{pmatrix} 1 & 0 \\ 0 & 1 \end{pmatrix}$.

Consider what happens in C when we multiply $\begin{pmatrix} 0 & 1 \\ -1 & 0 \end{pmatrix}$ by itself. We have

$$\begin{pmatrix} 0 & 1 \\ -1 & 0 \end{pmatrix} \cdot \begin{pmatrix} 0 & 1 \\ -1 & 0 \end{pmatrix} = \begin{pmatrix} -1 & 0 \\ 0 & -1 \end{pmatrix}.$$

We thus have an element in C whose square behaves like -1. In C we *can* solve the equation $x^2 = \begin{pmatrix} -1 & 0 \\ 0 & -1 \end{pmatrix}$ and this last matrix behaves like -1 within the system $(C, +, \cdot)$.

It is standard practice to call the members of C **complex numbers.** And it is a commonly accepted convention to write $\begin{pmatrix} a & b \\ -b & a \end{pmatrix}$ as $a + bi$. That is, we introduce "$a + bi$" as a name for the complex number $\begin{pmatrix} a & b \\ -b & a \end{pmatrix}$. Then we have $0 + 1i = \begin{pmatrix} 0 & 1 \\ -1 & 0 \end{pmatrix}$ and we usually condense "$0 + 1i$" to simply "i." We have $a + 0 \cdot i = \begin{pmatrix} a & 0 \\ 0 & a \end{pmatrix}$. And we usually write "$a + 0 \cdot i$" as simply "$a$." Thus, we have our usual situation of ambiguity when we are extending one number system to another. "3.15" might be a name for the real number 3.15 or for the complex number $3.15 + 0i$. But because of the isomorphism established above this ambiguity causes no difficulty.

From our calculations above, we see that we now have that $i^2 = -1$. Also $(a + bi) + (c + di) = (a + c) + (b + d)i$ and $(a + bi) \cdot (c + di) = (ac - bd) + (ad + bc)i$.

$0 + 0i$ is the additive identity in C. $-a - bi$ is the additive inverse for $a + bi$. $1 + 0i$ is the multiplicative identity. If $a + bi \neq 0$ then the multiplicative inverse of $a + bi$ is $(a + bi)^{-1} = a/(a^2 + b^2) - (b/(a^2 + b^2))i$ as may be checked by a simple calculation. C is closed under subtraction and under division by nonzero numbers. Multiplication distributes across addition in C as it does in any set of matrices which is closed under the two operations.

Any polynomial equation that we can solve in R, we can solve in C. And there are polynomial equations that we can solve in C that cannot be solved in R. For example, in C, the equation $x^2 + 1 = 0$ has the solutions i and $-i$.

In fact, it can be shown that in C any polynomial equation with real coefficients has at least one solution. This result is called the **fundamental theorem of algebra.** One way to realize the importance of this theorem is as follows. We found the system of natural numbers inadequate because, in that system, we could not always solve such a simple equation as $x + n = 0$. So we invented the integers to take care of this shortcoming. But in the system of integers we found we could not always solve such an equation as $px + q = 0$. So we invented the rational numbers. But in the system of rationals we found we could not solve such an equation as $x^2 - 2 = 0$. So we invented the reals. But even with the real number system we could not solve $x^2 + 1 = 0$. So we invented the complex numbers. Now what polynomial equations can't we solve in C so that we need to go on and invent a new, richer number system? The answer is that there aren't any. *Any* polynomial equation has a solution in C. The system $(C, +, \cdot)$ is up to the task of giving us a solution for *any* polynomial equation.

The complex numbers may seem a bit abstract. But it is perfectly possible to represent them geometrically in a concrete fashion. We simply use the point (a,b) in a rectangular coordinate system to represent the number $a + bi$. When we do this we refer to the resulting system as the **complex plane.** Notice then that the addition of two complex numbers is equivalent to vector addition of the corresponding points (vectors).

Despite their apparent abstractness, complex numbers have many uses in very practical affairs. They are especially useful to electrical engineers and physicists.

A complex number bi with $b \neq 0$ is sometimes called an **imaginary number** (or a **pure imaginary**). This terminology seems a bit unfortunate, especially when put alongside the practice of calling the numbers in R **real** numbers. It is well to remember that the words "real" and "imaginary" in these contexts are technical mathematical terms and are not to be taken in their everyday senses. Imaginary numbers are no more unreal than are real numbers.

PROBLEMS

1. Show that the system $(C^*, +, \cdot)$ of this section is isomorphic to $(R, +, \cdot)$.
2. Show that when $a + bi \neq 0$, then $(a + bi)^{-1}$ has the value claimed for it in this section.

3. Evaluate: (a) $(6 + 4i) + (-5 + 2i)$, (b) $(5 - 2i) + (5 + 2i)$, (c) $(-7 + 4i) + (7 + 4i)$, (d) $(6 + 4i)\cdot(-5 + 2i)$, (e) $(5 - 2i)\cdot(5 + 2i)$, (f) $(-7 + 4i)\cdot(7 + 4i)$, (g) $i(3 + 2i)$, (h) $6i\cdot 7i$, (i) $(1 + i)\cdot(1 - i)$, (j) $(1 + i) \div (1 - i)$ [NOTE: In C, as in F, we may divide by multiplying by the reciprocal], (k) $i \div (6 + 2i)$, (l) $(7 + i) \div i$, (m) $1 \div (\pi + ei)$.

4. Illustrate parts a, b, and c of Problem three by using vectors in the plane to represent the complex numbers.

5. Conjecture what happens to the vector for a complex number when that number is multiplied by i. To help in making the conjecture plot each of the following as vectors and then also plot the vector for their product with i. (a) 1, (b) i, (c) -1, (d) $1 + i$, (e) 7, (f) $7i$, (g) $1/\sqrt{2} - (1/\sqrt{2})i$.

6. What is the length of the vector that represents $a + bi$?

7. Show that if L_1 is the length of the vector for $a + bi$ and L_2 is the length of the vector for $c + di$, then L_1L_2 is the length of the vector for $(a + bi)(c + di)$.

8. The branch of mathematics called **complex analysis** considers functions which map subsets of C into C. Such functions are called **functions of a complex variable.** Taking z to be a complex number, diagram as a mapping, by showing a few corresponding points, the function f defined by (a) $f(z) = z$, (b) $f(z) = z + i$, (c) $f(z) = z + 1 - i$, (d) $f(z) = z^2$.

9. Solve the following equations. (a) $x^2 + 4 = 0$, (b) $ix + 6 + 2i = 0$, (c) $(1 + i)x + (1 - i) = 0$, (d) $x^4 - 1 = 0$. (There are four solutions to this.)

10. Show that -1, $1/2 + (\sqrt{3}/2)i$, and $1/2 - (\sqrt{3}/2)i$ are all solutions to $x^3 + 1 = 0$. Plot these three solutions in the complex plane.

11. Show that 1, $1/\sqrt{2} + (1/\sqrt{2})i$, i, $-1/\sqrt{2} + (1/\sqrt{2})i$, -1, $-1/\sqrt{2} - (1/\sqrt{2})i$, $-i$, and $1/\sqrt{2} - (1/\sqrt{2})i$ are all solutions to $x^8 - 1 = 0$. Plot these solutions in the complex plane.

12. The complex number $a - bi$ is called the **conjugate** of the number $a + bi$. (a) Plot several complex numbers and their conjugates in the complex plane. (b) What is the conjugate of the conjugate of a complex number? (c) Show that the product of a complex number and its conjugate is always a real number. (d) What is the relationship between the length of the vector for $a + bi$ and the product of $a + bi$ with its conjugate?

13. Show that $(C, +, \cdot)$ is a field.

14. Show that $(C, +, \cdot)$ is not an ordered field.

*12

Transfinite arithmetic

12.1 Denumerable sets

When we first dealt with the natural numbers in this book many pages ago, we started with some very basic ideas about sets and counting. We then gradually built up more and more useful systems of numbers. We looked carefully at several kinds of numbers which are useful to people everywhere on the earth—and on the moon and other extraterrestrial bodies.

We shall now turn back to our point of beginning and look again at basic notions about sets and counting. And in doing so, we shall discover another surprising and fascinating world of numbers—the realm of the transfinite.

We have said that a set A has exactly as many elements as a set B if it is possible to set up a one-to-one correspondence between A and B. And we defined the cardinal number of a set A to be the collection of all sets which have exactly as many elements as A. We shall now recognize the fact that these definitions make perfectly good sense when the sets involved are **infinite.**

In our work with the usual number systems, we have discovered (invented?) many infinite sets. N, I, F, and R are all infinite sets. Let us recall that we have said that a set is finite if its cardinal number is a natural number. Otherwise, the set is infinite.

Now, just as we gave a name to the cardinal number of the set $\{a,b\}$ by stating the definition $2 = \#(\{a,b\})$, let us give a name to the cardinal number of the infinite set N by stating the definition

$$\aleph_0 = \#(N).$$

The symbol $\aleph_0$ is read "aleph null." "$\aleph$" is the first letter of the Hebrew alphabet. And "$\aleph_0$" is recognized by mathematically sophisticated people all over the globe as a name for the cardinal number of N, just as "2" is recognized (by many more people) as a name for the cardinal number of $\{a,b\}$. We shall refer to $\aleph_0$ as a **transfinite cardinal (number).**

Let $E = \{0,2,4,6,8, \ldots\}$, the set of all even natural numbers. We claim that $\#(E) = \aleph_0$. To establish this, all we need to show is that N has exactly as many elements as E. We do this by displaying a one-to-one correspondence between N and E. And this is easy to do. The function f defined by $f(n) = 2n$ is such a correspondence. F is diagrammed as a mapping in Figure 12.1. There are exactly as many even natural numbers as there are natural numbers.

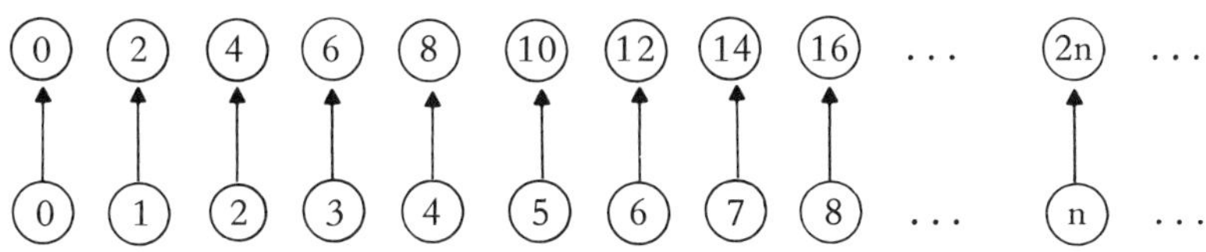

FIGURE 12.1

Since $\#(E) = \#(N) = \aleph_0$ and E is a proper subset of N we see that transfinite cardinals behave quite differently from finite cardinals (i.e. natural numbers). When A and B are finite sets and A is a proper subset of B we cannot have $\#(A) = \#(B)$. But when a set is infinite, it is evidently possible to remove many elements from it and yet make no change in its cardinal number.

Let $A = \{3,4,5,6, \ldots\}$. We claim that $\#(A) = \aleph_0$. This is proved by noting that if we define g by $g(n) = n + 3$, then g is a one-to-one correspondence between N and A. g is diagrammed as a mapping in Figure 12.2.

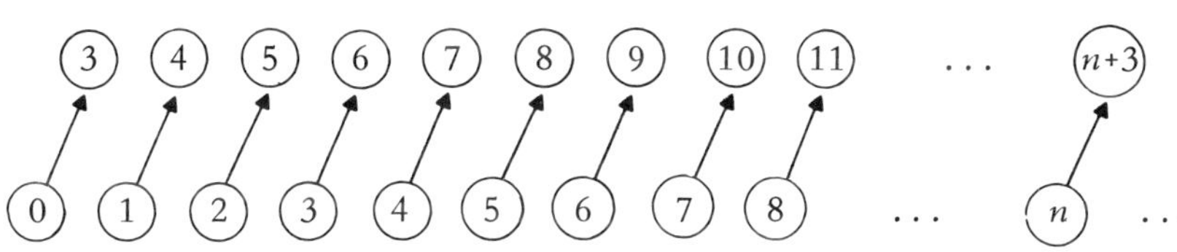

FIGURE 12.2

Let $B = \{1,3,5,7, \ldots\}$, the set of odd natural numbers. Then $\#(B) = \aleph_0$ since the function h defined by $h(n) = 2n + 1$ is a one-to-one correspondence between N and B. h is diagrammed as a mapping in Figure 12.3.

A set that has cardinal number $\aleph_0$ is said to be **denumerable** (or to be **denumerably infinite**). It is also common practice to say that a set is **countable** when the set is either finite or denumerable. These two technical terms and, indeed, $\aleph_0$ itself would be somewhat useless if it were not the case that there exist sets which are *not* countable. There do, as we shall see in the next section.

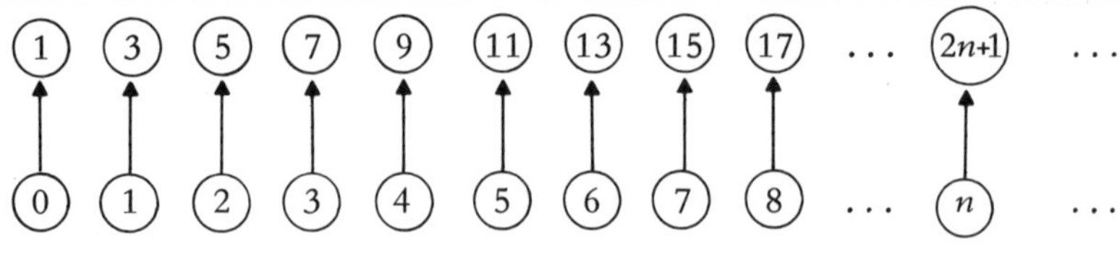

FIGURE 12.3

PROBLEMS

1. Find the cardinal number of each of the following sets. Justify your conclusion in each case. (a) $\{x \mid x \in N \text{ and } x + 1 > x\}$, (b) $\{x \mid x^2 = 4 \text{ and } x \in I\}$, (c) $\{x \mid x \in N \text{ and } x + 1 = x\}$, (d) $\{x \mid x \in N \text{ and } x \geq 6\}$, (e) $\{x \mid x = y^2 \text{ for some } y \text{ in } N\}$, (f) I (note that $I = \{0, 1, -1, 2, -2, 3, -3, \ldots\}$), (g) $\{1, 1/2, 1/3, 1/4, 1/5, \ldots\}$, (h) $\{(-1)^1, (-1)^2, (-1)^3, (-1)^4, (-1)^5, \ldots\}$, (i) $\{2^0, 2^1, 2^2, 2^3, \ldots\}$, (j) I^-, (k) I^+, (l) $\{0,1,10,11,100,101,110,111,1000,1001, \ldots\}$, (m) $\{.1,.01, .001,.0001, .00001, \ldots\}$, (n) $\{x \mid x = [(y^2 - y) \div y] + 1 - y \text{ for some } y \text{ in } N\}$.

2. Show that $N \times N$ is denumerable. [HINT: $N \times N = \{(0,0), (0,1), (1,0), (0,2), (1,1), (2,0), (0,3), (1,2), (2,1), (3,0), (0,4), \ldots\}$.]

12.2 A nondenumerable set

When we consider the number line, and think of the rational points on it and the points for natural numbers on it, it would seem that there are many more of the former than there are of the latter. But this is not so. There are exactly as many rational numbers as there are natural numbers.

To prove this we need only display a one-to-one correspondence between N and F. This may be done as follows. Consider the array

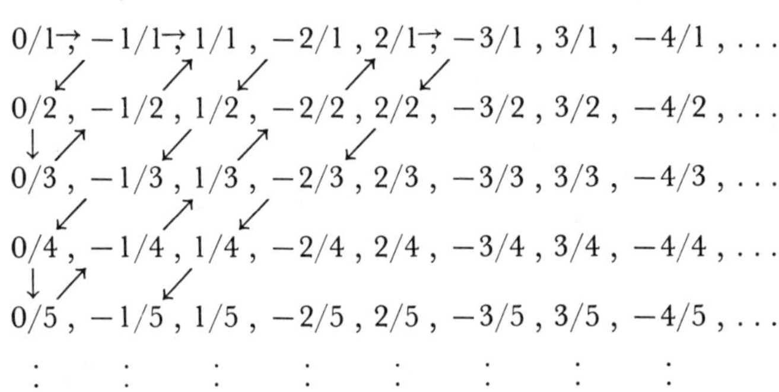

Every rational number appears at least once in this array. In fact, when q is positive p/q will be found in the qth row, and in the $2|p|$th column if p is negative, and in the $(2p + 1)$th column if p is nonnegative. But then we may set up a one-to-one correspondence between N and F by marching through this array as indicated by the arrows and assigning successive members of F to successive natural numbers. Each time we come to a rational number already assigned to a natural number we should omit it and move on to the next rational. We see then that $\#(F) = \#(N) = \aleph_0$ and the collection of all rational numbers is denumerable. If we had hoped to get an uncountable set by moving up to F, we must be disappointed. In fact, it almost begins to appear that any infinite set is denumerable and that this entire business of transfinite numbers is a tempest in a teapot. But when we take the step to the set of all real numbers, we find that we were not barking up an empty tree. The set of all reals *is* uncountable. There *does not* (*and cannot*) *exist* a function which is a one-to-one correspondence between N and R. In some basic sense there are *more* real numbers than there are natural numbers. There are different orders of infinity! Some infinite sets have more elements than other infinite sets!

To show that R is not denumerable, let us suppose that f is a function that maps N into R. We shall show that f *cannot* map N *onto* R and thus cannot be a one-to-one correspondence between N and R. Since f maps N into R, for any n in N, $f(n)$ is a real number. But then $f(n)$ can be written as an infinite decimal. Let $f(n) = b_n.a_{n0}a_{n1}a_{n2}\ldots$ where "b_n" names an integer and the "a's" are decimal digits. We shall now construct a real number y that is not the value of f at any n. This will show that f does not map N *onto* R and will complete the proof. Let $y = 0.d_0d_1d_2\ldots$ where $d_k = 1$ if $a_{k,k+1} = 2$ and $d_k = 2$ if $a_{k,k+1} \neq 2$. Then y cannot be $f(0)$ since it differs from $f(0)$ in the first decimal place. y cannot be $f(1)$ since it differs from $f(1)$ in the second decimal place. In fact, for any n in N, y differs from $f(n)$ in the $n + 1$st decimal place. Hence, for every n, $f(n) \neq y$ and the proof is finished.

We have shown that $\#(R) \neq \aleph_0$. The standard name for $\#(R)$ is "c" which is the initial letter of "continuum." c is a transfinite number and we have $c \neq \aleph_0$.

We have said that there are more real numbers than there are natural numbers. We shall now make this idea more precise.

If a and b are cardinal numbers we shall say that $a \leq b$ if for $\#(A) = a$ and $\#(B) = b$ we have that there exists a one-to-one correspondence between A and some subset of B. We say that $a < b$ if $a \leq b$ and $a \neq b$.

This definition agrees with our former one for $\leq$ when a and b are finite.

And if n is any finite cardinal we have $n < \aleph_0$.

In addition, we have $\aleph_0 < c$. This follows from the fact that N itself is a subset of R so that $\#(N) \leq \#(R)$ and we showed above that $\#(N) \neq \#(R)$.

This relation of $\leq$ for cardinal numbers has properties we might expect. It is transitive, and when $a \leq b$ and $b \leq a$ we may conclude that $a = b$. The former property is quite easy to prove, but the proof of the latter is quite deep and we shall omit it.

There are some very natural questions to ask in regard to this relation of $<$ in cardinal numbers.

Is there a cardinal number b such that $\aleph_0 < b < c$? This turns out to be an extremely subtle question. We shall return to it in Section 12.7.

Are there cardinal numbers which are larger than c? There are. We shall see how to construct them in Section 12.4.

All of the work that we have done so far with transfinite numbers was first done by George Cantor in about 1870. It is hardly *modern* mathematics. And yet it is of quite recent development when compared with the long history of the use of natural, integral, rational, and real numbers.

PROBLEMS

1. Show that if A and B are sets then (a) $\#(A) \leq \#(A \cup B)$, and (b) $\#(A \cap B) \leq \#(A)$.
2. Show that if x, y, and z are cardinal numbers and $x \leq y$ and $y \leq z$, then $x \leq z$.
3. Prove that if n is a finite cardinal then $n < \aleph_0$.
4. Show that if a and b are cardinal numbers and $a = b$, then $a \leq b$.
5. Suppose that in the proof of the nondenumerability of R we have

$$f(0) = .11567493\ldots$$

$$f(1) = .33333333\ldots$$

$$f(2) = .451451451\ldots$$

$$f(3) = .764295001\ldots$$

$$f(4) = .545445444\ldots$$

$$f(5) = .799799799\ldots$$

$$\vdots$$

What are the first few digits of the y constructed in the proof?

6. Which of the following sets are denumerable? (a) R^+, (b) R^-, (c) $\{x \mid x \in R$ and $0 \leq x \leq 1\}$. Explain. [HINT: Consider the proof given above that R itself is not denumerable.]

12.3 Operations on cardinal numbers

In Chapter 6 we defined the operations of addition and multiplication for cardinal numbers. In that chapter, we were not considering transfinite cardinals, but the definitions given apply to that case, nonetheless.

Suppose, for example, we wish to find $3 + \aleph_0$. To do so, we should find a set A such that $\#(A) = 3$, find a set B such that $\#(B) = \aleph_0$ and $A \cap B = \phi$, and then find $\#(A \cup B)$. Suppose we let $A = \{0,1,2\}$ and $B = \{3,4,5,6\ldots\}$. Then $A \cap B = \phi$ and, from our earlier work, $\#(A) = 3$ and $\#(B) = \aleph_0$. Hence, $3 + \aleph_0 = \#(A \cup B) = \#(N) = \aleph_0$.

More generally, we shall have for any finite cardinal n that

$$n + \aleph_0 = \aleph_0.$$

What about $\aleph_0 + \aleph_0$? We found earlier that $\#(\{0,2,4,6,8,\ldots\}) = \aleph_0$ and $\#(\{1,3,5,7,9,\ldots\}) = \aleph_0$. But $\{0,2,4,6,8,\ldots\} \cap \{1,3,5,7,9,\ldots\} = \phi$. So we have $\aleph_0 + \aleph_0 = \#(E \cup F) = \#(N) = \aleph_0$.

In other words, the union of two disjoint denumerable sets is denumerable.

These results of transfinite arithmetic again point up the striking difference between the situation with finite cardinals and that where infinite cardinals may be involved. We know that when x and y are finite cardinals (natural numbers) $y + x = x$ implies $y = 0$. We see that in the realm of transfinite numbers, $y + x = x$ does not imply $y = 0$.

We found that for all x in N we had $x + 1 > x$. But we have $1 + \aleph_0 = \aleph_0 + 1 = \aleph_0$ so a similar result does not hold for transfinite arithmetic.

A rereading of our work in Sections 6.1, 6.2, and 6.3 will show that our early arguments for the commutativity and associativity of addition and multiplication apply equally well to finite or infinite cardinals. Also, our earlier proofs that 0 and 1 are the identity elements for addition and multiplication, respectively, apply to the more general case. Our proof in Section 6.5 that multiplication distributes across addition applies equally well when infinite sets are involved.

Let us consider some cases of multiplication with transfinite numbers. We claim that $2 \cdot \aleph_0 = \aleph_0$. To prove this let $A = \{a,b\}$. Then $2 \cdot \aleph_0 = \#(A \times N) = \#(\{(a,0), (b,0), (a,1), (b,1), (a,2), (b,2), (a,3), (b,3), \ldots\})$. But the function k defined by $k(n) = (a,n/2)$ when n is even and $k(n) = (b,(n+1)/2)$ when n is odd is a one-to-one correspondence between N and $A \times N$. Hence, $\#(A \times N) = \aleph_0$. The function k is diagrammed as a mapping in Figure 12.4.

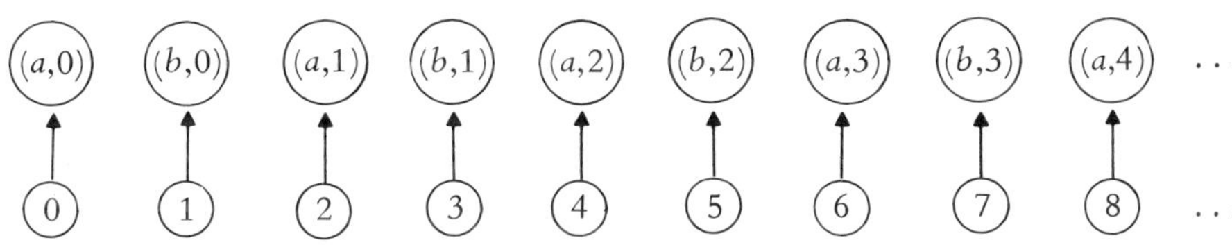

FIGURE 12.4

More generally, when n is any nonzero finite cardinal we have $n \cdot \aleph_0 = \aleph_0$. Also, from the result in Problem 2, Section 12.1, we may conclude that $\aleph_0 \cdot \aleph_0 = \aleph_0$.

When x, y, and z are natural numbers and $x \neq 0$, we know that $y \cdot x = z \cdot x$ implies that $y = z$. We see that this is not, in general, true for cardinal numbers since $2 \cdot \aleph_0 = 3 \cdot \aleph_0$ and, indeed, $2 \cdot \aleph_0 = \aleph_0 \cdot \aleph_0$. So we see again that transfinite arithmetic is quite different from ordinary arithmetic. And such results suggest, also, why it is proper to be very careful in giving proofs for the basic facts in any arithmetic. Results that we might claim to be obvious (in ordinary arithmetic) sometimes turn out to be false (in transfinite arithmetic).

Some further results of transfinite arithmetic, whose proofs we shall not give, are the following:

$\#(\{x \mid 0 \leq x \leq 1, x \in R\}) = c$. In other words, there are exactly as many points on the segment from 0 to 1 as there are on the entire real line.

If n is finite, $n + c = c$.

If n is finite and different from 0, $n \cdot c = c$.

$\aleph_0 + c = c$.

$\aleph_0 \cdot c = c$.

$c + c = c$.

$c \cdot c = c$.

Since $c \cdot c$ is the cardinal number of $R \times R$, this last result asserts that there are exactly as many points in a plane as there are on a line. Evidently, it is not possible to find the number of dimensions in a space by considering the cardinality of its collection of points.

Problems

1. Prove by direct application of the definitions of the operations that (a) $6 + \aleph_0 = \aleph_0$, (b) $1 \cdot \aleph_0 = \aleph_0$, (c) $3 \cdot \aleph_0 = \aleph_0$.

2. Accepting the fact that $1 + \aleph_0 = \aleph_0$, prove by induction that if n is a finite cardinal, then $n + \aleph_0 = \aleph_0$.

3. Accepting the facts that $1 \cdot \aleph_0 = \aleph_0$ and $\aleph_0 + \aleph_0 = \aleph_0$, use the distributive law and induction to prove that when n is any nonzero natural number, then $n \cdot \aleph_0 = \aleph_0$.

4. In Section 6.4, we proved that when x is a finite cardinal, we have $x + 1 \neq x$. Explain why this proof does not apply to cardinal numbers in general.

5. If x, y, z, and w are cardinal numbers prove that

   (a) $x \leq y$ implies $x + z \leq y + z$.
   (b) $x \leq y$ implies $xz \leq yz$.
   (c) $x \leq y$ and $z \leq w$ implies $xz \leq yw$.
   (d) $x \leq y$ and $z \leq w$ implies $x + z \leq y + w$.

6. Show that each of the following is false when x, y, and z are cardinal numbers.

   (a) $x < y$ implies $x + z < y + z$.
   (b) $x < y$ and $z \neq 0$ implies $xz < yz$.
   (c) $x \cdot z = x$ and $x \neq 0$ implies $z = 1$.
   (d) $x \neq 0$ implies there is a natural number y such that $y \neq x$ and $y + 1 = x$.

7. Suppose you are the manager of Hotel Aleph-Null which has rooms numbered 0, 1, 2, 3, ... Suppose your hotel is filled, with a guest in each room, and a new person arrives and wants a room. Tell how to arrange it so that he can get a room and yet each of your old guests also has a room. What fact from transfinite arithmetic does this illustrate?

8. Suppose you are the manager of the hotel in Problem 7 and there is a guest in each room. The Grand Order of the Negative Integers comes to town and each member wants a room. Show how you can put them all up and still house all your current guests. What fact from transfinite arithmetic does this illustrate?

9. In Figure 12.5 we have shown how to set up a geometric one-to-one correspondence between the set of all points on the interval from 0 to 1 and the set of points on the interval from 0 to 2. Find a geometric way to set up a one-to-one correspondence between the points x such that $0 \leq x < 1$ and the points y such that $0 \leq y$.

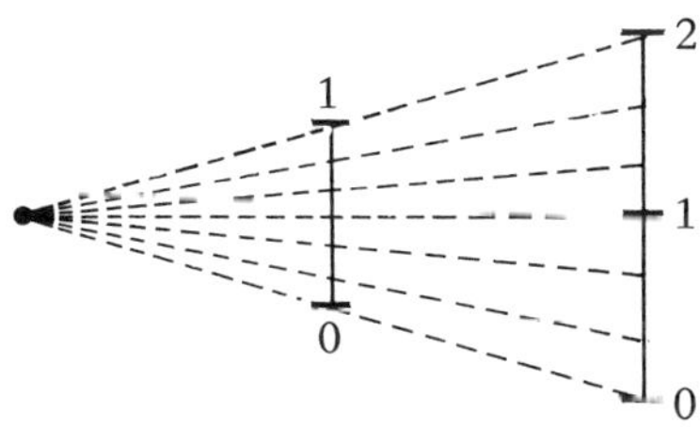

FIGURE 12.5

12.4 Exponents

In Section 6.22 we defined B^A to be the collection of all mappings of the set A into the set B. We discovered that when A and B are finite sets we have $\#(B^A) = [\#(B)]^{\#(A)}$. Therefore, when a and b are any cardinal numbers it is natural to define b^a to be the cardinal number of B^A where $\#(A) = a$ and $\#(B) = b$.

Hence, $\aleph_0{}^2$ is the cardinal number of the collection of all mappings from $\{0,1\}$ into N. The mapping K defined by $K(f) = (f(0), f(1))$ for each f in $N^{\{0,1\}}$ is clearly a one-to-one correspondence between $N^{\{0,1\}}$ and $N \times N$. Hence, $\aleph_0{}^2 = \aleph_0 \cdot \aleph_0$ as we would expect.

What is the value of $2^{\aleph_0}$? Consider the collection of mappings $\{0,1\}^N$. If f is in this collection then f assigns either 0 or 1 to each natural number N. Each real number in the set $B = \{x \mid x \in R, 0 \leq x \leq 1\}$ has an infinite binary decimal fraction that names it. And if we agree never to use a **terminating** binary fraction as a name,† we actually have it that each element of B has a unique binary

† So that we would write 1/2 as "$.01111\ldots_2$" instead of as "$.1_2$," for example.

fraction name. Suppose $x \in B$ and using the base 2 numeration system we have $x = .b_1b_2b_3 \ldots$ where the "b's" are binary digits. Let us pair up with x the function f defined by $f(0) = b_1$, $f(1) = b_2, \ldots, f(n) = b_{n+1}, \ldots$. This pairing sets up a one-to-one correspondence between B and a subset of $\{0,1\}^N$. Since $\#(B) = c$, this implies $c \leq 2^{\aleph_0}$. But it is very easy to show that $2^{\aleph_0} \leq c$. We may, therefore, conclude that $2^{\aleph_0} = c$.

This result is important in many ways. One thing we note from it is that we have performed an operation with the cardinal numbers 2 and $\aleph_0$ and succeeded in getting a larger transfinite cardinal number, c. Perhaps we have discovered a means for generating larger and larger cardinal numbers! We have:

> If a is any cardinal number then $2^a > a$.

This proof of this fact is not especially difficult but it will be omitted here and discussed in a problem.

As a consequence of the above theorem, we see that 2^c, the cardinal number of the collection of all mappings of R into $\{0,1\}$, is larger then c. If we let $f = 2^c$, then we have $f > c$. If we let $g = 2^f$, then $g > f$. If we let $h = 2^g$, then $h > g$. Et cetera. Thus, we may generate an infinite array of successively larger transfinite cardinal numbers.

Our arithmetic makes sense with these new numbers. As we might guess we get such results as $\aleph_0 \cdot f = f$, $c \cdot f = f$, $f \cdot f = f^2 = f$, $f + f = f$, etc.

It is perfectly legitimate to ask the question, "Of what earthly good are these wild things called transfinite numbers?" An answer could be put forth that they are simply fun and fascinating. But much more can be said. In doing such a down to earth (?) thing as planning a rocket trip to the moon the engineers and scientists get involved with such "practical" mathematics as, so-called, differential equations. But in dealing with differential equations and asking how their solutions behave, it is very helpful to make distinctions between denumerable and nondenumerable sets—that is, to use $\aleph_0$ and cardinal numbers larger than $\aleph_0$. In many of the most fruitful modern uses of the branch of mathematics called probability theory it turns out to be very helpful to distinguish between various transfinite cardinal numbers. There are many other "practical" applications of transfinite arithmetic.

Problems

1. Show that $2^{\aleph_0} \leq c$. [HINT: To each f in $\{0,1\}^N$ assign the real number whose *decimal* fraction name is "$.f(0)f(1)f(2) \ldots$."]

2. Let $F = \{0,1\}^R$. Give several examples of members of F. What is the nature of a member of $\{0,1\}^F$?

3. The **power set** of a set A is defined to be the collection of all possible subsets of A. Show that the power set of A has the same cardinal number as $\{0,1\}^A$.

4. Prove that if x, y, and z are cardinal numbers then (a) $x^y \cdot x^z = x^{y+z}$ and (b) $(x^y)^z = x^{yz}$.

5. Although in the system N we left 0^0 undefined, it has a natural value in the setting of cardinal numbers. What is it?

6. Fill in the details in the following proof that for any cardinal number a we have $2^a > a$. Proof: Let $\#(A) = a$. For x in A define the function f_x on A by taking $f_x(y) = 0$ if $y \neq x$ and $f_x(x) = 1$. The mapping which pairs f_x with x is a one-to-one correspondence between A and a subset of $\{0,1\}^A$. Hence, $a \leq 2^a$. Suppose K maps A into $\{0,1\}^A$. Let $K(x) = g_x$ for each x in A. Define the function h by letting $h(x) = 0$ if $g_x(x) = 1$, and $h(x) = 1$ if $g_x(x) = 0$. Then for every x in A, $h \neq g_x$. Therefore, K does not map A *onto* $\{0,1\}^A$. This completes the proof.

7. It can be shown that the set of all algebraic real numbers is denumerable. Explain why it follows from this that such things as transcendental numbers actually exist. What must be the cardinal number of the set of all transcendental real numbers?

12.5 Well-ordered sets

If we pursue in depth some of the ideas presented in Chapter 3 we are led to the arithmetic of so-called transfinite **ordinal** numbers.

Let us say that a set A is **simply ordered** by a relation R if (i) aRa for all a in A, (ii) aRb or bRa for all a and b in A, and (iii) aRb and bRc implies aRc for all a, b, and c in A. That is, A is simply ordered by R if R is reflexive in A, any two elements in A are comparable by R, and R is transitive in A. If A is simply ordered by R, then we say that R is a **simple order** relation on A.

If A is any collection of real numbers then A is simply ordered by $\leq$. Such an A is also simply ordered by $\geq$.

The collection K of all kings of England who were named "Henry" can be simply ordered by taking xRy to mean that x reigned no later than y.

If we assume that no two rivers in the world have exactly the same length, then we may simply order the set of rivers by saying that xRy means that x is at least as long as y.

Since no two states in the United States have exactly the same population, we may simply order the collection of all such states by saying that xRy means that x has at least as many residents as y.

On the other hand, if we let B be the collection of all items on sale in a supermarket and for x and y in B we take xRy to mean that x costs no more than y, then R is almost surely not a simple order relation. This is due to the fact that there are probably two different items that sell for the same price and it can be shown that in a simply ordered set aRb and bRa implies $a = b$.

If A is a finite set which is simply ordered by a relation R, then A is very much like a set of numbers with the relation $\leq$. More precisely, we have:

> If $\#(A) = n$ where n is a natural number and A is simply ordered by R, then R in A is isomorphic to the relation $\leq$ in $\{1,2, \ldots ,n\}$.

For example, the set of all states in the United States ordered by population, as described above, is isomorphic to the set $\{1,2, \ldots, 50\}$ ordered by $\leq$. The isomorphism between these two sets pairs California with the number 1 since California is the most populous state. It pairs New York with 2 since New York is the next most populous state. This is what is meant by saying that California is the *first* state in population rank and that New York is the *second* state in population rank.†

In general, when we say that a member of a collection is the Kth member, it is understood that the collection is simply ordered somehow and we are indicating that the member is paired with K by the isomorphism between the given collection and a set $\{1,2, \ldots, n\}$ ordered by $\leq$.

In the case of the kings of England named "Henry" ordered by the time of their reign, it is even customary to include the number assigned to a king as a part of his name. We have Henry the first, Henry the second, and so on.

In a simply ordered finite set, the kth member is sometimes said to have **ordinal number** k. In any sufficiently large simply ordered finite set we have a *first* member, a *second* member, a *third* member, and so on. On the other hand, if a finite set is not simply ordered, it may not make sense to ask for the first member of the set. If, for example, we ask someone to tell us who his best friend is, there may be no answer. The person may find it impossible to simply order his friends according to how much he likes them. Small children find such a fact hard to believe and will insist on your telling them, for example, what your *favorite* color is.

In any set A that *is* simply ordered by a relation R we may easily explain what it means to say that a certain element is the *first* element. We call x the *first element* of A if x is in A and xRy for all y in A. Similarly, we say that z is the *last element* in A if z is in A and yRz for all y in A.

Any nonempty finite simply ordered set must have both a first and a last element. However, the situation is different for *infinite* simply ordered sets. The collection of all integers ordered by $\leq$ is simply ordered, but has neither a first nor a last element.

Notice, nevertheless, that some infinite simply ordered sets *do* have first or last elements. For example, $\{x \mid x \in R \text{ and } 0 \leq x \leq 1\}$ ordered by $\leq$ has 0 as a first element and 1 as a last element. This points up the error in defining an infinite set to be one that does not have a last element as is sometimes done.

In Section 6.12 we observed that every nonempty set of natural numbers has a first element when we take the natural numbers to be ordered by $\leq$. More generally, if A is any set which is simply ordered by a relation R, we say that A is *well ordered* by R if every nonempty subset of A has a first element.

It is quite easy to show that any finite simply ordered set must be well

† These statements are based on 1969 population estimates. The 1960 census gave New York as the most populous state.

ordered. On the other hand, the set of integers is not well-ordered by $\leq$. The subset consisting of all the negative integers, for example, does not have a first element.

PROBLEMS

1. Which of the following sets is simply ordered by R if the meaning of xRy is given by the accompanying phrase?

(a) $\{a,b,c,d\}$, x comes no later in the alphabet than y does.
(b) $\{a,b,m,p,d\}$, x comes no later in the alphabet than y does.
(c) {Plato, Aristotle, Zeno}, x was born no later than y.
(d) $\{x \mid x$ is a student at your school$\}$, x is a sibling of y.
(e) $\{x \mid x$ is a student at your school$\}$, x earns no more than y does.
(f) $\{x \mid x$ is a state in the USA$\}$, x is no larger (in area) than y is.
(g) $\{x \mid x$ is or was a president of the USA$\}$, x served as president no later than y. (Be careful here.)

2. Diagram the relation given in Problem 1(a).
3. Which of the following sets is well ordered by $\leq$?

(a) $\{5,6,7,8,9,10, \ldots\}$
(b) $\{-4,-3,-2,-1,0,1,2,3, \ldots\}$
(c) $\{1,0,-1,-2,-3,-4,-5, \ldots\}$
(d) $\{1, 1/2, 1/3, 1/4, 1/5, \ldots\}$
(e) $\{1/2, 2/3, 3/4, 4/5, 5/6, \ldots\}$
(f) $\{x \mid x$ is rational and $0 \leq x \leq 1\}$
(g) $\{x \mid x$ is a rational number$\}$
(h) $\{1,2,4,8,16,32,64, \ldots\}$
(i) $\{1, 1/2, 1/4, 1/8, 1/16, 1/32, 1/64, \ldots\}$
(j) $\{\sqrt{1}, \sqrt{2}, \sqrt{3}, \sqrt{4}, \sqrt{5}, \ldots\}$.

4. Prove that any nonempty finite simply ordered set has a first member. [HINT: Use induction on the size of the set.]

5. Use induction and the result in Problem 4 to prove that if A is simply ordered by R and $\#(A) = n$ where $n \in N$, then R in A is isomorphic to $\leq$ in $\{1,2,3, \ldots, n\}$.

6. Prove that any finite simply ordered set is well ordered.

7. Diagram each possible relation that simply orders (i) $\{a,b\}$, (ii) $\{a,b,c\}$.

8. Which of the relations diagrammed in Problem 7 are isomorphic?

9. On the set $N \cup \{a\}$ define a relation R by saying that (i) when x and y are in N, then xRy if and only if $x \leq y$, (ii) when x is in N, then xRa, and (iii) aRa. Is $N \cup \{a\}$ simply ordered by R? Well ordered by R? Prove your conclusions are correct.

10. Define a relation R in I by saying that (i) when x and y are nonnegative integers, then xRy if and only if $x \leq y$, (ii) when x and y are negative integers,

then xRy if and only if $x \geq y$, and (iii) when x is negative and y is nonnegative, then xRy. Is I simply ordered by R? Well ordered by R? Justify your conclusions.

12.6 Ordinal numbers

When n is a natural number we have it that any n element simply ordered set is isomorphic to $\{1,2, \ldots ,n\}$ ordered by $\leq$. It follows that *any* two n element simply ordered sets are isomorphic and, thus, have similar structures as far as order goes. The standard terminology that is used in this and more general situations is as follows:

> When A and B are simply ordered sets we shall say that they have the same **order type** when their order relations are isomorphic.

For example, {Mount Everest, Mount McKinley, Mount Whittier} ordered by height has the same order type as $\{1,2,3\}$ ordered by $\leq$. N ordered by $\leq$ has the same order type as $\{1, 1/2, 1/3, 1/4, \ldots\}$ ordered by $\geq$.

On the other hand, N and I both ordered by $\leq$ are not of the same order type. We may prove this as follows. Suppose f is a one-to-one correspondence between N and I. We shall show that f cannot be an isomorphism with respect to the order relations given. Let $f(0) = k$. Clearly $k - 1 \in I$ and $k - 1 < k$. Let $n = f^{-1}(k - 1)$. We have $0 < n$ in N and yet in I we have $f(n) = k - 1 < k = f(0)$. Hence, f cannot be an isomorphism and N and I do not have the same order type.

The order types of well-ordered sets are especially interesting. When we consider them we are led to another kind of number. The standard definition in this regard is:

> If A is a well-ordered set, the *ordinal number* of A is the collection of all sets that have the same order type as A.

When n is a natural number, let us use "n" as a name for the collection of all sets that have the same order type as the set $\{1,2, \ldots ,n\}$ ordered by $\leq$. Then, for example, "3" will serve as a name for an ordinal number as well as for a cardinal number. And this dual meaning for "3" will cause no confusion because we may think of 3 as either the cardinal number of the (unordered) set $\{a,b,c\}$ or as the *ordinal number* of this same set when it is simply ordered (and, hence, well ordered).

Since any two finite simply ordered sets that have the same cardinal number are order isomorphic, they must have the same order type and, therefore, have the same ordinal number. The cardinal number of a finite simply ordered set completely determines the ordinal number of that set. It is, therefore, permissible to use a name of the cardinal number to identify the ordinal number.

To investigate the situation when we move up to infinite sets, let us agree, as is common practice, to use "ω" (pronounced "omega") as a name for the ordinal number of N ordered by $\leq$.

ω is a *transfinite* ordinal number. And, as we might expect from our work with cardinal numbers, there are many other transfinite ordinal numbers. In fact, there is an entire arithmetic of ordinal numbers in which the arithmetic of finite ordinal numbers (i.e. natural numbers) is simply a special case.

We shall not pursue this general arithmetic of ordinal numbers to any depth, but content ourselves with a single example. Consider the set $N \cup \{a\}$ ordered by R which is defined as follows. If x and y are in N then xRy if and only if $x \leq y$. And xRa for all x in $N \cup \{a\}$. Thus R orders the natural numbers in the usual way, but the element a comes after every natural number. a is the *last* element in this well ordered set. Since N does not have a last member it is not difficult to see that the ordinal number of $N \cup \{a\}$ ordered by R must be different from ω. The ordinal number of $N \cup \{a\}$ ordered by R is said to be $\omega + 1$.

Note that $\#(N \cup \{a\}) = \aleph_0$. Also $\omega + 1 \neq \omega$. Hence, when we consider infinite sets, we find that the cardinal number of a set does *not* determine its *ordinal number*.

PROBLEMS

1. Prove carefully that $N \cup \{a\}$ ordered by R as defined above is not of the same order type as N ordered by $\leq$.
2. Let us define the relation Q in $N \cup \{a\}$ by taking xQy whenever $x \leq y$ and x and y are in N, and taking aQx for all x in $N \cup \{a\}$. Show that $N \cup \{a\}$ ordered by Q has ordinal number ω.
3. The ordinal number of $N \cup \{a\}$ ordered by Q as defined in Problem 2 is said to be $1 + \omega$. Is addition of ordinal numbers commutative? Justify your conclusion.
4. Suggest a definition for the ordinal number (a) $\omega + 2$, (b) $2 + \omega$, (c) $\omega + \omega$, (d) $(\omega + \omega) + \omega$. (For part (c), see Problem 10, Section 12.5.)
5. If A is well ordered by R and $y \in A$, the *section* of A determined by y is the set $\{x \mid x \in A \text{ and } xRy\}$. If B has ordinal number b and C has ordinal number c, let us say that $b \leq c$ if B is order isomorphic to some section of C. We say $b < c$ if $b \leq c$ and $b \neq c$. Show then that (a) $n < \omega$ for every natural number n, (b) $0 < a$ for every nonzero ordinal number a.
6. Prove that the relation $\leq$ defined in Problem 5 is transitive.
7. If B ordered by Q has ordinal number b and C ordered by R has ordinal number c, let us define $b \cdot c$ to be the ordinal number of $B \times C$ ordered by S where $(x,y)S(z,w)$ if and only if xQz and $x \neq z$, or $x = z$ and yRw. Find (a) $2 \cdot \omega$ and (b) $\omega \cdot 2$.

12.7 Abstract set theory

Suppose that in a certain town there is a single barber. Suppose further that this barber shaves all (and only) those men in town who do not shave themselves. Who shaves the barber? If he shaves himself then it seems that he should not. And if he does not, then he should.

This strange situation is very closely related to a well-known difficulty in set theory that is known as Russell's paradox. Given any set it seems to make sense to ask whether or not that set is a member of itself. In most familiar cases, the sets we deal with do not happen to be members of themselves. So let us consider the collection S which contains *all* (and exactly) those sets which are not members of themselves. That is, let $S = \{A \mid A \text{ is a set and } A \notin A\}$. Let us ask if $S \in S$ or not. If $S \notin S$, then S is a set and S is not a member of itself; so it follows from the definition of S that $S \in S$. But this is a contradiction. That is, we see that $S \notin S$ implies $S \in S$.

If we suppose that $S \in S$, then by the definition of S we see that $S \notin S$. That is, $S \in S$ implies $S \notin S$. Another contradiction.

But surely we should have $S \in S$ or $S \notin S$. And, therein lies the paradox.

Mathematics, which is reputed to be the epitome of consistency, seems to harbor some contradictions at its very core—in the theory of sets on which we have built all these useful number systems.

Russell's paradox should be taken quite seriously. Something must be done to prevent our deriving contradictions from our basic notions. We shall discuss what can be done later. Let us first look at another difficulty.

Let T be the collection of *all* sets. Let z be the cardinal number of T. Let w be the cardinal number of $\{0,1\}^T$. We know that $z < w$. On the other hand, every element of $\{0,1\}^T$ is a function and, hence, is also a set. And T is the collection of all sets. So we have $\{0,1\}^T \subseteq T$. Therefore, $w \leq z$. And this is contrary to $z < w$.

Something is shaky at the foundations!

Perhaps it would be more accurate to say that something *was* shaky at the foundations. Mathematicians have been aware of such difficulties as the above for some years. And they realized that the source of the troubles seems to be the question of what exactly sets *are*. What kind of beasts are sets, anyway?

If you go back and reread Chapter 2, you will see that we never actually stated a definition for the word "set." We gave many examples of sets and we seemed to be saying implicitly that any fool knows what a set is. While this may be true in regard to fools, wise men, as usual, will be a bit uncertain.

How does a mathematician begin to come to grips with the question of what sets are? Using his simplest intuitive notions of what sets *should* be, he sets down a list of properties or assertions that should be true about any objects we would call sets. These assertions he may then take as axioms for an abstract *theory of sets*. He will then proceed to show that the usual properties of sets can be established as theorems from these axioms. And he will try to arrange it so that such contradictions as those above cannot be obtained in the system.

In recent times there have been several such axiomatic treatments of set theory that seem to be successful. The reader who wishes to pursue this matter further should consult one of the several text books now available for abstract set theory.

In an earlier section, we asked if there was a cardinal number b such that $\aleph_0 < b < c$. This basic question is one that has received much attention from logicians and mathematicians. The assertion that there is no cardinal number between $\aleph_0$ and c has come to be called the **continuum hypothesis.** In 1963 this question was settled in a rather strange way. Using results of Kurt Gödel, Paul Cohen showed that from the usually accepted axioms for set theory one could neither prove nor disprove the continuum hypothesis. There seem to be (at least) two *different* consistent set theories which fit most of our usual ideas about sets—in one the continuum hypothesis holds, while in the other it does not. The situation is analogous to the existence of two different kinds of geometry—euclidean and noneuclidean. Both the geometries are consistent and both are available for use in different situations.

The consequences of Cohen's results are still being investigated. We may take the recent date of these results as a reminder that, far from being stagnant, mathematics is a vibrant and developing vehicle for man's careful consideration of matters practical and abstract.

Problems

1. It can be shown that any ordinal number α is the order type of the collection of all the ordinal numbers less than α when this collection is ordered by $<$. Illustrate this conclusion with several examples.
2. Show that the idea of the set of *all* ordinal numbers leads to a contradiction.

Answers to Selected Problems

Section 1.1

1. (a) true (c) true (e) true (g) true (i) true (k) true (m) true (o) false
2. (i) (a) The square of a number is even if the number is even.
   (b) In order that the square of a number be even, it is sufficient that the number be even.
   (c) In order that a number be even, it is necessary that its square be even.
   (d) A number is even only if its square is even.
   (e) A number being even implies that its square is even.

   (iii) (a) $a + 2 = b + 1$, if $a + 1 = b$.
   (b) In order that $a + 2 = b + 1$, it is sufficient that $a + 1 = b$.
   (c) In order that $a + 1 = b$, it is necessary that $a + 2 = b + 1$.
   (d) $a + 1 = b$ only if $a + 2 = b + 1$.
   (e) $a + 1 = b$ implies $a + 2 = b + 1$.

   (iv) (a) You will get no ice cream if you don't eat your spinach.
   (b) In order that you get no ice cream, it is sufficient that you not eat your spinach.
   (c) In order that you not eat your spinach, it is necessary that you get no ice cream.
   (d) You haven't eaten your spinach only if you get no ice cream.
   (e) Your not eating your spinach implies that you get no ice cream.

   (v) (a) You don't understand the situation, if you're not confused.
   (b) In order that you not understand the situation, it is sufficient that you not be confused.
   (c) In order that you not be confused, it is necessary that you not understand the situation.

(d) You are not confused only if you don't understand the situation.
(e) Your not being confused implies that you do not understand the situation.

3. (a) (1) A necessary and sufficient condition that a triangle be equilateral is that it be equiangular.
(2) If a triangle is equilateral, then it is equiangular, and conversely.
(c) (1) Two sets are equal if and only if each is a subset of the other.
(2) If two sets are equal, then each is a subset of the other, and conversely.

4. (a) false (c) false (e) true (g) true (i) true

5. Equal, in the mathematical sense, means logical identity. In the statement, "All men are created equal", this would mean that all men are the same man.

7. (i) If the square of a number is even, then the number is even.
(iii) If $a + 2 = b + 1$, then $a + 1 = b$.
(v) If you don't understand, then you're not confused.

8. The original implication.

9. *Part 1:* Prove "If . . . , then ***" is true if "*** or not . . ." is true.
By hypothesis, "*** or not . . ."
If ***, then any implication with this as a conclusion, namely "If . . . , then ***", must be true.
If not . . . , then any implication with . . . as the hypothesis must be true.
Conclusion: By considering both possible cases, we find that the implication "If . . . , then ***" must be true if "*** or not . . ." is true.

Part 2: Prove "*** or not . . ." is true if "If . . . , then ***" is true.
By logic, either . . . or not . . . .
If . . . , by hypothesis we have ***, one of the choices of the "or" statement. Hence, "*** or not . . ." must be true.
If not . . . , then we again have one of the choices of the "or" statement.
Conclusion: By considering both possible cases, we find that the "or" statement, "*** or not . . .", must be true if "If . . . , then ***" is true.

Section 1.2

1. (a) True. 36 is a perfect square since $36 = 6^2$.
(c) False. Since $1^2 \neq 17$, $2^2 \neq 17$, $3^2 \neq 17$, $4^2 \neq 17$, and the square of any whole number greater than 4 will be greater than 17, there is no whole number whose square is 17. Hence, 17 is not a perfect square.
(e) True. 4 and 9 are perfect squares, since $4 = 2^2$ and $9 = 3^2$. We see that 9 is 5 greater than 4.
(g) True. Suppose N and M are perfect squares. Then there are whole numbers K and J such that $N = K^2$ and $M = J^2$. Then $N \cdot M = (K^2) \cdot (J^2) = K \cdot K \cdot J \cdot J = K \cdot J \cdot K \cdot J = (K \cdot J) \cdot (K \cdot J) = (K \cdot J)^2$. Hence, $N \cdot M$ is a perfect square. (The details of such an argument will be matters we shall consider later.)

2.

1	2	3	4	5	6	7	8	9	10
1	0	0	1	0	0	0	0	1	0
11	12	13	14	15	16	17	18	19	20
0	0	0	0	0	1	0	0	0	0
21	22	23	24	25	26	27	28	29	30
0	0	0	0	1	0	0	0	0	0
31	32	33	34	35	36	37	38	39	40
0	0	0	0	0	1	0	0	0	0
41	42	43	44	45	46	47	48	49	50
0	0	0	0	0	0	0	0	1	0

4. Let us call a number a **trio square** if it is the sum of three perfect squares. (There may well be other possible definitions that make just as much sense. There is no unique correct answer to such a problem as this.) Some trio squares according to the given definition would be 3, 6, 9, 14, 19, etc.

6. Since only conjectures are asked for here, any answers are correct for this problem. As a matter of fact, however, (a) and (c) are true while (b) and (d) are false.

Section 1.3

1. (a) properly constructed; true
   (c) properly constructed; true
   (e) properly constructed; true
   (g) not properly constructed
   (i) properly constructed; this is false because 4 is even and is a perfect square
   (k) not properly constructed

2. If $3 = 2 + 1$, then $5 + 3 = 5 + (2 + 1)$.
   If $a = b + 1$, then $a + a = a + (b + 1)$.
   (These are, of course, only two of many possible examples.)

3. (a) 3 is a number, but 3 is not even.
   (c) 7 is greater than 4, but 7 is not less than 6.
   (e) Every four years the day following February 28 is February 29.
   (g) "What is mathematics?" is a properly constructed English sentence, but it does not end with a period.
   (i) "$1 + 1 = 2$, or $1 + 2 = 4$" is true, since "$1 + 1 = 2$" is true, but "$1 + 1 = 2$ *and* $1 + 2 = 4$" is false, since the last part is false.
   (k) "$2 = 2$ *or* $3 = 4$" is true, but "$2 = 2$ *and* $3 = 4$" is false.

Section 1.4

1. As was pointed out in this section, there are often many different ways to perform a given task and, hence, many different correct flow charts for a task. We give on the next page one possible answer to the given problem, but there are others. This comment applies to all of the problems in this section.

(a)

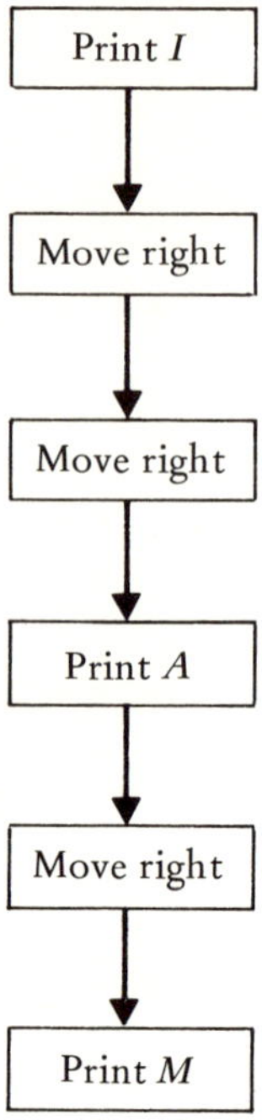
Print *I*
Move right
Move right
Print *A*
Move right
Print *M*

(b)

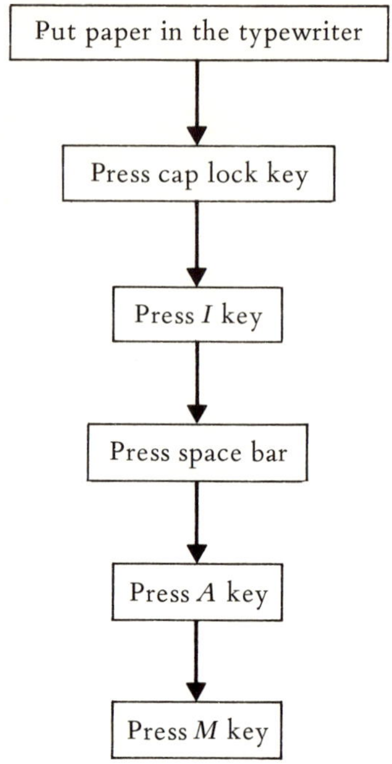
Put paper in the typewriter
Press cap lock key
Press *I* key
Press space bar
Press *A* key
Press *M* key

2. (c)

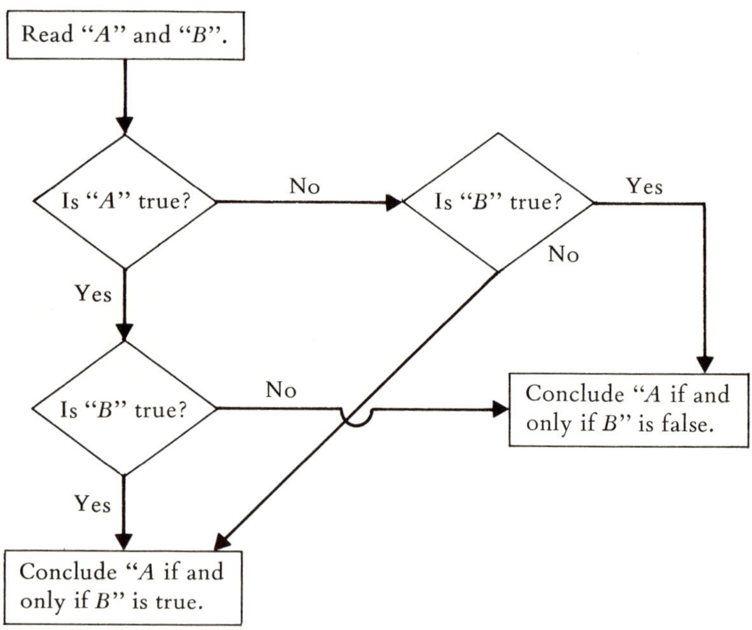
Read "*A*" and "*B*".
Is "*A*" true?
No
Is "*B*" true?
Yes
No
Yes
Is "*B*" true?
No
Conclude "*A* if and only if *B*" is false.
Yes
Conclude "*A* if and only if *B*" is true.

3.

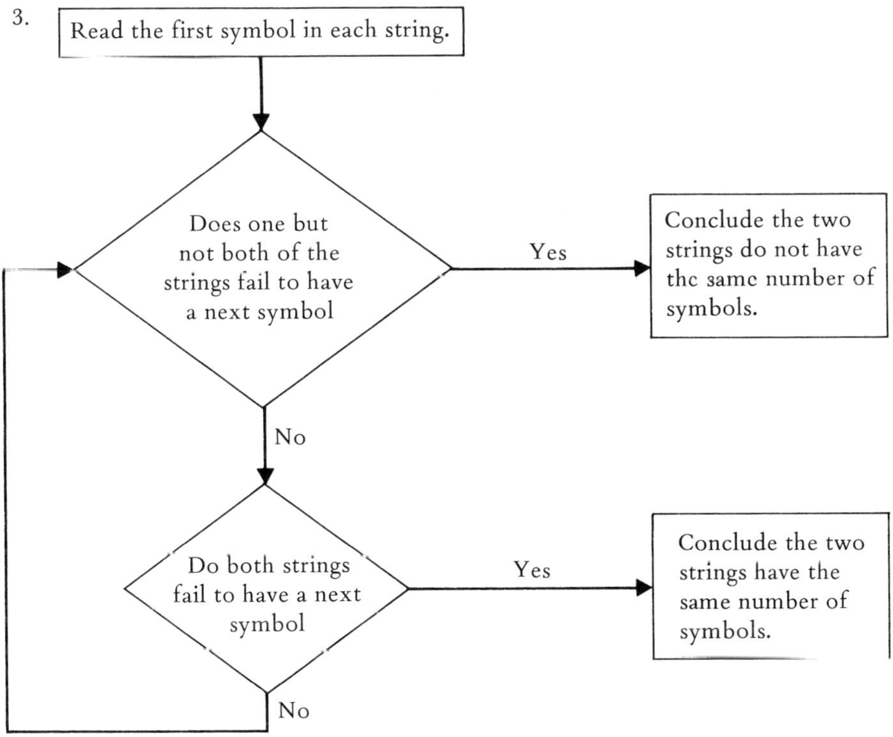
Read the first symbol in each string.
Does one but not both of the strings fail to have a next symbol
Yes
Conclude the two strings do not have the same number of symbols.
No
Do both strings fail to have a next symbol
Yes
Conclude the two strings have the same number of symbols.
No

5.

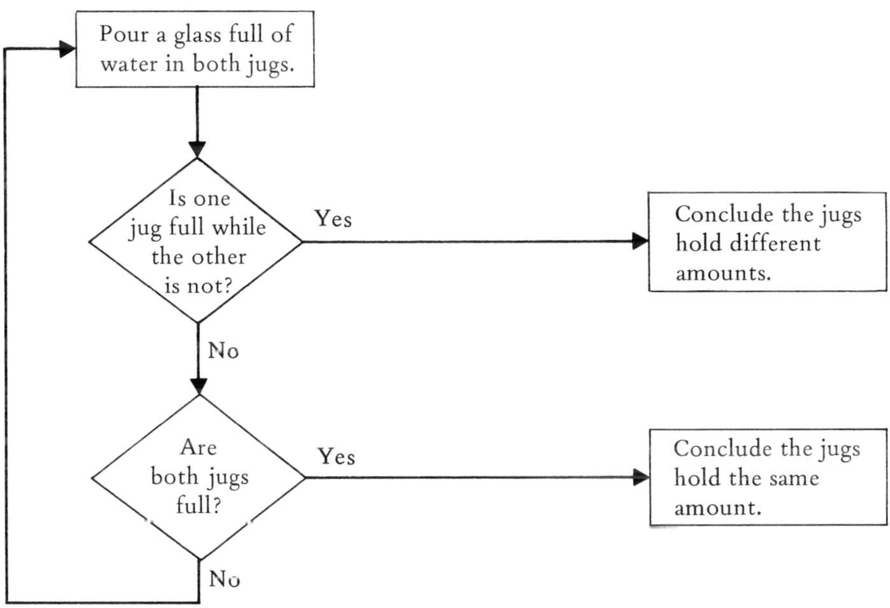
Pour a glass full of water in both jugs.
Is one jug full while the other is not?
Yes
Conclude the jugs hold different amounts.
No
Are both jugs full?
Yes
Conclude the jugs hold the same amount.
No

6.

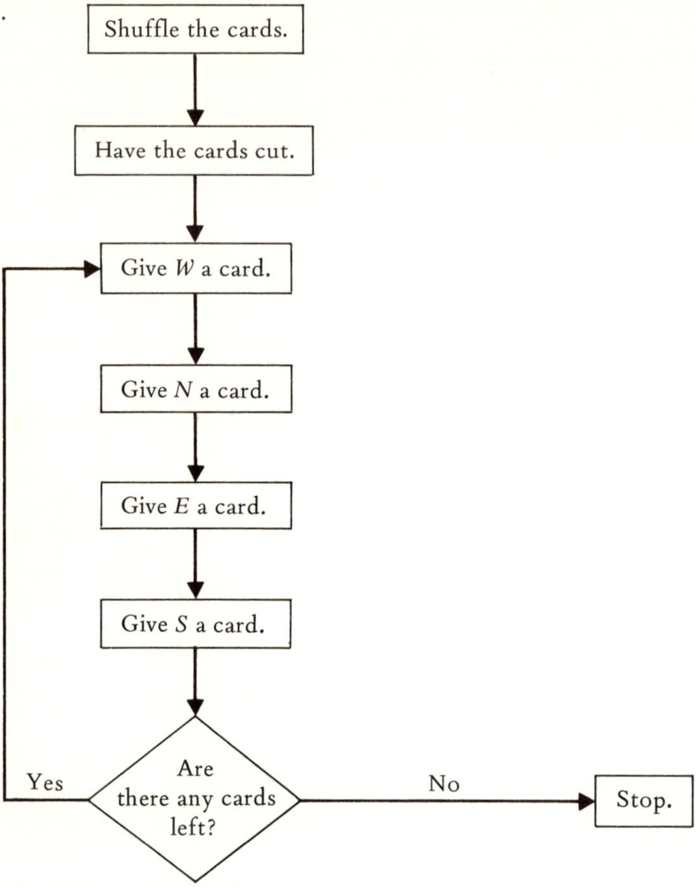

7. Insert a diamond with the suggested question in it on the "Yes path" from the diamond in the flow chart for Problem 4.

Section 2.1

1. (a) true (c) false (f) false (h) true (j) true

4. herd of elephants, deck of cards, grove of trees, crowd of people, set of silver, plates, cups, glasses, cord of wood, scale of notes, school of fish, group of people, string of pearls, fleet of ships, company of soldiers, clump of grass, mixture of chemicals, series of volumes

(There are, no doubt, many other examples.)

5. gathering, collection, aggregate, association, amalgamation, accumulation, confederation

(There are other possibilities also.)

7. (a) not properly constructed
   (c) properly constructed; false
   (e) not properly constructed
   (g) properly constructed; false
   (i) properly constructed; true

Section 2.2

1. (a) false (c) true (e) true (g) true (i) true
   (k) true (m) false (o) true (q) false

2. (a) ϕ
   (d) ϕ, $\{1\}$, $\{2\}$, $\{3\}$, $\{4\}$, $\{1,2\}$, $\{1,3\}$, $\{1,4\}$, $\{2,3\}$, $\{2,4\}$, $\{3,4\}$, $\{1,2,3\}$, $\{1,2,4\}$, $\{1,3,4\}$, $\{2,3,4\}$, $\{1,2,3,4\}$

3. (a) 32
   (b) 64
   (Of course, since conjectures are asked for, other answers are acceptable. The numbers given are the *correct* conjectures.)

5. (a) 0 (b) 0 (c) 1 (d) 3 (e) 6 (f) 10 (g) 15
   (In regard to parts f and g, see the comment at the end of Problem 3.)

6. (a) Let $A = \{x \mid x$ is a whole number between 1 and 2$\}$ and $S = \{x \mid x$ is a perfect square$\}$. Since $A = \phi$ we have $A \subseteq S$ and, therefore, the statement is true.
   (c) Let $C = \{y \mid y$ is a three-sided rectangle$\}$ and $G = \{w \mid w$ is green$\}$. Since $C = \phi$ we have $C \subseteq G$ and, therefore, the statement is true.

8. (a) True; since every element of A is an element of B and every element of B is an element of C, every element of A is an element of C and A is a subset of C. Since $B \subset C$, there exists an element in C that is not in B. Therefore this element cannot be an element of A and A is a *proper* subset of C.
   (c) True; if there were a proper subset of ϕ, then ϕ would have to contain an element that was not in the proper subset. But since ϕ contains no elements, this is impossible, and ϕ cannot have any proper subsets.
   (e) False; consider the example $A = \{1,2,3\}$ $B = \{3,4,5\}$.

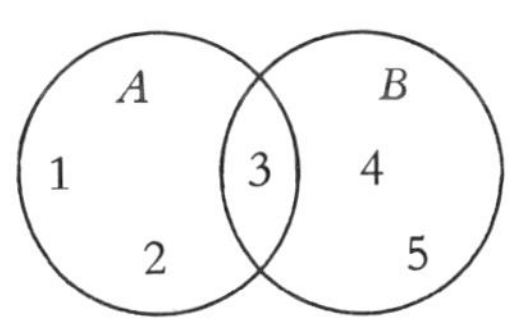

10. (a) *Theorem 2.2.7.* $A \subseteq B$ and $B \subseteq C$ implies $A \subseteq C$

Step	*Reason*
1. $A \subseteq B$	Hypothesis
2. $x \in A$ implies $x \in B$	Step 1 and Definition 2.2.1
3. $B \subseteq C$	Hypothesis
4. $x \in B$ implies $x \in C$	Step 3 and Definition 2.2.1
5. $x \in A$ implies $x \in C$	Steps 2 and 4
6. $A \subseteq C$	Step 5 and Definition 2.2.1

(c) If $A \subseteq B$ and $B \subset C$, then $A \subset C$.

Step	*Reason*
1. $A \subseteq B$	Hypothesis
2. $B \subset C$	Hypothesis
3. $B \subseteq C$	Step 2 and Theorem 2.2.6
4. $A \subseteq C$	Steps 1 and 3 and Theorem 2.2.7
*5. $A = C$	Assumption
*6. $C \subseteq A$	Step 5 and Theorem 2.2.5
*7. $C \subseteq B$	Steps 1 and 6 and Theorem 2.2.7
*8. $B = C$	Steps 3 and 7 and Theorem 2.2.5
9. $B \neq C$	Step 2 and Definition 2.2.3
10. $A \neq C$	Since steps 8 and 9 contradict each other, our assumption in Step 5 must be incorrect.
11. $A \subset C$	Steps 4 and 10 and Definition 2.2.3

11. (a) The rows for $n = 4$ and $n = 5$ are 1, 4, 6, 4, 1 and 1, 5, 10, 10, 5, 1 respectively.

Section 2.3

1. (a) $\{a,e\}$ (c) $\{e,a\ s\ y,t\}$ (e) $\{e,a,s,y,t\}$ (g) $\{y\}$ (i) ϕ
   (k) $\{e,a,s\}$ (m) $\{e,a,s\}$ (o) ϕ (q) $\{u\}$ (s) $\{a,e,i\}$

3.

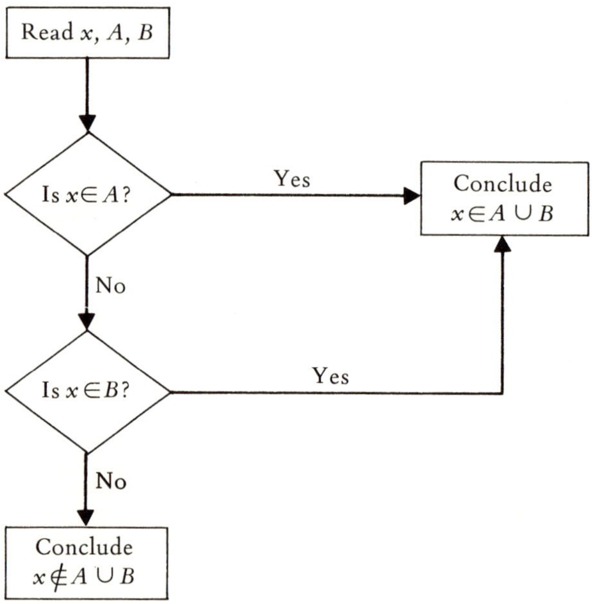

6. (a) ϕ is the only set for which $A \cap A = \phi$ is true.
   (b) ϕ is disjoint from every set, since $A \cap \phi = \phi$ for all sets A.

9. (a) $\{1,9,25\}$ (c) ϕ

10. PROOF. Suppose $B \subseteq A$. Let $x \in A \cup B$. By Definition 2.3.2, $x \in A$ or $x \in B$. If $x \in A$, $x \in A$. If $x \in B$, by hypothesis $x \in A$. Hence, $A \cup B \subseteq A$. By Theorem 2.3.11, $A \subseteq A \cup B$. Thus $A = A \cup B$ by Theorem 2.2.5.

    Next, suppose $A \cup B = A$. $B \subseteq A \cup B$ by Theorem 2.3.11. Therefore, $B \subseteq A$.

11. (a) True. PROOF: Suppose $x \in A \cup C$. Then by Definition 2.3.2, $x \in A$ or $x \in C$. If $x \in A$, then by hypothesis $x \in B$. Therefore $x \in B \cup C$. If $x \in C$, then $x \in B \cup C$. We conclude that $A \cup C \subseteq B \cup C$.

    (c) False. Counterexample: Let $A = \{1\}$, $B = \{1,2\}$, $C = \{1,2,3\}$. $A \subset B$ since $1 \in B$ but $2 \notin A$. However, $A \cup C = \{1,2,3\} = C$ and $B \cup C = \{1,2,3\} = C$. Therefore $A \cup C = B \cup C$ and *not* $A \cup C \subset B \cup C$.

    (e) False. Counterexample: Let $A = \{1,2\}$, $B = \{1\}$, $C = \{1,2,3\}$. $A \cup C = \{1,2,3\}$ and $B \cup C = \{1,2,3\}$, so $A \cup C \subseteq B \cup C$. But $2 \in A$ and $2 \notin B$, so *not* $A \subseteq B$.

    (g) True. If $A \cup B = \phi$, then $A = \phi$ or $B = \phi$. In fact, a *stronger* statement can be made. If $A \cup B = \phi$, then $A = \phi$ *and* $B = \phi$.

    PROOF: Suppose $A \neq \phi$. Then A contains some element x and, therefore, $x \in A \cup B$. This contradicts the hypothesis that $A \cup B = \phi$. Next, suppose $B \neq \phi$. A similar argument leads again to a contradiction. We conclude that if $A \cup B = \phi$, then $A = \phi$ *and* $B = \phi$.

    (i) True. By hypothesis $A = \phi$ or $B = \phi$. Assume the conclusion is false. That is, assume $A \cap B \neq \phi$. Then $A \cap B$ contains an element x, and by Definition 2.3.1, $x \in A$ and $x \in B$. Therefore, $A \neq \phi$ *and* $B \neq \phi$, a contradiction of the hypothesis. Thus, $A \cap B = \phi$.

Section 2.4

1. (a) $\{t\}$

   (c) $(\{t\})' = \{a,b,c,\ldots,r,s,u,v,w,x,y,z\}$ $\quad E \cup S' = (E')' \cup S' = (E' \cap S)'$

   (e) $\{g,h,i,j,k,l,m,n,o,p,q,r,w,t\}$

   (g) $\{e,a,s,y,t\}$ $\quad (E' \cap S')' = (E')' \cup (S')' = E \cup S$

2. $V' = \{x \mid x$ is a letter that is not a vowel$\}$
   $M' = \{y \mid y$ is a letter that does not appear in the word "mathematics"$\}$
   $S' = \{x \mid x$ is a letter that does not appear in the word "set"$\}$

3. (a) $\{2\}$ (c) $N = \{1,2,3,\ldots\}$ (e) ϕ (g) $\{1,3,5,7,\ldots\}$ (i) $\{2,4,6,8,\ldots\}$

4. (a) $\{i,o,u\}$ (c) $\{m,a,t,h,e,i,c,s\}$

5. (a) $(A \cup A)' \cap A' = (A)' \cap A'$ (Theorem 2.3.7)
   $= A'$ (Theorem 2.3.3)

   (c) $(A' \cup B') \cap (A \cup B)' = (A \cap B)' \cap (A \cup B)'$ (Theorem 2.4.5)
   $= [(A \cap B) \cup (A \cup B)]'$ (Theorem 2.4.5)
   $= (A \cup B)'$ (Theorem 2.3.13 says $A \cap B \subseteq A \cup B$. Then apply Theorem 2.3.12.)

6. True.

Statement	Reason
1. $A' = B'$	Hypothesis
2. $A' \subseteq B'$ and $B' \subseteq A'$	Theorem 2.2.5 and step 1
3. $(B')' \subseteq (A')'$ and $(A')' \subseteq (B')'$	Theorem 2.4.7 and step 2

4. $B \subseteq A$ and $A \subseteq B$	Theorem 2.4.6 and step 3
5. $A = B$	Theorem 2.2.5 and step 4

8. True.

Statement	*Reason*
1. $A \subseteq B$	Hypothesis
2. $B' \subseteq A'$	Theorem 2.4.7 and step 1
3. $C - B = C \cap B'$, $C - A = C \cap A'$	Definition from Problem 4
4. $C \cap B' \subseteq C \cap A'$	Step 2, Problem 11(b), Section 2.3, and Theorem 2.3.1
5. $C - B \subseteq C - A$	Steps 3 and 4

Section 2.5

1. (a) $\{(1,3), (2,3)\}$
   (c) $\{(1,2), (1,3), (2,2), (2,3)\}$
   (e) $\{(1,1), (1,2), (1,3), (2,1), (2,2), (2,3), (3,1), (3,2), (3,3)\}$
2. (a) Let h and t represent heads and tails on the penny. Let H and T represent heads and tails on the dime. Then $\{h,t\} \times \{H,T\} = \{(h,H), (h,T), (t,H), (t,T)\}$ represents all possible combinations.
   (c) Let $C = \{$all cards in the deck$\}$. Then find $C \times C$.
   (e) $\{$white, yellow, ivory$\} \times \{$white, yellow, ivory$\}$
3. (a) False. Counterexample: Let $A = \{1\}$, $B = \{1,2\}$, and $C = \{2\}$. Then $A \cup (B \times C) = \{1\} \cup \{(1,2), (2,2)\} = \{1, (1,2), (2,2)\}$ but $(A \cup B) \times (A \cup C) = \{1,2\} \times \{1,2\} = \{(1,1), (1,2), (2,1), (2,2)\}$.
   (c) False. Counterexample: Let $A = \{1\}$, $B = \{1,2\}$, and $C = \phi$. Then $A \subseteq B$. But $A \times C = A \times \phi = \phi$ and $B \times C = B \times \phi = \phi$. Therefore $A \times C = B \times C$, and *not* $A \times C \subset B \times C$.
   (e) False. Counterexample: Let $A = \{1,2\}$, $B = \{1\}$, and $C = \phi$. Then $A \times C = \phi$ and $B \times C = \phi$, so $A \times C \subseteq B \times C$. But $A \subseteq B$ is *false*.
   (g) True. PROOF: Suppose $(a,c) \in (A \times C)$. Then by Definition 2.5.1, $a \in A$ and $c \in C$. By hypothesis, then, $a \in B$ and $c \in D$. Thus $(a,c) \in B \times D$ by Definition 2.5.1. We conclude $A \times C \subseteq B \times D$.
4. No. Let $A = \{1\}$ and $B = \phi$. Then $A \times B = \phi = B \times A$ (Theorem 2.5.1). But $A \neq B$.

6. (i)

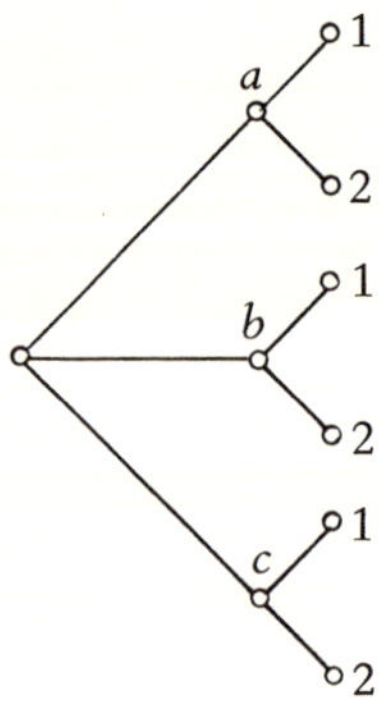

(ii)

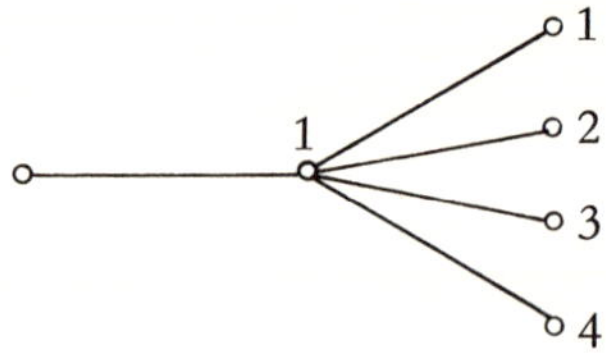

(iii)

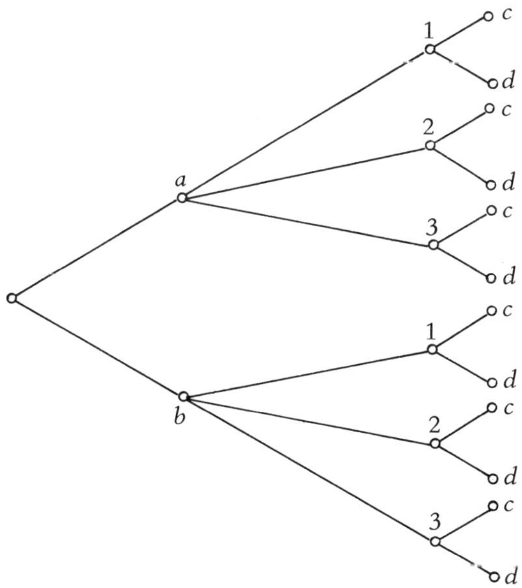

Section 3.1

1. (a) $L = \{(1,2), (1,3), (1,4), (1,5), (2,3), (2,4), (2,5), (3,4), (3,5), (4,5)\}$
2. $\{(1,1), (2,2), (3,3), (4,4), (5,5)\}$

3. (b) (The diagram for 3(a) is the same except that it includes an arrow from each subset back to that subset.)
4. (a)

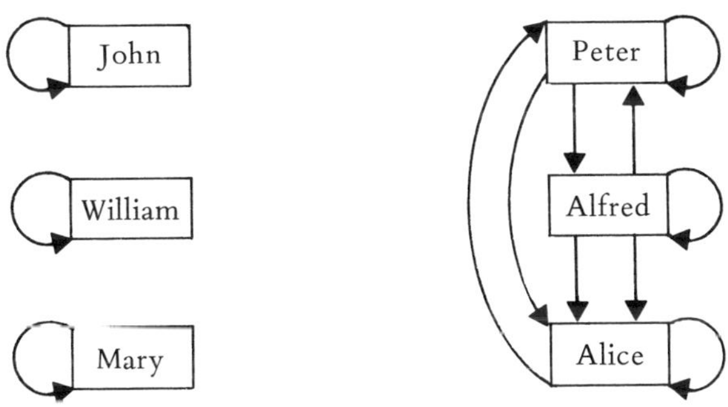

5.

	r	s	t
M			
G			
R			x
S		x	

	r	s	t
A	x	x	x
L	?	?	?
Y			x
H			x

	r	s	t
N	x	x	x
P	x	x	x
V	x	x	x
E	x	x	x

7. *M*, *G*, *Y*, *H*, *D*

9. *M*, *G*, *S*

10. (a) not reflexive since $(1,1) \notin \phi$;
symmetric since, if $(x,y) \in \phi$, then (y,x) is in ϕ;
(If you think this statement is false, try to find a counterexample.)
transitive since, if $(x,y) \in \phi$ and $(y,z) \in \phi$, then $(x,z) \in \phi$.

11. (a) We need all possible subsets of $\{1,2\} \times \{1,2\} = \{(1,1), (1,2), (2,1), (2,2)\}$.

ϕ	$\{(1,1), (1,2)\}$	$\{(1,1), (1,2), (2,1)\}$
$\{(1,1)\}$	$\{(1,1), (2,1)\}$	$\{(1,1), (1,2), (2,2)\}$
$\{(1,2)\}$	$\{(1,1), (2,2)\}$	$\{(1,2), (2,1), (2,2)\}$
$\{(2,1)\}$	$\{(1,2), (2,1)\}$	$\{(1,1), (2,1), (2,2)\}$
$\{(2,2)\}$	$\{(1,2), (2,2)\}$	$\{(1,1), (1,2), (2,1), (2,2)\}$
	$\{(2,1), (2,2)\}$	

12.

		Domain	*Range*
(a)	*M*:	{Mary}	{Peter, Alfred, Alice}
	G:	{John}	{Peter, Alfred, Alice}
	B:	{Peter, Alfred}	{Peter, Alfred, Alice}
	A:	*C*	*C*

13.

(1)
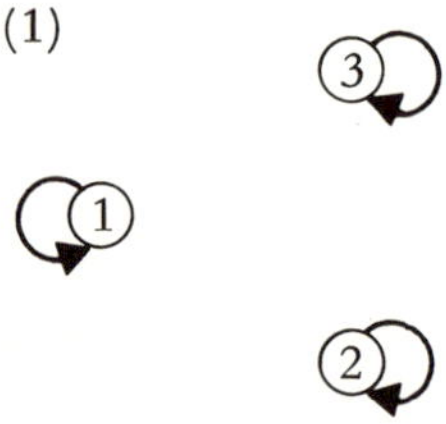

(2)
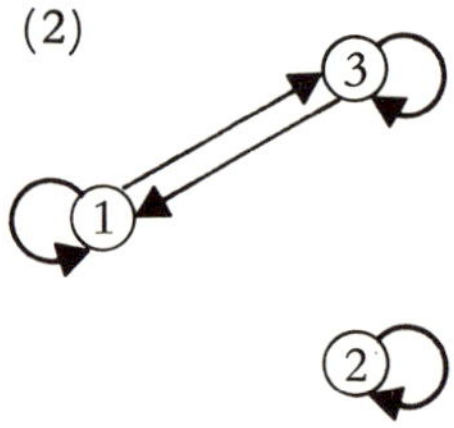

(3)
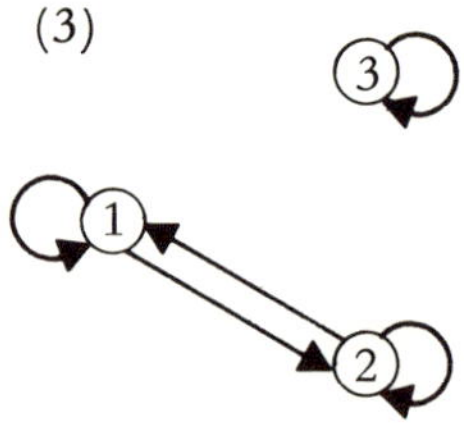

(4)
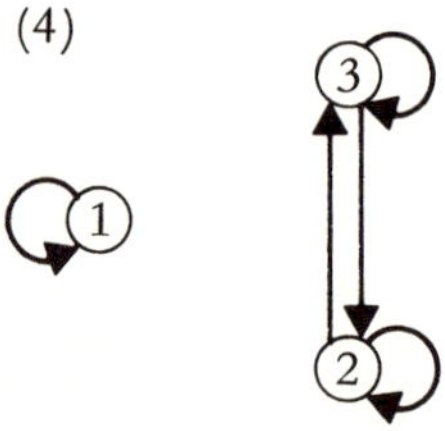

(5)
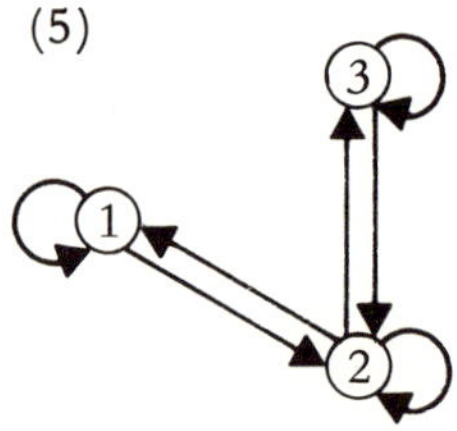

(6)
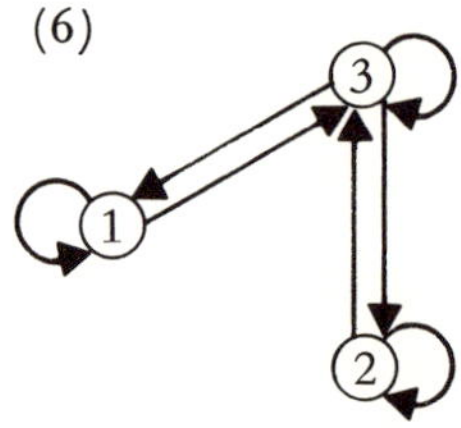

(7)
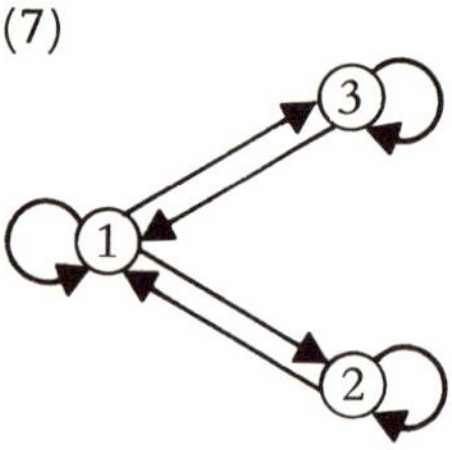

(8)
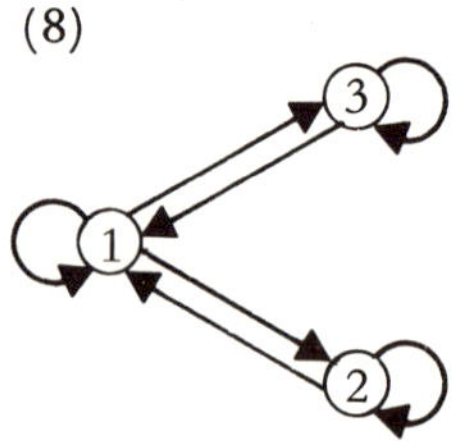

15. PROOF: To prove R is reflexive, we must show xRx, or $(x,x) \in R$ *for every* $x \in A$. Let x be any member of A. Then by hypothesis, x is in the domain of R and by definition of domain (see Problem 12) this means xRy for some $y \in A$. xRy implies yRx since R is symmetric and, therefore, xRx since R is transitive.

16. (a) True: If $A = \phi$, then the only relation on A is ϕ. In this case R and S must both equal ϕ, which is a reflexive relation on ϕ. (See Problem 10.)
    (c) False; counterexample: $A = \{1,2\}$, $R = \phi$ (this is transitive; see Problem 10). $S = \{(1,2), (2,1)\}$. S is not transitive since $(1,2)$ and $(2,1)$ are in S, but $(1,1) \notin S$. However $\phi = R \subseteq S$.

17. (a) True. PROOF: If $(x,y) \in R$, then $(y,x) \in R$ since R is symmetric on A. Thus, R is also symmetric in B.
    (c) True. PROOF: If $(x,y) \in R$ and $(y,z) \in R$, then since R is transitive in A, $(x,z) \in R$. Thus R is a transitive relation in B.

Section 3.2

1. (a) 3.1.2: $\{1\}, \{2\}, \{3\}, \{4\}, \{5\}, \{6\}$.
   (c) H: $[x] = \{y \mid y$ is a student with the same home town as $x\}$
   Y: $[x] = \{y \mid y$ is a student, the same year in school as $x\}$
   (e) $\{1,3,5\}, \{2,4,6\}$

2. S: $[x] = \{y \mid y$ is a student from the same state as $x\}$
   B: $[x] = \{y \mid y$ is a student the same as x or a sibling of $x\}$
   W: $[x] = \{y \mid y$ is a student whose weight rounded off to the nearest pound is the same as $x\}$
   E: $[x] = \{y \mid y$ is a student with the same color eyes as $x\}$
   (If x represents you in the above equivalence classes, what would these classes look like?)

4. Yes. PROOF: (x,x) is in $A \times A$ for every x so our relation is reflexive. If (x,y) is in $A \times A$, then so is (y,x) since $A \times A$ contains every ordered pair from A, hence our relation is symmetric. Likewise it will be transitive.
   There is only one equivalence class, and that is all of A. (Since given any x in A, $[x] = \{y \mid (x,y)$ is in $A \times A\}$ which is all of A.)

7. V: One equivalence class would consist of all Nixon voters; another of all Humphrey voters; etc.
   K: Not an equivalence relation.
   C: One equivalence class would consist of all people who own no cars; another would consist of all people who own one car; another of all those who own two cars; etc.
   T: Not an equivalence relation.
   F: One equivalence class would consist of all citizens with no children; another class of all citizens with one child; etc.
   W: Not an equivalence relation, since some citizens don't work in buildings.

8. (a) 5 (b) 15

10. (a) The relation ϕ is both kinds of relations if $A = \phi$. In general, for any A, the only relation that is both kinds of relations is the equality relation.

(c)

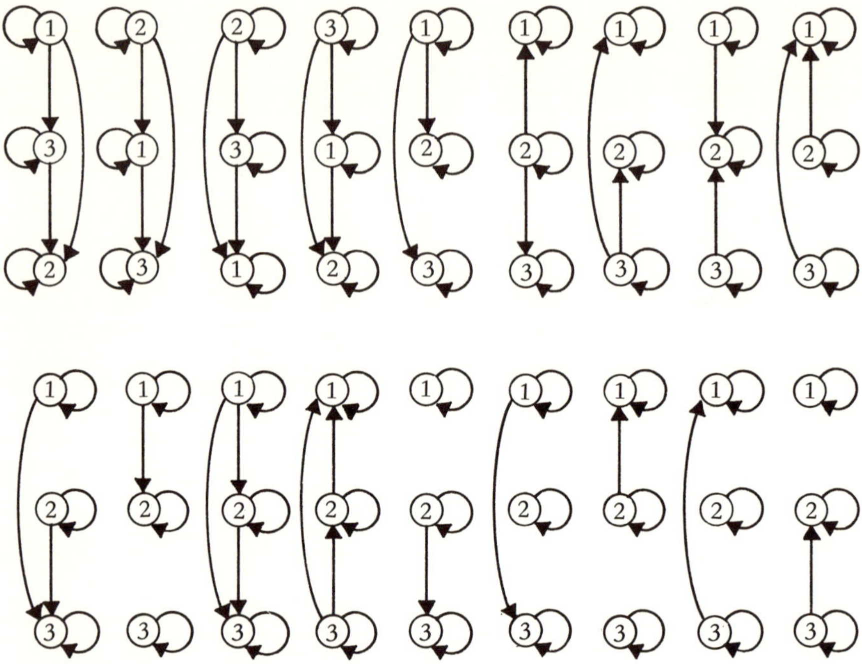

11. (a) We already know that set inclusion is a **partial** ordering, and so the only other property that we have to check to prove that it is a **linear** ordering is that for any x and y in A, that either $x \leq y$ or $y \leq x$. In the given collection of subsets, if we pick two, then one is a subset of the other, or vice versa.
    (c) For any x and y from the set $\{1,2,3,4,5,6\}$, either x is less than or equal to y, or y is less than or equal to x. Also for any number x from this set, x is less than or equal to x since $x = x$. Thus the relation is reflexive. If x and y are two numbers from this set and (x,y) and (y,x) are in the relation, then x is less than or equal to y, and y is less than or equal to x, which means $x = y$.
    Finally, if (x,y) and (y,z) are in the relation, then x is less than or equal to y and y is less than or equal to z, so that x is less than or equal to z, or (x,z) is in the relation; hence the relation is transitive. Thus the relation is a linear ordering.

Section 3.3

1. (a) Let us consider the relations given in the answer to Problem 11(a), Section 3.1, to be numbered from 1 to 16. Then the following sets indicate collections of isomorphic relations.
    $\{1\}$, $\{2,5\}$, $\{3,4\}$, $\{6,11\}$, $\{7,10\}$, $\{8\}$, $\{9\}$, $\{12,14\}$, $\{13,15\}$, $\{16\}$.
    (c) With the relations numbered as in the answer to Problem 11(d), Section 3.2, the following sets indicate isomorphic relations: $\{1,2\}$, $\{3,4,5,6,7,8\}$.
2. (ii) isomorphic to (viii). (iii), (v), (ix) are each isomorphic to each other. (i) isomorphic to (vii).
4. "Any two linear orderings on a fixed finite set are isomorphic."

6. No. The following diagram affords a quick counterexample:

R S

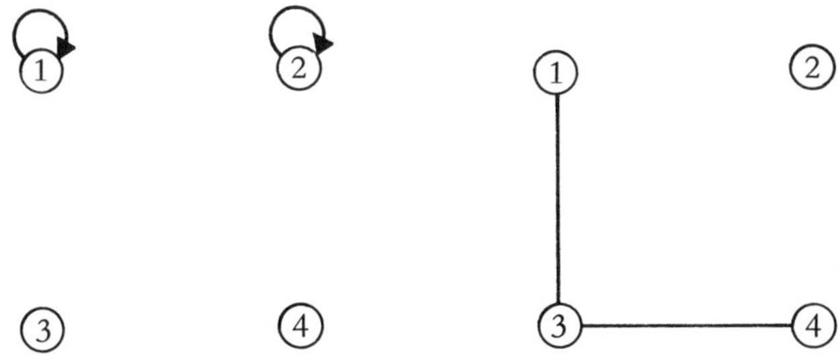

For the two relations R and S to have the same structure, S would have to have two elements related to themselves. But in S no element is related to itself.

8.

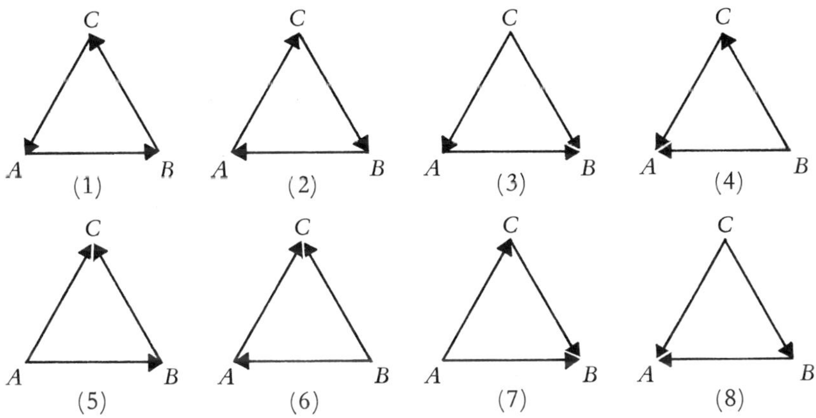

The following collections indicate which of these are isomorphic: $\{1,2\}$, $\{3,4,5,6,7,8\}$.

Section 4.1

1. M not a function
   G not a function
   B not a function
   S function

   domain: {William, Mary}
   range: {William, Mary}

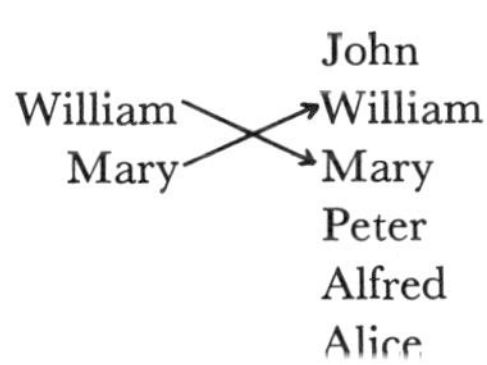

A is a function unless William and Mary are the same age or two of the three children are twins or they are triplets

L	not a function (we would hope)
Y	not a function
H	not a function
N	not a function
P	not a function
V	not a function
E	not a function
D	not a function

2. (a)

(i)

x	$f(x)$
1	1
2	1

(iii)

x	$h_1(x)$
1	1

x	$h_2(x)$
1	2

(v)

x	$n_1(x)$
1	1
2	1

x	$n_5(x)$
1	2
2	3

x	$n_2(x)$
1	1
2	2

x	$n_6(x)$
1	3
2	1

x	$n_3(x)$
1	1
2	3

x	$n_7(x)$
1	3
2	2

x	$n_4(x)$
1	2
2	1

x	$n_8(x)$
1	3
2	3

x	$n_9(x)$
1	2
2	2

(b)

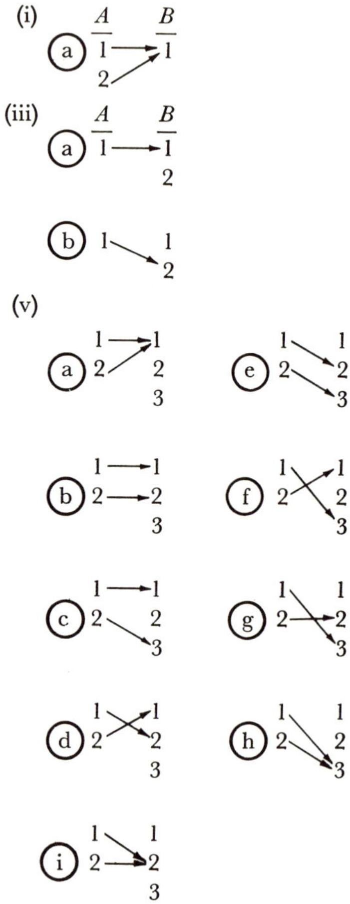

3. (a) $4 \times 4 \times 4 = 64$

4. (a) (i) Same as into.
   (iii) None of these functions is an onto mapping because *B* has more elements than *A*.
   (v) None is an onto mapping (see iii).
   (b) (a) None.

5. (a) yes (c) no (e) yes (g) yes (i) yes

6. (a) domain: $\{1,2,3,4,5\}$;
   range: $\{1\}$
   (f) domain: $\{x \mid x$ lives in Chicago and the mother of x lives in Chicago$\}$
   range: $\{y \mid y$ lives in Chicago and a child of y lives in Chicago$\}$
   (i) domain: $\{x \mid x$ is a student in your class$\}$
   range: $\{y \mid y$ is the number of sisters of at least one member of your class$\}$

7. $\{0,1\} \times \{0,1\} = \{(0,0), (0,1), (1,0), (1,1)\}$

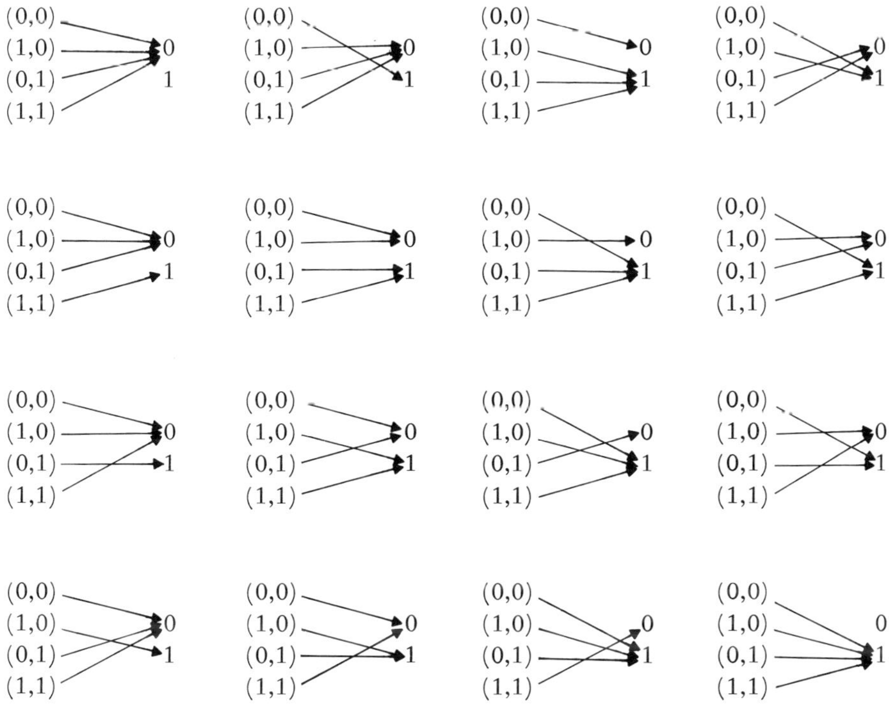

8. (a) 11 (c) 6 (e) 3

9. (a) 2 (c) none (e) 6

10. (a) (i) a (iii) a (v) a, b, c, g, i
    (c) none (Domain and range are two different types of things.)

11. (a) False. Counterexample: Let $f = \{(1,2), (2,1)\}$ and $g = \{(1,1), (2,2)\}$. Then $f \cup g = \{(1,1), (1,2), (2,1), (2,2)\}$ which is not a function since 1 is the first element in two different ordered pairs.
    (c) True. Assume $f \cap g$ is not a function. Then (x,y) and (x,z) are in $f \cup g$ where $y \neq z$. But this would mean that both these ordered pairs are in both f and g. Thus neither f nor g is a function, contradicting the hypothesis. Conclusion: $f \cap g$ is a function.
    (e) False. For counterexample see Problem 9.

12. (a) Let $g(x) = 1 - F_A(x)$. Suppose $x \in A$. Then $F_A(x) = 1$ and, hence, $g(x) = 1 - 1 = 0$. Suppose $x \in A'$. Then $x \notin A$ and $F_A(x) = 0$. Hence, $g(x) = 1 - 0 = 1$. We see then that $g(x) = 1$ if $x \in A'$ and $g(x) = 0$ if $x \notin A'$. That is, $g(x) = F_{A'}(x)$.
(c) Since $A \cup B = (A' \cap B')'$ we have from *a* and *b* of this problem that $F_{A \cup B}(x) = 1 - F_{A' \cap B'}(x) = 1 - F_{A'}(x)F_{B'}(x) = 1 - (1 - F_A(x))(1 - F_B(x))$.

13. Each subset determines a mapping (from $\{1,2,3,4,5,6\}$ into $\{0,1\}$) which is the characteristic function for that subset. (See Problem 12.) Conversely, any mapping from $\{1,2,3,4,5,6\}$ into $\{0,1\}$ determines a subset of which it is the characteristic function. Hence, there are exactly as many subsets as there are mappings.

14. (a) If x is true we have not-x false and also $T(x) = 1$. Hence, $1 - T(x) = 0$.
If x is false we have not-x true and also $T(x) = 0$. Hence, $1 - T(x) = 1$.
(c) Since (x or y) is equivalent to not-(not-x and not-y) this follows from parts (a) and (b).

16. (a)

B	ϕ	$\{a\}$	$\{b\}$	$\{c\}$	$\{d\}$
$f(B)$	ϕ	$\{c\}$	$\{d\}$	$\{b\}$	$\{c\}$

B	$\{a,b\}$	$\{a,c\}$	$\{a,d\}$	$\{b,c\}$	$\{b,d\}$
$f(B)$	$\{c,d\}$	$\{c,b\}$	$\{c\}$	$\{b,d\}$	$\{c,d\}$

B	$\{c,d\}$	$\{a,b,c\}$	$\{a,b,d\}$	$\{a,c,d\}$
$f(B)$	$\{b,c\}$	$\{b,c,d\}$	$\{c,d\}$	$\{b,c\}$

B	$\{b,c,d\}$	$\{a,b,c,d\}$
$f(B)$	$\{b,c,d\}$	$\{b,c,d\}$

Section 4.2

1. (a)

x	1	2	3	4	5	6
$(f \circ g)(x)$	1	1	1	2	2	2
$(g \circ f)(x)$	1	1	1	1	1	1

(c)

x	1	2	3	4	5	6	7
$(f \circ g)(x)$	2	2	2	2	2	2	2
$(g \circ f)(x)$	3	3	3	3	3	3	3

(e)

x	1	2	3	4	5
$(g \circ f)(x)$	4	1	4	3	4

y	a	b	c	d	e
$(f \circ g)(y)$	d	a	d	a	a

We give here only the diagrams for part (a). The first is the diagram for $g \circ f$.

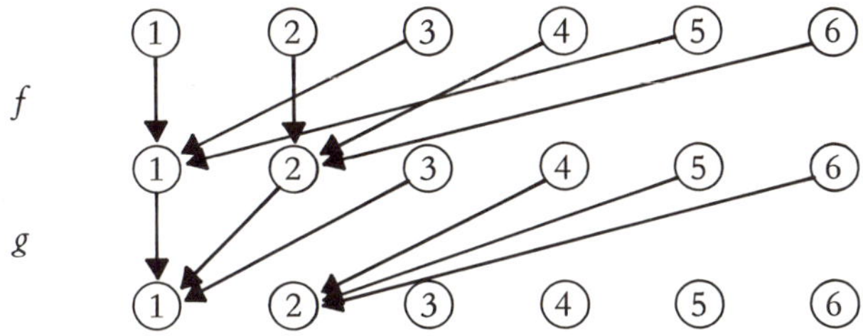

The diagram would also be acceptable with the points for 3, 4, 5, and 6 omitted from the second and third rows. These points are not involved in $g \circ f$.

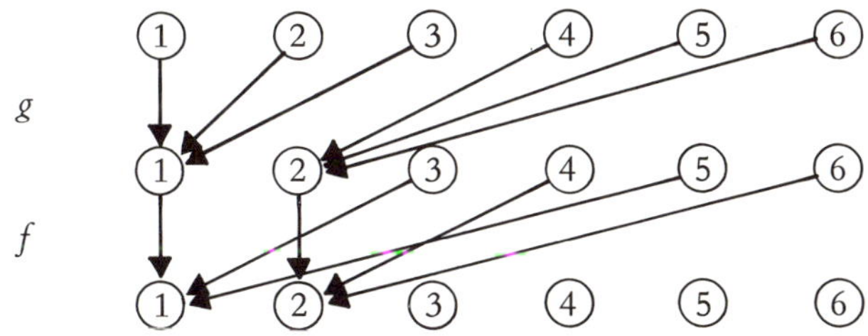

The points for 3, 4, 5, and 6 could again be omitted from the second and third rows.

2. (a)

x	a	b	c	d
$(F \circ F)(x)$	b	c	d	b
$(F \circ F \circ F)(x)$	d	b	c	d

3. $I \circ F = F$ and $F \circ I = F$.

5. (a) 3 (c) 4 (e) 5

6. (a) 4 (c) 6 (e) 2 (g) undefined

8. (a) True. PROOF: If y is in $F(B)$ then $y = F(x)$ for some x in B. But x must be in C also since by hypothesis, $B \subseteq C$. Thus y is in $F(C)$ also, and so $F(B) \subseteq F(C)$.
   (c) True. PROOF: To show $F(B \cup C) = F(B) \cup F(C)$. Let y be any member of $F(B \cup C)$. This means that $y = F(x)$ for some x in $B \cup C$. Thus x is in either B or C. If x is in B then y is in $F(B)$, and if x is in C then y is in $F(C)$. In either case we arrive at the conclusion that y is in $F(B) \cup F(C)$, and we have shown $F(B \cup C) \subseteq F(B) \cup F(C)$.

   Next let y be any member of $F(B) \cup F(C)$. This means that y is in $F(B)$ or $F(C)$. If y is in $F(B)$ then $y = F(x)$ for some x in B, and this x would also be in $B \cup C$, since $B \subseteq B \cup C$. Thus y is in $F(B \cup C)$. Also if y is in $F(C)$ we get, in a like fashion, that y is in $F(B \cup C)$. In either case we see that $F(B) \cup F(C) \subseteq F(B \cup C)$.

   We now have the desired conclusion: namely, $F(B \cup C) = F(B) \cup F(C)$.

9. The following groups of functions are isomorphic (see Problem 2(vi), Section 4.1):
   I: a,1,aa IV: j,o,t

II: *b,g,k,s,u,w* V: *c,d,e,m,p,x*

III: *n,r* VI: *h,i,q,v,y,z*

Section 4.3

1. (a) Neither has an inverse function.
   (c) Neither has an inverse function.
   (e) Neither has an inverse function.

2. (a) R

3. (b) niece or nephew of (d) sibbing of (f) predecessor of
   (h) shorter than (j) lighter than (l) smaller than
   (n) slimmer than (p) slower than (r) warmer than

4. (Answer in tabular form)

Function with an inverse:	A	B
ii—*b,c*	1,2	1,2
iii—*a*	1	1
iii—*b*	1	2
v—*b*	1,2	1,2
v—*c*	1,2	1,3
v—*d*	1,2	1,2
v—*e*	1,2	2,3
v—*f*	1,2	1,3
v—*g*	1,2	2,3
vi—*f,j,n,o,r,t*	1,2,3	1,2,3

6. Same as those for F.

7. (a) None (b) 24 (c) None

9. PROOF: First assume R is reflexive on A. We want to show R^{-1} is also reflexive on A. For any x in A, (x,x) is in R since R is a reflexive relation on A. But (x,x) is also in R^{-1} and so R^{-1} is also reflexive on A.

11. PROOF: Let (x,y) be any element of R^{-1}. Then (y,x) is in R, and thus by hypothesis (y,x) must also be in S, which means (x,y) is in S^{-1}. Hence $R^{-1} \subseteq S^{-1}$.

12. (a) No. The simple relation $\{(0,1)\}$ is not related to itself since it is not its own inverse.
    (c) No. Same counterexample as in a. $\{(0,1)\}I\{(1,0)\}$ and $\{(1,0)\}I\{(0,1)\}$ but not $\{(0,1)\}I\{(0,1)\}$.

14. F is a function whose domain is A and whose range is B. F^{-1} is a function whose domain is B and whose range is A. G is a function with domain B and range C. G^{-1} is a function with domain C and range B. Clearly, $G \circ F$ maps A *onto* C. But also $F^{-1} \circ G^{-1}$ maps C onto A and for any x in A we have

$$\begin{aligned}[(F^{-1} \circ G^{-1}) \circ (G \circ F)](x) &= [(F^{-1} \circ G^{-1}) \circ G](F(x)) \\ &= [F^{-1} \circ (G^{-1} \circ G)](F(x)) \\ &= F^{-1}([G^{-1} \circ G](F(x))) \\ &= F^{-1}(F(x)) \\ &= x.\end{aligned}$$

Hence, $F^{-1} \circ G^{-1}$ is an inverse function for $G \circ F$ by Theorem 4.3.3 and we see that $G \circ F$ is a one-to-one correspondence between A and C.

Section 4.4

1. (i) The possible isomorphisms between (b) and (c) are: $\{(1,2),(2,1),(3,3)\}$ and $\{(1,2),(2,3),(3,1)\}$.
   (iii) The possible isomorphisms between (c) and (d) are $\{(2,3),(1,1),(3,2)\}$ and $\{(2,3),(1,2),(3,1)\}$.
2. The constructed isomorphism is the only possible one. If g is to be an isomorphism between R and S, then clearly we must have $g(1) = d$ since 1 and d are the only points not connected to a different arrow. Also we must have $g(4) = c$ since 4 and c are the only points that have no arrows coming into them. But also we must have $g(5) = b$ since 5 and b are the only points with two arrows coming in and one leaving. Then since an arrow runs from 3 to 5 in R and from e to b in S we must take $g(3) = e$. Finally, to complete the one-to-one correspondence we shall have $g(2) = a$. Thus we are forced to construct the function obtained previously.
4. (a) The possible isomorphisms are: $\{(1,a),(2,b),(3,d),(4,c)\}$ and $\{(1,b),(2,a),(3,d),(4,c)\}$.
   (c) The possible isomorphisms are: $\{(1,b),(2,a),(3,c),(4,d)\}$, $\{(1,b),(2,d),(3,c),(4,a)\}$, and $\{(1,b),(2,c),(3,d),(4,a)\}$.
5. (a) Aside from R itself we have

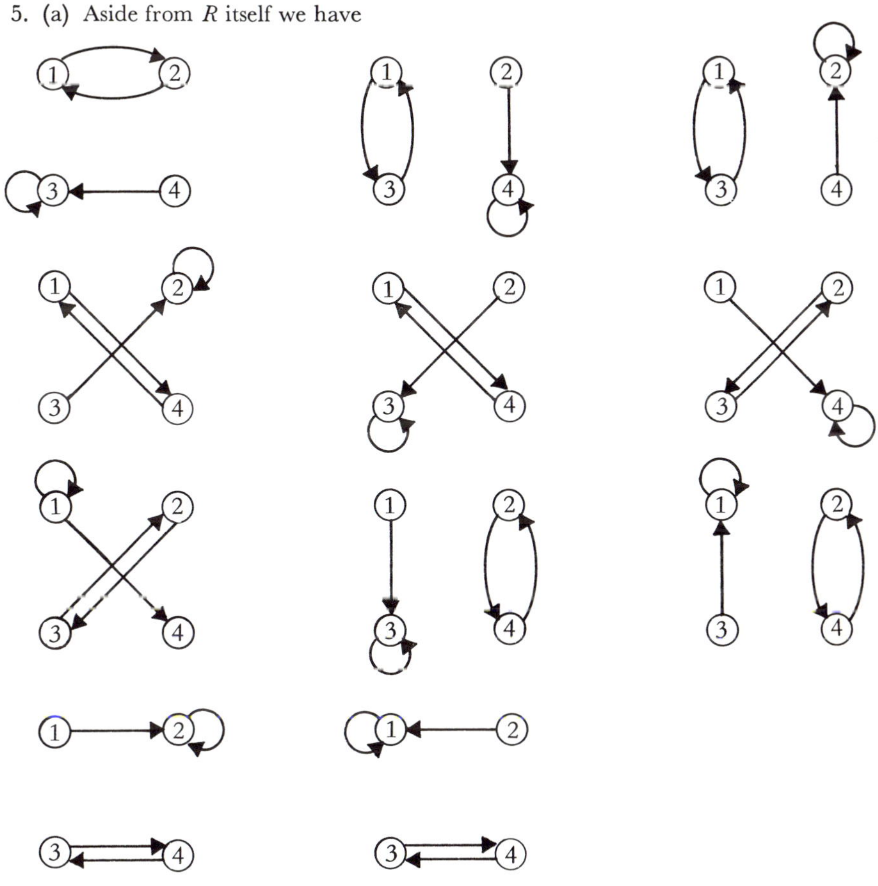

Section 5.1

1. (a) $1 \oplus 2 = 0$ (c) $2 \oplus 1 = 0$
   (e) $2 \odot (1 \oplus 2) = 2 \odot 0 = 0$ (g) $2 \oplus (2 \odot 2) = 2 \oplus 1 = 0$
   (i) $1 \oplus (1 \oplus 0) = 1 \oplus 1 = 0$ (k) $(1 \odot 0) \odot 1 = 0 \odot 1 = 0$
   (m) $1 \odot (1 \oplus 0) = 1 \odot 1 = 1$ (o) $1 \oplus (1 \odot 1) = 1 \oplus 1 = 2.$

2. (a) Yes, $\overset{+}{\sim}$ and $\overset{\cdot}{\sim}$ are both commutative operations. We can verify this by performing the operations on each pair of numbers given in the table, and finding that $x \overset{+}{\sim} y = y \overset{+}{\sim} x$ and $x \overset{\cdot}{\sim} y = y \overset{\cdot}{\sim} x$ for every x and y in $\{0,1,2,3\}$. For example, $0 \overset{+}{\sim} 1 = 1 \overset{+}{\sim} 0 = 1$, $2 \overset{\cdot}{\sim} 3 = 3 \overset{\cdot}{\sim} 2 = 2$, $1 \overset{+}{\sim} 3 = 3 \overset{+}{\sim} 1 = 0$, etc. But we can save time in verifying commutativity by checking to see that the tables are symmetrical about the diagonal from upper left to lower right. It is not difficult to make sure that if the table were flipped about this axis, in both cases zeros would replace zeros, ones would replace ones, and so forth.
   (d) $1 \overset{+}{\sim} (1 \overset{\cdot}{\sim} 2) = 1 \overset{+}{\sim} 2 = 3$ whereas $(1 \overset{+}{\sim} 1) \overset{\cdot}{\sim} (1 \overset{+}{\sim} 2) = 2 \overset{\cdot}{\sim} 2 = 0.$

3. (b) (i) For $\underset{+}{\sim}$ and $\underset{\cdot}{\sim}$, we may set up similar rules, using a two-hour clock with numerals 0 and 1 on its face. Then $x \underset{+}{\sim} y$ can be found by setting the clock at x and rotating it clockwise through y hours. We can define $\underset{\cdot}{\sim}$ by the following: $0 \underset{\cdot}{\sim} x = 0$, $1 \underset{\cdot}{\sim} x = x$, for all x in $\{0,1\}$.
   (ii) We can use a four-hour clock, numbered 0 through 3, to substitute for the $\overset{+}{\sim}$ and $\overset{\cdot}{\sim}$ tables. Again, $x \overset{+}{\sim} y$ is found by setting the clock to x and rotating it clockwise through y hours. In addition, $0 \overset{\cdot}{\sim} x = 0$, $1 \overset{\cdot}{\sim} x = x$, $2 \overset{\cdot}{\sim} x = x \overset{+}{\sim} x$, and $3 \overset{\cdot}{\sim} x = x \overset{+}{\sim} x \overset{+}{\sim} x$.

5. No. If $\oplus$ distributed across $\odot$, then $2 \oplus (2 \odot 2)$ should equal $(2 \oplus 2) \odot (2 \oplus 2)$; and comparing the answers from 1(g) and 1(h), we can see that it does not.

7. (0,0,0), (0,0,1), (0,1,0), (0,1,1), (0,0,2), (0,2,0), (0,1,2), (0,2,1), (0,2,2), (1,0,0), (1,0,1), (1,1,0), (1,1,1), (1,0,2), (1,2,0), (1,1,2), (1,2,1), (1,2,2), (2,0,0), (2,0,1), (2,1,0), (2,1,1), (2,0,2), (2,2,0), (2,1,2), (2,2,1), (2,2,2). The correct conjecture is $4 \times 4 \times 4 = 64$.

8. (b) No. (Compare with 2(a).)
   (d) No. For example, $7 - (3 - 2) = 7 - 1 = 6$, whereas $(7 - 3) - (7 - 2) = 4 - 5$, which is undefined.
   (f) Yes.

9. (a)

C	*a*	*b*	*c*	*d*
a	*aa*	*ab*	*ac*	*ad*
b	*ba*	*bb*	*bc*	*bd*
c	*ca*	*cb*	*cc*	*cd*
d	*da*	*db*	*dc*	*dd*

   (c) No. It doesn't fit the given definition because C is not an operation *in* $\{a,b,c,d\}$. Specifically, the range of C includes elements outside $\{a,b,c,d\}$, such as *aa*, *ba*, etc.

11. (a)

$\underset{+}{\sim}$	ϕ	$\{0\}$	$\{1\}$	$\{0,1\}$
ϕ	ϕ	ϕ	ϕ	ϕ
$\{0\}$	ϕ	$\{0\}$	$\{1\}$	$\{0,1\}$
$\{1\}$	ϕ	$\{1\}$	$\{0\}$	$\{0,1\}$
$\{0,1\}$	ϕ	$\{0,1\}$	$\{0,1\}$	$\{0,1\}$

(c) Yes.

12. (a) Yes.

(c) No, since 0 is in A and there is no x such that $0 \odot x = y$ when $y \in \{1,2\}$.

(e) No, since anything but $\{1,2\} \cap \{1,2\}$ yields results outside the proposed "complete set" of results.

13.
$$\begin{pmatrix}1&0&0\\0&0&0\end{pmatrix} \quad \begin{pmatrix}1&0&0\\1&0&0\end{pmatrix} \quad \begin{pmatrix}1&0&0\\0&1&0\end{pmatrix} \quad \begin{pmatrix}1&0&0\\1&1&0\end{pmatrix}$$
$$\begin{pmatrix}1&0&1\\0&0&0\end{pmatrix} \quad \begin{pmatrix}1&0&1\\1&0&0\end{pmatrix} \quad \begin{pmatrix}1&0&1\\0&1&0\end{pmatrix} \quad \begin{pmatrix}1&0&1\\1&1&0\end{pmatrix}$$
$$\begin{pmatrix}1&1&0\\0&0&0\end{pmatrix} \quad \begin{pmatrix}1&1&0\\1&0&0\end{pmatrix} \quad \begin{pmatrix}1&1&0\\0&1&0\end{pmatrix} \quad \begin{pmatrix}1&1&0\\1&1&0\end{pmatrix}$$
$$\begin{pmatrix}1&1&1\\0&0&0\end{pmatrix} \quad \begin{pmatrix}1&1&1\\1&0&0\end{pmatrix} \quad \begin{pmatrix}1&1&1\\0&1&0\end{pmatrix} \quad \begin{pmatrix}1&1&1\\1&1&0\end{pmatrix}$$

Section 5.2

1.

	a (identity)	b (inverses)
$\overset{+}{\sim}$	0	$0^{-1} = 1$, $1^{-1} = 3$, $2^{-1} = 2$, $3^{-1} = 1$
$\overset{\cdot}{\sim}$	1	$1^{-1} = 3$, $3^{-1} = 1$, 0 and 2 have no inverses
$\underset{+}{\sim}$	0	$0^{-1} = 0$, $1^{-1} = 1$
$\underset{\cdot}{\sim}$	1	$1^{-1} = 1$, but 0 has no inverse
$\dot{\cap}$	$\{1,2\}$	Only $\{1,2\}$ has an inverse: $\{1,2\}^{-1} = \{1,2\}$.

3. The empty symbol—that is, a blank.

5. M has 0 as its identity element, and only 0 has an inverse: $0^{-1} = 0$. This is because M gives us nothing to operate upon 1 with to get back the identity element 0.

7. As suggested, consider

$(aFb)F[(b^{-1})F(a^{-1})] = [aF(bFb^{-1})]F(a^{-1})$	Because F is associative
$= (aFe)F(a^{-1})$	By Definition 5.2.2, $bFb^{-1} = e$
$= (a)F(a^{-1})$	By Definition 5.2.1, $aFe = a$
$= e$	By Definition 5.2.2.

Now consider

$[(b^{-1})F(a^{-1})]F(aFb) = (b^{-1})F[(a^{-1})F(aFb)]$	By associativity of F
$= (b^{-1})F[(a^{-1}Fa)Fb]$	By associativity of F
$= (b^{-1})F(eFb)$	By Definition 5.2.2, $a^{-1}Fa = e$
$= (b^{-1})F(b)$	By Definition 5.2.1, $eFb = b$
$= e$	By Definition 5.2.2.

We now have shown that:

$(aFb)F[(b^{-1})F(a^{-1})] = [(b^{-1})F(a^{-1})]F(aFb) = e$. Thus by Definition 5.2.2, $(aFb)^{-1} = (b^{-1})F(a^{-1})$.

8. J has identity element a; $a^{-1} = a$ and $b^{-1} = b$.
K has identity element a; $a^{-1} = a$, $b^{-1} = c$, and $c^{-1} = b$.
L has identity element b; $b^{-1} = b$, $c^{-1} = c$, and $d^{-1} = d$. a has no inverse.
M has identity element a; $a^{-1} = a$, $b^{-1} = b$, $c^{-1} = c$, and $d^{-1} = d$.
N has identity element c; $c^{-1} = c$, and a, b, and d have no inverses.
V has identity element f; $f^{-1} = f$, and t has no inverse.
$\longrightarrow$ has no identity element, and hence neither t nor f has an inverse.
⊞ has identity element even; $(\text{odd})^{-1}$ = odd and $(\text{even})^{-1}$ = even.
⊡ has identity element odd; $(\text{odd})^{-1}$ = odd and even has no inverse.

9. (a) Identity elements exist in operations ⓔ, ⓗ, ⓘ, ⓛ.
(c) Remember that we can check for commutativity by looking to see if the operation table is symmetrical about the diagonal drawn from the upper left to lower right corners. In this way, we easily find that the operations that are commutative are: ⓐ, ⓑ, ⓔ, ⓗ, ⓘ, ⓛ, ⓞ, and ⓟ.

10. The associative operations are ⓐ, ⓔ, ⓗ, ⓘ, ⓙ, ⓚ, ⓛ, and ⓟ.

11. [NOTE: Names for the operations are arbitrary, and thus the names of our answers may differ without either of us, necessarily, being wrong. We should, however, have the same nine operations no matter what their names.]

ⓐ	1	2	3
1	1	2	3
2	2	1	1
3	3	1	1

ⓑ	1	2	3
1	1	2	3
2	2	1	2
3	3	2	1

ⓒ	1	2	3
1	1	2	3
2	2	1	3
3	3	3	1

ⓓ	1	2	3
1	1	2	3
2	2	1	2
3	3	1	1

ⓔ	1	2	3
1	1	2	3
2	2	1	1
3	3	2	1

ⓕ	1	2	3
1	1	2	3
2	2	1	3
3	3	1	1

ⓖ	1	2	3
1	1	2	3
2	2	1	1
3	3	3	1

ⓗ	1	2	3
1	1	2	3
2	2	1	3
3	3	2	1

ⓘ	1	2	3
1	1	2	3
2	2	1	2
3	3	3	1

(a) The commutative operations are ⓐ, ⓑ, ⓒ.

13. $(3W2)W2 = 1W2 = 2$, while $3W(2W2) = 3W1 = 3$. Since $2 \neq 3$, $(3W2)W2 \neq 3W(2W2)$ and W is not associative.

15. (a) 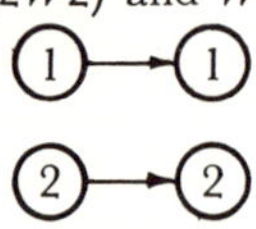
and 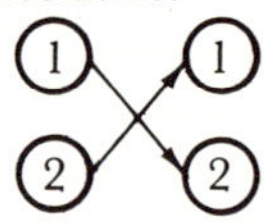
are the only mappings of $\{1,2\}$ *onto* $\{1,2\}$.

	Dave	Nancy
Dave	Dave	Nancy
Nancy	Nancy	Dave

Dave is again the identity element; and as before, $(\text{Dave})^{-1}$ = Dave and $(\text{Nancy})^{-1}$ = Nancy.

16. (a) $\{1,2\}$ can be made a semigroup 8 ways, as many as there are associative operations. (The semigroups are a, e, h, i, j, k, l, and p.)
(c) Yes.
(e) Yes. t and f are different names to call 1 and 2, respectively, which makes V the same operation as ⓔ in Problem 5.2.9. And as we found in Problem 16(a)

above, $\{1,2\}$ is a semigroup with respect to $\textcircled{e}$; hence, $\{\iota, f\}$ is a semigroup with respect to V.

17. (a) We can prove that an identity element is not necessary in A, where A is a semigroup with respect to F, by giving one counterexample. One possible choice for our counterexample is:

F	1	2
1	1	1
2	1	1 .

This A is closed under F and F is associative; hence it is a semigroup. But it has no identity element.

(c) (i) First let us assume $z = xFy$. Then $x^{-1}Fz = x^{-1}F(xFy)$. (We know that both sides of this new equation are in A because A is a semigroup with respect to F and is therefore closed under it.) Then, since F is associative, $x^{-1}Fz = (x^{-1}Fx)Fy = eFy = y$. Now to prove the "if" portion of the problem, assume $x^{-1}Fz = y$. Then $xF(x^{-1}Fz) = xFy$, and by the associativity of F, $(xFx^{-1})Fz = xFy$. By definition of inverse, $eFz = xFy$ and hence $z = xFy$. Taking the two parts of the proof together, $z = xFy$ if and only if $x^{-1}Fz = y$.

(ii) First, assume that $w = yFx$. Then

$$\begin{aligned} wFx^{-1} &= (yFx)Fx^{-1} & & \\ wFx^{-1} &= yF(xFx^{-1}) & &\text{Because } F \text{ is associative} \\ wFx^{-1} &= yFe & &\text{By definition of inverse} \\ wFx^{-1} &= y & &\text{By definition of identity.} \end{aligned}$$

Taking the proof in the other direction, assume that $wFx^{-1} = y$. Then

$$\begin{aligned} (wFx^{-1})Fx &= yFx \\ wF(x^{-1}Fx) &= yFx \\ wFe &= yFx \\ w &= yFx. \end{aligned}$$

Section 5.3

1. (b)

	B^{-1}	A	B	C
		$\overbrace{\qquad}^{y}$		
z	A	–	–	B
	B	–	B	C
	C	B	–	A

(c)

	$\odot^{-1}$	0	1	2
		$\overbrace{\qquad}^{y}$		
z	0	–	0	0
	1	–	1	2
	2	–	2	1

(e)

$\odot^{-1}$	1	2	3	4	5	6	7	8	9	10
1	1	–	–	–	–	–	–	–	–	–
2	2	1	–	–	–	–	–	–	–	–
3	3	–	1	–	–	–	–	–	–	–
4	4	2	–	1	–	–	–	–	–	–
5	5	–	–	–	1	–	–	–	–	–
6	6	3	2	–	–	1	–	–	–	–
7	7						1	–	–	–
8	8	4	–	2	–	–	–	1	–	–
9	9	–	3	–	–	–	–	–	1	
10	10	5	–	–	1	–	–	–	–	–

(g)

$\overset{+}{\sim}^{-1}$	0	1	2	3
0	0	3	2	1
1	1	0	3	2
2	2	1	0	3
3	3	2	1	0

(i)

w^{-1}	1	2	3
1	1	–	–
2	2	1	–
3	3	–	1

2. (a)

$\overset{+}{\sim}^{-1}$	0	1	2	3
0	0	3	2	1
1	1	0	3	2
2	2	1	0	3
3	3	2	1	0

3. (a) $x \oplus 1 = 2 \rightarrow x = 1$ (c) $x \overset{+}{\sim} 1 = 3 \rightarrow x = 2$
   (e) $3 \overset{+}{\sim} x = 2 \rightarrow x = 3$ (g) $x \sim 2 = 0 \rightarrow x = 0$ or 2
   (i) $x \overset{\cdot}{\sim} 3 = 1 \rightarrow x = 3$.

4. 

Consider a three-hour clock, numbered clockwise 0, 1, 2. We may think of — in this way: Given a problem $x - y$, we start our clock on the x-hour, and then move the hand y hours *counter-clockwise*. This, in effect, undoes the clockwise movements we worked out in 5.1.3 to explain the $\oplus$ operation.

6. No. Take as a counter example Problem 1b: A's inverse is B, but B's inverse operation is quite different from A.

A	A	B	C
A	C	A	B
B	C	B	C
C	B	A	A

B	A	B	C
A	–	–	C
B	C	B	A
C	–	–	B

B^{-1}	A	B	C
A	–	–	B
B	–	B	C
C	B	–	A

Section 5.4

1. ⓐ is isomorphic to ⓟ, ⓑ to ⓞ, ⓒ to ⓝ, ⓓ to ⓜ, ⓔ to ⓛ, ⓕ to ⓚ, ⓖ to ⓙ, and ⓗ to ⓘ, all by the isomorphism $f(1) = 2$ and $f(2) = 1$.

3. There is no way for us to set up an isomorphism between the two sets. $\{0,1,2\}$ has three elements while $\{\phi,\{1\},\{2\},\{1,2\}\}$ has four elements. There is no way to pair these in one-to-one correspondence.

5. The isomorphism f, where $f(0) = T$ and $f(1) = F$ makes $(\{0,1\},\sim)$ and $(\{T,F\},V)$ isomorphic. And the isomorphism g, such that $g(0) =$ even and $g(1) =$ odd, makes $(\{0,1\},\sim)$ and $(\{\text{odd, even}\},\boxdot)$ isomorphic. Thus all three systems are isomorphic to each other by Theorem 5.4.3.

6. (a) For any x, y, z in A, $(xFy)Fz = xF(yFz)$ because F is associative on A. Applying the isomorphism f between (A,F) and (B,G), we get:

$$[f(x)Gf(y)]Gf(z) = f(x)G[f(y)Gf(z)].$$

Now, since f is an isomorphism, we know that $f(x)$, $f(y)$, and $f(z)$ are all in B. And since f is one to one $f(x)$, $f(y)$, and $f(z)$ can be any elements in B. Therefore, G is associative on B, according to the displayed equation.

Let y and w be elements of B. Then there are elements x and y in A such that $z = f(x)$ and $w = f(y)$. Let $xFy = q$. Then q is in A since A is closed under F. We have $zGw = f(x)Gf(y) = f(xFy) = f(q)$. But $f(q)$ is in B. So B is closed under G.

(e) Each element x in A has an inverse element x^{-1} for F, so we know that $xFx^{-1} = x^{-1}Fx = e$ for every x in A. Applying our isomorphism f, we get: $f(x)Gf(x^{-1}) = f(x^{-1})Gf(x) = f(e)$ for every x in A. And we already know from part d that when e is an identity element for F on A, $f(e)$ is an identity element for G on B. So we have found that each $f(x)$ in B has an inverse element for G, namely $f(x^{-1})$, where x^{-1} is the inverse of x in A for F.

8. (Theorem 5.4.6) If F distributes across G in A, then, for all x, y, and z in A: $xF(yGz) = (xFy)G(xFz)$. Now, applying our isomorphism f, $f(x)H[f(y)Jf(z)] = [f(x)Hf(y)]J[f(x)Hf(z)]$, for all $f(x), f(y)$, and $f(z)$ in B. Thus, H distributes across J in B.

Section 6.1

1. $2 = \#(\{a,b\})$. $\#(\{a,b\}) = \#(\{x,y\})$ since if we define f by $f(a) = x$, and $f(b) = z$, f is a one-to-one correspondence between the two sets $\{a,b\}$ and $\{x,z\}$. Thus, since $2 = \#(\{a,b\})$, and $\#(\{a,b\}) = \#(\{x,z\})$ then $2 = \#(\{x,z\})$ by the transitive property of equality.

3. Suppose $0 = 1$ were true. Then by transitivity $\#(\phi) = \#(\{a\})$ would be true, which would imply that there was a one-to-one correspondence between the two sets ϕ and $\{a\}$. But such a correspondence could not possibly exist because there would never be an x in ϕ for which $f(x) = a$ since ϕ contains no members. Thus $0 = 1$ is impossible.

5. By definition, $8 = \#(\{a,b,c,d,e,f,g,h\})$. Consider the following function: $f(a) = 1$, $f(b) = 2, f(c) = 3, f(d) = 4, f(e) = 5, f(f) = 6, f(g) = 7, f(h) = 8$. Note that f is a one-to-one correspondence between the two sets $\{a,b,c,d,e,f,g,h\}$ and $\{1,2,3,4,5,6,7,8\}$. Therefore by Definition 6.1.1 $\#(\{a,b,c,d,e,f,g,h\}) = \#(\{1,2,3,4,5,6,7,8\})$ and so by transitivity $8 = \#(\{1,2,3,4,5,6,7,8\})$.

7. $f(a) = a$, $f(b) = b$, $f(c) = 1$, $f(d) = 2$, $f(e) = 3$, $f(f) = 4$ defines a one-to-one correspondence between the set $\{a,b,c,d,e,f\}$ and the set $(\{a,b\} \cup \{1,2,3,4\})$ which means they have the same cardinal number, which must be 6 by Definition 6.1.3.

9. The solution here is analogous to, if not a little more long-winded than, that to Problem 8. You could let f be the following one-to-one correspondence:

$(1,1) \to a$ $\quad (1,2) \to b$ $\quad (1,3) \to c$ $\quad (2,1) \to d$ $\quad (2,2) \to e$
$(2,3) \to f$ $\quad (3,1) \to g$ $\quad (3,2) \to h$ $\quad (3,3) \to i$.

Again, consult Definition 6.1.3.

Section 6.2

1. (a) Let $A = \{a,b,c\}$ and $B = \{d,e,f,g\}$. Then $\#(A) = 3$ and $\#(B) = 4$ and $A \cap B = \phi$. Thus $3 + 4 = \#(A \cup B) = \#(\{a,b,c,d,e,f,g\}) = 7$, by definition.

(c) Let $A = \{a,b\}$ so that $\#(A) = 2$. Let $B = \{c,d,e,f,g\}$ so that $\#(B) = 5$ and $A \cap B = \phi$. Thus $2 + 5 = \#(A \cup B) = \#(\{a,b,c,d,e,f,g\}) = 7$.

2. (i) Let $A = \phi$ and $B = \{a\}$. Then $\#(A) = 0$ and $\#(B) = 1$ and $A \cap B = \phi$. Thus $0 + 1 = \#(A \cup B)$ [Definition 6.2.1] $= \#(\{a\}) = 1$.
   (iii) Let $A = \{a,b\}$ and $B = \{c\}$ so that $\#(A) = 2$ and $\#(B) = 1$ and $A \cap B = \phi$. Thus $2 + 1 = \#(A \cup B) = \#(\{a,b,c\}) = 3$.
3. (a) 2 (c) 2
4. Disproof: Let $m = 0$, $n = 1$, $p = 1$, and $q = 0$. $0 + 1 = 1 = 1 + 0$ by Theorem 6.2.2. But $0 \neq 1$ and $1 \neq 0$ by Problem 3, Section 6.1.
6. $5 = \#(\{a,b,c,d,e\})$. $7 = \#(\{1,2,3,4,5,6,7\})$.
   (a) $5 + 7 = \#(\{a,b,c,d,e,1,2,3,4,5,6,7\})$.
   (c) $9 = \#(\{a,b,c,d,e,f,g,h,i\})$, and $4 = \#(\{1,2,3,4\})$. Thus $9 + 4 = \#(\{a,b,c,d,e,f,g,h,i,1,2,3,4\})$.

Section 6.3

1. (a) $A = B = \{a,b,c\}$ so that $A \times B = \{(a,a),(a,b),(a,c),(b,a),(b,b),(b,c),(c,a),(c,b),(c,c)\}$. $\#(A) = \#(B) = 3$. $3\cdot 3 = \#(A \times B) = 9$ since the following is a one-to-one correspondence between $A \times B$ and $\{a,b,c,d,e,f,g,h,i\}$ (that set whose cardinal number is 9 by definition):

   $(a,a) \leftrightarrow a$ $(a,b) \leftrightarrow b$ $(a,c) \leftrightarrow c$ $(b,a) \leftrightarrow d$ $(b,b) \leftrightarrow e$
   $(b,c) \leftrightarrow f$ $(c,a) \leftrightarrow g$ $(c,b) \leftrightarrow h$ $(c,c) \leftrightarrow i$

   (c) $\#(A) = 2$ if $A = \{a,b\}$.
   $\#(B) = 4$ if $B = \{a,b,c,d\}$.
   Thus $2\cdot 4 = \#(A \times B) = \#(\{(a,a),(a,b),(a,c),(a,d),(b,a),(b,b),(b,c),(b,d)\}) = \#(\{a,b,c,d,e,f,g,h\}) = 8$ as evidenced by this one-to-one correspondence:

   $(a,a) \leftrightarrow a$ $(a,b) \leftrightarrow b$ $(a,c) \leftrightarrow c$ $(a,d) \leftrightarrow d$
   $(b,a) \leftrightarrow e$ $(b,b) \leftrightarrow f$ $(b,c) \leftrightarrow g$ $(b,d) \leftrightarrow h$

3. To prove: If $n\cdot m = 0$, then $n = 0$ or $m = 0$.
   PROOF: Suppose $\#(A) = n$ and $\#(B) = m$ and $n\cdot m = \#(A \times B) = 0$. Then $A \times B = \phi$ in which case either A must be ϕ or B must be ϕ (or both).
5. $\#(\{a,b\}) = 2$ so $\#(\{c,d\}) = 2$ and $\#(\{e,f\}) = 2$ by the following correspondences:

   $a \leftrightarrow c$ (This shows $a \leftrightarrow e$ (This shows
   $b \leftrightarrow d$ $\#(\{c,d\}) = 2$.) $b \leftrightarrow f$ $\#(\{e,f\}) = 2$.)

   Thus $\#(\{a,b\}) + \#(\{c,d\}) + \#(\{e,f\}) = 2 + 2 + 2 = (2+ 2) + 2 = 4 + 2 = 6$. We must show $6 = \#(\{a,b\})\cdot 3$. $\#(A) = 3$ if $A = \{a,b,c\}$. Thus since $\#(\{a,b\}) = 2$ then $\#(\{a,b\})\cdot 3 = 2\cdot 3 = 6$ (see table in Section 6.3).
6. (b) Let a be the cardinal number of the first group, and b be the cardinal number of the second group; we will then have $a + b$ people at the party.
   (d) Let p = number of people. We have $p\cdot 10$ fingers.
   (f) Let a = number of water glasses,
   b = number of juice glasses, and
   c = number of ounces in each glass.
   We will need $c(a + b)$ ounces of beverage.
   (h) Let a = number of workers with smaller salary,
   b = number of workers with larger salary,
   c = number of dollars in smaller salary, and
   d = number of dollars in larger salary.
   The payroll amounts to $(a\cdot c) + (b\cdot d)$.

(j) Let a = number of shares of first stock,
b = number of dollars first share is worth,
c = number of shares of second stock, and
d = number of dollars second share is worth.
Total assets = $(a \cdot b) + (c \cdot d)$ dollars.

(l) Let n = number of shelves, and
p = number of copies on each shelf.
$n \cdot p$ = total number of copies.

Section 6.4

1. $0 + 1 = 1$ so $1 \in N$ by Definition 6.4.1 (ii) and since $0 \in N$ by (i). $1 + 1 = 2$ so $2 \in N$ by Definition 6.4.1. Similarly, $2 + 1 = 3$ implies $3 \in N$, $3 + 1 = 4$ implies $4 \in N$, $4 + 1 = 5$ implies $5 \in N$, $5 + 1 = 6$ implies $6 \in N$, $6 + 1 = 7$ implies $7 \in N$, $7 + 1 = 8$ implies $8 \in N$, $8 + 1 = 9$ implies $9 \in N$, $9 + 1 = 10$ implies $10 \in N$.

3. $B = \{k, k + 1, k + 2, k + 3, \ldots\}$ where you let k be any natural number other than 0. The set ϕ will also work.

5. Domain = N. Range = $\{x \mid x = N \text{ and } x \neq 0\} = B$.
Yes. S is a one-to-one correspondence between N and B.
PROOF: Onto: For any $x \in B$ there is a $y \in N$ such that $y + 1 = S(y) = x$, by Theorem 6.4.7.
One-to-one: If $S(x) = x + 1 = y + 1 = S(y)$, then $x = y$ by Theorem 6.4.4.; so S is a one-to-one correspondence.

7. If $0 \in A$ and the successor of k is in A whenever k is in A, then $A = N$.

9. $\{0,2,4,6,8, \ldots\} = \{x \mid x = 2k \text{ for some } k \in N\}$ is contained in A. But we may have $A \neq N$.

11. $\{0,1,2,3,4,5,6,7,8,9,10\}$ is contained in A.

13. No. As a counterexample, the empty set suffices. We may conclude that either $A = \phi$ or $A = N$.

15. If $A = \phi$ then the answer to the question, "Is S empty?", is yes, in which case the machine asserts that $\#(A) = 0$; the correct response. If $A = \{a\}$ (or in general contains one element), then the first time through the decision box the answer is "No" while the second time through the answer is "Yes." By this time the machine says $\#(A) = n = 1$ (since n is 1 by this point).

In general, for a finite set of m elements, the decision box will be answered no for the first m times through. Each time through the machine deletes an element from A. After the mth time through, the machine deletes the final element from A leaving A (or S) empty. At this point $n = m$ and the question answers "Yes" concluding correctly that $\#(A) = m = n$.

Section 6.5

1. (a) 10 (c) $8x + 5y$ (e) $8a + 4$ (g) 10 (i) $8w + 4a$

3. $$\begin{aligned} x + x + x + x + x &= (1 \cdot x + 1 \cdot x + 1 \cdot x + 1 \cdot x + 1 \cdot x) \\ &= (1 + 1 + 1 + 1 + 1)x \\ &= 5x \end{aligned}$$

4. (a) Disproof: $0 \cdot 2 = 0$ but $2 \neq 1$.

5. $(x + y)(z + w) = (x + y)(z) + (x + y)(w)$ [Dist.; Theorem 6.5.1]
$= x \cdot z + y \cdot z + x \cdot w + y \cdot w$ [Dist.; Theorem 6.5.2]

8.

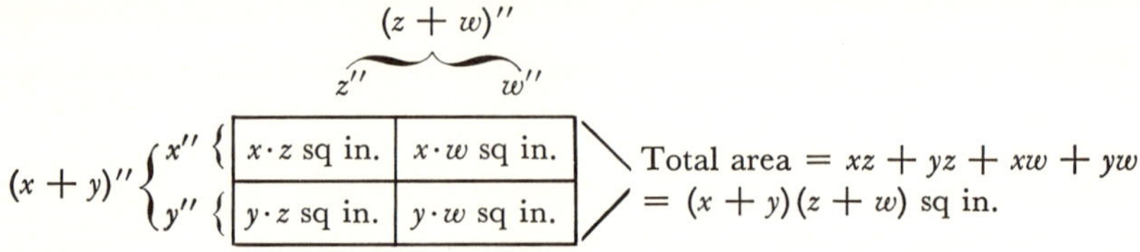

Section 6.6

1.

E	0	1	2	3	4	5
0	–	0	0	0	0	0
1	1	1	1	1	1	1
2	1	2	4	8	16	32
3	1	3	9	27	81	243
4	1	4	16	64	256	1024
5	1	5	25	125	625	3125

3. Range of $T = \{0,2,4,8,16,32, \ldots\}$. Yes.

5.

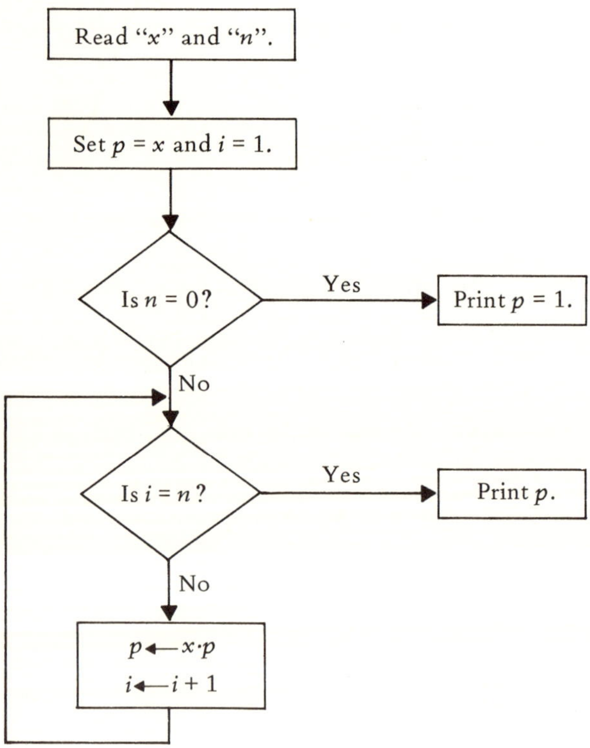

Section 6.7

1. (a) one hundred ninety-five
   (c) fifty-two thousand, seven hundred twenty-three
   (e) eight billion, nine hundred sixty-four million, thirty-two

3. Put a stack of ten cubes in each box on the table.

5. (a) Ambiguities would result. For example, 111 might be interpreted as $1 \cdot 10^2 + 1 \cdot 10 + 1$ or $(11) \cdot 10 + 1$ or $1 \cdot 10 + (11)$ which (in the decimal system) results in the three distinct numbers 111, 21, or 12.
   (c) Replacing a "blank space" for zero might result in this type of difficulty: Does 1 1 stand for eleven, one hundred one, or one hundred ten, or something else?

Section 6.8

1. (a) 17 (c) 156 (e) 1885

2. (a) 64 (c) 1727 (one less than 1000_{12} which is 12^3 or 1728)
   (e) 17,087 (g) 4432

3. (a) 1000000000 (c) 1000000000_{12}

4. If either number is zero the problem is trivial. If both numbers are nonzero, you receive the numeral for the sum by juxtaposing the numerals for the numbers added (e.g. 1111 + 111 = 1111111).

5. (a) 1001010_2 (c) 1110101001000_2

6. (a) 2201_3 (c) 101021122_3

7. (a) 62_{12} (c) $48t5_{12}$

9.

+	0	1	10	11	100
0	0	1	10	11	100
1	1	10	11	100	101
10	10	11	100	101	110
11	11	100	101	110	111
100	100	101	110	111	1000

·	0	1	10	11	100
0	0	0	0	0	0
1	0	1	10	11	100
10	0	10	100	110	1000
11	0	11	110	1001	1100
100	0	100	1000	1100	10000

11.

+	0	1	11	111
0	0	1	11	111
1	1	11	111	1111
11	11	111	1111	11111
111	111	1111	11111	111111

·	0	1	11	111
0	0	0	0	0
1	0	1	11	111
11	0	11	1111	111111
111	0	111	111111	111111111

13. The smallest base that would use a digit 3 (in the conventional notation) would be base 4 in which case you get the smallest possible value of the numeral "113"—namely 23_{10}.

    There is no largest such number because the numeral stands for a larger number in base $n + 1$ than it stood for in base n.

Section 6.9

1. (a) T (c) T (e) T (g) F (i) F (k) T (m) T

2. (a) $x = 0 + x$ so $x \geq 0$.
   (c) $x \geq y$ and $y \geq z$ implies there are natural numbers n and m such that $x = y + n$ and $y = z + m$ so $x = (z + m) + n = z + (m + n)$ and so $x \geq z$ since $m + n$ is also in N.
   (e) This is done in the appendix to this section.
   (g) $x \geq y$ implies that for some $n \in N$, $x = y + n$. This implies $x + z = (y + n) + z = y + (n + z) = y + (z + n) = (y + z) + n$. This implies $x + z \geq y + z$. $x + z \geq y + z$ implies for some $n \in N$ $x + z = (y + z) + n = (y + n) + z$, so that $x = y + n$ and finally $x \geq y$.
   (i) $x \geq y$ and $z \geq w$ implies that for some n and m in N $x = y + n$ and $z = w + n$. So $x + z = (y + n) + (w + m) = (y + w) + (m + n)$. This implies $x + z \geq y + w$ since $m + n \in N$.
   (k) $x \geq y$ implies $x = y + n$ for some $n \in N$. Then $xz = (y + n)z = yz + nz$. This implies $xz \geq yz$ since nz is in N.
   (m) $x \geq y$ implies $xy \geq y \cdot y = y^2$ and $x \geq y$ implies $x \cdot x \geq xy$. Thus $x^2 \geq xy \geq y^2$ and this implies (by Theorem 6.9.3) that $x^2 \geq y^2$.

3. (a) False. Let $x = 0$, $y = 1$, $z = 2$, and $w = 0$. Then $x < y < z$. But $xw = yw = zw = 0$.
   (c) False. Let $x = 0$, $y = 0$, $z = 0$, and $w = 1$. Then $x \leq y$ and $z < w$ but $xz = yw = 0$.
   (e) True. $x \leq y$ and $y \leq z$ implies $x \leq z$ (Theorem 6.9.3). Then $z \leq x$ implies $x = z$ (Theorem 6.9.6). Thus $x \leq y$ and $y \leq z$ implies $z \leq y$ and $y \leq z$. Then $y = z$, so $y = z = x$.

4. Yes. Let $A = \{6,7,8,9,\ldots\}$. $f : N \rightarrow A$ defined by $f(n) = n + 6$ is a one-to-one correspondence between A and N. If $n \geq m$ $(n,m \in N)$ then $f(n) = n + 6 \geq m + 6 = f(m)$ by Theorem 6.9.9.

6. No. Suppose f was a one-to-one correspondence from N to N that was an isomorphism. Then since $0 < x + 1$ for any x in N we must have $f(0) > f(x + 1)$ for any x in N which means there is a natural number that is greater than any other natural number, which contradicts Theorem 6.9.8.

7. (a) Follows from Theorem 6.9.20 except when $n = m = 0$ in which case $10^n = 10^0 = 1 \geq 1 = 10^0 = 10^m$.
   (c) Follows from Theorem 6.9.16.

8. This flow chart assumes the machine can decide if one of two numbers named by single digits is bigger than or equal to another. If $x = 10$ this chart works for decimal notation. If $x = 2$ this chart works for binary notation. (See page 531.)

Section 6.10

2. (a) (1,9): (0,8), (2,10), (3,11), (4,12), (5,13), (6,14), (7,15), (8,16), . . .
   (c) (5795,5798): (0,3), (1,4), (2,5), (3,6), (4,7), (5,8), (6,9), . . .

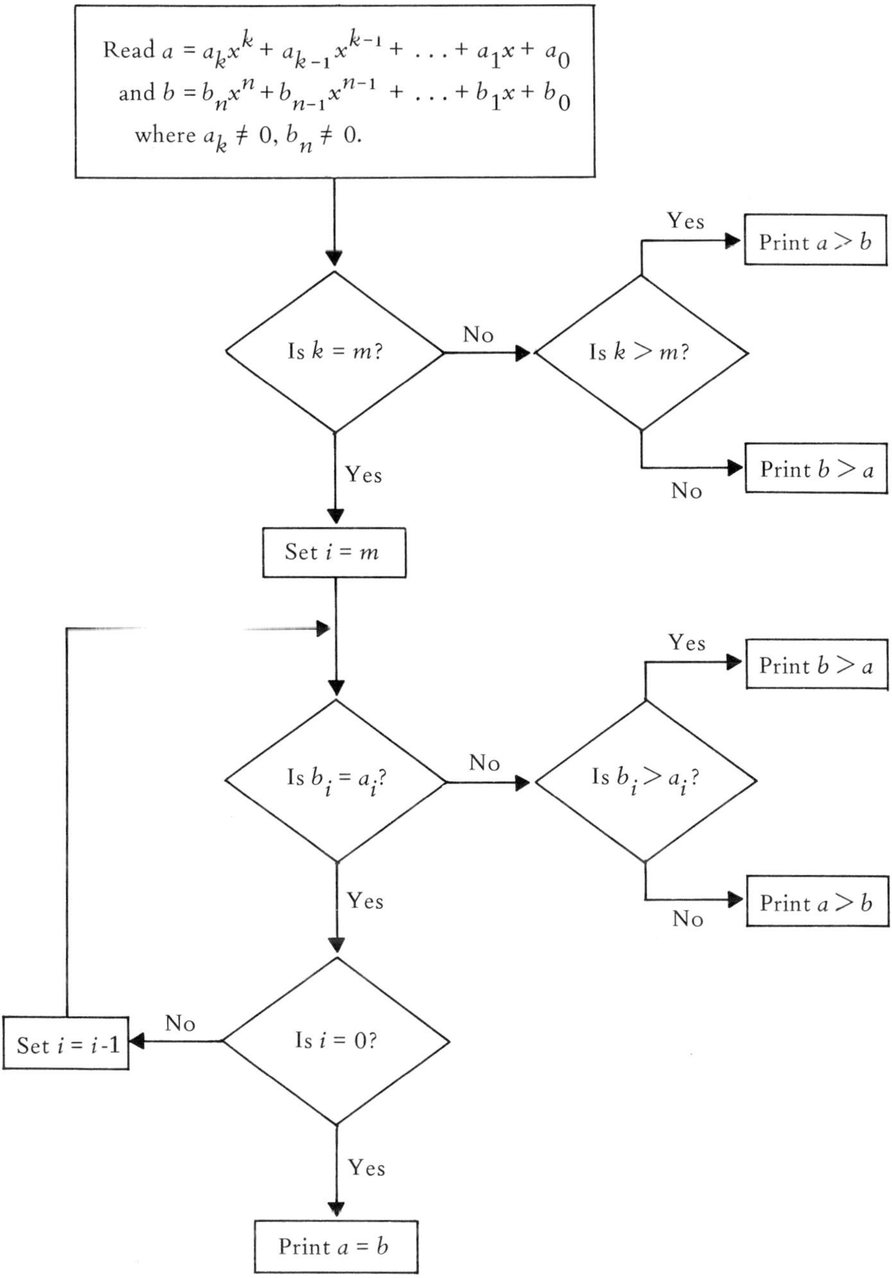

Flow chart for Problem 8, Section 6.9

4.

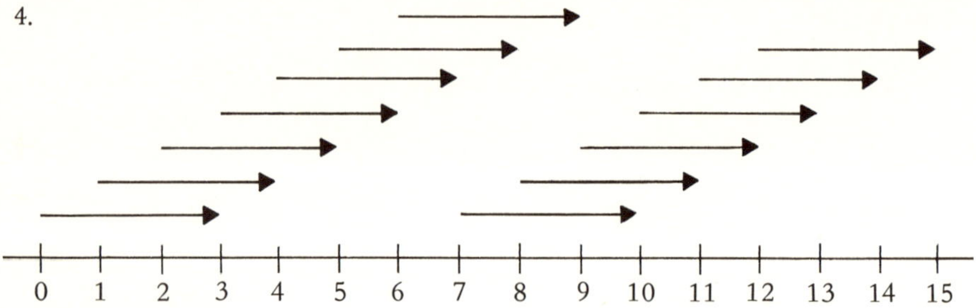

Section 6.11

3. We could add the straight lines passing through the origin and the points which are one unit to the right of the vertical axis and 7 and 8 units above the horizontal axis.
5. (a) 42 sq. in., (b) 48 sq. in., (c) 36 sq. in.

Section 6.12

3. (a), (b), (c), (e), (g), (i), (j), (m), (n), (o), (q), (s), and (t) are true. The rest are false.

Section 6.13

8. Write the tally name for the smaller number below that for the larger. Starting at the right, erase one tally from each name. Repeat this process until all the tallies in the shorter name are erased. What remains is the tally name for the difference.

Section 6.14

2. (a) $\{6\}$, (b) $\{4\}$, (c) $\{5\}$, (d) ϕ, (e) $\{2\}$, (f) ϕ, (g) ϕ, (h) N.
4. Suppose $y - x \leq z - x$ in N. Then $(y - x) + x \leq (z - x) + x$. But $(y - x) + x = y$ and $(z - x) + x = z$. So we have $y \leq z$.
5. Suppose that $y \leq z$. Then $z - y$ is in N. Hence, by the result in Problem 3, $x + y \leq z$ implies $(x + y) - y \leq z - y$. But this last gives $x \leq z - y$. On the other hand, if $x \leq z - y$, then $x + y \leq (z - y) + y = z$.
7. No. Although we have $x - 0 = x$ for all x in N, we do not have $0 - x = x$ for any x except $x = 0$.
9. Suppose $y \leq x$. Then there is an integer n such that $x = y + n$. We then have $n = x - y$ and, hence, $(x + y)(x - y) = (x + y)n = xn + yn = x(x - y) + y(x - y) = (x^2 - xy) + (yx - y^2) = [(x^2 - xy) + xy] - y^2 = x^2 - y^2$ where we have used Theorem 6.14.6 and Theorem 6.14.2.

Section 6.15

2. (a) 3, (c) 24, (e) 72, (g) 6, (i) 12, (k) 1, (m) 3, (o) 5, (q) 49, (s) 7, (u) 3, (w) 2, (y) undefined, (aa) 11.
4. (a) Let n be the number of people in the club and A be the amount donated. Then each person donated $A \div n$ dollars.
   (c) Let C be the number of coconuts and let R of them be rotten. If there are n people, each person will get $(C - R) \div n$ good coconuts.

(e) If there were F fighters then each of them cost $100{,}000{,}000 \div F$ dollars.
(g) Let x be the number. Then $(x + 5) \div 7 = 6$.
(i) Let A be Joan's age, and B be Joy's age. Then $A = 2B - 3$. It follows that $B = (A + 3) \div 2$.
(k) If A is the area of the rectangle and its length is L, then its width is $A \div L$.
(m) If Q is the number of quarts, then $Q \div 4$ is the number of gallons.
(o) If W is the number of ounces, then $W \div 16$ is the number of pounds.
(q) If D is the distance to Arcturus and S is the speed, the time for the trip will be $D \div S$.

6. (a) 1, (c) 125, (e) 100.

7. (a) 8, (c) 9, (e) 4, (g) 3.

9. If x is in N and $x \neq 0$, then $x = 1 \cdot x$ and from this it follows that $x \div x = 1$.

Section 6.16

1. (a) 10, (c) 2^4, (e) 1, (g) 3^2.

5. Suppose that x is a divisor of y, and x is a divisor of z, and $y \leq z$. Then there are natural numbers n and m such that $y = nx$ and $z = mx$. $y \leq z$ implies $nx \leq mx$. Since x cannot be zero this implies $n \leq m$. (For if $n > m$ then $nx > mx$ when $x \neq 0$.) Hence, $y \div x \leq z \div x$.

7. Suppose x divides y, w divides z, and u divides v. Then there are natural numbers n, m, and p such that $y = nx$, $z = mw$, and $v = pu$. Then

$$\begin{aligned}(ywu + zxu + xwv) \div (xwu) &= (nxwu + mwxu + xwpu) \div (xwu)\\ &= [(n + m + p)(xwu)] \div (xwu) = n + m + p\\ &= (y \div x) + (z \div w) + (v \div u).\end{aligned}$$

9. Suppose x divides y, w divides z, and $yw \geq xz$. Then there are natural numbers n and m such that $y = nx$ and $z = mw$. Since $yw \geq xz$, $yw - xz$ is defined and we have

$$\begin{aligned}(yw - xz) \div (xw) &= (nxw - xmw) \div (xw)\\ &= [(n - m)(xw)] \div (xw) = n - m\\ &= (y \div x) - (z \div w).\end{aligned}$$

Section 6.17

1. In each case we use only the more basic technique, leaving the application of the usual algorithm to the reader.
(a) $999 + 1 = 9 \cdot 10^2 + 9 \cdot 10 + 9 + 1 = 9 \cdot 10^2 + 9 \cdot 10 + 10 = 9 \cdot 10^2 + 9 \cdot 10 + 1 \cdot 10 = 9 \cdot 10^2 + (9 + 1) \cdot 10 = 9 \cdot 10^2 + 10^2 = 9 \cdot 10^2 + 1 \cdot 10^2 = (9 + 1) \cdot 10^2 = 10^3 = 1000$.
(c) $747 + 1049 = 7 \cdot 10^2 + 4 \cdot 10 + 7 + 10^3 + 4 \cdot 10 + 9 = 10^3 + 7 \cdot 10^2 + 4 \cdot 10 + 4 \cdot 10 + 7 + 9 = 10^3 + 7 \cdot 10^2 + 4 \cdot 10 + 4 \cdot 10 + 16 = 10^3 + 7 \cdot 10^2 + 4 \cdot 10 + 4 \cdot 10 + 1 \cdot 10 + 6 = 10^3 + 7 \cdot 10^2 + (4 + 4 + 1) \cdot 10 + 6 = 10^3 + 7 \cdot 10^2 + 9 \cdot 10 + 6 = 1796$.
(e) $736 + 554 + 1532 = 7 \cdot 10^2 + 3 \cdot 10 + 6 + 5 \cdot 10^2 + 5 \cdot 10 + 4 + 10^3 + 5 \cdot 10^2 + 3 \cdot 10 + 2 = 10^3 + 7 \cdot 10^2 + 5 \cdot 10^2 + 5 \cdot 10^2 + 3 \cdot 10 + 5 \cdot 10 + 3 \cdot 10 + 6 + 4 + 2 = 10^3 + (7 + 5 + 5) \cdot 10^2 + (3 + 5 + 3) \cdot 10 + 10 + 2 = 10^3 + (7 + 5 + 5) \cdot 10^2 + (3 + 5 + 3 + 1) \cdot 10 + 2 = 10^3 + (7 +$

$5 + 5) \cdot 10^2 + (10 + 2) \cdot 10 + 2 = 10^3 + (7 + 5 + 5 + 1) \cdot 10^2 + 2 \cdot 10 + 2 = 10^3 + (10 + 8) \cdot 10^2 + 2 \cdot 10 + 2 = 10^3 + 10^3 + 8 \cdot 10^2 + 2 \cdot 10 + 2 = 2 \cdot 10^3 + 8 \cdot 10^2 + 2 \cdot 10 + 2 = 2822.$

3. (a) 1111100_2, (c) 1617_8, (e) 100100_2.

Section 6.18

1. We show the calculation in terms of basic meanings and leave the application of the usual algorithm to the reader.
   (a) $47{,}963 - 36{,}462 = 4 \cdot 10^4 + 7 \cdot 10^3 + 9 \cdot 10^2 + 6 \cdot 10 + 3 - (3 \cdot 10^4 + 6 \cdot 10^3 + 4 \cdot 10^2 + 6 \cdot 10 + 2) = (4 - 3) \cdot 10^4 + (7 - 6) \cdot 10^3 + (9 - 4) \cdot 10^2 + (6 - 6) \cdot 10 + (3 - 2) = 10^4 + 10^3 + 5 \cdot 10^2 + 0 \cdot 10 + 1 = 11501.$
   (c) $7574 - (6943 - 427) = 7574 - [(6 \cdot 10^3 + 9 \cdot 10^2 + 4 \cdot 10 + 3) - (4 \cdot 10^2 + 2 \cdot 10 + 7] = 7574 - [(6 \cdot 10^3 + 9 \cdot 10^2 + 3 \cdot 10 + 10 + 3) - (4 \cdot 10^2 + 2 \cdot 10 + 7)] = 7574 - [6 \cdot 10^3 + (9 - 4) \cdot 10^2 + (3 - 2) \cdot 10 + (13 - 7)] = 7 \cdot 10^3 + 5 \cdot 10^2 + 7 \cdot 10 + 4 - (6 \cdot 10^3 + 5 \cdot 10^2 + 10 + 6) = 7 \cdot 10^3 + 5 \cdot 10^2 + 6 \cdot 10 + 14 - (6 \cdot 10^3 + 5 \cdot 10^2 + 10 + 6) = (7 - 6) \cdot 10^3 + (5 - 5) \cdot 10^2 + (6 - 1) \cdot 10 + (14 - 6) = 10^3 + 5 \cdot 10 + 8 = 1058.$
   (e) $4675 - 3966 - 709 = 4675 - (3966 + 709) = 4675 - 4675 = 0.$

3. (a) 1100110_2, (c) 112011_3.

6. (a) $1732t_{12}$, (c) $96e74_{12}$.

Section 6.19

1. We display the process using basic principles and leave the usual algorithm to the reader.
   (a) $749 \cdot 5 = (7 \cdot 10^2 + 4 \cdot 10 + 9) \cdot 5 = (5 \cdot 7) \cdot 10^2 + (5 \cdot 4) \cdot 10 + 5 \cdot 9 = (35) \cdot 10^2 + (20) \cdot 10 + 45 = (3 \cdot 10 + 5) \cdot 10^2 + (2 \cdot 10) \cdot 10 + 4 \cdot 10 + 5 = 3 \cdot 10^3 + 5 \cdot 10^2 + 2 \cdot 10^2 + 4 \cdot 10 + 5 = 3 \cdot 10^3 + (5 + 2) \cdot 10^2 + 4 \cdot 10 + 5 = 3745.$
   (c) $663 \cdot 94 = (6 \cdot 10^2 + 6 \cdot 10 + 3)(9 \cdot 10 + 4) = (9 \cdot 6) \cdot 10^3 + (9 \cdot 6) \cdot 10^2 + (9 \cdot 3) \cdot 10 + (4 \cdot 6) \cdot 10^2 + (4 \cdot 6) \cdot 10 + (4 \cdot 3) = (5 \cdot 10 + 4) \cdot 10^3 + (5 \cdot 10 + 4) \cdot 10^2 + (2 \cdot 10 + 7) \cdot 10 + (2 \cdot 10 + 4) \cdot 10^2 + (2 \cdot 10 + 4) \cdot 10 + 10 + 2 = 5 \cdot 10^4 + (4 + 5 + 2) \cdot 10^3 + (4 + 2 + 4 + 2) \cdot 10^2 + (7 + 4 + 1) \cdot 10 + 2 = 5 \cdot 10^4 + (10 + 1) \cdot 10^3 + (10 + 2) \cdot 10^2 + (10 + 2) \cdot 10 + 2 = (5 + 1) \cdot 10^4 + (1 + 1) \cdot 10^3 + (2 + 1) \cdot 10^2 + 2 \cdot 10 + 2 = 6 \cdot 10^4 + 2 \cdot 10^3 + 3 \cdot 10^2 + 2 \cdot 10 + 2 = 62{,}322.$
   (e) $555 \cdot 111 = (5 \cdot 10^2 + 5 \cdot 10 + 5) \cdot (10^2 + 10 + 1) = 5 \cdot 10^4 + 5 \cdot 10^3 + 5 \cdot 10^2 + 5 \cdot 10^3 + 5 \cdot 10^2 + 5 \cdot 10 + 5 \cdot 10^2 + 5 \cdot 10 + 5 = 5 \cdot 10^4 + (5 + 5) \cdot 10^3 + (5 + 5 + 5) \cdot 10^2 + (5 + 5) \cdot 10 + 5 = (5 + 1) \cdot 10^4 + (10 + 5 + 1) \cdot 10^2 + 5 = 6 \cdot 10^4 + 10^3 + 6 \cdot 10^2 + 5 = 61605.$

4. (a) 100011_2, (c) 1100100_2, (e) $1e7626_{12}$.

Section 6.20

1. (a) $40 - 10 = 30$, $40 - 2 \cdot 10 = 20$, $40 - 3 \cdot 10 = 10$, $40 - 4 \cdot 10 = 0$. Hence, $40 \div 10 = 4$.
   (c) $90 - 15 = 75$, $90 - 2 \cdot 15 = 60$, $90 - 3 \cdot 15 = 45$, $90 - 4 \cdot 15 = 30$, $90 - 5 \cdot 15 = 15$, $90 - 6 \cdot 15 = 0$. Hence, $90 \div 15 = 6$.

(e) $50{,}000 - 25{,}000 = 25{,}000$; $50{,}000 - 2 \cdot 25{,}000 = 0$. Hence, $50{,}000 \div 25{,}000 = 2$.

5. To check our results we should compute $xq + r$ and, of course, get y.

7. (a) 6, (c) 5342.

9. 5.

10. (a) 1,2,3,4,6,9,12,18,36.

11. 7.

13. (a) 111011_2, (c) 2012_3 with a remainder of 10_3, (e) 40_{12}, (g) 2_{12}, (i) 12_{12} and a remainder of 4_{12}.

Section 6.21

2. None of them.

3. They are 151, 157, and 163.

4. (a) 5^3, (c) $2^2 \cdot 5^2 \cdot 7^2$, (e) $37 \cdot 101$, (f) $2^5 \cdot 3 \cdot 5^2$.

5. Suppose x and y have the common divisor d which is different from 1. d may be written as a product of primes. Let p be one of the primes appearing in the product. Then x and y are both multiples of p and, hence, p is a prime divisor of both x and y.

7. 28 is also a perfect number.

9. They are 3 and 5, 11 and 13, 17 and 19, 29 and 31, 41 and 43, 59 and 61, 71 and 73, 101 and 103, 107 and 109, 137 and 139.

11. The only primary number is 1. Every natural number, except 0, can be written (uniquely) as a sum of primary numbers (i.e., as a sum of ones).

13. A number will be a perfect square if and only if all the powers appearing on its prime factors are even.

Section 6.22

3. $9! = 362{,}880$, $10! = 3{,}628{,}800$, $11! = 39{,}916{,}800$.

5. 24 ways

7. 3125 ways

9. 362,880 ways

11. n^n is larger than $n!$. There are n factors in each product, but each factor in n^n is equal to n, while in $n!$ almost all of the factors are smaller.

12. 365^k (or 366^k)

15. $365 \cdot 364 \cdot 363 \cdot \ldots \cdot (365 - k + 1)$

17. $g(4) = (((4!)!)!)! = ((24!)!)!$ and clearly, this is a very large number.

18. There are $52 \cdot 51 \cdot 50 \cdot 49 \cdot 48 = 311{,}875{,}200$ different hands.

Section 6.23

1. $\binom{6}{2} = 15$

2. $\binom{6}{4} = 15$

3. $\binom{52}{5} = 2{,}598{,}960$

5. $\binom{n}{k}$ is the number of different subsets of size k that can be chosen from a set of size n. But each time we pick a subset of size k we determine a subset of size $n - k$ (namely, those we leave behind) so that we must have $\binom{n}{k} = \binom{n}{n-k}$.

7. 256, 70.

9. $\binom{50}{4} = 230{,}300$

11. (a) $c^5 + 5c^4d + 10c^3d^2 + 10c^2d^3 + 5cd^4 + d^5$.
    (c) $2^7 + 7\cdot 2^6\cdot x + 21\cdot 2^5\cdot x^2 + 35\cdot 2^4\cdot x^3 + 35\cdot 2^3\cdot x^4 + 21\cdot 2^2\cdot x^5 + 7\cdot 2\cdot x^6 + x^7 = 128 + 448x + 672x^2 + 560x^3 + 280x^4 + 84x^5 + 14x^6 + x^7$.
    (e) $a^{10} - 10a^9b + 45a^8b^2 - 120a^7b^3 + 210a^6b^4 - 252a^5b^5 + 210a^4b^6 - 120a^3b^7 + 45a^2b^8 - 10ab^9 + b^{10}$.

13. We have $\sum_{k=0}^{n} \binom{n}{k} 1^k \cdot 1^{n-k} = \sum_{k=0}^{n} \binom{n}{k} = (1 + 1)^n = 2^n$.

Section 7.1

1. (a), (d), (e), (f), (g), (h).
2. In each case, we have given only 7 of the many pairs.
   (a) (0,6), (1,7), (2,8), (3,9), (4,10), (5,11), (6,12).
   (c) (0,22), (1,23), (2,24), (3,25), (4,26), (5,27), (6,28).
   (e) (1,0), (2,1), (3,2), (4,3), (5,4), (6,5), (7,6).
   (g) (0,0), (1,1), (2,2), (3,3), (4,4), (5,5), (6,6).
   (i) (693,342), (694,343), (695,344), (696,345), (697,346), (698,347), (699,348).
3. (a) $^+6$, (c) $^+22$, (e) $^-1$, (g) $\bar{0}$, (i) $^-351$.
4. (a) (0,33), (1,34), (2,35), (3,36), . . .
   (c) (5764,0), (5765,1), (5766,2), (5767,3), . . .
5. We collect into sets the pairs that are in the relation E to one another. $\{a,c,f,k\}$, $\{b,i,j\}$, $\{d,l\}$, $\{e,g\}$, $\{h\}$.
6. (a) $^+m = {}^+n$ implies $[(0,m)] = [(0,n)]$. From this we get $(0,m)\ E\ (0,n)$ and, hence, $m + 0 = n + 0$. But this implies $m = n$.

Section 7.2

1. (a)

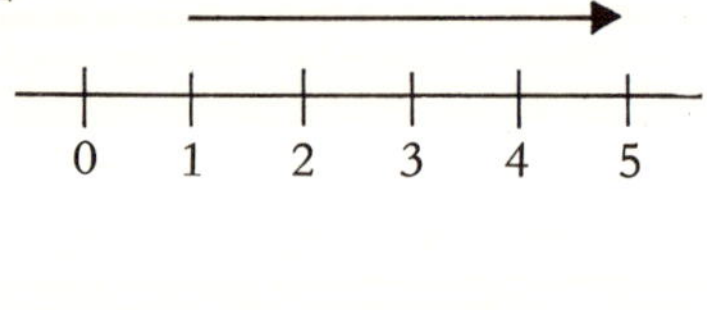

   (c)

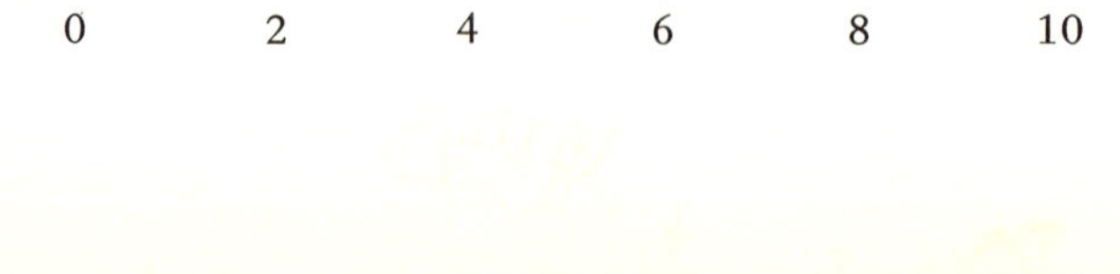

(e)

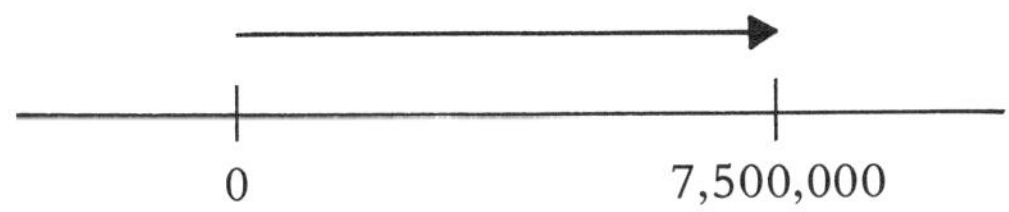

2. We can think of temperatures colder than 0 as losses of n degrees after the temperature was zero (i.e., as an "arrow" on the thermometer from 0 to n degrees below 0). We now label this point ^-n on the thermometer.

4. (a) We may think of $^-6$ as (i) the class $\{(6,0),(7,1),(8,2),\ldots\}$, (ii) the vector which is the collection of arrows running 6 units from right to left, (iii) the solution of the equations $x + 6 = 0$, $x + 7 = 1$, $x + 8 = 2$, ..., (iv) the result of the subtractions $0 - 6$, $1 - 7$, $2 - 8$, ..., (v) the additive inverse of 6, or (vi) the number corresponding to the point 6 units to the left of the origin on the number line.

   (c) We may think of $\bar{0}$ as (i) the class $\{(0,0),(1,1),(2,2),(3,3),\ldots\}$, (ii) the solution to the equations $x + 0 = 0$, $x + 1 = 1$, $x + 2 = 2$, ..., (iii) the result of the subtractions $0 - 0$, $1 - 1$, $2 - 2$, ..., (iv) the vector which is the collection of all arrows of 0 length (i.e. the collection of points), (v) the integer that behaves like the natural number 0, or (vi) the integer that is its own additive inverse.

5. (a) $^-27$, (c) $^-20{,}589$.

Section 7.3

1. (a) $[(0,55)]$, (c) $[(50,802)]$, (e) $[(3978,977)]$.
2. (a) 55, (c) 742, (e) 3001.
3. (a) $^+131$, (c) $^-4337$.
4. (a) $^+363$, (c) $^-779$, (e) $^+1548$.
6. (a) If x and y are positive, then there are natural numbers n and m such that $x = [(0,n)]$ and $y = [(0,m)]$. x and y have the same sign and $|x| = n$ and $|y| = m$. Then $x +_I y = [(0,n)] +_I [(0,m)] = [(0,m + n)] = {}^+(m + n) = {}^+(|x| + |y|)$.

   (c) If x is positive and y is negative, then there are natural numbers n and m such that $x = [(0,n)]$ and $y = [(m,0)]$. Then x and y differ in sign and $|x| = n$ and $|y| = m$. Suppose $|x| > |y|$. Then there is a nonzero natural number k such that $n = m + k$. We have $x +_I y = [(0,n)] +_I [(m,0)] = [(0,m + k)] +_I [(m,0)] = [(m,m + k)] = [(0,k)] = {}^+k = {}^+(n - m) = {}^+(|x| - |y|)$.

   The proof of the other parts of Problem 6 are similar and will not be reproduced here.
7. (a) $^+24$, (c) $^-16$.
8. (a) 74, (c) 2, (e) 12.
9. (a) $^+1$, (c) $^-1$, (c) $^+7$, (g) $^+1770$.
10. (i) If $y = {}^+n$: $x +_I y = x +_I ({}^+n) = z$ implies $(x +_I {}^+n) +_I {}^-n = z +_I {}^-n$. This implies $x +_I ({}^+n +_I {}^-n) = z +_I {}^-n$, which yields $x +_I (0) = x = z +_I {}^-n$.

    (iii) If $y = \bar{0}$: $x +_I y = x +_I \bar{0} = z$ implies $(x +_I \bar{0}) + \bar{0} = z +_I \bar{0}$. Then $x +_I (\bar{0} +_I \bar{0}) = x = z$.

11.

$+_I$	$^-6$	$^-5$	$^-4$	$^-3$	$^-2$	$^-1$	0	$^+1$	$^+2$	$^+3$	$^+4$	$^+5$	$^+6$
$^-6$	$^-12$	$^-11$	$^-10$	$^-9$	$^-8$	$^-7$	$^-6$	$^-5$	$^-4$	$^-3$	$^-2$	$^-1$	$\bar{0}$
$^-5$	$^-11$	$^-10$	$^-9$	$^-8$	$^-7$	$^-6$	$^-5$	$^-4$	$^-3$	$^-2$	$^-1$	$\bar{0}$	$^+1$
$^-4$	$^-10$	$^-9$	$^-8$	$^-7$	$^-6$	$^-5$	$^-4$	$^-3$	$^-2$	$^-1$	$\bar{0}$	$^+1$	$^+2$
$^-3$	$^-9$	$^-8$	$^-7$	$^-6$	$^-5$	$^-4$	$^-3$	$^-2$	$^-1$	$\bar{0}$	$^+1$	$^+2$	$^+3$
$^-2$	$^-8$	$^-7$	$^-6$	$^-5$	$^-4$	$^-3$	$^-2$	$^-1$	$\bar{0}$	$^+1$	$^+2$	$^+3$	$^+4$
$^-1$	$^-7$	$^-6$	$^-5$	$^-4$	$^-3$	$^-2$	$^-1$	$\bar{0}$	$^+1$	$^+2$	$^+3$	$^+4$	$^+5$
0	$^-6$	$^-5$	$^-4$	$^-3$	$^-2$	$^-1$	$\bar{0}$	$^+1$	$^+2$	$^+3$	$^+4$	$^+5$	$^+6$
$^+1$	$^-5$	$^-4$	$^-3$	$^-2$	$^-1$	$\bar{0}$	$^+1$	$^+2$	$^+3$	$^+4$	$^+5$	$^+6$	$^+7$
$^+2$	$^-4$	$^-3$	$^-2$	$^-1$	$\bar{0}$	$^+1$	$^+2$	$^+3$	$^+4$	$^+5$	$^+6$	$^+7$	$^+8$
$^+3$	$^-3$	$^-2$	$^-1$	$\bar{0}$	$^+1$	$^+2$	$^+3$	$^+4$	$^+5$	$^+6$	$^+7$	$^+8$	$^+9$
$^+4$	$^-2$	$^-1$	$\bar{0}$	$^+1$	$^+2$	$^+3$	$^+4$	$^+5$	$^+6$	$^+7$	$^+8$	$^+9$	$^+10$
$^+5$	$^-1$	$\bar{0}$	$^+1$	$^+2$	$^+3$	$^+4$	$^+5$	$^+6$	$^+7$	$^+8$	$^+9$	$^+10$	$^+11$
$^+6$	$\bar{0}$	$^+1$	$^+2$	$^+3$	$^+4$	$^+5$	$^+6$	$^+7$	$^+8$	$^+9$	$^+10$	$^+11$	$^+12$

Section 7.4

1. (a) [(8,8)], (c) [(0,30)], (e) [(0,1)], (g) [(658,1582)].
2. (a) $\bar{0}$, (c) $^+30$, (e) $^+1$, (g) $^+924$.
3. (a)
$$\begin{aligned}[(a,b)] \cdot_I [(c,d)] &= [(bc + ad,\ bd + ac)] \\ &= [(da + cb,\ db + ca)] \\ &= [(c,d)] \cdot_I [(a,b)].\end{aligned}$$
Here the second equality is justified by the commutativity of addition and multiplication in N.
(c) $[(a,b)] \cdot_I [(0,0)] = [(b\cdot 0 + a\cdot 0,\ b\cdot 0 + a\cdot 0)] = [(0,0)]$.
4. (a) $^-110100_2$, (c) $^-9648_{12}$.

Section 7.5

1. (a) Yes. Let $f(-n) = n$, for any nonzero $-n$ in Q and $f(\bar{0}) = 0$. Clearly f is a one-to-one correspondence between Q and N. Also, $f((-n) + (-m)) = f(-(n + m)) = n + m = f(-n) + f(-m)$ so f is the desired isomorphism.
2. (a) Yes. $f(x) = 2x$ is clearly a one-to-one correspondence between N and A. $f(x + y) = 2(x + y) = 2x + 2y = f(x) + f(y)$; so f is an isomorphism.
(c) No. For the reason, see part (b).
3. (a) Yes.
4. (a) Yes.
5. Yes; $(-n) + (-m) = -(n + m)$.
No; $(-1)\cdot(-1) = +1$ and $^+1$ is not in I^-.
9. Suppose f is a one-to-one correspondence between K and N and let $f(^-1) = a$ and $f(^+1) = b$. Then $a \neq b$. Now $f(^-1 \cdot_I {}^-1) = f(^+1) = b$ and $f(^-1)\cdot f(^-1) = a\cdot a = a^2$. Now if f is to be an isomorphism we must have $a^2 = b$ and also $b = 1$ since f must then map the multiplicative identity in I onto that in N. But in N, $a^2 = 1$ implies $a = 1$. To see this, note that if $a^2 = 1$, then clearly $a \neq 0$ and hence $a \geq 1$. But if $a \neq 1$ then $a > 1$ and $a^2 > 1$. It follows that f cannot be an isomorphism.

Section 7.6

1. (a) one withdrawal of 250 dollars.
   (c) one deposit of 110 dollars.
   (e) one withdrawal of 126 dollars.
   (g) one deposit of 3360 dollars.

2. (a) $W - R$ pounds force acting downward. (We take an upward force as negative here. The opposite choice is also possible in which case we get $R - W$.)
   (c) A net gain in dollars of $P - nC$ dollars. (A negative results in a loss.)
   (c) Let c = cost per item, n = number of items made, k = selling price per item, and s = number of items sold. The net profit is $k \cdot s - n \cdot c$.
   (g) Let D = number of tax dollars due, and W = number of tax dollars already withheld. Then the amount of refund you have coming is $W - D$. (A negative amount here would mean you must pay some more to the government.)

4. Notice the only difference between this operation and multiplication of integers is that a negative times a negative is now a negative, not a positive.
   We have: commutativity; $+1$ as identity element; and associativity.
   We don't have: distributivity and cancellation laws.

Section 7.7

1. (i) $4 \div 2 = 6 \div 3$ and $4 \cdot 3 = 6 \cdot 2$.
   $9 \div 3 = 12 \div 4$ and $9 \cdot 4 = 3 \cdot 12$.
   $7 \div 1 = 12 \div 2$ and $7 \cdot 2 = 1 \cdot 14$.
   $6 \div 2 = 12 \div 4$ and $6 \cdot 4 = 2 \cdot 12$.
   (iii) $6 \div 2 + 9 \div 3 = 3 + 3 = 6 = 36 \div 6 = (3 \cdot 6 + 2 \cdot 9) \div 6$
   $= (3 \cdot 6 + 2 \cdot 9) \div 2 \cdot 3$.
   $7 \div 1 + 2 \div 2 = 7 + 1 = (14 + 2) \div 2 = (2 \cdot 7 + 1 \cdot 2) \div 2 \cdot 1$.
   $8 \div 2 + 10 \div 5 = 4 + 2 = 6 = (40 + 20) \div 10 = (5 \cdot 8 + 2 \cdot 10) \div 5 \cdot 2$.
   $15 \div 3 + 14 \div 7 = 5 + 2 = 7 = (105 + 42) \div 21$
   $= (7 \cdot 15 + 3 \cdot 14) \div 7 \cdot 3$.

2. (a) Suppose x, y, z, and w are in I and y divides x and w divides z. Then there are integers p and q such that $x \div y = p$ and $z \div w = q$. These imply $x = py$ and $z = qw$. Now suppose that $p = q$. Then $xw = (py)w = (pw)y = (qw)y = zy$.

3. (a) -4, (c) $+4$, (e) $+74$.

4. (a) -16, (c) -16.

5. (a) 12, (c) -11.

8. (a) Let us first consider the natural numbers only. Let $A = \{x \mid x \in N$ and x is even or $x = 2k + 1$, for some natural $k\}$. $0 \in A$ since $0 \div 2 = 0$. Suppose $p \in A$. If p is even then there is a natural number k such that $p = 2k$. Thus $p + 1 = 2k + 1$ so $p + 1 \in A$. If $p = 2k + 1$ for some natural number k then $p + 1 = 2k + 1 + 1 = 2k + 2 = 2(k + 1)$ so $p + 1$ is even and $p + 1 \in A$. Thus $A = N$. Hence, any natural number (i.e. positive integer) has the desired property. Suppose $-n$ $(n \in N)$ is any negative integer. If n is even then so is $-n$. If n is not even then $n = 2k + 1$ for some positive integer k. Thus $-n = -(2k + 1) = -2k + (-1) = -2k - 2 + 1 = 2(-k - 1) + 1$ and is thus of the desired form.

Section 7.8

1. (a), (b), (c), (d), (e), (f), (i).

2. We give the set, then the minimum, and then the maximum.
   (a) $\{+8, +9, +10, +11, +12, +13\}$, $+8$, $+13$.
   (c) $\{0, +1, +2, +3, +4, +5\}$, 0, $+5$.
   (e) $\{-6, -5, -4, -3, -2, -1, 0, +1, +2\}$, -6, $+2$.
   (g) $\{+3, +4, +5, \ldots\}$, $+3$, none.
   (i) $\{-18, -17, -16, -15, \ldots\}$, -18, none.
   (k) I, none, none.
   (m) $\{\ldots, +9, +10, +11\}$, none, $+11$.

3. $x < 0$ implies there is a natural number (nonnegative integer) n such that $n \neq 0$ and $x + n = 0$. Thus $x = -n$ so x is negative. Conversely, if x is negative then $x = -n$ for some natural number n. Then $x + n = -n + n = 0$ so $x < 0$.

4. (a) Since $x + 0 = x$ for all x and 0 is a non negative integer, we have $x \geq x$ by Definition 7.8.1.
   (c) Suppose $x > y$ and $y > z$. Then by Theorem 7.8.5, $x - y$ and $y - z$ are positive. Then by Theorems 7.8.2 and 7.8.4 $(x - y) + (y - z) = x - z$ is positive. And, using Theorem 7.8.5 again, we conclude $x > z$.
   (e) Suppose $x \geq y$ and $y \geq x$. Then there are nonnegative integers n and m such that $x = y + n$ and $y = x + m$. Then $x = (x + m) + n$ and, hence, $m + n = 0$. But this implies $m = n = 0$.
   (g) Since 1 is a nonnegative integer we have $x + 1 \geq x$ immediately from Definition 7.8.1. And $x + 1 = x$ would imply $1 = 0$ which is impossible.
   (i) Suppose $x > y$. Then $x \geq y$ and by part (h), $x + z \geq y + z$. But if $x + z = y + z$ we have $x = y$ which is not the case. Therefore, we have $x + z > y + z$.
   Suppose $x + z > y + z$. Then $x + z \geq y + z$ and by part (h), $x \geq y$. But if $x = y$ then clearly $x + z = y + z$ and, since this is not the case, we must have $x > y$.
   (k) Suppose $x \geq y$ and $z > w$. Then $z \geq w$ and, by part (j), $x + z \geq y + w$. Suppose $x + z = y + w$. Then $z - w = y - x$. But $y - x$ is nonpositive since $x \geq y$. On the other hand, by Theorem 7.8.5, $z - w$ is positive since $z > w$. Hence, we cannot have $x + z = y + w$. It follows that $x + z > y + w$.

6. (a) True. If $y < x < z$, then $y + w < x + w < z + w$ by Theorem 7.8.6(i). The situation is similar if $z < x < y$.
   (c) False. Let $x = -1$. Then $x > 2x$ since $-1 > -2$.
   (e) True. If $x \leq 0$ and $y \leq 0$, then $x + y \leq 0 + 0 \leq 0$ by Theorem 7.8.6(j).
   (g) True. Suppose x is between y and z. Then $y < x < z$ or $z < x < y$. Then $y - x < 0 < z - x$ or $z - x < 0 < y - x$.

7. (a) $\{-4, -3, -2, -1, 0, +1, +2, +3, +4, +5, +6, +7, +8, +9, +10, +11\}$
   (c) $\{+1, +2, +3, +4, +5, +6, +7, +8, +9, +10\}$
   (e) ϕ

Section 7.9

1. (b) Let w be the wage in dollars per hour and let h be the number of hours to be worked. We want $w \cdot h \geq 2000$.

(d) Let i = income per month from annuity, j = income from fund, w = wage per hour, and h = the number of hours to be worked. We want $1000 \leq w \cdot h + i + j \leq 1500$.

(f) Let P_1, P_2, P_3 be the prices per unit respectively of the three articles. Let b_1, b_2, b_3 be the respective numbers of each to be bought. We want $P_1b_1 + P_2b_2 + P_3b_3 \leq 4700$.

(h) Let d_1, d_2, d_3 and v_1, v_2, v_3 be the distance and speed between A and B, B and C, and C and A, respectively. We want $d_1/v_1 + d_2/v_2 + d_3/v_3 < 15$.

(j) With n_i and p_i as in part (i), let c_i be the cost of producing one item of the ith kind. ($i = 1,2,3,4$.) We want all the inequalities of part (i) satisfied as well as $n_1c_1 + n_2c_2 + n_3c_3 + n_4c_4 \leq 1500$.

2. Let n be the number of times the game was played.
   (a) $n \geq 5$.
   (c) $n \leq 120$.
   (e) $n = 4,5,6,7$, or 8.
   (g) $n = 2,4,6$, or 8.

3. We give the set in roster notation, then its minimum, and then its maximum.
   (a) $\{\ldots, 0,1,2\}$, none, 2.
   (c) $\{\ldots, -6,-5,-4\}$, none, -4.
   (g) $\{\ldots, 0,1,2\}$, none, 2.
   (i) ϕ, none, none.

4. (a) $\{-1,0,1\}$, -1, 1.
   (c) $\{1,2,3,4,5\}$, 1, 5.
   (e) ϕ, none, none.
   (g) $\{\ldots, -4,-3,-2,2,3,4,\ldots\}$, none, none.
   (i) ϕ, none, none.

6. (i) All of the sets except that in (g) are finite.
   (ii) All of the sets are bounded except (g).

8. (a) True since $x = 0$ if and only if $|x| = 0$.
   (c) True. Suppose x is between y and z. Then either $y < x < z$ or $z < x < y$. If $w \neq 0$, then w is either positive or negative and it follows from Theorem 7.9.1 that we must have either $yw < xw < zw$ or $zw < xw < yw$.
   (e) False. Let $x = y = -1$. Then $xy = 1 > 0$ and yet $x < 0$ and $y < 0$.
   (g) False. Let $x = 0$ and $y = 1$. Then for all z we have $xz = 0 \cdot z = 0 = 1$.
   (i) True since $0^2 = 0$ and the product of integers of like sign is positive.
   (k) True. $\{x : x^2 \leq 9\} = \{-3,-2,-1,0,1,2,3\}$.

9. We consider several cases.
   (a) Suppose $x > 0$ and $y > 0$. Then $x + y > 0$ and we have $|x + y| = x + y = |x| + |y|$.
   (b) Suppose $x < 0$ and $y < 0$. Then $x + y < 0$ and $|x + y| = -(x + y) = (-x) + (-y) = |x| + |y|$.
   (c) Suppose $x < 0$ and $y > 0$ and $|x| > |y|$. Then $|x + y| = -(x + y) = -x - y = |x| - |y| < |x| + |y|$.
   (d) Suppose $x < 0$ and $y > 0$ and $|x| < |y|$. Then $|x + y| = x + y = -|x| + |y| < |x| + |y|$.

   The cases where $x > 0$ and $y < 0$ are similar to (c) and (d).

Section 7.10

2. (a)

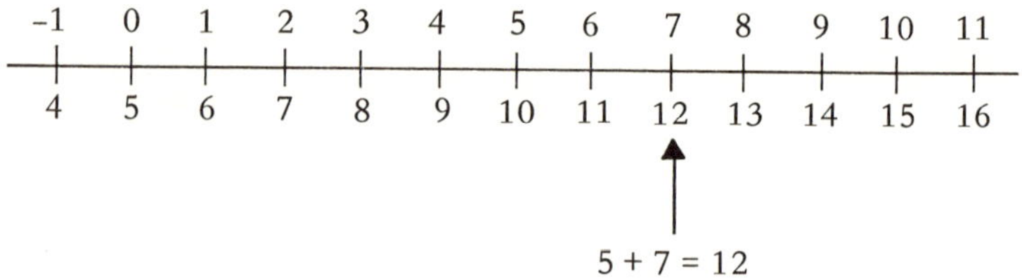

(c)

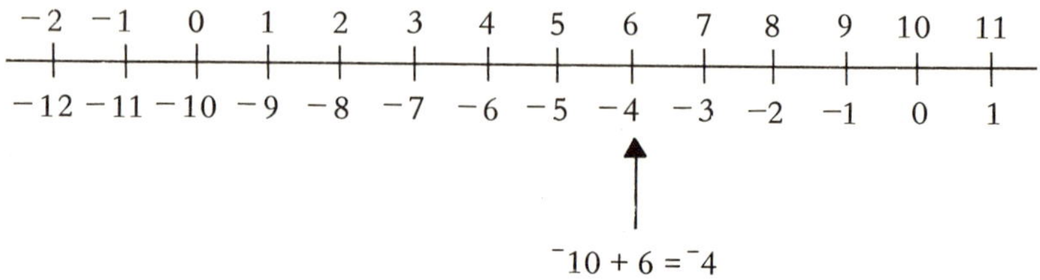

4. To find $z \div y$, draw the horizontal at z on the vertical axis, find the point where this horizontal hits the line labelled y, draw the vertical through this point, and note the coordinate x where this vertical hits the horizontal axis. We shall have $z \div y = x$. For example, the horizontal through $z = {}^{-}20$ hits the line $y = {}^{+}4$ at a point above $x = {}^{-}5$ on the horizontal axis, and we have ${}^{-}20 \div {}^{+}4 = {}^{-}5$.

6. To subtract y from z, move the top scale so that the point for z there is immediately above the point for y on the lower scale. Read the coordinate x on the bottom scale of the point below the origin on the upper scale. We shall have $x = z - y$.

Section 7.11

1. (a), (d), and (e) are true.

2. The equivalence classes are:
   (a) $\{\ldots,-3,0,3,6,9,\ldots\}$, $\{\ldots,-2,1,4,7,10,\ldots\}$, and $\{\ldots,-1,2,5,8,11,\ldots\}$.
   (c) $\{\ldots,-6,0,6,12,18,\ldots\}$, $\{\ldots,-5,1,7,13,19,\ldots\}$,
   $\{\ldots,-4,2,8,14,20,\ldots\}$, $\{\ldots,-3,3,9,15,21,\ldots\}$,
   $\{\ldots,-2,4,10,16,22,\ldots\}$, and $\{\ldots,-1,5,11,17,23,\ldots\}$.
   (e) $\{\ldots,-8,0,8,16,24,\ldots\}$, $\{\ldots,-7,1,9,17,25,\ldots\}$,
   $\{\ldots,-6,2,10,18,26,\ldots\}$, $\{\ldots,-5,3,11,19,27,\ldots\}$,
   $\{\ldots,-4,4,12,20,28,\ldots\}$, $\{\ldots,-3,5,13,21,29,\ldots\}$,
   $\{\ldots,-2,6,14,22,30,\ldots\}$, $\{\ldots,-1,7,15,23,31,\ldots\}$.

4. (a)

$+_7$	0	1	2	3	4	5	6
0	0	1	2	3	4	5	6
1	1	2	3	4	5	6	0
2	2	3	4	5	6	0	1
3	3	4	5	6	0	1	2
4	4	5	6	0	1	2	3
5	5	6	0	1	2	3	4
6	6	0	1	2	3	4	5

$\cdot_7$	0	1	2	3	4	5	6
0	0	0	0	0	0	0	0
1	0	1	2	3	4	5	6
2	0	2	4	6	1	3	5
3	0	3	6	2	5	1	4
4	0	4	1	5	2	6	3
5	0	5	3	1	6	4	2
6	0	6	5	4	3	2	1

6. (a) $x \equiv 1 \pmod{2}$,
   (c) $x \equiv 2 \pmod{5}$,
   (e) no solution,
   (g) no solution,
   (i) $x \equiv 2 \pmod{5}$.

7. (a) $-0 = 0$, $-1 = 2$, $-2 = 1$, $1^{-1} = 1$, $2^{-1} = 2$.
   (c) $-0 = 0$, $-1 = 4$, $-2 = 3$, $-3 = 2$, $-4 = 1$, $1^{-1} = 1$, $2^{-1} = 3$, $3^{-1} = 2$, $4^{-1} = 4$.
   (e) $-0 = 0$, $-1 = 6$, $-2 = 5$, $-3 = 4$, $-4 = 3$, $-5 = 2$, $-6 = 1$, $1^{-1} = 1$, $2^{-1} - 4$, $3^{-1} - 5$, $4^{-1} = 2$, $5^{-1} = 3$, $6^{-1} = 6$.

8. Suppose $x \equiv y \pmod{k}$ and $z \equiv w \pmod{k}$. Then there are integers n and m such that $x - y = nk$ and $z - w = mk$. Then $(x - y) + (z - w) = nk + mk = (n + m)k$. But also, $(x - y) + (z - w) = (x + z) - (y + w)$. So we see that $(x + z) - (y + w)$ is a multiple of k and, hence, $x + z \equiv y + w \pmod{k}$. In regard to multiplication we have $xz - yw = xz - zy + zy - yw = z(x - y) + y(z - w) = z(nk) + y(mk) = (zn + ym)k$. Hence, $xz - yw$ is a multiple of k and it follows that $xz \equiv yw \pmod{k}$.

Section 7.12

1. The systems given in (a) and (e) are groups.

3. It is a group when $k = 2,3,5$, or 7.

5. Only the system given in (c) is a group.

7. Yes. F is closed under composition since the composition of two one-to-one correspondences is a one-to-one correspondence. The identity map i defined by $i(x) = x$ is the identity element in F. Composition of functions is associative. And if $f \in F$ then f^{-1} is also. This group need not be commutative. For example, if $A = \{1,2,3\}$ and $f(1) = 2$, $f(2) = 1$, $f(3) = 3$, $g(1) = 1$, $g(2) = 3$, and $g(3) = 2$, then $f \circ g \neq g \circ f$.

10. The operation table for the system is

$\circ$	v_0	v_1	v_2	m_1	m_2	m_3
v_0	v_0	v_1	v_2	m_1	m_2	m_3
v_1	v_1	v_2	v_0	m_2	m_3	m_1
v_2	v_2	v_0	v_1	m_3	m_1	m_2
m_1	m_1	m_2	m_3	v_0	v_2	v_1
m_2	m_2	m_1	m_3	v_1	v_0	v_2
m_3	m_3	m_2	m_1	v_2	v_1	v_0

From the table we see that we have closure and v_0 for an identity element. We know that composition of mappings is associative so $\circ$ is associative. The inverse elements can be read off from the table as $-v_0 = v_0$, $-v_1 = v_2$, $-v_2 = v_1$, $-m_1 = m_1$, $-m_2 = m_2$, $-m_3 = m_3$.

11. The operation table for the system is:

$\circ$	r_0	r_1	r_2	r_3	h	v	d_1	d_2
r_0	r_0	r_1	r_2	r_3	h	v	d_1	d_2
r_1	r_1	r_2	r_3	r_0	d_2	d_1	h	v
r_2	r_2	r_3	r_0	r_1	v	h	d_2	d_1
r_3	r_3	r_0	r_1	r_2	d_1	d_2	v	h
h	h	d_1	v	d_2	r_0	r_2	r_1	r_3
v	v	d_2	h	d_1	r_2	r_0	r_3	r_1
d_1	d_1	v	d_2	h	r_1	r_3	r_0	r_2
d_2	d_2	h	d_1	v	r_3	r_1	r_2	r_0

From the table we see that we have closure and r_0 for an identity. Since the composition of mappings is associative, so is $\circ$. The inverse element can be read from the table as $-r_0 = r_0$, $-r_1 = r_3$, $-r_2 = r_2$, $-r_3 = r_1$, $-h = h$, $-v = v$, $-d_1 = d_1$, and $-d_2 = d_2$.

Section 7.13

1. Suppose $x = b \oplus (-a)$. Then $x \oplus a = (b \oplus (-a)) \oplus a = b \oplus ((-a) \oplus a) = b \oplus 0 = b$. Hence, $x = b \oplus (-a)$ is a solution for $x \oplus a = b$. Now suppose that y is a solution of the equation so that $y \oplus a = b$. Then $(y \oplus a) \oplus (-a) = b \oplus (-a)$. This implies $y \oplus (a \oplus (-a)) = b \oplus (-a)$, which gives $y \oplus 0 = b \oplus (-a)$ or $y = b \oplus (-a)$. This completes the proof.

2. We give the solution in each case. (a) 3, (c) 2, (e) 2, (g) 4, (i) 4.

4. If we let $f(a^n) = n$ (where we take $a^0 = 1$) then f is an isomorphism between $(\{1,a,a^2,a^3\},M)$ and $(\{0,1,2,3\},+_4)$. We may verify this by taking the table for M and replacing each a^j by j. We end up with the table for $+_4$.

5. (i) a is clearly a generator for this group since $a^2 = aMa$, $a^3 = aMaMa$, and $1 = aMaMaMa$. (Actually, a^3 is also a generator for this group.)
   (iii) 1 is a generator for $(\{0,1,2,3,4\},+_5)$.
   (v) 2 is a generator for $(\{1,2,3,4\},\cdot_5)$. (So is 3.)

7. We have the following table for the operation restricted to $H = \{0,2\}$ in $(\{0,1,2,3\},+_4)$:

$+_4$	0	2
0	0	2
2	2	0

This system is isomorphic to the group $(\{0,1\},+_2)$.
(a) $(\{0\},+_6)$, $(\{0,2,4\},+_6)$, $(\{0,3\},+_6)$, and $(\{0,1,2,3,4,5\},+_6)$ are all subgroups of the given group.
(c) $(\{r_0,r_1,r_3\},\circ)$, $(\{r_0,m_1\},\circ)$, $(\{r_0,m_2\},\circ)$, and $(\{r_0,m_3\},\circ)$ are all non-trivial subgroups.

Section 8.1

1. (a), (b), (c), (e), and (f) are true.

2. Some of the many possibilities are:
   (a) $(14,2), (21,3), (28,4), (^-7,^-1), (^-14,^-2), (^-21,^-3)$.
   (c) $(9,1), (^-9,^-1), (18,2), (^-18,^-2), (27,3), (^-27,^-3)$.
   (e) $(0,1), (0,2), (0,3), (0,4), (0,^-1), (0,^-2), (0,^-3)$.
   (g) $(1,^-7), (^-1,7), (^-2,14), (3,^-21), (^-3,21), (4,^-28)$.
   (i) $(16,9), (32,18), (^-32,^-18), (48,27), (^-48,^-27), (64,36)$.

3. (a), (c), (e), (f), (i), (j), (k).

5. (a), (b), (d), (f), (g), and (h) are true.

6. (a) $5/9 = 70/126$ since $126 \cdot 5 = 630$.
   (c) $144/12 = {}^-48/{}^-4$ since $(144)(^-4) = {}^-576 = (12)(^-48)$.
   (e) If $a/b = c/d$ and $c/d = e/f$ then $a/b = e/f$ by the transitivity of the equality relation.

Section 8.2

1. (a) $^-1/2$ is the number we invented to be the solution of the equation $2x = {}^-1$. It is also the number we invented to be the result of dividing $^-1$ by 2. It is the number corresponding to the point obtained by dividing the interval from $^-1$ to 0 into two equal pieces. It is the equivalence class $\{^-1/2, 1/\ 2, {}^-2/4, 2/^-4, {}^-3/6, \ldots\}$.
   (c) $^-4/5$ is the solution of $5x = {}^-4$, the result of dividing $^-4$ by 5, the point obtained by dividing the interval from $^-1$ to 0 into 5 equal parts and taking the first division point to the right of $^-1$, the class $\{(^-4,5),(4,^-5),(^-8,10),(8,^-10),(^-12,15),\ldots\}$.
   (e) $37/2$ is the solution of $2x = 37$, the result of dividing 37 by 2, the class $\{(37,2),(^-37,^-2),(54,4),(^-54,^-4),(111,6),\ldots\}$, and the point that divides the interval from 18 to 19 into two equal pieces.
   (g) $^-10/2$ is the solution of $2x = {}^-10$, the result of dividing $^-10$ by 2, the class $\{(^-5,1),(5,^-1),(^-10,2),(10,^-2),(^-15,3),\ldots\}$ and the point obtained by dividing the interval from 0 to 1 into two equal pieces and then moving 10 times as far to the left of the origin as the division point is to the right of the origin.

2. (a)

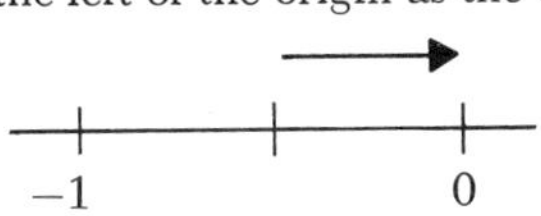

(c)

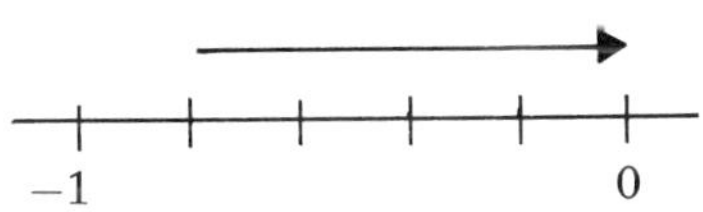

(e)

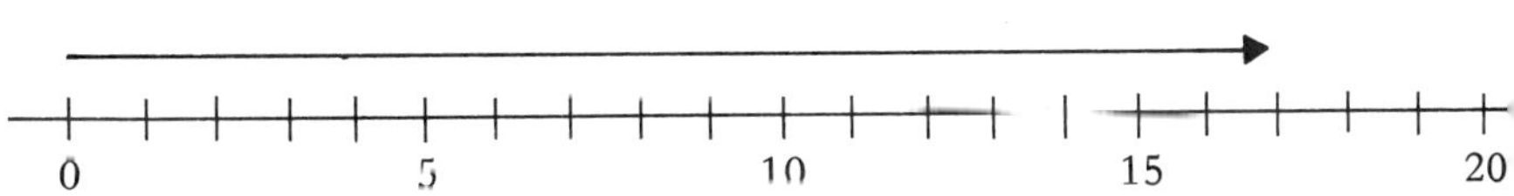

(g)

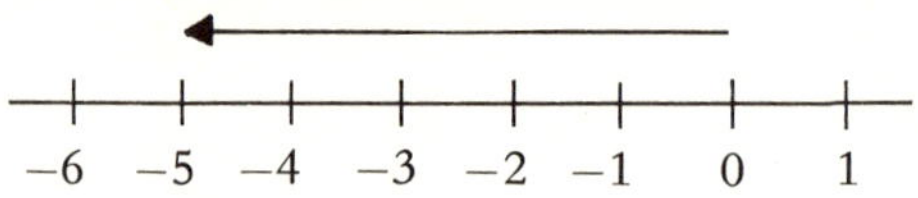
−6
−5
−4
−3
−2
−1
0
1

3. (a)

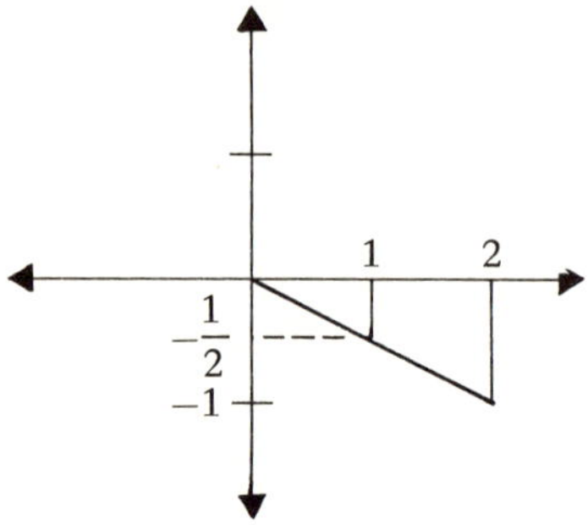
1
2
$-\frac{1}{2}$
−1

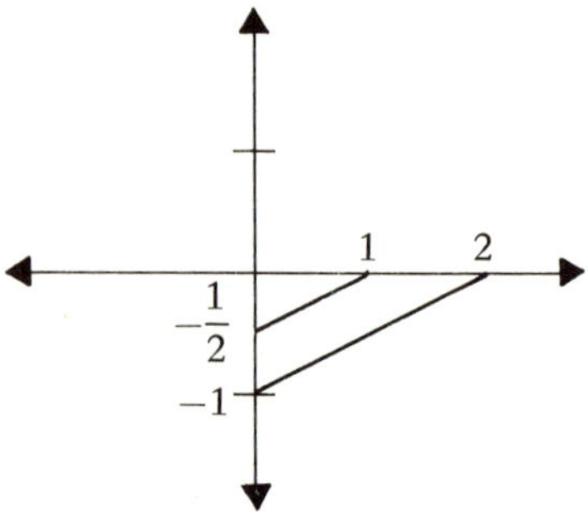
1
2
$-\frac{1}{2}$
−1

(e)

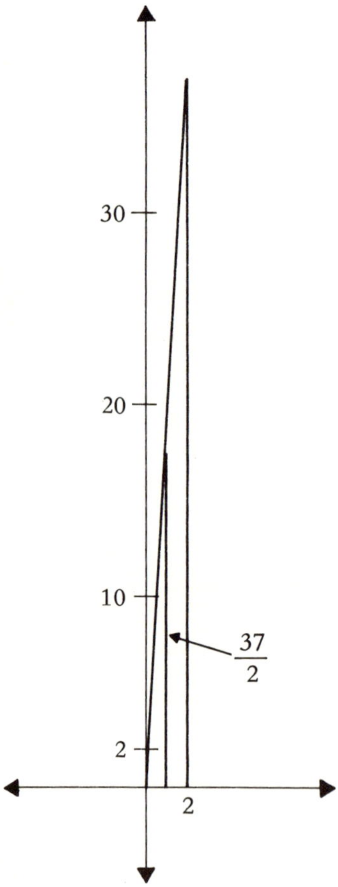
30
20
10
2
2
$\frac{37}{2}$

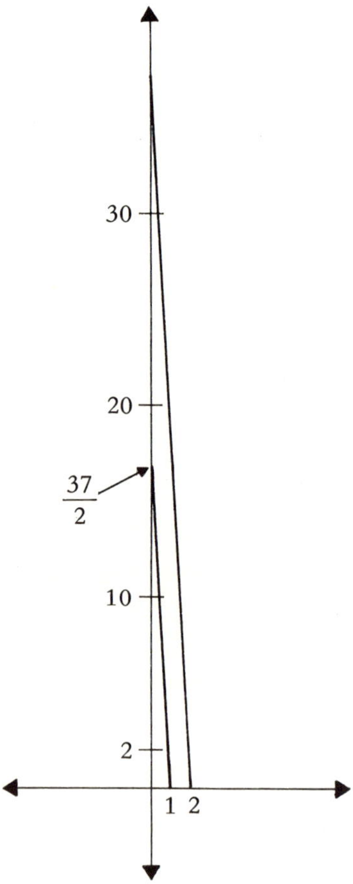
30
20
10
2
$\frac{37}{2}$
1 2

(g)

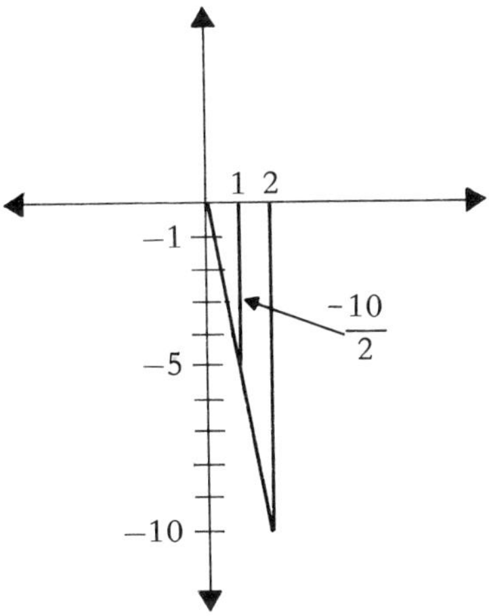

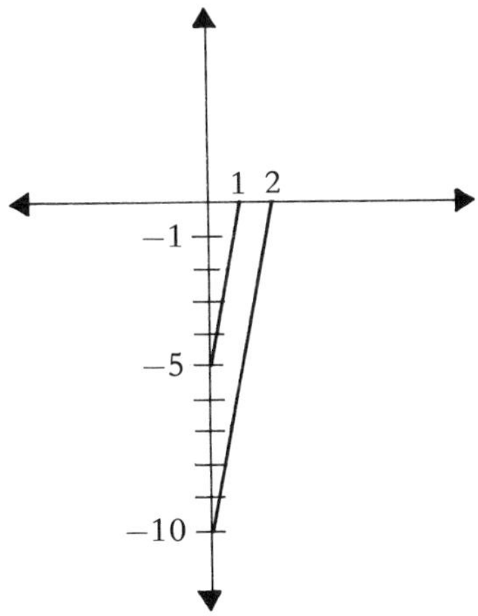

4. Some of the many possible answers are:
   (a) "14/2," "21/3," "28/4," "35/5," "42/6," "49/7."
   (c) "$^{-}729/^{-}10{,}935$," "1458/21,870," "1458/$^{-}21{,}870$," "5832/43,740," "$^{-}5832/^{-}43{,}470$," "7290/109,350."
   (e) "$^{-}7/1$," "$7/^{-}1$," "$14/^{-}2$," "$28/^{-}4$," "$^{-}28/4$," "$^{-}7000/1000$."

5. (a)

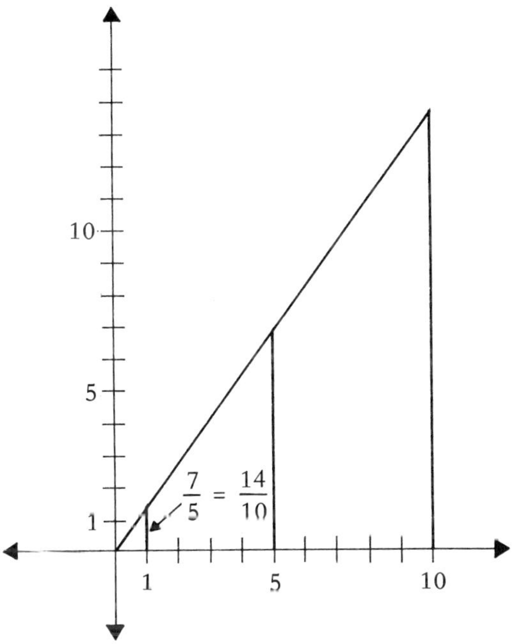

6. (a) To split the segment from A to B into 4 pieces of equal length, construct a line at A perpendicular to AB. Along this line lay off 4 equally spaced points P_1, P_2, P_3, and P_4. Connect P_4 and B by a line L_4. Through P_1, P_2, and P_3 draw

lines L_1, L_2, and L_3 parallel to L_4. These lines will intersect AB in the points Q_1, Q_2, and Q_3 which will divide AB into 4 pieces of equal length.

Section 8.3

1. (a) 4/1, (c) $-2/3$, (e) 0/1, (g) $-6/1$, (i) 5/6, (k) 6857/1092, (m) 10361/51, (o) 33/70.
   (Note: The problem does not insist that the fractions we give as answers be in the simplest form. So other answers are possible. But the fractions given should name the same rational numbers as those above.)
2. By the definition of addition in F $a/b +_F c/b = (ab + cb)/b^2$. But we may use the distributive law and Theorem 8.1.3 to get $(ab + cb)/b^2 = [(a + c)b]/b \cdot b = (a + c)/b$. Here, of course, we assume that $b \neq 0$ since if $b = 0$ then both a/b and c/b are undefined.
3. (a) $-1/2$, (c) $-24/1$, (e) $-1/9$, (g) $-1/25$, (i) 23/41.
5. Yes. F is closed under $+_F$, $+_F$ is associative, 0/1 is an identity element for $+_F$, and $-a/b$ is an inverse element for a/b with respect to $+_F$.
7. Yes. S is closed under $+_F$ since $a/2 +_F b/2 = (a + b)/2$. $+_F$ is clearly associative. 0/2 is the identity and $-a/2$ is the inverse for $a/2$ with respect to $+_F$.

Section 8.4

1. (a) 7/6, (c) $-4/3$, (e) 25/3, (g) 0/1, (i) $-1/1$, (k) 24/1.
   (Note: The problem does not *require* the simplest form for the fraction, so other (equivalent) answers are possible.)
2. (a) $1/2 \div_F 2/1 = 1/2 \cdot_F 1/2 = 1/4$ while $2/1 \div_F 1/2 = 2/1 \cdot_F 2/1 = 4/1$ and $1/4 \neq 4/1$.
   (Note: These are, of course, only examples among many possible ones.)
3. Yes. $(a/b) \cdot_F (c/d) = (ac)/(bd)$. I is closed under multiplication so we know that ac and bd are in I. Also $bd \neq 0$ when $b \neq 0$ and $d \neq 0$. Hence $(ac)/(bd)$ is in F.
5. No. A single example (among many possible ones) will suffice. We have $2/1 \div_F (2/1 +_F 3/1) = 2/1 \div 5/1 = 2/5$ whereas $(2/1 \div_F 2/1) +_F (2/1 \div 3/1) = 1/1 +_F 2/3 = 5/3$.
7. (a) Yes. Q is closed under $\cdot_F$ since $x \cdot_F y = 0$ implies $x = 0$ or $y = 0$ in F. $\cdot_F$ is associative and 1 is the identity element. If a/b is in Q, then so is b/a which is its inverse with respect to $\cdot_F$.

Section 8.5

1. (a) Yes. h is clearly one-to-one between T and I, and we have $h(a/1 +_F b/1) = h((a + b)/1) = -(a + b) = {}^-a + {}^-b = h(a/1) + h(b/1)$.
3. (a) 5 is the collection of all sets which can have exactly as many members as $\{a,b,c,d,e\}$.
   (c) 5 is the collection $\{(5,1),(^-5,^-1),(10,2),(^-10,^-2),(15,3),(^-15,^-3),\ldots\}$.
4. (a) Yes, since $a/1 +_F b/1 = (a + b)/1$ and I is closed under addition so that $(a + b)/1$ is in T.
   (c) Yes, since $a/1 -_F b/1 = (a - b)/1$ and I is closed under subtraction.

5. No. G is not a one-to-one correspondence between F and itself since $G(0/1)$ is not in F.
7. (a) 85/9, (c) 195/148, (e) 700/351.

Section 8.6

1. Yes. The equivalence classes are $\{0\}$, F^+, and F^-.
2. (a) No. For example $2/1 -_F 4/1$ is not in F^+.
3. (a) Yes. Suppose x and y are in F^-. Then there are negative integers a and c and positive integers b and d such that $x = a/b$ and $y = c/d$. Then $x + y = a/b + c/d = (ad + cb)/bd$. But ad and cb are negative integers and I^- is closed under addition. So $ad + cb$ is in I^-. Since I^+ is closed under multiplication bd is in I^+. Hence, $(ad + cb)/bd$ is in F^-.
   (c) No. For example $(-1/1)(-2/1)$ is not in F^-.
4. Suppose x and y are in F^+. Then there are positive integers a, b, c, and d such that $x = a/b$ and $y = c/d$. Then $x + y = a/b + c/d = (ad + cb)/bd$. But I^+ is closed under addition and multiplication so that $ad + cb$ and bd are in I^+. Hence, $x + y$ is in F^+ and we see that F^+ is closed under multiplication. We also have $x \cdot y = (a/b) \cdot (c/d) = (ac)/(bd)$. But I^+ is closed under multiplication so both ac and bd are in I^+. Hence, $x \cdot y$ is in F^+ and we see that F^+ is closed under multiplication.
6. (a) 144/5, (c) 324.
7. (a) 160,000/3, (c) 4,300,000/3.
8. (a) 1/10, (c) 200.

Section 8.7

1. (a), (d), (e), (h), (i), and (j) are true.
2. (a) max = 1/2, no min. (c) min = −1/2, no max.
   (e) min = −1/17, max = 1/17. (g) min = 1/3, max = 3/2.
3. (a) Let P be the selling price. Then we want $520P \geq 1780$. That is, we must have $P \geq 1780/520$ dollars per item.
   (c) Let V be the number of people in the town who are of voting age. Then there are $7500 - V$ who are below voting age. We know that $(7500 - V)/7500 \geq 1/3$. Then $7500 - V \geq 2500$ and, hence, $V \leq 5000$.
4. Let a, b, and c be positive integers. By Theorem 8.7.3 $a/b \geq a/c$ when and only when $ac \geq ba$. But $ba = ab$ so $a/b \geq a/c$ when and only when $ac \geq ab$. But by Theorem 7.9.1(a) this is true if and only if $c \geq b$.
7. Let f be the function defined on Q by $f(a/1) = a$; then f is the desired isomorphism.

Section 8.8

1. (a) no max, min, lub, or glb,
   (c) max = min = glb = lub = 1,
   (e) min = glb = $^-1$, max = lub = $^+1$,
   (g) no min, glb = $^-1$, max = lub = 2,
   (i) no min, glb = $^-1$, no max, lub = 2,
   (k) no min, no glb, no max, lub = $^-5$,
   (m) min = glb = $^-17$, no max, no lub,

(o) min = glb = max = lub = 13/5,
(q) min = glb = $^-1$, max = lub = 5,
(s) no min, glb = 0, max = lub = 1/10,
(u) no min, no glb, no max, no lub.

2. Many values for n will do the job. Some that will are: (a) 101, (b) (679)(97,453) + 1, (c) $30 \times 10^{12} + 1$, (d) 1.

3. (a) Suppose $x = \max(A)$. Then $x \in A$ and $x \geq y$ for all y in A. Hence x is an upper bound for A. But we also see that no number smaller than x can be an upper bound for A, since $x \in A$. Hence, $x = \text{lub}(A)$.

5. (i) From the description of the set, it is clear that $^-1$ is a lower bound for the set. To show that $^-1$ is the glb suppose that $z > {}^-1$. We know that there is a y such that $^-1 < y < z$. But then y is in the set and, hence, z cannot be a lower bound. Hence, $^-1$ is the glb. By Theorem 8.8.1, if the set has a min, it must be $^-1$. But $^-1$ is not in the set, so there cannot be a min. Similar arguments apply at the right end.
(s) The set is $\{1/10, 1/100, 1/1000, \ldots\}$. Clearly, 1/10 is the max and the lub of this set. Since 10^n is positive for all n, so is $1/10^n$ and, therefore, 0 is a lower bound for the set. Suppose $z > 0$. We claim that we can find an n such that $1/10^n < z$. This inequality is equivalent to $10^n > 1/z$. Now $1/z$ can be written as a mixed number, say as $m\frac{p}{q}$ where $0 < p/q < 1$. m has a decimal name. Let k be the number of digits in this name. Take $n > k + 1$. Then $10^n > m + p/q$ and, hence, for this n we have $10^n > 1/z$ or $1/10^n < z$. It follows that no positive number can be a lower bound for the set, and hence, 0 is the glb. Since 0 is not in the set, Theorem 8.8.1 tells us that the set cannot have a min.

6. No, a set may be unbounded and yet have a glb (or a lub). An example would be F^+ (F^-). It can be said, however, that if S is unbounded, then either glb(S) or lub(S) must fail to exist.

7. (i) The sets in (c), (d), (e), (f), (g), (h), (i), (o), (q), (r), (s), and (v) are bounded.

8. glb(S) = 0 since $k/10^n$ can be made arbitrarily close to 0 by taking n big enough.

Section 8.9

1. (a) $4\frac{7}{10}$, (c) $64\frac{17}{74}$, (e) $-4\frac{67}{101}$, (g) $-20\frac{3}{47}$.

2. (a) 46/7, (c) 4357/100, (e) 374,940/1000, (g) 275/16. (Note that some of these fractions can be simplified to obtain other answers that are, of course, also correct.)

3. (a) $13\frac{5}{7}$, (c) $-91/102$, (e) $32\frac{25}{51}$, (g) 76/3, (i) 140/289, (k) $-147/17$. (Note that it is a subjective matter as to which form is more convenient so that equivalent expressions are surely acceptable as answers here.)

4. The areas from left to right with the top row first are 1/2, 1/4, 1/8, 1/2, 1/2, 39/20, and 133/32 square inches.

6. (a) 3/4 sq in. (c) 19/16 sq in.

7. (a) 9/8 sq in.

8. (a) 27/64 sq in., (c) 63/64 sq in., (e) 153/128 sq in.

Section 8.10

1. (a) .4, (c) 5.27, (e) 7, (g) .0000364, (i) .0000436.
2. (a) 2/5, (c) 33/100, (e) −81/10, (g) 111/20,000.
3. We give only the final results except in part (a) where the method asked for is displayed. (a) .3 + .5 = 3/10 + 5/10 = 8/10 = .8, (c) 3.604, (e) .1575, (g) .02, (i) 2000, (k) 50, (m) 50, (o) 5,347,600,000.
5. When we look at the decimal representations for the products $1 \cdot 7$, $2 \cdot 7$, $3 \cdot 7$, ..., $9 \cdot 7$, we see that none of them ends in a "0". If 1/7 can be written as a decimal fraction then there must be integers k and n such that $1/7 = k/10^n$ and k is not a multiple of 10. For this k and n we would have $7 \cdot k = 10^n$. But k does not have a "0" in the units place in its decimal representation, hence $7 \cdot k$ does not either, and this is impossible if $7k = 10^n$.
6. (a) 1/7,000,000.
7. (a) Suppose that $x = 1/2^n$, then $x = 5^n/2^n \cdot 5^n = 5^n/10^n$ and, hence, x can be written as a decimal fraction.

Section 8.11

1. We give only the final result except in case (a) where a method is displayed. (a) 6.43 + 2.59 = 643/100 + 259/100 = 902/100 = 9.02, (c) 37.1000552, (e) −326.82, (g) 6.593, (i) .42, (k) −51.75, (m) 375, (o) 33330, (q) 390,000, (s) 724,000, (u) 500,000.
2. All except (c) and (g). (a) 3/25 = .12, (c) 2/15 = (2/3)(1/5). So if $x = 2/15$ and x can be expressed as a decimal fraction then so can $5x = 2/3$. But 2/3 cannot be so expressed. (Another justification would be to carry out the long division and show that the process does not terminate.) (e) 3/8 = .375, (g) the division algorithm for 4/9 yields .444 . . . which does not terminate.
3. (a), (b), and (c) are true as may be established by the division algorithm. (d) is false since $3/31 = .096\frac{24}{31}$.

Section 8.12

1.

x \ y	−3	−2	−1	0	1	2	3
−2	−1/8	1/4	−1/2	1	−2	4	−8
−1	−1	1	−1	1	−1	1	−1
−1/2	−8	4	−2	1	−1/2	1/4	−1/8
1/10	−1000	100	−10	1	−1/10	1/100	−1/1000
0	–	–	–	–	0	0	0
1/10	1000	100	10	1	1/10	1/100	1/1000
2/10	1000/8	25	5	1	2/10	1/25	1/125
1/2	8	4	2	1	1/2	1/4	1/8
1	1	1	1	1	1	1	1
2	1/8	1/4	1/2	1	2	4	8

3. (a) −1, (c) 1, (e) 1, (g) −1.

5. We shall do an induction on n. Let $A = \{n | n$ is a positive integer and $x^n = 0$ implies $x = 0\}$. We see that 1 is in A since $x^1 = x$ and $x^1 = x = 0$ clearly implies $x = 0$. Suppose k is in A. Assume $x^{k+1} = 0$ for some x in F. We know that $x^{k+1} = x^k \cdot x$. By Theorem 8.4.7, $x^k \cdot x = 0$ implies $x^k = 0$ or $x = 0$. But since k is in A, $x^k = 0$ implies $x = 0$. Hence, in any case, $x^{k+1} = 0$ implies $x = 0$. Therefore, $k + 1$ is in A. We see then that $A = I^+$ and this completes the proof.

6. (a) 1/3, (c) .1849, (e) 1/2, (g) 100, (i) 5/3, (k) 4.

7. (a) \$125.

8. (a) \$150.

9. $(50{,}000)/(1.05)^5 \cong 39{,}176.31$.

11. (a) $3 + 2(3/4) \cdot 3 = 7\frac{1}{2}$ ft, (c) 429/32 ft, (e) 8565/512 ft.

12. (a) 3.6 ft, (c) 3.666 ft, (e) 3.66666 ft.

13. The members of the solution set for each equation are: (a) 1/2, −1/2, (c) .1, (e) 1, −1, (g) 0, 1, (i) 1/2, (k) 1.

15. Let x and y be rational numbers.
    (a) We have $(x/y)^1 = x/y = x^1/y^1$. If $(x/y)^k = x^k/y^k$ for some positive integer k, then we have $(x/y)^{k+1} = (x/y)^k(x/y) = (x^k/y^k)(x/y) = (x^kx)/(y^ky) = x^{k+1}/y^{k+1}$. Hence, by induction $(x/y)^n = x^n/y^n$ for all positive integers y.
    (c) Suppose $n \geq m$. If $n = m$, then $x^n/x^m = 1 = x^0 = x^{n-m}$. If $n > m$, then there is a positive integer q such that $n = m + q$. Hence, by part (b) of this problem, $x^n/x^m = x^{m+q}/x^m = (x^m \cdot x^q)/x^m = x^q = x^{n-m}$.

16. Let x and y be rational numbers. Let n and m be positive integers.
    (a) We have $(xy)^{-n} = 1/(xy)^n = 1/(x^ny^n) = (1/x^n)(1/y^n) = x^{-n} \cdot y^{-n}$ where the second equality is justified by part (b) of the preceding problem.

17. Let x be a rational number and suppose that n is a positive integer and m is a negative integer. Let $k = -m$. Then k is positive.
    (a) We have $x^n \cdot x^m = x^n \cdot x^{-k} = x^n(1/x^k) = x^n/x^k$. If $n \geq k$, then $x^n/x^k = x^{n-k} = x^{n+m}$ by part (c) of Problem 15. If $n < k$, then there is a positive integer q such that $n + q = k$ and we have $x^n/x^k = x^n/x^{n+q} = x^n/(x^n \cdot x^q) = 1/x^q = x^{-q} = x^{-(k-n)} = x^{-k+n} = x^{n+m}$.

Section 8.13

1. (a) 5, (c) −1, (e) −2, (g) 2, (i) 10, (k) 10, (m) 3, (o) .9.

2. (a) $25^{1/2} = 5$, (c) $(-1)^{1/3} = -1$, (e) $(-8)^{1/3} = -2$, (g) $64^{1/6} = 2$, (i) $10{,}000^{1/4} = 10$, (k) $(100{,}000{,}000)^{1/8} = 10$, (m) $81^{1/4} = 3$, (o) $(.81)^{1/2} = .9$.

3. (a) 0, (c) 343, (e) 1000, (g) 100,000, (i) 3, (k) 1000, (m) undefined, (o) 10,000/81.

4. (a) $\{1,-1\}$, (c) $\{-1\}$, (e) ϕ, (g) $\{-1\}$, (i) $\{2\}$, (k) $\{10,-10\}$, (m) $\{1/2\}$, (o) $\{2,-2\}$, (q) $\{1/2,-3/2\}$, (s) $\{1,5/2\}$, (u) $\{-3\}$.

5. The following examples are some of the many possible ones. (i) Let $x = y = -1$ and $z = 1/2$. (ii) Same as for (i). (iii) $x = -1$, $y = 1/2$, $z = -1/2$. (v) $x = -1$, $y = 1/2$, $z = 2$.

7. Let x, y, and z be rational numbers and suppose that $(x^y)^z$ is defined. Let a, b, c, and d be integers such that $y = a/b$ and $z = c/d$. Let $x^y = u$ and $(x^y)^z = u^z = v$. From $x^y = x^{a/b} = u$ we get that $x^a = u^b$. From $u^z = u^{c/d} = v$ we get $u^c = v^d$. Now $x^z = u^b$ implies $(x^a)^c = (u^b)^c$ or $x^{ac} = u^{bc}$ since we may apply Theorem 8.12.1(e). Using this same theorem again we see that $u^c = v^d$ implies $(u^c)^b = (v^d)^b$ or $u^{bc} = v^{bd}$. It follows that $x^{ac} = v^{bd}$. Note that if either b or d is even, then v is positive. And bd will be even only if b or d is even. Hence, we have it that $v = x^{ac/bd} = x^{yz}$.

8. In each case we give values for x, y, and z that make the hypothesis true and the conclusion false. Of course, the values we give are only particular ones among many possible. (a) $x = -2$, $y = 1$, $z = 2$; (b) $x = 2$, $y = 3$, $z = -2$; (c) $x = 1$, $y = 2$, $z = -1$; (d) $x = 1$, $y = 2$, $z = 0$; (e) $x = -2$; (f) $x = y = 1$, $z = 2$; (g) $x = 9$, $y = 16$.

Section 8.14

1. (a) 5/8, (c) 341/32, (e) 2032/81, (g) 15,251/20,736.
2. (a) $.001_2$, (c) cannot be done, (e) 101.111_2.
3. (a) $.01_3$, (c) 1.11_3, (e) cannot be done.
4. (a) $.01_{12}$, (c) $.109_{12}$, (e) $.1594_{12}$.

Section 8.15

1. (a) 4.8, (c) -1.7, (e) -7.1.
4. (a)

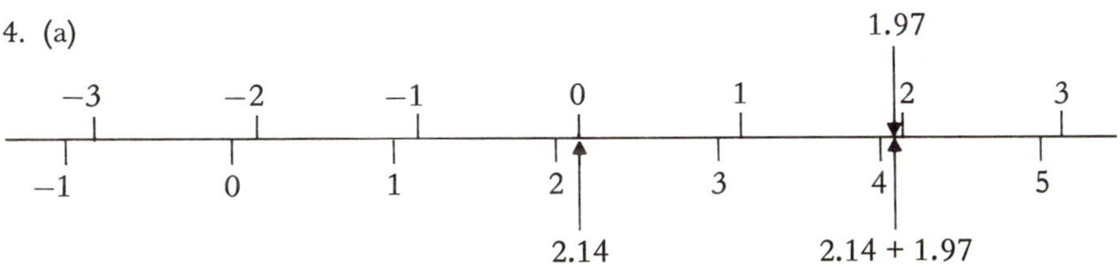

5. (a) 3, (c) $-3/2$, (e) .9, (g) $-.9$.
6. To find the result x of dividing z by y, draw a horizontal line through z on the vertical axis, find the point where this line hits the line labelled y, draw a vertical line through this point, and observe the point x where this vertical hits the horizontal axis. For example, the horizontal line through $z = 3$ hits the line $y = 2$ directly above the point $x = 3/2$. The horizontal through $z = -4$ hits the line $y = -2$ at a point directly below $x = 2$. Et cetera.
8. (a) On the two scales lay off equally spaced points for 1/2048, 1/1024, 1/512, . . . , 512, 1024, 2048.
9. Yes. Since $3^n \cdot 3^m = 3^{n+m}$ and I is closed under addition, T is closed with respect to multiplication. Multiplication is clearly associative in T. $3^0 = 1$ is the identity. 3^{-n} is the inverse element for 3^n.

11. (a) B is closed under multiplication since $b^n \cdot b^m = b^{n+m}$ and I is closed under addition. Multiplication is associative in B since it is in F. $b^0 = 1$ is the identity element. b^{-n} is the inverse element for b^n.

Section 8.16

1. (a) $2^4 \cdot 3$, (c) $13 \cdot 17$, (e) 7^4, (g) $2^6 \cdot 5^6$, (i) $2^6 \cdot 3^3$, (k) 101, (m) $41 \cdot 271$.
2. (a) 8, (c) 1, (e) 49, (g) 101.
3. (a) 6/125, (c) 987/250, (e) 17/49, (g) 1/12, (i) 17/391, (k) 52/9.
4. (a) $2^4 \cdot 3 \cdot 5^3$, (c) $2^3 \cdot 3 \cdot 5^3 \cdot 7 \cdot 47$, (e) $7^4 \cdot 17$, (g) 11211.
5. (a) 25/49, (c) 31/100, (e) 17/72, (g) 6982/169, (i) 221/5328, (k) 245/13,824.
6. (a) 25/12, (c) 3/4, (e) 15/16, (g) 21/16, (i) 422/243.
8. If p/q is a fraction in lowest terms, it can be expressed as a base b fraction if and only if the only prime factors that occur in q also occur in b. In particular, if b is a prime, p/q can be expressed as a base b fraction if and only if q is a power of b.

Section 8.17

1. The systems given in (b) and (d) are fields. In (a), $\{1\}$ has no inverse element with respect to U. In (c), L does not distribute across M.

2. (a) Let x be a member of S. We know that $1 \boxdot x = x$ and, therefore, $(1 \boxdot x) \boxplus ((-1) \boxdot x) = x \boxplus ((-1) \boxdot x)$. Since $\boxdot$ distributes across $\boxplus$, this implies $(1 \boxplus (-1)) \boxdot x = x \boxplus ((-1) \boxdot x)$. But $1 \boxplus (-1) = 0$ and, also, $0 \boxdot x = 0$ by part (b) above. Hence, $x \boxplus ((-1) \boxdot x) = 0$ and since addition is commutative, $((-1) \boxdot x) \boxplus x = 0$. Then by the definition of $-x$, the additive inverse for x, is equal to $(-1) \boxdot x$.
   (c) If x is in S we have $x \boxplus (-x) = 0$ and, hence, $(x \boxplus (-x)) \boxplus (-(-x)) = 0 \boxplus (-(-x)) = -(-x)$. Since addition is associative, this implies $x \boxplus ((-x) \boxplus (-(-x))) = -(-x)$. But $(-x) \boxplus (-(-x)) = 0$ and, therefore, $x \boxplus 0 = x = -(-x)$.
   (e) Suppose x is in S and $x \neq 0$. Then $x \boxdot x^{-1} = 1$ and $x^{-1} \boxdot x = 1$. By the definition of the multiplicative inverse, it follows that $(x^{-1})^{-1} = x$.

4. Suppose x is in S. If $0 = 1$ in S, we have $x = 1 \boxdot x = 0 \boxdot x = 0$. Hence, if $1 = 0$ in a field every element in S is equal to 0. This proves the contention.

6. Suppose k is not a prime. Then there are positive integers p and q less than k such that $k = pq$. But then $[p] \cdot_k [q] = [k] = [0]$ and yet $[p] \neq [0]$ and $[q] \neq [0]$. But in view of the result in Problem 2(f) this is impossible if the system is a field.

7. (a) Suppose $ij = ik \pmod{p}$. Then $ij - ik$ is divisible by p. But $ij - ik = i(j - k)$. Both i and $|j - k|$ are less than p. Now if $i(j - k) \neq 0$ and it is written as a product of primes all the prime factors that occur must be less than p. But then $i(j - k)$ cannot be divisible by p unless it is 0. This implies $j = k$.
   (c) Since $[i]$, $[2i]$, . . . , $[(p - 1)i]$ are all distinct and different from $[0]$, one of them must be the class $[1]$. Hence, $[i]$ has a multiplicative inverse.

Section 9.1

Sequence	*Problem 1*	*Problem 2*	*Problem 3*
(a)	.33333, .333333	$(1 - 10^{-n})/3$	not null
(c)	1, −1	$(-1)^{n+1}$	not null
(e)	1/6, 1/7	$1/n$	null
(g)	13, 15	$2n + 1$	not null
(i)	25, 20	$45 - 5(n - 1)$	not null
(k)	6/7, 7/8	$n/(n + 1)$	not null
(m)	60/256, 60/1024	$60/4^{n-1}$	null
(o)	10^5, 10^6	10^n	not null
(q)	1/4 − .249999, 1/4 − .2499999	$(.1)^{n+1}$	null
(s)	.00001, −.00001	$\begin{cases} (.1)^{n/2} \text{ for even } n \\ -(.1)^{(n-1)/2} \text{ for odd } n \end{cases}$	null
(u)	−1/18, −1/21	$-1/2n$	null
(w)	$497^{674}/32$, $497^{674}/64$	$497^{674}/2^n$	null

4. (b), (g), (h), and (i).

6. (f) and (j) (only) insure that the sequence is null. (b), (e), (g), and (k) insure that the sequence is not null.

7. (a) Let $z_n = n$ and $w_n = -n$.
(This, of course, is only one among many possible examples.)

8. (a) Let $z_n = 1/n^2$ and $w_n = 1/n$. Many other examples are possible.

10. Let $(z_1,z_2,z_3, \ldots)$ be a null sequence. Find N such that $n > N$ implies $|z_n| < 1$. Let $A = \max\{|z_1|,|z_2|, \ldots, |z_n|\}$. Let $B = \max\{A,1\}$. Then $|z_n| < B$ for all n. That is, $(z_1,z_2,z_3, \ldots)$ is bounded.

Section 9.2

1. (a) 1/3, (c) no limit, (e) 0, (g) no limit, (i) no limit, (k) 1, (m) 0, (o) no limit, (q) 0, (s) 0, (u) 0, (w) 0.

2. The following are some of the many possible examples. (a) Let $z_n = n + (1/n)$, $w_n = (1/n) - n$; (c) Let $(z_1,z_2,z_3, \ldots) = (1,1,2,1/2,3,1/3,4,1/4,5,1/5, \ldots)$ and $(w_1,w_2,w_3, \ldots) = (2,1,5/4,2,10/9,3,17/16,4,26/25,5, \ldots)$.

3. (a), (c), (d), (e), (h), (i), (m), and (o) insure convergence. (f), (k), and (p) insure divergence.

4. (a) No. E fails to be reflexive on the set since, for example, $\lim_{n \to \infty} n$ does not exist and we cannot, therefore, assert that $(1,2,3, \ldots)$ is in relation E to itself.

5. (a) No. If $(z_1,z_2,z_3, \ldots)$ and $(z_1 + w_1,\ z_2 + w_2,\ z_3 + w_3, \ldots)$ are both convergent, then so is $((z_1 + w_1) - z_1,\ (z_2 + w_2) - z_2,\ (z_3 + w_3) - z_3, \ldots) = (w_1,w_2,w_3, \ldots)$.
   (c) No. If $(z_1,z_2,z_3, \ldots)$ and $(w_1/z_1,w_2/z_2,w_3/z_3, \ldots)$ are both convergent, then so is $((w_1/z_1)z_1,(w_2/z_2)z_2,(w_3/z_3)z_3, \ldots) = (w_1,w_2,w_3, \ldots)$.

Section 9.3

1. (i) (a), (b), (g), (h), (j), (k), (o), (r), (u), (v), and (x).
   (iii) All in part (i) except (b).

4. (a) Impossible since the z_n's cannot get close to 2 if $z_n \leq 1$ for all n.
   (c) Let $z_n = -n$.
   (e) Impossible. If $z_4 = 5$ and the sequence is strictly increasing then $z_{5+k} > z_5 > z_4 = 5$ for all k. Hence, z_n cannot be close to 5 far out in the sequence.
   (g) Impossible. Such a sequence must have terms larger than any given fixed number.
   (i) Impossible. If $z_{n+1} \geq z_n$ for all n or $z_{n+1} \leq z_n$ for all n, then $-z_{n+1} \leq -z_n$ for all n or $-z_{n+1} \geq -z_n$ for all n.

Section 9.4

1. In each case we display two subsequences that converge to different limits. (a) (0,0,0, . . .) and (1,1,1, . . .), (c) (1/2,1/3,1/4,1/5, . . .) and (2/1,3/2,4/3,5/4,6/5, . . .), (e) (1/2,3/4,5/6,7/8, . . .) and (−2/3,−4/5,−6/7,−8/9, . . .). (Note that other choices are possible in all parts.)

3. There are, of course, many possible examples. Perhaps the simplest is (1,2,3,1,2,3,1,2,3,1,2,3, . . .).

4. There are many possible answers in each case. We give only one for each case. (a) (1/2,1/3,1/4,1/5, . . .), (c) (1/2,2/4,3/6,4/8, . . .), (e) (2/72,3/144,4/216,5/288, . . .), (g) Every positive rational number appears as a term in the sequence. In fact, every positive rational number appears infinitely often in the sequence.

6. We give only one of many possible examples. (a) (1,−1,1/2,−1/2,1/3,−1/3,1/4, −1/4,1/5, . . .).

7. One of the many possible examples would be (1,0,1,0,1,0,1,0, . . .) and (0,1,0,1,0,1, 0,1, . . .).

Section 9.5

1. In each case, we give the sequence of partial sums and then the limit.
   (a) (1,1,2,2,3,3, . . .), none.
   (c) (1,1.2,1.24,1.248,1.2496, . . .), 1.25.
   (e) (.31,.3131,.313131, . . .), 31/99.
   (g) (.9,1.71,2.439,3.0951, . . .), 9.
   (i) (.357,.357357,.357357357, . . .), 357/999.
   (k) (.572,.5721,.57211,.572111, . . .), 5149/9000.
   (m) (.07,.0707,.070707, . . .), 7/99.

3. Since $S_{n+1} = S_n + a_{n+1}$ when $S_k = a_1 + a_2 + \ldots + a_k$, we have $S_{n+1} > S_n$ when $a_n > 0$.

4. The series (i), (ii) and (iii) have the sequences of partial sums $(S_1 + T_1, S_2 + T_2, S_3 + T_3, \ldots)$, $(S_1 - T_1, S_2 - T_2, S_3 - T_3, \ldots)$, and $(kS_1,kS_2,kS_3, \ldots)$ respectively. But by our results in Section 9.2, these three sequences converge if $(S_1,S_2, \ldots)$ and $(T_1, T_2, \ldots)$ do.

7. The table displays the value of r^n correct to 9 decimal places.

n \ r	.9	.99	.9999
5	.59049	.950990050	.99950010
10	.348678440	.904382075	.999000450
50	.005153775	.605006067	.995012230
100	.000026561	.366032341	.990049339

9. (a) $S_{2n+2} - S_{2n} = b_{2n+2} + b_{2n+1} = -a_{2n+2} + a_{2n+1}$. But $a_{2n+2} \leq a_{2n+1}$ so $-a_{2n+2} + a_{2n+1} \geq 0$ and the desired result follows.

10. If $(S_1,S_2, \ldots)$ converges to S then every subsequence of $(S_1,S_2, \ldots)$ must converge to S. The subsequences $(S_1,S_3,S_5, \ldots)$ and $(S_2,S_4,S_6, \ldots)$ are monotonic. Since they converge to S we have $S_{2n} \leq S \leq S_{2n+1}$ for all n. Then S cannot differ from S_{2n} and S_{2n+1} by more than $S_{2n+1} - S_{2n} = a_n$. That is, $|S_n - S| \leq a_n$ for all n.

12. The time required for Achilles to catch the tortoise is $1/10 + \left(5 \cdot \frac{1}{10}\right)\Big/500 + \left(5 \cdot \frac{1}{1000}\right)\Big/500 + \left(5 \cdot \frac{1}{10^5}\right)\Big/500 + \left(5 \cdot \frac{1}{10^7}\right)\Big/500 + \cdots = 1/10 + 1/10^3 + 1/10^5 + 1/10^7 + \ldots = (.1)/(1 - .01) = 10/99$ minutes.

Section 9.6

1. (a) 56/99, (c) 2/3, (e) 537,084/99,900, (g) 572,802/999.
2. (a) $6.\overline{0}$, (c) $.375\overline{0}$, (e) $.\overline{428571}$, (g) $58.\overline{1}$.
4. Yes, with this restriction each rational number has exactly one infinite decimal associated with it and each repeating decimal is associated with a rational number.

Section 10.1

1. The sequences in (c), (d), (e), (f), (g), (h), (j), (k), (l), (m), and (n) are in S. Grouping together those sequences in S that are in the relation E to one another we have the classes {c}, {d,g}, {e}, {f,k}, {h}, {j,l,m}, {n}.
3. $b_1 = 1.7$, $b_2 = 1.73$, $b_3 = 1.732$, $b_4 = 1.7320$. If $b = [(b_1,b_2,b_3,\ldots)]$ we would expect to have $b^2 = 3$.
5. The numbers given in (a), (c), (e), and (g) are rational numbers.
7. Let "p/q" be in lowest terms. Then if we write p and q as products of primes, there must be no prime that appears in both factorizations. (We are assuming that we will allow no prime to appear with a 0 exponent.) But the prime factorizations of p^n and q^n can be obtained from those for p and q by simply multiplying each exponent on each prime by n. We see then that p^n and q^n will have no prime factor in common—and, hence, no factor at all in common. That is, "p^n/q^n" is in lowest terms.
9. We may take as an example the sequence given in Problem 1, part (m).
11. By Problem 10, if $(x_1,x_2, \ldots)E(0,0,0, \ldots)$ and some x_k is positive, then some term in $(0,0,0, \ldots)$ must be positive. And this is clearly not the case.

Section 10.2

1. (a) Let $x = [(.01,.01001,.010010001, \ldots)]$. Then x is the number we invented to be the limit of the sequence $(.01,.01001,.010010001, \ldots)$, or to be the lub of

the set $\{.01,.01001,.010010001,\ldots\}$, or to be the number named by ".010010001 ...", or to be the coordinate of the point approached by the rational points .01, .01001, .010010001, ...

(c) Let $z = [(1.2,1.25,1.259,1.2599,\ldots)]$ as defined in Problem 4 in Section 10.1. Then z is the number we invented to be the limit of the sequence $(1.2,1.25,1.259,1.2599,\ldots)$, the lub of the set $\{1.2,1.25,1.259,1.2599,\ldots\}$, the number named by the infinite decimal "1.2599 ...", the coordinate of the point approached by the rational points $1.2,1.25,1.259,1.2599,\ldots$, the $\sqrt[3]{2}$.

(e) The real number $[(.6,.60,.600,\ldots)]$ is the number we invented to be the limit of the sequence $(.6,.60,.600,\ldots)$, the lub of the set $\{.6\}$, the number named by the infinite decimal ".6000 ...," the coordinate of the point approached by the points $.6,.6,.6,\ldots$, the real number that behaves like the rational number .6.

2. The first five terms of the sequence are .04,.0119,.00064,.00003836,.000010076.

4. Let $x = [(x_1,x_2,x_3,\ldots)]$ and let $y = [(x_{k+1},x_{k+2},x_{k+3},\ldots)]$. We may think of x as the lub of $\{x_1,x_2,x_3,\ldots\}$ and y as the lub of $\{x_{k+1},x_{k+2},x_{k+3},\ldots\}$. The first set contains the numbers $x_1,x_2,\ldots,x_k$ that may not be in the second set. But $x_1 \leq x_{k+1}$, $x_2 \leq x_{k+1},\ldots$, and $x_k \leq x_{k+1}$ so the removal of the numbers $x_1,x_2,\ldots,x_k$ from the set cannot change the lub of the set. That is, we must have $x = y$.

6. Yes, E is an equivalence relation on T. The proofs given for E on S apply equally well for T. However, aside from the equivalence classes of bounded sequences which coincide with the real numbers, there are also many different equivalence classes corresponding to unbounded sequences. For example, if $s_1 = (1,2,3,4,\ldots)$, $s_2 = (1 + 1,2 + 1/2,3 + 1/3,4 + 1/4,\ldots)$, $s_3 = (3,4,5,6,\ldots)$, and $s_4 = (1,4,9,16,\ldots)$ then s_1Es_2 and yet it is false that s_1Es_3, s_3Es_4, or s_1Es_4.

Section 10.3

1. (a) $y +_R z = [(.9,.98,.989,.9898,.98988,.989889,.9898898,\ldots)]$.
   (c) $x +_R z = [(.9,.99,.999,.9999,\ldots)]$.

2. Let $x = [(x_1,x_2,x_3,\ldots)]$ and $y = [(y_1,y_2,y_3,\ldots)]$ and $z = [(z_1,z_2,z_3,\ldots)]$. Then $x +_R y = [(x_1 + y_1,x_2 + y_2,x_3 + y_3,\ldots)] = [(y_1 + x_1,y_2 + x_2,y_3 + x_3,\ldots)] = y +_R x$ where the middle equality is justified by the commutativity of addition in F. Also $x +_R (y +_R z) = [(x_1,x_2,x_3,\ldots)] +_R [(y_1 + z_1,y_2 + z_2,y_3 + z_3,\ldots)] = [(x_1 + (y_1 + z_1),\ x_2 + (y_2 + z_2),\ x_3 + (y_3 + z_3),\ldots)] = [((x_1 + y_1) + z_1,\ (x_2 + y_2) + z_2,\ (x_3 + y_3) + z_3,\ldots)] = [(x_1 + y_1,\ x_2 + y_2,\ x_3 + y_3,\ldots)] +_R z = (x +_R y) +_R z$ where the middle equality is justified by the associativity of addition in F.

4. The sum of the given real numbers is $[(-.1,-.01,-.001,-.0001,\ldots)] = [(0,0,0,0,\ldots)]$.

5. (a) "1.0000 ..." is a name for the real number $[(1.0,1.00,1.000,\ldots)]$. ".9999 ..." is a name for the real number $[(.9,.99,.999,\ldots)]$. The two numbers are the same if $(1.0,1.00,1.000,\ldots)E(.9,.99,.999,\ldots)$. But this is so since $(1 - .9,\ 1 - .99,\ 1 - .999,\ldots) = (.1,.01,.001,\ldots)$ is clearly a null sequence.

6. Suppose $x = [(x_1,x_2,x_3,\ldots)]$ and $y = [(y_1,y_2,y_3,\ldots)]$ are both positive real numbers. Then there is an x_k and a y_m such that $x_k > 0$ and $y_m > 0$. Let $n = \max\{k,m\}$. Because of the monotonicity of the sequences $j > n$ implies $x_j \geq x_k > 0$ and $y_j \geq y_k > 0$. Hence, $x_j + y_j > 0$. It follows that $x +_R y = [(x_1 + y_1,\ x_2 + y_2,\ldots)]$ is positive.

8. Let $x = [(x_1,x_2,\ldots)]$ be a negative real number. Then $x_n < 0$ for each n. Let d_n be the largest integral multiple of 10^{-n} such that $d_n \leq x_n$ for each n. Then $d_n \leq d_{n+1}$ for every n and $(d_1,d_2,\ldots)$ is bounded by the same upper bound as that for $(x_1,x_2,\ldots)$. We have $(d_1,d_2,\ldots)E(x_1,x_2,\ldots)$ since $x_n - d_n \leq 10^{-n}$ for all n. Hence, $x = [(d_1,d_2,\ldots)]$. Let $y = [(-d_1 - 10^{-1}, -d_2 - 10^{-2},\ldots)]$. Then, $x +_R y = [(-10^{-1}, -10^{-2},\ldots)] = [(0,0,0,\ldots)]$ and, therefore, $y = -x$. If y is not positive then $-d_n - 10^{-n} < 0$ for all n and, hence $-10^{-n} < d_n < 0$ for all n. But then $x = [(d_1,d_2,\ldots)] = [(0,0,0,\ldots)]$ which cannot be since x is negative. We see then that y is positive. To show that the inverse is unique, suppose that $z = [(z_1,z_2,\ldots)]$ is another real number such that $x +_R z = [(0,0,0,\ldots)]$. Let $x_n + y_n = a_n$ and $x_n + z_n = b_n$ for all n. Then $(a_1,a_2,\ldots)$ and $(b_1,b_2,\ldots)$ are both null sequences. Then $(a_1 - b_1, a_2 - b_2,\ldots) = (y_1 - z_1, y_2 - z_2,\ldots)$ is also a null sequence. Hence, $(y_1,y_2,\ldots)E(z_1,z_2,\ldots)$ and, therefore, $y = z$.

Section 10.4

1. (a) $[(.15,.15,.15,.15,\ldots)]$
   (c) $[(.3,.303,.303003,.3030030003,\ldots)]$
   (e) $[(.5293,.53993893,.540045854,\ldots)]$
   (g) $[(-.6,-.55,-.541,-.5401,-.54005,-.540047,-.5400470,\ldots)]$
   (i) $[(.0,.02,.029,.0293,.02939,.029395,.0293950,.02939505,\ldots)]$

2. Let $x = [(x_1,x_2,\ldots)]$, $y = [(y_1,y_2,\ldots)]$, and $z = [(z_1,z_2,\ldots)]$ be nonnegative real numbers.
   (a) $x \cdot_R y = [(x_1y_1,x_2y_2,\ldots)] = [(y_1x_1,y_2x_2,\ldots)] = y \cdot_R x$ where the middle equality is justified by the commutativity of multiplication in F.
   (c) $x \cdot_R (y +_R z) + x \cdot_R [(y_1 + z_1, y_2 + z_2,\ldots)] = [(x_1(y_1 + z_1), x_2(y_2 + z_2),\ldots)] = [(x_1y_1 + x_1z_1, x_2y_2 + x_2z_2,\ldots)] = [(x_1y_1, x_2y_2,\ldots)] +_R [(x_1z_1, x_2z_2,\ldots)] = (x \cdot_R y) +_R (x \cdot_R z)$ where the middle equality is justified by the distributive law in F.

4. Suppose $x = [(x_1,x_2,\ldots)]$ and $y = [(y_1,y_2,\ldots)]$ are positive real numbers. Then there is an x_k and a y_m such that x_k and y_m are positive rationals. Let $n = \max\{k,m\}$. If $j > n$ we see that $x_j \geq x_k > 0$ and $y_j \geq y_m > 0$ from the monotonicity of the sequences. Hence, $x_jy_j > 0$. It follows that $x \cdot_R y = [(x_1y_1,x_2y_2,\ldots)]$ is positive.

Section 10.5

1. When "3" is taken as a name for a natural number, then 3 is the collection of all sets which can be put into a one-to-one correspondence with $\{a,b,c\}$. When "3" is taken as a name for an integer it refers to $^+3 = \{(0,3),(1,4),(2,5),(3,6),\ldots\}$. When "3" is taken as a name for a rational number it refers to $3/1 = \{(3,1),(-3,-1),(6,2),(-6,-2),(9,3),(-9,-3),\ldots\}$. When "3" is taken as a name for a real number it refers to $[(3,3,3,\ldots)]$ which is the class of all bounded monotone increasing sequences of rationals $(x_1,x_2,x_3,\ldots)$ such that $(x_1-3, x_2-3, x_3-3,\ldots)$ is a null sequence.

3. Yes, since $(R^*,+_R,\cdot_R)$ and $(F,+,\cdot)$ are isomorphic, and $-_R$ and $\div_R$ are defined in terms of $+_R$ and $\cdot_R$, and $-$ and $\div$ are defined in terms of $+$ and $\cdot$.

4. (a) No. For example $\sqrt{2} + (-\sqrt{2}) = 0$ which is rational.

5. No. For suppose x is rational and y is irrational. $x = p/q$ for some integers p and q

and if $x + y$ is rational we would have $p/q + y = r/s$ for some integers r and s. This would imply $y = (r/s) - (p/q)$ which tells us that y is rational which is impossible.

7. (a) Suppose a and b are rational numbers and $a \neq b$. Let $c = a - b$. Then $c \neq 0$. Hence, $(a - b, a - b, a - b, \ldots) = (c, c, c, \ldots)$ and this sequence is not null. That is, it is false that $(a, a, a, \ldots)E(b, b, b, \ldots)$. Hence, $[(a, a, a, \ldots)] \neq [(b, b, b, \ldots)]$.

Section 10.6

1. (a) .101001000100001 . . . , (c) 1, (e) $\sqrt{3}$, (g) none.
2. Yes. If a non-empty set of natural numbers is bounded above, then it is finite and, thus, has a maximum. This maximum is the lub for the set.
4. Let A be a non-empty set of real numbers which is bounded below. Let b be a lower bound for A and $B = \{y | y = -x \text{ for some } x \text{ in } A\}$. Then $-b$ is an upper bound for B. If A is nonempty, so is B. Hence B has a lub. Call it u. We claim that $-u$ is a glb for A. First we see that $-u$ is a lower bound for A since for all y in B we have $y \leq u$ and hence $-y \geq -u$ so that $x \geq -u$ for all x in A. Second, to see that $-u = \text{glb}(A)$, suppose $-u < z$. Then $u > -z$ and, hence, there is y_0 in B such that $u \geq y_0 > -z$. This implies $-u \leq -y_0 < z$. But $-y_0$ is in A. So this shows that no number larger than $-u$ can be a lower bound for A and, hence, $-u$ is the glb for A.
5. We give, in each case, one of the many possible rational numbers and then one of the many possible irrational numbers.
   (a) $3/2, \sqrt{2}$ (c) impossible (e) $1.5, \sqrt{5/2}$
   (g) $-.47695335, -.47695335050050005\ldots$ (i) $6.001, \sqrt{73/2}$.
6. Let x and y be distinct real numbers. There is no loss in generality if we suppose $y > x$. Let $d = y - x$. Then $d > 0$. There is an n such that $10^{-n} < d$. If x is rational then $x + 10^{-n}$ is a rational number between x and y, and $x + (10^{-n}/\sqrt{2})$ is an irrational number between x and y. If x is irrational, then $x + 10^{-n}$ is an irrational number between x and y, and z will be a rational number between x and y if we let z be the largest integral multiple of 10^{-n} that exceeds x.
8. The sequence of partial sums for the series as given is bounded and is also monotone increasing since $a_n \geq 0$. Hence, it has a limit by Problem 7 and, therefore, the series converges.

Section 10.7

1. (a) $2^2 = 4$ and $3^2 = 9$. $4 < 5 < 9$. $(2.2)^2 = 4.84 < 5 < 5.29 = (2.3)^2$. $(2.23)^2 = 4.9729 < 5 < 5.0176 = (2.24)^2$. Hence $2.23 \cong \sqrt{5}$.
   Similar procedures yield (c) $\sqrt[3]{7} \cong 1.91$, (e) $\sqrt[3]{5} \cong 1.71$, (g) $\sqrt{.15} \cong .387$, (i) $\sqrt[4]{.0001} = .1$.
2. (a) 4.123106, (c) 164.3168, (e) 3.162278, (g) .08366600, (i) .05830952.
3. (a) Since $\sqrt{2} \cong 1.4142$, we have $2^{\sqrt{2}} \cong 2^{1.4142} = 2^{14,142/10.000} = 2^{7071/5000} = 2 \cdot 2^{3071/5000}$. And we could attempt to find the 5000th root of 2^{2071}. This is clearly not very practical. A more practical procedure is to make use of logarithms which

are discussed in Section 10.10. Another device could be used in case a calculator is available that takes square roots automatically. It makes use of binary fractions. We first find a binary fraction approximation for the square root of 2. We have $\sqrt{2} \cong 1.01101010_2$. Then $2^{\sqrt{2}} \cong 2^{1.01101010_2} = 2^{(1+1/4+1/8+1/32+1/128)} = 2\cdot\sqrt[4]{2}\cdot\sqrt[8]{2}\cdot\sqrt[32]{2}\cdot\sqrt[128]{2}$ and each of these roots is easy to find by taking repeated square roots. (For example, $\sqrt[8]{2} = \sqrt{\sqrt{\sqrt{2}}}$.)

4. The first few terms of the sequence are approximately: 1.4142135, 1.4422495, 1.4142135, 1.3797296, 1.3480061. The limit of the sequence is 1. It is difficult to see this if we just have these first few terms, but we can find that $\sqrt[32]{32} \cong 1.1143867$, $\sqrt[128]{128} \cong 1.0386341$, and $\sqrt[1024]{1024} \cong 1.0067920$. And these values strongly suggest that the limit is 1. These decimal values are quite easily found on a calculator that takes square roots automatically.

5. (a) 1,1,1,1, . . .
(c) 10000, 100, 10, 3.1622777, . . .
(e) .01, .1, .31622777, .56234132, . . .
The limit of the sequence is 1 if $k > 0$, 0 if $k = 0$, and undefined if $k < 0$.

7. (i) 1, (iii) 3.

8. Suppose x and y are in R. We have $f(x+y) = 10^{x+y} = 10^x \cdot 10^y = f(x)\cdot f(y)$. Hence, f is an isomorphism between $(R,+)$ and $(R^+,\cdot)$.

Section 10.8

1. (b) $3x - 2 = 0$, (d) $x^2 - 77 = 0$, (f) $x^6 - 4 = 0$, (h) $x^6 - 4x^3 - 2 = 0$, (j) $x^4 - 10x^2 + 1 = 0$. (Other answers are, of course, possible in all these cases.)

2. Suppose e^n is algebraic. Then there are integers $a_1, a_2, a_3, \ldots, a_k$ such that e^n is a zero of $a_kx^k + \cdots + a_1x + a_0$. That is, we would have $a_k(e^n)^k + \cdots + a_1(e^n) + a_0 = 0$. Then $a_ke^{nk} + \cdots + ae^n + a_0 = 0$. That is, e would be a zero of a polynomial with integral coefficients. This is impossible since e is transcendental. Hence, our assumption that e^n is algebraic must be wrong.

4. By Problem 3, every rational number is algebraic. Hence, every transcendental number is irrational. But there are algebraic numbers that are irrational. An example is $\sqrt{2}$. Hence, the collection of transcendental numbers is a proper subset of the set of irrationals.

6. Yes. For if $-x$ is algebraic then so is $-(-x) = x$ by Problem 5. And if $1/x$ is algebraic, then $1/(1/x) = x$ is also by Problem 5.

Section 10.9

1. (a) 5.002×10^3, (c) 7.5×10^6, (e) 6.95×10^2, (g) 6.37692×10^2, (i) 7.976×10^{-7}.

2. (a) 4, (c) undetermined, (e) 3, (g) 6, (i) 4.

3. (a) 5475, (c) 1.37×10^7, (e) 6.996×10^{-3}, (g) 48.0, (i) 5.21×10^4, (k) .37, (m) -3×10^{-5}.

4. Between $(.495)(6.425) = 3.180375$ and $(.505)(6.435) = 3.249675$ square miles.

Section 10.10

1. (a) 4, (c) 9, (e) -6, (g) $2/5$.

2. (a) $\log_{10} 21 = \log_{10} 7 + \log_{10} 3 = 1.32222$
   (c) $\log_{10} 77 = \log_{10} 7 + \log_{10} 11 = 1.88649$
   (e) $\log_{10} 42 = \log_{10} 2 + \log_{10} 3 + \log_{10} 7 = 1.62325$
   (g) $\log_{10} 11{,}000 = 3 + \log_{10} 11 = 4.04139$
   (i) $\log_{10} \sqrt[5]{11} = (1/5) \log_{10} 11 = .20828$
   (k) $\log_{10} .0003 = -4 + \log_{10} 3 = .47712 - 4$.

3. $\log 4 = 2 \log 2$, $\log 5 = 1 - \log 2$, $\log 6 = \log 2 + \log 3$, $\log 8 = 3 \log 2$, $\log 9 = 2 \log 3$, $\log 10 = 1$, $\log 12 = \log 3 + \log 4$.

4. (a) .17609, (c) 2.17667, (e) $.17667 - 3$, (g) 3.86770, (i) $.04437 - 4$, (k) 4.65992.

5. (a) 10, (c) .0001, (e) 5.6.

6. (a) 2, (c) 5,630, (e) 2.790, (g) 5.757, (i) 2.6955, (k) .05630, (m) .0003634.

7. (a) 1.3205, (c) 4.3267, (e) 10.746.

8. (a) 1, (c) $1/2$, (e) $2/5$.

Section 10.11

1. We know that in this system $-1 = 2$ and $-2 = 1$. The only sets which do not contain 0 are $\{1\}$, $\{2\}$, and $\{1,2\}$. We cannot take $P = \{1,2\}$ since this would violate condition (iv) of the definition. We cannot take $P = \{1\}$ since $1 + 1 = 2$ so $\{1\}$ is not closed under addition. And the set $\{2\}$ does not contain the multiplicative identity. Hence, no set P in $\{0,1,2\}$ can have the desired property.

3. Let $(S,+,\cdot)$ be an ordered field with P its set of positive elements. Then 1 is in P. We have $0 + 1 = 1$. Hence, there is a number y (namely 1) in P such that $0 + y = 1$. Hence, $0 \leq 1$. It remains to prove that $0 \neq 1$. But by condition (iv), not both 0 and 1 can be in P. So we must have $0 \neq 1$.

4. (a) Since 0 is in $P \cup \{0\}$ and $x + 0 = x$ we have $x \geq x$.
   (c) If $x \geq y$ and $y \geq z$, then there are numbers k and n such that $x = y + k$, $y = z + n$ and k and n are in P. Then $x = y + k = (z + n) + k = z + (n + k)$. But P is closed under addition so $n + k$ is in P. Hence, $x \geq z$.
   (e) Suppose $x \geq y$ and $z \in P$. There is a number k such that $x = y + k$ and $k \in P \cup \{0\}$. Then $x \cdot z = (y + k) \cdot z = yz + kz$. If $k = 0$ then $k \cdot z = 0$. If $k \in P$, then $k \cdot z \in P$ since P is closed under multiplication. Hence, $k \cdot z \in P \cup \{0\}$. It follows that $xz \geq yz$.

Section 11.1

1. (i) Since triangle DEF is similar to triangle $D^*E^*F^*$ we know that $d/d^* = e/e^* = f/f^*$. It is given that $d = e = f = d^* = 10$. Therefore $d/d^* = 10/10 = 1$. Therefore $1 = e/e^* = 10/e^*$ which means $e^* = 10$. Also $f/f^* = 10/f^* = 1$ so $f^* = 10$.
   (iii) If, instead, we are given that $d = e = 20$, $f = 10$, and $d^* = 50$ then $d/d^* = 20/50 = e/e^* = 20/e^*$ so $e^* = 50$; similarly $20/50 = f/f^* = 10/f^*$ so $f^* = 25$.

(v) If $d = 5$, $e = 12$, $f = 13$, and $d^* = 1$, then $d/d^* = 5/1 = e/e^* = 12/e^*$ so $e^* = 12/5$; similarly $5/1 = f/f^* = 13/f^*$ so $f^* = 13/5$.

(vii) If $d = 4$, $e = \sqrt{7}$, $f = \sqrt{9} = 3$, $d^* = 16$, then $d/d^* = 4/16 = e/e^* = \sqrt{7}/e^*$ so $e^* = 4\sqrt{7}$. Also $4/16 = f/f^* = 3/f^*$ so $f^* = 12$.

2.

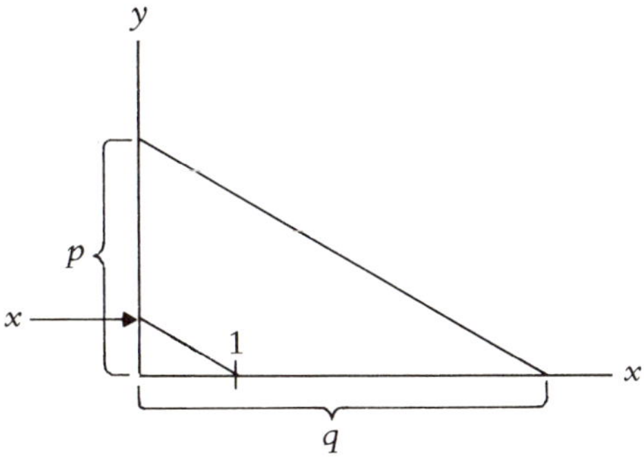

To show that the two triangles are similar, we need to make use of a theorem in plane geometry which has not been mentioned. The theorem can be stated as follows with reference to the following diagram: If the lines l and m are parallel then $\measuredangle A = \measuredangle B$.

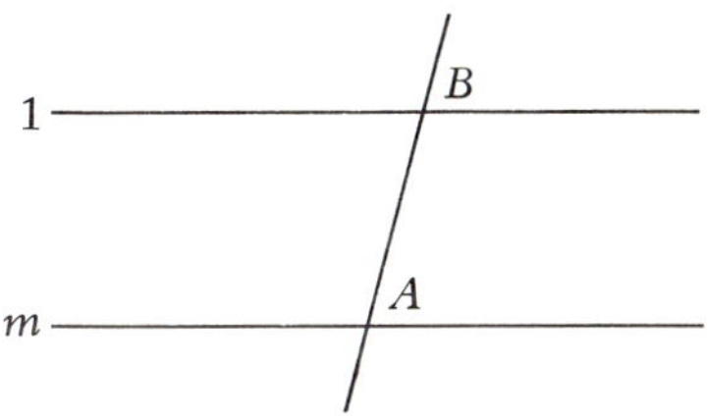

Using this theorem, it is easy to see that the two triangles are similar. Therefore, $p/x = q/1$ or $x = p/q$ which was to be shown.

4. (i) If $a = 3$ and $b = 4$, then $3^2 + 4^2 = c^2$ or $c^2 = 25$ so $c = 5$.

(iii) If $a = 1$, $b = \sqrt{3}$ then $c^2 = 1^2 + (\sqrt{3})^2 = 4$ so $c = 2$.

(v) If $a = 12$, $c = 13$ then $b^2 = 13^2 - 12^2 = 25$ so $b = 5$.

(vii) From $a^2 + b^2 = c^2$ we see c^2 is always greater than a^2 or b^2 and so c is greater than a or b. Hence, we cannot have a right triangle with $a = 10$, $c = 9$ if c is the hypotenuse.

5. Only those given in (iv), (v), (vi), and (vii).

7. The area of a triangle is 1/2 (base) (height). For a right triangle we can choose either leg as the base and the other as the height. In each case, we give the area of the unstarred and then of the starred triangle. (iv) 30, 120, (v) 30, 6/5, (vi) $\sqrt{15}/2$, $(3\sqrt{15})/2$, (vii) $(3\sqrt{7})/2$, $24\sqrt{7}$.

9. No. For example, there are many different shapes possible for each rectangle where *all* the angles are equal to 90°.

Section 11.2

1.

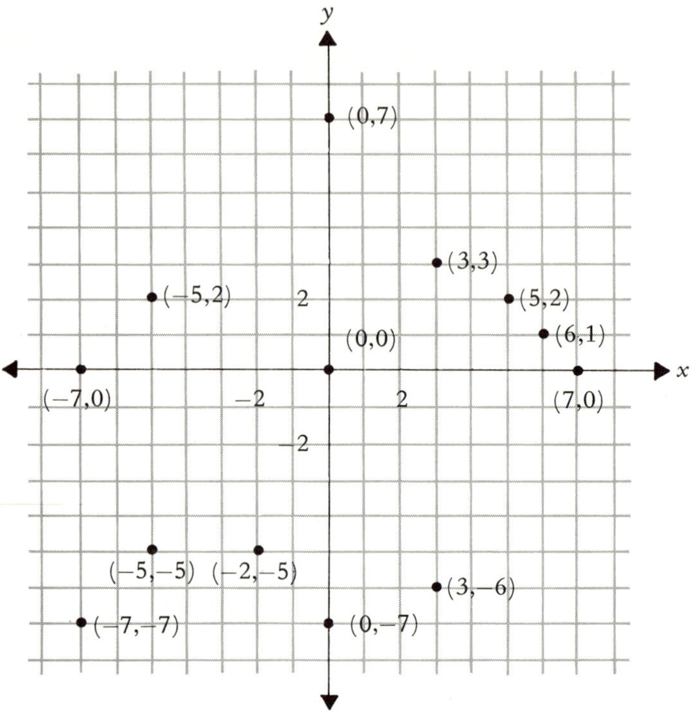

2. (a)

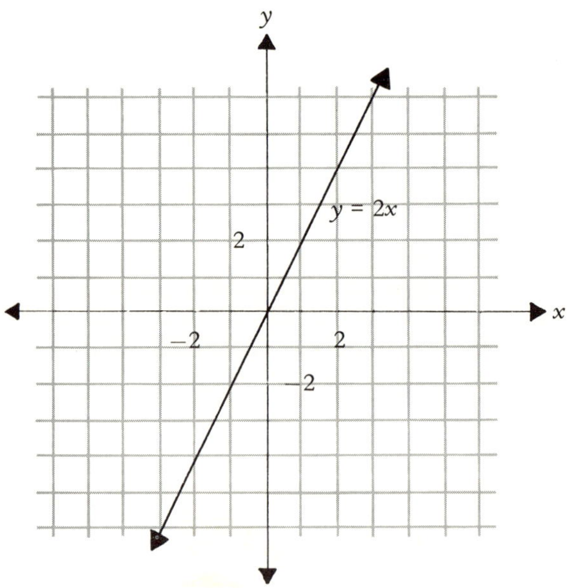

(c)

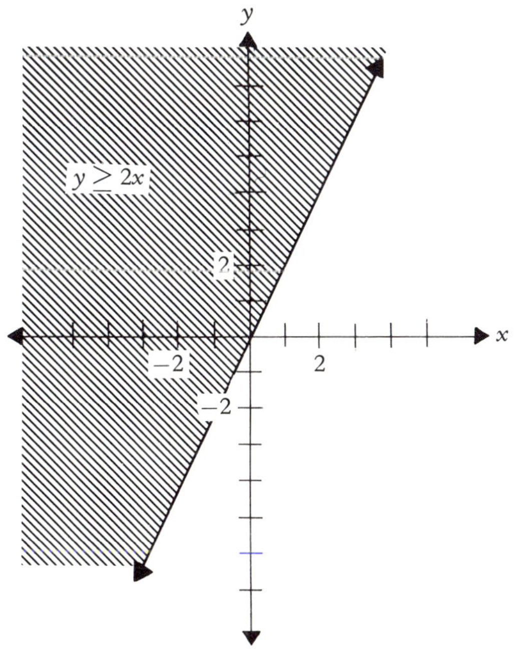

(e)

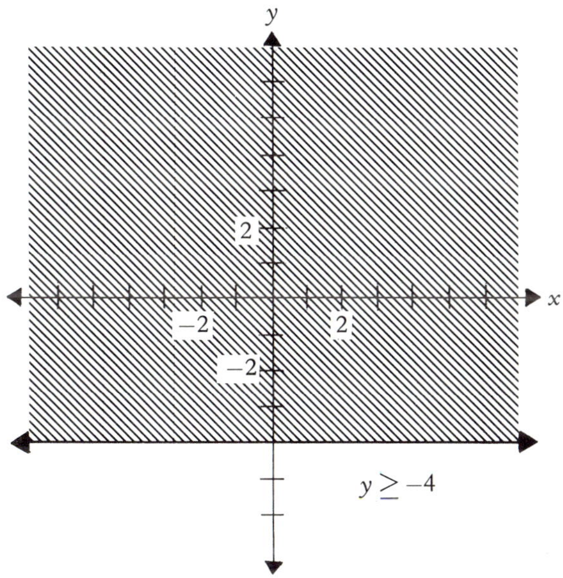

(g)

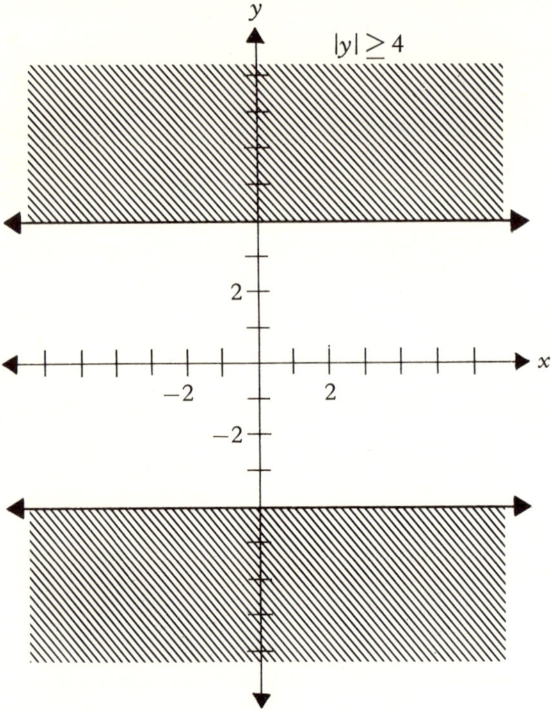

3. (a)

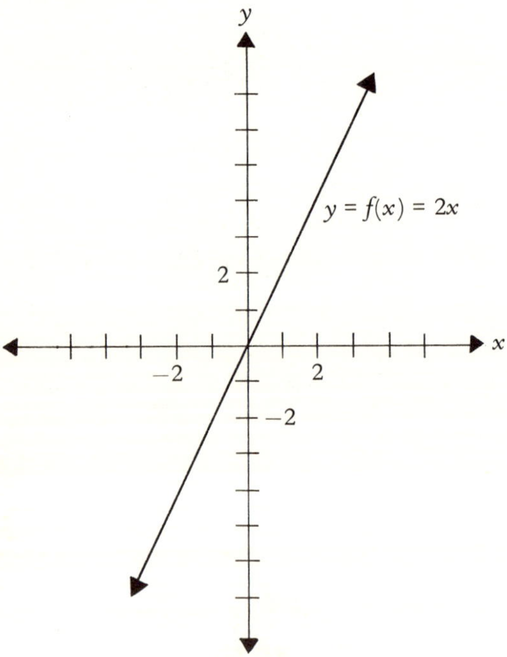

(c)

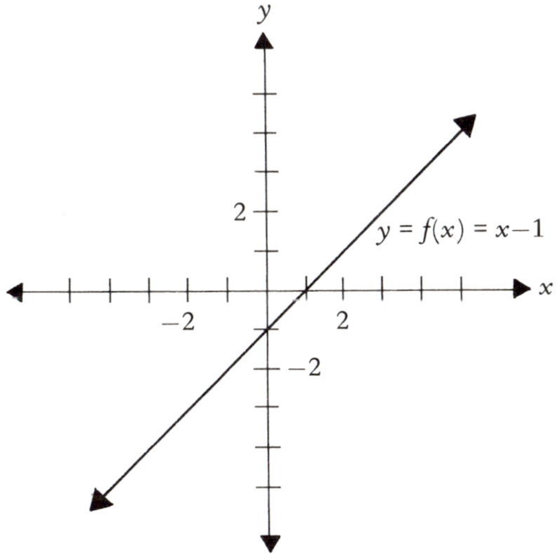

4. (a) The height of the ball at time $t = \frac{1}{2}$ is the y coordinate of the intersection of the graph with the line $t = \frac{1}{2}$. It looks like the intersection occurs at s somewhere between 3 and 4, call it $3\frac{1}{2}$. So the height of the ball at $t = \frac{1}{2}$ is almost $3\frac{1}{2}$ yards.
   (c) The total time the ball is in flight is 2 seconds.

5. (a) The ball is moving upwards whenever $v > 0$. From the graph, $v > 0$ from $t = 0$ to 1 and thereafter $v < 0$. The ball is therefore moving upwards for times between 0 and 1.
   (c) If you stop and think about it you will realize that the ball reaches its maximum height just as it stops going up and starts going down. This occurs when the velocity, v, changes from positive to negative. From the graph, this is at $t = 1$.

6. (a) From the graph, the maximum height attained is about 72 inches. So we want to see at what age the person reaches height 36 inches. From the graph this seems to be at about age 4.

7. (a) The barometer can be said to be rising whenever the graph is rising. This appears to be between the following times (approximately): 0 to 17, 37 to 45, 55 to 66, 91 to 107.
   (c) Between the following times (approximately): 36 to 37, 45 to 55, 80 to 91.

8. (a) From the graph, the maximum temperature is about 32°, occurring at time 101.
   (c) The temperature is warming up whenever the graph is rising. This occurs between the following times: 0 to 8, 32 to 38, 51 to 68, 82 to 101.

10. (a) Some of the other points on the line are: $(3,3)$, $(3\frac{1}{2},1)$, $(5, -5)$, $(6, -9)$, $(1, 11)$, $(4\frac{1}{2}, -3)$ and $(1\frac{1}{2}, 9)$. All of these points have coordinates (x, y) which satisfy the equation $y = -4x + 15$.
    (c) From the graph we notice that the line connecting the points $(0, -3)$ and $(-1, -5)$ also contains the point $(1, -1)$. Since a line is determined by any two points on it we know that this line must be the same as the line for part 10(b). A check of the points on the two lines verifies this. Hence the points on this line must also satisfy the equation $y = 2x - 3$.

11. We leave all but one of these graphs to the reader since they contain personal information. We give an example for part (c) however. (See graph.) One may collect quite a bit of information about the author of the graph for part 11(e). He was the second of five children, his younger siblings were born when he was about 4, 6, and 8. One of his siblings died when he was 42. Note, also, that conceivably when he was, say, 18, his mother gave birth to a baby on the same day his brother died and so the graph does not change.

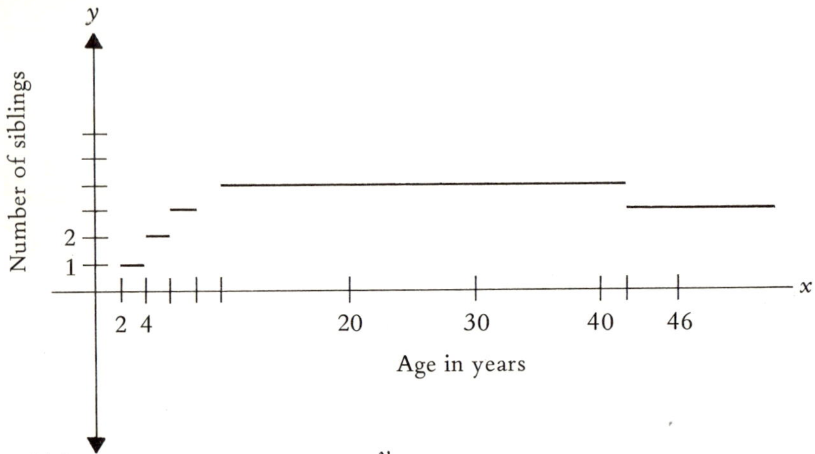

Section 11.3

For Problems 1 and 3 see the graphs.

1.

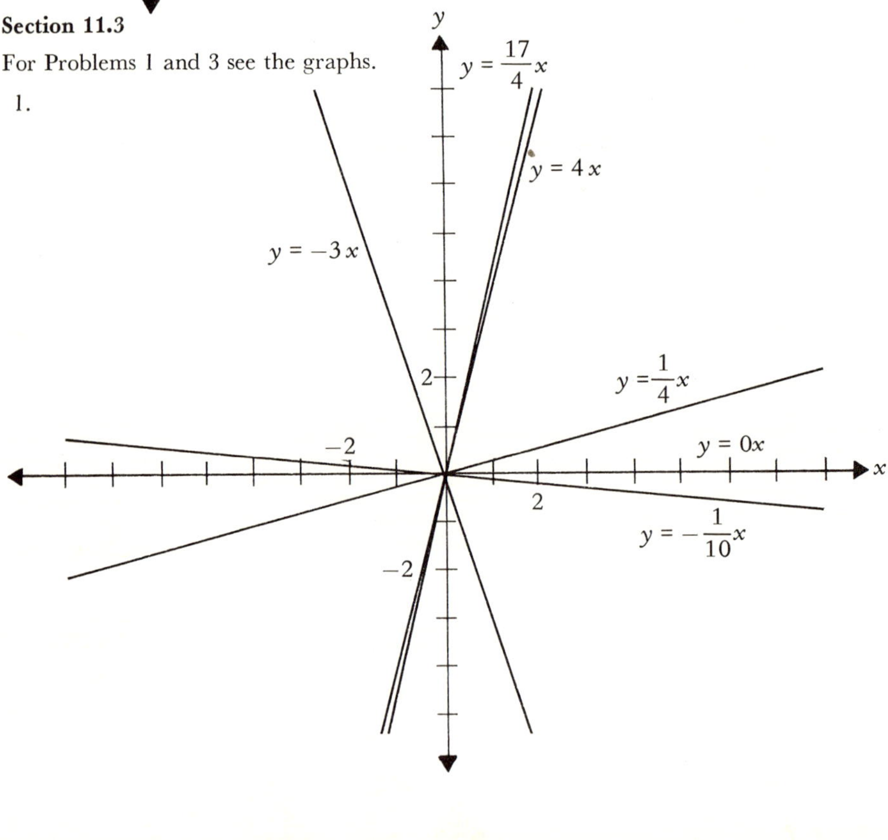

3.

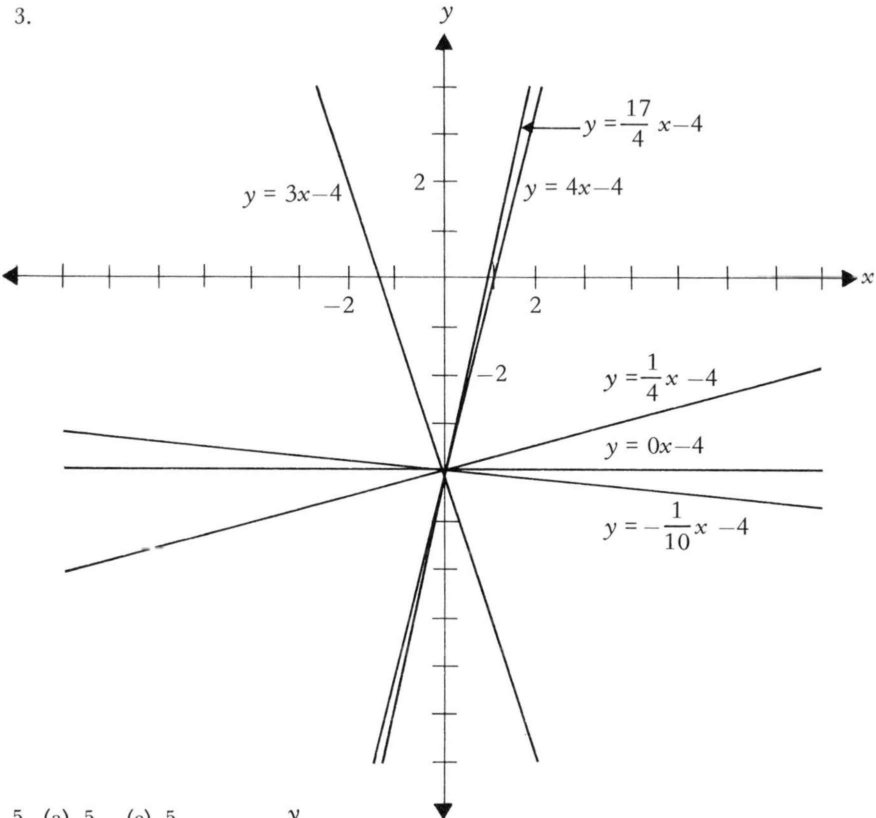

y
x
2
−2
2
−2
$y = \frac{17}{4}x - 4$
$y = 3x - 4$
$y = 4x - 4$
$y = \frac{1}{4}x - 4$
$y = 0x - 4$
$y = -\frac{1}{10}x - 4$

5. (a) 5, (c) 5.

6.

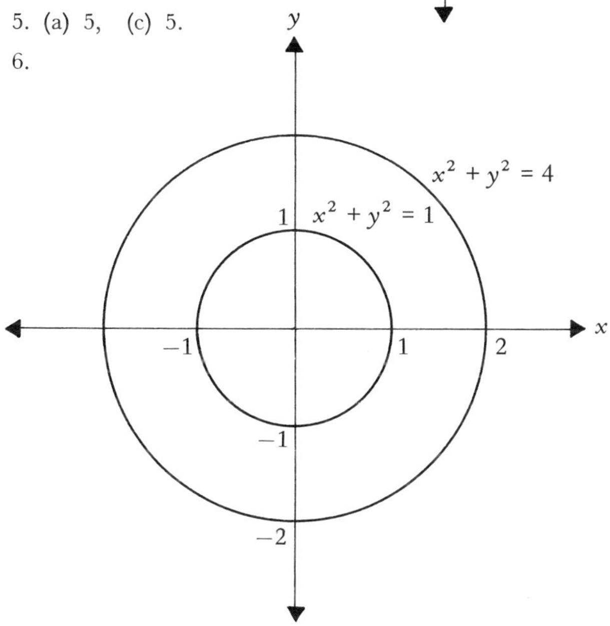

y
x
$x^2 + y^2 = 4$
$x^2 + y^2 = 1$
1
−1
1
2
−1
−2

8. (a) $y = f(x)$. We note that y is proportional to x means that there is a k such that $f(x) = kx$ or $y = kx$. We note also here that if we are given one point (x_0, y_0) which is on the graph of f, then we can find k since $k = y_0/x_0$. $f(x) = kx$ and $(1, 5)$ is on the graph. Hence $k = 5/1 = 5$. So $y = f(x) = 5x$.
   (c) $y = k_1z$ and $z = k_2x$. $y = 15$ when $z = 3$ implies $k_1 = 15/3$ so $y = (15/3)z$. $z = 4$ when $x = 2$ implies $k_2 = 4/2 = 2$ so $z = 2x$. We substitute the quantity $(2x)$ for z in the equation $y = (15/3)z$. So $y = (15/3)(2x) = 10x$. Hence $f(x) = 10x$.
   (e) $y - b = kx$. Hence, $y = kx + b$. When $x = 0$, $y = b$. We are given that when $x = 0$, $y = 32$ so $b = 32$. Hence $y - 32 = kx$. Now $y = 212$ when $x = 100$ so $k = (212 - 32)/100 = 9/5$. We might note here that if x is the temperature in degrees Celsius then y is the temperature in degrees Fahrenheit. $f(x) = (9/5)x + 32$ which is perhaps more familiar as $F = 9/5\ C + 32$.

10. Suppose $f(x) = kx$ for all real x.
    Then
$$\begin{aligned} f(ax_1 + bx_2) &= k(ax_1 + bx_2) \\ &= akx_1 + bkx_2 \\ &= af(x_1) + bf(x_2). \end{aligned}$$

11. (c) The mapping takes the interval $2 \leq x \leq 4$ to the interval $4 \leq f(x) \leq 8$.

12. See graphs. (a) $y = 6x$, (c) $y = x + 3$.

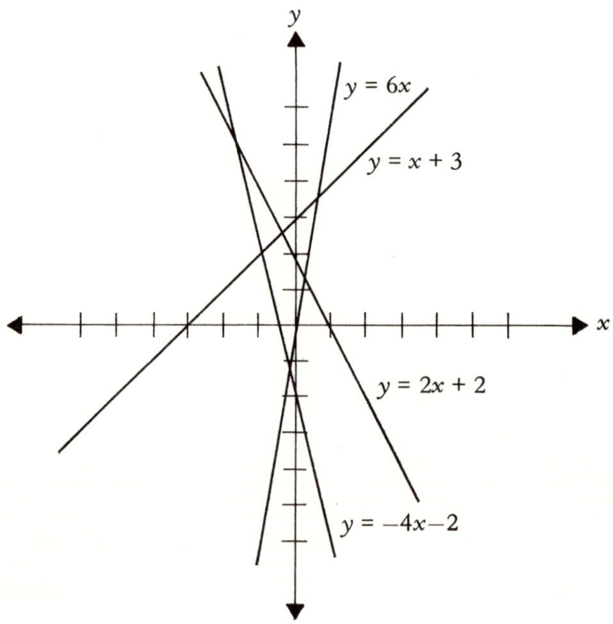

Section 11.4

1. Suppose f is a constant, then $f(x) = k$ for some real number k and all real x. Let $h_1, h_2, h_3, \ldots$ be a null sequence with no $k_n = 0$. Then

$$f'(x) = \lim_{n\to\infty} \frac{f(x + h_n) - f(x)}{h_n} = \lim_{n\to\infty} \frac{k - k}{h_n} = 0.$$

3. (a)

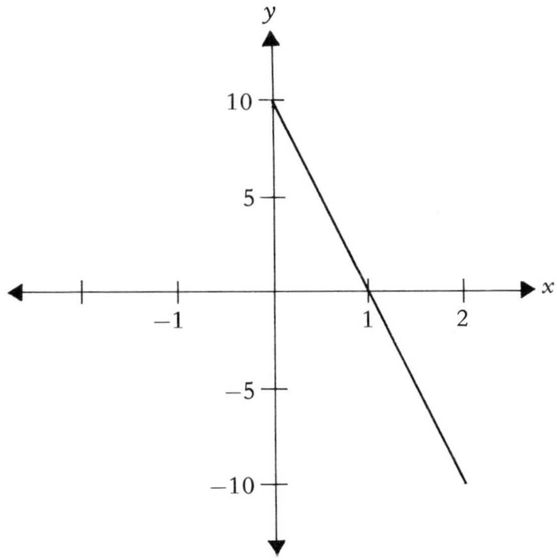

4. Suppose f is differentiable at a and that $f'(a) > 0$. We know from the definition of f' that $f'(a) > 0$ means that for h near 0 $\dfrac{f(a+h) - f(a)}{h} \simeq f'(a) > 0$.

   Therefore for h near zero $\dfrac{f(a+h) - f(a)}{h} > 0$. But this implies that

   (a) if h is positive, then $f(a+h) - f(a) > 0$ or $f(a+h) > f(a)$, and
   (b) if h is negative, then $f(a+h) - f(a) < 0$ or $f(a+h) < f(a)$ when h is near 0.

6. See Figure 11.11(a). $y = f(t)$ is the distance of the ball above ground at time t. It should be clear from the chapter that $f'(t)$ is the velocity of the ball (either up or down) at time t. Clearly the ball has stopped rising and begun falling right when it reaches its maximum height. This occurs when the velocity is zero or when $f'(t) = 0$.

7. We use the formulas given in the chapter to find the derivative of f and then the derivative of f'.
   (a) $f'(x) = 0; f''(x) = 0$
   (c) $f'(x) = 2x; f''(x) = 2$
   (e) $f'(x) = 7x^6; f''(x) = 42x^5$

8. (a) $1/(2\sqrt{x})$, (c) $1/(10\sqrt[10]{x^9})$, (e) $-1/(4\sqrt[4]{x^5})$.

Section 11.5

1. (a) 25, (c) 0.
2. (a) 1, (c) 12, (e) $2 + 3\pi - 8$.
   See the graph on page 572.
4. (a) $\displaystyle\int_0^x 2dt = 2(x - 0) = 2x$. (See Problem 3).

   If $G(x) = 2x$ then $G'(x) = 2$ which is what the fundamental theorem says $G'(x)$ should be.

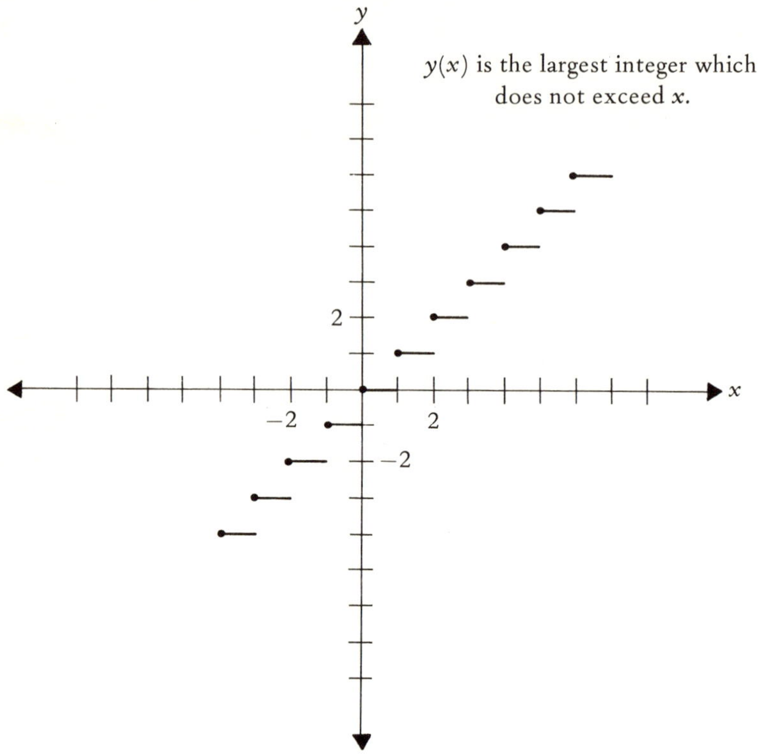

Graph for Problem 2, Section 11.5

4. (c) $\int_0^x t\,dt = \frac{1}{2}x^2$ (since it is the area of a triangle with base x and height x). If $G(x) = \frac{1}{2}x^2$ then $G'(x) = x$, which is again in agreement with the fundamental theorem.

5. (a) The distance traveled in the first 3 seconds is $\int_0^3 v(t)dt = 144$ feet. (c) 256 ft.

6. If $y = kt$ then the area under the graph from times $t = 0$ to $t = t$ is the area of a triangle with base t and height kt. This area is $\frac{1}{2}(t)(kt) = (\frac{1}{2})kt^2$ which is proportional to t^2.

8. (a) $\int_1^5 (1 + 3)dx = 16$; $\int_1^5 1dx = 4$; $\int_1^5 3dx = 12$.

   (c) $\int_1^5 2xdx = 24$; $\int_1^5 xdx = 12$; $\int_1^5 xdx = 12$.

Section 11.6

1. Let's call the plants 1, 2, and 3. If P_i represents the production at plant i for $i = 1,2,$ and 3, we could look at the vector (P_1,P_2,P_3) as representing the production for that item for that time period. If plant i has a labor cost L_i for a certain time period, it would be natural to think of (L_1,L_2,L_3) as the labor cost vector for that item for that

time period. If M_i is the material cost in plant i, then (M_1,M_2,M_3) would be the material cost vector.

3. (a) $(5,4) + (-5,3) = (0,7)$.
   (c) $(5,2) + (6)$ has no meaning. Our definition does not tell us how to add vectors of different dimension.
   (e) $(5,6,7,8,9,10) + (10,9,8,7,6,5) = (15,15,15,15,15,15)$.
4. (a)

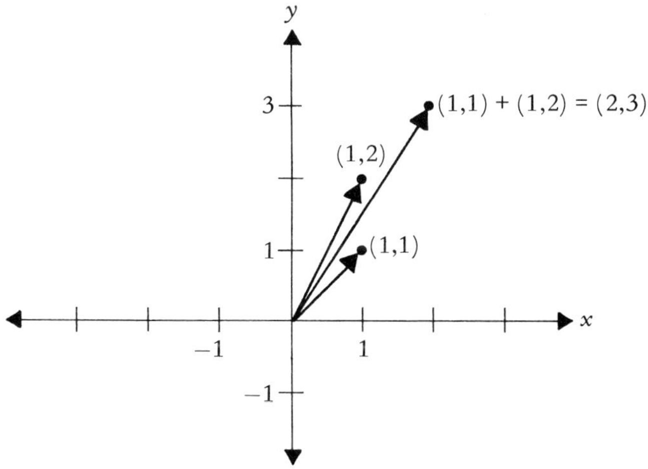

(d)

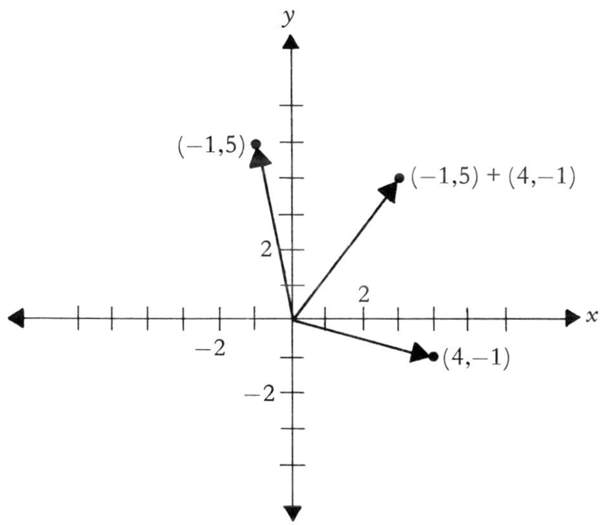

5. (a) Prove that R^n addition is commutative. What is necessary is to show that: $(x_1,x_2, \ldots ,x_n) + (y_1,y_2, \ldots ,y_n) = (y_1,y_2, \ldots ,y_n) + (x_1,x_2, \ldots ,x_n)$ but $(x_1,x_2, \ldots ,x_n) + (y_1,y_2, \ldots ,y_n) = (x_1 + y_1, x_2 + y_2, \ldots , x_n + y_n) = (y_1 + x_1, y_2 + x_2, \ldots , y_n + x_n)$ (since addition of reals is commutative) $= (y_1,y_2, \ldots , y_n) + (x_1,x_2, \ldots ,x_n)$.

6. It is easy to see that the additive identity for R^n is the vector $(0,0,0, \ldots ,0)$ since $(x_1,x_2, \ldots ,x_n) + (0,0, \ldots ,0) = (x_1,x_2, \ldots ,x_n)$. It is also easy to see that the additive inverse of $(x_1,x_2, \ldots ,x_n)$ is $(-x_1,-x_2, \ldots ,-x_n)$ since $(x_1,x_2, \ldots ,x_n) + (-x_1,-x_2, \ldots ,-x_n) = (0,0, \ldots ,0)$. If we define subtraction to be the inverse operation for addition, we have $(x_1,x_2, \ldots ,x_n) - (y_1,y_2, \ldots ,y_n) = (x_1,x_2, \ldots ,x_n) + (-(y_1,y_2, \ldots ,y_n))$ where $-(y_1,y_2, \ldots ,y_n)$ is the additive inverse of $(y_1,y_2, \ldots ,y_n)$. Hence, $(x_1,x_2, \ldots ,x_n) - (y_1,y_2, \ldots ,y_n) = (x_1 - y_1,x_2 - y_2, \ldots ,x_n - y_n)$ and we see that R^n is closed under subtraction since R is closed under subtraction.

7. In Section 6.22, $R^{\{1,2,\ldots,n\}}$ was defined to be the collection of all mappings from $\{1,2, \ldots ,n\}$ into R. If f is such a mapping then it may be displayed in a table as

k	1	2	3	. . .	n
$f(k)$	x_1	x_2	x_3	. . .	x_n

We may identify such a function by simply specifying the numbers $x_1,x_2, \ldots ,$ and x_n. That is, we may think of the vector $(x_1,x_2, \ldots ,x_n)$ as corresponding to the indicated member of $R^{\{1,2,\ldots,n\}}$ and vice versa.

Section 11.7

1. If we drive for x_i hours on the ith day and we drive at an average speed of a_i miles per hour on the ith day, then the total distance we shall drive for the 5 days is $a_1x_1 + a_2x_2 + a_3x_3 + a_4x_4 + a_5x_5$. Representing this number as a vector product we have

$$a_1x_1 + a_1x_2 + \ldots + a_5x_5 = (a_1\ a_2\ a_3\ a_4\ a_5)\begin{pmatrix} x_1 \\ x_2 \\ x_3 \\ x_4 \\ x_5 \end{pmatrix}.$$

2. As an extension of the notation in Problem 1, let us call b_i the average speed on the ith day according to the second speed policy. Then

$$\begin{pmatrix} a_1 & a_2 & a_3 & a_4 & a_5 \\ b_1 & b_2 & b_3 & b_4 & b_5 \end{pmatrix}\begin{pmatrix} x_1 \\ x_2 \\ x_3 \\ x_4 \\ x_5 \end{pmatrix} = \begin{pmatrix} a_1x_1 + a_2x_2 + a_3x_3 + a_4x_4 + a_5x_5 \\ b_1x_1 + b_2x_2 + b_3x_3 + b_4x_4 + b_5x_5 \end{pmatrix}.$$

We see that the product matrix is 2×1 and that the ith row in it is the distance traveled according to the ith speed policy.

4. Suppose you had two speed policies $(a_1, \ldots ,a_5)$ and $(b_1, \ldots ,b_5)$ and two time vectors $(x_1, \ldots ,x_5)$ and $(y_1, \ldots ,y_5)$. Then

$$\begin{pmatrix} a_1 & a_2 & a_3 & a_4 & a_5 \\ b_1 & b_2 & b_3 & b_4 & b_5 \end{pmatrix}\begin{pmatrix} x_1 & y_1 \\ x_2 & y_2 \\ x_3 & y_3 \\ x_4 & y_4 \\ x_5 & x_5 \end{pmatrix}$$

$$= \begin{pmatrix} a_1x_1 + a_2x_2 + \ldots + a_5x_5 & a_1y_1 + a_2y_2 + \ldots + a_5y_5 \\ b_1x_1 + b_2x_2 + \ldots + b_5x_5 & b_1y_1 + b_2y_2 + \ldots + b_5y_5 \end{pmatrix}.$$

The number in the ith row and the jth column of the product matrix represents the distance traveled according to the ith speed policy and the jth time vector.

5. (a) Suppose a person had a calorie vector $(c_1,c_2, \ldots ,c_n)$ and a diet vector $(x_1,x_2, \ldots, x_n)$. Then his calorie intake for the week would be

$$(x_1,x_2, \ldots ,x_n)\begin{pmatrix} c_1 \\ c_2 \\ \cdot \\ \cdot \\ \cdot \\ c_n \end{pmatrix} = c_1x_1 + c_2x_2 + \ldots + c_nx_n.$$

6. (a) (20), (b) not defined, (c) $\begin{pmatrix} 10 & 25 \\ 4 & 10 \end{pmatrix}$,
(e) $\begin{pmatrix} 3 & 0 \\ 0 & -4 \end{pmatrix}$, (g) $(18 \quad -6 \quad 30)$,
(i) $\begin{pmatrix} 14 & 0 \\ 4 & 8 \end{pmatrix}$, (k) $\begin{pmatrix} 24 & 2 \\ 12 & 16 \end{pmatrix}$, (m) $\begin{pmatrix} 0 & 2 \\ -2 & 0 \end{pmatrix}$,
(o) $\begin{pmatrix} 2 & 8 & 36 \\ 3 & 3 & 18 \\ -6 & 2 & -4 \end{pmatrix}$, (q) $\begin{pmatrix} 0 & 1 & 0 \\ 0 & 2 & 1 \\ 1 & 2 & 0 \end{pmatrix}$,
(s) $\begin{pmatrix} 0 \\ 5 \end{pmatrix}$, (u) $\begin{pmatrix} 1 & 0 \\ 0 & 1 \end{pmatrix}$.

7. (a), (b) Yes, M_{nm} is closed under addition; addition is commutative and associative in M_{nm}. These facts follow from the definition of matrix addition and the facts that for real numbers we have closure under addition, commutativity, and associativity, respectively.

   (d) Each element in M_{nm} has an additive inverse. If the inverse of M is denoted by $-M$ and if the element in the ith row and jth column of M is m_{ij}, then the element in the ith row, jth column of $-M$ is $-m_{ij}$.

8. (a) Suppose $\vec{x} = (x_1,x_2, \ldots ,x_n)$ and $\vec{y} = (y_1,y_2, \ldots ,y_n)$. Then $c\vec{x} = (cx_1,cx_2, \ldots ,cx_n)$ and $c\vec{y} = (cy_1,cy_2, \ldots ,cy_n)$. So $c\vec{x} + c\vec{y} = (cx_1,cx_2, \ldots ,cx_n) + (cy_1,cy_2, \ldots ,cy_n)$ $= (cx_1 + cy_1,cx_2 + cy_2, \ldots ,cx_n + cy_n) = (c(x_1 + y_1),c(x_2 + y_2), \ldots ,c(x_n + y_n)) = c(x_1 + y_1,x_2 + y_2, \ldots ,x_n + y_n) = c[(x_1,x_2, \ldots ,x_n) + (y_1,y_2, \ldots ,y_n)]$ $= c(\vec{x} + \vec{y})$.

9. We note first that this operation is commutative since $(x_1,x_2, \ldots ,x_n) \cdot (y_1,y_2, \ldots ,y_n)$ $= x_1y_1 + x_2y_2 + \ldots + x_ny_n = y_1x_1 + y_2x_2 + \ldots + y_nx_n = (y_1,y_2, \ldots ,y_n) \cdot (x_1,x_2, \ldots ,x_n)$.
   It is not associative since $(\vec{x} \cdot \vec{y}) \cdot \vec{z}$ is not defined, the first part in parentheses being a real number. We note that it is distributive over addition. That is, $\vec{x} \cdot (\vec{y} + \vec{z}) = \vec{x} \cdot \vec{y} + \vec{x} \cdot \vec{z}$. Another property is that $(a\vec{x}) \cdot \vec{y} = a(\vec{x} \cdot \vec{y})$. The proof of this is left to the reader. In view of what we discovered about distance in the plane it would make sense to define the length of $\vec{x} = (x_1,x_2, \ldots ,x_n)$ to be the square root $\vec{x} \cdot \vec{x}$. This is indeed what is done when the geometry of n-dimensional space is studied.

Section 11.8

2. $\begin{pmatrix} d/e & -b/e \\ -c/e & a/e \end{pmatrix}\begin{pmatrix} a & b \\ c & d \end{pmatrix} = \begin{pmatrix} a & b \\ c & d \end{pmatrix}\begin{pmatrix} d/e & -b/e \\ -c/e & a/e \end{pmatrix} = \begin{pmatrix} 1 & 0 \\ 0 & 1 \end{pmatrix}$.

We note that this fact gives us a way to find the inverse for any two by two matrix if $ad - cb \neq 0$.

3. (a) We write the system as $\begin{pmatrix} 6 & 0 \\ 0 & -10 \end{pmatrix}\begin{pmatrix} x_1 \\ x_2 \end{pmatrix} = \begin{pmatrix} 30 \\ 30 \end{pmatrix}$. We multiply both sides by the inverse of the first matrix [see 1(a)] to obtain

$$\begin{pmatrix} 1/6 & 0 \\ 0 & -1/10 \end{pmatrix}\begin{pmatrix} 6 & 0 \\ 0 & -10 \end{pmatrix}\begin{pmatrix} x_1 \\ x_2 \end{pmatrix} = \begin{pmatrix} 1 & 0 \\ 0 & 1 \end{pmatrix}\begin{pmatrix} x_1 \\ x_2 \end{pmatrix} = \begin{pmatrix} x_1 \\ x_2 \end{pmatrix}$$

$$= \begin{pmatrix} 1/6 & 0 \\ 0 & -1/10 \end{pmatrix}\begin{pmatrix} 30 \\ 30 \end{pmatrix} \quad \text{or} \quad \begin{pmatrix} x_1 \\ x_2 \end{pmatrix} = \begin{pmatrix} 5 \\ -3 \end{pmatrix}.$$

This gives us the solution $x_1 = 5$, $x_2 = -3$. Of course, this is the hard way to do this problem. But the *method* applies in cases where other approaches are difficult. In the subsequent cases we shall simply give the solution vector.

(c) (0, 0), (e) (71, 64), (g) (7, 9, 29), (i) $(-41/5, 116/15, 4/5)$, (k) there is no solution.

4. (a) The 2×2 permutation matrices are $\begin{pmatrix} 1 & 0 \\ 0 & 1 \end{pmatrix}$ and $\begin{pmatrix} 0 & 1 \\ 1 & 0 \end{pmatrix}$. They are each their own inverses.

(b) The 3×3 permutation matrices are $\begin{pmatrix} 1 & 0 & 0 \\ 0 & 1 & 0 \\ 0 & 0 & 1 \end{pmatrix}$, $\begin{pmatrix} 1 & 0 & 0 \\ 0 & 0 & 1 \\ 0 & 1 & 0 \end{pmatrix}$, $\begin{pmatrix} 0 & 1 & 0 \\ 1 & 0 & 0 \\ 0 & 0 & 1 \end{pmatrix}$,

$$\begin{pmatrix} 0 & 1 & 0 \\ 0 & 0 & 1 \\ 1 & 0 & 0 \end{pmatrix}, \begin{pmatrix} 0 & 0 & 1 \\ 1 & 0 & 0 \\ 0 & 1 & 0 \end{pmatrix}, \begin{pmatrix} 0 & 0 & 1 \\ 0 & 1 & 0 \\ 1 & 0 & 0 \end{pmatrix}.$$

The inverse of any one of these matrices is obtained by taking the transpose. That is, by interchanging the rows and the columns.

5. (b) Suppose A has an inverse matrix A^{-1}. Then we have $A\vec{x} = \vec{b}$ if and only if $A^{-1}A\vec{x} = A^{-1}\vec{b}$. And this is true if and only if $\vec{x} = \vec{b}$.

6. (a) We must show that $\begin{pmatrix} 10 & 4 \\ 3 & 11 \end{pmatrix} - \begin{pmatrix} 7 & 0 \\ 0 & 7 \end{pmatrix} = \begin{pmatrix} 3 & 4 \\ 3 & 4 \end{pmatrix}$ has no inverse. Suppose $\begin{pmatrix} a & b \\ c & d \end{pmatrix}$ is the inverse of this matrix. Then $\begin{pmatrix} 3 & 4 \\ 3 & 4 \end{pmatrix}\begin{pmatrix} a & b \\ c & d \end{pmatrix} = \begin{pmatrix} 1 & 0 \\ 0 & 1 \end{pmatrix}$. But this implies $3a + 4c = 1$ and $3a + 4c = 0$ which is impossible.

(c) We need to show that $\begin{pmatrix} 1 & 1 \\ 0 & 1 \end{pmatrix} - \begin{pmatrix} 1 & 0 \\ 0 & 1 \end{pmatrix} = \begin{pmatrix} 0 & 1 \\ 0 & 0 \end{pmatrix}$ has no inverse. If $\begin{pmatrix} 0 & 1 \\ 0 & 0 \end{pmatrix}\begin{pmatrix} a & b \\ c & d \end{pmatrix} = \begin{pmatrix} 1 & 0 \\ 0 & 1 \end{pmatrix}$ then we must have $0 \cdot b + 0 \cdot d = 1$ and this is impossible. Hence, the given matrix has no inverse.

Section 11.9

1. First of all we can establish a one-to-one correspondence between the elements of C^* and R. If a is in $(R,+,\cdot)$ then let $f(a) = \begin{pmatrix} a & 0 \\ 0 & a \end{pmatrix}$. Clearly f is a one-to-one

correspondence. What we have left to show is that $f(ab) = f(a)\cdot f(b)$ and $f(a+b) = f(a)+f(b)$. But $f(ab) = \begin{pmatrix} ab & 0 \\ 0 & ab \end{pmatrix} = \begin{pmatrix} a & 0 \\ 0 & a \end{pmatrix}\begin{pmatrix} b & 0 \\ 0 & b \end{pmatrix} = f(a)\cdot f(b)$ and $f(a+b) = \begin{pmatrix} a+b & 0 \\ 0 & a+b \end{pmatrix} = \begin{pmatrix} a & 0 \\ 0 & a \end{pmatrix} + \begin{pmatrix} b & 0 \\ 0 & b \end{pmatrix} = f(a) + f(b)$.

3. (a) $(1 + 6i)$, (c) $8i$, (e) 29, (g) $-2 + 3i$, (i) 2, (k) $1/20 + (3/20)i$, (m) $\pi/(\pi^2 + e^2) - ei/(\pi^2 + e^2)$.

4. (a)

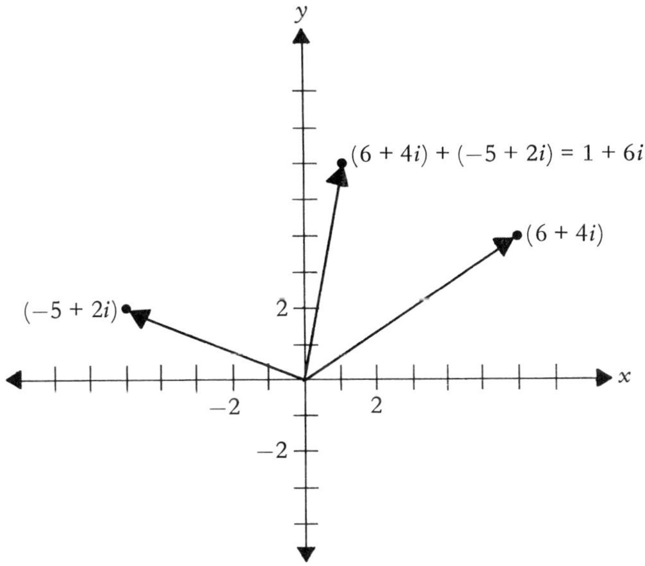

(b)

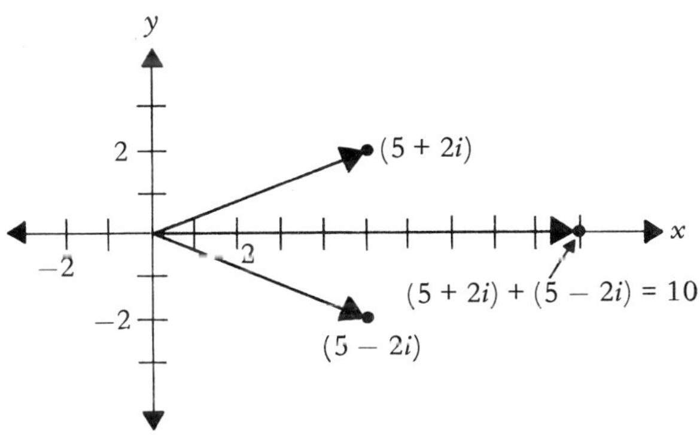

(c)

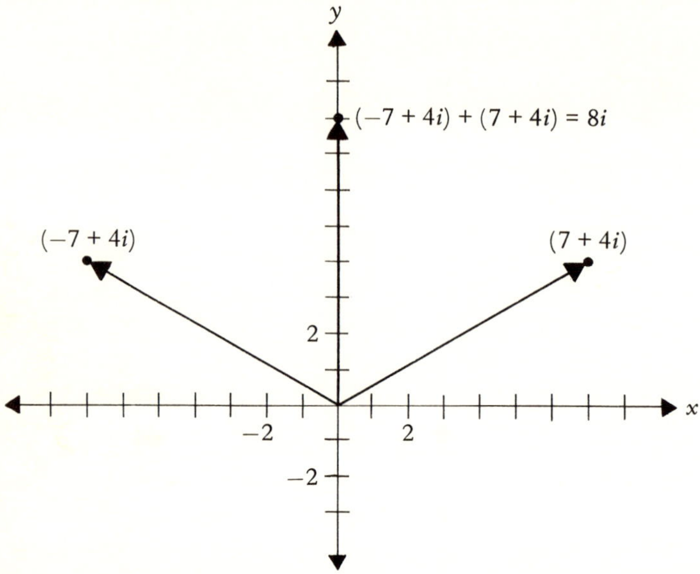

5. Consideration of what happens to these complex numbers when multiplied by i leads us to the conjecture that the vector which represents a complex number is rotated 90° counterclockwise when the vector is multiplied by i.
6. The length of the vector representing $a + bi$ is $\sqrt{a^2 + b^2}$.
9. (a) $x = \pm 2i$, (c) $x = i$.
10.

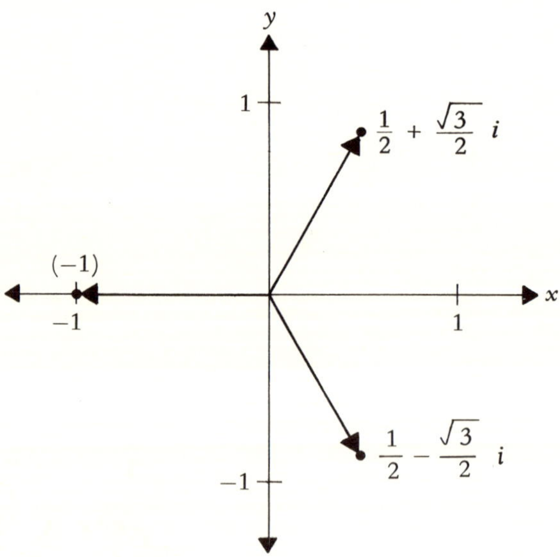

11.

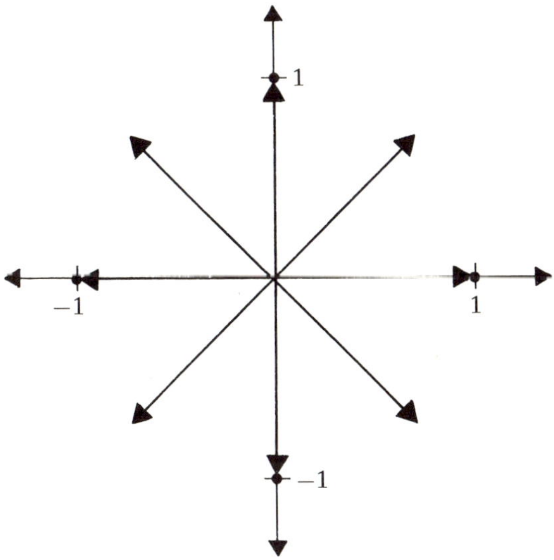

12. (a) We denote the conjugate of $a + bi$ by $\overline{a + bi}$. We have then $\overline{a + bi} = a - bi$. Note that the conjugate of a real number x is x. See graph.

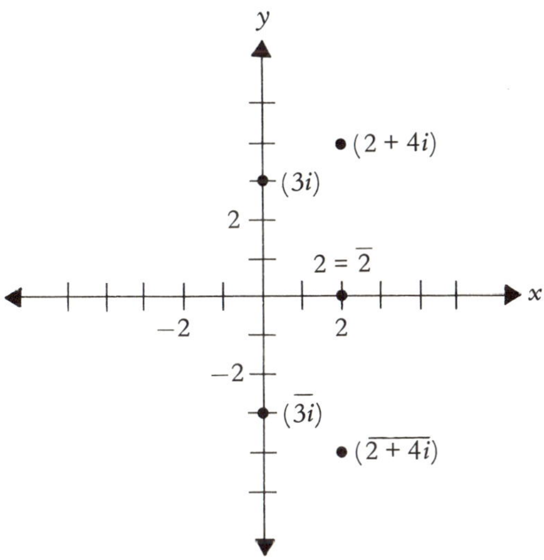

(b) $\overline{(\overline{a + bi})} = \overline{(a - bi)} = a - (-bi) = a + bi$.
(c) $(a + bi)(\overline{a + bi}) = (a + bi)(a - bi) = a^2 + b^2$ which is real.
(d) The length of the vector $a + bi$ is $\sqrt{a^2 + b^2}$; this is the square root of the product of $a + bi$ and its conjugate.

14. Let P be a set in C such that $1 + 0i$ is in P, $0 + 0i$ is not in P, and P is closed under addition and multiplication. If i is in P then $i^2 = -1 + 0i = -(1 + 0i)$

is also in P. If $-i$ is in P then $(-i)^2 = -1 + 0i = -(1 + 0i)$ is also in P. Hence P cannot have property (iv) of the definition given in Section 10.11.

Section 12.1

1. (a) $\aleph_0$ since the given set is simply N and $\aleph_0 = \#(N)$.
   (c) 0 since the given set is ϕ.
   (e) The given set is $B = \{0,1,4,9,16, \ldots\}$. If we let $f(n) = n^2$, then f is a one-to-one correspondence between N and B and, hence $\#(B) = \#(N) = \aleph_0$.
   (g) Let $f(n) = 1/(n + 1)$. Then f is a one-to-one correspondence between N and the given set which, thus, has the cardinal number $\aleph_0$.
   (i) Let $f(n) = 2^n$. Then f is a one-to-one correspondence between N and the given set which, thus, has cardinal number $\aleph_0$.
   (k) Let $f(n) = n + 1$. Then f is a one-to-one correspondence between N and I^+. Thus, $\#(I^+) = \#(N) = \aleph_0$.
   (m) Let $f(n) = 10^{-n}$. Then f is a one-to-one correspondence between N and the given set which, thus, has cardinal number $\aleph_0$.

2. If we let f be defined by $f((0,0)) = 0$, $f((0,1)) = 1$, $f((1,0)) = 2$, $f((1,1)) = 3$, etc., then f is a one-to-one correspondence between $N \times N$ and N. Hence $\#(N \times N) = \#(N) = \aleph_0$. The idea here is that each pair in $N \times N$ will eventually appear some place in the proposed list and, indeed, will occur exactly once in the list. This is supposed to be evident even though we have not actually given an algebraic formula for f. (Incidentally, it is *possible* to do so.)

Section 12.2

1. We first note that when A and B are sets we have $A \subseteq A \cup B$ and $A \cap B \subseteq A$. If we let f be the identity mapping given by $f(x) = x$ on A, we see that f is a one-to-one correspondence between A and a subset (A itself) of $A \cup B$. Hence, $\#(A) \leq \#(A \cup B)$. Similarly, if we let g be the identity mapping in $A \cap B$, then g is a one-to-one correspondence between $A \cap B$ and a subset (namely $A \cap B$) of A. Hence, $\#(A \cap B) \leq \#(A)$.

3. Let $A = \{1,2, \ldots ,n\}$. Then $\#(A) = n$. Let f be a mapping from A to N. We shall show that f cannot be a one-to-one correspondence between A and N and, hence, we cannot have $n = \aleph_0$. The set $\{f(1),f(2), \ldots ,f(n)\}$ is finite and, hence, has a maximum. Call this maximum M. Choose a natural number k such that $k > M$. Then k can not be a value of f on A. Hence, f does not map A *onto* N. Since $A \subseteq N$ we see that $\#(A) \leq \#(N)$ and since $\#(A) \neq \#(N)$, we have $\#(A) = n < \aleph_0 = \#(N)$.

5. $y = .222122 \ldots$.

Section 12.3

1. (a) Let $A = \{0,1,2,3,4,5\}$. Then $\#(A) = 6$. Let $f(n) = n + 6$. Then f is a one-to-one correspondence between N and $B = \{6,7,8,9, \ldots\}$. Thus, $\#(B) = \aleph_0$. We note that $A \cap B = \phi$ and, hence, $6 + \aleph_0 = \#(A) + \#(B) = \#(A \cup B) = \#(N) = \aleph_0$.
   (c) Let $C = \{a,b,c\}$. Then $\#(C) = 3$. And $C \times N = \{(a,0),(b,0),(c,0),(a,1),(b,1),(c,1),(a,2), \ldots\}$. Let $f(n) = (a,n)$ if $n \equiv 0 (\text{mod } 3)$, $f(n) = (b,n)$ if $n \equiv 1 (\text{mod } 3)$, and $f(c,n)$ if $n \equiv 2 (\text{mod } 3)$. Then f is a one-to-one correspondence between N and $C \times N$. Hence, $3 \cdot \aleph_0 = \#(C \times N) = \#(N) = \aleph_0$.

3. Let $A = \{n \mid n \in N$ and $n \neq 0$ and $n \cdot \aleph_0 = \aleph_0\}$. Since we know that $1 \cdot \aleph_0 = \aleph_0$ we see that 1 is in A. Suppose k is in A. Then $(k + 1) \cdot \aleph_0 = k \cdot \aleph_0 + 1 \cdot \aleph_0 = \aleph_0 + \aleph_0 = \aleph_0$. Thus, k in A implies $k + 1$ is in A and this completes the proof.

5. Let $x = \#(A)$, $y = \#(B)$, $z = \#(C)$, and $w = \#(D)$. Suppose that $A \cap C = B \cap C = \phi$.
   (a) If $x \leq y$, then there is a one-to-one correspondence f between A and some subset of B. Let g be the mapping defined on $A \cup C$ by $g(x) = f(x)$ for x in A and $g(x) = x$ for x in C. Then g is a one-to-one correspondence between $A \cup C$ and a subset of $B \cup C$. Hence, $\#(A \cup C) \leq \#(B \cup C)$. That is, $x + z \leq y + z$.
   (c) Suppose $x \leq y$ and f is as in part (a). Suppose $z \leq w$. Then there is a one-to-one correspondence k between C and some subset of D. Define a mapping j on $A \times C$ by letting $j((a,c)) = (f(a),k(c))$. Then j is a one-to-one correspondence between $A \times C$ and a subset of $B \times D$. Hence, $\#(A \times C) \leq \#(B \times D)$. That is, $xz \leq yw$.

6. (a) Let $x = 1$, $y = 2$, and $z = \aleph_0$. Then we have $x < y$ and yet $x + z = y + z$.
   (c) Let $x = \aleph_0$ and $z = 2$. Then $xz = x$ and yet $z \neq 1$.

7. Send the new arrival to room 0 and ask the guest in room n to move to room $n + 1$ for $n = 0,1,2, \ldots$. This illustrates $\aleph_0 + 1 = \aleph_0$.

Section 12.4

1. Define the function G on $\{0,1\}^N$ by taking $G(f) = .f(0)f(1)f(2) \ldots$ where we interpret the expression on the right as the decimal name for a real number. Since $f(n)$ is 0 or 1 for each n, $G(f)$ names a real number. And if $G(f) = G(g)$ then $f(n) = g(n)$ for every n so that $f = g$. Thus G is a one-to-one correspondence between $\{0,1\}^N$ and a subset of R and it follows that $\#(\{0,1\}^N) \leq \#(R)$. That is, $2^{\aleph_0} \leq c$.

3. Let P be the power set of A. Define a function G on P as follows: For S in P let $G(S) = f_S$ where $f_S(x) = 1$ if $x \in S$ and $f_S(x) = 0$ if $x \notin S$. Then G is a one-to-one correspondence between P and $\{0,1\}^A$. Hence, $\#(P) = \#(\{0,1\}^A)$.

4. Let $x = \#(A)$, $y = \#(B)$, and $z = \#(C)$ and suppose $B \cap C = \phi$.
   (a) Define a mapping G on $A^B \times A^C$ as follows. For $f \in A^B$ and $g \in A^C$, let $G((f,g)) = f \cup g$. (Here we treat the functions f and g as sets of ordered pairs.) Then G is a one-to-one correspondence between $A^B \times A^C$ and $A^{B \cup C}$. Hence, $\#(A^B \times A^C) = \#(A^{B \cup C})$ and, therefore, $x^y \cdot x^z = x^{y+z}$.

5. 0^0 should be the cardinal number of ϕ^ϕ, the collection of all mappings from ϕ into ϕ. There is exactly one such mapping—the empty map. So we should have $0^0 = 1$.

7. Let A be the set of all algebraic real numbers and T be the set of all transcendental real numbers. We are told that $\#(A) = \aleph_0$. If $\#(T) = \aleph_0$, then $\#(A \cup T) = \#(A) + \#(T) = \aleph_0 + \aleph_0 = \aleph_0$. But $A \cup T = R$ and $\#(R) = c \neq \aleph_0$. Hence, T cannot be denumerable. We must, in fact, have $\#(T) = c$.

Section 12.5

1. The relations in (d), (probably) (e), and (g) are not simple order relations. The rest are.

3. (a), (b), (e), (h), and (j) are situations where the set is well ordered by $\leq$.

5. Let $A = \{n | n \in I^+$ and $\#(S) = n$ and S simply ordered by R implies R in S is isomorphic to $\leq$ in $\{1,2,\ldots,n\}\}$. 1 is in A since a simply ordered set $\{a\}$ is obviously isomorphic to $\{1\}$ ordered by $\leq$. Suppose k is in A. Let $\#(S) = k + 1$ and suppose S is simply ordered by R. By Problem 4, S has a first member x_1. There are k members in $S - \{x_1\}$ so $S - \{x_1\}$ is isomorphic to $\{1,2,\ldots,k\}$ ordered by $\leq$. Let f be an isomorphism that establishes this fact. Define a mapping g on f by letting $g(x_1) = 1$ and $g(x) = f(x) + 1$. Then g is an isomorphism between S ordered by R and $\{1,2,\ldots,k+1\}$ ordered by $\leq$. Hence $k + 1$ is in A. Therefore, by induction, $A = I^+$ and th,s completes the proof.
8. All of them in (i) are isomorphic and all of them in (ii) are isomorphic.
10. If we take it that xRy means $x = y$ or x is to the left of y in our list, we may display the given situation as $\{-1,-2,-3,\ldots;0,1,2,3,\ldots\}$. I is both simply ordered and well ordered here. To see that R is reflexive we note that x has the same sign as x and hence xRx for all x in I. We see that if x and y are in I, then either xRy or yRx because we must have x and y both negative, or both nonnegative, or different in sign. To show that R is transitive, suppose that x and y and z are in I and xRy and yRz. If x is nonnegative we see that we have y and z nonnegative and $x \leq y$ and $y \leq z$ and, hence, $x \leq z$ so that xRz. If x is negative and z is negative, then we must have y also negative and $x \geq y$ and $y \geq z$ so that $x \geq z$ and, hence, xRz. If x is negative and z is positive, then we have immediately that xRz. Thus, in every possible case, xRy and yRz implies xRz. Since R is reflexive and transitive in I and any two elements in I are comparable by R, we see that R simply orders I.

    To show that I is well ordered by R, suppose A is a nonempty subset of I. We consider two different cases. If $A \subseteq N$, then A has a smallest element (since N is well ordered by $\leq$) and this element will be a first element in A with respect to the relation R. If A is not contained in N, then $B = A \cap I^-$ is nonempty. Since I^- is well ordered by $\geq$, B will contain a first member with respect to $\geq$. This element is in the relation R to all elements of B, but it also is negative so that it is in the relation R to all elements in $A \cap N$. Hence, this element is a first element of A with respect to R. We have shown that in each case A has a first element and, therefore, I is well ordered by R.

Section 12.6

1. Let f be a one-to-one correspondence between $N \cup \{a\}$ and N. We shall show that f cannot be an isomorphism. Let $f(a) = k$. Then k is a natural number. Hence, $k + 1$ is in N and also $k < k + 1$. Let $f^{-1}(k + 1) = b$. If f were an order preserving mapping we would need to have aRb and $a \neq b$ in $N \cup \{a\}$ and this is impossible since bRa for all b in $N \cup \{a\}$. Hence, f cannot be an isomorphism and we see that $N \cup \{a\}$ ordered by R and N ordered by $\leq$ are of different order types.
3. No, addition of ordinal numbers is not commutative since by Problem 1 we have $\omega + 1 \neq \omega$ and by Problem 2 we have $1 + \omega = \omega$.
4. (a) $\omega + 2$ is the ordinal number of the set $\{0,1,2,\ldots;a,b\}$ ordered from left to right.
   (c) $\omega + \omega$ is the order type of the set $\{-1,-2,-3,\ldots;0,1,2,3,\ldots\}$ ordered from left to right.
5. (a) ω is the order type of $\{0,1,2,3,\ldots\}$ ordered by $\leq$. n is the order type of $\{0,1,2,$

$\ldots, n\}$ ordered by $\leq$. Clearly the latter set is order isomorphic (in fact, identical with) the section of N determined by n. Hence, $n \leq \omega$. On the other hand, it is impossible to set up a one-to-one correspondence between $\{0,1,2, \ldots, n\}$ and N (let alone an *order preserving* one-to-one correspondence) so that $n \neq \omega$. Hence, $n < \omega$.

6. Let a, b, and c be the ordinal numbers of well ordered sets A, B, and C respectively, and suppose $a \leq b$ and $b \leq c$. Then A is order isomorphic to some section of B and B is order isomorphic to some section of C. Let f be the isomorphism between A and the section of B determined by the element k. Let g be the isomorphism between B and a section of C. We claim that $g \circ f$ is an order isomorphism between A and the section of C determined by the element $g(k)$. Hence, $a \leq c$.

7. (a) $2 \cdot \omega$ is the ordinal number of the set $\{(a,0),(b,0),(a,1),(b,1),(a,2),(b,2), \ldots\}$ ordered from left to right.

Section 12.7

1. 1 is the ordinal number of $\{0\}$ which is the set of ordinals less than 1. 6 is the ordinal number of $\{0,1,2,3,4,5\}$. ω is the ordinal number of $\{0,1,2, \ldots\}$ which is the set of all ordinal numbers which are less than ω. $\omega + 2$ is the ordinal number of $\{0,1,2, \ldots; \omega, \omega + 1\}$. $\omega + \omega$ is the ordinal number of $\{0,1,2, \ldots; \omega, \omega + 1, \omega + 2, \ldots\}$. (Note that we are assuming here that all the sets are ordered from left to right as given.)

Index

A B C D E F G H I J 9 8 7 6 5 4 3 2 1